EURO-CERAMICS

Volume 1: Processing of Ceramics

The Proceedings of the First European Ceramic Society Conference (ECerS '89) held on 18–23 June 1989 at the Maastricht Exhibition & Congress Centre, The Netherlands.

Volume 1: Processing of Ceramics

Synthesis
Forming
Densification
Phases and Reactions

Volume 2: Properties of Ceramics

Characterisation
Electronic Ceramics
Superconducting Ceramics
Traditional Ceramics

Volume 3: Engineering Ceramics

Bioceramics
Standardisation
Fracture
Deformation and Reinforcement
Composites
Surface-Related Phenomena
Applications

EURO-CERAMICS

Volume 1

Processing of Ceramics

Edited by

G. de WITH

Philips Research Laboratories, Eindhoven, The Netherlands

R. A. TERPSTRA

*Netherlands Energy Research Foundation, ECN,
Petten, The Netherlands*

and

R. METSELAAR

*Centre for Technical Ceramics,
Eindhoven University of Technology, The Netherlands*

ELSEVIER APPLIED SCIENCE
LONDON and NEW YORK

ELSEVIER SCIENCE PUBLISHERS LTD
Crown House, Linton Road, Barking, Essex IG11 8JU, England

Sole Distributor in the USA and Canada
ELSEVIER SCIENCE PUBLISHING CO., INC.
655 Avenue of the Americas, New York, NY 10010, USA

WITH 97 TABLES AND 437 ILLUSTRATIONS

British Library Cataloguing in Publication Data

Euro-Ceramics.
1. Ceramics
I. With, G. de II. Terpstra, R. A. III. Metselaar, R.
666

ISBN 1-85166-426-2 (set)
ISBN 1-85166-430-0 (Vol. 1)
ISBN 1-85166-431-9 (Vol. 2)
ISBN 1-85166-432-7 (Vol. 3)

Library of Congress CIP data applied for

Printed in Northern Ireland by The Universities Press (Belfast) Ltd.

FOREWORD

In September 1987, during the 14th 'Science of Ceramics' conference, the European Ceramic Society was founded. The primary goal of this new society is to improve and facilitate communications both between European countries and between Europe and the rest of the world. Eleven countries have now joined the new society: Belgium, Federal Republic of Germany, France, Italy, The Netherlands, Norway, Portugal, Spain, Sweden, the United Kingdom and Yugoslavia.

To achieve its goal, the Society has decided to organize a set of conferences. This proceedings volume is a record of the first, ECerS '89, which took place in Maastricht, The Netherlands, in June 1989. The aim was to bring together engineers, scientists, technicians and all others with an interest in ceramics. The meeting covered the whole field of ceramics, from traditional clay-based materials to advanced materials with applications in the fields of engineering, electronics, magnetism and medicine. The 30 years tradition of the 'Science of Ceramics' meetings has been incorporated into the ECerS conferences under the heading of Basic Science. In 1991 the next conference of the European Ceramic Society, ECerS '91, will be held in the Federal Republic of Germany, probably near Munich. Directly after this conference the exhibition 'Ceramitec' will be organized. This will further enlarge the scope of the meeting, hopefully in a way that is appreciated by the European ceramic community.

It has been decided to present at each ECerS conference the A. L. Stuijts Memorial Award. As a research scientist at Philips Research Laboratories, Professor Stuijts contributed considerably to the development of magneto- and electro-ceramics, in particular to a better understanding of the relationship between the science and technology of these materials. To commemorate his contributions to the scientific community, an award with his name will be presented to a ceramist noted for his or her outstanding contributions to ceramic science or technology. The awardee will present a memorial lecture. During ECerS '89 the first Stuijts Memorial Award was presented to Professor R. J. Brook, who gave his Memorial Lecture entitled 'The Mastery of Microstructure'. Hopefully with this first memorial lecture a tradition has been started that will last for many years.

As mentioned by the editors, the space available to the authors in these proceedings is limited. I am pleased to announce that the ECerS now has its own *Journal of the European Ceramic Society*. This journal offers the possibility of publishing full-length papers of high scientific quality.

Organizing a conference is an elaborate task which is impossible without the help of many persons. This was true for ECerS '89 as well. It is impossible to mention all who contributed. An exception must be made for all collaborators of the Maastricht Exhibition & Congress Centre, MECC. Their enthusiastic cooperation has contributed considerably to the success of the conference.

I hope that the new series of conferences will be as successful as its predecessor, 'Science of Ceramics', and will fulfil all the needs of ceramists in Europe and the rest of the world.

R. Metselaar
President of the European Ceramic Society, 1987–1989,
Centre for Technical Ceramics, Eindhoven, The Netherlands.

PREFACE

This proceedings volume contains, besides the text of six key-note lectures and an extended abstract text of the first Stuijts Memorial Lecture, a representative selection of the more than 450 papers presented during ECerS '89.

The division between the various topics was approximately as follows: basic science 40%, engineering ceramics including bioceramics 30%, electronic ceramics including high T_c superconductors 20%, and traditional ceramics 10%. As was to be expected, the division is almost the same as in the American Ceramic Society Annual Meetings during the last few years. The fact that proper processing is the key to good properties is reflected here by the large number of contributions dealing with this topic. Moreover, there was a considerable number of papers discussing the processing and properties of composite materials.

Because of the large number of contributions, the number of pages for each manuscript had to be limited. This may seem a handicap to authors. The usefulness of proceedings is, in our opinion, mainly found in an overview of activities going on, rapidly published after the conference. Detailed theories or extensive descriptions of experimental work are out of place in such proceedings. Therefore, apart from practical reasons, in our view a limited number of pages is sufficient. For reasons of speed only limited reviewing was carried out. This procedure limited itself mainly to format and obvious errors. Opinions expressed remain, however, entirely the responsibility of the authors.

The editing of the proceedings would not have been possible without the help of many individuals. Without mentioning them all personally, this is the proper place to thank all those who contributed. One exception, however, must be made: Miss E. Ramaekers performed many secretarial duties, and without her help the editing of these proceedings could not have been completed in so timely a manner.

G. de With
Philips Research Laboratories and Centre for Technical Ceramics, Eindhoven, The Netherlands

R. A. Terpstra
Netherlands Energy Research Foundation, ECN, Petten, The Netherlands

R. Metselaar
Centre for Technical Ceramics, Eindhoven, The Netherlands

Honorary Committee ECerS '89

J. Kremers	Queen's Commissioner of Limburg
Ph. J. I. M. Houben	Mayor of Maastricht
R. W. de Korte	Minister of Economic Affairs
W. J. Deetman	Minister of Education and Science
R. E. Selman	Board of Management DSM
S. van Houten	Board of Management Nederlandse Philips Bedrijven BV
M. C. van Veen	Board of Management Hoogovens Groep BV
G. Kaptein	Board of Management NV Koninklijke Sphinx
G. J. M. Teeuwen	Board of Management Redland Bouwprodukten BV

Advisory Committee ECerS '89

A. McDonald (*Australia*)
F. Cambier (*Belgium*)
P. H. Duvigneaud (*Belgium*)
D. Yen (*China*)
F. Aiazzi (*Italy*)
D. T. Beruto (*Italy*)
G. Ziegler (*FRG*)
G. König (*FRG*)
P. Boch (*France*)
W. Simonet (*France*)
S. Saito (*Japan*)
A. J. Burggraaf (*The Netherlands*)
R. Pampuch (*Poland*)
J. L. Baptista (*Portugal*)
A. E. Benloch (*Spain*)
S. de Aza Pendas (*Spain*)
R. Pompe (*Sweden*)
G. J. Gittens (*UK*)
D. P. Thompson (*UK*)
D. E. Niesz (*USA*)
D. Kolar (*Yugoslavia*)

Scientific Programme Committee ECerS '89

G. de With (*Chairman*)
R. A. Terpstra (*Secretary*)
R. Metselaar
R. Wernicke
R. Nelissen
K. de Groot
K. Keizer
H. J. Timmers

Council of the European Ceramic Society

Belgium:	P. Fierens F. J. Cambier
FRG:	H. Hausner (*President Elect*) M. Blumenberg
France:	R. Cauville D. Broussaud
Italy:	G. N. Babini C. Palmonari
The Netherlands:	J. T. van Konijnenburg R. Metselaar (*President*)
Norway:	Th. Møller J. F. Haslund
Portugal:	A. F. Faria Frasco J. L. Baptista
Spain:	A. G. Verduch M. A. Delgado Méndez
Sweden:	R. Carlsson R. Ahlgren
United Kingdom:	G. J. Gittens D. Taylor
Yugoslavia:	D. Kolar

Sponsors of ECerS '89

City of Maastricht
Commission of the European Communities
Dutch Ministry of Economic Affairs
Elsevier Science Publishers Ltd
NV Gouda Vuurvast
KLM Royal Dutch Airlines
Koninklijke Mosa BV
NV Koninklijke Porceleyne Fles Ao 1653
NV Koninklijke Sphinx
Netherlands Energy Research Foundation (ECN)
D. van Oord & Co. Clay Tiles BV
Philips International BV
Philips Research Laboratories
Redland Bouwprodukten
Strategic Analysis Inc. SAI (Brussels)
Teewen BV
Tichelaar's Koninklijke Makkumer Aardwerk en Tegelfabriek BV
TNO Institute for Applied Physics

CONTENTS

Volume I: PROCESSING OF CERAMICS

I.4 PHASES AND REACTIONS

THE STUIJTS AWARD

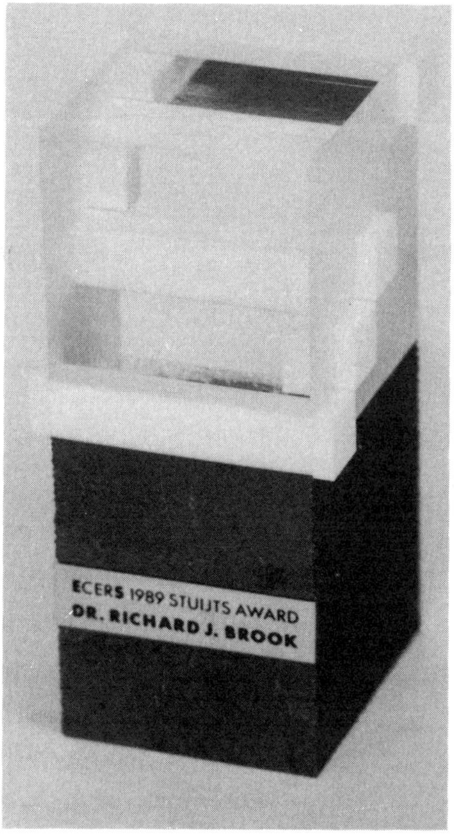

At the founding of the European Ceramic Society it was decided to present at each ECerS conference the A. L. Stuijts Memorial Award. As a research scientist at Philips Research Laboratories, Eindhoven, The Netherlands, Professor A. L. Stuijts (1922–1982) contributed considerably to the development of magneto- and electro-ceramic materials, in particular to a better understanding of the relationship between the microstructure and processing of these materials. He was also a Professor Extraordinary at the Eindhoven University of Technology. In 1976 he was elected Fellow of the American Ceramic Society and in 1977 he delivered the Sosman Memorial Lecture for this society. To commemorate his contributions to the ceramic community, during each conference of the European Ceramic Society an award bearing his name will be presented to a ceramist noted for outstanding contributions to ceramic science or technology. Professor R. J. Brook was the first to receive this award during ECerS '89 in Maastricht, The Netherlands.

Richard J. Brook (1938–) was born in Leeds, UK, and started his career in ceramics at the Massachusetts Institute of Technology, Boston, USA, where he obtained his PhD for research on ferrite thin films in 1966. From 1966 to 1970 he was Assistant Professor of Materials Science at the University of Southern Carolina where he did research on the development of microstructures in ceramics and on electrical properties. After 4 years at the AERE, Harwell, UK, as principal scientist and leader of the Electronic Ceramics Group, where he mainly researched electronic ceramics, Richard Brook joined the University of Leeds as Professor and Head of the Department of Ceramics. Apart from researching ceramic fabrication, solid electrolytes and ceramics processing, he has taken part in many university and external committees. Since 1988 Professor Brook has been one of the directors at the Max Planck Institute for Metals Research, Stuttgart, FRG.

The Stuijts award was presented by Dr F. M. A. Carpay, Executive Director, Philips Research Laboratories, after which Professor Brook delivered the Memorial Lecture entitled 'The Mastery of Microstructure'.

Stuijts Memorial Lecture 1989
"The Mastery of Microstructure"

R.J. Brook
Max Planck Institute for Metals Research
Stuttgart, FRG

In the light of his experience in the successful development of ceramic microstructures to match specific electromagnetic applications, Professor Stuijts proposed a number of themes in 1972 as being ripe for progress in the general task of exploiting ceramics for advanced engineering use. Three of these topics are reviewed with emphasis on the work that has been done and on current attitudes. The three themes are powder preparation, the importance of technological factors, and the need for valid generalised pictures of complex processes.

In powder preparation, the dramatic influence of chemical preparation methods on the quality of powders available for ceramic fabrication is one of the clearest developments of recent years. Major advances in processing convenience and in product reliability have resulted. In respect to technological factors, considerations of cost and of the need for reproducilbility are now seen as critical in determining the rate of adoption of ceramics into new applications; introduction of ceramic components into heat engines which has long been seen as a target in respect to the full and active use of ceramics on the large volume scale is largely determined by such aspects. The very great complexity of ceramics processing science and the resulting risk that detailed explanations of phenomena cannot then be extended for more general use has made the search for precise atomistic interpretation of microstructural change less persuasive than it once was.

Three sectors that have proved fruitful in the intervening years in tackling the issues that were identified as being of concern are the use of mapping techniques, the use of computer modelling and the use of detailed microstructural characterisation notably in the form of electron microscopy.

These methods have allowed the development of a systematic approach to ceramic processing along the lines that Professor Stuijts was seeking. The approach has shown striking success in the development of the set of high strength and toughness ceramics based on the transformation toughening mechanism; it has also proved effective in the work on high temperature materials based on silicon nitride and the sialons.

The path to the mastery of microstructure that was advocated by Professor Stuijts has been found to be a productive one. The early view that detailed knowledge of precise atomistic mechanisms might one day underlie our approach to ceramic processing has if anything been seen as excessively and unnecessarily ambitious; in its place, a pragmatic pattern of development based on good understanding of the possible contributions of generalised mechanisms has developed.

A framework for the successful processing of ceramics is now in place; the need, however, for imagination in the exploitation of this framework remains as true today as at the time of its recognition by Professor Stuijts.

CHEMICAL SYNTHESIS OF CERAMIC POWDERS

R. F. Hill, J. Lee, L. C. Montgomery, W. M. Shen
Union Carbide Advanced Ceramics
Union Carbide Coatings Service Corporation
Cleveland, Ohio 44101

ABSTRACT

A review of powder synthesis developments at Union Carbide Advanced Ceramics is presented. High purity alumina and cordierite have been synthesized by hydrolysis of metal alkoxide solutions. Titanium diboride and boron carbide powders made by a continuous carbothermic process are characterized by high purity and small primary particle size. Pressureless sinterable TiB_2 is introduced. Extremely fine high purity silicon nitride powders have been made by thermal decomposition of silane gas.

INTRODUCTION

In the ceramic industry there is good agreement that powder raw materials have a major effect on the ultimate attainable physical, chemical, and structural properties of advanced ceramic components. In order to meet the ever increasing demand of ceramic component producers, the role of the powder suppliers must be expanded. A variety of new ceramic powders have been made by three different chemical synthesis techniques to provide the purity, particle size, consistency, and homogeneity required by parts producers. Included are alumina and cordierite powder made from organometallics; titanium diboride and boron carbide made by a continuous carbothermic process; and silicon nitride made by a gas phase process. The characteristics of these powders which were produced on the commercial and pilot plant scale are presented.

ROLE OF THE POWDER PRODUCER

It has become evident that producing powders of higher purity than users are accustomed to presents some unusual problems. In many cases the impurities in powders play a major role in the densification behavior and physical properties of ceramics made from them. In some circumstances "impurities" are required in powders. In these cases, impurities are no longer benign contaminants, but are really essential ingredients of the powder. Therefore, it is not enough for a powder supplier to provide only high purity powder. The supplier must synthesize powders complete with the appropriate additives, so that they will perform up to the expectations of the ceramic component producer. Today's powder suppliers must increasingly rely on continuous powder production to give the reproducibility the component producer needs. This reproducibility is improved when starting from high purity powders which are modified with specific additives rather than letting random impurities affect powder characteristics.

I. SYNTHESIS OF CERAMIC POWDERS FROM ORGANOMETALLICS

It is well known that ceramic powders can be synthesized quite readily from organometallics. However, what is generally not taken advantage of or widely practiced is the fact that chemical processes can be utilized to recycle all of the byproducts of the reactions involved, thus making the process both economical and environmentally acceptable. (1,2) Figure 1 shows the outline of a typical process which has been employed for the manufacture of ultra high purity aluminum oxide. Of particular importance the fact that the only raw materials introduced into the process are aluminum, metal, and water. In Step 1, aluminum alkoxide is synthesized from aluminum metal and alcohol. Depending on the alcohol chosen, some purification can take place during this step, by preferential reaction or distillation. The alkoxide formed is hydrolized with water in Step 2 to form a hydrated aluminum oxide-alcohol slurry. In Step 3, the alcohol is removed from the slurry by distillation, leaving a dilute slurry product which can be dried and calcined to alpha alumina. In Step 4, the alcohol which was removed in Step 3 is dehydrated and then recycled to make more alkoxide in Step 1. The organic materials remain in the process and are continually refined and reused. The only effluents are hydrated oxide products, water, and hydrogen. The resulting powder products are of exceptionally high purity, Table 1. For the hydrated alumina, or boehmite made by this process, no metallic impurities were detected by ICP emission spectroscopy. The only impurities found in the alumina calcined to the alpha phase were a few parts per million of iron and silicon. Indeed this high purity has been demonstrated by what is considered to be the ultimate test of purity: that is, the powder has been used to grow single crystal sapphire.

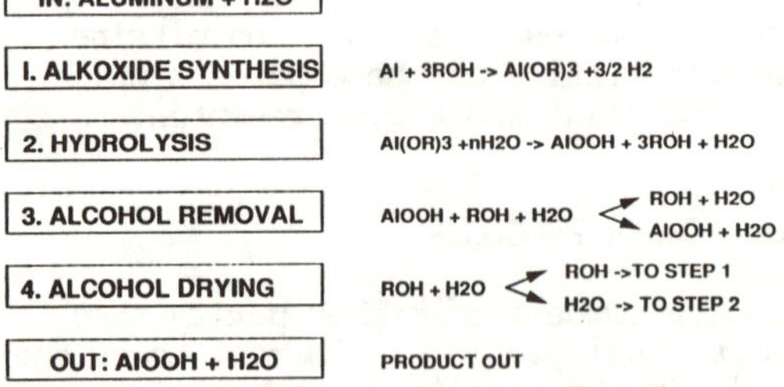

Figure 1. Hydrated Aluminum Oxide Process

	BOEHMITE	ALPHA Al2O3
B	<5	<6
Ba	<1	<1
Ca	<3	<4
Co	<3	<4
Cr	<5	<6
Cu	<3	<4
Fe	<3	5
K	<11	<12
Mg	<2	<2
Mn	<2	<2
Na	<11	<12
Si	<5	14
Ti	<3	<4

Table 1. ICP Em. Spec. Trace Elemental Analysis, ppm

Today, high purity alumina derived from alkoxides is produced commercially. The use of inexpensive aluminum metal raw materials, in combination with novel chemical processing methods, leads to economically viable production of high purity alumina powders.

Cordierite powder has also been produced in pilot plant quantities using the same basic organometallic process. (3) The processing characteristics of this powder are an excellent example of the improved densification behavior which can be realized by utilizing the organometallic process. Careful processing of this powder according to the recipe on Figure 2, including pre-calcination, deagglomeration, compaction, and sintering has resulted in hexagonal cordierite ceramics which can achieve full density at temperatures below 875°C. This is well below the normal processing temperatures for cordierite made from conventional oxide powders which sinter at 1360°C or from glass which sinters at 1150°C.

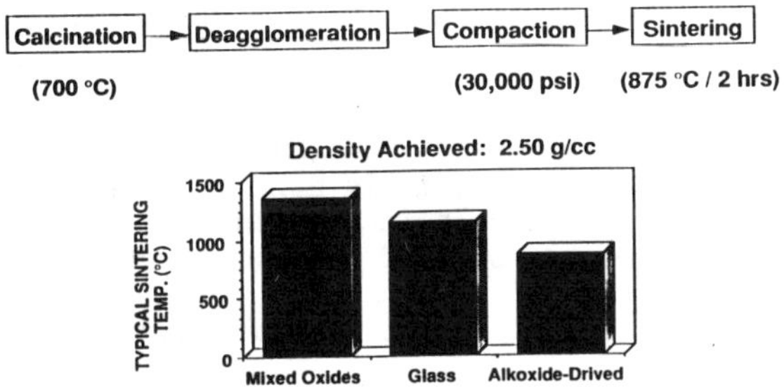

Figure 2. Cordierite Densification Process

II. POWDER SYNTHESIS BY CARBOTHERMIC REDUCTION

Union Carbide Advanced Ceramics utilizes the carbothermic reduction process to manufacture titanium diboride and boron carbide on a commercial scale. Powders made by the traditional Holmantube batch process Union Carbide used in the 1960's and 70's have been significantly improved by making the production process continuous. The new method has been carefully developed so that control of particle size during continuous synthesis eliminates the need for extensive milling of the product, thereby avoiding contamination which is characteristic of some milling processes. Only two primary steps are required for the production of high purity, fine particle size titanium diboride powder. These steps include blending of the raw materials, followed by continuous heat treatment at 1800-2000°C. The morphology of the "as synthesized" powder particles shown on Figure 3 is one of flat hexagonal platelets about 10 microns in diameter by 1 to 2 microns thick. Table 2 shows that the typical carbon, oxygen, and nitrogen content of this powder is quite low. Trace elemental analysis by ICP emission spectroscopy of this powder indicates the exceptionally high purity which can be achieved using the continuous process. Most elements are undetectable, except for 200 ppm iron and minor amounts of calcium and silicon.

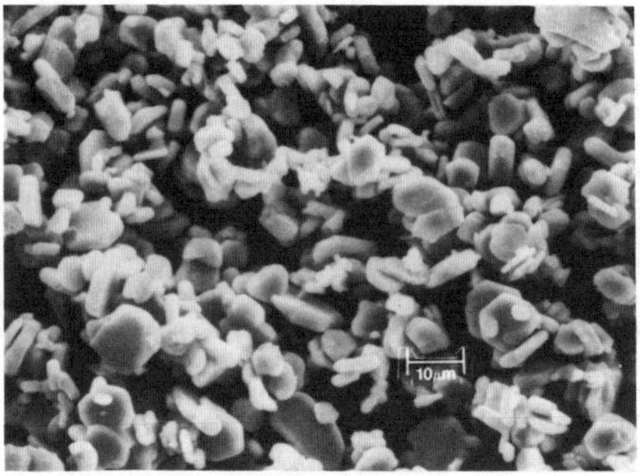

Figure 3. Titanium Diboride "As Synthesized" Powder Morphology

Carbon	0.2%
Oxygen	0.3%
Nitrogen	0.1%
Al (ppm)	<155
Ca	20
Co	<16
Cu	<38
Fe	204
K	<33
Ni	<16
Si	25
W	<17
Zn	<10

Table 2. Analysis of Continuously Produced TiB_2

When titanium diboride is manufactured by this continuous process, and it is made with very few impurities, it is not possible to densify even by rigorous hot pressing conditions. In fact, even at 2000°C and 2000 psi, the maximum density attained was only 73 percent of theoretical. The sintering aids required for densification simply are not present. (4,5) Reducing the particle size to 4 to 5 microns only increased the density to about 85 percent.

To a powder supplier, this situation can be quite advantageous. By carefully controlling the amount of sintering additives incorporated into the powder, the densification behavior can be reliably controlled and reproduced. For example, if the high purity 10 micron size powder which was previously not densifiable is prepared with just 0.38 volume percent of cobalt, iron, nickel, and tungsten sintering aids, it is possible to achieve above 99% of theoretical density by hot pressing. Since the additives are controlled, and not left to chance, the reproducibility of densification performance is excellent.

If in addition to controlling the sintering additives, the particle size distribution is adjusted and a grain growth inhibitor is added, the powder can be further improved. A total of 0.4 volume percent of Co, Ni, Fe, and W densification aids and 2.5 volume percent of a chrome diboride grain growth inhibitor were added to the powder. The average particle size of the powder was controlled to be about 5 microns. The powder can now be sintered without the application of pressure according to the process which is outlined on Figure 4. The pressureless sinterable powder*, which contains a Poly-vinyl-acetate binder was compacted at 30,000 psi in a uniaxial die of 1.25 inch diameter. The shapes were embedded in loose TiB_2 powder, and heated to 2050°C in an argon atmosphere. Results show that this powder which previously could not be densified by hot-pressing at 2000°C can now achieve above 98% density by pressureless sintering.

*Union Carbide Advanced Ceramics Titanium Diboride, Grade HCT-SB

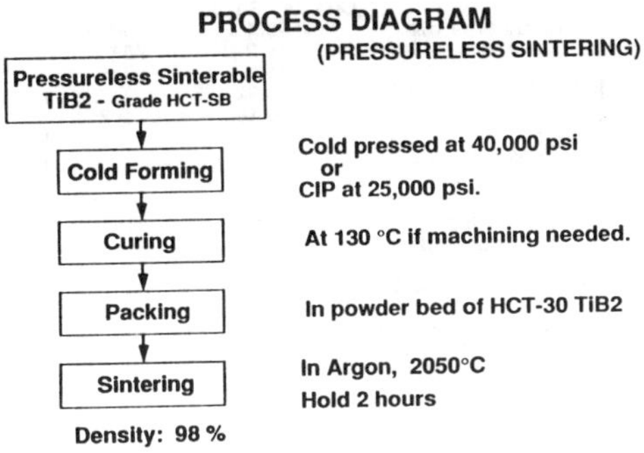

Figure 4. Process for Pressureless Sintering HCT-SB TiB_2

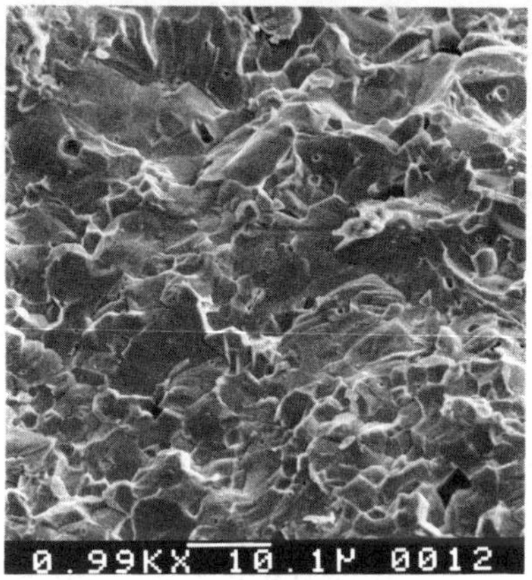

Figure 5. Fracture Surface of Pressureless Sintered TiB_2

A fracture surface of sintered TiB_2 is shown on Figure 5. The microstructure is composed of small grains of approximately 5 microns in size. The grain size has not increased much over the average size of the starting powder, which is an indication that growth has been controlled.

This commercially available pressureless sinterable titanium diboride has some interesting physical properties for applications where a combination of strength, corrosion resistance and hardness are required. Preliminary physical properties (Table 3) of ceramics made from it recorded 305 MPa for four point bend strength. The resistance of this electrically conductive powder was measured at room temperature to be 0.19 micro ohm-meters. The sonic modulus was measured to be 51.7 GPa.

For comparison purposes, data for hot pressed TiB_2 taken from the technical literature of a commercial manufacturer is listed on the same table. The structure properties of hot pressed ceramic properties can now be approached using a pressureless sintering process. It is anticipated that many new applications will develop for titanium diboride components, since this material is no longer limited by the economic or shape constraints of the hot press process.

Many non-oxide powders can be prepared by the same continuous carbothermic process which has been used to make TiB_2. Table 4 shows the properties of boron carbide made by the continuous carbothermic process. The data in the left column represents powder which has been synthesized without milling to a particle size of about 30 microns. The only impurities in this "as furnaced" B_4C included 350 ppm iron and minor amounts of other elements shown in the elemental analysis table. Starting from this 30 micron base powder, it is possible to reduce the particle size without contaminating the powder. The data in the right hand column is representative of high purity milled powder with a particle size of about 5 microns. Hot pressing this powder at 2000°C and 2000 psi results in densities of about 99 percent.

	Pressureless Sintered (HCT-SB)	Hot Pressed *
4 Point Bend MOR	305 MPa	345 MPa
Electrical Resistivity	0.19 μ-ohm-m	
Sonic Modulus	51.7 GPa	>50 GPa

* Source: Data Sheet of Commercial TiB2 Hot Presser

Table 3. Comparison of Physical Properties of Pressureless Sintered HCT-SB TiB_2 with Commercially Hot Pressed TiB_2

PROPERTIES OF B4C

	Coarse - 30 μm	Fine - 5μm
Composition, w/o	TS-5030	TS-5031
B	78.2	78.0
C	21.0	20.9
O	0.52	0.92
N	0.18	0.2
Trace Metals, w/o		
Fe	0.03	0.04
Si	0.02	0.02
Al,Ca,Cr,Mg, Ni,W,Zn,Zr, ppm	< 100	<100
Surface Area, m2/g	0.25	0.9
Hot Pressed Density, % (2200 °C/2000 psi)	87.0	98.8

Table 4. Properties of Boron Carbide Made by the Continuous Carbothermic Process

III. GAS PHASE POWDER PRODUCTION

Gas phase powder synthesis has been utilized for the pilot plant production of ultra high purity silicon nitride. (6) This process utilizes a commercially available high purity silane gas as the raw material to produce silicon nitride powder of extremely high purity and fine particle size. Figure 6 shows an outline of the basic process. Silane is reacted with nitrogen or ammonia in a heated reactor. By adjusting the process controls, it is possible to produce either amorphous or crystalline silicon nitride powder. The very fine powder is collected using special filters.

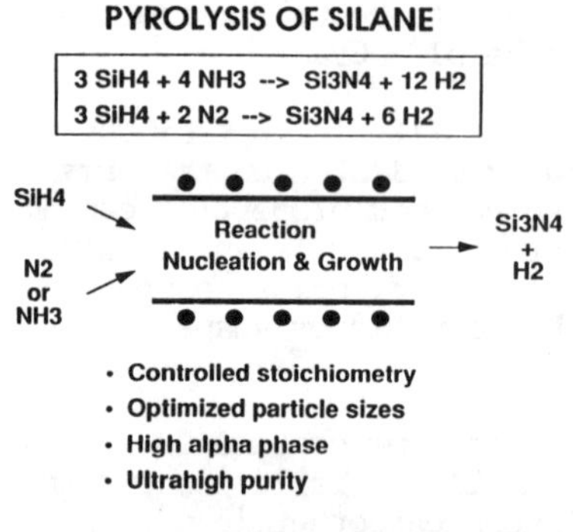

Figure 6. Gas Phase Process for Silicon Nitride Powder

PROPERTIES OF TS-1929 SI3N4

Alpha Phase:	93.1%
Oxygen:	0.79%
Carbon:	0.01%
Surface Area:	4.6 m2/g
Tap Density:	0.33 g/cc
Trace Elements:	
Al	<4
B	<2
Ca	<1
Cr	<4
Fe	<4
Na	<8
Ti	<2
Other metals also undetectable.	

Table 5. Properties of Si_3N_4 Powder Made by Thermal Decomposition of Silane Gas

The particle size shown by SEM micrograph (Figure 7) is approximately 0.2 microns. The powder has a morphology which is predominantly spherical.

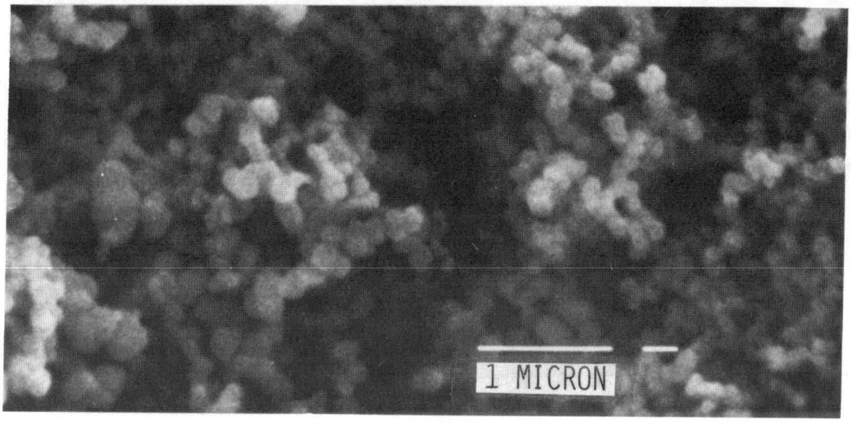

Figure 7. SEM of Si_3N_4 Made by Gas Phase Process

Table 5 shows the properties of silicon nitride made by this process. Trace elemental analysis by ICP emission spectroscopy indicates that the chemical purity of the powder is exceptionally high. Most elements are not detectable even at very low levels. The oxygen content is typically about 0.8 percent. Hot pressing this powder with aluminum oxide and yttrium oxide sintering aids results in high density ceramics with four point bend strengths of 700 to 1000 MPa. Current efforts are directed towards optimizing the powder in order to maximize the physical properties of ceramics made from it.

REFERENCES

1. Weber, W.W., Caglione, A.J., and Frost, A.C., Process for Producing Alumina, U.S. Patent 4,202,870, 1980.

2. Weber, W.W., and Hill, R.F., Apparatus for Preparing Aluminum Alkoxides, U.S. Patent 4,288,410, 1981.

3. Hill, R.F., and Terbot, J.F., Cordierite Composition and Methods of Production, European Patent Application 87.201829.6, 1988.

4. Watanabe, T., and Tokunaga, Y., "Recent Studies of Metal Diborides," Bulletin Japan Institute Met., 1986, 25, (21), 1018-1025.

5. Finch, C.B., Becher, P.F., Angelini, P., Baik, S., Bamberger, C.E., and Brynstad, J., "Effect of Impurities on the Densification of Submicrometer TiB_2 Powders," Advanced Ceramics 1 (1), 50-54, 1988.

6. Shen, W.M., and Sara, R.V., "Characteristics and Processability of Gas Phase Synthesized and Direct Nitridation Si_3N_4 Powders," Ceramic Transactions, Vol. 1, Part B, Ceramic Powder Science, The American Ceramic Society Corp., 1986, pp. 1042-1050.

SYNTHESIS AND PROPERTIES OF NANOPHASE CERAMICS AND COMPOSITES

A.J. BURGGRAAF, G.S.A.M. THEUNISSEN, A.J.A. WINNUBST
University of Twente, Faculty of Chemical Technology, Lab. of Inorganic Chemistry, Materials Science and Catalysis, P.O. Box 217, 7500 AE Enschede, The Netherlands

ABSTRACT

A review is given on recent developments. It is shown that dense ceramics with a mean grain diameter of 50 nm can be synthesized. They exhibit very interesting (low temperature) sintering behaviour, enhanced superplastic deformation and low temperature deformability. Porous composite ceramic membrane structures has been synthesized with mean pore diameters ranging from 2.5-50 nm. By modification of these pore structures the pore diameter can be decreased to molecular sieve dimensions (0.5-2.0 nm) and nanoscale substructures can be produced giving rise to a change in the chemical nature of the internal surface. The membranes show large permeabilities, they have permselective properties and can be used in separation of mixtures and in (catalytic) membrane reactors.

INTRODUCTION AND GENERAL CONSIDERATIONS

Nanocrystalline materials can be thought of as a polycrystalline state of matter in which the size of the individual crystallites is of the order of 1-50 nanometer. In some cases these crystallites form the elementary building units of the microstructure, in other cases they form a substructure of ceramic grains.
Characteristic is the large volume fraction of the atoms located in grain boundaries or interfacial regions. In grains with a size of 5 nm and a thickness of the grain boundary region of 1 nm already 50% of the atoms is located in that grain boundary region. In this region the composition, structure (coordination) and chemical binding is different from that of the bulk. Consequently, quite different properties must be expected for nano-crystalline materials. This will especially be the case when the microstructural scale is comparable to the physical dimensions that characterise the phenomena (domain size, Debye length in surface region, dimensions of plastic zone around a crack tip), when the property is dependent on a large volume fraction of the interface or on the very small radius of curvature.

A number of cases are reported in literature [1-3] and the field is reviewed by Burggraaf et al. [4].
The materials can be made in different ways: i) from solids by spinodal demixing or micro(nano) domain formation, ii) from gases or melts by e.g. evaporation-condensatiaon or laser/plasma melting and splat cooling followed by powder compaction, iii) from sols or gels by gelation of colloidal syspension or (partial) crystallisation of inorganic polymeric gels and iv) by modification technologies like impregnation-precipitation in xerogels or porous ceramics or by ion implantation. These methods are briefly reviewed [4].
Three main groups of nanoscale ceramics can be distinguished: i) porous ceramic bodies (xerogel monolith, ceramic membranes), ii) dense ultra fine grained material and iii) nanophase surface layers (e.g. by ion implantation).

CERAMIC MEMBRANES

Ceramic membranes can be produced in a variety of ways and with different morphologies. A review is presented elsewhere [5]. On the moment the most important group of ceramic membranes is produced by slibcasting of a colloidal suspension onto a porous support. In this process a thin gel layer is formed. During carefull drying, capillary forces causes compaction of the gel layer and a regular packing of particles (size 5-10 nm) is formed. Finally calcining transforms the gel material into a layer of oxidic particles and a stable membrane is formed with a sharp pore size distribution. An essential feature is that the membrane is crack and defect free. A schematic representation of a typical asymmetric three layer membrane configuration is given in Figure 1.

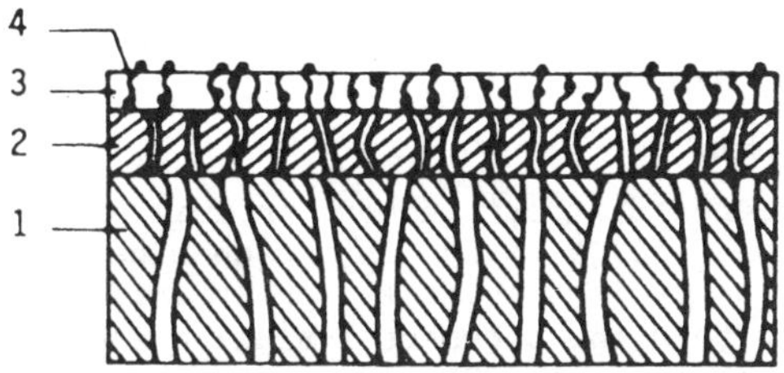

1. Porous support (1-15 μm pores)
2. Intermediate layer(s) (100-1500 nm)
3. Separation (top)layer (3-100 nm)
4. Modification of separation layer

1+2 is microfiltration or 'primary' membrane
1+2+3 is ultrafiltration or 'secondary' membrane
1+2+3+4 is hyperfiltration and/or gas separation membrane

Figure 1. Schematic representation of an asymmetric three layer membrane structure.

The thin (1-10µm) toplayer can be produced with porosities of 30-50% and diameters as small as of 2.5-4 nm from e.g. γ-alumina, titania, zirconia and mixtures of them on (mainly) α-alumina supports [5]. Details of the nanostructure development for γ-alumina membranes are given by Leenaars et al [6].

The pores of a toplayer of an existing membrane system can be modified by deposition of a new phase from liquids or gases. Several morphologic situations are given in Figure 2.

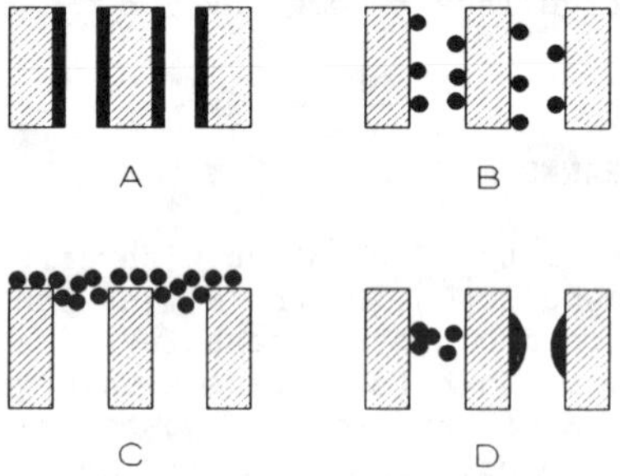

A = mono or multilayer deposit
B = micro(nano) particles within pore, partial plugging
C = plugs in the pore entrance, film on top of the membrane
D = plugs or constrictions on certain sites within the pore

Figure 2. Schematic representation of the microstructure of a modified (catalytic) membrane toplayer.

All these situations are realised in our current program and result in i) a decrease of the pore size or ii) a change in the chemical nature of the internal pore surface. In this way iron, vanadium or Zr oxide (mono)layers have been produced within the pores of γ-alumina and titania membranes, MgO/Mg(OH) or Ag particles (size 5-20nm) are obtained within γ-alumina and titania membranes (pore diameter 4-8 nm) filling up to 50% of the pore volume. Most interesting is the deposition of silica or titania porous plugs or thin films within the pore entrance resulting in an effective pore diameter between 5-15 Å [5,7]. In this way molecular sieving properties are obtained.

Membrane systems as described have (potential) application in separation of mixtures of liquids, vapors or gases and as active components of membrane reactors (catalytic membranes, biotechnology). These are all based on the fact that the transport properties (permeability) of the membrane are different for different substances and dependent on pore size and/or interaction with the internal solid surface (permselectivity) and/or on interaction between transport properties and reactivity. Some typical problems and properties and the potential of membranes are discussed elsewhere by Burggraaf et al [8].

DENSE NANOSCALE CERAMICS

The synthesis of bulk materials with a density larger than 95% of the theoretical density and a grain size below 100 nm is a new field which can yield materials with very interesting properties. Some results has been published by Birringer et al [1] and by Hahn et al [1,9] for TiO_2 ceramics. The material was prepared by evaporation-condensation of Ti metal particles followed by oxidation to TiO_2 particles with a size of 8-12 nm. The powder was compacted with very high pressures (1.5-5 GPa) at room temperature. Plastic deformation of the ceramic material at 180°C in a short time (seconds) was shown but densities were not reported. Grain growth does not occur untill a temperature of about 550°C and becomes rapid at about 800°C. Winnubst and Burggraaf et al [10] synthesised ZrO_2-Y_2O_3 ceramic solid solutions with 6 at % Y (ZY6) with a grain size of 50 nm. The material was prepared with pressureless sintering at a (very) low temperature (1040°C) using a very reactive powder obtained by gel-precipitation techniques. The powder is characterised by crystallite "chains" with a crystallite size of about 8 nm. These chains form small cluster "aggregates" with a size of about 13 nm and very weak agglomerates which can be completely destroyed by isostatic pressing. The green compact obtained in this way has a very regular microstructure consisting of very small particles (8-13 nm) and a sharp pore size distribution around a pore diameter of 6 nm before sintering and 12 nm after sintering at 650°C.
Note that this is a structure similar to that of ceramic membranes. Sintering at 1317 K results in 6 hours in a relative density of 97,5 % and a grain size of 53 nm. The microstructure develops in three steps. The main characteristics are given in Table 1. Complete analysis shows a number of interesting features. Due to the very rapid sintering and recrystallisation of the present aggregates between 550°C-650°C, the starting size of the grains in the grain growth temperature curve during densification between 650 and 850°C is approximately the aggregate size. The main densification mechanism is grain boundary slip assisted by grain boundary diffusion. This points already to the possibility of very rapid superplastic deformations of the final ceramic. Superplastic deformation (120%) of ZY6 with a grain size of 0.5 µm has been reported by Wakai [11] at 1450°C. Grain boundary sliding with enhanced grain boundary diffusion was believed to be the dominant mechanism. This was also the determining mechanism during the main part of the sintering process of our ceramics (Table 1). The deformation (strain) rate $\mathring{\varepsilon}$ of nanoscale ceramic can now be estimated with the help of the relation (1) and the results of Wakai:

$$\frac{1}{\mathring{\varepsilon}} = \frac{1}{\mathring{\varepsilon}_d} + \frac{1}{\mathring{\varepsilon}_i} \qquad (1)$$

where $\mathring{\varepsilon}_i = \text{const}\,\frac{\sigma^2}{d}$ (creep by interface controlled diffusion) and $\mathring{\varepsilon}_d = \text{const.}\,\frac{\sigma}{d^2}$ (creep by lattice diffusion) with σ = applied stress. It is assumed that the two processes are sequential (the first term is dominant). For nanoscale ceramics with d=50 nm at 1040°C and stress levels of about 10MPa, the calculated value amounts $\mathring{\varepsilon} \approx 10^{-2} sec^{-1}$.
This value has to be compared with the value of $\mathring{\varepsilon} = 2 \times 10^{-4} sec^{-1}$ at 1450°C found by Wakai and indicates high deformation rates at relatively low temperature. This is experimentally verified now in our program on hot forging of ceramic (composite) materials. It is anticipated that this can result in near net shaping, defect elimination and improved bonding of toughened materials. A brief review of obtainable effects and possibilities is given elsewhere [12].

TABLE I

Microstructure development and mechanisms during sintering of ultra fine grained Y-TZP with 6 at% Y.

temperature	mechanisms:
- 550-650°C	- aggregate densification - recrystallisation to single crystallite
- 650-850°C	- grain boundary slip (E_a= 100 kJ mol^{-1}) - E_a= 110 kJ mol^{-1} for grain growth
- 990-1200°C	- grain boundary diffusion and slip (E_a=320 kJ mol^{-1}) - E_a = 150 kJ mol^{-1} for grain growth

- large internal sinterpressure: 12 MPa (20 nm grain size).
- grain growth rate exponent n≈1 → non anomal growth!
- grain boundary drag by Y segregation → retarded growth.

REFERENCES

1. Birringer, R., Hahn, H., Höfler, H., Karch, J. and Gleiter, H., Defect and Diffusion Forum, 1988, **59**, 17.
2. Bender, H.A., Ingel, R.P., Mc. Donough, W.J. and Spain, J.R., Advanced Ceramic Materials, 1986, **1**, 137.
3. Komarneni, S., Breval, E. and Roy, R., J. Mater. Sci., 1986, **21**, 737.
4. Burggraaf, A.J., Keizer, K. and van Hassel, B.. Solid State Ionics, 1989, **32/33**, 771.
5. Burggraaf, A.J. and Keizer, K., Synthesis of Inorganic Membranes, in Inorganic Membranes; Synthesis, Characteristics and Applications, ed. R. Bhave, Nostrand Reinhold, New York , 1989, in press.
6. Leenaars, A.F.M., Keizer, K. and Burggraaf, A.J., J. Mater. Sci., 1984, **19**, 1077.
7. a) Huis in 't Veld, M.H.B.J. et al.; this conference proceedings
   b) Praag, W. van, et al.; this conference proceedings.
8. Burggraaf, A.J. and Keizer, K., Synthesis, Microstructure and Properties of new and modified ceramic Membranes, Proc. of 1st Int. Conf. on Inorganic Membranes, 3-7 July 1989, Montpellier, France.
9. Hahn, H., in Ceramic Powder Processing ed. by Messing, G, Fuller, E., Hausner, H., American Ceramic Soc. Inc., 1988, p.1115.
10. Winnubst, A.J.A., Theunissen, G.S.A.M. and Burggraaf, A.J., Compaction and sintering Behaviour of Nanocryst. Y-TZP; this conference proceedings.
11. Wakai, F. and Nagono, T., J. Mat. Sci. Letters , 1988, **7**, 607-609.
12. Burggraaf, A.J., Nanoscale or ultra fine grained Ceramics, Internal Report, University of Twente no. CT89/093/128.

ULTRAFINE Mn,Fe AND Fe,Co SPINEL OXIDES PREPARED FROM MIXED CARBONATES

J.M. JIMENEZ MATEOS, J. MORALES AND J.L. TIRADO
Departamento de Química Inorgánica e Ingeniería Química,
Facultad de Ciencias, Universidad de Córdoba.
14004 Córdoba, Spain.

ABSTRACT

Mn,Fe and Fe,Co spinel oxides were prepared by the thermal decomposition of mixed carbonate precursors at 400 and 300°C, respectively. These oxides are cation-deficient phases and occur as fine particles of *ca.* 10 nm, which are scarcely distorted single crystalline domains with highly homogeneous composition. The specific surface of the samples was rather high (40-120 m^2g^{-1}) as a result of the small size of the particles and the presence of an interparticle porous system.

INTRODUCTION

Utrafine mixed oxides of first-row transition metals with spinel structures can be prepared by two straightforward procedures: direct precipitation reactions (1-3) and low-temperature thermal decomposition of precursor compounds (4-7). Recently, the potential ceramic interest of these highly disperse materials was evidenced by their sintering behaviour to almost theoretical densities (3), resulting from the absence of intergranular porosity in the pelletized material.

Some of the precursor compounds used more recently in the preparation of fine-particle oxides include hydrazinecarboxylate hydrates (4), carbonates (5) and oxalates (6,7). Mn, Fe and Co carbonates crystallize with a calcite structure and can be prepared over a wide compositional range of solid solutions. This paper reports a study on the applicability of carbonate precursors in the preparation of highly disperse Mn,Fe and Fe, Co oxides with spinel structures. The properties of the spinels prepared from carbonates and by other wet and ceramic methods are compared.

MATERIALS AND METHODS

Mixed carbonates were prepared by direct precipitation from divalent metal solutions with 1 M HCO_3^- while continuously passing CO_2. Fe/(Fe+Co) and Fe/(Fe+Mn) ratios were determined by atomic absorption spectroscopy and electron microprobe ED analysis. Metal/Oxygen ratios were evaluated by dissolving the sample in the presence of Fe(II) and back-titrating it with $K_2Cr_2O_7$.

Powder X-ray diffraction patterns were obtained on a Siemens D500 diffractometer using CuK $_\alpha$ radiation and a graphite monochromator. Co_3O_4 was used as standard in the determination of unit cell parameters and broadening analysis. Differential Scanning Calorimetry (DSC) curves were recorded on a Mettler TA3000 instrument. N_2 adsorption isotherms were obtained at 77 K on a standard high-vacuum apparatus. Electron microscopy studies were carried out on a JEOL 200CX microscope.

RESULTS AND DISCUSSION

The mixed Mn/Fe and Fe/Co carbonates were obtained as powdered monophasic solids with a calcite structure as shown by X-ray diffraction. The Fe/M ratios determined by atomic absorption spectrometry are listed in Table 1 and show the possibility of preparing carbonate precursors throughout the interval of cation substitution. Electron microprobe X-ray ED microanalyses showed the distribution of transition metal ions in the samples to be highly homogeneous.

The DSC curves of selected compositions are shown in Fig. 1. The endothermal decomposition reaction of the mixed carbonates took place in several steps, as found in previous work (8). The complex decomposition endotherms were followed by exothermal effects of variable intensity. This effect occurs at lower temperatures for Co-containing samples (Fig. 1 a-c), and is more marked for Mn-containing products. On the other hand, the effect was not shown in the DSC traces recorded under a dynamic Ar atmosphere. The exotherm can be ascribed to oxidation phenomena of the initial product of decomposition of the divalent-metal carbonates.

Cation-deficient spinel oxide phases were obtained after 1 hour of low-temperature isothermal treatment of the various mixed carbonates at 300 and 400°C for the Fe/Co and Fe/Mn systems respectively. These products showed X-ray diffraction patterns ascribable to cubic spinel oxides except for Mn-containing samples with Fe/(Fe+Mn) ratios below 0.33. These samples showed a tetragonal spinel structure similar to that of Mn_3O_4 due to the effect of the Jahn-Teller distortion of the octahedrally coordinated Mn(III) ions.

The isothermal decomposition of the carbonate precursors at 600°C for 1 h led to more complex products. Samples with Fe/(Fe+Co) ratios between 0.36 and 0.47 led to a biphasic product with unit cell parameters ascribable to mixtures of Co_3O_4 and $CoFe_2O_4$ in proportions related to the initial Fe/(Fe+Co) ratios. Similarly, the high-temperature products of the thermal decomposition of Mn-containing carbonates were mixtures of phases structurally related to Mn_2O_3, α-Mn_2O_3 and α-Fe_2O_3, as shown by X-ray diffraction patterns.

The results obtained in the characterization of the monophasic spinel products are shown in Table 1. We should note that the O/M ratios were frequently higher than 1.333 (corresponding to an M_3O_4 stoichiometry), due to the cation-deficient nature of the products obtained after the oxidation processes. This effect was specially evident in Mn-containing samples, where O/M ratios higher than 1.5 indicate the presence of Fe(III), Mn(III) and Mn(IV) in the oxides.

On the other hand, for cobalt ferrites, previous studies (1,9) showed that the unit cell parameter _a_ is highly dependent on the method of preparation and the particle size. According to Burriesci _et al_. (9), ceramic samples have lower _a_ values than wet samples. The results given in

TABLE 1
Chemical composition, structural paremeters and specific surface area of mixed Mn,Fe and Fe,Co spinel oxides.

System	Fe/M	O/M	a (nm)	S_{BET} (m^2g^{-1})	D_{220} (nm)	$\tilde{e}_{220}$ (10^3)
	0.875	1.433	0.8365(7)	54	13.2	0.87
	0.684	1.380	0.8375(9)	40	12.7	0.54
Co,Fe	0.623	1.382	0.8376(4)	56	9.2	2.01
prepared	0.467	1.372	0.8243(7)	118	6.7	11.68
at 300°C	0.396	1.359	0.8203(2)	61	7.8	14.95
	0.354	1.354	0.8200(8)	75	10.8	10.06
	0.153	1.336	0.8132(5)	76	12.2	3.59
	0.856	1.569	0.8364(6)	53	23.5	3.11
	0.650	1.607	0.8376(4)	71	12.6	0.86
Mn,Fe	0.577	1.600	0.8396(5)	60	10.3	4.34
prepared	0.502	1.593	0.8434(3)	64	11.1	6.02
at 400°C	0.360	1.599	0.8442(0)	90	12.3	7.18
	0.330	1.609	0.8448(0)	40	13.6	7.85
	0.134	1.600	tetragonal	37	-	-

S_{BET}: BET surface   D: Crystallite size $\tilde{e}$: microstrain content

Table 1 are closer to those expected for ceramic samples. This could be initially assigned to a lower particle size for wet samples (1). However, the results of the evaluation of N_2 adsorption isotherms (Fig. 2) and electron microscopy do not support this argument. Thus, the values of BET surface shown in Table 1 are somewhat high and hence indicative of high sample dispersion. In addition, Fig. 3a reveals a small particle size of Co-containing spinels. These particles consist of few lattice planes as shown by the reduced number of lattice fringes. The average particle size is ca. 10 nm, which is close to previously reported data for samples prepared by wet procedures (2,3). The particle size of these finely divided materials was also determined by X-ray diffraction line broadening analysis (Table 1). The size of the coherently diffracting domains is consistent with the electron micrographs and shows particles to consist of single domains. In addition, lattice distortions are low as shown by the microstrain contents ($\tilde{e}$), except for those samples undergoing a redistribution of cations at higher temperatures (Fe/(Fe+Co) = 0.36-0.47). The above values of the structural and X-ray line broadening parameters can be considered to result from the homogeneous composition of the parent carbonates, which is preserved after the decomposition reactions.

The values of the unit cell parameter of the Mn-containing phases (Table 1) are smaller than those reported for Mn,Fe mixed spinels prepared by the thermal decomposition of oxalate precursors (6). However, the O/M ratios shown in Table 1 indicate that the level of cation vacancies is higher in ex-carbonate spinels. This may determine the smaller unit cell dimensions of the sample. On the other hand, the values of BET surface shown in Table 1 are larger than those reported in ref. (6) thus indicating a higher dispersion of the samples obtained from carbonate precursors. The specific surface area of Mn-containing spinels is in some cases lower than in Co,Fe spinels. However, the average particle size obtained from the micrographs (Fig. 3b) and the broadening analysis (Table 1) is also indicative of a high dispersion in mixed Mn,Fe spinels.

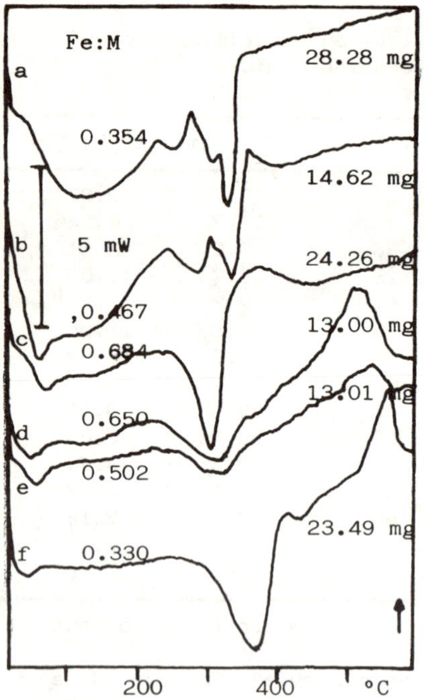

Figure 1. DSC traces of mixed Fe,Co (a–c) and Mn,Fe (d–f) carbonates.

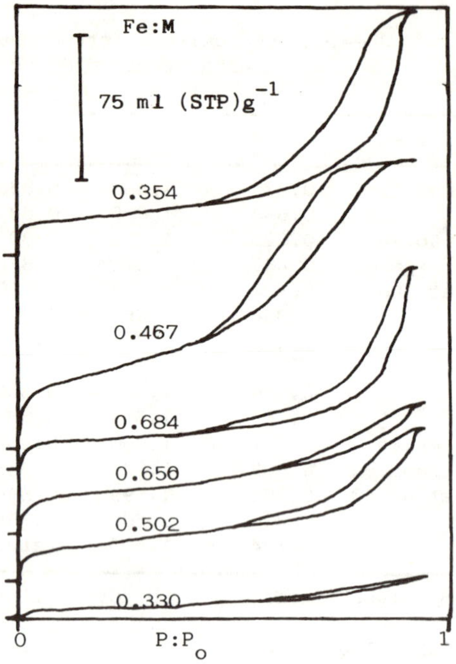

Figure 2. N_2 adsorption isotherms of decomposed Fe,Co and Mn,Fe carbonates.

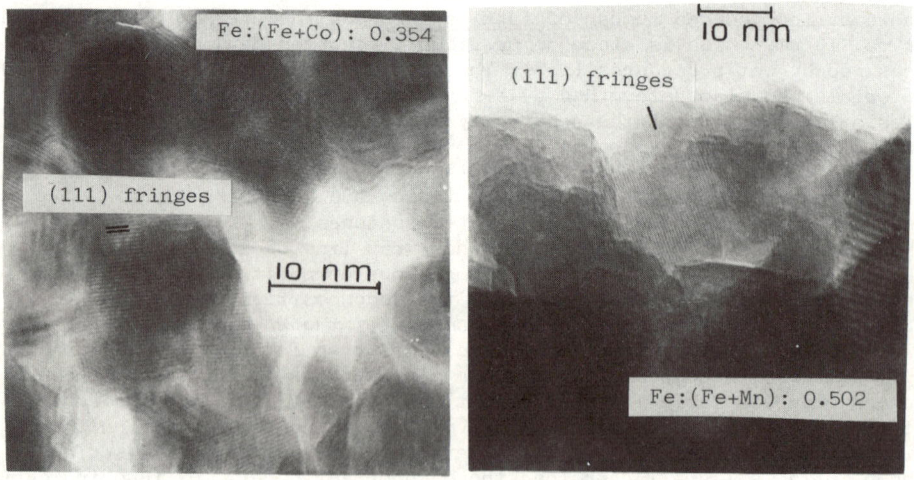

Figure 3. Electron micrographs of spinel oxides:
(a) Fe, Co system (b) Mn, Fe system

A more detailed analysis of the adsorption data (Fig. 2) revealed the occurrence of hysteresis phenomena, especially in Co-containing samples. It can be interpreted in terms of interparticle mesoporosity, on the basis of the electron micrographs. The absence of intraparticle porosity is of interest in the preparation of pelletized aggregates of these materials. However, the porous volume defined by the interparticle space should be minimized to obtain high-density pellets as reported in ref. (3).

ACKNOWLEDGEMENTS

The authors acknowledge CICYT and PFPI for financial support.

REFERENCES

1. Vanderberghe, R.E., Vanleerberghe, R., De Grave, E. and Robbrecht, G., Preparation and magnetic properties of ultra-fine cobalt ferrites. J. Magn. Magn. Mat., 1980, 15-18, 1117-1118.

2. Tirado, J.L., Thomas, J.M., Jefferson, D.A., Millward, G.R. and Charles, S.W., On the ultrastructure and morphology of colloidal cobalt ferrite. J. Chem. Soc., Chem. Commun. 1987, 365-368.

3. Komarneni, S., Fregeau, E., Breval, E. and Roy, R., Hydrothermal preparation of ultrafine ferrites and their sintering. J. Am. Ceram. Soc., 1988, 71(1), C26-C28.

4. Ravindranathan, P., Mahesh, G.V. and Patil, K.C., Low-temperature preparation of fine-particle cobaltites. J. Solid State Chem., 1987, 66, 20-25.

5. Rao, C.N.R. and Gopalakrishnan, J., Synthesis of metal oxides by novel routes. Acc. Chem. Res., 1987, 20, 228-235.

6. Gillot, B., El Guendouzi, M., Rousset, A. and Tailhades, P., Electrical conductivity study of oxidation process in manganese-substituted magnetites. J. Mater. Sci., 1986, 21, 2926-2932.

7. Peshev, P., Toshev, A. and Gyurov, G., Preparation of high-dispersity MCo O (M = Mg, Ni, Zn) spinels by thermal dissociation of coprecipitated oxalates. Mat. Res. Bull. 1989, 24, 33-40.

8. Webb, T.L. and Kruger, J.E., in Differential Thermal Analysis, ed. R.C. Mackenzie, Academic Press, New York, 1982, pp. 326-332.

9. Burriesci, N., Garbassi, F. and Pizzini, S., A comparison between the structural, magnetic and surface properties of cobalt ferrites prepared by wet and ceramic methods. J. Magn. Magn. Mat. 1987, 7, 52-57.

GROWTH OF SiC WHISKERS AND PLATELETS FROM THE GAS PHASE

GABRIELA GÓRNY and ROMAN PAMPUCH
Institute of Materials Science,AGH,Cracow,al.Mickiewicza 30,PL

ABSTRACT

SiC whiskers of different growth axes and polytype composition grown by using CVD from methyltrichlorosilane and $SiCl_4$ + CCl_4 ,respectively, at different temperatures, were studied by X-ray and electron diffraction. A rational explanation of changes of the growth forms with temperature and gas phase composition is given on the basis of an atomistic-statistical nucleation model which permits also to understand the growth of SiC whiskers and platelets in methods involving sublimation-deposition.

INTRODUCTION

Much attention is being paid at present to SiC whiskers, layers and platelets grown from the gas phase, by chemical vapour deposition (CVD) and by various methods utilising sublimation-deposition.For an efficient process control better understanding of the growth mechanism is needed because many details of SiC growth from the gas phase, especially that of whiskers,remain unclear in spite of much earlier work (1) (2).Structural investigations of CVD-grown SiC whiskers shall be discussed in the present paper on the basis of an atomistic-statistical nucleation model which is also applicable to growth of SiC whiskers and platelets from the gas phase in methods other than CVD, e.g. in methods involving sublimation-deposition.

EXPERIMENTAL

Using a modified van Arkel-deBoer method (2) ,SiC whiskers were grown by CVD from highly purified $SiCl_4$ + CCl_4 + H_2 and CH_3SiCl_3 + H_2 gaseous mixtures, respectively, on inductively-heated graphite substrates,covered by a SiC layer in order to exclude additional sources of C. Flow rates from 6E-4 to 1.86E-3 mole/h of $SiCl_4$ and 5E-4 to 1.6E-3 mole/h of CCl_4 were used in proportions ensuring a Si/C molar ratio of 1.2.The CH_3SiCl_3 and H_2 flow rates were 3.19E-2 to 6.72E-2 and 2 mole/h ,respectively.

In each of the two gaseous mixtures the whiskers were grown for 2 hours at 1673, 1723, and 1773 K, respectively. The conditions chosen ensure deposition of SiC as a single solid phase.Basing on ASTM cards and the known d_{hkl} and (hkl) data for SiC polytypes, the overall polytype composition of whiskers from each growth series was determined on the basis of X-ray patterns. It was found that whiskers grown from the silicon and carbon chlorides belong to 2H or 3C SiC polytypes and the ones grown from methyltrichlorosilane to 15R and 6H polytypes only.A presence of Si,C,and SiO_2 was not detected.About 500 whiskers were next investigated by using TEM. Bright and dark field images as well as electronograms from selected areas were registered at least at two whisker orientations with respect to the electron beam. Typical results are exemplified in Figures 1a-1b.The electronograms were evaluated by using a computer-aided analytical methhod.A use was made of the following programs: CRYSTAL for a standard solution of the electronograms, FACTOR in which the structure factors F_{hkl} are used for optimalisation of the solution,and PATTERN which enables to devise theoretical electronograms of defect-free polytypes at different orientations with respect to the electron beam.The last program served for a differentiation of diffraction effects from the (hkl) planes of similar polytypes,e.g.of 3C and 6H SiC.In this way the polytype composition and the whisker growth-axis (using the analysis of traces) were determined. In order to ascertain the position and orientation of microstructural defects an evaluation of diffraction contrasts in dark and bright field images, as described in (3) , was made in conjuction with the above analysis.

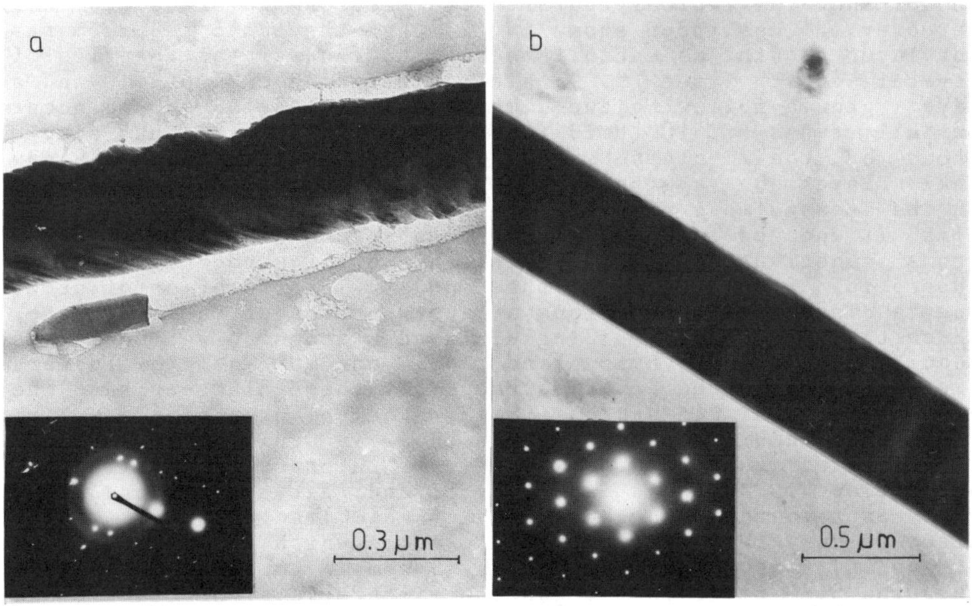

Figure 1.Electron image and diffraction pattern of: (a) 3C SiC whisker grown from $SiCl_4$+ CCl_4 at 1773 K, growth axis [111] ,microtwins in (111); (b) 6H SiC whisker grown from CH_3SiCl_3 at 1773K, growth axis [10$\bar{1}$0] , stacking faults in (10$\bar{1}$0)

TABLE 1
Polytype composition, growth directions and structural defects of SiC whiskers grown by CVD

Gas mixture	Temperature deg.K	Polytype	Growth axis	Structural defects
$SiCl_4$+ CCl_4	1673	2H	0001	Stacking faults in (0001) planes
$SiCl_4$+ CCl_4	1723	2H	0001	
		(2H + 3C)	(111)	
$SiCl_4$+ CCl_4	1773	3C	111	Microtwins in (111)
CH_3SiCl_3	1673	15R	$11\bar{2}0$	Stacking faults normal to growth axis
		(6H)	($10\bar{1}0$)	
CH_3SiCl_3	1723	15R	$10\bar{1}0$	
		(6H)	($10\bar{1}0$)	
CH_3SiCl_3	1773	6H	$10\bar{1}0$	

Whiskers observed in minor amounts are indicated in parentheses

DISCUSSION

The results shall be discussed in terms of an atomistic model of heterogeneous nucleation from the gas phase by assuming layer growth.The vapour-liquid-solid (VLS) mechanism,often considered to be the very mechanism of whisker growth is not taken into account because most SiC whiskers,including rice-hull-derived ones which grow in presence of liquid- forming impurities, do not show the typical features of VLS-grown whiskers (2) (4). Moreover,it has been shown in (2) that the basic requirements for layer growth, as established in the fundamental theories of crystal growth (see (5)), are satisfied in case of SiC and a layer growth following two-dimensional nucleation should occur on all types of SiC (hkl) faces.It can be assumed that single atoms or their clusters may give rise to a nucleus of a new (hkl) layer on a given (hkl) plane of the SiC substrate when on the average more than 1/2 of the valence electrons of the atoms is engaged in bonds with the substrate and with the other atoms constituting the cluster.This applies both to physical adsorption from the gas phase and to incorporation of atoms due to dissociation reactions at active sites of the substrate, which take place in CVD. Single Si and C atoms do not meet this requirement on any (hkl) plane of SiC and more-atomic clusters are to be considered.It is assumed that these clusters have the composition $(Si_2C)m$ and/or (SiC)m, m being an integer equal to or higher than 1. Namely,the gas phase in equilibrium with SiC at elevated temperatures,which is of interest in sublimation-deposition, is composed mainly by Si and,next,by Si_2C and SiC_2, what is favourable to generation of $(Si_2C)m$ clusters.It may be also assumed that single Si and C atoms which are liberated at active surface sites in CVD may assemble to $(Si_2C)m$ clusters when the Si/C ratio in the gaseous mixture is higher than 1 and to (SiC)m ones when this ratio is equal to 1.$(Si_2C)m$ or (SiC)m clusters having the lowest value of m among the clusters in which more than 1/2 of valence electrons is engaged are considered as the critical, nuclei-forming ones and shown in Table 2.

TABLE 2
Values of m of critical (SiC)m and (Si_2C)m clusters, and average number, per atom, of valence electrons not engaged in bonds by the constituent atoms (in parentheses) for given (hkl) planes

Plane:	(0001)	$(10\bar{1}0)$	$(11\bar{2}0)$	(111)	(112)	(110)
(SiC)m	3(1.5)	3(1.33)	3(1.33)	3(1.5)	3(1.66)	3(1.5)
(Si_2C)m	3(1.33)	2(1.66)	x	3(1.33)	2(1.66)	2(1.5)

Note: Si_2C clusters cannot be attached directly to $(11\bar{2}0)$ faces

Only the main low-index (hkl) SiC planes, which constitute the faces of SiC whiskers and platelets in macro-and microscale (2) (4), have been taken into account. On layer growth, two-dimensional nucleation of a new layer should be the rate-determining step and the growth rate in the direction normal to a given (hkl) plane proportional to saturation density of the critical clusters on that plane, which is given, according to (6), by:

$$(N)_i = \left(\frac{N_{hkl}}{\nu^{i-1}}\right)^{1/i} J^{(i-1)/i} \exp\left[\left(\frac{i-1}{i}\right)\left(\frac{E_d+e(4-n)}{kT}\right)\right] \qquad (1)$$

in case of physical adsorption, and in case of CVD by:

$$(N)_i = \left(\frac{N_{hkl}A}{\nu^{i-1}}\right)^{1/i} J_x^{q(i-1)/i} \exp\left[\left(\frac{i-1}{i}\right)\left(\frac{E_d-E_p+e(4-n)}{kT}\right)\right] \qquad (2)$$

A is a constant; N_{hkl} is the surface density of atoms in a given (hkl) plane; q is the order of dissociation reactions at active sites; ν is the vibration frequency of atoms; J and J_x are the flux densities of gaseous species on physical adsorption and CVD, respectively; i is the number of atoms in the critical cluster; E_p and E_d are the activation energies of dissociation at active sites and of surface diffusion per atom, respectively; the last term in the exponent is the dissociation energy of the critical cluster, per atom, which is equal to e(4-n) in case of Si-C clusters and where e is the dissociation energy of a Si-C bond, n is the average number of not-engaged valence electrons. Figures 2a and b show plots of $\ln(N)_i$ vs. 1/T calculated from equations (1) and (2) by using the known N_{hkl} values and the following typical data: J_x = 1.E-8; J = 1E-18 mole/cm^2s; q = 2; E_p = 1.64E-19 J; E_d = 2.8E-19 J; e = 4.7E-19 J; ν = 1E13 s^{-1}; A = 1. The plot in Fig.2a, calculated from equation (2) for critical (SiC)m clusters, applies to CVD from methyltrichlorosilane. It indicates that a preferential growth in $[11\bar{2}0]$ direction below about 1673 K and in $[10\bar{1}0]$ direction above this temperature should take place. Moreover, the higher values of $(N)_i$ on $(10\bar{1}0)$ or $(11\bar{2}0)$ planes, as compared with the ones for (0001) planes, indicate the tendency for unidirectional growth, i.e. growth of whiskers. These results are in a good agreement with experimental data. In the plots in Fig.2b, calculated for critical (Si_2C)m clusters, solid lines indicate results obtained by using Eqn.(2) which apply to CVD from $SiCl_4$+ CCl_4 mixtures of a Si/C ratio of 1.2, while dashed lines indicate results obtained by using Eqn.1 which apply to SiC growth by sublimation-deposition. They indicate a tendency for unidirectional growth in [0001] direction over a broad range of temperatures.[0001] growth axis is indeed observed on whisker growth ,not only in CVD from non-stoichiometric chloride mixtures but also in the Lely method involving sublimation-deposition and in growth from rice hulls (4).It is also well known that in the first two methods SiC platelets may be grown at temperatures above 2000 K.Growth of platelets at these temperatures is also suggested by Fig.2b,which indicates similar values of $(N)_i$ and, hence,similar growth rates normal to (0001) and (10$\bar{1}$0) planes at these high temperatures.Besides the growth forms also the polytype composition of SiC whiskers may be rationally explained by the model presented here.The model assumes a predominance of $(Si_2C)m$ clusters in CVD from non-stoichiometric chloride mixtures and under sublimation-deposition conditions.Under these conditions only whiskers having the metastable 2H structure were found to grow below 1700 K and the ones having the 3C structure above this temperature. 2H SiC is believed to have indeed a small C-deficit (7) and known to transform irreversibly to 3C SiC above 1700 K (8). The model also rationally explains the growth of whiskers of the rather stoichiometric 15R and 6H SiC polytypes from CH_3SiCl_3 , where a predominance of (SiC)m clusters has been assumed.

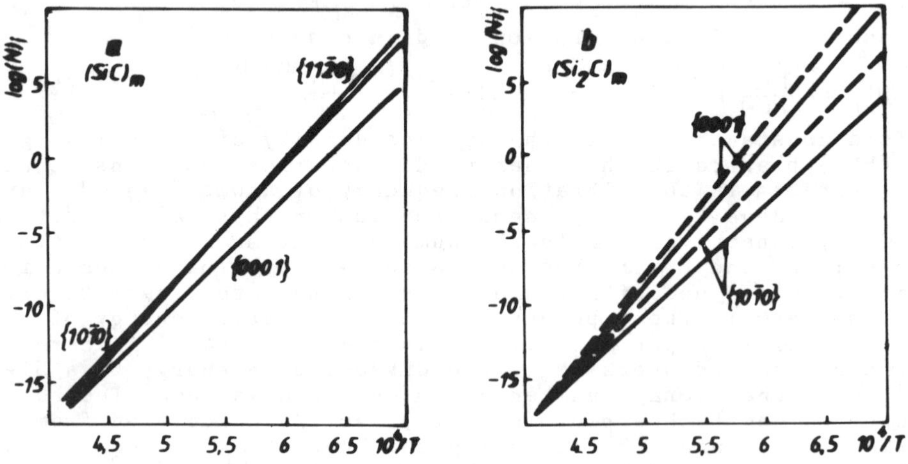

Figure 2. Plot of $(N)_i$ vs. 1/T (a) for $(SiC)_m$ and (b) for $(Si_2C)_m$ critical clusters

REFERENCES

1. Knippenberg, W.F. and Verspui, G. In Proceed.3rd Int.Conf. on Silicon Carbide, Miami, September 1973, pp. 109-122
2. Pampuch,R. and Stobierski,L.,Ceram.Int. 1977, 3, 43-52.
3. Hirsch,P.B.,Howie,A.,Nicholson,R.,Pashley,D.V.and Wheelan M. Electron Microscopy of Thin Crystals,Butterworth,London,1961
4. Nutt,S.R., J.Am.Ceram.Soc.,1988, 71, 149-56.
5. Bennema,I., J.Cryst.Growth, 1974 ,24/25,76-81.
6. Palatnik,L.S., Epitaksyalnye plenki,Nauka,Moskva, 1971.
7. Berman,I. and Ryan,C.E., J.Cryst. Growth,1971, 9, 314-316.
8. Krishna,P.and Marshall,R.C, J.Cryst.Growth, 1971,8 ,129-131.

PREPARATION OF FINE PARTICLES OF $SrTiO_3$ BY ALKOXIDE METHOD AND THEIR SINTERABILITY

JUNICHI HOJO AND HIDEYOSHI KIMURA
Department of Applied Chemistry, Faculty of Engineering, Kyushu University,
6-10-1 Hakozaki, Higashi-Ku, Fukuoka-Shi, 812 Japan

ABSTRACT

Fine crystalline particles of $SrTiO_3$ having the size of 0.01μm were produced by hydrolysis of Sr-Ti mixed alkoxide at 80°C by using a large excess of H_2O. However, the powder included large agglomerates and needed a high sintering temperature of 1300°C to give a dense sintered body. The freeze-drying and the treatment with NH_3 aq. solution were effective to disperse $SrTiO_3$ particles. The dispersed powder showed a high sinterability and was densified to above 95% of theoretical at 1200°C. The addition of V_2O_5 to $SrTiO_3$ powder also lowered the sintering temperature.

INTRODUCTION

A metal alkoxide decomposition method has been receiving a great attention to produce high-purity submicron oxide powder which makes possible very low temperature sintering [1]. $SrTiO_3$-based ceramics are potential dielectrics with high dielectric constant and low dissipation factor. The particular application is a grain boundary layer capacitor but a high sintering temperature around 1400°C is required to obtain a well-sintered microstructure from conventional $SrTiO_3$ powder produced by a solid state reaction method [2]. Fine $SrTiO_3$ powder has been prepared by hydrolysis of Sr-Ti mixed alkoxide [3,4]. However, the sintering behavior has been less studied. In the present work, the sinterability of alkoxy-derived $SrTiO_3$ powder was investigated with emphasis on the effects of powder properties and sintering aid.

MATERIALS AND METHODS

Metallic strontium flakes were mixed with an equal mole of titanium isopropoxide in isopropanol and the solution was refluxed for 2hr at 83°C in nitrogen atmosphere to form a mixed alkoxide of $Sr(i\text{-}OPr)_2$ and $Ti(i\text{-}OPr)_4$. Since a part of mixed alkoxide was precipitated in isopropanol, p-xylene was added to the solvent to increase the solubility. The concentration of mixed alkoxide, $SrTi(i\text{-}OPr)_6$, was 0.1mol/ℓ. Water diluted in isopropanol was dropped into the solution at 30°C or 80°C to hydrolyze the alkoxide. Since three moles of H_2O are required for one mole of $SrTi(i\text{-}OPr)_6$, the molar ratio, $[H_2O]/[SrTi(i\text{-}OPr)_6]$, was varied in the range of 3 to 30. Powder produced was washed with ethanol and dried at 70°C in vacuum. In some experiments, powder was treated with NH_3(15%) aq. solution at 70°C before drying, or freeze-dried after dispersed in water. The calcination was done for 0.5hr at 800°C in air. V_2O_5 was added to $SrTiO_3$ powder as sintering aid by the precipitation of NH_4VO_3 in $SrTiO_3$-isopropanol suspension and the subsequent decomposition at 600°C. Powder sample was analyzed with TEM, XRD, TG-DTA and IR spectra.

Powder was pressed into pellets of 10mm in diameter and about 2.5mm thick. The green density was about 50%. The pellets were heated for 1hr or 4hr at 1000°~1400°C in air. The bulk density of sintered body was determined from the size and weight or by an Archimedes method. The fracture surface was observed with SEM.

RESULTS AND DISCUSSION

Powder Properties

X-ray diffraction patterns of as-produced powders and calcined one are shown in Fig. 1. When the hydrolysis temperature is 30°C, powders produced are totally amorphous. On the other hand, when the hydrolysis temperature is 80°C, powder produced is amorphous at $[H_2O]/[SrTi(i\text{-}OPr)_6]=3\sim6$ but crystalline (cubic phase) at $[H_2O]/[SrTi(i\text{-}OPr)_6]=30$. The X-ray diffraction peaks are a little broad compared to that of calcined powder. The crystalline size calculated from the half-width of peak was about 0.01μm in agreement with the size of primary particles observed with TEM.

TG and DTA curves of amorphous $SrTiO_3$ powder are shown in Fig. 2. The large weight loss from room temperature to 600°C seems mainly due to the evaporation of water from hydrous amorphous powder. The exothermic reaction at 260°C may be assigned to the oxidation of organic residue. The X-ray diffraction and IR spectra suggested that the crystallization of

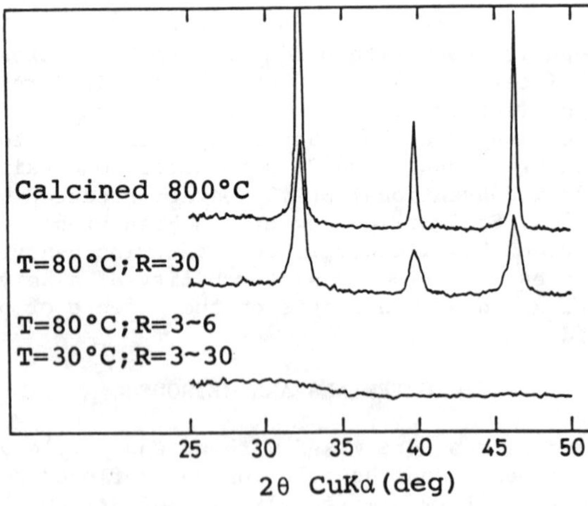

Figure 1. X-ray diffraction patterns of $SrTiO_3$ powders produced. T=hydrolysis temp., $R=[H_2O]/[SrTi(i\text{-}OPr)_6]$.

amorphous $SrTiO_3$ occurred around 400°C and $SrTiO_3$ was also formed by the reaction of $SrCO_3$ with TiO_2 around 600°C. The inclusion of $SrCO_3$ and TiO_2 may be caused by the reaction of amorphous $SrTiO_3$ with CO_2 in air.

Electron micrographs of as-produced and freeze-dried crystalline $SrTiO_3$ particles are shown in Fig. 3. As-produced powder consists of very fine particles having the size of about 0.01μm but forms large agglomerates. The calcination at 800°C increased the size of primary particles to about 0.05μm and the calcined powder also included large agglomerates. On the other hand, freeze-dried powder consists of well-dispersed particles. The treatment with NH_3 aq. solution was also effective to disperse $SrTiO_3$ particles. It has been reported that alkoxy-derived TiO_2 particles were dispersed in NH_3 aq. solution [5].

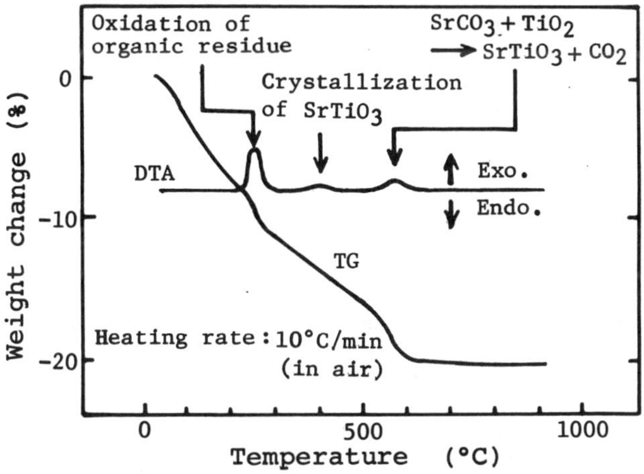

Figure 2. TG and DTA curves of amorphous $SrTiO_3$ powder.

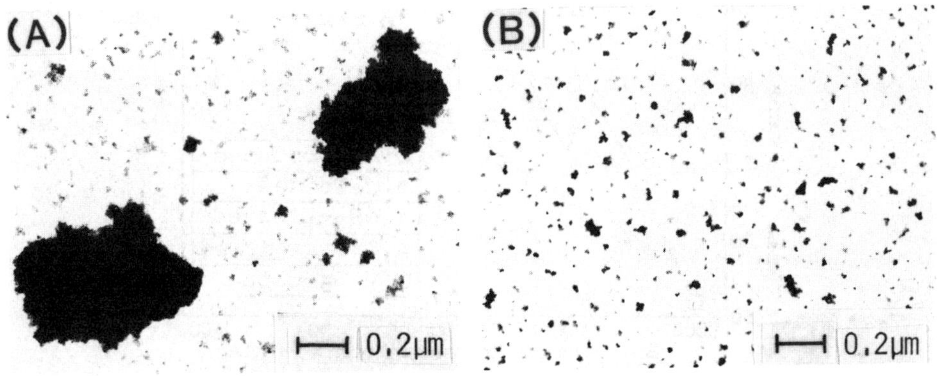

Figure 3. TEM photographs of crystalline $SrTiO_3$ particles.
(A) as-produced powder (T=80°C, R=30), (B) freeze-dried powder.

Sinterability of $SrTiO_3$ Particles

The bulk density of sintered body of crystalline $SrTiO_3$ powder is shown as a function of sintering temperature in Fig. 4. As-produced powder shows a gradual increase in bulk density with raising temperature and the relative density reaches 97% at 1300°C. The sinterability of calcined powder is similar to that of as-produced powder. On the other hand, freeze-dried and NH_3-treated powders show a higher sinterability at 1100°~1200°C and the relative density reaches 98% at 1200°C in freeze-dried powder.

The microstructures of sintered bodies obtained from as-produced and freeze-dried powders at 1300°C are shown in Fig. 5. The sintered body of as-produced powder has a wide distribution of grain size from 0.5 to 1.6μm, whereas that of freeze-dried powder has a uniform structure consisting of about 0.8μm-sized grains. NH_3-treated powder also gave a fine-grained structure at 1200°C, although the grain growth became significant at 1300 °C. The heterogeneous grain growth in as-produced powder may be caused by the agglomeration of particles, as seen in Fig. 3, and large agglomerates may reduce the sinterability.

Amorphous $SrTiO_3$ powder showed the sinterability similar to that of as-produced crystalline powder but many cracks were generated in the sintered body.

Effect of V_2O_5 as Sintering Aid

It has been reported that V_2O_5 aided the sintering of $SrTiO_3$ particles [6]. The bulk density of sintered body of V_2O_5-doped $SrTiO_3$ powder is shown in Fig. 4. The densification becomes remarkable at 1100°C, where the relative density reaches 80% by 1hr-sintering and 91% by 4hr-sintering. Since the melting point of V_2O_5 is 690°C, the sintering may be accelerated by a liquid phase mechanism. V_2O_5 showed the large effect at a lower temperature than in the literature [6], because the present powder was very fine. However, the effect of V_2O_5 is reduced above 1100°C. It is possible that the liquid phase disappeared by the dissolution of V_2O_5 into $SrTiO_3$ during the sintering at high temperatures.

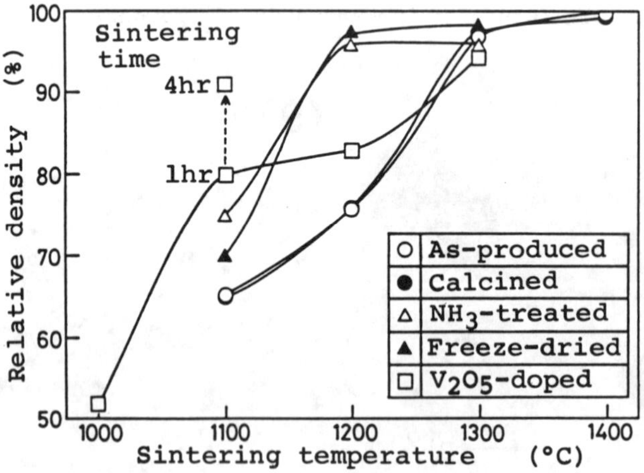

Figure 4. Variation of bulk density of sintered body with sintering temperature in crystalline $SrTiO_3$ powders.
Sintering time; 1hr, V_2O_5 added: 2wt%.

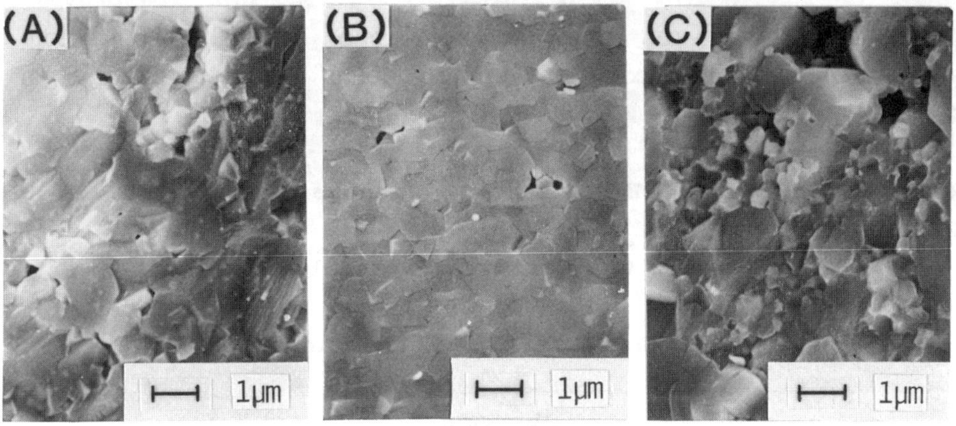

Figure 5. SEM photographs of fracture surface of sintered body obtained at 1300°C (see Fig. 4).
Powder sample: (A) as-produced, (B) freeze-dried, (C) V_2O_5-doped.

The microstructure of sintered body obtained at 1300°C is shown in Fig. 5. The structure is not uniform, having a wide distribution of grain size. In the V_2O_5-doped sample, the local densification was observed in sintered body at 1100°C and 1200°C. This may be caused by the heterogeneous distribution of V_2O_5 in the mixing with $SrTiO_3$ particles.

CONCLUSION

Fine crystalline particles of $SrTiO_3$ having the size of 0.01μm are produced by hydrolysis of Sr-Ti mixed alkoxide at a high temperature by using a large excess of H_2O. The fine powder is very sinterable at a low temperature but the dispersion of particles is important to enhance the sintering. V_2O_5 is also effective to lower the sintering temperature. The sintering may be accelerated by a liquid phase mechanism.

REFERENCES

1. Mazdiyasni, K.S., Fine Particle Perovskite Processing. Am. Ceram. Soc. Bull., 1984, 63, 591-94.
2. Fujimoto, M. and Kingery, W.D., Microstructures of $SrTiO_3$ Internal Boundary Layer Capacitors During and After Processing and Resultant Electrical Properties. J. Am. Ceram. Soc., 1985, **68**, 169-73.
3. Smith II, J.S., Dolloff, R.T. and Mazdiyasni, K.S., Preparation and Characterization of Alkoxy-Derived $SrZrO_3$ and $SrTiO_3$. J. Am. Ceram. Soc., 1970, 53, 91-95.
4. Kasai, T., Ozaki, Y. and Yamamoto, S., Preparation of $BaTiO_3$ and $SrTiO_3$ from Metal Alkoxides. Yogyo-Kyokai-Shi, 1987, 95, 1000-06.
5. Ikemoto, T., Uematsu, K., Mizutani, N. and Kato, M., Synthesis of Monodispersed Titania Fine Particles by Hydrolysis of $Ti(OC_2H_5)_4$. Yogyo-Kyokai-Shi, 1985, **93**, 261-66.
6. Chen, J-S., Young, R-J. and Wu, T-B., Densification and Microstructural Development of $SrTiO_3$ Sintered with V_2O_5. J. Am. Ceram. Soc., 1987, **70**, C260-64.

LASER-SYNTHESIS OF LOW AGGLOMERATED AND NARROW SIZED SUBMICROMETER SILICON NITRIDE POWDERS FROM CHLORINATED SILANES.

Robert A. Bauer, F.Einar Kruis, Johanna G.M. Becht, Joop Schoonman

Delft University of Technology
P.O. Box 5045
2600 GA Delft, The Netherlands.

Abstract

The Laser Chemical Vapor Precipitation (L-CVP) of Si_3N_4 powder from mixtures of SiH_2Cl_2 and NH_3 has been studied. The reactant gases were mixed at different heights with respect to the laser beam, thus preventing low temperature reactions and agglomeration of the produced particles. Particle diameters are typically between 15 and 110 nm depending on the injection height.

Introduction

The quality of sintered structural ceramics such as Si_3N_4 is improved by the application of sintering aids and high pressure at the expense of decreased high-temperature strength and high cost of the finished parts. A constant high quality of the products could be obtained if green bodies were formed using ultrafine, non-agglomerated, monodisperse powders. Our research program is based on producing such powders using laser chemical vapor precipitation (LCVP).[1,2]

In the laser-CVP of Si, SiC and Si_3N_4, SiH_4 is invariably used as the silicon reactant[3,4]. We have developed a new reactor to be able to use the relatively cheap chlorinated silanes in the synthesis of silicon nitride with LCVP.[1,2] It is possible to produce ultrafine, narrow-sized powder with LCVP. However the high number density and extremely small diameters of the particles in the hot reaction zone makes it difficult to obtain a sufficiently low agglomeration degree of the powder. Work of Aoki e.a.[5]

injection of the reactants. This result is obtained however at the cost of increased particle size and particle size distribution. The initial reaction step was the laser-excited formation of silicon from SiH_4. At high reaction temperature, higher than 1683 K, the silicon particles are liquid. The liquid particles collide and larger but still spherical particles will be formed. Slightly above the first reaction zone NH_3 is injected and the liquid silicon particles are nitrided. Then the particles should stop growing. Turbulence may cause circulation of the particles and keep the particles growing. A wide particle size distribution will be obtained in the powders.

Here we will show that by gas flow control it is possible to form low agglomerated, ultrafine silicon nitride powders and still maintain a narrow particle size distribution using dichlorosilane as a silicon source.

Experimental Procedure

A schematic drawing of the stainless steel reactor is presented in Figure 1. The used reactor has been described

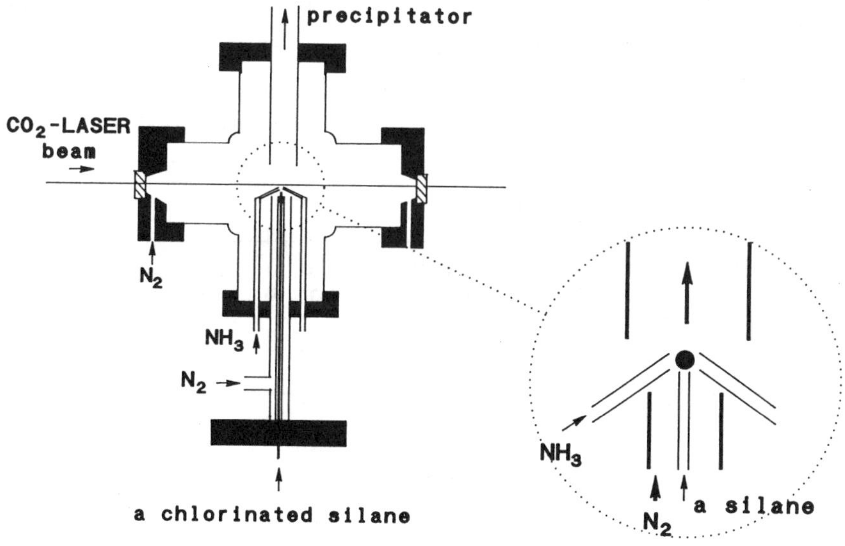

Fig.1 Laser reactor

previously.[1] The gas inlet nozzle had four separated concentric outlets for NH_3 pointed towards the middle, and a centered outlet for the chlorinated silanes. The height of the injection of NH_3 relative to the laser beam could be adjusted. The flow of SiH_2Cl_2 was 33 sccm and the NH_3 flow was 115 sccm.

The produced powders were collected in a electrostatic precipitator. Afterwards the powders were heated at 770 K in vacuum during 15 h to remove NH_4Cl.

The powders were studied with IR, EDX, XRD and TEM.

Results and discussion

When NH_3 was injected into the SiH_2Cl_2 stream below the laser beam, heavily agglomerated, amorphous particles are formed as shown in figure 2. IR-spectroscopy showed that the particles consist of Si_3N_4 with a high level of imide impurities. The presence of chlorine has been determined with EDX. When NH_3 was injected in the laser beam, the temperature was raised there, because of the exothermic nitride formation. The synthesized particles were somewhat larger and less agglomerated than in the previous experiment. With TEM several of the particles appeared to contain crystalline silicon cores. IR-spectroscopy showed the presence of Si-N bonds (950 cm^{-1}) and a small amount of N-H bonds. Therefore the powders consisted of mainly silicon nitride with a low contamination of imides. With EDX some chlorine contamination was detected.

TABLE I Physical characteristics of Si_3N_4 as function of the NH_3-injection height

h*	part. d (nm)	agglom. d (μm)	core	outside	whiskers	bulges
-1.5	5-40	0.05-2	SiN	SiN	+	
-0.5	20-30	0.3-3	SiN	SiN	++	
+0.3	10-30	0.1-2	SiN/Si	SiN	+++	
+0.5	±50	0.2-1.4	Si	SiN		
+1.2	50-80	0.1-0.7	Si	SiN		
+1.5	50-100	0.1-3	Si	SiN		+++
+2.9	90-110	0.8-1.2	Si	SiN/Si		++

* Injection height of NH_3 in the central SiH_2Cl_2 flow

Besides spherical particles a large number of whiskers was observed with TEM. The whiskers were conically shaped with a spherical top. Similar whiskers were reported by others[6] and attributed to growth by a VLS-mechanism.

When ammonia was injected above the laser beam the highest temperature region was the injection area of NH_3 in the upper part of the reaction flame. Once even a second flame was observed above the ammonia injection. This increased temperature could be related to the exothermic nitriding of silicon. In these experiments larger spherical and egg-shaped particles were formed as shown in figure 3. The degree of agglomeration as observed with TEM was relatively low. Most of the particles had large crystalline silicon cores and an amorphous exterior. In the experiment with the highest NH_3 injection position this outside layer was very small. On the outside sometimes bulges were observed with TEM. They were probably related to a strong nitrification process.

With EDX no chlorine could be detected. The relative number of Si-N bonds compared to that of Si-H was very high. This indicates that the outside was amorphous silicon nitride. In the experiments with the high injection position little nitrification has occurred. In the same experiment extreme small particles with diameter of approximately 5 nm were observed. Chemical analyses of this particles were not possible.

These result support a reaction mechanism, where Si_3N_4 is formed by the nitridation of Si particles.

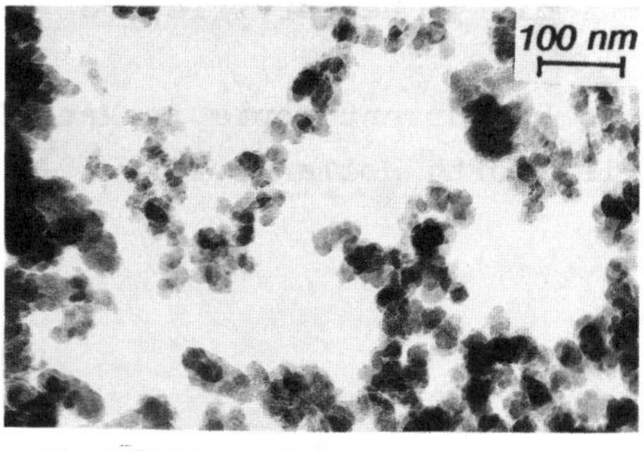

Fig.2 Si_3N_4 particles

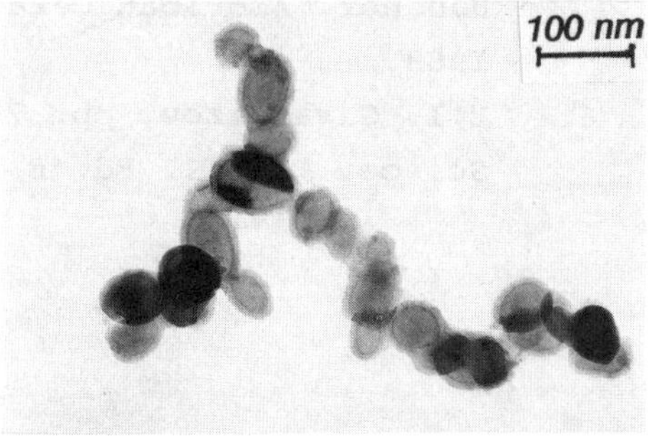

Fig.3 Si_3N_4 particles with Si core

Conclusion

The particle formation mechanism of silicon nitride with Laser-CVP can be changed considerably by altering the height of the NH_3 injection position. By injecting NH_3 a little above the laser beam loosely agglomerated nitrided silicon particles with a diameter around 100nm can be formed. Experiments to optimize the nitrification process are currently in progress.

Acknowledgment:

This research is financed by the Netherlands Ministry of Economic Affairs in the framework of the Innovation Directed Research Programs (IOP).

References

1. R.A. Bauer, R. Smulders, J.G.M. Becht, P.J. van der Put and J. Schoonman, to be published in J. Am. Cer. Soc..
2. R.A. Bauer, R. Smulders, E. Geus, P.J. van der Put, J.G.M. Becht and J. Schoonman, Ceram. Eng. Sci. Proc., 9, [7-8], 949-956 (1988).
3. W.R. Cannon, S.C. Danforth, J.H. Flint, J.S. Haggerty and R.A. Marra, J. Am. Ceram. Soc., 65 [7], 324-30 (1982).
4. W.R. Cannon, S.C. Danforth, J.S. Haggerty and R.A. Marra, J. Am. Ceram. Soc., 65 [7], 330-35 (1982).
5. M. Aoki, T.N. Kogyo, J.H. Flint and J.S. Haggerty, pp. 250-260 in Ceramic Transactions, Vol. 1A, Ceramic Science II. Ed. G.L. Messing, E.R. Fuller and H. Hausner. American Ceramic Society, Westerville, OH, USA, 1988.
6. E.I. Givargizov, pp. 79-145 in Current Topic in Material Sciece, Vol. 1. Ed. E.Kalis, North Holland, 1978.

PYROLYTIC PREPARATION OF CERAMIC POWDERS

D. VOLLATH
Kernforschungszentrum Karlsruhe GmbH,
Institut für Material- und Festkörperforschung III,
P.O.Box 3640, D-7500 Karlsruhe, Germany

ABSTRACT

Methods for pyrolytic fabrication of ceramic powders are described, based on aqueous solutions of nitrates, chlorides or other water soluble compounds. Using a two phase nozzle, the solutions are sprayed into a stream of hot carrier gas. After evaporation of the water, pyrolysis is obtained by heating the dried particles with light. In a second process the energy is introduced by a hydrogen flame, where hydrogen is propellant gas for the nozzle and heating gas at the same time.

Using a proper selection of process parameters hollow spheres can be largely avoided. In a few systems it is possible to obtain powders with interesting properties within an amorphous region. The production rate is about 100 - 150 g ceramic powder per hour.

1. INTRODUCTION

Starting from the state of the arts a facility for spray calcination of ceramic powders was developed. The following targets were reached:

- Very homogeneous product.
- Very fast process in order to avoid separation of multiphase systems.
- Highly sinterable powders with small particle sizes.
- Production rate of about 100 - 150 g powder per hour.
- Suitable for pyrolysis of nitrates, chlorides or other source compounds into the desired oxides.

This process avoids the problems of the standard processes which are:

Either mechanically mixing and subsequent grinding of the source powders or coprecipitating of the hydroxides from salt solutions. In the first case, several hours of grinding are required and in the second case the various constituents of a system are frequently precipitated, at greatly differing pH-values, leading to problems with the powder homogeneity. The problems are similar in preparation techniques based on the hydrolysis of alcoxides. Fully homogeneous products are obtained if - as possible with some of the ZrO_2-Al_2O_3 ceramics - organic compounds which contain already the metallic constituents at the desired ratio are used as the source material /1/.

2. DEVICE FOR PYROLYTIC PREPARATION OF POWDERS

2.1 SPRAYING OF THE SOLUTIONS OR SUSPENSIONS

To overcome the problems described above methods for pyrolytic preparation of powders were developed, e.g., solutions consisting of nitrates or acetates are sprayed into a furnace and decomposed thermally to oxides. These processes have been described already for fabrication of electroceramics /2/. Using this technique spherical powder particles with diameters in the range of 2 - 50 µm are obtained. In this case about 0.2 molar solutions are sprayed in a furnace kept at temperatures between 500 and 600 °C. Powders with particle sizes from 10 to 20 nm are obtained if nitrate solutions are sprayed and the fine fraction of the sprayed mist is heated in an ICP (inductively coupled plasma) flame. In this way, small amounts of powders have been fabricated /3, 4, 5/.

The device for pyrolytic powder fabrication developed, is characterized by the following features:

- The aqueous solution or a suspension is finely sprayed through a two phase nozzle. The mean size of droplets furnished by the nozzle is between 8 and 9 µm with a spraying capacity of 1.5 l/h. The maximum spraying capacity of the nozzle is about 6 l/h.

- A ceramic frit surrounds the spray nozzle concentrically and is passed by a laminar, gas stream heated up to 800 °C. The gas stream serves for two purposes: it ensures the transport of the droplets generated in the spray nozzle, and it is to heat and evaporate them.

2.2 DRYING AND CALCINATION

Two methods for drying of the mist emering from the spray nozzle we developed:

- Introducing the energy by light /6/.
- Introducing the energy by a H_2-flame /7/.

Each of these two processes has its own problems:

In the case of calcination by light water of the droplets is evaporated by contact with the hot carrier gas stream and the molten nitrate or chloride residues are heated by radiation heating until they decompose Halogen filament lamps are used as sources of radiation heating. A high efficiency of radiation heating requires that no powder deposits in that part of the reaction vessel where the lamp is located. This can be achieved by selection of a suitable flow velocity of the carrier gas. The powders formed in the process described are so fine that as result of thermophoresis they would preferably deposit at the colder parts of the walls of the reaction chambers. This is avoided by thermostating.

For flame calcination hydrogen is used as spray gas and as propellant for the two phase nozzle. In this case the main problem is to ignite and to avoid disruptions of the flame. This problem is solved using a high voltage spark direcly at the nozzle.

3. PRACTICAL APPLICATION

Among the oxide ceramic materials those based on zirconium oxide offer a particularly high potential of application, especially in the partially stabilized tetragonal modification. The following compositions were produced.

97 mole % ZrO_2 + 3 mole% Y_2O_3,

85 mole% ZrO_2 + 15 mole% MgO,

78 mole% ZrO_2 + 2 mole% Y_2O_3 + 20 mole% Al_2O_3.

The source products used for zirconia were zirconium nitrate and zirconyl chloride, respectively, and for all the other oxides the nitrates or chlorides. In contrast to the nitrates, the water containing chlorides do not melt during heating, they rather hydrolyse liberating hydrochloric acid. Based on this difference, the powders obtained from chloride solutions tends to form hollow spheres. This tendency can be suppressed by details of the experimental setup. This difference is important in the case of calcination by light. In the case of flame calcination hollow particles are rarely observed, they tend to desintegrate. A mixture of chlorides and nitrates has a detrimental effect as observed in all cases. As far as one has a complete solution and no effects of gelation, the concentration of the solution to be sprayed is not significant. In the case of zirconia based ceramics, it is possible to produce up to 150g powder per liter solution.

Figs. 1 and 2 exhibit micrographs of partially stabilized zirconia powder with 20% Al_2O_3 prepared by both methods.

All powder containing no alumina showed crystallisation. Due to the broad diffraction lines it is not possible to decide if the powder is crystallized in the cubic or the tetragonal modification.

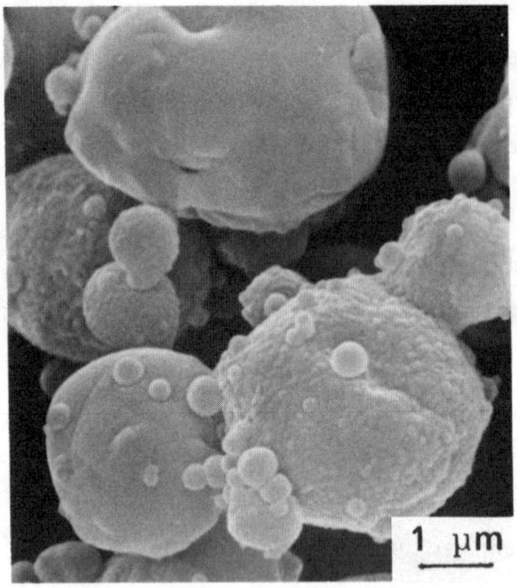

Figure. 1: Micrograph a powder obtained by light calcination ($78\ ^{m}/_{0}\ ZrO_2$, $2\ ^{m}/_{0}\ Y_2O_3$, $20\ ^{m}/_{0}\ Al_2O_3$)

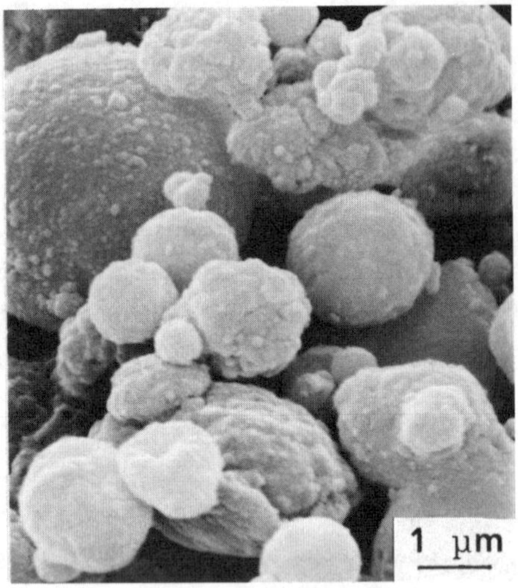

Figure 2: Appearance of a powder obtained of flame calcination ($78\ ^{m}/_{0}\ ZrO_2$, $2\ ^{m}/_{0}\ Y_2O_3$, $20\ ^{m}/_{0}\ Al_2O_3$)

From DTA measurements one can assume, that the powder is according to Ostwald´s rule crystallized in the cubic phase. Crystallisation of the alumina starts in any case during sintering.

Besides the powders for zirconia based ceramics, powders for the aluminium-magnesium-spinell and for the silicates and aluminates of lithium were produced. In the latter case the powder was produced from suspensions prepared by an organic route /7, 8, 9/.

4. CONCLUSIONS

A facility for preparing ceramic powders with a production capacity of about 100-150 g/h has been presented. By application of the technique described also those systems are allowed to be prepared as very homogeneous powders, which can be prepared only with great difficulties by conventional techniques.

By experimental results it has been demonstrated that also suspensions can be profitably processed.

Dependent on the systems used, the product fabricated is either poorly crystallized or amorphous. Solutions prepared, either of nitrates or chlorides can be used. Mixtures of nitrates with chlorides has been proved to be disadvantageous.

LITERATURE:

/1/ H. Yoshimatsu, H. Kawasaka, A. Osaka
STABILITY OF TETRAGONAL ZrO_2 PHASE IN Al_2O_3 PREPARED FROM Zr-Al ORGANOMETALLIC COMPOUNDS
J. Mat. Sci. 23 (1988) 322-336.

/2/ E.R. Ivers-Tiffée, V. Magori
German Patent DE 0534 09 815 (1985).

/3/ M. Kagawa, M. Kikuchi, R. Ohno, T. Nagae
PREPARATION OF ULTRAFINE MgO BY THE SPRAY-ICP TECHNIQUE
J. Am. Ceram. Soc. 64 (1981) C7-C8.

/4/ M. Kagawa, F. Honda, H. Onodera, T. Nagae
THE FORMATION OF ULTRAFINE Al_2O_3, ZrO_2 AND Fe_2O_3 BY THE SPRAY-ICP TECHNIQUE
Mat. Res. Bull. 18 (1983) 1081-1087.

/5/ N. Kagawa, Y. Imamura, S. Usui, Y. Syono
DISTRIBUTION OF ZrO_2 AND Al_2O_3 IN ULTRAFINE ZrO_2-Al_2O_3 POWDERS FORMED BY THE SPRAY-ICP TECHNIQUE
J. Mat. Sci. Lett. 3 (1984) 699-702.

/6/ D. Vollath, B. Dörzapf
German Patent Application P 3719825.4 (1987).

/7/ D. Vollath, B. Dörzapf
German Patent Application P 3840316.1 (1988).

/8/ D. Vollath, H. Wedemeyer, E. Günther
IMPROVED METHODS FOR FABRICATION OF $LiAlO_2$ AND Li_2SiO_2 PELLETS AND SPHERES AS BREEDER MATERIALS IN FUSION REACTORS
Proc. of the 13th Symp. on Fusion Technology (1984) Vol. 2, 967-972.

/9/ D. Vollath, H. Wedemeyer
TECHNIQUES FOR SYNTHESIZING LITHIUM SILICATES AND LITHIUM ALUMINATES
Advances in Ceramics Vol. 25 (1989) in the press.

/10/ D. Vollath, H. Wedemeyer
ON THE PREPARATION OF THE LITHIUM SILICATES SERIES FROM Li_2SiO_3 TO Li_8SiO_6 IN ALCOHOLIC MEDIA
J. Nucl. Mat. 141-143 (1986) 334-338.

SYNTHESIS OF AlN FROM AMINOALANES

M.-A. Einarsrud, Institute of Inorganic Chemistry, Norwegian Institute of Technology, University of Trondheim, 7034 Trondheim, Norway;
W.E. Rhine and M.J. Cima, Ceramics Processing Research Laboratory, MIT, Cambridge, MA 02139, USA.

Aluminium nitride precursors were synthesized by reacting dimethylaminoalane with ammonia in hydrocarbon solvents. Various reaction conditions were investigated in an attempt to control particle size and morphology. Using an excess of ammonia gave submicron, high-purity aluminium nitride powder having large surface area (118 m^2/g) after calcination at 1000°C. After calcination at 1600°C the surface area decreased to 1.3 m^2/g. The oxygen content of samples calcined at 1200°C was 0.5%. A colloidally processed sample densified to >95% of the theoretical density at 1700-1750°C.

INTRODUCTION

Aluminium nitride (AlN) has the potential application as an electronic substrate material for high-power, high-density electronic circuits because of its high theoretical thermal conductivity (320 W/mK) (Slack,1973). The thermal conductivity decreases, however, if oxide impurities are present in ceramics (Slack,1973), so a ceramic powder of high purity is desired.The commercial procedures for preparing AlN powder involve the carbothermal reduction of alumina in the presence of nitrogen (Ste d'Electrochimie, 1957) or the reaction of aluminum with either ammonia or nitrogen (Yamashita *et al.*, 1986). The powders produced by these methods often contain high concentrations of oxygen and other impurities and are grey in color.

Several other chemical synthesis routes have been investigated in the literature to produce high-purity AlN powder. For example, AlN can be prepared by reacting ammonia with either $AlCl_3$ (Lewis, 1970) or $(NH_3)_6AlF_6$ (Huseby, 1983), though these reactions require high temperatures and yield highly corrosive byproducts. Several other researchers have alternatively proposed an organoaluminum compound as a good starting material for synthesizing AlN powder. Interante *et al.* (1986) reacted triethylaluminum with ammonia to form a precursor of high-purity AlN powder. Bolt and Tebbe (1987) also used triethylaluminum as a starting material to synthesize a thermoplastic precursor of AlN. This precursor could be drawn into fibers that densified between 1600°C and 1800°C without the use of sintering additives.

Wiberg and May (1955) reported that AlH_3 reacts with ammonia in THF solution to form various compounds in which amino or imido groups replace one or more hydrogens in AlH_3. These compounds were shown to release NH_3 or H_2 upon heating and approach the stoichiometry of AlN. However, these researchers did not describe the conditions required to form AlN. Maya (1986) investigated these reactions, reacting $AlH_3{\cdot}Et_2O$ with NH_3 to form a product with the empirical formula $[Al(NH_2)NH]_n$, and converting this intermediate product into AlN. Ochi, Rhine, and Bowen (1988) studied the reaction between AlH_3 and NH_3 in a tetrahydrofuran (THF) solution. When excess NH_3 was used, a white, high-powder AlN powder was obtained; when a stoichiometric amount of NH_3 was used, however, the synthesized AlN contained a carbon residue, probably due to THF decomposition during heat-treatment through association with the precursor.

In the present study, H_2AlNMe_2 was investigated as a starting material for synthesizing a precursor for AlN powder. The compound H_2AlNMe_2 is soluble in hydrocarbons, which should facilitate solvent removal. Two synthesis routes were tried, in efforts to control the morphology of the powder produced.

EXPERIMENTAL

Due to the reactivity of aluminum hydride compounds with oxygen and/or moisture, handling of the chemicals and all the experimental work were performed in an inert atmosphere or under vacuum.

Dimethylamine hydrochloride, $HNMe_2{\cdot}HCl$ (Aldrich Chemical Company, Milwaukee, WI, USA 98%; dried under vacuum at room temperature for 24 h), was reacted with excess $LiAlH_4$ (Alfa Products, Danvers, MA, USA; 1.5 M solution in THF, used as received) in 150 ml of THF according to the following equation:

$$LiAlH_4 + HNMe_2{\cdot}HCl \rightarrow H_2AlNMe_2 + LiCl + 2H_2 \qquad (1)$$

The resulting solution was filtered and the solvent removed under vacuum. The H_2AlNMe_2 sublimed easily at 50-60°C (-70°C at the cold spot); the product was sublimed twice to give a total yield of 77%.

The reaction scheme below was used to synthesize AlN powder. This paper discusses AlN powder production following the two synthesis routes shown in Table 1.

$$H_2AlNMe_2 + n\,NH_3 \rightarrow \text{precursors} \rightarrow AlN + \text{gaseous byproducts} \qquad (2)$$

Table 1. Reaction conditions for the synthesis of AlN precursors.

	Temperature (°C)	Solvent	Amount NH_3	Atm. for pyrolysis
Precursor I	-70°C	toluene	excess	N_2
Precursor II	25°C	hexane	slight excess	NH_3/N_2

AlN via Precursor I

H_2AlNMe_2 (3.6 g) was dissolved in toluene and reacted with excess NH_3 at -70°C. At this temperature, ammonia condensed into the reaction flask. Upon heating, gas evolved, beginning at -60°C, and a voluminous white precipitate formed. The solvent was removed under vacuum at room temperature. The precipitate (Precursor I) was placed in alumina or molybdenum boat and heat-treated for 2 h under nitrogen at a rate of 10 K/min to various selected temperatures to form AlN.

AlN via Precursor II

In the second synthesis, H_2AlNMe_2 (3.6 g) was dissolved in 150 ml hexane, and this solution was reacted with a slight excess of ammonia at room temperature. The reaction took place at the interface between the solution and the gas phase as ammonia flowed over the solution. A white layer formed on the solution surface and the powder gradually sank to the bottom of the reaction flask; gas evolved from the product at the bottom of the reaction flask. The solvent was removed under vacuum and a white precipitate obtained. The precipitate (Precursor II) was pyrolyzed under ammonia at temperatures up to 200°C, then heat-treated under nitrogen to 1000°C.

Powder Characterization

Fourier transform infrared (FTIR) spectra were recorded using a closed diffuse-reflectance cell with KBr windows (IR44, IBM Instruments, Danbury, CT, USA). Surface areas of the powders obtained by both synthesis routes were measured by the one-point BET method using N_2 gas (Quantachrome, Syosset, NY, USA). X-ray diffraction (XRD) patterns were measured using Cu Kα radiation (RU300, Rotaflex, Rigaku/USA, Inc., Danvers, MA, USA). The powder morphologies were studied by scanning electron microscopy (SEM) (S-530, Hitachi Scientific Instruments, Rockville, MD, USA). Chemical analyses were performed by commercial laboratories (Galbraith Laboratories, Inc., Knoxville, TN, and IRT Corporation, San Diego, CA, USA).

Sintering

Colloidal pressing produced green compacts with higher green densities than were obtained by dry pressing. The AlN powder was dispersed in a hexane solution using 4 wt% OLOA™1200 (Chevron Chemical Company, San Fransisco, CA, USA) as the dispersant. The resulting AlN compact was pressed at 10 000 psi and sintered under 1 atm nitrogen in a molybdenum crucible at 1700-1800°C. Final density was measured using Archimedes′ method.

RESULTS AND DISCUSSION

Precursor Characterization

The FTIR spectra obtained for Precursors I and II after pyrolysis under ammonia /nitrogen are shown in Figure 1. The broad band at ~750 cm^{-1} is due to Al-N stretching vibrations; the broad bands in the region 3000-3300 cm^{-1} are due to the N-H stretching vibrations; the band at ~1800 cm^{-1} is due to Al-H stretching vibrations; and the rather sharp bands in the region 1500-1600 cm^{-1} are N-H bending modes. For Precursors I and II after pyrolysis under ammonia, no absorption is apparent in the Al-H stretching region (1700-1800 cm^{-1}) and only weak features can be detected in the C-H stretching region (2850-3050 cm^{-1}). From these observations, we conclude that both the hydrogens and almost all the NMe_2 groups on the aluminum atom of the starting compound must have been replaced by ammonia, and that both precursors were probably an imide-amide, $[AlNH(NH_2)]_n$. Maya (1986) also observed absorption bands at the frequencies 3200, 1505, 900, and 680 cm^{-1} for a precursor with empirical formula $[Al(NH)NH_2)]_n$. Also included in Figure 1 is an FTIR spectrum of Precursor II before pyrolysis under ammonia. In addition to the bands observed for the precursor after pyrolysis, a band at ~1800 cm^{-1} due to Al-H stretching vibrations and bands between 1200 and 1600 cm^{-1} probably due to C-H and N-H bending modes are observed. This spectrum suggests that the precursor is $H(NH_2)AlNMe_2$.

Characterization of AlN Powders

The AlN powders derived from heat-treatment of the precursors white or slightly grey. Chemical analysis of AlN obtained from Precursor I calcined at 1200°C showed it to contain 33.10% nitrogen and 65.24% aluminum (theoretical: 34.17% N, 65.83% Al), which gives an Al/N ratio of 1.02. The sample contained 0.547±0.022 wt% oxygen (neutron activation analysis); The oxygen source was presumably the atmosphere during heat-treatment of the precursor.

Figure 2 shows the development of X-ray diffraction patterns of the AlN powder derived from Precursor I after heat-treatment at 1000, 1200, and 1500°C. Figure 3 shows the XRD pattern of AlN powder produced from Precursor II; the same pattern was observed for AlN produced from Precursor I at the same temperature.

The surface areas of powders obtained from Precursor I when calcined at different temperatures are given in Figure 4. At 1000°C, the surface area was rather high (118 m^2/g), indicating high reactivity of the powder; the surface area decreased to 1.3 m^2/g at 1600°C. The surface area of an AlN powder obtained from Precursor II, when calcined at 1000°C, was measured to be 109 m^2/g, slightly lower than that of the powder obtained from Precursor I.

Figures 5a-b show scanning electron micrographs (SEMs) of the AlN powders obtained from Precursor I and II, respectively, after calcination to 1000°C consisted of very fine, agglomerated particles. From the SEMs, the particle size can be estimated to be less then 0.1µm.

Sintering behavior

Heat-treatment to 1500°C of the powder produced from Precursor I coarsened the particles. At this temperature, the surface area was about 10% of its value after calcination at 1000°C. At 1600°, the surface area had decreased to 1.3 m^2/g and sintering was observed. In some samples, only partial sintering was observed: some sections were almost 100% dense, others were more porous. This variability in sintering may be due to differences in impurities and oxygen content, a suggestion offered by other authors (Bolt and Tebbe, 1987).

Dry-pressing the powder (10 000 psi) gave a very low green density and a final density only 67% of theoretical. Colloidal pressing produced a higher-density green compact that sintered to an average of 95% of theoretical without the use of sintering aids.

Scanning electron microscopy of the section of this sample that was topmost during sintering showed it to be almost 100% dense (Fig. 6a). The average grain size was 1.5 µm, with only a few pores apparent in the microstructure. This part of the sample, ~100% of the theoretical density, was slightly grey in color. The bottom section of the sample during sintering was porous (Fig. 6b). The difference in density might be an artifact of the colloidal pressing, producing different green densities in the two sections of the sample.

CONCLUSIONS

- Two precursors for AlN have been synthesized from H_2AlNMe_2 and NH_3 in a hydrocarbon solution.
- Upon pyrolysis under NH_3 and/or N_2, high-purity AlN powder has been formed.
- The powder has a high surface area at low calcination temperatures (1000°C) and consists of very fine (0.1 µm) particles.
- From 1000°C to 1500°C, the surface area decreases almost linearly with the temperature, and at 1500°C particle coarsening occurs.

- At 1600°C, the surface area decreases to 1.3 m^2/g and sintering occurs. Partial sintering has been observed in which some sections of samples were almost 100% dense while other sections were more porous.
- A sample formed by colloidal pressing sintered to an average density of 95% of theoretical without the use of sintering aids.

ACKNOWLEDGEMENT

A scholarship from the Royal Norwegian Council for Scientific and Industrial Research (NTNF) to M.A. Einarsrud and funding from the Air Force Office of Scientific Research (Contact no. F49620-84-C-0097) is gratefully acknowledged.

REFERENCES

American Society for Testing and Materials, "X-Ray Powder Data File," Philadelphia, PA.
J.D. Bolt and F.N. Tebbe, Proceedings of the American Ceramics Society Electronics Division Meeting, Denver, CO, 1987.
I.C. Huseby, *J. Am. Ceram. Soc.*, **66**, 217-20 (1983).
L.V. Interante, L.E. Carpenter, C. Withmarsh, and W. Lee, Mat. Res. Soc, Symp. Proc., **73**, 359-66 (1986).
D.W. Lewis, *J. Electrochem. Soc. Electrochem. Thecnol.*, **117**, 978-82 (1970).
L. Maya, *Adv. Ceram. Mater.*, **1**, 150-53 (1986).
A. Ochi, W.E. Rhine, and H.K. Bowen, Proceedings of the Materials Research Society Spring Meeting, Reno, NV, 1988.
G.A, Slack, "Nonmetallic Crystals with High Thermal Conductivity," *J. Phys. Chem. Solids,* **34** [2] 321-35 (1973).
Ste d'Electrochimie: French Patent 1.149.539 (1957).
E. Wiberg and A. May, Z. *Naturforch,* **10B**, 229-32 (1955).
H. Yamashita, Y. Imai, R. Oguma, and T. Hayashi, Pat. Appl. 86/205606, Jpn. Kokai Tokkyo Koho JP61/205606 (1986).

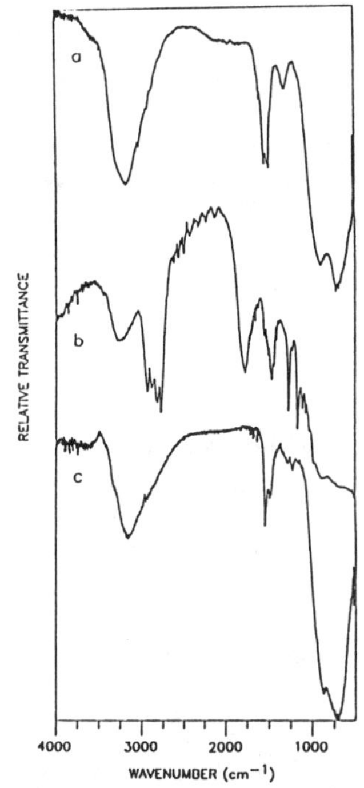

Figure 1. FTIR diffuse reflectance spectra of the intermediate compounds (AlN precursors): a) Precursor I, b) Precursor II before pyrolysis under ammonia, and c) Precursor II after pyrolysis under ammonia.

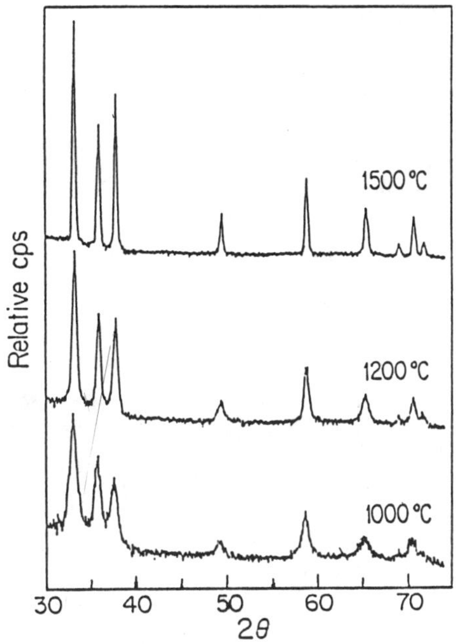

Figure 2. XRD patterns of AlN powder from Precursor I and heat-treated at a) 1000°C, b) 1200°C, and c) 1500°C.

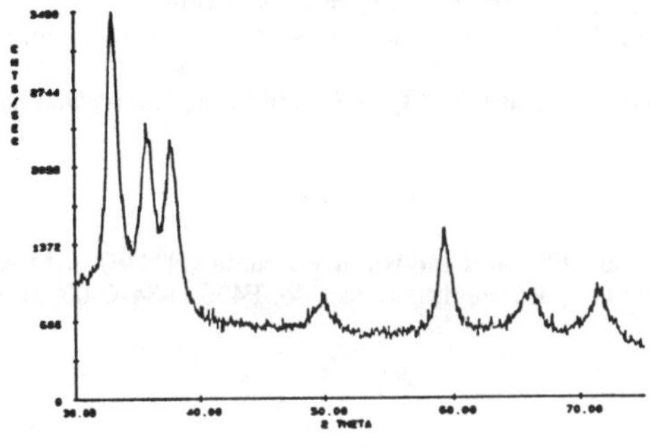

Figure 3. XRD pattern of AlN powder prepared from Precursor II after pyrolysis under ammonia to 200°C and subsequent heat treatment under nitrogen to 1000°C.

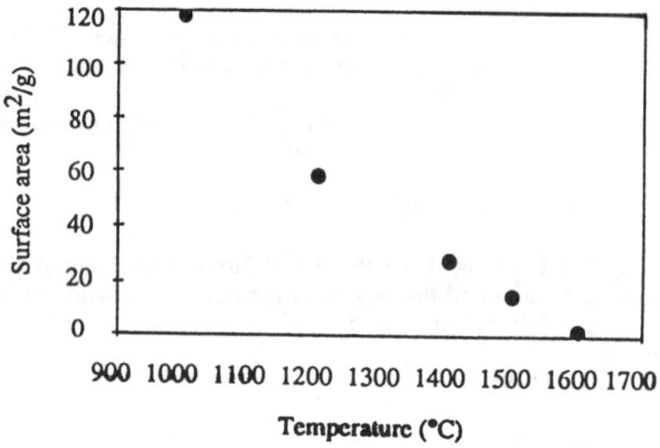

Figure 4. Surface areas of AlN powders prepared from Precursor I.

a b

Figure 5. Scanning electron micrographs of AlN powder produced at 1000°C from a) Precursor I and b) Precursor III.

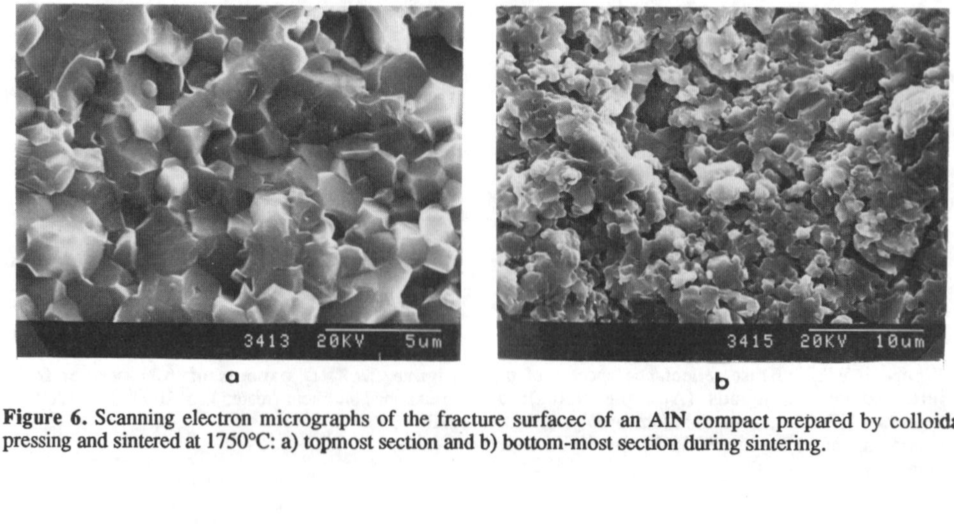

a b

Figure 6. Scanning electron micrographs of the fracture surfacec of an AlN compact prepared by colloidal pressing and sintered at 1750°C: a) topmost section and b) bottom-most section during sintering.

STUDY OF NEW POLYSILAZANE PRECURSORS TO Si-C-N-O CERAMICS

C. COLOMBIER
ATOCHEM, Centre de Recherche Rhône-Alpes,
Boîte Postale n° 63, F-69310 PIERRE-BENITE, France

ABSTRACT

Polysilylhydrazines were synthetised by reacting hydrazine on chlorosilanes. The influence of the nature of the chlorosilane on the polymer structures and pyrolysis results (ceramic yields, formed gases during pyrolysis) was studied. The use of these precursors to manufacture ceramic coatings and ceramic matrices for composites is briefly discussed.

INTRODUCTION

The precursors of silicon derived ceramics (SiC, Si_3N_4, Si_2N_2O) have been the object of intense research efforts during the past twenty years. Such products are being found applications principally for the manufacture of ceramic fibers (1), ceramic coatings (2), ceramic matrices for composites (3) and as binders in green body fabrication (4). The precursors must have several properties, namely high ceramic yields, application adapted rheological behaviour, adapted derived ceramic (relevant composition and microstructure).

The research on polysilazanes, precursors of silicon nitride based ceramics have resulted in extended works involving ammonolysis, transamination, thermolysis or catalytic approaches (5). The present article describes a new route for the synthesis of polysilazanes, the originality of which is the use of hydrazine (H_2N-NH_2)* as nitrogen donor. The polysilazanes are obtained by reacting hydrazine on chlorosilanes (mainly dichlorosilanes of formula $RSiCH_3Cl_2$). Polysiloxazanes, precursors of silicon oxynitride ceramics, can be easily obtained using the same process by reacting liquid mixtures N_2H_4/H_2O** on chlorosilanes. The basic properties of these products will be described with simple polysilazanes

* Hydrazine : liquid, bp : 113,5°C
** N_2H_4 and H_2O are miscibles in all proportions

involving only one chlorosilane during the synthesis. Commercially available precursors* are more complex mixtures based on this chemistry.

EXPERIMENTAL

The reaction is carried out in a stirred reactor in the presence of triethylamine in order to reduce the formation of hydrazine hydrochloride. Hence, the basic reaction of the process is :

$$RSiCH_3Cl_2 + N_2H_4 + 2\ Et_3N \longrightarrow \left[SiRCH_3N_2H_2\right]_n + 2\ \underline{Et_3N.HCl}\downarrow$$

Hydrazine is introduced in one hour at room temperature in a mixture of toluene (solvent), chlorosilane and triethylamine. The temperature of the reactor is then increased to 70°C during 6 hours. The formed salt (by-product) is separated by filtration and the solvent is evaporated under vacuum at 60°C in a rotating apparatus.

RESULTS AND DISCUSSION

Structures and properties of the polymers

Their physical state was depending strongly on the nature of the R group bonded to the silicon atom (see table 1). If R is $-CH_3$ or $-CH{=}CH_2$, oils of low molecular weight are obtained. When R is a hydrogen atom, an only partially soluble and infusible solid results. By means of elemental analysis and of a chemical method** of determination of the amount of N-N bonds in the polymer it can be stated that the N/Si mol. ratio is close to 1. It must then be assumed that the structures of our polymers are somewhat different from that described by Niederprüm and Wannagat (6) : $\left[Si(CH_3)_2{-}NH{-}NH\right]_n$. Furthermore ^1H NMR revealed that, in average the amount of hydrogen atoms bonded to the nitrogen ones was nearly 1.2 H/N-N when R is $-CH_3$ or $-CH{=}CH_2$ and only 0.7 H/N-N when R is H. From these results, it must be assumed that hydrazine is trifunctional toward $RSiCH_3Cl_2$ when R is $-CH_3$ or $-CH{=}CH_2$ and partially tetrafunctional when R is H. Thus, the tendency of hydrazine to react with dichlorosilanes increases when the mass of the groups bonded to the silicon atoms decreases. This can be further confirmed by studying the reaction of hydrazine with diphenyldichlorosilane : in the same conditions of synthesis as for the above described polymers, about 10 % wt of chlorine remained bonded to silicon in the end-product (the chlorine con-

* Pyrofine PV and Pyrofine PM registered trade mark of ATOCHEM
**This method involves hydrolysis of the polymer followed by the quantitative oxydation of hydrazine by iodine

TABLE 1
Structure of the polymers

Nature of R in $RSiCH_3Cl_2$	Polymer properties	$\overline{MW}$ (GPC)	Elemental mol. ratio Si	C	N	O	H	N-N/Si mol. ratio	H_R/CH_3 mol. ratio (^1H NMR)	H_{NH+NH_2}/CH_3 mol. ratio (^1H NMR)	H_{NH+NH_2}/N-N
-H	partial. soluble nonmeltable	6 600	1	1.15	1.27	0.04	4.5	0.61	0.8	0.44	0.72
$-CH_3$	oil (≃6 p at 20°C)	640	1	2.17	1.32	0.06	7.0	0.64		0.77	1.20
$-CH=CH_2$	oil (227 p at 20°C)	500	1	3.0	1.39	0.05	6.7	0.65	2.9	0.81	1.25

tent was < 200 ppm in case of the polymers of table 1). Steric factors are probably responsible for such differences. As an indication we propose as representative structure :

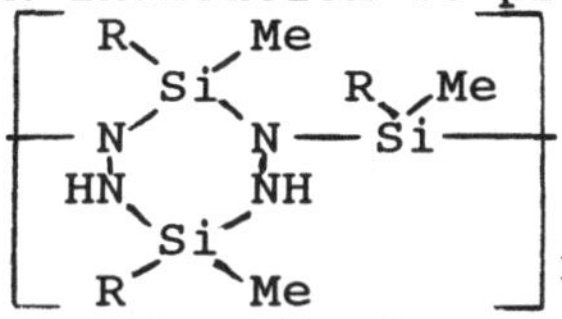

when R is CH_3 or $CH=CH_2$.

When R is a hydrogen atom, the product cross-linked partially during evaporation of the solvent, most probably through -SiH and NH bonds with formation of hydrogen. This is revealed through ^1H NMR (table 1) by a loss of 20 mol. % of the SiH bonds of the starting chlorosilanes and by the extremely low quantity of hydrogen atoms on the nitrogen ones (0.72 H/N-N). This supplementary reaction contributes further to explaining the higher molecular weight of the CH_3SiHCl_2 derivate.

Pyrolysis of the polymers under nitrogen

The composition of the ceramics (table 2) reflected the composition of the polymers without apparent influence on the ceramic yield (mass % residue of pyrolysis at 1000°C under N_2. Heating rate : 100°C/hr). In particular, the overall carbon content as well as the free carbon content can be directly related to the carbon content of the polymer. On the other hand, the ceramic yield was more dependent on the functionality of the R group bonded to the silicon atoms. The three studied products can be seen as representative of three different cases :

R is H: Mass-spectrometer analysis of the gases evolved during pyrolysis (see fig. 1) revealed the presence first of hydrazine, then $MeSiH_3$ (near 300°C), NH_3 (250 to 500°C), CH_4 (400 to 1000°C) and H_2 (maximum near 800°C). The formation of

TABLE 2

Pyrolysis results (up to 1000°C under N_2, 100°C/hr)

Nature of R in $RSiCH_3Cl_2$	Ceramic yield (wt %)	Elemental analysis of the ceramic (mol. ratio) Si	N	C	O	Approximate composition (wt %) (by calculation from the elemental analysis) Si_3N_4	SiC	C_{free}	SiO_2
-H	72	1	1.15	0.51	0.053	79.5	8.7	9.4	3.2
$-CH_3$	5	1	0.82	1.01	0.13	52.3	23.6	15.2	7.0
$-CH=CH_2$	59	1	1.03	1.36	0.06	59.9	13.1	23.2	2.8

methylsilane at 300°C corresponds to a major change of structure in the product, since an important heat generation was stated by differential scanning calorimetry. The N-N bond is probably implicated in this structure change since the ammonolysis product of CH_3SiHCl_2 [low molecular weight oligomers generally considered as impracticable as ceramic precursors because of their low ceramic yield (20-40 %)] doesn't generate such heat in this range of temperature. Absence of hydrogen in the low temperature range can be surprising if the cross-linking reaction responsible for the high ceramic yield (72 %) is taken into consideration : $\equiv SiH + HN= \longrightarrow \equiv Si-N= + H_2$. Nevertherless, this can be seen as a consequence of the already cross-linked structure of the analysed sample and perhaps of a limit of sensitivity of the mass-spectrometer. Note that the NMR study (table 1) was necessarily limited to the soluble fraction of the sample which was then less cross-linked.

R is CH_3: In spite of the very low ceramic yield (5 %), no molecule containing silicon could be found in the gases of pyrolysis. Most probably, since the product exhibited a low molecular weight (≃ 640) it distillates on heating and condensates prior to reaching the mass-spectrometer. In this case, no cross-linking reaction took place at the beginning of the pyrolysis to hinder volatilization of the oligomers. Absence of any thermic signal was also stated by DSC up to 600°C. Nevertheless, methane and ammonia were evolved during pyrolysis from the decomposition of the higher MW fraction of the product.

R is $CH=CH_2$: This is an intermediate case of the two former. The polymer became very fluid on heating (< 0.5 poises at 150°C) but cross-linked rapidly between 230 and 270°C. Obviously, a vinylic polymerization (which doesn't generate a gas) took place with heat formation (DSC). Up to 300°C, no gas could be detected : only low MW oligomers volatilized and condensed prior to reaching the mass-spectrometer. From 400 to 800°C, NH_3, CH_4, $H_2C=CH_2$, $H_3C-CH=CH_2$ and $H_3C-CH_2-CH=CH_2$ are

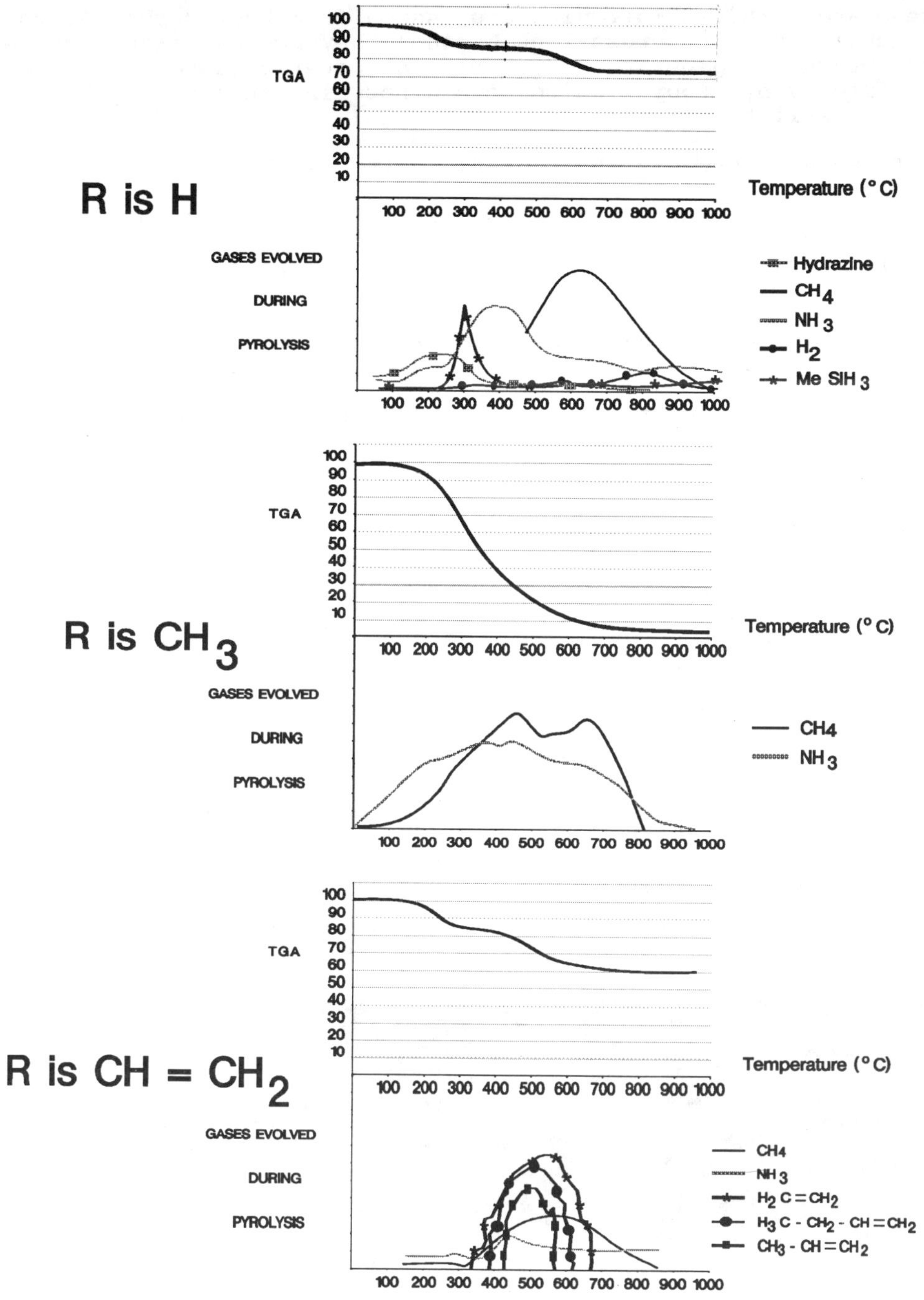

Figure 1 . Pyrolysis of the polymers under N_2. Thermogravimetric analysis and mass-spectrometer analysis of the gases evolved during pyrolysis.

generated. These molecules are issued from hydrocarbon groups which could not cross-link because they are saturated, as Me groups which generate CH_4, or for steric factors, such as isolated vinylgroups which could not find another vinylgroup to cross-link.

Influence of the gas environment during pyrolysis

N_2, Ar and NH_3 were studied as carrier gases during pyrolysis. Equivalent compositions of the ceramic and equal ceramic yields were found for N_2 and Ar which can be seen as inert toward the precursors. On the contrary, under ammonia the ceramic yields were slightly lower (71 % in state of 72 % when R is H, and 54 % in state of 59 % when R is $CH{=}CH_2$) and the carbon content of the ceramics was drastically reduced.

Complete removal of carbon was not reached for a product containing vinyl groups : 2 wt % carbon remained in the final ceramic (in state of 23 wt % for N_2 or Ar). It then appeared black as is normally the case for all above described products when they are pyrolysed under N_2 or Ar. When R is hydrogen, the very low carbon content (0.2 % in state of 13 % for N_2 or Ar) resulted in a white appearance. Such a reactivity of ammonia toward organosilicon precursors was already demonstrated in case of polycarbosilanes (7), polysilanes and polysilazanes (8). It was proposed that the process of removal of carbon occurs via the attack of NH_2 radicals, generated during the decomposition of ammonia, on silicon (8).

Applications

The best ceramic yields are obtained when cross-linking reactions take place at the beginning of the pyrolysis. This could be observed for the hydrazine derivative of CH_3SiHCl_2 and $CH_3SiCH{=}CH_2Cl_2$. Both are the base of commercially available liquid precursors (respectively Pyrofine PM and Pyrofine PV).

A major problem arising in using ceramic precursors is the control of the shrinkage during pyrolysis. Since the

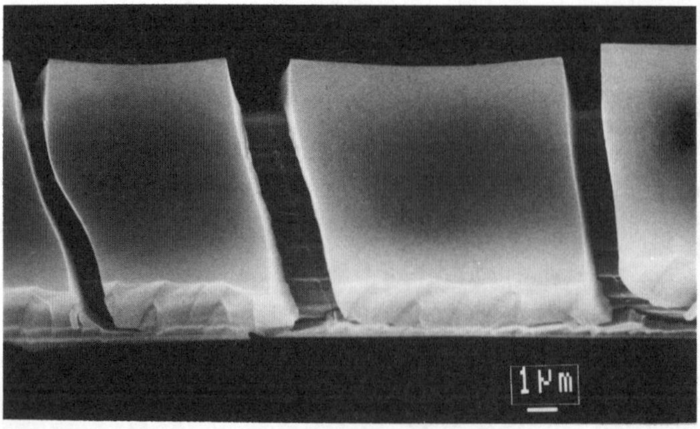

Figure 2 . Si-C-N-O ceramic on carbon fibre.

density of the precursor is about 1 and the density of the ceramic is ≃ 2.5, the volume ratio of the ceramic to the corresponding polymer is ≃ 24 % in case of a 60 wt % ceramic yield. Fig. 2 shows a Si-C-N-O coating on a carbon fibre. The formed cracks illustrate the problems of compatibility between the precursor material whose volume decreases drastically during pyrolysis, and the carbon substrate which tends on the contrary to expand on heating.

Coatings: The manufacture of ceramic coatings by dipping in a liquid followed by a pyrolysis is a particularly simple process in comparison with some vapour phase deposition techniques. Furthermore, the usual problems of diffusion of gases to the surface are eliminated and a more homogeneous coating thickness can be attempted without influence of the macroscopic shape of the substrate. Coatings on stainless steel substrates were obtained (Fig. 3) by dipping them in toluenic solutions of Pyrofine PV or PM followed by a heat treatment in nitrogen between 850 and 1000°C (heating rate : 100°C/hr). The concentration of the precursor was the determinant factor for the adhesion of the coating. Too high concentrations (50 wt %) resulted in the formation of scales showing a poor adhesion. Adherent coatings were obtained with 10 and 20 wt % solutions of precursor. As a consequence of the low concentration, the coating thickness was low, typically 0.35 µm for a 10 wt % solution and 0.7 µm for a 20 wt % solution. The adhesion could be markedly improved by roughening the surface of the substrate with a corundum jet. In this case, SEM investigations of the cross-section of the coating (Fig. 3b) revealed the tendency of this process to smooth the surface : because of its liquid state and its good wetting to metallic surfaces, the precursor tends to fill the cavities. Such pretreated surfaces retained more precursor than smooth surfaces and so received a thicker coating (1.2 µm in spite of 0.7 µm for non-roughened

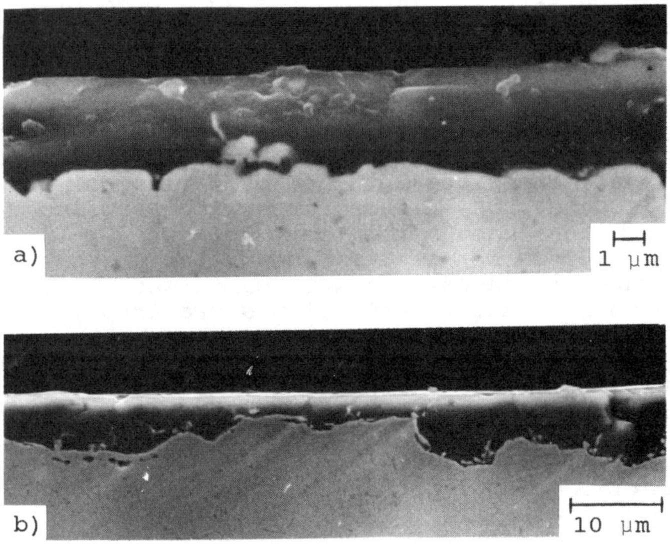

Figure 3 . Si-C-N-O coating on stainless steel. The substrates were dipped in a 20 wt % toluenic solution of precursor and pyrolysed under N_2. Five cycles of dipping/pyrolysis were subsequently performed.
a) coating of ceramic ex-Pyrofine PV on a non-roughened substrate.
b) coating of ceramic ex-Pyrofine PM on a roughened substrate.

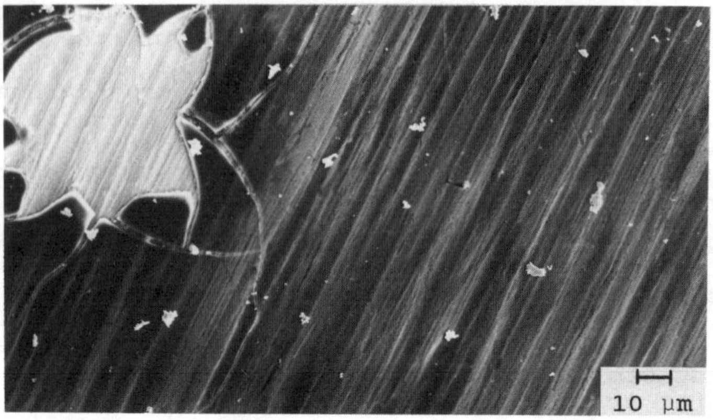

Figure 4 . Si-C-N-O coating on molybdenum. The substrate was dipped in a 10 wt % toluenic solution of Pyrofine PV and heated to 1000°C under N_2.

surfaces in case of a 20 wt % toluenic solution). Coatings on metals with lower thermal expansion coefficients, closer to that of silicon nitride (3.10^{-6} $°C^{-1}$), as molybdenun or zirconium ($5.5\ 10^{-6}$ $°C^{-1}$) were generally of better quality. Flaking in some areas of the substrate was difficult to avoid in the absence of corundum jet pretreatment (see fig. 4). Nevertheless, large areas without apparent cracks could be obtained. Efforts are necessary to improve the adherence of such coatings and particularly an increase of interdiffusion between the coating and substrate could have significant beneficial effects.

<u>Ceramic</u> <u>matrix</u> <u>composites</u>: Liquid precursors of sufficently low viscosity can be injected in fibrous preforms. Fig. 5 shows unidirectional composites obtained by cycles of impregnation/pyrolysis (1000°C) of carbon fibre with Pyrofine PV. The process starts with a vacuum treatment followed by injection of the liquid precursor at 100°C under low pressure (5 bars). The subsequent impregnations were performed by simply dipping the composite in a hot (150°C) liquid precursor. Fig.5 shows the excellent wetting of the carbon fibres by the precursor in spite of the high density of fibres in the composite. Cracks resulting from the shrinkage are necessarily present and perfectly dense materials cannot be realized in this way. Nevertheless, since 10 % porosity can be reached in only 3 cycles of impregnation/pyrolysis (for 70 vol % of fibres), it may be a potential competitor for the Chemical Vapour Infiltration (CVI) technique (9).

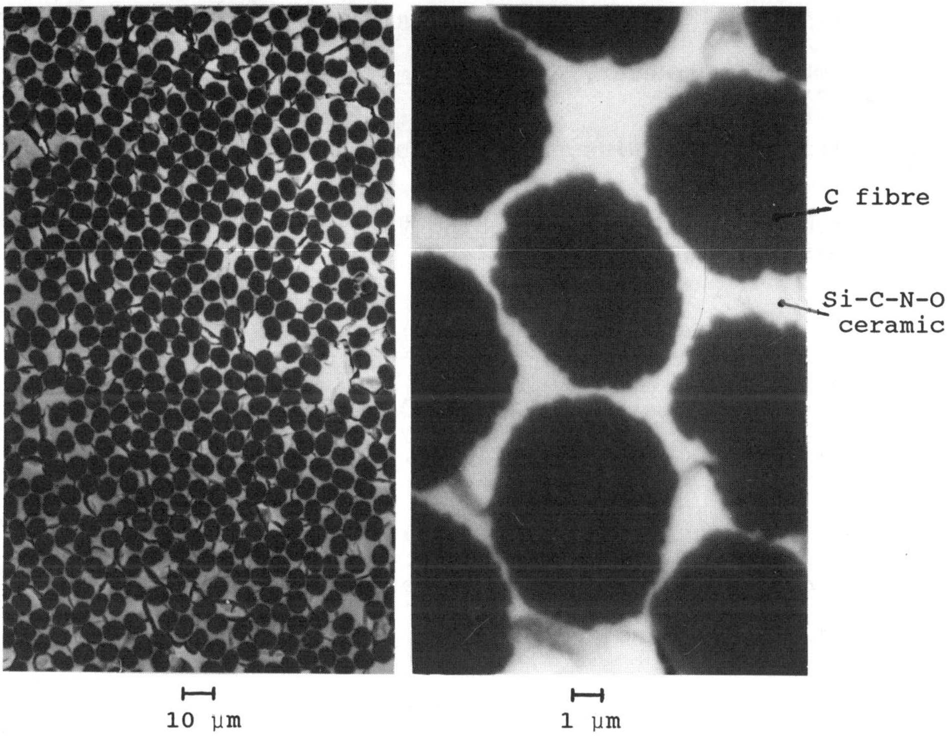

Figure 5 . Carbon fibre (T 300, Torray) reinforced unidirectional composite.

CONCLUSIONS

Depending on the nature of the chlorosilane as reaction partner with hydrazine, polysilazanes with very different properties (viscosity, thermal decomposition ...) are obtained. Two dichlorosilanes can be selected, $CH_3SiCH=CH_2Cl_2$ and CH_3SiHCl_2 as a base for hydrazine derived ceramic precursors. These chlorosilanes allow the polymer to cross-link at the early stage of the pyrolysis resulting in increased ceramic yields. These precursors can be used to prepare ceramic coatings for example on metals. Roughening the substrate prior coating is beneficial to improve the adhesion. In this case, the liquid precursor fills the cavities of the surface and tends to diminish the roughness. Deposition of the precursor from an organic solution is necessary in order to control adhesion and coating thickness. Ceramic matrices can be easily manufactured by the injection of liquid precursors in dense fibrous preforms.

REFERENCES

1 . Chaim, R. and Heuer, A.H., J. am. ceram. soc., 71 11 960 (1988)

2 . Coblenz, W.S., Wiseman, G.H., Davis, P.B. and Rice, R.W., Mat. Sci. Res. 17, 271 (1984)

3 . Diefendorf, R.J., Proceeding Verbundwerk 1988, 13, Demat exposition managing, Frankfurt

4 . Schwab, S.T. and Blanchard-Ardid, C.R., Mat. Res. Soc. Symp. Proc. vol. 121, 581 (1988)

5 . Laine, R.M., Blum, Y.D., Tse, D. and Glaser, R., Inorganic and Organometallic Polymers, ACS Symposium Series 360, edited by M. Zeldin, K.J. Wynne and H.R. Allcock (ACS, Washington, DC, 1988), p. 124

6 . Niederprüm, H. and Wannagat, U., Z. f. anorganische u. allgemeine Chemie 311, 270 (1961)

7 . Okamura, K., Sato, M. and Hasegawa, Y., EP 0167230 (1985)

8 . Burns, G.T., and Chandra, G., J. am. ceram. soc., 72 2. 333 (1989)

9 . Naslain, R. MRS-Europe, Nov. 1985, p. 99

PREPARATION AND PROPERTIES OF A HIGH SURFACE AREA SILICON OXYNITRIDE POWDER

P.W. LEDNOR* AND R. DE RUITER
KONINKLIJKE/SHELL-LABORATORIUM, AMSTERDAM
(Shell Research B.V.)
P.O. Box 3003, 1003 AA Amsterdam, The Netherlands

ABSTRACT

The synthesis of a silicon oxynitride powder with a high surface area by reaction of a particular silica phase with ammonia is described. The stability of the surface nitrogen towards hydrogen, air or steam has been determined.

INTRODUCTION

Our primary objective is to extend the range of inorganic materials for use in heterogeneous catalysis as catalysts or catalyst supports. We are currently investigating whether compositions other than silica, alumina or alumina-silicates such as zeolites might lead to different catalytic properties in terms of surface chemistry, stability or texture (surface area and pore system).

Candidate materials should have a high surface area, i.e. greater than 50 $m^2.g^{-1}$, and preferably be available as porous materials or as fine powders. Thus the secondary objective is the synthesis of high surface area materials.

In this paper we provide (i) a summary of our recent work on the synthesis of high surface area silicon (oxy)nitrides [1,2], and (ii) new data on the surface stability of this material. Silicon nitride was selected for this study since it is a simple system, and one for which extensive work on silica provides a good reference. Since the surface of a nitride will often contain some oxygen, it is appropriate to refer to the surface chemistry as that of an oxynitride. Furthermore, our synthetic work has led to a high surface area material, which is also a bulk oxynitride.

RESULTS AND DISCUSSION

Synthesis

The synthesis of non-oxides in high surface area forms is not a trivial problem: e.g. there is no readily available solution chemistry analogous to the sol-gel preparations of porous oxides.

We have described various approaches to potentially high surface area silicon (oxy)nitrides using solution chemistry [1]. One of these routes led to a material with a surface area of about 100 $m^2.g^{-1}$. Further analysis showed that this was a non-porous powder. An interesting question is whether the solution chemistry could be adapted to give a porous gel.

In a separate approach [1,2] we have explored the reaction of various silica phases with ammonia at high temperature, in which the crystallinity and porosity of the silica were systematically varied. The main results are shown in Tables I and II.

TABLE 1

Types of silica used for reaction with ammonia

	Porous	Non-porous
Crystalline	Silicalite[a]	Quartz wool, Quartz powder
Amorphous	Silica gel[b]	Aerosil-300[b]

a. An Al-free structural analogue of zeolite ZSM-5
b. S.A. ca. 300 $m^2.g^{-1}$

TABLE 2

Reaction of various types of SiO_2 with flowing NH_3 (30 $l.h^{-1}$) at 1100 °C[a]

SiO_2	Bulk N, %w	Surface N, %w	Surface N, % at.	Surface area, $m^2.g^{-1}$
Quartz wool	<1	14	20	
Quartz powder	2	10	14	
Silicalite/15 h	2	7	10	
Silicalite	4	7	11	
Silica gel (10 g)	10	13	19	34
Aerosil	22	24	34	
Aerosil/12 h	25	28	40	132
Aerosil/24 h	27	28	40	150

a. All reactions carried out for 4 h using 1-5 g silica, unless otherwise stated. The products obtained from Aerosil and silica gel are amorphous, as determined by X-ray powder diffraction.

It can be seen that the reaction of Aerosil silica with ammonia can give a material whose surface and bulk nitrogen contents correspond closely to the 28 %w N calculated for Si_2N_2O. Furthermore, the product of the reaction has a high surface area. Additional characterisation has been described [1,2]. It has recently been noted that silicon oxynitride has attractive material properties but that a convenient synthesis is lacking [3].

Stability

The stability of the silicon oxynitride surface under conditions relevant to catalysis was determined by treating samples with flowing gases, and subsequently re-measuring the surface nitrogen concentration by X-ray photoelectron spectroscopy. Results are given in Table III.

TABLE 3

Stability of silicon oxynitride surfaces to air, steam and hydrogen[a]

	Weight % N (surface)	Weight % N (bulk)	Surface area, $m^2.g^{-1}$
Silicon oxynitride (A)	28	25	132
After steam at 680 °C	19	21	140
After hydrogen at 750 °C	20		
Silicon oxynitride (B)	22	18	152
After steam at 750 °C	15	17	
After hydrogen at 750 °C	15		
After air at 750 °C	17		

a. All treatments for 4 hours at 1 bar using a gas flow of 5 l/h (steam = nitrogen + 2 % H_2O) and 0.1 - 1.0 g SiON.

Clearly, the high surface area silicon oxynitrides have good stability, illustrated in particular by the retention of significant amounts of surface nitrogen under both reducing and oxidising conditions.

CONCLUSIONS

1. A high surface area amorphous silicon oxynitride powder has been synthesized by reaction of Aerosil silica (ex Degussa) with gaseous ammonia at 1100 °C. This material is of interest in the context of exploring the surface chemistry, relevant to catalysis, of a non-oxide, and may also be of value as a ceramic precursor.
2. The surface of the silicon oxynitride powder is stable with respect to treatment with steam, air or hydrogen at elevated temperature. Further investigation of this and related materials [4] is in progress.

REFERENCES

1. Lednor, P.W. and De Ruiter, R., in "Better Ceramics Through Chemistry III", ed. C.J. Brinker, D.E. Clark and D.R. Ulrich., Materials Research Society Symposium Proceedings, Vol. 121, 1988, pp. 497–502.
2. Lednor, P.W. and De Ruiter, R., J.C.S. Chem. Commun., 1989, 320.
3. Yu, Y.-F. and Mah, T-I., in "Better Ceramics Through Chemistry II", ed. C.J. Brinker, D.E. Clark, and D.R. Ulrich, Materials Research Society Symposium Proceedings, Vol. 73, 1986, pp. 559–564.
4. Lednor, P.W. and De Ruiter, R., to be presented at the 5th Internat. Symposium on Inorganic and Organometallic Polymers, American Chemistry Society Fall Meeting, Miami, Sept. 10–15th, 1989.

WET CHEMICAL PREPARATION AND CHARACTERISTICS OF TETRAGONAL $ZrO_2(-Y_2O_3)-CeO_2$ CERAMICS

G.S.A.M. THEUNISSEN, R.S.A. DE LANGE, A.J.A. WINNUBST, and A.J. BURGGRAAF
University of Twente, Faculty of Chemical Technology, Laboratory of Inorganic Chemistry, Materials Science and Catalysis,
P.O. Box 217, 7500 AE Enschede, The Netherlands

ABSTRACT

The microstructure development is followed for Zirconia materials doped with various amounts of Ceria and/or Yttria. With the Yttria containing compacts very small grainsize at low temperatures can be reached (53 nm at 1044°C). This is interesting for post sintering heat treatments and superplastic deformation. The chemical stability however is poor. Ceria containing compacts have good chemical stability but large grain sizes which makes it difficult to stabilize the tetragonal phase at room temperature. It will be shown that both effects can be combined in one ceramic. In that case reduction of the grain growth is caused by Yttrium and improved chemical stability by Cerium.

INTRODUCTION

Polycrystalline 100% tetragonal ZrO_2-Y_2O_3 ceramics (Y-TZP) are regarded as materials exhibiting high strength and toughness [1]. However, when exposed to humid atmospheres, the critical grainsize to preserve the tetragonal phase reduces drastically [2,3]. In case of 3 mol% Y_2O_3 - ZrO_2 the grainsize should be less than 0.25 μm to preserve the tetragonal phase [2]. In most cases however the resulting grain size is larger. Besides Yttrium, a lot of other stabilizing agents has been used to stabilize the tetragonal phase like Ca, Mg, Ce or Ti. Especially in the last few years, increasing attention has been paid to Ceria doped tetragonal zirconia (Ce-TZP). A very high value for the fracture toughness (30 MPa$\sqrt{m}$) has been reported for a Ce-TZP ceramic doped with 12 at% Ce [4]. This material shows also a good ageing behaviour [5]. Besides this, Ce-TZP shows very interesting plastic deformation behaviour [6]. A disadvantage of Ce-TZP is that larger dopant concentrations are needed for stabilization and higher temperatures and/or longer sintering times are needed for densification of the material. This results in a large grain size after sintering which is a disadvantage for plastic deformation processes and to obtain high strength. On the contrary it is possible to obtain a dense 6 at% Y-TZP ceramic (ρ > 97%) with a very small grain size (< 50 nm) [7]. In these ceramics Yttrium is believed to inhibit the grain growth by means of an impurity drag mechanism [7,8]. A small grain size can be interesting for post sintering heattreatments where the grain growth can be controlled very well. In this way one can construct a TTT-diagram as has been done by Leach [9] and Lange [10]. The second advantage is that a small grain size is favourable for superplastic deformation of the ceramic as has been shown by Wakai et al. [11] and Burggraaf et al. [12].

It would be very favourable when a good chemical stability (Ce) can be combined with a small grainsize (Y) in one ceramic. Some work performed by Sato et al. [13] on Y,Ce-TZP shows an improved chemical stability compared to ZY. Duh et al. [14] showed that doping of Ce-TZP with Yttrium gives rise to smaller grainsizes.

Before studying the combined effect of Y and Ce their specific effects as dopant are investigated with compositions having the same dopant concentration. Attention in this study has been mainly focussed on ceramics with a total dopant concentration of 6 at%. The microstructure development is studied in order to identify grain growth parameters.

EXPERIMENTAL PROCEDURE

Zirconia ceramics with different Yttrium and Cerium concentrations has been synthesized with an overall composition of $(100\text{-}x\text{-}y)ZrO_2.xYO_{1.5}.yCeO_2$ (denoted as ZYxCy) with $0 \leq x \leq 6$ and $0 \leq y \leq 12$. The powders were prepared by a gel-precipitation technique. This implies the hydrolysis of a diluted metalchloride solution ("chloride" method") in an excess of 25 wt% ammonia. [15]. During hydrolysis the pH was kept at a value of 11 or more. After completion of the precipitation, the gel was washed with a water/ammonia mixture in order to remove the chlorides. Finally the gel was washed with ethanol. After drying at 120°C the gel was calcined for two hours at 550°C. Powder compacts were isostatically pressed at a pressure of 400 MPa. Sintering took place at temperatures between 600 and 1500°C in either a furnace or a Netzsch 402E dilatometer. X-ray fluorescence spectometry, using a Philips PW 1410 spectrometer, was used for the analysis of the overall composition. Nitrogen adsorption/desorption isotherms were measured at 77 K by means of a Carlo Erba Sorptomatic Series 1800 instrument. Pore size distributions were calculated from the desorption branch according to the method of Dollimore et al. [16]. Crystallite sizes were determined by means of X-ray line broadening measurements in the region < 100 nm, using a Philips PW 1370 X-ray diffractometer with CuKα-radiation. A Jeol 200CT transmission electron microscope was used for direct observations of grain sizes in the region < 100nm. Grain sizes > 100 nm were calculated by the method of Oel [17], using a Jeol JSM-35CF scanning electron microscope.

RESULTS AND DISCUSSION

The chloride synthesis resulted for the Yttrium containing compositions in a powder with crystallite "chains" with a primary crystallite size of 8 nm. These chains form very small cluster-"aggregates" (with a size of about 13 nm) at the intersections of the chains and weak agglomerates which can be completely destroyed by isostatic pressing [15]. For the powders containing only Ceria the primary crystallite size was 11-12 nm (XRLB). For both the ZY6 as the ZY4C2 powder a BET-surface of about 120 m^2/g was measured. The BET-surface of ZC6 was 87 m^2/g. According to the relation $d = 6/\rho * S_{BET}$ the crystallite size calculated from S_{BET} corresponds with those mentioned above and one can conclude that all three powders are not aggregated [15]. This means that, except for the primary crystallite size, the microstructures are similar.

The dilatometer curves for a ZY6 and ZC6 compact are shown in fig. 1 (the curve of ZY4C2 is similar to the curve of ZY6). As can be seen from this figure, densification starts for both compacts close to 900°C. At this temperature is the ZY6 compact 3% more dense than the ZC6 compact (in terms of relative densities). At 1000°C the compacts are equal dense. The ZC6 compact has reached his final density at 1170°C (92%). However at this point the ZY6 compact is 5% less dense. This means that the overall densification rate of ZC6 in the 1000-1170°C range ($\Delta l/l_o$ per min) is larger than the densification rate of ZY6. By increasing the temperature the ZC6 compact does not densify anymore whereas the ZY6 compact densifies to more than 97% of the relative density (which is reached at a temperature of about 1370°C).

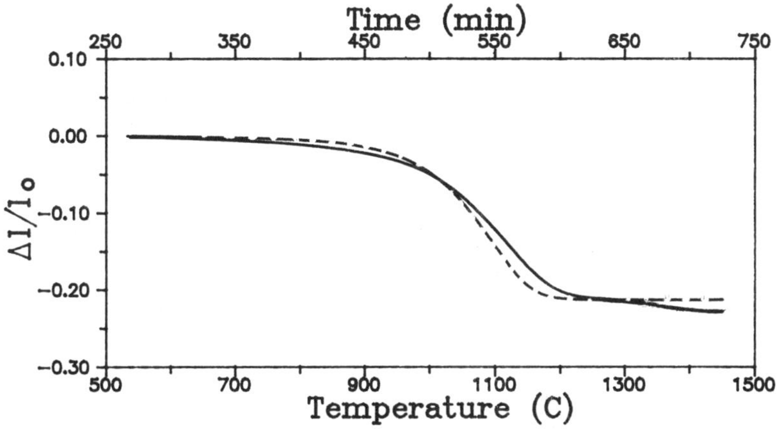

Figure 1: Dilatometer curves of an ZY6 compact (solid line) and a ZC6 compact (dashed line). The curve for a ZY4C2 compact is identical to the ZY6 curve. Heating rate: 2 K/min.

The fast but incomplete densification of ZC6 is probably caused by the fast grain growth (at 1400°C the resulting grainsize is 17.5 times larger than the grain size in the ZY6 compact) which reduces the driving force in the final sintering stage. In the initial stage of sintering (from 550-900°C) no difference in pore size and pore distribution seems to be present between the three compacts. In all three cases the poresize at 600°C amounts about 6-8 nm and 18-20 nm at 850°C. However some differences are observed in crystallite size between the compacts (table 1). At 600°C, the Yttria containing compacts have crystallite sizes which are about 10 nm. The ZC6 compact however has a crystallite size of 13 nm.
When increasing the temperature rapid grain growth takes place. This has been described more in detail in [7]. In the ZC6 compact the grain growth is more rapidly (see table 1). This suggests that the grain growth is not inhibited with Cerium as a dopant. The differences in resulting grainsize become larger with increasing temperature (table 1). The restraining effect of Yttrium in the ZY4C2 compact is also clear (especially compared to ZC6) although the inhibiting effect is not so strong as in ZY6.

TABLE 1
Grainsizes of Yttria/Ceria containing zirconia ceramics after sintering at various temperatures.

Specimen	Cryst. size*(nm) $600^{o}C/0hr$	Grain size (nm) $850^{o}C/0hr$	Grain size (μm) $1400^{o}C/2hr$
ZY6	10	19	0.4
ZY4C2	10	20	1.5
ZC6	13	29	7.0

* The grains are denoted as crystallites because they represent the primary crystallites in an aggregate. Above $650^{o}C$ the crystallites in the aggregates have recrystallized to one new crystallite with the size of a densified aggregate and are denoted as grains.

Starting from a general thermodynamical equation a relation has been derived by Garvie and Swain [18], which describes the correlation between the critical grain size (d_c) and the temperature where the transformation from the tetragonal to the monoclinic phase takes place (M_s-temperature).

$$1/d_c = C_1 + C_2 * M_s \qquad (1)$$

In this equation are C_1 and C_2 constants. The M_s-temperature is determined from the dilatometer curves (temperature where sudden dilatation is present which corresponds with the phase transformation).

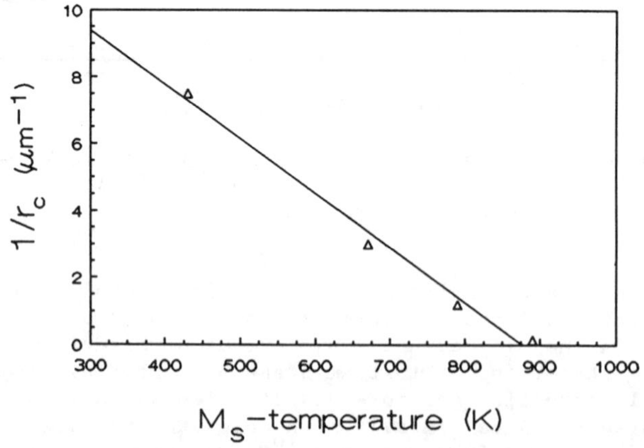

Figure 2: Critical grainsize vs. M_s-temperature for a ZC6 compact.

In case of ZC6, d_c decreases drastically with decreasing M_s-temperature (fig. 2). The following constants are derived: $C_1 = 3.28*10^{7}$ and $C_2 = -3.64*10^{4}$ (these values depend on the composition). These values are close to those derived by Garvie [18] for 8.4 mol% Ca-PSZ (resp. $3.27*10^{7}$ and $-3.65*10^{4}$). This suggests that Ca and Ce should have similar effects on M_s. It can also be seen from figure 2 that (for this composition) d_c should be about 80-90 nm to preserve the tetragonal phase at roomtemperature.

The results suggest that there are two ways to preserve the tetragonal phase i) increasing the dopant concentration (an increase in dopant concentration will especially decrease the value of C_1) and ii) the usage of a grain growth inhibitor like Yttrium).

Finally preliminary ageing experiments indicate that the stability of the tetragonal phase is improved when a certain amount of Ceria is added. This shows that the positive effects of Ceria and Yttria can be combined in one ceramic.

ACKNOWLEDGEMENT

This research was partly supported by the Innovative Research Program on Technical Ceramics (IOP-TK) with financial aid of the Dutch Ministry of Economic Affairs.

REFERENCES

1. Claussen N., in Science and Technology of Zirconia II" ed. by N. Claussen, M. Rühle and A.H. Heuer, American Ceramic Society, Columbus, Ohio, 1984 p.325.
2. Winnubst A.J.A., Burggraaf. A.J., The aging behavior of ultrafine-grained Y-TZP in hot water, in Advances in Ceramics, Vol. 24: Science and Technology of Zirconia III ed. T. Noma, M. Yoshimura, S. Somiya, The American Ceramic Society, Inc. 1988 pp. 39-47.
3. Sato T., Shimada M., J. Am. Ceram. Soc., 1985 **68** 356.
4. Tsukuma K., Am. Ceram. Soc. Bull., 1986 **65** 1386-1389.
5. Sato T., Shimada M., Am. Ceram. Soc. Bull., 1985 **64** 1382-1384.
6. Reyes-Morel P.E., Chen I-W., J. Am. Ceram. Soc., 1988 **71** 343-353.
7. Theunissen G.S.A.M., Winnubst A.J.A., Burggraaf A.J., in preparation.
8. Theunissen G.S.A.M., Winnubst A.J.A., Burggraaf A.J., J. Mater. Sci. Letters, 1989 **8** pp. 55-57.
9. Leach C.A., J. Mater. Sci. Letters, 1987 **6** pp. 303-304.
10. Lange F.F., Marshall D.B., Porter J.R., "Controlling microstructures through phase partitioning from metastable precursors: The ZrO_2-Y_2O_3 system" in Ultrastructure processing of advanced ceramics ed. J.D. MacKenzie, D.R. Ulrich, Wiley and Sons, New York, Chichester, (1988).
11. Wakai F., Nagono T., J. Mater. Sci. Letters, 1988 **64** pp. 607-609
12. Burggraaf A.J., Theunissen G.S.A.M., Winnubst A.J.A., Synthesis and properties of nanoscale ceramics and composites, Proc. 1st. Eur. Cera. Soc. Conf. (this issue).
13. Sato T., Ohtaki S., Endo T., Shimada M., Int. J. High. Technology Ceramics, 1986 **2** pp. 167-177.
14. Duh J-G., Dai H-T., Hsu W-Y., J. Mater. Sci. 1988 **23** pp. 2786-2791.
15. Groot Zevert W.F.M., Winnubst A.J.A., Theunissen G.S.A.M., Burggraaf A.J., accepted for publication in J. Mater. Sci..
16. Dollimore D., Heal G.R., J. Appl. Chem. 1964 **14** pp. 109.
17. Oel H.J., Ber. Dtsch. Keram. Ges. 1966 **43** pp. 624-631.
18. Garvie R.C., Swain M.V., J. Mater. Sci., 1985 **20** pp. 1193-1200.

SOL-GEL PREPARATION OF COMPLEX PEROVSKITE CERAMICS

FREDERIC CHAPUT and JEAN-PIERRE BOILOT
Groupe de Chimie du Solide, Laboratoire de Physique de la Matière Condensée,
Ecole Polytechnique, 91128 Palaiseau Cédex, France

ABSTRACT

Using an alkoxide-hydroxyde route, in which the precursors are titanium, zirconium, hafnium isopropoxides and hydrated alkaline earth (Ba, Sr) hydroxides in a solid state, we have prepared a homogeneous solution which only advances by condensation of the $Ti(OH)_6^{2-}$ Ba^{2+} type species and leads to a controlled stoichiometry powder. Dried powders exhibit the cubic perovskite structure.
A direct sintering at 1150°C, without calcination, leads to highly dense $BaTiO_3$ bodies with fine-grained uniform microstructure (1µm) which exhibit high permittivity value at room temperature (K=5400).

INTRODUCTION

Ferroelectric properties of $BaTiO_3$ ceramics strongly depend on the average grain size. At 25°C, an optimal value of the permittivity (K≈5000-6000) is observed for fine-grained bodies (0.7-1 µm) (1). Consequently, there is an increasing interest in the chemical techniques to prepare fine barium titanate powders. Coprecipitation from inorganic salt solutions is a well known technique to prepare fine powders. However, concerning barium titanate, the rapid condensation reactions between Ti-OH groups involve the growth of TiO_2,xH_2O dense particles corresponding to the formation of an highly cross-linked Ti-O network. A mixture of hydrated titanium dioxide and barium carbonate is then produced by coprecipitation (2) and, during calcination near 1000°C, other titanates, such as Ba_2TiO_4 and $BaTi_4O_9$, are formed to be converted to $BaTiO_3$.

In order to avoid or to control the Ti-O-Ti cross-linking formations, chelating organic ligands such as glycols, organic acids or β-diketones (3-5) can be used. Other methods are known of obtaining barium titanate powders: atomization of salt solution (use of unidentate ligands) (6), hydrolysis of alkoxides (7-8), aging of TiO_2 gels in the presence of $Ba(OH)_2$, hydrolysis of titanic acid esters in aqueous and basic solutions, containing barium ions (9).

We have recently investigated (10-11) an alkoxide-hydroxide sol-gel route to prepare perovskite powders with the formula $Ba_xSr_{1-x}(Ti_uZr_vHf_w)O_3$ (with $0 \leq x \leq 1$ and u+v+w=1). This method of processing of perovskite powders makes possible the synthesis of the perovskite structure at very low temperature (100°C) and the preparation of dense bodies by sintering at low temperature (1200°C) without prior calcination. We present here new results concerning the alkoxide-hydroxide route in which the barium precursor is the hydrated barium hydroxyde ($Ba(OH)_2.8H_2O$) in the solid state.

EXPERIMENTAL PROCEDURE

The first reaction which takes place during the alkoxide-hydroxide method is nucleophilic. The alkoxide, such as $Ti(OR)_4$, is hydrolyzed by water molecules and hydroxyl ions, leading to $Ti(OH)_6^{2-}$ species which are neutralized by Ba^{2+} or Sr^{2+} ions. The unstable transition compound goes through a condensation which leads directly to the perovskite end product. The difficult thing is to completely avoid the formation of hydrated titanium oxide.

We begin with a mixture containing a high concentration of the alkaline earth hydroxide to increase the basicity, and consequently the kinetics of the condensation of $Ti(OH)_6^{2-}$ Ba^{2+}. The barium precursor is the hydrated barium hydroxide ($Ba(OH)_2.8H_2O$) in the solid state. We employ an alkoxide precursor with a relatively long organic group (titanium isopropoxide) to decrease the hydrolysis rate, the molecular associations (Ti-O-Ti bonds) and the volatility. This allows, through vigorous stirring, the creation of a homogeneous Ba-Ti solution which can be heated at 60-80°C before the condensation. Gelling occurs usually within three minutes after a homogeneous solution was obtained. The IR spectrum exhibits bands at 380 and 570 cm^{-1} clearly demonstrating that the gel presents the perovskite organization at least at the molecular scale.The resulting slurry is then either oven-dried in air for 2hours at 150°C. Titanium can be also partially or entirely replaced by zirconium or hafnium atoms. Chemical analysis by X-ray fluorescence shows that all these powders exhibit the starting solution stoichiometry (Ti/Ba= 1 ± 0.005 for the pure barium titanate) indicating that our experimental procedure allows complete formation of the perovskite phase.

The dried powders are passed through a 100 µm nylon sieve, then are cold isostatically pressed at 200MPa. The stoichiometric barium titanate green samples are sintered without calcination, in air, for 2h at different temperatures (between 1100 and 1450°C).

RESULTS (stoichiometric barium titanate)

Electron micrograph (figure 1a) of the alkoxide-hydroxide derived powders show familiar structures consisting of agglomerates of 80nm in size. X-ray patterns exhibit peaks corresponding to the cubic perovskite structure and the specific

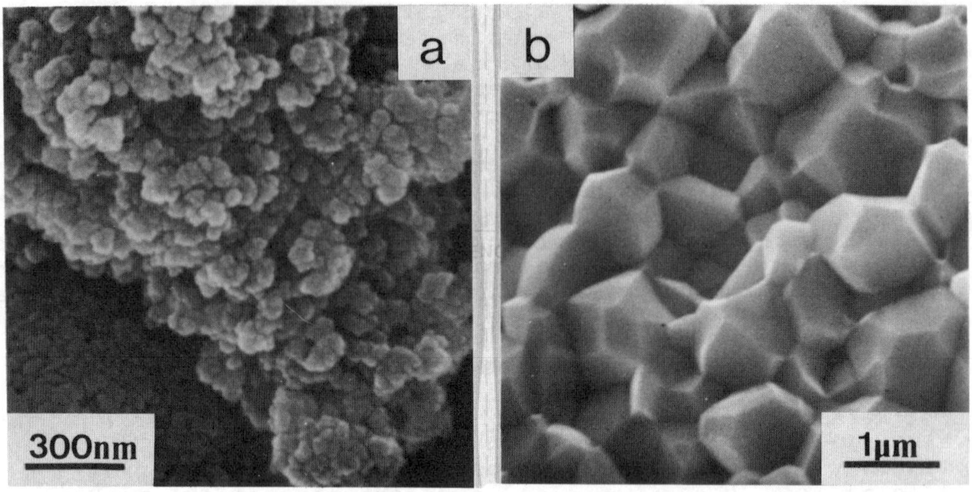

Figure 1. a) Morphology of the pure barium titanate powders after conventional oven-drying. b) Microstructure of the pure barium titanate ceramics sintered at 1150°C with fine-grained uniform microstructure (1µm).

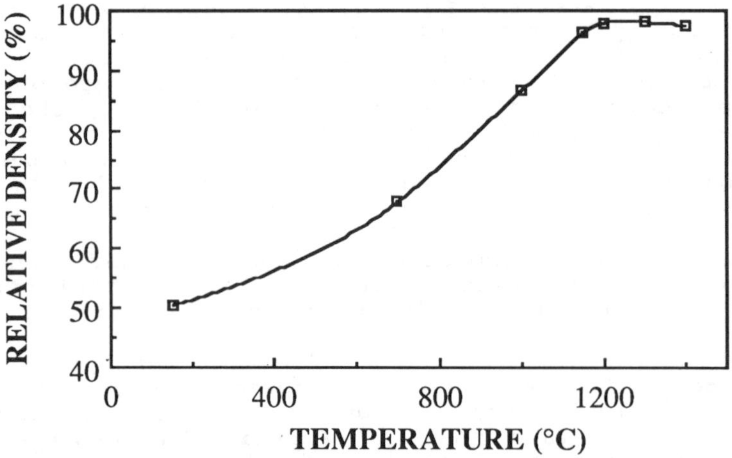

Figure 2. Evolution of the sintering ratio (% of the theoretical density) as a function of the sintering temperature for barium titanate ceramics.

surface area progressively decreases from $50m^2/g$ after drying to $1m^2/g$ at 1050-1100°C.

A value of 50-55% of the theoretical density is observed for isostatic pressed samples and a dilatometric curve indicates that the sintering takes place between 800 and 1150°C. Figure 1b shows the microstructure of barium titanate ceramics sintered at 1150°C. Firing at this temperature is sufficient to reach 97% theoretical density and a fine-grained uniform microstructure is observed with a 1-1.5 µm grain size. Above 1200°C, abnormal grain growth is observed and at 1450°C, the whole matrix is transformed into a large-grained microstructure with 25-35 µm grain size. Due to the formation of intergranular porosity, a slight decrease of densification is noted above 1200°C (figure 2).

Table 1 presents the dielectric characteristics at 20°C and 1kHz for samples sintered at different temperatures. The dielectric constant is greatest (about 5400) for a sintering temperature of 1150°C corresponding to fine-grained ceramics (<1 µm). As expected (12), the permittivity sharply decreases ($K_{20°C}$≈1500-2000) for coarse-grained materials (>5 µm). The dissipation factor remains very low (2.9 % for the fine-grained ceramic) but increases when the sintering temperature rises.

TABLE 1.
Electrical characteristics (1kHz) at 20°C for pure barium titanate ceramics sintered at different temperatures.

Sintering temperature (°C)	Dielectric constant	Dissipation factor (%)
1100	3900	1.9
1150	5400	2.9
1200	4100	2.6
1250	3700	3.3
1300	3500	4.2
1350	3300	5.1
1400	3100	4.8
1450	2050	4.7

CONCLUSION

A way of manufacturing $BaTiO_3$ fine powders is to mix barium and titanium precursors in a strongly basic medium containing KOH or $Ba(OH)_2$ (which can result from the hydrolysis of an alkoxide). In this method, the condensation of the $Ti(OH)_6^{2-}$ Ba^{2+} species directly leads to the perovkite compound. Unfortunately, the formation of hydrated titanium oxide by self-condensation between Ti-OH groups is generally not completely prevented and stoichiometry discrepancies result.

Using an alkoxide-hydroxyde route, in which the precursors are the titanium isopropoxide and the hydrated barium hydroxide in a solid state, we have prepared a homogeneous Ba-Ti solution which only advances by condensation of the $Ti(OH)_6^{2-}$ Ba^{2+} species and leads to a controlled stoichiometry powder. The same route can be used to prepare perovskite powders with the complex formula: $Ba_xSr_{1-x}(Ti_uZr_vHf_w)O_3$ (with $0 \leq x \leq 1$ and u+v+w=1).

Concerning pure barium titanate, dried powders exhibit the cubic perovskite structure. By using a direct sintering at

1150°C, highly dense $BaTiO_3$ bodies with fine-grained uniform microstructure (1μm) were prepared. Electrical property measurements on these bodies (1kHz) show high permittivity value at room temperature ($K_{20°C}$ =5400).

REFERENCES

1. Hennings, D., Barium titanate based ceramic materials for dielectric use. Int. J. High Technology Ceramics, 1987, **3**, 91-111.

2. Savos'kina, A.I.,Limar', T.F. and Kisel', N.G., Formation of $BaTiO_3$ from coprecipitation products. Izv. Akad. Nauk. SSSR, Neorgan. Mat. ,1975, **11** (12), 2245.

3. Rehspringer, J.L. and Bernier, J.C., Chemical processing of $BaTiO_3$ based ceramics. Mat. Res. Soc. Symp. Proc.,1986, **72**, 67-72.

4. Hennings, D. and Mayr, W., Thermal decomposition of (BaTi) citrates into barium titanate. J. Solid State Chem.,1978, **26**, 329.

5. Ali, N.J. and Milne, S.J., Synthesis and properties of barium titanate powder derived from a catechol complex. Br. Ceram. Trans. J., 1987, **86** (4), 113-117.

6. Shaikh, A.S. and Vest, G.M. Kinetics of $BaTiO_3$ and $PbTiO_3$ formation from metallo-organic precursors. J. Am. Ceram. Soc., 1986, **69**(9), 682-688.

7. Mazdiyasni, K.S., Dolloff, R.T. and Smith II, J.S., Preparation of high-purity submicron barium titanate powders. J. Am. Ceram. Soc. , 1969, **52**(10), 523.

8. Kasai, T., Ozaki, Y. and Yamamoto, S., Preparation of $BaTiO_3$ and $SrTiO_3$ from metal alkoxides. Yogyo-Kyokai-Shi, 1987, **95** [10], 1000.

9. Flaschen, S.S., Aqueous Synthesis of Barium Titanate. J. Am. Chem. Soc., 1955, **77**(12), 6194.

10. Chaput, F. and Boilot, J.P., Chemical processing of (Ba,Sr)(Ti,Zr)O_3 perovskite ceramics. J. Mater. Sci. Lett.,1987, **6**, 1110-1112.

11. Chaput, F. and Boilot, J.P., Ferroelectric ceramics prepared from a sol/gel route. Electroceramics, King College, London 16th-18th December 1987 p.21.

12. Arlt, G., Hennings, D. and Dewith, G., Dielectric properties of fine-grained barium titanate ceramics. J. Appl. Phys., 1985, **58**, 1619.

BARIUM TITANATE MONOLITHIC GELS

FREDERIC CHAPUT and JEAN-PIERRE BOILOT
Groupe de Chimie du Solide, Laboratoire de Physique de la Matière Condensée,
Ecole Polytechnique, 91128 Palaiseau Cédex, France

ABSTRACT

Pure and doped barium titanate monolithic gels have been prepared by hydrolysis of a mixture of titanium alkoxide and acetic acid with an aqueous solution of barium acetate. As previously observed by S. Doeuff, I.R. experiments show that the titanium alkoxide is chemically modified by the bridging acetate ligands which partially substitute the OR groups and increase the coordination number of titanium atoms. This allows to stabilize the titanium precursor and consequently to avoid the precipitation of titanium oxide. By using the monomeric $Ti(O^{i}Pr)_4$ precursor, barium titanate films have been prepared by the dip-coating technique and the optically clear gel turns in the perovskite phase at 800°C.

INTRODUCTION

The sol-gel process is a metal organic processing route which can be adapted to give thin films, monoliths (1) or powders by direct precipitation (2). For multicomponent oxides, it is generally admitted that the sol-gel process leads to a great homogeneity (3). This is rather unexpected, because alkoxides may exhibit very different rate of hydrolysis. Moreover, the degree of oligomerization of alkoxides can induce differences in the processing and in the resulting products, for instance, affecting homogeneity at a molecular level. In this study sol-gel route has been used for the deposition of pure and doped barium titanate thin film. Changes induced by the molecular association of titanium alkoxides on the process are discussed (final composition of multicomponent systems, reaction kinetics and physical properties of the solutions).

EXPERIMENTAL PROCEDURE

Barium titanate gels were obtained through hydrolysis and polycondensation of modified titanium alkoxides (4). Hydrolysis was performed with a concentrated aqueous solution of barium acetate. Titanium precursors were modified with acetic acid in order to get less moisture sensitive compounds. Carboxylic acid react with the precursors at a molecular level and therefore change the hydrolysis-condensation reactions (5). We have employed alkoxides with different organic groups. In one case (titanium isopropoxide) the steric hindrance prevents the molecular associations between Ti atoms, in the other case (titanium ethoxide) the titanium alkoxide presents a trimeric structure (6). Barium titanate films have been prepared by the dip-coating technique.

RESULTS AND DISCUSSION

Gel formation

Figure 1 gives the infra-red spectrum recorded from the clear solution freshly prepared with titanium isopropoxide and glacial acetic acid ($CH_3COOH/Ti(O^iPr)_4$ molar ratio = 4). The absorption bands of vibrations typical of the pure $Ti(O^iPr)_4$ are observed (7). Isopropyl alcohol (3300 cm^{-1}), free acetic acid (1720 cm^{-1}) and isopropyl acetate (1745 cm^{-1}) can be also displayed. As previously reported by S.Doeuff, I.R. experiments show that acetate groups (1430 and 1550 cm^{-1}) act as bridging ligands ($\Delta\nu(C\text{-}O) \approx 120 cm^{-1}$) (5,8). They react chemically with the titanium precursor by substituting the OR groups. This step leads to a new precursor $Ti(OR)_x(OAc)_{6-x}$ in which the coordination number of titanium atoms is increased (6). There are no significant differences between the IR spectra of samples prepared with isopropoxide and ethoxide.

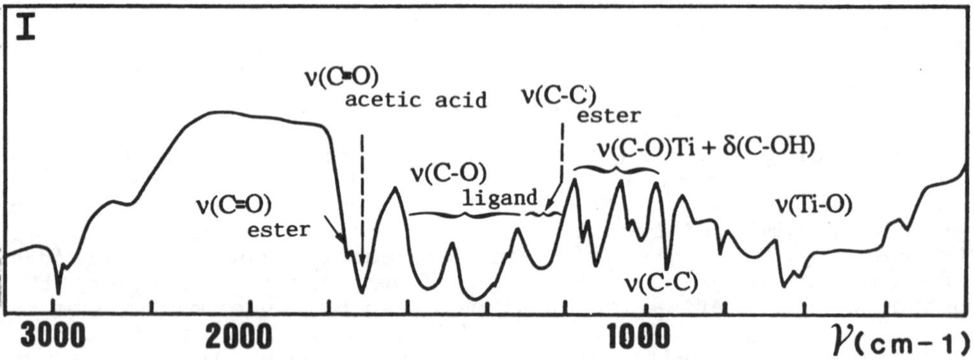

figure 1. Infra-red spectrum of the clear solution freshly prepared with $Ti(O^iPr)_4$ and AcOH ($AcOH/Ti(O^iPr)_4$ molar ratio=4).

After ageing for several hours, a colloidal precipitate appears in these solutions, it may be attributed to the formation of Ti-O-Ti bonds. In fact, water which is produced by slow esterification reactions gives rise to the hydrolysis of OR groups and the condensation of Ti-OH species. The corresponding infra-red spectrum exhibits a large absorption band at 470-650

cm^{-1} due to the Ti-O-Ti bonds (9).

In order to avoid the precipitation of titanium oxide, it is necessary to pour the barium solution immediately after that the acetic acid - titanium alkoxide - alcohol mixture has been made. After acidification, hydrolysis is performed by adding a saturated aqueous solution of barium acetate (H_2O/Ti molar ratio = 22.5). Gelation time and gel aspect strongly depend on the concentration of different reagents and on the nature of the titanium alkoxide (Table 1). The infra-red spectrum of the gels indicates that most of the OR groups are removed. However, acetate groups are not easily hydrolyzed, which allows to control the condensation. So, formation of monolithic gels is easier.

Table 1.
Mixture and gel aspects and gelation time as a function of the amount of acetic acid (X=AcOH/$Ti(OR)_4$ molar ratio) and of alcohol (Y=ROH /$Ti(OR)_4$ molar ratio)

a) $Ti(OEt)_4$-AcOH-EtOH-H_2O-$Ba(OAc)_2$ system

X or **Y**		mixture aspect	gelation time (min)	gel aspect
Y=6	0.5	whitish slurry	--	heterogenous
	1	clear	7	transparent
	6	clear	197	transparent
X=2.5	**0**	colloidal suspension	--	heterogenous
	1	colloidal suspension	--	heterogenous
	3	clear	12	transparent
	10	clear	46	transparent

b) $Ti(O^iPr)_4$-AcOH-^iPrOH-H_2O-$Ba(OAc)_2$ system

X or **Y**		mixture aspect	gelation time (min)	gel aspect
Y=6	1	whitish slurry	--	heterogenous
	3	whitish slurry	--	heterogenous
	4	clear	10	transparent
	6	clear	40	transparent
X=4	**0**	colloidal suspension	--	heterogenous
	1	colloidal suspension	--	heterogenous
	3	clear	7	transparent
	10	clear	18	transparent

Thermal behaviour of gels

The resulting transparent monolithic gels, after ageing and drying at 60°C for several hours, were heated at different temperatures. During thermal treatment of gels, several peaks are noted on the differential thermal analysis and thermogravimetric anlysis curves. The first exothermic peak near 350°C is due to the pyrolysis of acetyl groups and the second, near 680°C may be attributed to the crystallization of the perovskite phase. The last exotherm is also associated to the decomposition of small amounts of barium carbonate. $BaCO_3$ is formed during the combustion of organic ligands. It is very stable and slows down the perovskite crystallization considerably. The weight loss is 40% between 100 and 700°C.

The development of the perovskite phase was also revealed by X-ray diffraction (XRD) and infra-red spectroscopy (KBr pellets) taken at room temperature from samples heated from 400 to 1200°C. The xerogels appear to be amorphous by XRD and the perovskite formation strongly depends on the nature of titanium alkoxides. $BaCO_3$ is present in the gel between 450 and 750°C (absorption bands at 1400, 830 and 660 cm^{-1}) and the obtained mixture is completely transformed into the perovskite structure above 750°C in the case of xerogel prepared from titanium isopropoxide. In the other case (titanium ethoxide), we still observed the $BaCO_3$ phase above 800°C. The oligomerization of $Ti(OEt)_4$ seems to involve a greater molecular heterogeneity between Ti and Ba atoms. Consequently during calcination other titanates, such as Ba_2TiO_4 are formed to be converted to $BaTiO_3$.

To sum up, the gel structure can be seen as a titanium oxide polymer (figure 2a) with localization of the Baryum salt in pores (20-30 nm in size) (10). A careful control of the rheological properties of the organometallic solution ($Ti(O^iPr)_4$-AcOH-^iPrOH-H_2O-$Ba(OAc)_2$ system) allows us to deposit $BaTiO_3$ thin film onto fused quartz substrate by using dip coating technique (figure 2b).

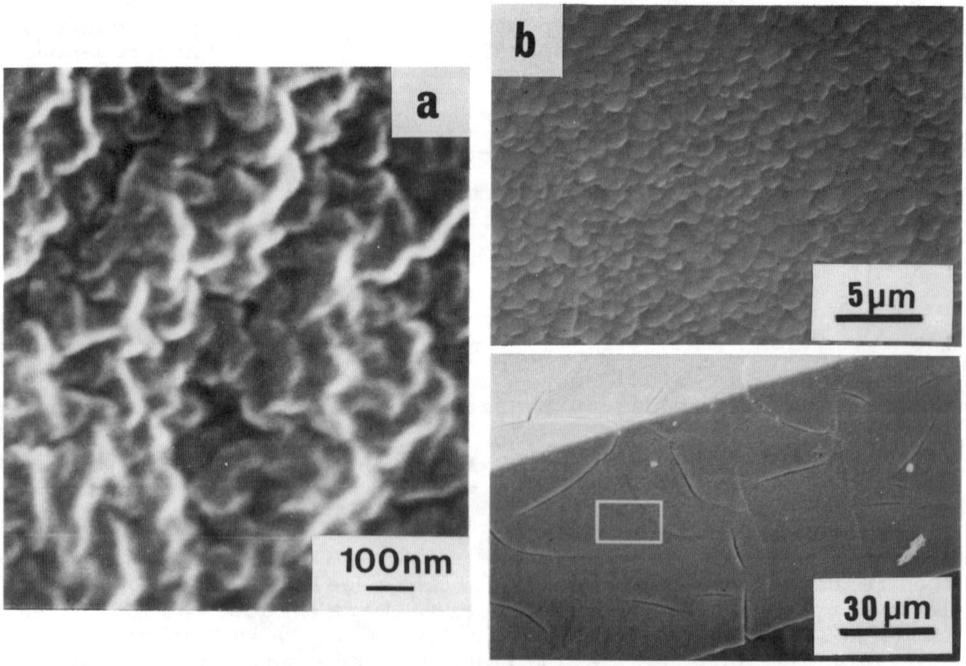

figue 2. SEM micrographs of : a) $BaTiO_3$ gel dried at 60°C; b) $BaTiO_3$ film calcined at 1200°C (isopropoxide route).

CONCLUSION

The sol-gel technique has been sucessfully used to prepare $BaTiO_3$ gels from several titanium alkoxides and nonalkoxide Ba precursor. The introduction of a complexing agent in the organic solution reduces the functionality of titanium atoms. This allows to avoid the direct precipitation of titanium oxide. The formation of the perovskite phase from these gels is highly sensitive to the nature of the starting precursor. The use of an oligomeric alkoxide leads to a multiphase product above 1000°C. In contrast, by using the monomeric titanium isopropoxide, thin fims of pure barium titanate can be prepared at 800°C.

REFERENCES

1. Pouxviel, J.C. and Boilot, J.P., Gels from a double alkoxide : $(BuO)_2$-Al-O-Si-$(OEt)_3$. J. Mater. Sci., 1989, **24**, 321-327.

2. Chaput, F., Boilot, J.P. and Beauger, A., Alkoxide-hydroxide route to synthetize barium titanate based powders. J. Am. Ceram. Soc., to be published.

3. Chaput, F., Boilot, J.P., Lejeune, M., Papiernik, R. and Hubert-Pfalzgraf, L.G., Low-temperature route to lead magnesium niobate. J. Am. Ceram. Soc.,in press.

4. Chaput, F. and Boilot, J.P., Chemically derived Ba titanate gels and ceramics. In High Tech. Ceramics, ed. P. Vincenzini, Elsevier Science Publishers B.V., Amsterdam, 1987, pp 1459-68.

5. Doeuff, S., Henry, M., Sanchez, C. and Livage, J., Hydrolysis of titanium alkoxides : modification of the molecular precursor by acetic acid. J. Non-Cryst. Solids, 1987, **89**, 206-216.

6. Doeuff, S., Elaboration de TiO_2 par voie sol-gel : du précurseur au matériau. Thesis, University Paris VI, France, 1988.

7. Kriegsmann, H. and Licht, K., Spectroscopic investigations of esters of titanic and silicic acid. Z. Electrochem., 1958, **62**, 1163-74.

8. Thiele, K.H. and Panse, M., Preparation of titanium acetates from tetramethyl and tetrabenzyl titanium. Z. anorg. allg. Chem., 1978, **441**, 23-28.

9. McDevitt, N.T. and Baun, W.L., Infrared absorption study of metal oxides in the low-frequency region (700-240 cm^{-1}). Spectrochim. Acta, 1964, **20**, 799-808.

10. Hennings,D., Review of chemical preparation routes for barium titanate. Electroceramics, King College, London 16th-18th December 1987 p.1-10.

BARIUM TITANATE PROCESSING: EFFECT OF RAW MATERIAL NATURE ON THEIR SYNTHESIS AND SINTERING BEHAVIOUR

J.F.FERNANDEZ, P.DURAN and C.MOURE
Instituto de Cerámica y Vidrio (CSIC), Electroceramics Department,Arganda del Rey, 28500 Madrid, Spain.

ABSTRACT

The solid state formation of barium titanate have been studied with respect to the physical and chemical characteristics of the raw materials. Special attention has been paid to the impurity levels, particle size distribution and particle morphology of the $BaCO_3$ and TiO_2. The effect of the processing parameters such as milling media, calcining temperature and time, and sintering schedule have been also determined.

In a general mode, small particle sizes of starting materials lead to a lower synthesis temperatures and times, but a strong difference in the sizes of TiO_2 and $BaCO_3$ causes an inhomogeneous reaction, with formation of agglomerates and other phases as Ba_2TiO_4. As the purity of $BaCO_3$ increases, the synthesis temperature decreases which leading to lower particle sizes of the obtained $BaTiO_3$.

INTRODUCTION

The solid state synthesis of $BaTiO_3$ is commonly carried out by calcining mixtures of $BaCO_3$ and TiO_2. Because of the reaction nature (1-2), the final particles of the obtained $BaTiO_3$ maintain the shape of the TiO_2 particles, with a diameter increase of 27% (3). Therefore the reaction occurs by diffusion of Ba cation througth the titania crystal. Owing to this fact, the aggregation of TiO_2 play a very important role in the synthesis reaction process (4).The main goal of the present work is to determine the influence of the raw material characteristics on the final features of $BaTiO_3$ powder synthesized from them.

EXPERIMENTAL METHODS.

Three $BaCO_3$ of different purity and morphology are chosen.

Table.1 shows the chemical analysis of them. The higher impurity levels correspond to isovalent cations as Sr^{2+}, and Ca^{2+}. A and A* are different lots of the same product. $BaCO_3$-A shows the smaller mean grain size, and needle-shaped particles. The other carbonates show rounded particles and a higher aggregation state. The ball milling during 24 hours causes an increase in grain size homogeneity and lowers the mean size.

Three TiO_2 of different chemical and granulometry nature were used, (Table.2). TiO_2-1 is a pigment grade material with 0,37 μm mean grain size. TiO_2-2 is a modified pigment grade material without Nb_2O_5 and slighly higher mean grain size (0,45 μm). TiO_2-3 is a coarse grained material (~13μm) with the highest purity level. Their crystalline structure is mainly anatase with 5, 2 and 45% rutile content respectively.

TABLE 1
Characteristic of $BaCO_3$ studied

	% Total Imp.	SrO	CaO	Na_2O	K_2O	size (μm)
BaCO3-A	0,66	0,63	0,004	-	-	1,8
BaCO3-A*	0,32	0,12	0,07	0,125	-	2,4
BaCO3-B	1,65	1,38	0,06	0,15	-	2,8
BaCO3-C	0,32	0,16	0,075	0,008	0,05	4,6

TABLE 2
Main impurities detected in TiO_2 (% weight).

	Al_2O_3	SiO_2	Nb_2O_5	CaO	K_2O	P_2O_5	SO_4
TiO_2-1	0,005	0,005	0,35	0,001	0,05	0,38	0,065
TiO_2-2	0,18	0,12	0,06	0,012	0,16	0,38	0,009
TiO_2-3	0,013	0,005	0,01	0,006	0,003	0,025	0,05

The raw materials were weighted, blended, and homogeneized by two different methods: wet ball milling and high speed turbine stirring (6000 rpm). The calcining temperature was 1150ºC for all the samples. The synthesis process was followed by TGA and DTA. The sintering temperatures ranged between 1175 and 1375ºC. The reacted powders and the sintered materials were identified by XRD and observed by SEM.

RESULTS AND DISCUSSION

The final blend homogeneity depends strongly of the agglomeration degree of the raw materials. So, in the A*-2 mixture the TiO_2 particles are agglomerated between the carbonate particles (Fig.1a). When the blending is made by stirring a lower agglomeration state is shown (Fig.1b). The granulometrical distribution is similar for both blend methods. Fig.2 shows the DTA curves for differents mixtures. The weight loss end coincides with a sharp increasing in the base line. Shuch

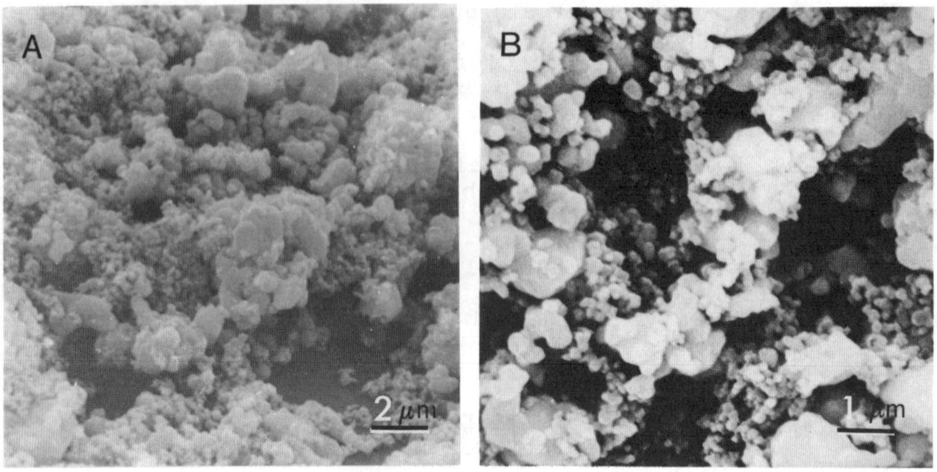

Figure 1. SEM micrographs of A*-2 mixtured by blending (a) or by stirring (b).

increasing can be attributed to the sintering beginning of the $BaTiO_3$-synthesized particles. It appears at lower temperatures when the particle size of the raw materials is smaller. For the same particle size, the temperature is lower when the $BaCO_3$ purity is higher.

In the powders with small particle size of titanium dioxide, the calcination at 1150ºC leads to a barium titanate powder which is, according to the XRD analysis, very pure except for the small traces of unreacted $BaCO_3$. When this last compound is used with large agglomerates or with large grain sizes, formation of barium ortotitanate Ba_2TiO_4 is observed (A milling-1,C-1). If coarse grained TiO_2 is used, the formed Ba_2TiO_4 remains, even for temperatures above 1225ºC (A-3). The rutile phase present in the titanium dioxide labeled TiO_2-3 plays a role less important than their high grain size, according with the author's experience exposed elsewhere (5).

At the same time that the synthesis process takes place a grain growth occurs by partial sintering between the emerging $BaTiO_3$ grains. These grains can be broken by milling, being attrition milling the most effective milling procedure.

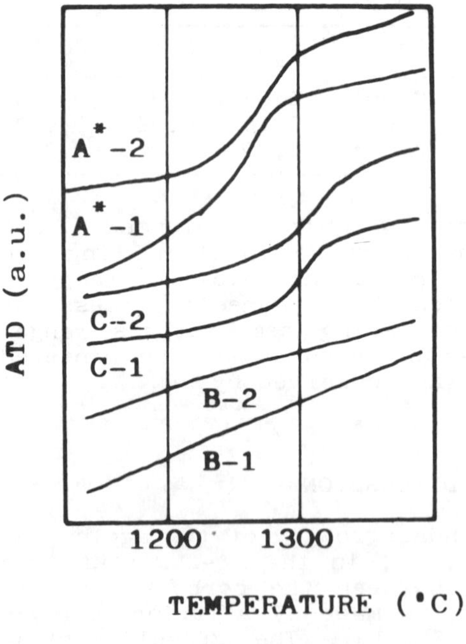

Figure 2. DTA curves for different mixtures.

After the milling step, the grain size of the different reaction powders was studied. Table 3 shows some of the mean grain sizes determined by the Laser Granulometric technique.

TABLA 3
Grain size distribution for different mixtures (µm).The time is refered to the attrition milling process duration.

	A*-2	6h	4h	3h	A*-1	3h	A-2	2h
d 10%		0.32	0.34	0.34		0.32		0.35
d 50%		0.57	0.61	0.66		0.59		0.75
d 100%		1.10	1.28	1.46		1.50		1.75

The highest grain growth with regard to the initial size corresponds to the material with the most fine-grained TiO_2. On the other hand, higher purity of $BaCO_3$ leads to a more homogeneous and smaller grain size of final calcined powder. In any case, the materials obtained from the mixing of raw materials with similar sizes and morphologies between they, show rounded particles with narrow size distributions (fig.3). When the raw materials are coarsed or have disimilar grain sizes, the final powders shows bimodal grain size distribution.

The sintering behaviour (fig.4), of the synthesized powders is straightly related with the initial particle size, size distribution and the morphology. The powders with the smaller grain size, which have been obtained at the lower temperatures and times, reached the highest densities for low sintering temperatures (powder A*-2, sintering temperature of 1200ºC). On the other hand, powders prepared with the same synthesis schedule, and similar final particle size have different sintering behaviour (A*-2 and A*-1 powders). The most important difference between they is the nature and percentage of the impurities. The TiO_2-2 material shows higher

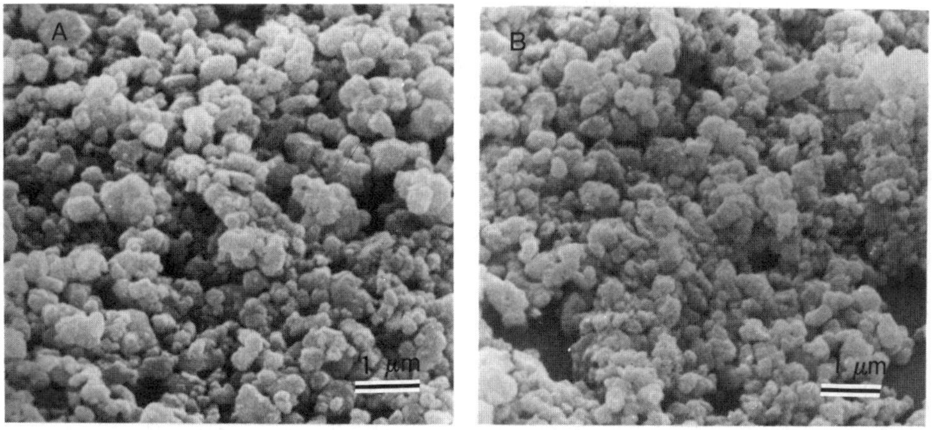

Figure 3. Morphology of $BaTiO_3$ synthesized powders a) A*-1, b) A*-2.

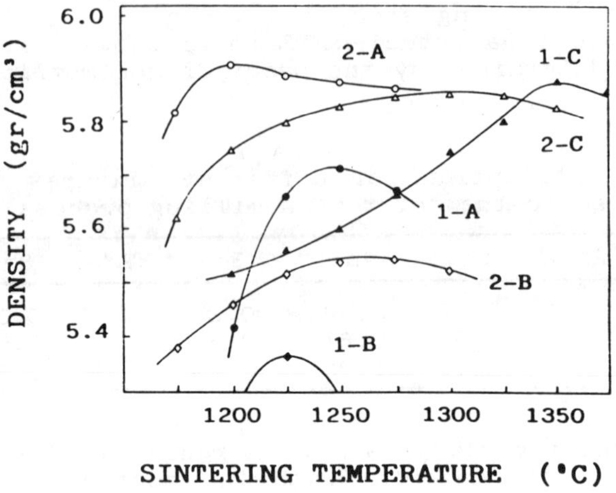

Figure 4. Sintering curves versus temperature.

contents of glass-forming oxides SiO_2, Al_2O_3, K_2O than the TiO_2-1 one, whereas this last have higher amount of Nb_2O_5. For the A*-2 powder, small amounts of liquid phase homogeneously distributed between the titanate grains promotes densification (6), without exaggerated grain growth appearance. On the contrary, the A*-1 powder have a relatively high amount of Nb_2O_5 and very small amounts of glass-forming oxides. The Nb_2O_5 acts as grain growth inhibitor (sintering inhibitor). The density reaches values lower than the A*-2 powders.

ACKNOWLEDGEMENTS

The authors are grateful to RIO RODANO S.A. for finantial support of this work under contract.

REFERENCES

1. Beauger, A., Mutin, J.C., Niepce, J.P., "Synthesis Reaction of Metatitanate $BaTiO_3$. Part 1: Effect of the Gaseous Atmosphere upon the Thermal Evolution of the $BaCO_3$-TiO_2" J.Mater.Sci. 1983,18,3041-46.
2. Idem "Part 2: Study of Solid-Solid Reaction Interfaces" 1983,18,3543-50.
3. Hertl, W., "Kinetic of Barium Titanate Synthesis" J.Am.Cer.Soc. 1988,71,10,879-83.
4. Nomura, T., Yamaguchi, T., "TiO_2 Aggregation and Sintering of $BaTiO_3$ Ceramics" J.Am.Ceram.Soc. 1980,59,4,453-55,458.
5. Fernandez, J.F., Duran, P., Moure, C., To be published.
6. Matsuo, Y., Sasaki, H., "Exaggerated Grain Growth in Liquid Phase Sintering" J.Am.Ceram.Soc. 1971,54,471.

SYNTHESIS OF N-DOPED BARIUM TITANATE BY PRECIPITATION FROM COMPLEXES.

P.H. DUVIGNEAUD, M. GILBERT, M. PROMEL.
Laboratoire de Chimie Industrielle et Analytique - CP 165
Université Libre de Bruxelles, 50 Av. F. Roosevelt, 1050
Bruxelles, Belgium.

ABSTRACT.

La-doped $BaTiO_3$ has been synthesized by a wet chemical route involving the decomposition of Ti-H_2O_2 and Ba-EDTA complexes in an oxidizing medium. Kinetics experiments show that i) the rate of Ti precipitation is greater than that of Ba; ii) H_2O_2/EDTA > 10 are necessary for complete precipitation of the Ba in the temperature range 35-60°c. The peroxidic precursor consists of amorphous micronic spherical particles. The concentrations investigated (2,5 to 5 10^{-2} m/l) tend to agglomerate the particles. However, calcination results in enhancing the submicronic fraction . A fine-grained microstructure (<5μ) is obtained after sintering giving a PTC characteristic of 4 orders of magnitude.

INTRODUCTION

Many wet chemical routes for preparing fine, homogeneous and high-purity $BaTiO_3$ powders are known such as thermal decomposition of oxalates, citrates, carbonates or metallo-organic compounds. Despite their advantages over the mixed oxide way, the chemical routes are not always suited for producing non-agglomerated particles of uniform sizes and sha-pes. Moreover, organic residues and carbonates may remain in the product and complete mixing of the Ti and Ba on an atomic scale seems often doubtful.

A less known method, described by a few authors (1) (2) consists of the precipitation of a peroxidic precursor of $BaTiO_3$ from a Ti-H_2O_2 complex and $BaCl_2$ in an aqueous alkaline solution. As shown in a previous work (3), non-carbonated powders can be obtained by this route but the sinterability is limited by the undefinite morphology of the particles. Recently, it has been demonstrated that such a peroxidic route could give rise to spherical, monodisperse and non-agglomerated particles of the precursor in diluted solutions (10^{-3} to 10^{-2} m/l Ba & Ti) (4).

In this process, Ti(IV) and Ba(II) are dissolved in the form of stable complexes at basic pH values. The heating of the solution in an excess of H_2O_2 causes the decomposition of the Ti and Ba complexes with concomitant homogeneous precipitation of the peroxidic precursor.

The present work is aimed at investigating the applicability of the above process to the synthesis of PTC materials. In particular, emphasis is placed on : i) the possibility of working at higher Ti and Ba concentrations, ii) the possibility of incorporating the usual dopants of the PTC thermistors, iii) the kinetic study of the precipitation giving valuable information on the powder homogeneity.

EXPERIMENTAL

Titanium tetraisopropylate (TiPT) was dissolved in water containing H_2O_2 and ammonia. On the other hand, Ba acetate was dissolved in water containing ethylenediaminotetraacetic acid (EDTA) and ammonia. Both solutions were mixed together so that : i) pH was close to 10, ii) EDTA/Ba = 2, iii) H_2O_2/EDTA = 10 or 20, iiii) Ti and Ba concentrations ranged between 10^{-2} and 10^{-1} m/l. Lanthanum as donor dopant was added to the Ba solution so that La/Ba = 3 10^{-3} (in at%).

The reactor was heated at temperatures ranging between 35°c and 60°c. The Ba, Ti and La contents of the solution were analysed as a function of the reaction time. Liquid samples were removed during the process, evaporated on a filter paper and analysed by X-Ray Fluorescence. The size and morphology of the particles were studied by SEM (Jeol JXM-35) and by LASER Particle Sizer (Fritsch analysette 22). The granulated powders were pressed into pellets, heated at 1400°c for 1 hour and allowed to cool in several steps. The structure parameters were investigted by XRD, IR spectroscopy and EDX. The measure of the PTC characteristic has been described elsewhere (5).

RESULTS

Kinetic aspects of the peroxide precipitation.

Fig. 1 &2 emphasize the influence of temperature and H_2O_2/EDTA ratio on the rate of precipitation of Ti, Ba initially present at 2,5 10^{-2} m/l and La .

The precipitation of titanium is quantitative in the temperature range 35-60°c. The rate of precipitation of this element is only temperature dependent. It is independent of the H_2O_2/EDTA ratio.

On the other hand, the precipitation of barium is strongly dependent on the H_2O_2/EDTA ratio. When H_2O_2/EDTA = 10, only 90% of the Ba is incorporated in the precipitate. When H_2O_2/EDTA=20, more than 99% of the barium is incorporated in the precipitate at a rate growing with the temperature.

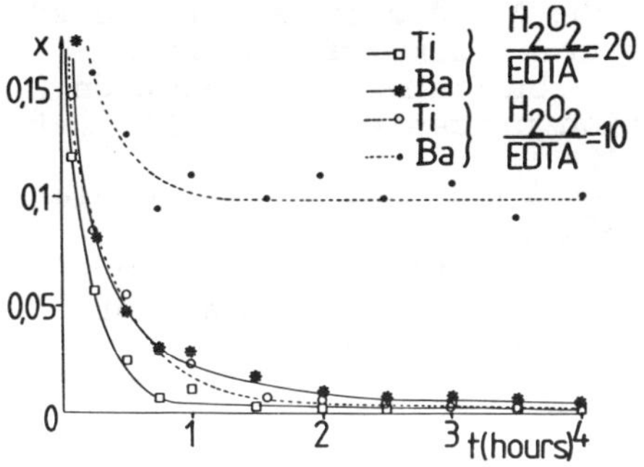

Fig.1: Evolution of the fraction (x) of Ti & Ba remaining in solution (T = 60°c)

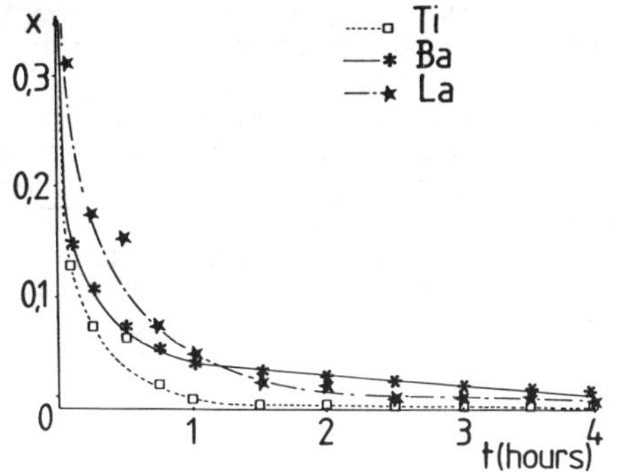

Fig.2: Evolution of the fraction (x) of Ti, Ba & La remaining in solution (T = 50°c)

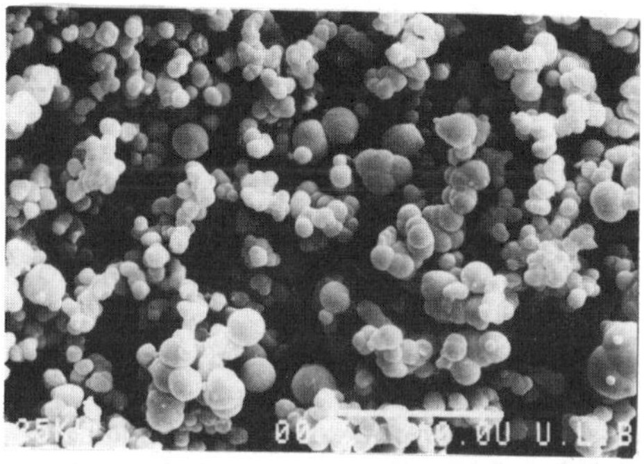

Fig.3: SEM image of powder dried at 110°c

Fig.4: SEM image of powder calcined at 900°c

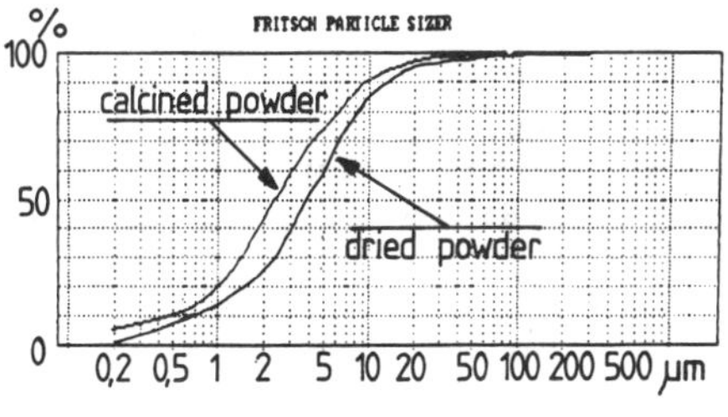

Fig.5: Particle size distribution of powders

Fig.6: Microstructure of La-doped $BaTiO_3$ sintered 1h. at 1400°c

Fig.7: R-T characteristic of La-doped $BaTiO_3$ sample

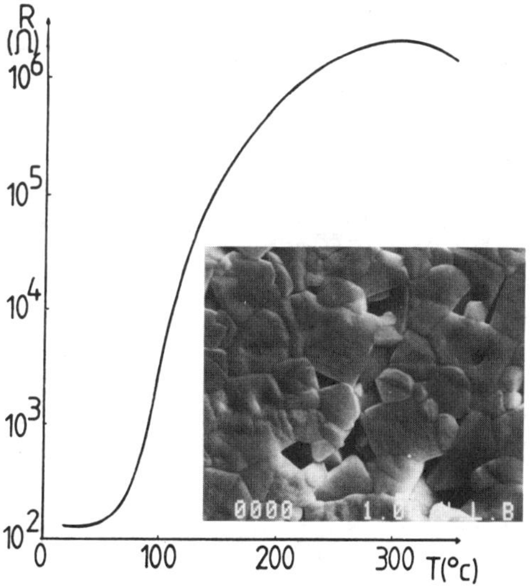

According to Fig. 2, most of the lanthanum is introduced as well in the solid phase after a reaction time of 4 hours.

Thus, the analysis of the solution during the reaction shows evidence that the precipitation processes of Ti(IV) and Ba(II) occur at different rates.

Typical particle sizes and shapes obtained after complete reaction (4 hours) and subsequent drying at 110°c are shown in Fig. 3 and 4. The particles have grown to form micronic spheres.

The particle size and shape are slightly dependent of the temperature and more dependent of the H_2O_2/EDTA ratio. Increasing H_2O_2 contents in the solution tend to increase the particle size and sphericity.

In the concentration range investigated here (2,5 to 5 10^{-2} m/l), the particles are polydispersed and agglomerated. The size distribution of the particles (Fig. 4) centered on a mean diameter of 5 µm confirms that coalescence occurs in the early stages of the process.

Nevertheless, the monomodal character of the size distribution shown in Fig. 4 suggests that only one type of particles is formed.

Nature and processing of the precursor.

The amorphous character of the precursor dried at 110°c is evidenced by XRD. The IR spectrum of the precursor reveals the absorption bands of water at 1670cm^{-1} & 3400 cm^{-1}, H_2O_2 (1400cm^{-1} & 860cm^{-1}) and Ti-O (600cm^{-1}). The CO_3^{2-} bands which interfere with the H_2O_2 bands are weak so that the barium seems only slightly carbonated. Alkyl and ammonia groups are not detected.

The thermal decomposition of the precursor is characterized by a mean weight loss of 16,9%, which is achieved at 850°c. Although the $BaTiO_3$ cubic phase is already formed at 600°c, the calcination of all samples was performed at 900°c for 4 hours. From the micrographs and sizes distributions of Fig.4 & 5, it turns out that: i) the morphology of the particles is unaltered, ii) shrinkage occurs owing to gaz evolution, iii) fragmentation occurs in connection with the cristallisation of submicronic $BaTiO_3$ crystallites in the particles (Fig.4).

After sintering, the relative densities of doped samples reach 0,9. The microstructure shows grain sizes ranging from 1 to 5 µm (Fig.6).

The PTC effect of a sample doped with 0,3at% La (Fig.7) reaches 4 orders in magnitude.

DISCUSSION.

The rate-controlling steps of the precipitation of Ba & Ti are of different nature. The destabilisation of the Ti-complex is due to the thermal effect rather than the oxidizing character of the solution. It turns out from Fig. 1 & 2 that the process of Ti elimination is of the second order, i.e.: $-d[Ti]/dt = K[Ti]^2$. It would correspond to the formation of dimers according to: $\equiv Ti\text{-}OH + HO\text{-}Ti\equiv \rightarrow \equiv Ti\text{-}O\text{-}Ti\equiv + H_2O$.

The rate of the Ba precipitation is limited to a step of the EDTA degradation which is strongly dependent on the H_2O_2 concentration. A solid of composition $BaO_2.xH_2O$ is expected.

Despite the above differences, the similar rates of Ba and Ti decrease in the solution enable the coprecipitation of the Ti polymer and the Ba hydrated peroxide in agreement with a monomodal size distribution. However, complete mixing of Ti and Ba on an atomic scale is doubtful. The surface layers of the particles are probably enriched in barium. In fact, the samples dried at room temperature are particulary sensitive to carbonatation. Drying of the powders at 110°c stabilizes a precursor whose formula is close to $BaO_2.TiO_2.2H_2O$, according to the weight loss at 900°c.

The starting concentrations experienced in this study causes growth and agglomeration of the spherical particles up to the micron scale. This feature is not necessarily a handicap for the attainement of a dense and regular microstructure. In fact, the calcination increases the submicronic fraction of the size distribution, owing to the formation of $BaTiO_3$ crystallites having sizes less than 0,3µ. Although the microstructure of the sintered samples is fine-grained (< 5µ), the residual porosity and the size distribution are still apprecia-ble. These two parameters would be reduced by a better control of the calcination followed by an adequate desagglomerating treatment of the $BaTiO_3$ crystallites.

The processing at high Ba and Ti concentrations ($>10^{-2}$ m/l) is interesting in that it favors the incorporation of small amounts of donor dopants in the solid. As evidenced by the high PTC effect shown in Fig.7, lanthanum is completely introduced in the material. Further investigations are made in order to incorporate other dopants (Mn, Sr, Pb) able to modify the PTC characteristic.

CONCLUSION.

The precipitation of a peroxidic precursor of $BaTiO_3$ from Ti and Ba complexes is a promising route for producing homogeneous n-doped materials. Precipitation, filtering and washing are easy to perform; particle shapes, sizes and stoichiometry can be controlled from the solution; carbonates and organic residues are easily eliminated upon heating. However, the maximum amounts of Ba and Ti in the solution are limited to < 0,1 m/l due to: i) the high H_2O_2 concentrations necessary for complete precipitation, ii) growth and agglomeration of the spherical particles preventing the obtaining of dense and regular microstructures.

REFERENCES.

1. Murata, M., Kitao,A., U.S. Pat., 4,061,583 (1977).
2. Ahn, Kim, Hwang, kyu, J. Kor. Ceram. Soc., 1984,**21**,3
3. Michel, A., TFE, Lab. de Chimie Industr.et Anal.,ULB,(1987)
4. Gherardi, P., Matijevic, E., Europ. Pat., 0,212,935 (1987)
5. Duvigneaud, P.H., Gilbert, M., Sc. of Ceram. XIV,979-984 (1987)

MODELLING CHEMICAL COMPOSITION AND MICROSTRUCTURE OF Mg-Al HYDROXIDES COPRECIPITATED FOR Mg-Al SPINEL FORMATION

DARIO BERUTO, RODOLFO BOTTER, MARINO GIORDANI
Centro Interdipartimentale di Ingegneria dei Materiali,
Universita' di Genova,
Piazzale J.F.Kennedy, Pad. D, I 16129 Genova, Italy

GUALTIERO GUSMANO, ENRICO TRAVERSA
Dipartimento di Scienze e Tecnologie Chimiche,
Universita' di Roma "Tor Vergata",
Via Orazio Raimondo, I 00173 Roma, Italy

EDOARDO PROVERBIO
Dipartimento di Ingegneria Chimica, dei Materiali, Materie Prime e Metallurgia, Universita' di Roma "La Sapienza",
Via Eudossiana 18, I 00184 Roma, Italy

ABSTRACT

In a recent paper we have proved that Mg-Al spinel formation from solid state is strongly dependent upon the mixing degree Coprecipitation of Mg-Al hydroxides is a promising process to get such mixing at molecular scale.

In this paper we study the coprecipitation reaction to get ultrafine particles of Mg-Al mixed hydroxides from different salts solutions (chloride, nitrate, sulphate). Chemical composition and microstructure of the precipitates have been investigated as a function of reaction parameters and comparisons have been made with mixed hydroxides obtained previously under the influence of non uniform electric field (NUEF).

INTRODUCTION

Traditional processing method of magnesium aluminium spinel ($MgAl_2O_4$) powders is based upon solid state reaction between the corresponding oxides at temperature as high as 1600-1700 K [1-3]. These temperatures are necessary to enhance solid state diffusion and reaction.

In a previous paper we have proved that the spinel formation temperature is strongly dependent upon the mixing degree of the precursors [4]. Due to this observation, it seems reasonable that the formation temperature can be sharply decreased if the most intimate mixing degree is obtained. This result can be achieved by means of Mg-Al mixed hydroxides coprecipitation [5-7].

The aim of the present work is to study the influence of different parameters on the mixed Mg-Al hydroxides characteristics, in order to facilitate the spinel formation.

EXPERIMENTAL

Precipitation Procedure

Standard precipitation of mixed hydroxides was obtained using 75 ml of magnesium and aluminium chlorides aqueous solution 1.5 M, with Mg/Al molar ratio equal to 1/2, added drop by drop to 225 ml of stirred NH_4OH solution. During the reaction the pH was kept constant at the value of 10 by continuous addition of 30 % NH_4OH solution. The total volume of NH_4OH added during the precipitation was about 600 ml. The temperature of precipitation was kept constant at the value of 333 K.

At the end of the reaction the precipitate was allowed to age for a total time of 2 hours at the same temperature, under moderate stirring. The precipitate was filtered under vacuum and washed with a NH_4OH solution in distilled water at pH=9 until the disappearance of the salts anions. Finally, it was dried in a oven at 378 K up to constant weight and milled in a vibratory mill.

Further samples were prepared from Mg and Al nitrate and sulphate solutions using the same procedure above described, in order to verify the anion influence during the precipitation.

To test the different degree of reactivity of $Al(OH)_3$ and $Mg(OH)_2$, a two steps salt solution addition was performed for a number of experiments.

The influence of the precipitation temperature and of the pH has been preliminarily investigated.

Chemical purity of the reagents was indicated by Carlo Erba RPE-ACS.

Characterization Techniques

The precipitates were characterized by:
- X-Rays Diffractometry (XRD), using Cu $k\alpha$ radiation ($\lambda = 1.542 \times 10^{-1}$ nm) (Siemens mod. Kristalloflex 810);
- Scanning Electron Microscopy (SEM) (Philips mod.505);
- Specific surface area measurements using B.E.T. method (Carlo Erba Strumentazione mod. Sorpty).

RESULTS AND DISCUSSION

Standard Precipitation

XRD pattern of the powder obtained by standard precipitation is shown in fig. 1a. Three different crystalline phases can be identified: mixed Mg-Al hydroxide, α-$Al(OH)_3$ (gibbsite) and β-$Al(OH)_3$ (bayerite).

These results are in agreement with those obtained by Mortland et al. [8], while Bratton [5] did not observe the presence of bayerite. As far as the identification of the mixed hydroxide is concerned, three different mixed hydroxides are reported on JCPDS [9], that are $Mg_6Al_2(OH)_{18}\cdot 4.5H_2O$ and $Mg_4Al_2(OH)_{14}\cdot 3H_2O$ from a work of Mascolo et al [10], and $MgAl_2(OH)_8$ from a work of Ketchik et al [11], who reports also a $Mg_2Al(OH)_7$ mixed hydroxide. Other mixed hydroxides, with different Mg/Al ratio included between 2 and 4, are reported in literature [5, 8, 12]. The peaks of the mixed hydroxide got by standard precipitation can be identified as those corresponding to a mixed hydroxide with molar ratio Mg/Al of 2/1.

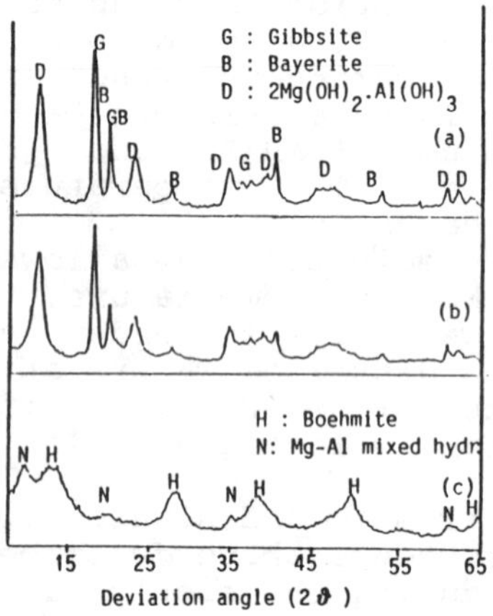

Fig.1: XRD patterns of powders got from chloride (a), nitrate (b) and sulphate (c) solutions.

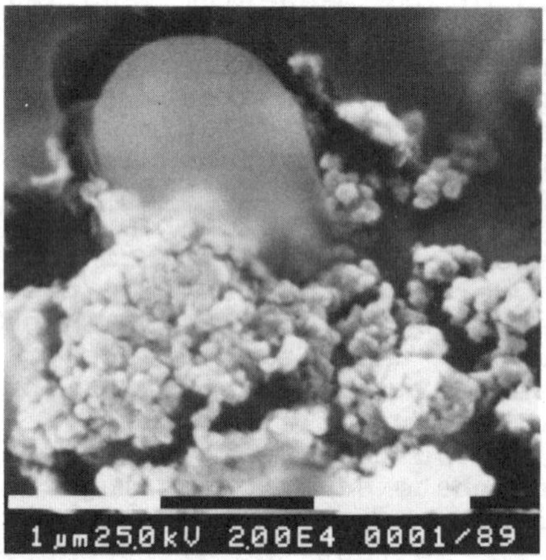

Fig.2: SEM micrograph of the product obtained from Mg-Al chloride solution.

SEM observations (fig. 2) showed that the powder was formed by homogeneous aggregates of particles with an average diameter of about 0.1 μm, and by larger spherical particles with an average diameter of about 1 μm. EDS analysis showed that these particles were made of aluminium hydroxide. The measured specific surface area was 100 m^2/g.

Influence of Anions

Fig. 1b shows the XRD pattern of the precipitate obtained from nitrates solution. No difference was observed with respect to the reference precipitate, except for a slight reduction in bayerite content. The SEM examination too, gave similar results for the two powders. The measured specific surface area was 114 m^2/g.

Fig. 1c illustrates the XRD pattern for the precipitate obtained from sulphate aqueous solution. By comparison with the previous ones, it is quite evident that Mg and Al from a sulphate aqueous solution precipitate giving completely different phases, with a poor degree of cristallinity. The latter might be due to extremely fine grain size of some components.

Even if a clear identification of the nature of the products is not easy, boehmite can be identified. The remaining other peaks, not related to boehmite, are close to those of the mixed hydroxides given in literature, except for the main reflection peak, which corresponds to a crystallographic distance d included between 8.5 and 9 x 10^{-1} nm.

At the present we have not enough data to clarify the detailed steps and the chemical nature of the products, however the ultrafine grain size can be related to the behaviour of aluminium sulphate. Indeed we have precipitated Al hydroxide from the corresponding sulphate under the same conditions. Results show that boehmite - gel can be formed, suggesting that the quite amorphous feature of the product might be related to the mechanism of boehmite formation.

Two Steps Precipitation

Precipitation of Al hydroxide followed by precipitation of Mg hydroxide: XRD pattern of the product (fig. 3a) shows, by comparison with fig. 1a, that $Al(OH)_3$ is present mostly in bayerite form; no evidence of $Mg(OH)_2$ presence was found, due to the fact that Mg^{2+} reacts fastly with bayerite to give the mixed hydroxide, instead of forming the corresponding $Mg(OH)_2$.

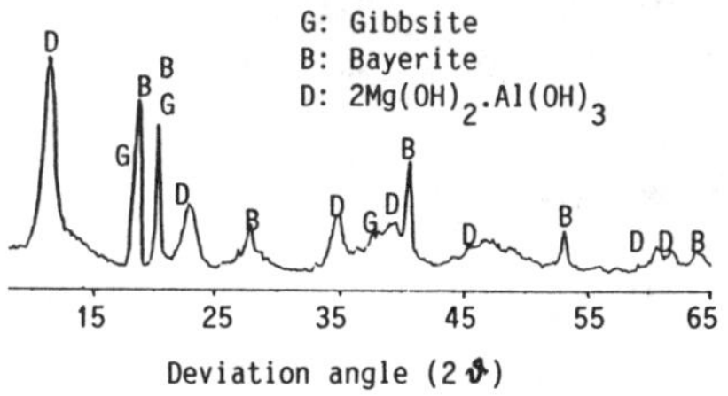

a

b

Fig.3 - XRD pattern (a) and SEM micrograph (b) of the powder obtained from precipitation of $Al(OH)_3$ followed by recipitation of $Mg(OH)_2$

SEM observations (fig. 3b) showed in this case unhomogeneous and large aggregates with average particle diameter of 0.5 µm; however particles of about 0.1 µm are present. The measured specific surface area was 97 m^2/g.

Precipitation of Mg hydroxide followed by precipitation of Al hydroxide: Precipitation of Mg hydroxide followed by addition of Al^{3+} yields to $Al(OH)_3$ and $Mg(OH)_2$. The mixed hydroxides are forming later during the aging period. To confirm this hypothesis, samples were taken off at different aging times and washed by centrifugation, in order to minimize the time required for the complete Cl^- removal.

Fig. 4 shows the XRD patterns of powders without aging and after 2, 4 and 8 hours of aging. At the beginning the precipitate was a mixture of brucite and bayerite; the aging allowed the mixed hydroxide formation: increasing the aging time, the brucite peaks underwent reduction, while the mixed hydroxide peaks increased. Moreover, with the aging the mixed hydroxide, which initially was closer to a composition with Mg/Al ratio 3/1, evolved to 2/1. At the end of 8 hours aging, the powder was a mixture of bayerite, brucite and mixed hydroxide in a smaller amount with respect to the reference powder, which means that only a part of $Mg(OH)_2$ was able to react with $Al(OH)_3$ in order to form the mixed hydroxide.

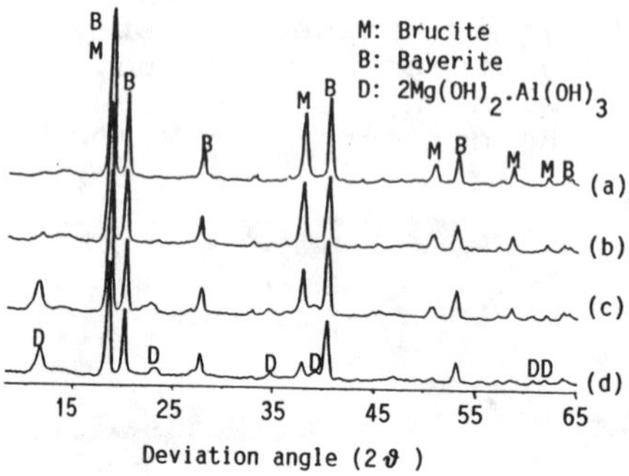

Fig.4 - XRD pattern of the product obtained from the precipitation of $Mg(OH)_2$ followed by addition of Al^{3+}. a - no aging; b - 2 h aging; c - 4 h aging; d - 8 h aging.

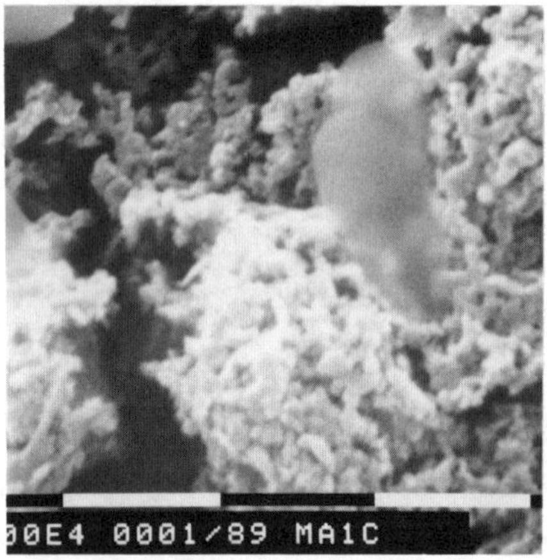

Fig.5 - SEM observation of Mg-Al two steps precipitation after 8 hours of aging.

SEM observations (fig. 5) gave results similar to that obtained for the reference powder, with ultrafine aggregates (about 0.1 µm) and large spheres (1 µm). The measured specific surface area was very high, thus 190 m^2/g.

The main conclusion that can be drawn from these results is relevant to a different reactivity of $Al(OH)_3$ with Mg^{2+} and of $Mg(OH)_2$ with Al^{3+}. The particles of $Al(OH)_3$ have a faster kinetics rate to react with the other cations in order to give the mixed hydroxide.

Influence of NUEF

The above described experiments show that the nature and microstructure of the precipitate can be greatly influenced by modifying the kind of anions as well as the precipitation procedure.

Although the detailed mechanism of these processes is quite difficult to understand, it seems reasonable to state that the kinetics of the process is affected by changing the solvation process of the ions as well as the double electrical layer around the formed solid particles.

These steps can be varied with the application of NUEF during the precipitation. In a previous communication [13], some of the authors have shown that the products obtained from nitrate solution under NUEF application are different in nature and microstructure from those got without NUEF.

The NUEF seems to act both on the microstructure as well as on the chemical nature of the precipitate. Coprecipitation under NUEF gives finer and non porous particles, probably due to the increase of the repulsive forces among dispersed particles [13].

It is interesting to observe that the results here obtained from the sulphate solution give some features comparable with those obtained from nitrate solutions under the influence of NUEF. Particularly the formation of a Mg-rich mixed hydroxide phase with a very similar XRD pattern has been observed in both the experiments. It was suggested that bayerite is a possible precursor of this phase; the fact that precipitation from sulphate yields a product where bayerite is not present leaves open this question and it requires further investigations.

CONCLUSIONS

Preparation of mixed Al-Mg hydroxides can be influenced by those parameters which affect the precipitation of $Al(OH)_3$ and modify the step of its nucleation. The above given results suggest that the precipitation of poorly crystalline phases sharply influences the crystallographic distance d of the mixed hydroxide.

Due to the fact that the increase of d is too high to be explained with an increase of the Mg/Al ratio in the mixed hydroxide, the following hypothesis can be drawn:

- the precipitation conditions introduce a high degree of disorder into the lattice of the mixed hydroxide;
- a new phase is obtained in such conditions.

More investigations are in progress in order to clarify this point.

Furthermore, while the specific surface area of particles obtained with NUEF is in agreement with the observed particles diameters [14], in the other cases the measured specific surface areas are larger than the ones calculated from the particles diameters. This suggest that the particles obtained with NUEF are not porous, while the other precipitations give porous particles.

REFERENCES

1. Ryskhewitch, E., Oxide Ceramics, Academic Press, New York and London, 1960, p. 271.

2. Totsenko, S.B., Degtyareva, E.V. and Kainarskii, I.S., Kinetics of magnesium spinel synthesis, Izv. Akad. Nauk SSSR, Neorg. Mater., 1968, **4**, 563-7.

3. West, A.R., Preparative methods. Solid State Chemistry and its Applications, J. Wiley & Sons, New York, 1984, 5-18.

4. Beruto, D., Botter, R., Giordani, M., Grazioli, F., Gusmano, G., Carassiti, F., Chiocchio, G. and Proverbio, E., Effect of MgO precursors microstructure on Mg-Al spinel formation from gamma-alumina ultrafine particles, II Int. Conf. on Eng. Mat., Bologna, June 20-23, 1988.

5. Bratton, R.J., Coprecipitates yielding $MgAl_2O_4$ spinel powders, Cer. Bull., 1969, **46**, 759-62.

6. Chalyi, V.P. and Makarova, Z.Ya., Isothermal treatment of system of metal hydroxides, Ukr. Khim. Zh., 1969, **35**, 497-9

7. Usov, P.G., Amelina, V.A. and Zhdanova, N.S., Influence of the particle size and different ratio between the components on the synthesis of aluminium-magnesium spinel (magnesium aluminate), Sb. Tr. Molodykh Uch. Tomsk. Politekh. Inst., 1973, **1**, 103-6.

8. Mortland, M.M. and Gastuche, M.C., Cristallisation d'hydroxides mixtes de magnesium et d'aluminium en milieu dialyse', C. R. Acad. Sci., 1962, **255**,2131-3.

9. Joint Committee on Powder Diffraction Standards, 1986.

10. Mascolo, G. and Marino, O., A new synthesis and characterization of magnesium-aluminium hydroxides, Miner. Mag., 1980, **43**, 619-21.

11. Ketchik, S.V., Plyasova, L.M. and Ivanova, A.S., X-ray diffraction investigation of the structure of aluminium-magnesium compounds, Inorg. Mater., 1980, **16**, 1356-9.

12. Feitknecht, W. and Gerber, M., Zur Kenntnis der Doppelhydroxyde und basischen Doppelsalze III uber Magnesium-Aluminiumdoppelhydroxyd, Helv. Chim. Acta, 1942, **25**, 131-7.

13. Beruto, D., Botter, R., Giordani, M. and Giannetti, G., Influence of non uniform electric field on the coprecipitation of fine particles of Mg-Al hydroxides, 2nd International Conference on Ceramic Powder Processing Science, Berchtesgaden, October 12-14, 1988.

14. Beruto, D. and Searcy, A.W., High-reactivity calcium oxides, Nature, 1976, **263**, 221-2.

PREPARATION AND CHARACTERISTICS OF ZIRCONIA/ALUMINA CERAMIC POWDERS

P. DEN EXTER, T.H.P. LEUWERINK, A. BOS, A.J.A. WINNUBST,
E.B.M. DOESBURG, A.J. BURGGRAAF
University of Twente, Faculty of Chemical Technology, Laboratory
of Inorganic Chemistry, Materials Science and Catalysis,
P.O. Box 217, 7500 AE Enschede, The Netherlands.

ABSTRACT

Alumina/zirconia ceramic composites have been produced by both a coprecipitation and a sol-gel technique. The influence of calcination temperature and of alumina phase on sintering behaviour has been investigated. The coprecipitation derived powder compacts containing θ-alumina could be sintered to dense (>95 %) ceramics at 1600°C. The sol-gel derived powder compacts could not be sintered to densities above 91 %. The formation of α-alumina in the powder during calcination must be avoided.

INTRODUCTION

It has been demonstrated that the fracture toughness and bending strength of alumina ceramics can be enhanced by dispersing fine zirconia particles in an alumina matrix [1]. The mechanisms associated with zirconia toughened alumina ceramics include stress-induced transformation toughening, stress-induced microcrack toughening and spontaneous microcrack toughening [2, 3]. These mechanisms are related to the stress-induced martensitic phase transformation from metastable tetragonal zirconia to the monoclinic form during cooling or mechanical testing. Several of these mechanisms can act simultaneously resulting in either additional or opposite influences on the fracture toughness and bending strength of alumina. For instance the combination of transformation toughening and microcrack toughening can operate counteracting on fracture toughness because the presence of microcracks decreases the rigidity of the process-zone around a propagating crack [4].

For unstabilized tetragonal zirconia dispersed in an alumina matrix the transformation to monoclinic zirconia rather easily takes place during the sintering process. An increase of the amount of monoclinic zirconia results in a larger amount of microcracking, making transformation toughening less effective. Transformation toughening may be optimized by stabilizing tetragonal zirconia with yttria and by reduction of its grain size. The latter requires a homogeneous distribution of the zirconia grains preventing the grains from getting together and to grow. Furthermore a very small zirconia grain size distribution is required to allow the largest amount of tetragonal zirconia to transform during mechanical testing. Finally a small average grain size and grain growth rate is wanted to enhance plastic deformation behaviour in hot and sinter forging experiments. For these reasons zirconia toughened alumina ceramics have been prepared by gel-precipitation and sol-gel techniques.

EXPERIMENTAL PROCEDURE

Powder Preparation
Zirconia (+ 3 mol% yttria)/ 85 wt% alumina ceramic powders have been prepared by both a coprecipitation and a sol-gel technique using commercially available reagent-quality chemicals. The coprecipitation derived powders have been prepared by the hydrolysis of a solution of aluminium chloride, yttrium chloride and zirconium chloride by adding it dropwise to a large excess of 25 % NH_4OH. After the coprecipitation the gel has been washed thoroughly with a water/ammonia mixture in order to remove all chlorine and has been washed subsequently with ethanol to remove free water.

The sol-gel derived powders have been prepared by the hydrolysis of aluminium secondary butoxide in hot water followed by the peptization of the hydroxide with a solution of zirconyl nitrate and yttrium nitrate. The clear sol has been added dropwise to a dilute solution of ammonia (pH = 9.5). After the precipitation the gel has been washed thoroughly with water and ethanol.

The gels have been dried at 120°C and calcined in air at temperatures ranging from 550 to 1300°C for 2 hours.

Characterization
X-ray fluorescence spectrometry has been used for the analysis of the chemical composition of the powders whereas x-ray diffraction has been used for the analysis of the phase composition. Specific surface areas of the powders have been determined by the BET-method using the equipment as described by Bosch et al. [5]. Powder compaction behaviour has been investigated by measuring the changes in densification with increasing isostatic pressure as described by Van de Graaf et al. [6]. The pore-size distribution in the green compacts have been calculated from nitrogen adsorption/desorption isotherms while pores with radii larger than 15 nm have been determined by mercury-intrusion porosimetry. After isostatically pressing (400 MPa) the sintering behaviour of the powder compacts has been studied by pressureless sintering for 2 hours at 1600°C and densification curves have been obtained using a dilatometer. Bulk densities of the sintered compacts have been measured by a mercury displacement method while microstructures of polished and thermally etched compacts have been examined using a scanning electron microscope. Ceramic grain sizes have been measured using the Lineal Intercept Technique for measuring grain sizes in two-phase polycrystalline ceramics as described by Wurst et al. [7].

RESULTS AND DISCUSSION

X-ray fluorescence measurements demonstrate that both coprecipitation and sol-gel technique are suitable for obtaining a ceramic powder with the proper chemical composition. However depending on the experimental procedure and calcination temperature a ceramic powder is obtained consisting of tetragonal zirconia and one of the transition aluminas or α-alumina. The sintering behaviour of the isostatically pressed powders (400 MPa) as a function of calcination temperature and related alumina phase composition is given in table 1.

TABLE 1
Calcination temperature, alumina phase composition, specific surface area and sintering behaviour of the ceramic powders

Synthesis	T. calc. (°C)	alumina phase	S_{BET} (m^2/g)	density* (%)
sol-gel	500	γ	254	89
,,	850	γ/δ	187	91
,,	1000	θ	136	70
,,	1200	α/θ	20	53
coprecip.	700	γ	138	93
,,	1100	θ	55	98
,,	1200	θ	32	98
,,	1300	α	6	86

* % theoretical density after sintering for 2 hours at 1600°C

As can be seen from table 1 the sol-gel derived ceramic powders can not be sintered to dense ceramic (> 95%). The coprecipitation derived ceramic powders can be sintered to dense ceramic on the condition that the calcination temperature is high enough but no α-alumina is formed yet. Obviously the microstructure of the calcined powder is changed when the θ to α phase transformation takes place during calcination, giving rise to a decrease in sinterability.

When bulk densities of powder compacts are plotted as a function of the logarithm of the applied isostatic pressure two linear parts are obtained. According to Van de Graaf [6] the intersection of these linear parts represents the strength of the powder agglomerates. For the coprecipitation derived powders a similar compaction behaviour is observed implying that the powders contain agglomerates. Both linear parts are not sharply separated indicating that the maximum strength for the agglomerates of each powder is not very close to its average strength. For the coprecipitation derived powders the mean agglomerate strength increases from 96 MPa for the powder containing θ-alumina to 210 MPa for the powder containing α-alumina. After isostatic pressing at 400 MPa the agglomerates are broken in both cases. The pore radius increases from 29 nm for the green compact (400 MPa) containing θ-alumina to 94 nm for the green compact containing α-alumina.

After sintering of the coprecipitation derived powder compact containing θ-alumina a dense ceramic is obtained with tetragonal zirconia particles with a mean grain size of ≅ 1 μm homogeneously dispersed between α-alumina grains with a mean grain size of ≅ 2 μm (figure 1). After sintering of the coprecipitation derived powder compact containing α-alumina a porous ceramic is obtained with irregular shaped pores which are much larger than the alumina grain size, i.e. pores with a large coordination number.

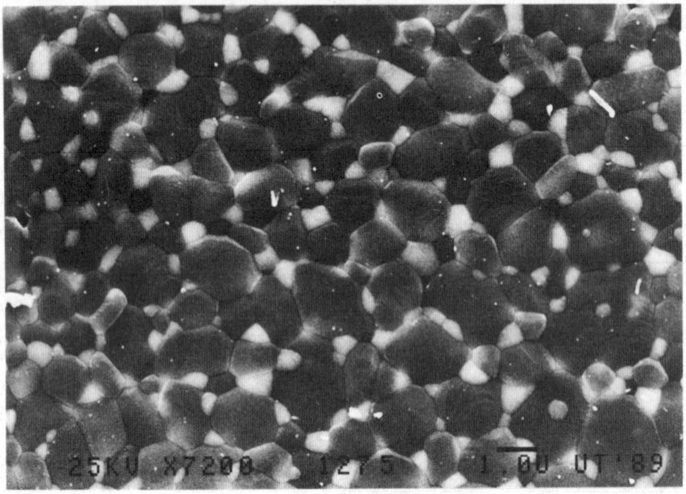

Figure 1. Dense ceramic obtained from coprecipitation derived powder containing θ-alumina.

In the coprecipitation derived powder compact containing θ-alumina densification begins during heating up (2.5 K/min.) at 1100°C. At 1240°C when the densification rate is maximal the θ to α phase transformation takes place and further densification is inhibited. At still higher temperatures densification of the powder compact slowly proceeds having its highest densification rate at 1470°C.

The sol-gel derived powders show a similar compaction behaviour indicating they also contain agglomerates. The powder containing α-alumina exhibit the highest agglomerate strength (150 MPa). The agglomerates are broken during compaction at 400 MPa and a bimodal pore size distribution remains with pore radii of 30 and 349 nm. Assuming the latter being inter-aggregate pores of close packed spherical aggregates, the aggregates must be 3 μm in size.

After sintering of the sol-gel derived powder compacts, irrespective of whether they contain θ or α-alumina, a porous ceramic is obtained consisting of a packing of dense polycrystalline aggregates (3 μm) with large inter-aggregate pores. Within these aggregates tetragonal zirconia (≅ 0.4 μm) is homogeneously dispersed between α-alumina grains with a mean grain size somewhat smaller than 1 μm. Because the aggregate size after sintering is the same as calculated from the pore size in the green compact these denser domains are not the result of the sintering proces.

The poor sinterability of the coprecipitation and sol-gel derived powders containing α-alumina can reasonably be described by a change in the microstructure of the powders, probably by a change in the aggregate structure. The only difference between the coprecipitation and sol-gel derived powders is that probably even the sol-gel derived powder containing θ-alumina contains very large aggregates. If the agglomerates are fractured during compaction an aggregate packing with very large inter-aggregate pores remains. According to Lange et al. [8] grain growth reduces the coordination number of inter-aggregate pores allowing them to disappear. Because of the presence of well-dispersed intergranular zirconia grain growth within the aggregates is inhibited and inter-aggregate pores can not disappear.

Nevertheless the sol-gel derived powders are very interesting because of their small grain size after sintering. When the coordination number of the inter-aggregate pores can be reduced by appling stresses by hot-pressing an interesting material results.

ACKNOWLEDGEMENT

This research was partly supported by the Innovative Research Program on Technical Ceramics (IOP-TK) with financial aid of the Dutch Ministry of Economic Affairs.

REFERENCES

1. Claussen, N., Fracture toughness of Al_2O_3 with an unstabilized ZrO_2 dispersed phase. J. Am. Ceram. Soc., 59 [1] 49 - 51 (1976).

2. Rühle, M., Claussen, N. and Heuer, A.H., Transformation and Microcrack Toughening as Complementary Processes in ZrO_2-Toughened Al_2O_3. J. Am. Ceram. Soc., 69 [3] 195 - 197 (1986).

3. Wang, J. and Stevens, R., Toughening mechanisms in duplex alumina - zirconia ceramics. J. Mater. Sci., 23 (1988) 804 - 808.

4. Clarke, D.R., A Simple Calculation of Process-Zone Toughening by Microcracking. J. Am. Ceram. Soc., 67 [1] C15 - C16 (1984).

5. Bosch, H. and Peppelenbos, A., Automatic and low-cost determination of BET surface areas. J. Phys. E: Sci. Instr. 10 (1977) 605 - 608.

6. Van de Graaf, M.A.C.G., Ter Maat, J.H.H. and Burggraaf, A.J., Microstructure and sintering kinetics of highly reactive ZrO_2-Y_2O_3 ceramics. J. Mater. Sci. 20, (1985), 1407 - 1418.

7. Wurst, J.C. and Nelson, J.A., Lineal Intercept Technique for Measuring Grain Size in Two-Phase Polycrystalline ceramics. J. Am. Ceram. Soc., 55 [2] 109 (1972).

8. Lange, F.F. and Davis, B.I., Sinterability of ZrO_2 and Al_2O_3 Powders: The Role of Pore Coordination Number Distribution. In Advances in Ceramics, Vol 12, ed. N. Claussen, M. Rühle and A. Heuer, The American Ceramic Society, Inc., pp. 699 - 713.

THE PRODUCTION OF FINE-GRAINED PURE ALUMINA AND ZIRCONIA-ALUMINA COMPOSITES.

A. BOS, P. DEN EXTER, A. J. A. WINNUBST, E. B. M. DOESBURG,
A. J. BURGGRAAF, H. J. VERINGA[1].

Centre for Ceramics and Catalyst Research CKKO*,
Faculty of Chemical Technology, University of Twente,
P.O.Box 217, 7500AE Enschede, The Netherlands

[1]The Netherlands Energy Research Foundation ECN
P.O.Box 1, 1755ZG Petten, The Netherlands

ABSTRACT

Al_2O_3 and Al_2O_3-ZrO_2 (ZTA) ceramics are produced using powders, which are prepared by hydrolysis of mixed metal-salt solutions in ammonia. The ZrO_2 (15 wt%) is added as unstabilized and as yttria(5.8 mole%)-stabilized ZrO_2. The effect of the addition of MgO on the sintering behaviour of the different compositions has been studied. Other parameters which were varied during the powder preparation were the drying procedure of the hydrolysis precipitate (drying after alcohol-washing or freeze-drying) and the calcination temperature.
MgO additions had a notable effect on the sintering behaviour of the compacts, producing dense Al_2O_3-ZrO_2 ceramics at temperatures as low as 1450°C.

INTRODUCTION

Al_2O_3 ceramic matrices may exhibit a considerable increase in material toughness, if fine-grained ZrO_2 is dispersed in the ceramic matrix [1]. The toughness increase is commonly ascribed to the martensitic t->m phase transformation of ZrO_2, which absorbs the strain energy. The transformation temperature is influenced by the grain size of the ZrO_2, or by the addition of phase-stabilising dopants, e.g. yttria [2]. Several toughening mechanisms based on this transformation are mentioned in the litterature. Microcracking and stress-induced phase transformation are considered to be the most important mechanisms in the toughening of ZrO_2-dispersed ceramics [3].
The addition of ZrO_2 however inhibits densification of the composite during sintering, making high sintering temperatures necessary. A high sintering temperature causes undesirable grain growth. The effect of MgO as a sintering additive to Al_2O_3 is well-known and has been studied extensively [4,5]. MgO inhibits grain growth and promotes densification, depending on the MgO-concentration. The effect of MgO on the sintering of Al_2O_3-ZrO_2 composites was studied by Kosmac et al. [6]. At ca 1000ppm MgO they observed a maximum in both relative density and grain size. A maximum in strength and fracture toughness was observed at ca 2000ppm MgO,

---

*The CKKO is jointly operated by ECN Petten and the University of Twente for fundamental research into ceramics and catalytic materials.

where a high relative density and small grain size were combined with a low MgO content. The effect of MgO on the sintering of Al_2O_3-ZrO_2 composites during HIPping has been studied by Min and Kim [7]. Their results show a decrease in the grain growth rate and an increase in the densification rate relative to MgO-doped Al_2O_3.
In this study it is aimed to produce fine-grained, dense Al_2O_3-ZrO_2 composites at low sintering temperatures by doping with MgO. The initial results on the effect of MgO-doping on the sintering behaviour and the microstructure of the Al_2O_3 and ZrO_2-Al_2O_3 composites are presented.

EXPERIMENTAL PROCEDURES

Ceramic powders were produced having the following compositions: pure Al_2O_3, Al_2O_3 with 15 wt% unstabilized ZrO_2 and Al_2O_3 with 15 wt% Y_2O_3(5.8 mole%)-stabilized ZrO_2. The powders were produced by coprecipitation of metal-nitrate and -chloride solutions in concentrated (25%) ammonia at pH=11-12. The precipitated hydroxides were washed with water to remove nitrates and chlorides. To prevent agglomeration during drying, two procedures were applied: 1.- drying at 120°C after washing the powder with alcohol to remove water and 2.- freeze-drying. The powders were calcined at different temperatures between 700°C and 1300°C. For standard sintering procedures the calcined powders were cold-isostically pressed at 400 MPa. The pressed compacts were sintered pressureless in air at 1600°C for 2 hours. For dilatometer experiments the powders were mixed with binders and plasticizers to improve the pressing behaviour (further referred to as pressing additives) and cold-pressed uniaxially into tablets, followed by isostatic pressing at 400 MPa. The tablets were finally sawn into bars.
MgO addition was performed after powder preparation, using two methods. The first method (method A) was to ball mill the calcined powder with the required amount of $Mg(NO_3)_2.6H_2O$ and pressing additives in alcohol, followed by drying and pressing. In the second method method B) the pressed calcined powder was ball milled with pressing additives in alcohol, dried and pressed. After sawing into bars, the green compacts were calcined at 600°C to remove the additives. The porous bars were then soaked in aqueous Mg-nitrate solutions and dried overnight at 80°C .
The powders were characterized with X-ray diffraction (XRD), tap density and S_{BET}. Green and sintered densities were measured with the Archimedes method.

RESULTS

The phases produced after calcination of the Al_2O_3 and Al_2O_3-ZrO_2 powders were determined

TABLE 1
Phases after calcination of the alumina and zirconia-alumina powders

Calc.T (°C)	Drying	Phases alumina	composite
--	ALC	bayerite	bayerite/amorphous zirconia
700	ALC	eta-alumina	
950	ALC	theta-alumina	theta-alumina/tetragonal zirconia
1000	ALC	theta-alumina	theta-alumina/tetragonal zirconia
1200	ALC	alfa-alumina	
1300	ALC		alfa-alumina/tetragonal zirconia
700	FD	eta-alumina	
950	FD	theta-alumina	theta-alumina/tetragonal + monoclinic zirconia
1000	FD		theta-alumina/tetragonal + monoclinic zirconia
1200	FD	alfa-alumina	
1300	FD		alfa-alumina/tetragonal zirconia

with X-ray diffraction (XRD) and are listed in Table 1. Also is indicated whether the powders were freeze-dried (FD) or alcohol-washed followed by drying at 120°C (ALC) prior to calcination. To obtain a complete transformation of the theta-Al_2O_3 to alfa-Al_2O_3 in the presence of ZrO_2, a higher calcination temperature was required than for the pure Al_2O_3 compositions.

Tap densities, green and sintered densities for the freeze-dried powders are normally slightly lower than for the powders dried after alcohol-washing. Sintered densities are highest for the theta-Al_2O_3 powders (more than 98% relative density). For the alfa-Al_2O_3 powders the relative densities are less than 95% for the pure Al_2O_3 and less than 60% for the ZrO_2-Al_2O_3 composites. The eta-Al_2O_3 was sintered only as pure Al_2O_3 and reached sintered densities of 90% or lower.

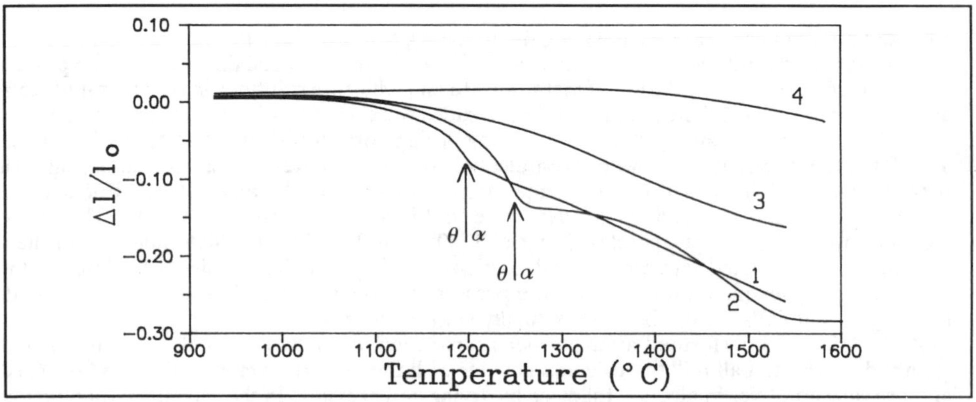

Figure 1. Typical densification curves for the green compacts of different compositions: 1: theta-Al_2O_3; 2: theta-Al_2O_3-ZrO_2; 3: alfa-Al_2O_3; 4: alfa-Al_2O_3-ZrO_2. Heating rate is 2.5°C/min. for all experiments. For the theta-Al_2O_3 compacts (curves 1 and 2) the presumed theta- to alfa-Al_2O_3 transformation is indicated with arrows.

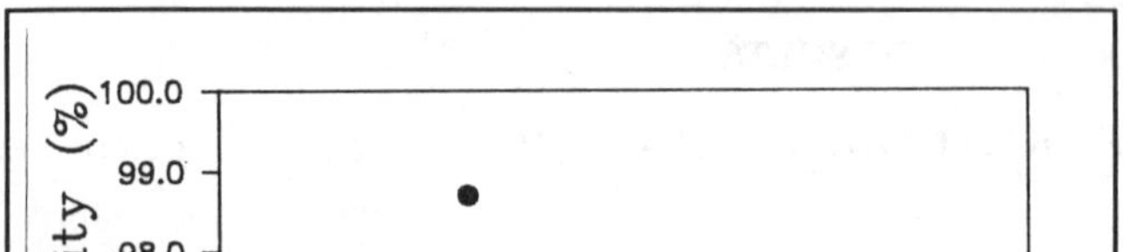

Figure 2. The effect of MgO additon on the sintering behaviour of a theta-Al_2O_3-ZrO_2 compact. Sintering conditions: 1450°C for 2 hours in air.

Typical densification curves for the undoped Al_2O_3 and Al_2O_3-ZrO_2 compacts were obtained with a Netzsch dilatometer type 402E and are shown in Fig.1. The sudden decrease in sintering rate at ca 1200°C for the pure theta-Al_2O_3 and at ca 1250°C for the theta-Al_2O_3-ZrO_2 compact is assumed to be induced by the theta- to alfa-Al_2O_3 transition. The temperature of 1250°C correlates well with the theta- to alfa-Al_2O_3 transformation in Al_2O_3-ZrO_2 powders, observed by den Exter et al. [8]. The transition coincides with a maximum in the

densification rate dl/dt.
The influence of MgO doping was studied with different addition techniques and at concentrations up till 2500ppm. In Fig.2 the effect of MgO-doping on the sintered density of an unstabilized ZrO_2-Al_2O_3 compact is illustrated. The effect of different addition techniques of the MgO is shown in Fig.3, where the densification of an undoped, stabilized ZrO_2-Al_2O_3 is compared with compacts doped with MgO following the two methods described above.

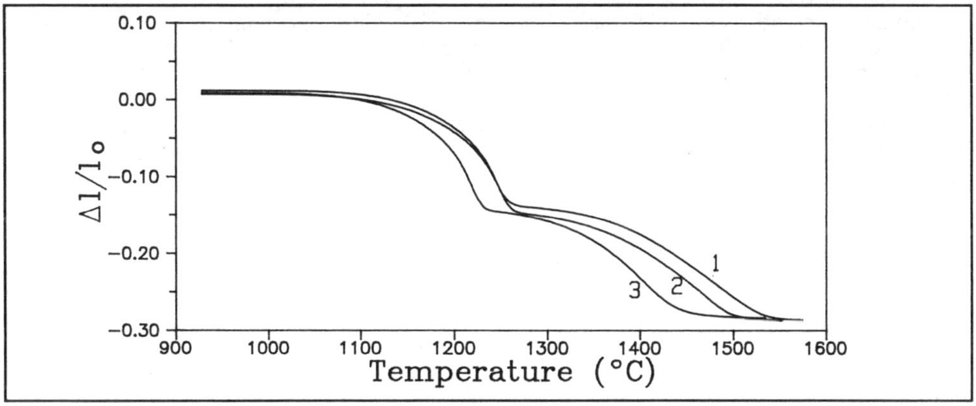

Figure 3. The influence of the method of MgO-addition on the sintering behaviour of a theta-Al_2O_3-ZrO_2 compact: 1: no MgO addition; 2: addition of 500 ppm MgO by method A; 3: addition of 500 ppm MgO by method B. The heating rate was 2.5°C/min for all experiments.

DISCUSSION

The addition of ZrO_2 has a rather substantial influence on the sintering behaviour of Al_2O_3. In the pure theta-Al_2O_3 compact the theta- to alfa-Al_2O_3 transition takes place at ca 1200°C after 10% linear shrinkage (curve 1, Fig.1). The transformation is associated with a negative volume change of ca 10% and decreases the transformation rate due to an increase in the porosity of the compact (no linear shrinkage is observed) and thus an increase in the diffusional distances. After the transformation the densification proceeds almost linearly with temperature.
In the theta-Al_2O_3-ZrO_2 compact (curve 2, Fig.1) the transition temperature is shifted to ca 1250°C. Consequently the theta-Al_2O_3 sinters to higher temperatures and thus the transition to alfa-Al_2O_3 takes place at a higher densification level. The sudden decrease in densification suggests a fast transformation. After the transition temperature the porosity of the compactis increased, but because of the higher densification level at this temperature the diffusional distances after transition will be smaller compared to the pure Al_2O_3 compact. Therefore, at higher temperatures the densification increases to a higher rate and the maximum densification level is attained at lower temperatures.
The sintering behaviour of the alfa-Al_2O_3 compacts (curves 3 and 4, Fig.1) is influenced rather dramatically by the presence of ZrO_2. Without ZrO_2 a sintered density of ca 90% is reached at 1550°C. The maximum densification level has not been reached within the sintering trajectory. This may be caused by the presence of hard agglomerates, introduced during calcination of the powder. In the presence of ZrO_2 there is almost no densification at all. A reason may be that the rather high calcination temperature (1300°C) causes sintering of both the Al_2O_3 and the ZrO_2, forming large, unreactive agglomerates which inhibit densification during the sintering stage.
A microstructural study of the different compacts at the different stages during the sintering trajectory will be performed to reveal more information regarding the sintering mechanisms.
The effect of MgO-addition on the sintering behaviour of a theta-Al_2O_3-ZrO_2 compact (Fig.2)

agrees very well with the results of Kosmac et al. [11], who prepared their composites with commercial Al_2O_3 powders. From Fig.2 it is obvious that by the addition of 1000 ppm MgO a dense zirconia-alumina ceramic can be obtained at a sintering temperature as low as 1450°C. Preliminary microstructural observations with SEM indicate a very fine grained microstructure, with a maximum grain size smaller than 1 μm. The addition method of the MgO (Fig.3) demonstrates that ball milling the powder with Mg-nitrate before pressing has less effect on the densification than soaking a pressed and porous compact in a Mg-nitrate solution.
From the observations of the sintering behaviour of the different compacts it is clear that in all cases the theta-Al_2O_3 containing compacts produce dense products at relatively low temperatures, which are still decreased by the addition of MgO. A disadvantage is that the linear shrinkage of these compacts is large (ca 25 to 30 %) and that it is ca 5% larger than for the alfa-Al_2O_3 containing compacts. A better control of processing variables during powder and green compact preparation must produce denser green compacts, in which the linear shrinkage during sintering is considerably reduced.

CONCLUSIONS

- Freeze-drying produces slightly less dense ceramics than drying after alcohol-washing.
- The Al_2O_3 phase present after calcination has a strong influence on the sintering behaviour.
- The presence of ZrO_2 increases the transformation temperature of theta-Al_2O_3, but also increases the relative density after sintering of the theta-Al_2O_3-ZrO_2 compact, compared with the pure theta-Al_2O_3 compact.
- The presence of ZrO_2 in the alfa-Al_2O_3 compacts inhibits densification. drastically.
- The method of MgO-addition has a large influence on the densification temperature.
- The addition of 1000 ppm MgO to a theta-Al_2O_3-ZrO_2 produces a dense ceramic at a temperature of 1450°C.

REFERENCES

1. Claussen, N., Fracture toughness of Al_2O_3 with an unstabilized ZrO_2 dispersed phase. J. Amer. Ceram. Soc., 1976, **59**, 49-51.
2. Heuer, A. H., Claussen, N., Kriven, W. M., and Rühle, M., Stability of tetragonal particles in ceramic matrices. J. Amer. Ceram. Soc., 1982, **65**, 642-650.
3. Rühle, M., Claussen, M., and Heuer, A. H., Transformation and microcrack toughening as complementary processes in ZrO_2-toughened Al_2O_3. J. Amer. Ceram. Soc., 1986, **69**, 195-197.
4. Harmer, M. P., and Brook, R. J., The effect of MgO additions on the kinetics of hot pressing in Al_2O_3. J. Mater. Sci., 1980, **15**, 3017-3024.
5. Ikegami, T., Kotani, K., and Eguchi, K., Some roles of MgO and TiO_2 in densification of a sinterable Al_2O_3. J. Amer. Ceram. Soc., 1987, **70**, 885-890.
6. Kosmac, T., Wallace, J. S., and Claussen, N., Influence of MgO additions on the microstructure and mechanical properties of Al_2O_3-ZrO_2 composites. J. Amer. Ceram. Soc., 1982, **65**, C66-67.
7. Min, J.-W., and Kim, D.-Y., Effect of pretreatment sintering temperature on the densification of Al_2O_3 and Al_2O_3-ZrO_2 ceramics by sinter plus HIP. Ceram. Int., 1988, **14**, 191-194.
8. Den Exter, P., Leuwerink, T. H. P., Bos, A., Winnubst, A. J. A., Doesburg, E. B. M., and Burggraaf, A. J., Preparation and characteristics of zirconia/alumina ceramic powders. This volume.

INFLUENCE OF ADDITIVES ON THE PRESSING AND SINTERING BEHAVIOUR OF SPRAY-DRIED SUBMICRON Al_2O_3

G.J.J. BECKERS, K. ZEILSTRA
Netherlands Energy Research Foundation, ECN
P.O. Box 1, 1755 ZG PETTEN, the Netherlands

To improve the pressing behaviour of Alcoa A16SG, a submicron Al_2O_3 powder, it has been spray-dried with addition of binders and plasticizers. The resulting product was pressed into pellets and sintered. Quality of the pellets as well as green and fired density were determined. The results of experiments with 13 binders and 5 plasticizers, added in concentration up to 3% weight, will be presented.

INTRODUCTION

The powders used for the production of high quality ceramics are normally in the size range of a few tenths of a micron upto a few micron. From point of view of sinterability, these powders have excellent properties. However, in this particle size range, interparticle forces are large compared to external forces, resulting in the formation of large, loose agglomerates of primary particles and poor flow properties. In a pressing application both effects are undesirable, since they will lead to large internal stresses and inhomogeneities in the product.

One of the possibilities to overcome these problems is to spray-dry the powder. In a spray-drying operation, a suspension of the powder is atomised in a hot gas stream, resulting in a dry powder, consisting of large, dense agglomerates of controlled size. Binders and plasticizers are added to the suspension to control the strength of the agglomerates and to diminish the internal friction between particles. A number of binders can be used, while as a plasticizer glycerol, polyethene glycol [1] and, sometimes, water [2] is recommended.

MATERIALS AND METHODS

All spray-drying experiments are performed with a Büchi-190 Mini spray-dryer, capable of atomizing and drying upto 1 liter of suspension per hour. Suspensions, based on water, containing 60 weight percent of Alcoa A16SG were spray-dried with and without additives. To disperse the powder the suspension was either brought at pH = 3 with nitric acid, or an organic deflocculant (Dolapix CE 64) was used. The exit temperature of the spray-dryer was kept just over 100°C to ensure a dry product. The binders and plasticizers used are summarised in table 1.

TABLE 1

Polyvinyl alcohol	PA-4G	binder
Polyvinyl alcohol	PA-20G	binder
Polyvinyl alcohol	PAF 35	binder
Polysacharide	PS-13	binder
Cellulose ether	AM-20	binder
Cellulose ether	AM-100	binder
Methylcellulose	A15LV	binder
Polyacrylate	Uramul	binder
Polyetheneglycol	PEG 100	plasticizer
"	PEG 400	plasticizer/binder
"	PEG 4000	plasticizer
"	PEG 1M	binder
"	F68	binder
"	F108	binder
Glycerol		plasticizer
Water		plasticizer

The concentration of the binders is varied between 0 and 1.25 weight %, of the plasticizers between 0 and 3 weight %.
Scanning Electron microscopy is used to determine the shape of the agglomerates and whether they stayed intact in the spray-dryer. The particle size distribution is measured by a Malvern Mastersizer.
The resulting powder is uniaxially pressed into pellets of 19 mm diameter at a pressure of 150 MPa. The filling height of the die is 30 mm.
From each powder 5 pellets were fired in air during 1 hour at 1500°C.
Green and sintered density of the pellets were determined gravimetrically.

RESULTS AND DISCUSSION

Büchi-190 mini spraydryer

There are a few limitations on the type and concentration of the additives, which are set by the operating principle of the spray-dryer. The first limitation is due to the feed pump, which cannot transport suspensions of a viscosity over 800 mPas. For this reason the cellulose-ether AM-100 was unsuitable.

The second limitation is set by the strength of the agglomerates, which should be high enough to keep them intact in the spray-dryer. For this reason PAF-35 and all polyethene glycol could not be used as binders.

The other binders yielded strong, donut-shaped, agglomerates (fig. 1).

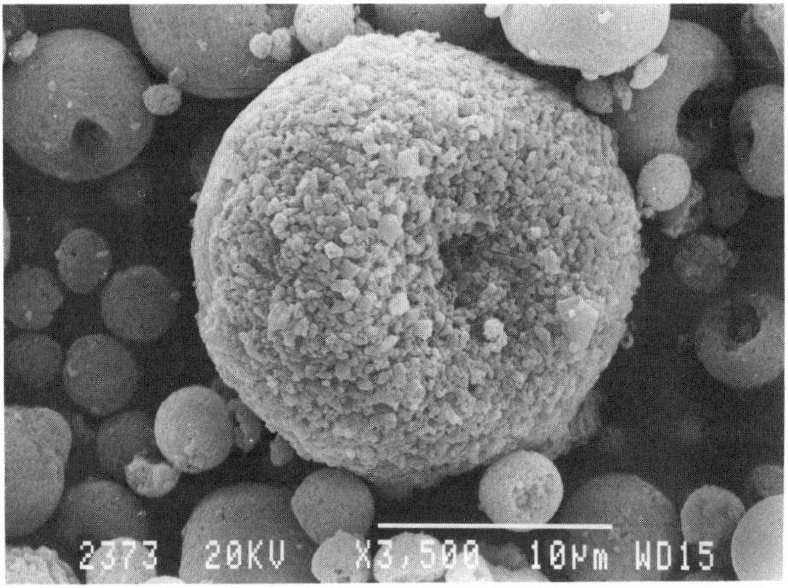

Figure 1. Typical spraydried agglomerate, showing primary particles of Alcoa A16SG.

The volume averaged diameter of the agglomerates ranged from 9 to 19 µm in all experiments. The parameter that affected this diameter the most was the viscosity of the suspension, with high viscosities yielding large diameters. The particle size distribution was log normal with a standard deviation of 1.7.

Type and concentration of binders

The pressing behaviour of the powder, which normally cannot be pressed, is greatly improved by the addition of the other binders. When added in a concentration of 0.5 weight percent on dry powder basis, green density of the compact was at a maximum of 56.1% for PA-4G and at a minimum of 55.1% for PS-13. When sintered in air for 1 hour at 1500°C, the density of the pellets ranged from 94.1% for PA-4G and PA-20G to 93.1% for PS-13.

The effect of binder concentration is qualitatively the same for all binders used (fig. 2). The green density decreases as binder concentration increases, probably caused by the increasing strength of the agglomerates. Sintered density initially increases somewhat, but later decreases again. The former effect is caused by the better homogeneity of the product containing less large pores, the latter by the decrease in green density.

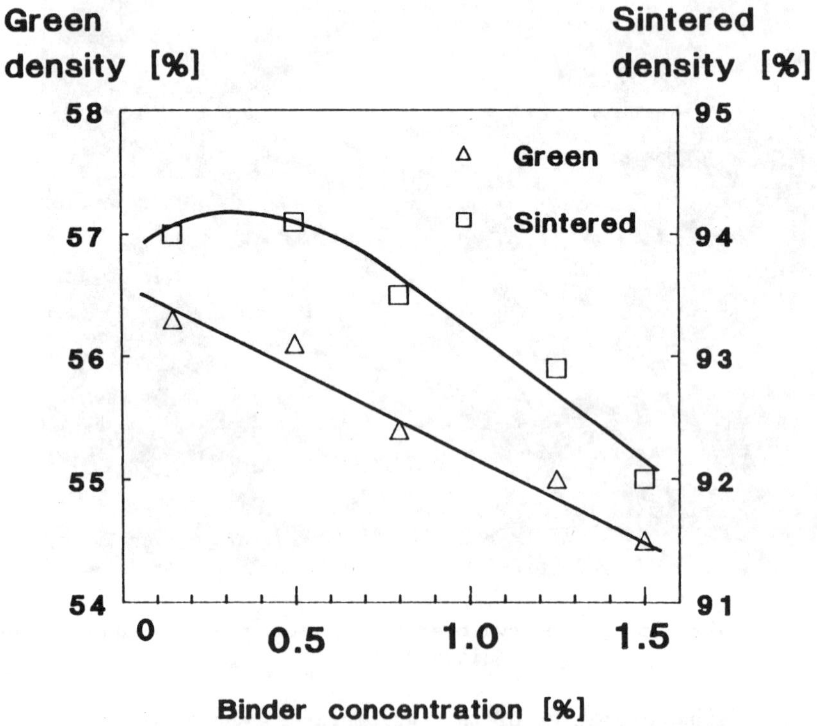

Figure 2. Influence of binder concentration on green and sintered density (for polyvinyl alcohol PA-4G).

Since PA-4G showed slightly better results than the other binders, all experiments with the plasticizers were performed with this binder added in

a concentration of 0.5 weight percent.
Addition of glycerol showed no effect on green and sintered density. A slight increase in green and sintered density could be noticed, when adding the polyethene glycol. The best results were obtained when adding two weight percent of PEG-400, yielding a green density of 58.7% and a sintered density of 95.2%.
The addition of water showed a marked increase in green density from 56.1% for a water content of less than 0.1% to 59.8% for 3.9 weight percent of water. However, when more than 1% of water is present, the powder tends to stick to the pressing die. This problem gets more serious as the water content is increased. At the highest water content large cracks can be observed in the green pellets. Probably some large internal voids exist also at a lower water content, leading to a maximal sintered density of 95.0% with 1.2% of water, which is appreciably lower than could be expected from green density.

CONCLUSIONS

The agglomerate size produced by the Büchi-190 mini spray-drier is too small to yield an optimum pressing powder.
However, the pressing behaviour of Alcoa A16SG, which cannot be pressed as such, can be improved enormously. In a 60 weight percent suspension of A16SG, dispersed with Dolapix CE64 or nitric acid, the addition of 0.5 weight percent polyvinyl alcohol PA-4G and two weight percent polyethene glycol PEG-400 is optimum.
When pressed into pellets at a pressure of 150 MPa a green density of 58.7% is achieved. After sintering for 1 hour at 1500°C the sintered density is 95.2%.

REFERENCES

[1] Danforth, S.C., Ceramic Powder Processing, ACerS Course, november 1987.

[2] Dimilia, R.A. and Reed, J.S., Effect of humidity on the pressing characteristics of spray-dried alumina, in Advances in Ceramics, Vol. 9.

SOL-GEL-SYNTHESIS FOR PREPARATION OF YTTRIUM ALUMINIUM GARNET

J.W.G.A. Vrolijk*, J.W.M.M. Willems and R. Metselaar
Centre for Technical Ceramics,
P.O.Box 513, 5600 MB Eindhoven, The Netherlands

ABSTRACT

Sol-gel-synthesis for preparation of pure yttrium aluminium garnet powder with small grain size is subject of this ongoing study. Starting materials were sulfates and chlorides of yttrium and aluminium. To obtain pure YAG ($Y_3Al_5O_{12}$) pH during hydrolysis as well as temperature during calcination and conversion into YAG must be carefully chosen. Presence of water has an adverse effect on the degree of agglomeration. To avoid this the influence of dispersion liquids was studied and this led to optimized conditions for precipitation. Sintering behaviour is discussed.

INTRODUCTION

Sintering behaviour is determined strongly by precursor powder characteristics. Those depend on the conversion steps in the process. In a wet chemical route synthesis the way of precipitation, calcination and conversion into garnet will contribute to sintering behaviour. Agglomerates formed during precipitation and remaining in the powder compact after calcination are disastrous for sintering to full density. To optimize process conditions the influence of dispersion medium on the agglomeration of YAG precursor particles during precipitation has been studied. To obtain a stoichiometric precipitate in the quasi-binary system furthermore it is important to take into account the dependence on pH for precipitation of Y and Al. The present paper describes the circumstances to reach the goals mentioned.

EXPERIMENTAL

Titration of aqueous solutions of yttrium sulfate and aluminium sulfate resp. with sodium hydroxide and simultaneous measuring of pH provides the graph of fig. 1. Precipitation of aluminium is completed at pH 5, on the

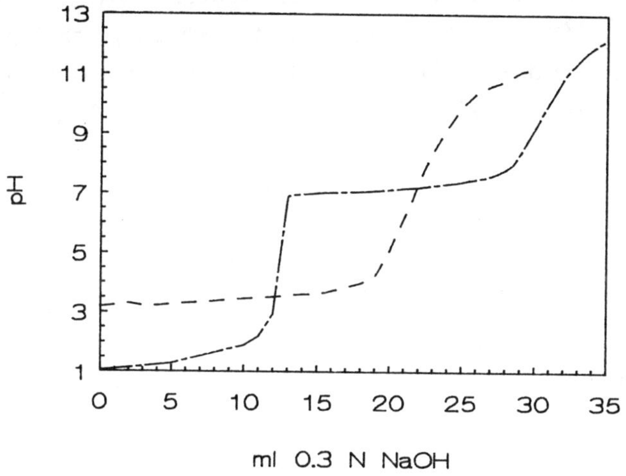

Fig. 1. Titration curves of Al sulfate (– – –) and Y sulfate (— - —).

other hand precipitation of yttrium occurs at pH >7. It is evident that Al precipitates before Y, so the reactions occur after each other and a homogeneous mixture of the precipitates can not be expected. To examine the Y and Al content in the precipitate as a function of pH a mixture of Y and

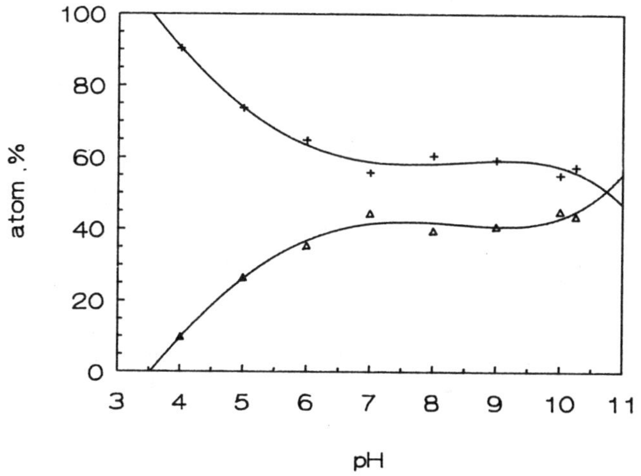

Fig. 2. Y (Δ) and Al (+) content in precipitate.

Al sulfate (Y/Al = 0.6), was titrated to certain pH values, filtered, dried and analyzed by SEM. Fig.2 shows that the ratio Y/Al = 0.6 is reached between pH 7 and pH 11. After substituting sulfuric acid by hydrochloric acid the procedure described gave identical results. Adding the salt solution drop wise into the alkaline medium has the advantage that possible inhomogeneous precipitation of yttrium and aluminium hydroxide will be limited to the volume of one drop. To avoid the relative high temperature of decomposition the sulfates are replaced by chlorides. Furthermore, instead of sodium hydroxide, ammonia has been used for precipitation. Ammonium chloride, that is left after washing, will be removed simply by heating as the temperature of sublimation is 340°C. The precipitate was separated by centrifugation, washed with a mixture of distilled water/ ammonia (pH 9) until no chlorides could be demonstrated. Subsequently the precipitate is washed several times with ethanol to remove the attached free water and dried in a rotating evaporator. The powder was heated at 350°C to remove ammonium chloride, calcined at 650°C and next fired at 800°C to obtain YAG.

AGGLOMERATES

Due to the presence of water the particles in the precipitate will agglomerate. To avoid this the precipitate is usually washed with ethanol [1], propanol-2 [2] or acetone [3], prior to drying. This gave rise to

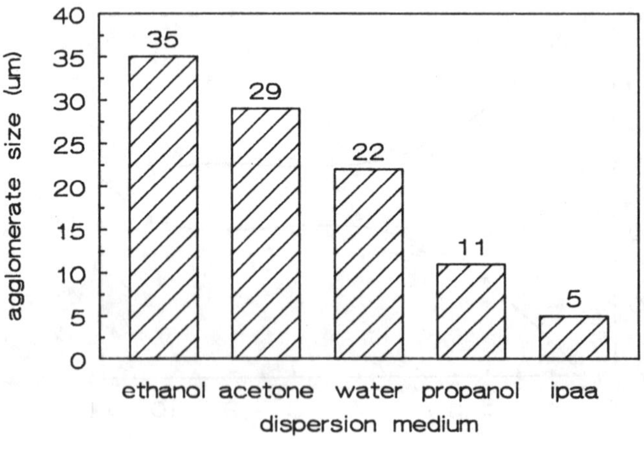

Fig.3. Precursor particle sizes.

carry out precipitations in the organic liquids mentioned after raising pH by means of ammonia to pH 9, making the restriction that pH is not well defined in organic substances. To diminish the water content born from the aqueous salt solution the precipitate has been redissolved in acetic acid and flocculated again (referred to ipaa) in the organic solutions prepared as described above. The precursor particle sizes obtained in this way were optical analyzed (Malvern). The decrease of volume mean diameter of agglomerates in the order of ethanol, acetone and isopropanol, as dispersion medium and of the method "ipaa" is convincing (fig.3). SEM analysis showed the same image. This holds too for the sizes of primary crystallites analyzed by XRD line broadening.

CONVERSION INTO YAG

The calcined powder was fired at 800, 900 and 1000°C to study the conversion into YAG. After firing at 800°C only pure YAG (fig.4) was detectable by means of XRD. At 900°C and at 1000°C besides $Y_3Al_5O_{12}$ (YAG, garnet

Fig.4. SEM photograph of YAG powder fired at 800°C.

structure) $YAlO_3$ (YAP, perovskite structure) was formed initially. Within 25 hrs. at 900°C and 1 hr. at 1000°C resp. all YAP was converted into YAG.

SINTERING

The YAG powder without any dope was uniaxially pressed to 176 MPa. The compact was subsequently sintered for 4 hours in air (1450°C) and vacuum (1750°C). To avoid grain growth the temperature was raised fast to the desired temperature. At 1450°C the relative density reached was 80% and at 1750°C, as first translucency appeared, the density was >99% .

CONCLUSIONS

- Stoichiometric YAG precursor can be obtained by precipitation from salt solutions at pH 9.
- To obtain small precursor particles the best dispersion liquid during flocculation is an isopropanol/ammonia mixture.
- To diminish the amount and size of the agglomerates the water content of the liquids used is decreased by redissolving the precipitate in an isopropanol/acetic acid mixture. Subsequently YAG precursor is flocculated again in an isopropanol/ammonia mixture.
- Sintering at 1750°C of the YAG powder compact without any dope resulted in a fully densified (>99%) and slightly translucent material.

ACKNOWLEDGEMENTS

The authors wish to thank H. de Jonge Baas for performing the XRD measurements and G.J. Bezemer for performing the TG/DTA analyses.

REFERENCES

1. Winnubst, A.J.A., Theunissen, G.S.A.M., Groot Zevert, W.F.M., and Burggraaf, A.J., The sintering behaviour of fine-grained ZrO_2-Y_2O_3 ceramics. In Science of Ceramics 14, ed. D. Taylor, The Institute of Ceramics, Stoke-on-Trent, 1988, pp. 309-14.

2. Van de Graaff, M.A.C.G., Ter Maat, J.H.H. and Burggraaf, A.J., Microstructural development during pressing and sintering of ultra fine zirconia powders. In Ceramic Powders, ed. P. Vincenzini, Elsevier Scientific Publishing Company, Amsterdam, 1983, pp. 783-94.

3. Dole, S.L., Scheidecker, R.W., Shiers, L.E., Berard, M.F. and Hunter Jr., O., Technique for preparing highly-sinterable oxide powders. Materials Science and Engineering, 1978, **32**, 277-81.

YAG POWDER SYNTHESIS FROM ALKOXIDES

M. Steinmann*, and G. de With**
Centre for Technical Ceramics
POB 513, 5600 MB Eindhoven, The Netherlands.

* : At present employed by Philips Lighting, Eindhoven
** : Also affiliated with Philips Research Laboratories, Eindhoven, The Netherlands.

ABSTRACT

Translucent yttrium-aluminum-garnet (YAG) ceramic is of interest because of its use in optical engineering. We investigated the feasibility of YAG powder synthesis from metal alkoxides. Based on the hydrolysis of a stoichiometric mixture of $Al(OC_3H_7{}^i)_3$ and $Y(OC_3H_7{}^i)_3$ we developed a procedure with which ultra-fine mixed oxide particles can be prepared. EDX analysis shows a homogeneous distribution of aluminum and yttrium in the stoichiometric ratio inside individual submicrometer particles. Between 800 and 900°C a phase transformation to the garnet structure takes place. The sinter activity of YAG powders depends on the powder microstructure and the phase transformation temperature. The highest relative density obtained after sintering at 1700°C reaches only 95 %. An optimized sintering schedule and/or better deagglomeration of the powder are therefore necessary.

INTRODUCTION

A number of ceramic materials can be sintered to translucency or even transparancy. The most well-known material, alumina, was first sintered to translucency by Coble [1]. More recently Gazza [2] claimed promising feasibilities of hot-pressed translucent Yttrium Aluminum Garnet ($Y_3Al_5O_{12}$ or YAG) for optical engineering. The interesting properties of YAG ceramics are based on the cubic lattice structure. As compared with translucent alumina the main virtue of YAG ceramics is a low optical absorption. Moreover, for various applications the better mechanical properties are also relevant [3].

The classical synthesis of YAG powders consists of a reaction of alumina- and yttria powder [4]. Therefore a homogeneous mixture of both powders must be heated at a temperature at which a solid state diffusion process can take place. Because of the long exposure times required at these temperatures, grain growth occurs. Consequently the resulting YAG powders have to be milled intensively afterwards thus obtaining an acceptable particle size distribution. Inhomogeneous composition and contamination are the main disadvantages of this preparation method.

Wet-chemical synthesis of mixed oxidic powders through simultaneous precipitation of metal hydroxides starting from alkoxides, eliminates the above mentioned disadvantages [5].

In the present study the synthesis of YAG powders from aluminum- and yttrium isopropoxides is investigated. YAG was synthesized by the hydrolysis of a mixture of both alkoxides and subsequent decomposition of the hydroxide precipitate. The main characteristics of powders and powder compacts related to the sintering behaviour of these powders are discussed.

MATERIALS AND METHODS

Powder preparation

Aluminum isopropoxide $Al(OC_3H_7)_3$ was synthesized by heating aluminum metal (99.9999%) in an excess of analytical grade 2-propanol with mercury(II)chloride (10^{-4} mol/mol of metal) as a catalyst for 15 hours at 82°C. After vacuüm distillation aluminum isopropoxide was dissolved in 2-propanol. Yttrium isopropoxide was commercially supplied and directly mixed with the aluminum isopropoxide solution. In this way sufficient yttrium isopropoxide could be dissolved because of the formation of the double alkoxide $Y[Al(OPr)_4]_3$ [6]. The concentration, 0.27 M and proper stoichiometry (Y:Al=3:5) of both metals was set by means of EDTA titration. No premature hydrolysis could be observed during storage. The mixed alkoxides were hydrolysed by adding the solution dropwise to a mixture of 2-propanol and double distilled water at room temperature (molar ratio $H_2O/Me=10$). Immediately a gelatineous hydroxide mixture precitated. In order to achieve weak agglomerates, a maximum degree of dispersion had to be accomplished during hydrolysis and free water had to be totally removed from the surface of the gelatinous precipitate prior to drying and thermal decomposition. Therefore after hydrolysis the precipitate was separated and washed with 2-propanol. The concentration of Yttrium and Aluminum in the remaining liquid phase was negligible. The separated gel was ball-milled prior to each washing step. After the final washing step the precipitate was treated ultrasonically at different power levels. Afterwards the precipitate was dried, ball-milled in a plastic jar and calcined at various temperatures.

Characterization of powders and compacts

After hydrolysis the atomic ratio Y/Al inside individual precipitated submicrometer particles was determined by means of energy dispersive x-ray analysis (EDX). The particle size distribution after ball-milling and ultrasonic dispersion was measured by means of light scattering using a Malvern Particle Sizer. Thermogravimetry (TG) and differential thermal analysis (DTA) were performed in a nitrogen flow at a heating rate of 10°C/min. X-ray diffraction was used for the analysis of phase composition and crystal structure of powders before and after heat treatment. Specific surface areas of the dried and calcined powders were determined according to the standard (BET) N_2 adsorption technique. Powder compaction behaviour was investigated by measuring changes in compact density with increasing uniaxial pressure as described by Niesz et al. [7]. The microstructure of powder compacts was examined by scanning electron microscopy.

Sintering experiments on powder compacts were performed in a furnace with lanthanum chromite heating elements. Exposure time for each temperature was 24 hours. Prior to compaction, 500 ppm MgO was added to the powders as sintering aid.

RESULTS AND DISCUSSION

Powder characteristics

EDX analyses of separate precipitated particles show a homogeneous and approximately stoichiometric distribution of Yttrium and Aluminum inside individual particles.

Ball-milling of the separated gel prior to each washing step did not significantly change the agglomerate size distribution. Ultrasonic treatment at different intensity levels initially revealed a small decrease in size. Longer exposures caused a coarsening of the agglomerates. Consequently, the powders still were agglomerated after drying. Nevertheless the agglomerates were very weak as will be described in the next section. The specific surface of the powders was 270 m^2/g.

Fig.1 shows TG and DTA curves of the dried powder. The sample weight decreases as a result of endothermic vaporization of residual isopropanol and endothermic conversion of yttrium- and aluminum hydroxide into amorphous oxides.

The dehydration is essentially completed at 700°C. The DTA curve shows one exothermic peak at 900°C due to the phase transformation to YAG as is confirmed by X-ray diffraction. The sharp peak demonstrates the homogeneity of the yttrium- and aluminum oxide mixture. XRD analysis excluded the formation of intermediate compounds, prior to or simultaneously with the YAG formation.

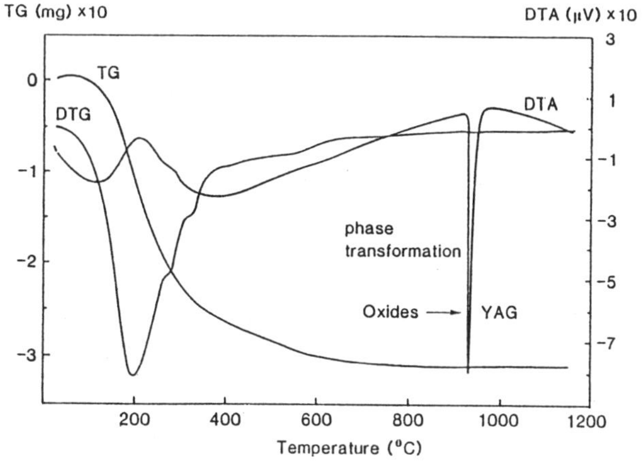

Figure 1. Thermal analysis of the dried powder.

Other samples were heated at 800°C during 1,5 and 15 hours. Exposure times between 5 and 15 hours resulted in complete transformation into the garnet structure. During heat treatment the specific surface of the powders strongly decreased due to particle growth. Table 1 shows the decrease of specific surface after calcination at 800 and 900°C.

As a result of high reactivity, adjacent particles already show neck-formation after calcination at 900°C, hence turning weak agglomerates into strong aggregates.

TABLE 1
Specific surface area according to the BET method and derived particle sizes

heat treatment	specific surface area (m^2/g)	derived particle size (nm)
none	271	7
800°C/1h	70	19
900°C/1h	3.6	370

Powder compaction
The compaction behaviour of powders can be illustrated by plotting the relative density of the powder compact as a function of compaction pressure.

Stepwise compaction of the non-crystallized oxide mixture, and two crystallized oxide mixtures, YAG(800) and YAG(900) respectively, shows different densification behaviour as illustrated in fig.2. The intersections between the two linear parts of the oxide- and the YAG(800) curve represent the average compression strenght (10 MPa) of the agglomerates. At pressures exceeding 10 MPa, the agglomerate fragments are being compressed, thus changing their internal microstructure. At a pressure of 350 MPa both compact microstructures are extremely homogeneous, possessing a uniform pore size of less then 10 nm as revealed by mercury intrusion porosimetry.

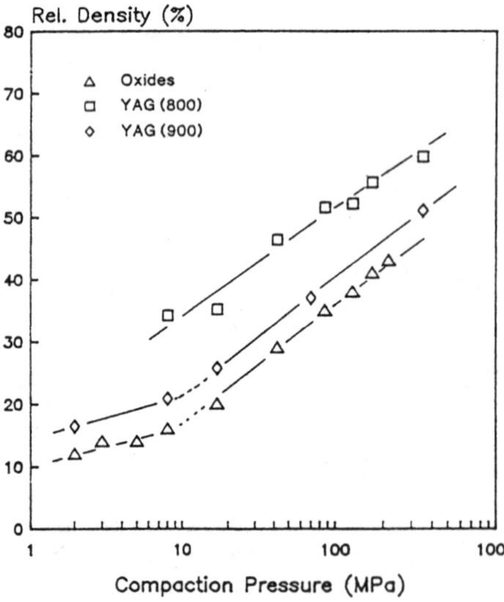

Figure 2. Compaction curve of oxides, YAG(800) and YAG(900)

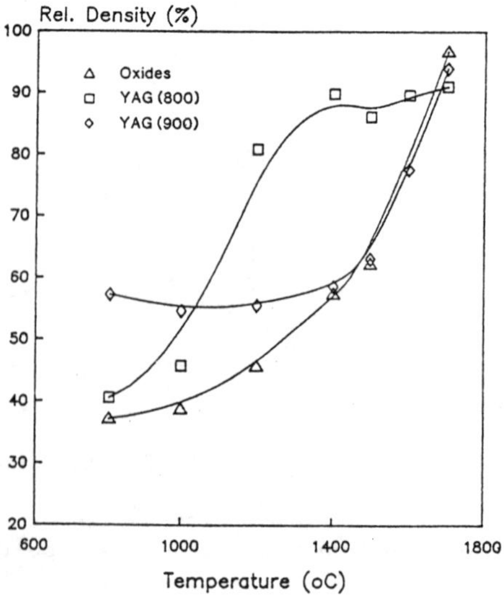

Figure 3. Sinter curve of oxides, YAG(800) and YAG(900)

The compaction curve of YAG(900), however, only illustrates the rearrangement of the strong aggregates. The inability of compressing the aggregates, impedes the formation of a homogeneous powder compact even at high compaction pressures.

Sintering behaviour

The sintering behaviour of the oxide mixture and both YAG powders, compacted at 170 MPa, is illustrated in fig.3. The YAG(800) compact reveals the highest sinter activity and starts shrinking at temperatures beyond 800°C. The oxide mixture at first transforms into YAG between 800 and 900°C. This phase transformation presumably has a detrimental effect on the microstructure of the powder compact hence decreasing the sinter activity. The YAG(900) compact starts shrinking just beyond 1200°C due to the inactive aggregates and inhomogeneous microstructure. Consequently, none of the compacts reaches the theoretical density of the garnet structure at 1700°C.

CONCLUSIONS

YAG powder synthesis from alkoxides was investigated.
Homogeneous and weakly agglomerated powders of nanometer particle size containing yttrium- and aluminum oxide were prepared. The phase transformation to the YAG structure occurs between 800 and 900 °C, excluding the formation of intermediate compounds.
From powders calcined at 800°C, not yet converted to YAG, homogeneous compacts can be prepared, but the homogeneity probably deteriorates during the phase transformation. For powders calcined at 900°C, which are converted to YAG, no homogeneous compacts could be pressed. Hence in spite of the favourable powder characteristics full density after sintering could not be obtained. The presently used deagglomeration method and/or sintering scheme should be improved.

REFERENCES

1. R.L. Coble, U.S. Patent 3026210 (1962)

2. G.E. Gazza et al., U.S. Patent 4029755 (1977)

3. G. de With, "Preparation, Microstructure and Properties of $Y_3Al_5O_{12}$ Ceramics", Philips J. Res. Vol.42, p. 119, (1987)

4. V.B. Glushkova et al., Translated from Izvestiya Akademii Nauk SSSR, Neorganicheskie Materialy, Vol. 19, No. 1, pp. 95-99, (1983)

5. K.S. Mazdiyasni, "Powder Synthesis from Metal-Organic Precursors", Ceramics International, Vol.8, No.2, p.42,(1982).

6. D.C. Bradley, R.C. Mehrotra and D.P. Gaur, Metal Alkoxides. Academic Press Inc., London, (1978).

7. D.E. Niesz, R.B. Bennett and M.J. Snyder, "Strength Characterization of Powder Aggregates", Ceramic Bulletin, Vol. 51, No. 9, p.677, (1972).

DCCAs IN SOL-GEL PROCESSING OF GLASS MONOLITHS

L. Nikolic and L. Radonjic
Faculty of Technology, 21000 Novi Sad
Yugoslavia

ABSTRACT

One of possible ways of solving the cracking problem is sol-gel processing of gel monoliths is using of drying control chemical additive (DCCA). Gelation and drying behavior of DCCA modified silica gels prepared from the TEOS solutions was investigated. Formamide, glycerol and oxalic acid as DCCA have different influence and the geling time ands stability against the cracking. As DCCA, only formamide demonstradet the positive effect.

INTRODUCTION

During the past decade many studies have been deminstrated the possibilitty of using sol-gel route to prepare glass and ceramics However, the difficulties of preparing large monolitic glass, crack free are not overcome. To reduce very long drying time, gel strenght need to be increased to resist cracking and the drying stress reduced.

Several approcaches have been suggesting ways of obtaining monolithic gels by: (a) increasing the mechanical strenght of gels by aging (1), (b) deminishing the magnitude of capillary force by using surfactants (2) "drying control chemical additives, DCCA (3, 4, 5), (c) reducing the rate of evaporation of the solvent from the pores by using semipermeabile membrane during drying (6) and (d) eliminating the pore liquid-solid interface by percritical evaporation (7,8).

The objective of this study is to demonstrate the possibility to produce monolitic gel using different DCCA in TEOS ethanol solutions. University of Florida group (3,4,9,10) have demostrated the use of formamide and oxalic acid as DCCA in producint crackfree gel monoliths from TMOS solutions. It was proved that DCCA controle: the rate of hydrolysis and condesation, pore size distribution, pore liquid vapor pressure and dryng stresses. The net effect is to enhance the rate of drying without inducing crack in the gel monoliths.

EXPERIMENTAL PROCEDURE

Besically the two types of samples were prepared: either without (type T,H1) or with DCCA (formamide-F, glycerol-G, oxalic acid O). As avariable the type and ratio of DCCA to ethanol (Et), the pH and the type of acid catalyst (HCl or HNO_3) were used, Table 1.

Desired amount of water, ethanol and DCCA were mixed for 5 min (room temperature) followed by addition to the solution of TEOS and mixed for 30 min. under reflux to accelerate hydrolysis of the TEOS. Homogenous solutions are cast into molds and mantained at 60°C until gelation. After gelation the temperature was raised.

Table 1.
Gelation time to different sols (T,E=3,R=15; H,E=4, R=10)

	T	H	T_F	T_G	T_{G2}	T_o	H1	$H1_G$	$H1_G$	$H1G_2$	$H2_F$	$H2_G$
PH	2,5	1	4	3	5	1,5	4	4	4	4	4	4
A	1	1	1	1	2	1	1	1	1	2	2	1
Gelation time(days)	8	2	1	11	14	21	1	3	1	1	1	1

A,R,E are the molar ratio of: DCCA/Et,Et/ TEOS and water /TEOS rexpectively.

The temperature was futhure raised to 150°C in 8 days and held there for 24 h. Densification can be achived by firing in a ambient atmosphere up to given temperature and kept there for 1h, depending on the desired degree of densification.

Gels as prepared and heat treated where characterized by:differential scanning calorimetry (DSC), thermal gravimetric (TGA), IR absorption spectrometry, and scanning electron microscopi (SEM).

RESULTS AND DISCUSSION

Table 1 shows the influence of the type and quantity of DCCA on gelation time the two type of cilica gels H and T. Formamide accelerates the gelation of the both gels T and H, but more T gels. Glycerol can accelerate the gelation only in higher concentration and oxalic acid on the contrary, increases gelation time. Both, F and G change the rate of hydrolysis and condensation. This is confirmed by the observation the rate of shrincage during aging due to resultant rapid Si-O bond formation leading to the stronger gel in the shorter time. Formamide changes the both pH and viscosity (Figure 1) and glycerol does not change viscosity but only pH in sol HI. On the contrary, oxalic acid decreases the both pH and viscosity increasing the geling time due to very slow the hydrolysis rate.

The sharp decrease in the weight of the gels and 150-300^{0}C (Figure 2) due to desorption of water is more pronounced for gels with F and G than witout. But O type of gels as do not loose the water or do not contained at all. The endothermic feature on DSC (Figure 3) is atributed to desorption of water. The difference in the temperature of water desorption for the F, G and O the type of gels is the resalt of the different nature of water bonding in these gels, which is in agrrement with water los (Figure 2).

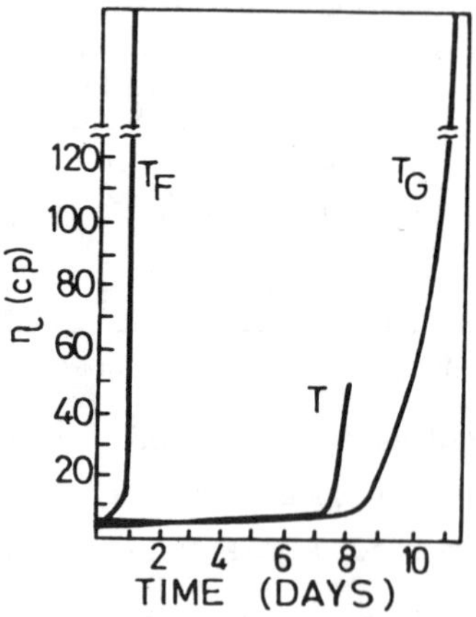

Figure 1. Viscosity change by time for T,T_F,T_G and T_o gels

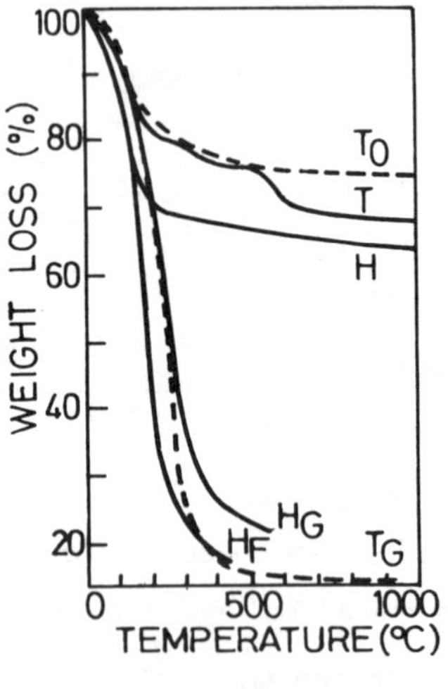

Figure 2. Weight loss of the gels with and without DCCA

The structure was monitored using IR absorption spectra. The IR spectra of as-prepared gels without and with DCCA, dryed and the gels heated in an microprocessor controlled furnace very slowely up to 550^{0}C are shown in Figure 4. The three characteristics bands of amporphous silica S (1080 cm^{-1}), D (455 cm^{-1}) and R (800 cm^{-1}) are to the vibration of Si-O or Si-O-Si bonds (11). Three other bands appear at 3500, 1620 cm^{-1} are attributed to molecular water. The band at 950 cm^{-1} corresponds to the vibration of Si-O (OH) bonds. The IR spectra of T, T_F, T_G and T_O are rather different (Fig. 4) showing the different the degree of hydrolysis and condensation. After drying (Fig. 4 b.d), the absorption

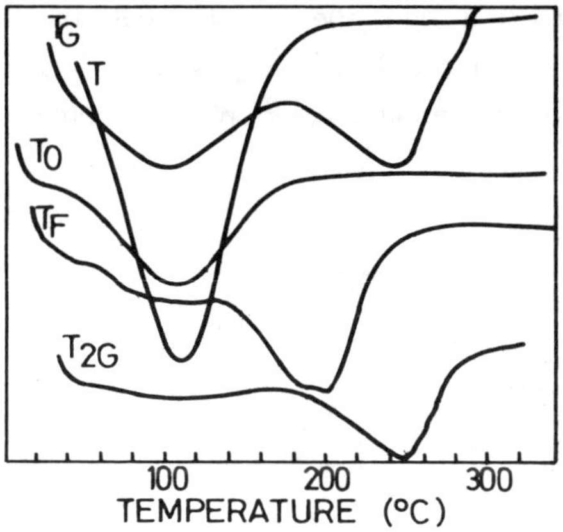

Figure 3. DSC analysis of the gels T, T_F, T_G, T_o

bands corresponding to molecular water are deminishing and the band 110-950 cm^{-1} and R,S,D are more pronounced in all gels. This is due to increasing the number of Si-O or Si-O-Si bonds after condensation. Futhure heating have similar tendency to the changing IR spectra. Series of H gels have similar features, only H sols need higher concentration of G (even pH is the same) in solution to increase the rate of hydrolysis. The bands S, R and D are too more pronaunced than in T, showing higher the condensation rate ih H than in T.

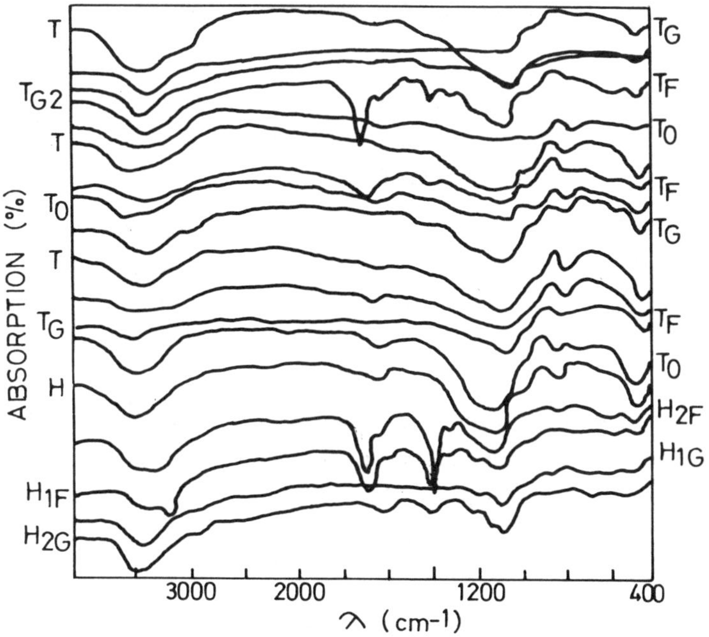

Figure 4. Infrared spectra of as-prepared and heated gels: without and with DCCA

Figure 5 showes the different microstructures of DCCA modified, acid cathalysed gels. T_F-modified gels in respect to the T_G-modified has particulate microstructure on somewhot coarese scale. The same is true for the T and H gels (Figure 5 a.b.d). Even F is more of efective as DCCA in both the T and H gels than G, the degree of coarseining the microstructure of the T and H gels with F is not high enough. It could be still the result of the small resolution of scanning microscope for this fine scale changes.

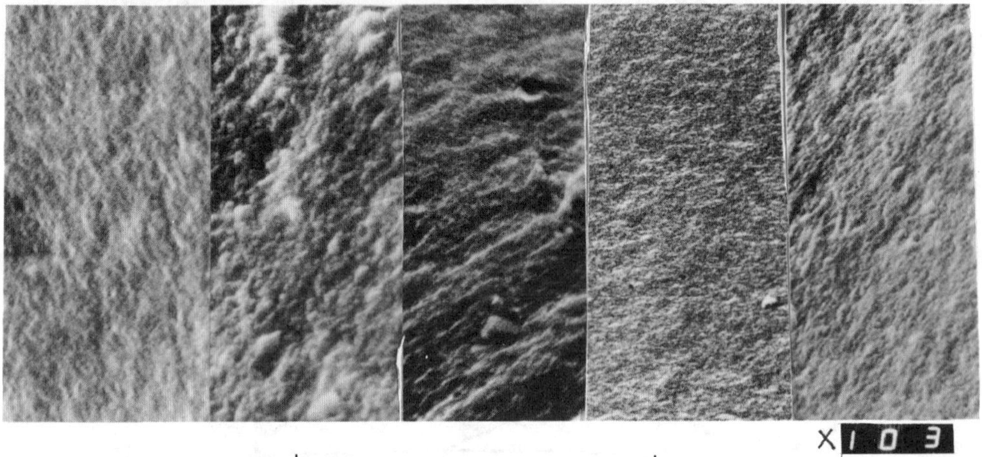

a. b. c. d. e.

Figure 5. Electron micrograph of the gels: (a) T, (b) T_F, (c) T_G, (d) H1, (e) H1

CONCLUSIONS

Formamide is only of DCCA used in this study which could be used for improving the processing silica gels from TEOS solutions. It reduces the geling, aging and drying times, coarsener the microstructure and increases the size of gel monoliths crack free. The acid catalyst and correct the type and volume of DCCA increases the hydrolysis rate allowing a stronger gel to be formed. Glycerol is less effective as DCCA, showing some influence on drying time only for high volume ratio to ethanol. Oxalic acid by decreasing the hydrolysis rate in TEOS solution can not be use as the effective DCCA.

REFERENCES

1. Iler, R.K., The Chemistry of Silica, Wiley - Interscience, New York, 1979, Ch. 1-6.
2. Zarzycki, J., Monolitic Xero-and Aerogels for Gel Glass Processing in Ultrastructure Processsing of Ceramics, Glasses and Composites, L.L. Hench and R.D. Ulrich, eds., John Wiley & Sons, New York, 1984.
3. Hench, L.L., Science of Ceramic Chemical Processing. L.L. Hench D.R. Ulrich,

eds., John Wiley & Sons, New York, 1986, p. 52

4. Hench, L.L., "SiO_2 Gel Glasses, Proc. of SPIEs Ann.Intern.Techn.Symposium on Optic and Electro-Optic, San Diego, 1984, p. 24
5. Wang S.H. and Hench, L.L., Science of Ceramic Chemical Processing, L.Hench and D.R. Ulrich eds., John Wiley & Sons, New York, 1986, p. 201
6. M. Prassas, Hench, L.L., and Zarzycki, J., Preparation of Na_2O-SiO_2 Gels, for the Gel Glass Process, J.Non Cryst.Soll., 1982, 48, 79
7. Zarzycki, J., Prassas M., and Phalippou, J., Synthesis of Glasses from Gels-the Problem on Monolithe Gels, J. Mater. Sci., 1982, 17, 3371
8. Phalippou, J. and Zarzicki, J., Behavior of Monolithic Silica Aerogels at Temperature above 1000^{0}C, in Ultrastructure Processing of Ceramics, Glasses and Composites, L.L. Hench ang R.D. Ulrich eds., John Wiley & Sons, New York, 1984.
9. Wallace, S. and Hench, L.L., The Processing and Characterization of DCCA modified Gel-Derived Silica in Better Ceramics Through Chemixtry, C.J. Brinker, D.E. Clark and D.R. Ulrich, eds., North-Holland, New York, 1984, p. 47.
10. Wang, S.H. and Heach, L.L., Processing and Properties of Sol-Gel Derived Glasses, in Better Ceramics through Chemistry, C.J. Brinker, D.E. Clark and R.D. Ulrich, eds., North-Holland New York, 1984, p.71.
11. Hair, M.L., Infrared Spectroscopy in Surface Chemistry, Marsel Dekker, New York, 1967.

SYNTHESIS OF MULLITE AND CORDIERITE POWDERS BY SOL-GEL PROCESS AND THEIR SINTERING BEHAVIOUR

D. SPORN AND H. SCHMIDT
Fraunhofer-Institut für Silicatforschung (ISC)
Neunerplatz 2, 8700 Würzburg, FRG

INTRODUCTION

In the field of high temperature applications the number of suitable oxidic, ceramic systems is limited. Mullite and cordierite are promising candidates, since they show excellent thermal properties. In the recent years an increasing interest in this systems could be observed, including new powders with high purity and good stoichiometry (e. g. by coprecipitation, spray pyrolysis, sol-gel etc.) [1 - 3]. The potential of the sol-gel process is to prepare high purity powders with good stoichiometry, and homogeneity to achieve pure mullite or cordierite phases at temperatures below 1000 °C [4, 5].

The homogeneity of sols and gels in multicomponent systems is strongly influenced by different rates of hydrolysis [6]. The hydrolysis of tetraethylorthosilicate (TEOS) in combination with high reactivity systems like $Al(OR)_3$ is too slow to prepare homogeneous amorphous solids if no special precaution is applied. This can be the prehydrolysis of fast reacting species by addition of understoichiometric amounts of water and their subsequent fixation by condensation to the slow reacting component. Another mean is the prehydrolysis of the slowly reacting species and the fixation of the highly reacting compound to residual OH groups.

In this paper, special preparation routes, including chelated Al-precursors for the systems of fine cordierite and mullite powders are investigated and the sintering behaviour of mono- and diphasic gels are compared.

EXPERIMENTAL

A scheme of applied sol-gel techniques for mullite is shown in fig. 1 and 2. Five series of sols having the mullite com-

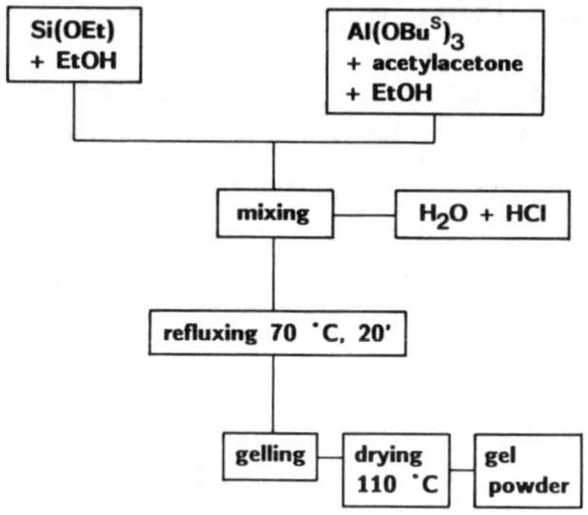

Fig. 1. Flow chart of mullite synthesis (monophasic)

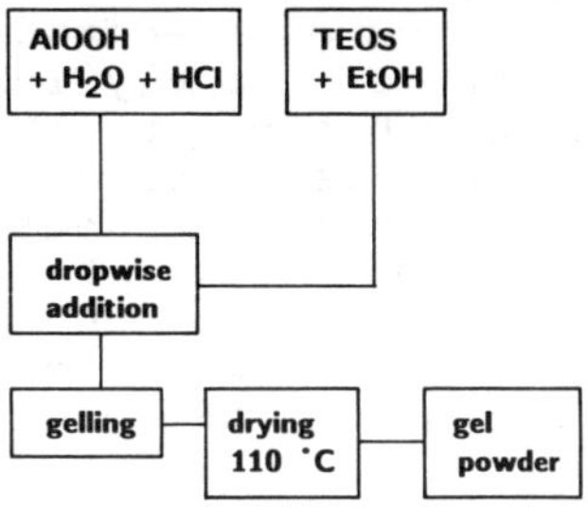

Fig. 2. Flow chart of mullite synthesis (diphasic)

position (molar ratio Si:Al = 1:3) were prepared for the study of the influence of prehydrolysis of TEOS with different amounts of water in the TEOS/ EtOH solution (in molar ratios H_2O/TEOS = 1, 2, 3, 4, and 60). Catalysis was carried out generally with 0,005 mole/l HCl in the hydrolyzing water. Concentration of TEOS was 0,5 mole/l. The partial hydrolyzed TEOS solutions were mixed with the ethanolic solutions of Al-nitrates, Al(OBu^s)$_3$ or chelated compounds, and transparent solutions were obtained. These solutions were refluxed for 20 min at 70 °C for solidifying to a gel. All the gels were dried at 110 °C and calcined stepwise to 1200 °C. By using diphasic sols from AlOOH and TEOS (fig. 2) the Al-source was peptized in 0,15 m aqueous HCl in concentrations up to 15 mass %. TEOS was added dropwise and gelation was carried out by heating up to 70 °C. All gel powders calcined at each temperature for 2 hrs were analyzed by XRD. A differential thermal analysis (DTA) was carried out in air up to 1350 °C.

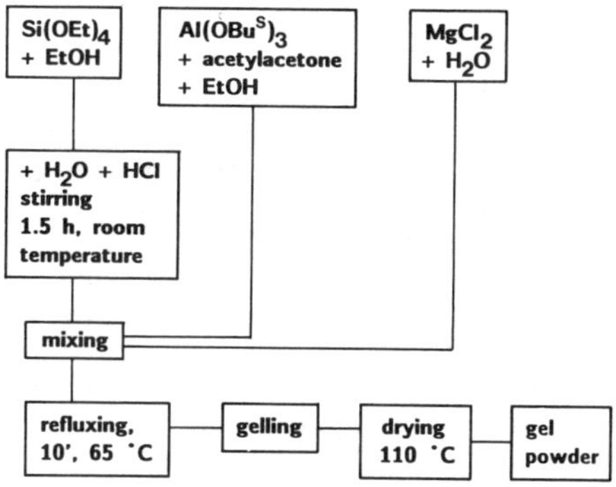

Fig. 3. Flow chart of cordierite synthesis (monophasic)

Bars from calcined gel powders were dry-milled for 10 min in a ball mill and pressed isostatically at 500 MPa. The linear shrinkage was investigated by dilatometry. Sintering density was determined by the Archimedes method and microstructure was investigated by SEM.

Cordierite powders were prepared by the way shown in fig. 3. The influence of prehydrolysis of TEOS and of the Al-source, powder characterization and sintering behaviour were investigated as described for mullite.

RESULTS AND DISCUSSION

Monophasic mullite gels crystallize at 970 °C (DTA data) in dependence on preparation techniques. The crystallization of mullite is strongly influenced by the Al-precursors, as XRD-analysis has shown. It is remarkable that the chelated reaction path leads directly to mullite without α-Al_2O_3 intermediates. Chelating seems to prevent the formation of crystalline Al_2O_3 phases at lower temperatures. The diphasic gel forms γ-Al_2O_3 at 500 °C and mullite at 1300 °C without α-Al_2O_3 as intermediate phase. The influence of $Si(OR)_4$ prehydrolysis in the monophasic mullite gel is shown in fig. 4. Obviously, prehydrolysis does not lead to an improved crystallization behaviour. This can be interpreted with the low diffusion coefficient of mullite: The mobility is better, if SiO_2 glass phases are present. In the homogeneous (prehydrolysed) case, the formation of SiO_2-glass phases is prevented. In opposition to this, cordierite shows a different behaviour. The optimum of the cordierite phase formation is shown at a TEOS prehydrolysis with 1 and 3 moles water per mole TEOS. In the case of cordierite, the differences between the mono- and the diphasic route are not so significant. The X-ray data indicate a strong influence of the Al-precursors.

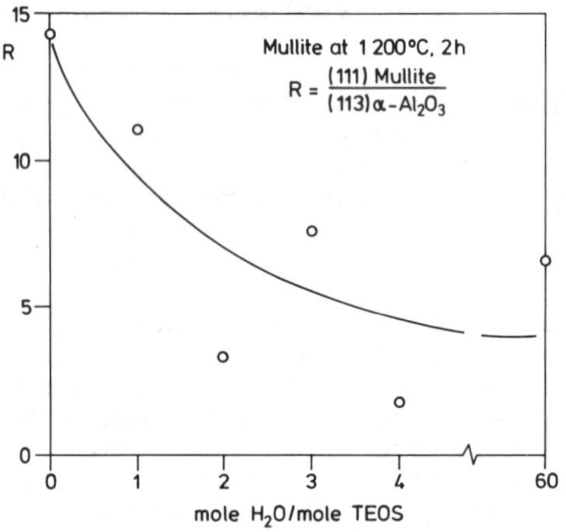

Fig. 4. Influence of prehydrolysis of TEOS on crystallization behaviour of monophasic mullite sols

The best results (pure cordierite phase at low temperature) are obtained from chelated Al-butylate as starting material.

The densification behaviour of monophasic and diphasic gel powders is very different (fig. 5). In monophasic mullite powders the sintering process starts at 850 °C combined with a strong linear shrinkage of 13 % up to 970 °C. The final density was 80 % T.D. at samples sintered at 1050 °C. The sintering step includes the formation of mullite and therefore the full densifying at low temperatures was inhibited because the diffusion coefficient of Si^{4+} in the mullite lattice is very low. By avoiding the mullite formation at low temperatures the densification behaviour can be improved. Gel powders from diphasic sols crystallize to mullite at 1300 °C in 2 hrs. Before mullite formation starts pressed gel powders can be densified at 1250 °C to nearly theoretical density (> 99 % T.D.). After heating up to 1300 °C the sample exhibits an open porosity of 0,2 %. Fig. 6 shows the microstructure of such a mullite ceramic with particle size ranging from 100 to 200 nm. Therefore, the use of diphasic mullite gel powders lead to dense ceramics at sintering temperatures which are 300 - 400 K lower compared to conventional mullite powders.

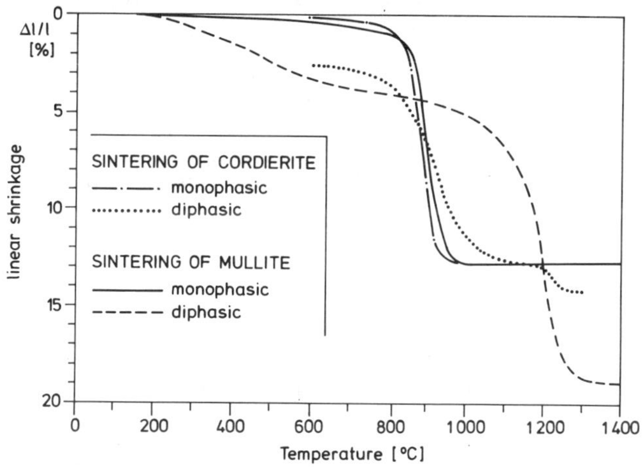

Fig. 5. Comparison of linear shrinskages of monophasic and diphasic gel powders

At 800 °C calcined monophasic pressed cordierite gel powders show a strong linear shrinkage between 820 °C and 920 °C combined with cordierite crystallization (fig. 5). At 1050 °C sintered samples attained > 95 % (T.D.). Fig. 7 shows the microstructure of such a sample which exhibits particle sizes in the range of 100 to 250 nm. Gel powders derived from diphasic gels showed a different densification behaviour with two sintering steps and a lower densification rate. The final density obtained at 1280 °C was 90 % (T.D.).

In the case of cordierite monophasic powders with high homogeneity in the amorphous state exhibit a better densification behaviour, the sintering temperature could be lowered by about 400 K compared to conventional powders.

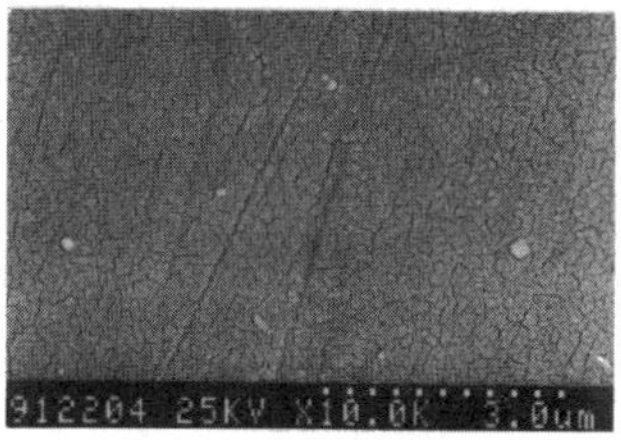

Fig. 6. Microstructure of sintered mullite ceramic from diphasic gel powders, 1300 °C, 2 hrs, heating rate 2 K/min, bar 3 μm

Fig. 7. Microstructure of sintered cordierite ceramic from monophasic gel powders, 1050 °C, 2 hrs, heating rate 2 K/min, bar 1 μm

CONCLUSIONS

Two sol-gel processing methods were developed to get dense cordierite and mullite ceramics at temperatures as low as with conventional powders. In the case of cordierite the use of chelated Al-alkoxide and prehydrolysis of TEOS to achieve very homogeneous and amorphous powders from monophasic sols was successful. Such powders can be densified at a temperature of 1050 °C to densities > 95 % (T.D.).

Mullite ceramics from diphasic gels showed a better densification behaviour in comparison to homogeneous gel powders. By intimate mixing of boehmite and TEOS the γ-Al_2O_3 phase is stabilized, this prevents the crystallization to mullite and makes the viscous-phase sintering to near full density at 1250 °C possible.

REFERENCES

[1] Takeuchi, K. and Tsuzuki, H., Synthesis of mullite by coprecipitation method using aluminum sulfate and sodium silicate. First Intern. Workshop on Mullite, Tokyo, Japan, November 1987.

[2] Nakagawa, Z., Sato, T., Sawabe, Y. and Hamano, K., Crystallization and characterization of mullite from a spray dried amorphous material. First Intern. Workshop on Mullite, Tokyo, Japan, November 1987.

[3] Prochazka, S. and Klug, F.J., Infrared transparent mullite ceramic. J. Am. Cer. Soc., 1983, **66** [12], 874- 80.

[4] Hoffmann, D.W., Roy, R. and Komarneni, S., Diphasic xerogels, a new class of materials: Phases in the system Al_2O_3-SiO_2. J. Am. Cer. Soc., 1984, **67** [7], 468-71.

[5] Bernier, J.C., Sol-gel processing for the synthesis of powders for dielectrics, 1986, Pow. met. intern. **18**, 164-68

[6] Suzuki, H., Ota, K. and Saito, H., Preparation of cordierite ceramics from metal alkoxides. I. Preparation and characterization of powder. J. Ceram. Soc. Jpn. Int. Ed., 1987, 95, 148-53

FACTORS INFLUENCING THE SINTERABILITY OF β-SiC POWDERS

VARUŽAN KEVORKIJAN, MILOŠ KOMAC, DRAGO KOLAR
"Jožef Stefan" Institute, Department of Ceramics, Jamova 39,
61111 Ljubljana, Yugoslavia

ABSTRACT

The sinterability of beta-SiC powder compacts depends strongly on the chemical and morphological characteristics of the starting powder. In order to evaluate the critical characteristics, the sinterability of different beta-SiC powders was compared.

It was found that once the level of impurities was sufficiently low, the sinterability of beta-SiC powders apparently depended mainly on particle size distribution rather than on average particle size, provided that the particles are mainly in the sub-micrometre range. A narrow particle size distribution and the absence of particles greater that 1 μm enhance sinterability. Oxygen pick-up can take place during the storage of SiC powders, resulting in the impairment of densification.

INTRODUCTION

Silicon carbide powders must meet stringent requirements to be sintered to high density. In general, the average particle size should be in the sub-micrometre range and the concentration of metallic impurities and in particular of oxygen must be minimized.

Such powders cannot be prepared directly by chemical synthesis. Further processing, mainly involving comminution, classification and chemical purification is required in order to convert crude powders into sinterable quality.

Depending on the selection and the intensity of post-synthesis treatment, the resulting SiC powders which may be regarded as sub- micrometre and acceptably pure, exhibit different particle size distributions and contain different amounts of impurities. Moreover, due to their fineness they are susceptible to reoxidation. Their densification behaviour can thus be affected, which appears to be rather critical when a density exceeding 95% T.D. and a uniform microstructure consisting of equiaxed grains is the target.

This paper describes the results of a study of the sintering of β-SiC powders of different origin, either commercial or experimental, the latter being synthesized by carbothermal reduction. Special emphasis was placed on the effect of particle size distribution and oxygen pick-up during storage on their sinterability.

EXPERIMENTAL PROCEDURE

Four commercial β-SiC powders were used in this study: Ibiden UF 0090, Superior Graphite HSC 100 GL, H.C.Starck B-10 and Mitsui Toatsu MSC-20. An experimental β-SiC powder (in subsequent text denoted as IJS powder) was prepared by carbothermal reduction of colloidal SiO_2. The procedure, which enables a powder with a high fraction of submicrometre particles to be prepared, is discussed in more detail elsewhere[1].

Chemical purification involved the removal of excess carbon and SiO_2 by heating in air and exposing the powders to a flow of dry gaseous HF, respectively.

Slurries for sedimentation were prepared by dispersing β-SiC powders in water at pH 11 with addition of a wetting agent.

Powders for sintering studies were doped with constant additions of 0.6 wt.% B and 4 wt.% C. Boron was added in the elemental form while phenolformaldehyde resin was used as a carbon precursor. Sintering was performed at 2000 - 2200°C for 15 to 60 minutes in a static argon atmosphere.

RESULTS AND DISCUSSION

In order to reveal the influence of particle size distribution on the sinterability, different commercial β-SiC powders were sintered at normal pressure employing an identical sintering regime.

Particle size distributions of the respective powders are shown in Fig. 1. A broad distribution with approximately 40% particles having diameters in the range 1 - 10 μm is characteristic of B-10 and HSC 100 GL powders. In contrast, MSC 20 and UF-0090 powders exhibit a rather narrow size distribution, predominantly in the sub-micrometre range. The particle size distribution of IJS β-SiC powder differs significantly, the main characteristics being comparable proportions of ultrafine particles (< 0.2 μm) and uniformly sized aggregates (~ 1.5 μm).

Due to the differences in oxygen and free carbon contents of the powders in the as-received condition, they were chemically treated immediately before sintering. Thus it is believed that the differences in the sinterability result predominantly from the different particle size distributions of the respective powders. Since it was shown that during storage of powders in humid air oxygen pick-up is possible (2), powders were exposed to an air atmosphere after HF treatment for different periods of time and then sintered. Results are presented in Figs 2 and 3, showing the relative sintered density as a function of sintering temperature and time of storage. As expected, higher densities were obtained in the case of finer powders with a narrower particle size distribution. Under the same conditions IJS powder sinters only to 80% T.D. in spite of the relatively high fraction of very fine particles. Consequently, in the case of β-SiC powders, particles with diameters above approximately 1 μm are detrimental for densification. The low sinterability of IJS powder may be explained by the peculiar particle size distribution, which obviously favours exaggerated grain growth and the formation of particle assemblies with pores of high coordination number which will not disappear on sintering.

The final density depends on the sintering temperature. More or less expressed density maxima exist in all cases. By increasing the sintering temperature, abnormal grain growth is promoted which in turn inhibits densification.

The influence of storage time is also evident. The highest densities are obtained if sintering is performed immediately after purification with HF. Samples which were stored before sintering for 100 hours or more sinter to lower densities, and the density maximum moves towards a higher temperature in the case of the coarser B-10 and HSC 100 GL powders. These rather pronounced effects are somewhat surprising, especially with regard to the role which free carbon plays in the sintering of SiC. It seems that at very low oxygen concentrations, which was achieved with a repeated HF treatment, other factors that are also known to control the densification of β-SiC become less critical, resulting in improved sinterability, provided of course that the general requirements for sinterable SiC powders are fulfilled.

In contrast to expectations, UF-0090 powder does not exhibit the best densification behaviour. It is hypothesized that a rather rapid oxygen pick-up occurs after cleaning with HF due the high fineness of that powder, which in turn is reflected in its reduced sinterability. This asumption, however, still needs verification.

Obviously, β−SiC powders can be sintered to high final densities, provided that very strict requirements regarding powder morphology and chemistry are fulfilled. Tedious processing of powders prior to sintering is thus unavoidable.

If low reaction temperatures can be used for the preparation of β-SiC, as exemplified by carbothermic synthesis, powders which contain a high fraction of submicrometre particles can be obtained(1). Processing consisting only of classification and removal of excess carbon and oxygen may be sufficient for producing powders with high sinterability. To test this, the IJS β-SiC powders were processed by sedimentation and chemical purification. Upon sintering at 2060°C in argon, densities >98.5 T.D. were obtained.

CONCLUSIONS

β-SiC powders, consisting predominantly of sub-micrometre particles can be pressurelessly sintered to densities above 95% T.D. at temperatures around 2050°C. The sinterability of the powders is enhanced by a narrow particle size distribution. Removal of particles with diameters in excess of 1 μm and deoxidation of powders immediately prior to sintering further improves densification.

REFERENCES

1. Kevorkijan, .V., Komac, M. and Kolar, D., The influence of preparation conditions on the properties of beta-SiC powders synthesized by carbothermic reduction, Proc. Int. Conf. "Ceramic Powder Processing Science", Berchtesgaden, 1988 (in press).

2. Matje, P. and Schwetz, A.K., Surface oxygen pick-up in submicron SiC and B_4C sintering powders at room temperature. Proc. Int. Conf. "Ceramic Powder Processing Science", Berchtesgaden, 1988 (in press).

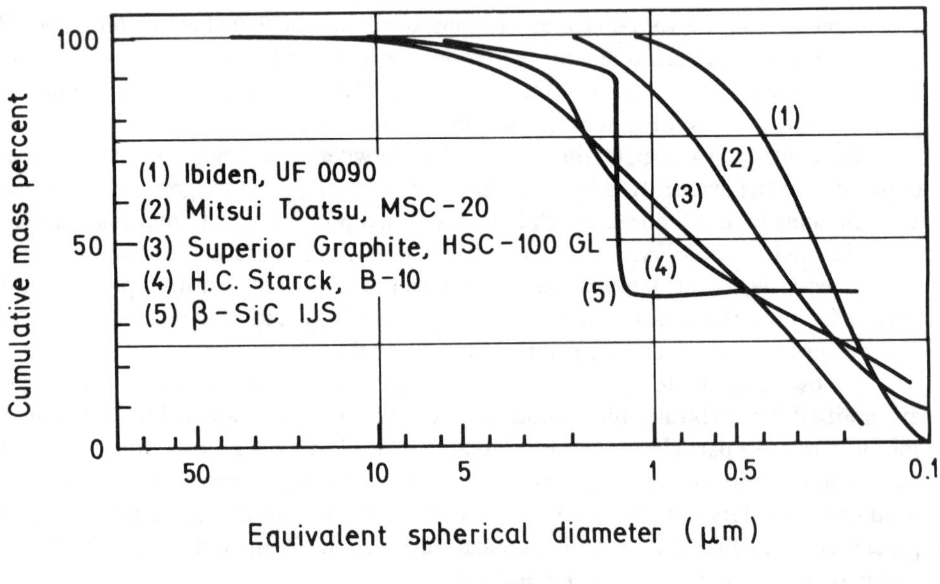

Figure 1. Particle size distribution of investigated β-SiC powders.

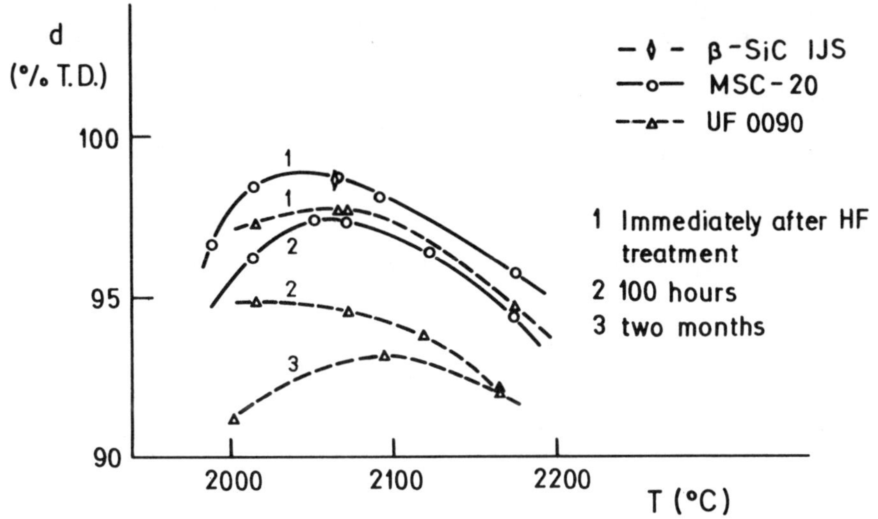

Figure 2. Effect of temperature and time of storage after HF treatment on the sinter-ed density of MSC 20 and UF - 0090 β-SiC

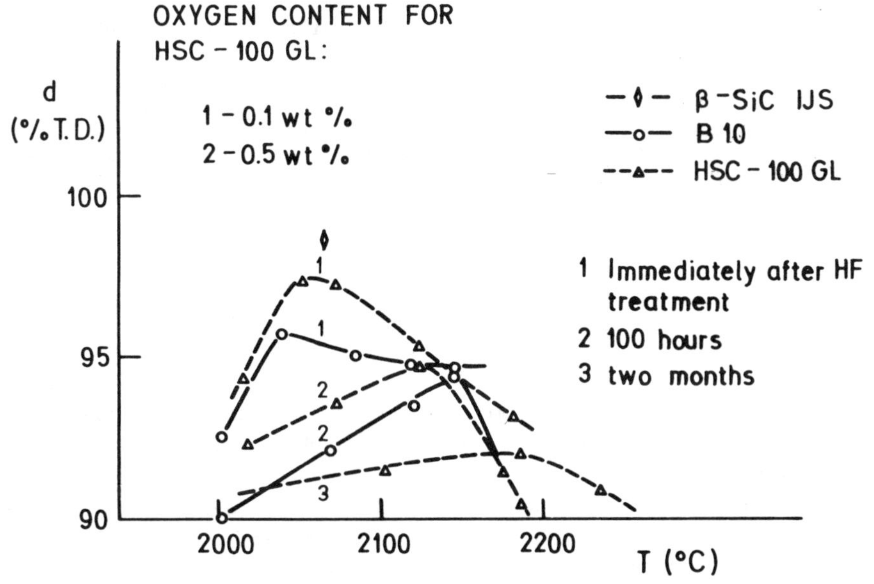

Figure 3. Effect of temperature and time of storage after HF treatment on the sintered density of B-10 and HSC 100 GL β-SiC.

LASER SYNTHESIS AND CHARACTERIZATION OF ULTRAFINE Si/C/N POWDERS

M. CAUCHETIER, O. CROIX, M. LANCE, C. ROBERT and M. LUCE
CEA - IRDI/DESICP - Département d'Etude des Lasers et
de la Physico-Chimie,
CEN Saclay, 91191 Gif-sur-Yvette Cedex (France)

ABSTRACT

Ultrafine Si/C/N composite particles in the submicronic range (<0.1 µm) were synthesized by a CO_2 laser induced reaction either between silane (SiH_4) and amine (i.e. CH_3NH_2) or between silane and a mixture of ammonia (NH_3) and hydrocarbon (i.e. C_2H_2). By varying the laser power, the flow rates and the ratios of the reactant gases, amorphous Si/C/N particles or mixed Si/SiC/Si_3N_4 powders were obtained. The powders were characterized by several methods : IR spectroscopy, X-ray diffraction, SEM, chemical analysis. Heating treatments by thermogravimetric analysis were performed under air, argon or nitrogen.

INTRODUCTION

It is now well established that CO_2 lasers are useful tools in the production of ultrafine ceramic powders with ideal characteristics for sintering. Laser gas phase reactions lead to the formation of fine, monodispersed, non-agglomerated and pure particles. The synthesis of Si_3N_4 [1] and SiC [2] powders in the range 10-100 nm are attractive examples of this new process. On the other hand with the use of fine particles of SiC-Si_3N_4 or homogeneous Si/C/N powder precursors it is expected an improvement in the thermomechanical properties of the sintered bodies. Such powders have been already synthesized in the vapor phase by using a furnace reactor[3], a plasma flame [4] and a CO_2 laser [1, 5]. In [1] the SiH_4-CH_3NH_2 mixture has been used only in the aim to prepare Si_3N_4 powders. We present here results obtained with a 1 kW CO_2 laser (CILAS CI 1000) in the synthesis of Si/C/N particles from SiH_4-CH_3NH_2 and SiH_4-C_2H_2-NH_3 mixtures.

EXPERIMENTAL

Synthesis of the powders

The experimental device used for the synthesis has been described previously [2]. The power of the unfocused laser beam was kept mainly constant at 600 W and the cell pressure was 10^5 Pa (1 atm) in all experiments. Pure gases purchased by L'Air Liquide were irradiated without purification. Gaseous phase exiting the collection chamber was in-line analyzed by IR spectroscopy.

Characterization of the powders

The powders were handled under argon from the collection chamber to a glove box to prevent contamination by air or water vapor. The particle morphology is determined by scanning electron microscopy (SEM), the surface area by the BET method, the molecular structure by IR spectroscopy and the crystalline phase by X-ray diffraction (XRD). Chemical analysis led to the Si-C-N contents and to the amount of oxygen as main impurity.

Heat treatment

The as-synthesized powders were heat treated in an alumina vessel by thermogravimetric analysis from room temperature to 1 500 °C under different atmospheres of flowing argon, air or nitrogen.

RESULTS AND DISCUSSION

Synthesis reactions and characterizations

Table 1 summarizes the experimental conditions for the synthesis and gives some characteristics of the powders. The yield in term of silane dissociation from IR absorption measurements is greater than 95 %, very similar to the yields obtained in the SiC synthesis [2]. On the IR spectrum appears an absorption band at 712 cm^{-1} which is related to the formation of HCN as impurity.

Figure 1 gives the C/N atomic ratio of the greenish-grey powders which are collected during the experiments with an initial C/N atomic ratio of 1 (from CH_3NH_2 or from 2 SiH_4 - 1 C_2H_2 mixtures). Below a reaction temperature of 1 500 °C the powder remains essentially amorphous and nitrogen rich. An explanation can lies in the fact that the NH_2 radical resulting from the dissociation of the methylamine is more reactive than the CH_3 radical. At higher temperatures (> 1 500 °C) nitrogen evolution occurs due to the instability of the Si/C/N powder or to the decomposition of Si_3N_4.

With ternary mixtures by varying the initial C/N atomic ratio it is possible to obtain either N rich (C/N = 0.72 in run 6) or C rich powder (C/N = 2.04 in run 7). The size of the particles are given by SEM (20 nm in run 0) or by the BET method (41 m^2/g in run 0 and 54 m^2/g in run 5).

TABLE 1
Si/C/N powders production

Run	Flow rates (cm^3/min) SiH_4		CH_3NH_2	T (°C)	Production rate (g/h)	Crystalline phases (XRD)
0	350		200	1495	32.6	n.i.(a)
1	275		275	1300	27.3	n.i.(a)
2	525		300	1400	49.0	n.i.(a)
3	188		107	1590	17.5	Si, β-SiC,
4(b)	187		108	1615	18.4	α-Si_3N_4
		C_2H_2	NH_3			
5	175	50	100	1680	16.4	Si, β-SiC
6	140	35	100	1700	13.3	Si, β-SiC, Si_3N_4
7	158	52	70	1740	15.0	Si, β-SiC

(a) : Amorphous phase not identified, (b) : laser power 800 W.

TABLE 2
Heat treatment of Si/C/N powders

Run	Heat treatment	Weight Change (%)	Phase composition (XRD)
0	1500°C, 7 min, argon	- 4.7	β-SiC, α, β-Si_3N_4 (α > β)
	1500°C, 50 min, nitrogen	+ 2.6	β-SiC, α, β-Si_3N_4 (α < β)
	1500°C, 30 min, air	+ 39.6	am. cristoballite
1	1500°C, 50 min, air	+ 38.7	" "
	1500°C, 63 min, argon	- 30.0	Si, β-SiC, traces α-Si_3N_4
2	2050°C, 30 min, argon(a)	- 34.6	β-SiC, traces α-SiC
	1500°C, 54 min, air	+ 40.1	am. cristoballite
3	1500°C, 33 min, air	+ 38.9	" "
	2050°C, 30 min, argon(a)	- 33.4	β-SiC, traces α-SiC

(a) during pressureless sintering of SiC powders.

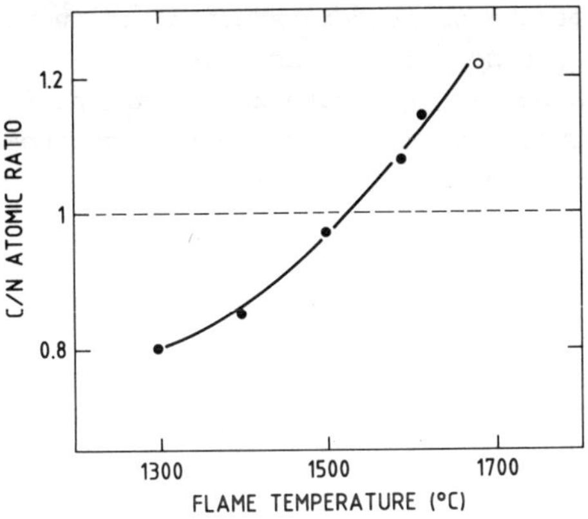

Figure 1. The C/N atomic ratio in powders versus the reaction temperature (• : binary mixture, ○: ternary mixture with C/N = 1).

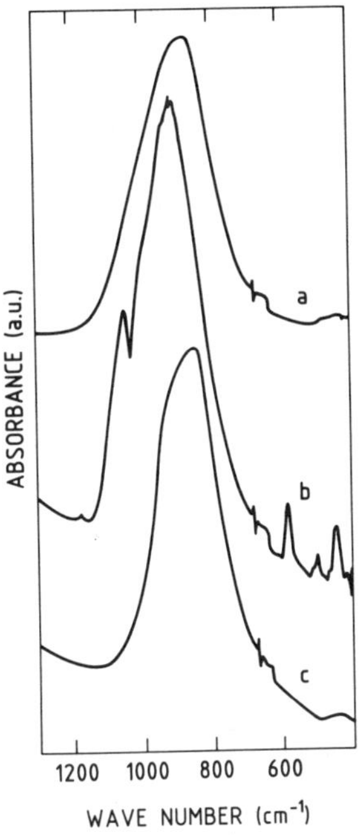

Figure 2. Infrared spectra of as-synthesized Si/C/N powders (a : run 1, b : run 6, c : run 7).

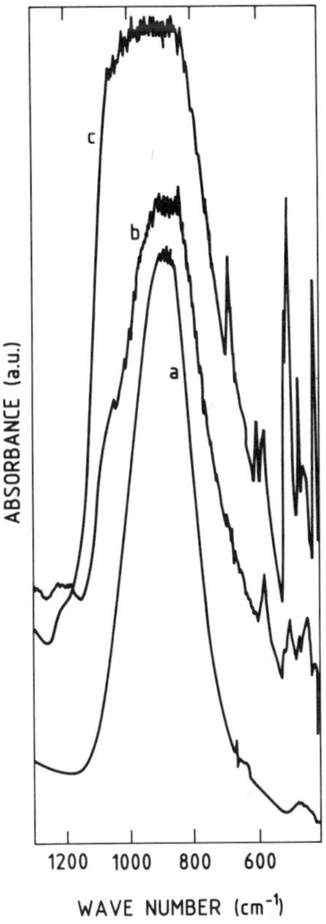

Figure 3. Crystallization of a Si/C/N powder (a : run 0) by heat treatment at 1 500 °C under argon (b) and under nitrogen (c).

Figure 2 shows IR absorption spectra of powders synthesized in runs 1(2a), 6(2b) and 7(2c). The broad absorption bands are assigned to stretching vibrations of Si-C and Si-N bonds. In Figure 2b appear the fine peaks related to α-Si_3N_4. The Figure 2c is very similar to the IR absorption spectrum of ultrafine SiC powder [2].

Heat treatment

The different heat treatments are given on the Table 2. The weight changes are related to the powder oxidation into silica under air with a good agreement between weight increases which are obtained by TGA and calculated from Si content. Under argon the weight loss increases with the heating time and with the temperature and corresponds to the vanishing of Si_3N_4. Under pure nitrogen the nitridation of amorphous powders leads to the formation of α- and β-Si_3N_4 as shown on Figure 3 : the IR absorption peak of the as-synthesized powder of run 0(3a) is broadened after heat treatment under argon (3b) and nitrogen (3c).

CONCLUSION

The use of CO_2 laser has permitted to obtain large quantities of Si/C/N powders. Further experiments will be concerning with the synthesis of such composites from ternary mixtures (SiH_4-NH_3- hydrocarbons) and the processing of the powders with oxide additives.

REFERENCES

1. Aoki M., Flint J.H. and Haggerty J.S., Synthesis of non-agglomerated Si_3N_4 powder. In Ceramic Powder Science II, A, ed. G. L. Messing, E. R. Fuller Jr, H. Hausner, The American Ceramic Society, Inc. , Westerville, Ohio, 1988, pp. 253-260.

2. Cauchetier M., Croix O. and Luce M., Laser synthesis of silicon carbide powders from silane and hydrocarbon mixtures. Adv. Ceram. Mat., 1988, 3 (6), pp. 548-52.

3. Hojo J., Maeda H. and Kato A., Preparation of composite particles of SiC-Si_3N_4 system with the vapor phase reaction method. J. Ceram. Soc. Jpn Intern. Ed. , 1987, 95, pp. 40-45.

4. Hojo J., Nakamine M. and Kato A., Preparation of SiC-Si_3N_4 composite particles by plasma gas phase reaction, in ref. 1 , pp. 277-284.

5. Rice G.W., Laser synthesis of Si/C/N powders from 1,1,1,3,3,3 - hexamethyldisilazane. J. Am. Ceram. Soc., 1986, 69(8), pp. C-183/C-185.

ALUMINUM PHOSPHATE $AlPO_4$ VIA SOLUTION ROUTES.

LAURE COURY*, FLORENCE BABONNEAU AND JACQUES LIVAGE.
Chimie de la Matière Condensée, Université Paris 6, 4 place Jussieu, 75005 Paris, France.

ABSTRACT

Aluminum phosphate $AlPO_4$ is a refractory material with a low chemical reactivity that could have potential applications as ceramic matrix for composites. $AlPO_4$ has been prepared via solution routes, using mineral precursors. The precursor solutions were characterized by ^{27}Al and ^{31}P NMR that already showed the presence of Al-O-P bonds. The transformation process of these solutions into the final crystalline ceramics was essentially followed by XRD and ^{27}Al and ^{31}P MAS-NMR. The influence of the precursors on the transformation mechanism is discussed in this paper.

INTRODUCTION

The solution routes have been widely used to produce oxide-based glasses and ceramics. These routes are especially interesting for the preparation of multicomponent systems: the mixture at a molecular level of the different precursors can lead to the formation of bonds in solution, that exist in the final network. However, the success of such preparations depends on the ability of the precursors to react together. In the case of phosphates, the high complexing ability of the phosphate anion allows an easy formation of metal-phosphate bonds. This paper presents a study on aluminum phosphate prepared via solution routes. Mineral salts (chloride and nitrate) have been used for aluminum precursors while phosphoric acid was the phosphorus precursor for all the preparations. $AlPO_4$ has already been prepared from $AlCl_3$ and H_3PO_4

[1-2]. The present study has been focused on the characterization of all the steps of the transformation, from the precursor solution to the final ceramics, using ^{27}Al and ^{31}P liquid and solid state NMR and XRD. $AlPO_4$ is a refractory material with a low chemical reactivity that can have potential applications as ceramic matrix for composites. For such applications, the solution route is quite promising: it can be used to densify a fibrous reinforcement using impregnation techniques.

EXPERIMENTAL

The precursor solutions have been prepared by mixing an aqueous solution of aluminum chloride or nitrate with phosphoric acid, in a molar ratio Al/P of 1/1. The solvent is then removed by evaporation, and the resulting powder dried at 90°C. NMR spectra were recorded with AM250 and MSL400 Bruker equipments. Solutions of aluminum nitrate in water and of phosphoric acid (85%) were taken as references. XRD patterns were recorded with a Philips equipment using Cu Kα radiation.

RESULTS

Characterization of the precursor solutions.

The solutions prepared from aluminum chloride or nitrate have been characterized by ^{27}Al and ^{31}P NMR. The results are similar for both salts and are represented in Figure 1 for solutions prepared from aluminum chloride at two different concentrations, 2 and 22 weight% in $AlPO_4$. A solution of aluminum chloride in water exhibits one peak in the ^{27}Al NMR spectrum (δ = 0 ppm) due to the species $Al(H_2O)_6^{3+}$. After the addition of phosphoric acid, two new peaks appear at -3.4 and -7.9 ppm assigned to species in which the aluminum ion, in octahedral environment, is bonded to phosphate groups [3-4]. The same solution exhibits several peaks in the ^{31}P NMR. The peak at 0 ppm is due to free phosphoric acid, while the peaks at -6.7, -7.4 and -12.1 ppm are assigned to condensed species in which phosphate groups are certainly bonded to aluminum ions [3]. At this low concentration, the reaction between the aluminum precursor and phosphoric acid is far from complete. An increase of concentration modifies the NMR spectra. The intensities of the ^{27}Al NMR peaks due to complexes between aluminum ions and phosphate ligands increase. The ^{31}P NMR spectrum show only a small peak due to free phosphoric acid. On the other hand, new signals appear at -16.6, -17.3 and -23.6 ppm. They reflect the condensation reactions occuring between the aluminum and the phosphate ions. The NMR study clearly shows that Al-O-P bonds are formed in solution. The evaporation of the solvent and thus the concentration of the solution lead to the formation of more condensed species, precursors of the aluminum phosphate network.

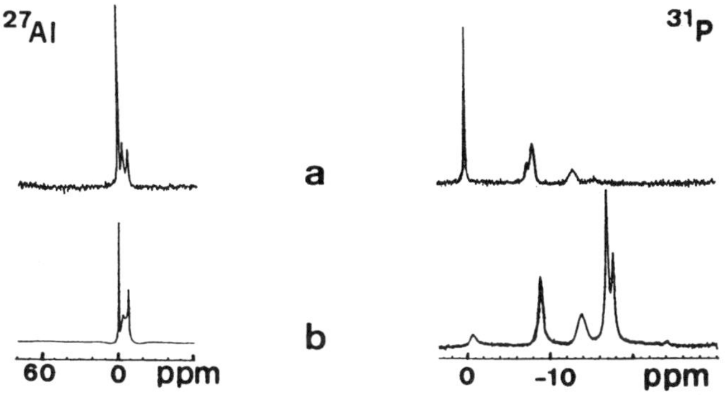

Figure 1 : ^{27}Al and ^{31}P NMR spectra for diluted (a) and concentrated (b) solutions.

Characterization of the fired samples.

The two powders obtained from aluminum nitrate (noted AN) and aluminum chloride (noted AC) have been characterized by TGA. AN exhibits only a weight loss of 1% up to 500°C while AC shows a total weight loss of 12% below 250°C. The difference of behaviors should correspond to a better decomposition of the nitrate ions during the evaporation process done by heat treating the solution.

The XRD pattern of the AN powder reveals the presence of two well crystallized phases, tridymite and cristobalite. The proportion of cristobalite regurlarly increases during the heat treatment. The AC powder is less crystalline than the AN powder (Figure 2a). Peaks due to tridymite and cristobalite are present in the XRD pattern, but together with a broad signal due to the presence of an amorphous phase. The system then crystallizes by increasing the temperature. At 1000°C, the cyrstalline phase is mainly the cristobalite with a small amount of tridymite.

In the case of the AC system in which a large amount of amorphous phase is present in the starting powder, the local environment of the aluminum and phosphorus atoms have been followed by ^{27}Al and ^{31}P MAS-NMR during the crystallization process (Figure 2b and 2c). The ^{27}Al MAS-NMR spectrum of the AC powder exhibits two signals due to aluminum ions bonded to phosphate groups. The main peak at 39.1 ppm is due to aluminum in a tetrahedral environment. The chemical shift is close to the chemical shift of $AlPO_4$, around 39 ppm depending on the phases [5-7]. The broad peak at -10.9 ppm is due to octahedral sites. These sites should be related to the amorphous phase revealed by XRD. In fact, the crystallization of the material occurs with a disappearance of this peak and a sharpening of the peak due to tetrahedral sites.

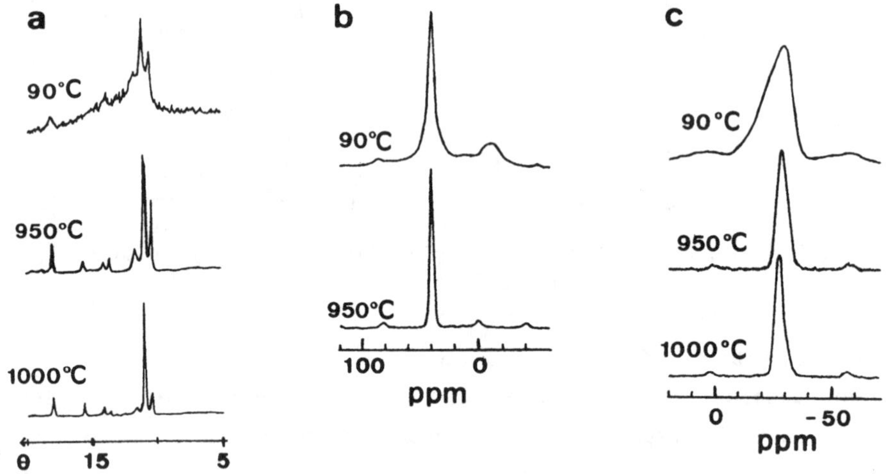

Figure 2 : XRD patterns (a), ^{27}Al (b) and ^{31}P (c) MAS-NMR spectra of the AC powder after heat treatments.

The ^{31}P MAS-NMR spectrum of the AC powder shows a rather broad peak centered on -28.8 ppm. The chemical shift values for the various crystalline forms of $AlPO_4$ are -25.6 ppm for berlinite, -29.5 ppm for tridymite and -27.1 ppm for cristobalite [6].The signal of the AC powder should thus correspond to phosphorus sites in an environment close to $AlPO_4$ phases. The broadness of the peak confirm the poor crystallinity of the sample. At 950°C, the peak is sharper and centered on -28.3 ppm, an intermediate value between the chemical shifts of tridymite and cristobalite. At 1000°C, the value of chemical shift, $\delta = -27.5$ ppm, is close to the value for cristobalite, but the non symmetry of the peak upfield reveals the presence of tridymite.

DISCUSSION AND CONCLUSION

Aluminum phosphate can be easily prepared via a solution route, using aluminum chloride or nitrate and phosphoric acid. The present study has clearly shown that Al-O-P bonds are formed as soon as the precursors are mixed in aqueous solutions. The evaporation of the solvent helps the formation of more condensed species that appear precursors for the $AlPO_4$ network. However, in solution, all the aluminum ions are in octahedral environment ; during the drying process, the symmetry changes from octahedral to tetrahedral. In the case of nitrates, this reaction is complete : all the tetrahedral aluminum atoms are in a crystalline $AlPO_4$ network. In the case of chlorides, octahedral aluminum atoms remain in an amorphous phase. These different behaviors can be due to a better decomposition of the nitrates during the heat treatment of the initial solution. The residual chloride anions certainly prevent a good organization of the network. In this study, NMR in both liquid and solid state appeared a powerful technique to follow the conversion process of a solution into a ceramics. A better understanding of the structural evolution during the process is necessary to obtain better control of the final ceramics.

REFERENCES

1. Cassidy J.E., Jarvis J.A.J. and Rothon R.N., J. Chem. Soc. Dalton Trans., 1974, 1497.

2. Rothon R.N., Thin Solid Films, 1981, **77**, 149.

3. Akitt J.W., Greenwood N.N. and Lester G.D., J. Chem. Soc. (A), 1971, 2450.

4. Karlik S.J., Elgavish G.A., Pillai R.P. and Eichhorn G.L., J. Magn. Reson., 1982, **49**, 164.

5. Müller D., Grunze I., Hallas E. and Ladwig G., Z. anorg. allg. Chem., 1983, **500**, 80.

6. Müller D., Jahn E., Ladwig G. and Haubenreisser U., Chem. Phys. Letters, 1984, **109** [4], 332.

7. Blackwell C.S. and Patton R.L., J. Phys. Chem., 1984, **88**, 6135.

COLLOIDAL PROCESSING THROUGH SURFACE MODIFICATION

E. BOSTEDT, M. PERSSON and R. CARLSSON
Swedish Ceramic Institute
Box 5403
S-402 29 Göteborg, Sweden

ABSTRACT

Particle surfaces of different ceramic powders have been transformed to "alumina like" surfaces by a surface modification technique, using Al-alkoxide. These new and uniform particle surfaces simplify the mixing and forming of ceramic multicomponent systems, since all modified powders have the same surface charge behaviour and can be treated as if they were Al_2O_3. By controlling the hydrolysis and polymerization reactions of the alkoxide molecules the surface modification technique can be optimized. Electrophoretic measurements, BET surface area measurements and X-ray fluorescence analysis were used to analyze the modified powders.

INTRODUCTION

Colloidal processing is necessary to produce high performance ceramics but to be able to obtain a stable, highly concentrated suspension it is important to have knowledge of the surface characteristics of the ceramic powder. Since the surface charge behaviour of different ceramic powders varies with type of powder, producer and sometimes even between batches it is often necessary to develop a processing agent for each separate powder. When producing multicomponent ceramics, where more than one kind of ceramic powder has to be dispersed, there can be difficulties in obtaining high stability and homogeneity of the system.

In this paper a surface modification technique is described by which these problems can be solved. With this technique the surfaces of different powders can be transformed to "alumina like" surfaces (expressed as an isoelectric point at pH 8-9) by reacting Al-alkoxide with the hydroxyl groups on the particle surfaces as is schematically shown in figure 1. With this new and uniform particle surface it becomes much easier to perform mixing and forming of different ceramic systems in both water and organic solvents. Different metal alkoxides can be used but the advantage of using Al-alkoxide is that the dispersing and stabilizing of Al_2O_3 is well known from the literature and the presence of Al_2O_3 is not a disadvantage in many ceramic systems.

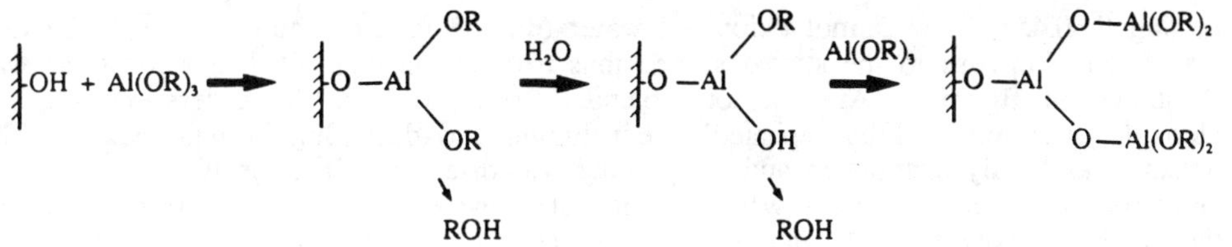

Figure 1. Schematic drawing of the reaction of an Al-alkoxide with an oxide surface.

The addition of alkoxides to ceramic powders has recently been studied by several different authors. In most cases alkoxides have been used to prepare composite powders with uniform composition (1, 2, 3) or to homogeneously disperse small amounts of sintering agents in a ceramic system (4, 5). Due to their ability to form strong covalent bonds with the particle surfaces, metal alkoxides have been used as coupling agents when dispersing ceramic powders in nonaqueous media (6, 7). However, alkoxide-coating of ceramic powders as a tool for changing the surface charge behaviour of the particles has not been reported.

To take advantage of the surface modification technique it is important that the amount of adsorbed Al-alkoxide is minimized, since the goal is to change the surface charge behaviour without increasing the specific surface area of the powder. A very high specific surface area will give problems during the dispersing and forming steps especially when complex shaped components with high and uniform green densities are to be produced.

The aim of the present study was to describe the idea and potential of the surface modification technique and this paper reports the optimization of the surface modification by control of the hydrolysis and polymerization reactions of the alkoxide.

MATERIALS AND METHODS

The starting materials used in this study were silicon nitride (SN-E10, UBE Industries, Japan), silicon carbide (Carbogran, UF-15, Lonza Werke, Fed. Rep. of Germany) and aluminium-tri-iso-propylate ($Al(OC_3H_7)_3$), referred to as Al-alkoxide, which was supplied by Merck-Scuchardt in Germany.

To be able to optimize the surface modification, controlled conditions were needed to avoid the alkoxide molecules reacting with each other before they reacted with the hydroxylated particle surface. Since any small amount of water will cause uncontrolled hydrolysis reactions, the experiments were performed in dry N_2-atmosphere in a glove box and the Al-alkoxide was vacuum distilled prior to use. The solvent, n-hexane, was dried by using molecular sieves and the powders were dried at 140 °C for 12 h.

40g powder was dispersed in 500 ml n-hexane, using a high shear mixer to break up hard agglomerates. Then 6 wt-% distilled Al-alkoxide (equivalent to 0.8 wt-% Al) diluted in hexane was slowly added to the suspensions during stirring. After 1 h of mixing 0, 0.45, 1 and 5 mol deionized water (diluted in 2-propanol) per mol alkoxide was added dropwise to the stirred suspensions to start the controlled hydrolysis, as was illustrated in figure 1. After a total blending time of 3 h the excess, non-reacted alkoxide, was removed by repeatedly centrifuging and decanting in pure hexane. The hexane was finally evaporated and the powder was dried at 130°C for 10 h.

Electrophoretic measurements were used to determine the change in isoelectric point of the modified powders. The amounts of reacted Al-alkoxide were studied by X-ray fluorescence analysis and the specific surface areas of the modified Si_3N_4 powder were measured using a single point BET method.

RESULTS

Surface modification performed in dry N_2-atmosphere with freshly distilled Al-alkoxide showed that only 0.12 wt-% Al had reacted with the Si_3N_4-particle surface and 0.18 wt-% had reacted with the SiC-surface, even when excess Al-alkoxide was added. This means that the same amount of Al-alkoxide (4.2μmol Al/m^2 particle surface) had reacted with both the Si_3N_4 and SiC particle surfaces, indicating that the amount of silanol groups on these powders was about the same. However, the zeta potential measurements did show that this "monolayer" of Al-alkoxide was not enough to change the surface charge characteristics of the powders.

To increase the amount of reacted alkoxide water had to be added to start the hydrolysis and polymerization reactions. Figure 2 shows that an increased amount of water added during the reaction increased the amount of alkoxide bound to the particle surfaces.

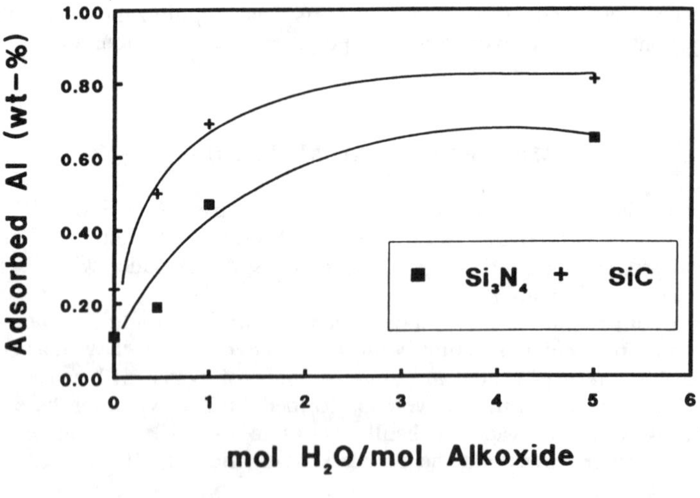

Figure 2. Amount of aluminium adsorbed vs amount water added during hydrolysis.

When more Al was bound to the particle surfaces the isoelectric point was moved to a higher pH and the particles became more and more "alumina like" as can be seen in figure 3. The results showed that 0.5 wt-% Al was required to modify the Si_3N_4 surface and 0.7 wt-% Al was required for SiC. This means that 0.5 mg Al/m² is the smallest amount needed to obtain a modification of Si_3N_4 and SiC particle surfaces.

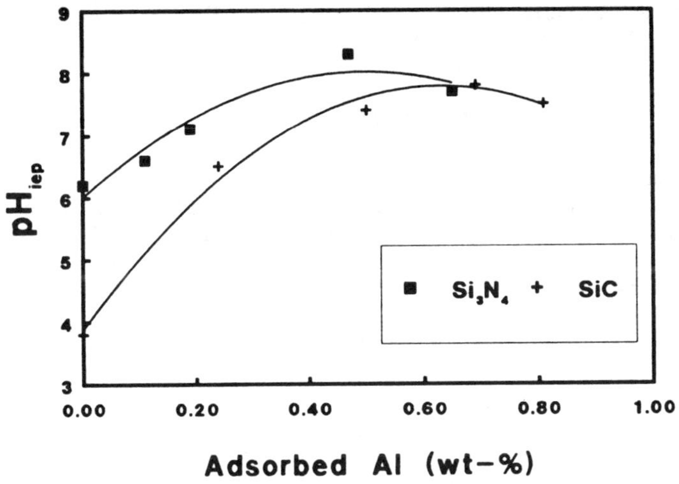

Figure 3. Change in isoelectric point vs adsorbed amount of aluminium.

When the amount of adsorbed Al-alkoxide was increased the specific surface area of the powder was increased. However, the increase in specific surface area was relatively small when excess non-reacted alkoxide was removed and distilled alkoxide had been used. The smallest amount of alkoxide needed to obtain the surface modification (0.5 wt-% Al for Si_3N_4) increased the specific surface area by about 10-15 % and this should not give any problems in the dispersing steps prior to forming.

DISCUSSION

The surface modification technique described in this paper has also been applied to several other ceramic powders, as for example Si, ZrO_2 and Y_2O_3. Ceramic systems containing three or more components with completely different surface charge behaviour have been modified with Al-alkoxide so that all powders had "alumina like" surfaces and could be treated as if they were Al_2O_3. These surface modified multicomponent systems have been stabilized in water by simply adding 1 wt-% of a commercial polyacrylic dispersing agent which is a convenient way to prepare Al_2O_3 slips.

CONCLUSIONS

A surface modification technique, using Al-alkoxide has demonstrated that the surfaces of different ceramic powders can be transformed to an "alumina like" surface The surface modification can be optimized by using controlled hydrolysis and polymerization of the alkoxides. The smallest amount of Al needed to modify Si_3N_4 and SiC, without increasing the specific surface area or changing the powder morphology was 0.5 mg Al/m^2 particle surface. Very promising results have been obtained from stabilizing surface modified multicomponent systems, which without surface modification would have been very difficult to disperse.

ACKNOWLEDGEMENTS

The authors wish to thank L. Bergström at the Institute for Surface Chemistry for help with the electrophoretic measurements and for fruitful discussions. The authors also want to thank B. Engström at KemaNord Industrikemi AB for help with X-ray fluorescence analysis. Finally the authors wish to thank Dr. I. Brown for revision of the english text. This work was sponsored by the Swedish Board for Technical Development.

REFERENCES

1. Fegley, B., White, P. and Bowen, H.K., Preparation of Zirconia-Alumina Powders by Zirconium Alkoxide Hydrolysis. J. Am. Ceram. Soc., 1985, **68**, C60-62.

2. Kishi, K., Umebayashi, S., Tani, E and Kobayashi, K., Synthesis of ß- Sialon from Si_3N_4 and Aluminium-iso-Propoxide. Yogyo-Kyokai-Shi, 1985, **93**, 629-635.

3. Okamura, H., Barringer, E.A. and Bowen, H.K., Preparation and Sintering of Monosized Al_2O_3-TiO_2-Composite Powders. J. Am. Ceram. Soc., 1986, **69**, C22-24.

4. Takahashi, Y. and Chizaki, J., Titania Coatings of Alumina Powders. J. Ceram. Soc. Jpn. Inter. Ed., 1988, **96**, 238-244.

5. Mah, T., Mazdiayasni, K.S. and Ruh, R., Characterization and Properties of Hot-Pressed Si_3N_4 with Alkoxy-Derived CeO_2 or Y_2O_3 as Sintering Agents. Ceram Bull., 1979, **58**, 840-844.

6. Kishi, K., Umebayashi, S., Pompe, R. and Persson, M., Interaction of Aluminium-Isopropoxide Solution and Si_3N_4 powder. J. Ceram. Soc. Jpn. Inter. Ed., 1988, **96**, 687-689.

7. Landham, R.R., Parish, M.V., Bowen, H.K. and Calvert, P.D. Organotitanate dispersants for $BaTiO_3$ and Al_2O_3. J. Mater. Sci., 1987, **22**, 1677-1681.

DEFLOCCULATION OF AQUEOUS α-Al_2O_3 SUSPENSIONS IN THE PRESENCE OF ADSORBING POLYMER AND POLYELECTROLYTE

B.C. BONEKAMP, W.H. VAN 'T VEEN, M.J. SCHOUTE and H.J. VERINGA
Netherlands Energy Research Foundation, ECN,
P.O. Box 1, 1755 ZG PETTEN, the Netherlands

ABSTRACT

Deagglomeration of aqueous α-Al_2O_3 suspensions around the iso-electric point (pH 8.5-9) of the oxide is discussed in relation to the adsorption from solution of anionic polyelectrolyte and the competitive adsorption of both polyelectrolyte and a non-ionic polymer (polyvinylalcohol PVA). Results of adsorpion-, electrophoresis-, turbidity- and sediment volume measurements are reported. In the presence of PVA the polyelectrolyte induced deagglomeration transition is shifted to higher polyelectrolyte concentration, which is important when partial deagglomeration is the goal as in the forming of porous ceramics.

INTRODUCTION

Due to their adsorption properties polyelectrolytes strongly affect the agglomeration/deagglomeration properties of aqueous colloidal dispersions [1,2]. These properties make polyelectrolytes important processing tools for wet processing techniques as colloidal filtration and tape casting [3]. In addition to a polyelectrolyte deflocculant, ceramic dispersions often contain a non ionic binder polymer such as polyvinylalcohol (PVA). The primary purpose of the latter is to enhance the green strength of a dried particle compact. However such a polymer may also affect the properties of the suspension. This aspect is often neglected and not well studied.
In this communication we discuss the adsorption of a polyanionic deflocculant/stabilizer in the presence and absence of PVA. In the presence of PVA competitive adsorption occurs as we will show.
This competition alters the deflocculating action of the polyanion, which is important for the forming of porous ceramic structures where a controlled state of agglomeration and not a priory complete deagglomeration is the goal.

EXPERIMENTAL

Alumina A16SG from Alcoa was used in the experiments. The specific surface area as determined by us with N_2 adsorption amounts 10 m^2/g. The mean particle size, as measured with static light scattering (Malvern Master Sizer) in optimally deagglomerated dilute suspensions is ≈ 0.7 μm.
Stock suspensions were prepared by mixing in a ballmill.
Polymers used were obtained from Zschimmer & Schwartz. The polyelectrolyte was a polycarboxylic ammonium salt with the commercial name Dolapix CE64 (DPE from now). The molecular mass of the polysalt is estimated to be about 5000 on the basis of capillary viscosimetry. The PVA was optapix 4G.
Adsorbed amounts were measured by the usual depletion technique [4]. Polyelectrolyte concentrations were determined by acid/base titrations. PVA concentrations were determined by a colorimetric method [5].
Turbidity (optical density) measurements to estimate the degree of deflocculation were performed in 1 cm cuvettes at a wavelength of 500 nm using a Beckman DU40 spectrophotometer. Diluted samples from the adsorption measurements were used in the experiments. The turbidity was recorded after 120 minutes for the quiescent samples. A relative decrease in turbidity (due to sedimentation) indicates the state of agglomeration.
Sediment volumes of suspensions, with the same composition as the suspensions used in the adsorption experiments, were determined in 10 ml measuring cylinders. Gravity settling was allowed for at least one week.
Electrophoretic mobilities were measured with a Zetasizer IIC (Malvern Ind.). Samples were prepared in polyethylene flasks. The average of 5 measurements at both stationary levels in a cylindrical cell was taken.

RESULTS AND DISCUSSION

Technical grade macromolecules are used in this study. These compounds are less well characterized than other commercially available polymers. However the compounds used by us are also applied in industrial ceramic powder pro-

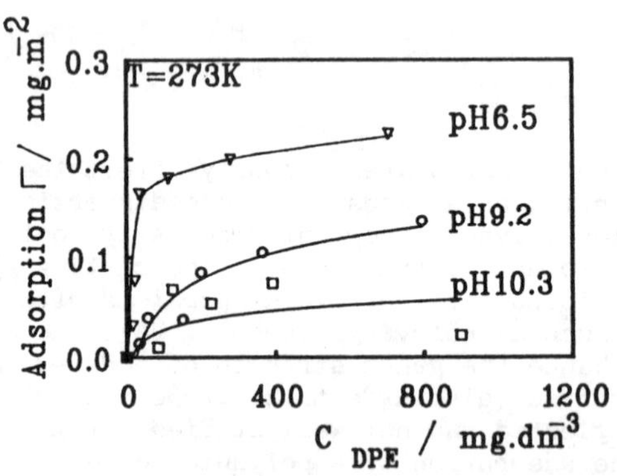

Figure 1. Adsorbed amount polysalt DPE on α-alumina as a function of the equilibrium concentration for three pH values indicated.

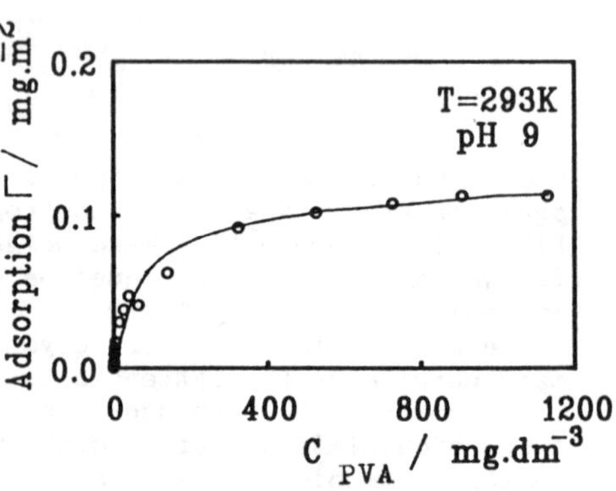

Figure 2. Adsorbed amount PVA-4G on alumina as a function of the equilibrium concentration.

cessing situations which makes our results of direct qualitative and quantitative relevance. As shown in fig. 1 and 2 the ammonium polyanion (DPE), as well as the PVA, adsorb (individually) on the alumina surface.
The isotherms are rather rounded and the affinity seems to be not very high. The roundedness may be indicative for the weak acid character and low molecular weight of DPE [2, 4, 7]. In the case of PVA the roundedness is indicative for a broad molecular weight distribution. But also weak polymer segment adsorption and low molecular weight may cause this effect [4].

The adsorbed amount DPE decreases with increasing pH due to the increasing ionization of DPE. Lateral repulsions between charged DPE segments in the adsorbed layer are dominating the adsorption behaviour. The adsorption behaviour of DPE is very similar to that found by Aksay [6] for polymethacrylic acid and to that of polyelectrolytes in general [7] and need not to be discussed here further.
In the presence of a constant total amount of PVA the adsorption of PVA and DPE are competitive to some extent (see fig.3.) but DPE does not displace all of the PVA and a mixed adsorption layer is still present at high DPE concentration.

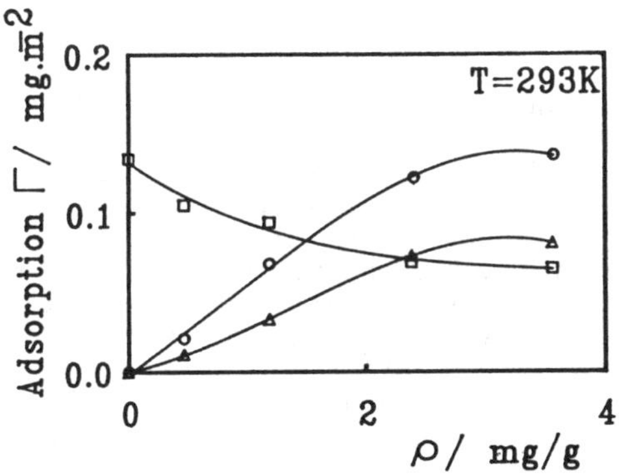

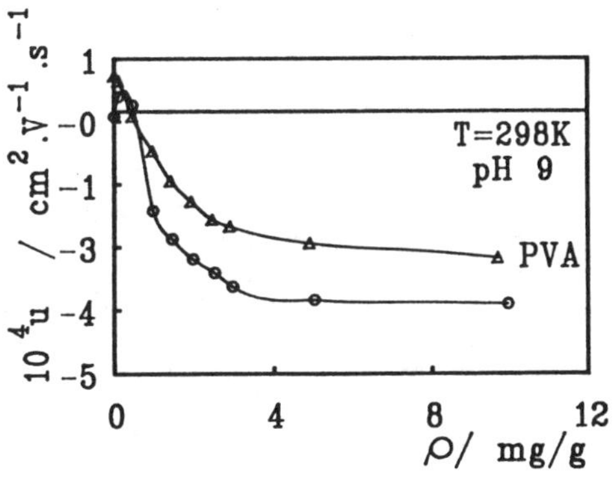

Figure 3. Adsorption of DPE and PVA (11.6 mg/g) as a function of the DPE-content. □-□ adsorbed amount PVA; Δ-Δ Adsorbed amount DPE in the presence and o-o absence of PVA.

Figure 4. Electrophoretic mobility as a function of the DPE content ρ in the absence (o-o) and presence (Δ-Δ) of PVA (11.6 mg/g). The initial pH is slightly below the i.e.p.

The plateau value of DPE decreases due to the presence of PVA and a lower colloidal stability is expected under those conditions. It is interesting to note here that Kawaguchi et al. [8] found that polyethylene oxide displaces polycations from silica surfaces.
The adsorption behaviour from DPE/PVA binary solutions is reflected in the electrophoretic mobility u_e of the suspended alumina particles (see Fig. 4). The effect of PVA on u_e as function of the DPE content ρ is clear: the DPE content ρ where charge reversal occurs ρ_0 is shifted to a higher value.

Further the mobility becomes less negative in the presence of PVA and hence electrostatic repulsion between the suspended particles is less in the presence of PVA.

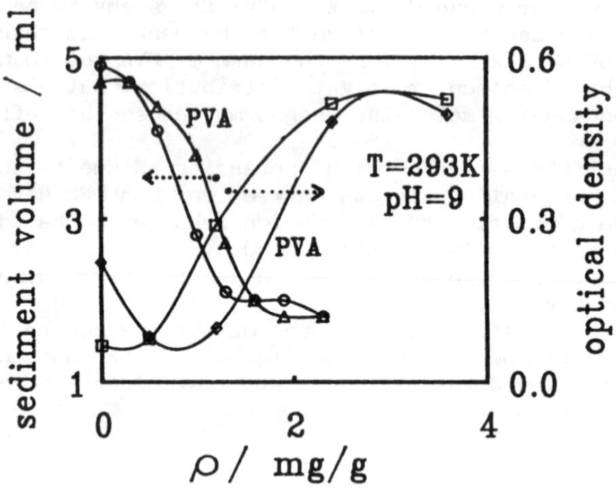

Figure 5. Turbidity ($O.D._{1\ cm}^{500\ nm}$) of dilute suspensions and sediment volume of 10 ml suspension (5.6% v/v) as a function of the polyelectrolyte content ρ in the absence and presence of PVA (11.6 mg/g).

Fig. 5 shows that DPE has the ability to deflocculate dilute and more concentrated alumina suspensions around the i.e.p of the oxide. The midpoint of this deagglomeration transition is shifted to a higher ρ in the presence of PVA. Comparison with the u_e results in fig 4 suggests that the midpoint of the transition in both cases occurs at about the same zeta potential and hence that an increasing electrostatic repulsion is the main cause of deflocculation both in the absence and presence of PVA.
In the transition region the effect of PVA in the sediment volume is substantial and hence it is expected that in this region large differences in the wet pore microstructure of the particle compacts exist. It depends on the microstructural changes during drying and the extent of control of this process whether these differences are effectuated in the green form.

ACKNOWLEDGEMENT

Financial support from the National program on technical ceramics IOP-TK is gratefully acknowledged.

REFERENCES

1. Gregory, J., Flocculation by polymers and polyelectrolytes in solid/liquid dispersions ed. Th.F. Tadros, Academic. Press, 1987, CH 8, pp 163.

2. Lyklema, J. and Fleer, G.J., Electrical contributions to the effect of macromolecules on colloid stability, Colloids and Surfaces, 1987, 25, 357-368.

3. Lange, F.F., Powder processing science and technology for increased reliability, J. Am. Ceram. Soc., 1989, **72**, 3-15.

4. Cohen Stuart, M.A., Cosgrove, T. and Vincent, B., Experimental aspects of polymer adsorption at solid/solution interfaces. Adv. Colloid Interface Sci., 1986, **24**, 143-239.

5. Horácek, J., Kolorimetrické Stanoveni polyvinylalkoholu, Chem.Prûmysl, 1962, **12**, 385-387.

6. Cesarano III, J., Aksay, I.A. and Bleier, A., Stability of aqueous α-Al_2O_3 suspensions with poly(methacrylic acid). polyelectrolyte, J. Am. Ceram. Soc.,1988, **71**, 250-255.

7. Cohen Stuart, M.A., Polyelectrolyte adsorption, J. Phys. France, 1988, **49**, 1001-1008.

8. Kawaguchi, M., Kawaguchi, H. and Takahashi, A., Competitive and displacement adsorption of polyelctrolyte and water soluble non-ionic polymer at the silica surface, J. Colloid Interface Sci., 1988, **124**, 57-62.

INCORPORATION OF SINTERING AIDS IN SILICON NITRIDE BY MEANS OF SURFACTANTS

REMCO de JONG, RONALD A. McCAULEY
Center for Ceramic Research, Rutgers University, Piscataway, NJ, USA.

ABSTRACT

Silicon nitride powders were coated with organometallic surfactants. Powder-surfactant interactions were investigated in an aqueous medium. Samples were slip cast, pressureless sintered, and hot-isostatically pressed to show the advantages of these coated systems over conventionally ball-milled systems. Fully dense materials ($E=315\pm3$ GNm^{-2}) were obtained after 1 hour at 1950 °C under 200 MPa nitrogen using 0.45±0.05 w% metal oxide as additive.

INTRODUCTION

It is difficult to obtain dense, high-strength Si_3N_4 ceramics because of the largely covalent bonding and limited self-diffusion. Eventhough pure Si_3N_4 compacts have been sintered to high density, the pressures required have been as high as 1.5 GPa [1]. Conventional hot-isostatic pressing has not resulted in a fully dense material [2]. Dense Si_3N_4 materials, however, can be obtained with the use of additives. These additives provide a liquid phase and promote densification, but they also introduce secondary phases into the densified material, which generally deteriorate the high-temperature mechanical properties.

The primary methods for sintering Si_3N_4 materials are pressureless sintering, high-pressure sintering, hot-pressing and hot-isostatic pressing [2-4]. The amount of additives used ranges from 5-20$^+$ mol%, being lowest for the latter two methods. The most common way of adding and mixing the additives into the Si_3N_4 powder is by ball-milling. This may affect the particle size distribution, powder surface characteristics, and purity of the system, but will not allow for control over the distribution of the additives. Prolonged milling does not improve the distribution. The kinetics for densification and transformation to the β-phase will, therefore, vary over the microstructure during sintering. Precipitation from a dispersion containing the Si_3N_4 powder introduces the additive more uniformly but has the disadvantage of excessive shrinkage [5].

The purpose of this study was to incorporate the additives in a controlled, uniform, reproducible manner. This should provide a liquid phase at the sintering temperature in intimate contact with the Si_3N_4 grains, thus improving the kinetics for densification and transformation. The use of a surfactant system containing the additive was expected to do this. The system used should be stable against processing conditions. This requires a strong interaction with the powder surface. The surfactants investigated were water soluble metal complexes with the hydrated metal ions as an anchor group. This was unlike most commercially available organometallics that have the metal fully coordinated by the organic ligands. These are more likely to give hydrogen bonding with the powder, which is relatively weak and affected by changes in the aqueous environment during processing. Processing in an aqueous medium with a minimum amount of additives was investigated. The powders were slipcast and the resulting materials were sintered under various conditions. Sintering characteristics were assessed by density, E-modulus, hardness, K_c and X-ray diffraction (XRD).

EXPERIMENTAL

Two Si_3N_4 powders were used, designated as A (KemaNord P95) and B (UBE E-10), with BET surface areas of ~10 m^2/gr. The powders were characterized for size distribution, morphology and bulk and surface chemistry. The powders were dispersed in triple-deionized, distilled water and sonicated to brake up weak agglomerates. The water soluble surfactants were a methacrylato chromic chloride hydroxide, a chromium acetate, and an aluminum lactate (labeled S1, S2 and S3). These were dissolved and introduced into the agitated dispersion. The pH of the dispersion was then adjusted with 10^{-2} M HNO_3 or NH_4OH to a value close to the iso-electric point (IEP) value. The unstable system was centrifuged at 4000 rpm and the supernatant with excess surfactant decanted. The powder was redispersed in water and the pH adjusted away from the IEP value. The powders from slips before and after centrifuging were used for mobility measurments in 10^{-3} M KCl aqueous systems and for TEM, ESCA, FTIR, and thermal analysis (TA: TGA/DSC) . To check for the stability of the system centrifuging and redispersing of the powders was repeated up to five times and the pH was varied between 3 and 11. Potentiometric titration was used to monitor the changes in the surface potential of the powder while the surfactant solution was introduced. The slip was then cast, after a short time to allow large agglomerates to settle, into discs 25x10 mm. The discs were oxidized at 500 and 675°C up to 36 hours and analyzed by ESCA and SEM. The discs were pressureless sintered at 1730 °C up to 6 hrs under 0.1 MPa ultra-high purity nitrogen in a tungsten crucible and tungsten lined furnace, using a Si_3N_4 powder bed. HIPing was also used at 1950 C for 1 hour under 200 MPa in BN powder. The sintered discs were analyzed by SEM, ESCA and XRD. The density (water immersion), microhardness (Vickers indenter), E-modulus (sonic velocity delay) and K_c (indentation) were determined. A few discs prepared from dry-pressing ball-milled mixtures of powder A, both with and without surfactant, and high-purity magnesia (SA=35 m^2/gr) and alumina powder (SA=10 m^2/gr) were also sintered under 0.1 MPa nitrogen for 8 minutes at 1730 °C for comparison.

RESULTS

The powder characteristics were reported in an earlier paper [6]. Most of the data was collected on powder A. Powder B contained a small amount of Si_3N_4 whiskers. Both powders behaved acidic in water and stable dispersions resulted. Addition of the pure surfactants lowered the pH considerably due to initial polymerization of the surfactant molecules. Powder-surfactant interactions were obvious from potentiometric titration and acoustophoretic mobility data that showed a change in the sign of the powder mobility and the phase of the monitored signal upon addition of the surfactant, lowering the pH. However, increased interaction was obtained with surfactants in solution. The pH of these solutions was raised to 6.5 with NH_4OH to promote polymerization due to the formation of hydroxyl bridges These solutions added to the powder dispersion showed maximum interaction as indicated by rapid mobility and phase changes. This was confirmed by ESCA data; the latter process gave powders with consistent, reproducible amounts of additive. Precipitation of surfactant molecules due to the addition of excess base had to be avoided because precipitates would act as imperfections in the cast microstructure. Precipitation occurred around pH 8.5-9 for all surfactants. The IEP of the surfactants was around pH 7.5-8 and the IEP of a coated powder dispersion shifted from the values reported in Ref.[6] to these values. Exposure to excess acid or base during mobility measurements and repeated centrifuging and redispersing did not alter the IEP values nor the ESCA values, indicating that once proper processing put the surfactants onto the powder surface, they were stable against processing conditions.

A possible model for the surfactants used and their interaction with the Si_3N_4 surface is given in Figure 1. Additional data is presented in Table 1.

TABLE 1
Data on surfactants (M[organic]) investigated

Surfactant	Label	M	#M	organic	IEP	mgr M_2O_3/gr Si_3N_4
Methacrylic acid	S1	Cr	2	$C_4H_5O_2$	8.0	4.5 ± 0.5
Formate	S2	Cr	3	$C_6H_6O_{13}$	7.5	4.0 ± 0.8
Lactate	S3	Al	3	$C_3H_5O_3$	8.1	4.0 ± 0.8

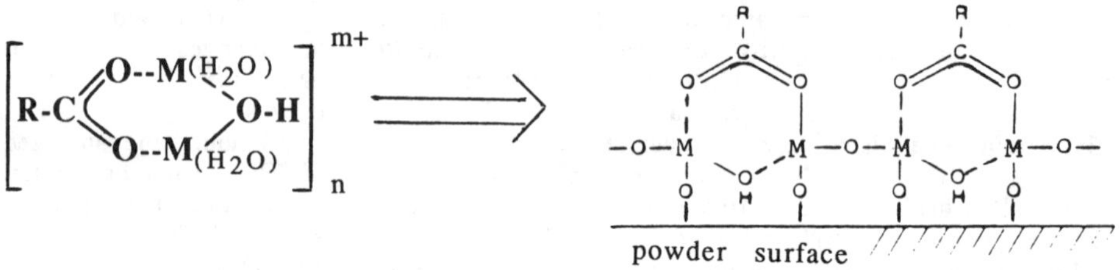

Figure 1. A possible model for the interaction with surface hydroxyl groups.

System 1 was used for initial experiments because chromium(III)complexes demonstrate a relative kinetic inertness in aqueous solution. Much of the classical complex chemistry involved chromium and a large number of complexes can be made. System S1 was commercially available. Surfactant S2 was a modification of S1. S3 was added to incorporate aluminum. The potentiometric titration data indicated that formation of hydroxyl bridges increased the net effective charge on a surfactant molecule. That resulted in the higher degree of interaction with the powder surface upon addition of base. These data also showed a saturation concentration of surfactant added, in excess of which the mobility did not further increase. This concentration was used to minimize the amount of excess surfactant to be removed by centrifuging and was used by the calculation of the amount of surfactant at the powder surface. The coated powders had higher surface potentials than as-received powders and stable dispersions were obtained containing up to 50 w% solids. Rheology measurements showed that slips had lower viscosity than electrostatically stabilized slips of Si_3N_4 without additives. The presence of the surfactant at the powder surface was shown by FTIR, ESCA and TEM analysis. FTIR showed adsorption bands for coated powders that corresponded to a few strong bands (1600-1200 cm^{-1}) for the pure surfactant solutions, after subtraction of the solvent. The OH-stretching band at 3743 cm^{-1} disappeared and a new, weak band appeared around 3010 cm^{-1}. This indicated interaction of surfactant (metal) ions with surface silanols. The weak band resulted probably from perturbed surface hydroxyl groups.

The TEM showed a less agglomerated dispersion for the coated systems than for the pure powders. Electron diffraction indicated the presence of a crystalline phase, not present in as-received powders. Identification of the surfactants was possible using ESCA binding energy data. The coated systems had the additive as a metal hydroxide on the surface. For S1, for example, the O1s values were 532.6 with a shoulder at 531.1 eV, indicating SiO_2 and $Cr(OH)_3$, respectively. The Cr2p values were 577.3, from $Cr(OH)_3$, or 577.7 eV, indicating the interaction with the organic group. The Si2p peak at 101.9 eV had a shoulder at 103.3 eV, from Si_3N_4 and SiO_2, respectively. The oxidized and sintered system showed the shoulder for Si2p at 102.7 eV, probably due to silicon oxynitride. The O1s values were 532.5 and 531.6 eV and the Cr2p value was 577.1 eV. The latter two correspond to values for Cr_2O_3. The conversion to the metal oxides due to oxidation was confirmed by FTIR analysis of oxidized, pure surfactants. Thermal analysis showed a weight loss and exothermic transition for the coated powders below 400 °C, indicating the combustion of the organic. No further weight loss occurred after 24 hours at 500/675 °C, indicating the completion of the oxidation treatment. The amount of residual carbon in the oxidized discs was less than 1 w%.

The amount of additive introduced into the system was calculated for the various surfactants using the saturation concentrations obtained from potentiometric titration and repeated centrifuging as well as using ESCA and TA data. The calculated amounts were in close agreement and are presented in Table 1 as mgr M_2O_3 per gram of Si_3N_4 powder. The number of surface sites needed to accommodate the amounts of additive listed would be 2.8-3.5 sites nm^{-2}.

Various systems were sintered at 1670-1950 °C. The results of these experiments are summarized in Tables 2 and 3. The green densities were 52-58% of theoretical.

TABLE 2
Data on discs sintered for 8 min. at 1730 °C under 0.1 MPa Nitrogen

System	2 w% MgO	2 w% MgO+S1	4 w% MgO	4 w% MgO+S1	4 w% MgO+S3
Density (% theor).	78.5	85.8	84.1	93.2	92.9
E-mod. (GNm^{-2})		175±15		240±10	265±9

The rather small amount of additive that could be incorporated (Table 1) did not provide enough densification. The coated systems did transform to the β-phase but the final density was limited to 62 %theo. after 6 hrs at 1750 °C. Systems S2 and S3, which were designed so as to introduce three metal atoms per monomer did not give better results. The microstructures showed rather coarse elongated β-grains and mostly open porosity. However, substituting the surfactant for a fraction of the additives conventionally incorporated by ball-milling only improved sintering results. Two well-studied systems, MgO and $MgO-Al_2O_3$, were selected and sintering cycles were taken from data in literature for these systems. The results in Table 2 show the increase in density with the addition of S1. Densification was limited both because of the decomposition and evaporation of MgO and because the total amount of additives was limited. Increasing the amount of MgO ball-milled into the coated system to 4 w% resembles systems reported in literature with maximum densities of 86 %theo. Again addition of S1 clearly gave better results. To see whether the increase in the amount of additive by 0.45 w% from S1 caused this improvement, discs with 0.5 w% Cr_2O_3 and 4 w% MgO were sintered as well. Densities were comparable with the 4 w% MgO system. The use of S3 did also give better results than a 1 w% Al_2O_3-4 w% MgO ball-milled system . Comparing S1 and S3 showed that their effect on the MgO system was similar. In addition, the final density under the sintering conditions listed was higher with 2 w% MgO+S1 than with 4w% MgO. The phase equilibrium diagrams did not predict an increased amount of liquid formation at 1730 °C for the S1 system.

TABLE 3
Data on discs HIPed for 60 min. at 1950 °C under 200 MPa. Nitrogen

System	Density %theo	H_V (GNm^{-2})	E (GNm^{-2})	K_C ($MNm^{-3/2}$)
pwd A+S1	100	19.2±.6	315±3	4.8
pwd B+S1	100	19.1±.5	305±5	4.8
pwd A+S2	100	18.4±.4	300±5	4.1
pwd A+S3	100	19.5±.5	297±2	5.3

The data in Table 3 shows that dense materials were obtained using the conditions and systems listed. System A+S3 sintered to only 95.6 %theo when the pH of the surfactant solution was not raised, causing a decrease in the amount of additive introduced. This indicated that for a ~10 m^2/gr powder the amounts listed in Table 1 approach the minimum amount of additive needed, if uniformly distributed.

CONCLUSIONS

Silicon nitride powders were coated with surfactants, thus incorporating ~0.5 w% additive in the microstructure. Proper selection of surfactant systems and control over processing conditions resulted in stable aqueous systems. Cast discs containing less additive were pressureless sintered to densities higher than obtained with conventional, ball-milled only systems. Hiping with coated only systems gave excellent results using a fraction of the amount of additives in systems reported in literature.

REFERENCES

1. Yamada, T., Shimada, M. and Koizumi, M., Densification of Si_3N_4 by High Pressure Hot-Pressing. Cer. Bull., 1981, **60**, 1281-1283.
2. Uchida, N., Koizumi, M. and Shimada, M., Fabrication of Si_3N_4 ceramics with Metal Nitride Additives by Isostatic Hot-Pressing. J. Am. Cer. Soc., 1985, **68**, C38.
3. Loehman, R. E. and Rowcliffe, D. J., Sintering of Si_3N_4-Y_2O_3-Al_2O_3. J. Am. Cer. Soc., 1980, **63**, 144-148.
4. Sanders, W. A. and Herbell, T. P., Characteristics of Si_3N_4-SiO_2-Ce_2O_3 Compositions sintered in High-Pressure Nitrogen. J. Am. Cer. Soc., 1983, **66**, 835.
5. Shaw, T. M. and Pethica, B. A., Preparation and Sintering of Homogeneous Silicon Nitride Green Compacts. J. Am. Cer. Soc., 1986, **69**, 88-93.
6. De Jong, R. and McCauley, R. A., Dispersion of Silicon Nitride in an Aqueous Medium. In Proc. 1st Int. Conf. on Powder Proc. Sci., Nov 1987, Orlando,Fl.

RHEOLOGICAL BEHAVIOUR OF SiC-Si AQUEOUS MIXTURES

Germán J. Piderit, Dept. of Materials Science,
Patricio Toro, Dept. of Basic Chemistry,
Ernesto Croquevielle, Dept. of Materials Science,
Universidad de Chile, Casilla 1420, Santiago-Chile

ABSTRACT

The rheological behaviour of aqueous Si and SiC and of mixtures thereof was studied. Potentiometric measurements of individual suspensions of Si and SiC showed the zero point of charge (zpc) at pH 7.0-8.5 and pH 3.5-4.0, respectively. The zpc of mixtures (containing up to 60% Si) is modified to a greater pH range. Si particles sizing was coarser than SiC particles with 90% under 20μ and 35% under 5μ. BET specific area of SiC was found to be 1.1 m^2/g. Viscosity variations (60% solids) decreased proportionally with increasing alkalinity in the pH range of 3-12 while the negative zeta potential did not change appreciably. Zeta potential zero was not reached. The weak acid nature (pKs) of SiC inhibits a stronger effect of pH on zeta potential.

INTRODUCTION

Interactions between particles in suspension is studied to the end of ascertaining the mechanism and structure of said particles deposits and more specially the defects originating during said formation.
This work aims at clarifying therefor the potentiometric and rheological behaviours of SiC and Si components. This pair of components along with the studies on alumina compounds and zirconia would allow to gain knowledge and experience for the production of improved parts of ceramics in the near future.

MATERIALS AND METHODS

SiC from the ISESA Chilean firm and Si from Matheson Products were used as the principal components in the present work. The grain size analysis of both solids gave the following result:

SIZE under:	40μ	30μ	20μ	10μ	5μ	3μ	2μ
% Si	65.9	59.4	54.8	38.1	22.7	15.1	12.3
% SiC	98.1	97.7	90.7	65.9	35.3	19.8	15.4

The electrolyte used were analytical grade KNO_3 and KOH (Analar, Trade Mark). The suspensions preparation, acid-base titration and Zeta potential measurements were done as described in previous papers (6).

RESULTS AND DISCUSSION

Potentiometric titrations

Surface charge on SiC, Si and mixtures thereof is variable and depends on the pH of the corresponding aqueous suspensions.

In keeping with the appraisals for obtaining the concentration of groups having an electric charge (1) (ionic, conjugated base or acid), and the density of surface charge (2), the same can be determined through potentiometric titration, and calculated from net proton consumption on the surface of these solids. Both computations are showing that the typical pH values allowing to ascertain the prevailing sort of charge corresponding to the zero point of charge (zpc) or to equal and opposed electrical charges, are 4.0 and 8.0 for SiC and Si suspensions, respectively. In the case of the suspensions containing a mixture of both solids a zpc change is observed in accordance with the increasing Si content of the mixture.

The zpc, which is different on every solid, reveals the nature of its surface. Accordingly, surface charge on SiC and Si becomes null under acid and basic conditions of the suspension, respectively. The solids whose structure does not contain oxygen will have an electrical charge in keeping with the pH of the medium and with the chemical species included in their impurities. Owing to SiC used, among these impurities there would be present valence-3 elements giving rize to a surface with the predominance of acid species as a result of an hydrolysis and/or neutralization of more acid sites (3). For solid Si there would be expected a behaviour similar to that of silica, whose surface has shown to have weak acid species preferently (1). Thus Figure 1 is showing that at value pH=10 the conjugated basic groups present on Si are larger than those on SiC (factor ≥2). In the wole pH range the charged surface groups on SiC are amaller than those on Si, thus showing that the groups charged with high pKa and larger than those on Si would prevail on SiC surface.

Electric charge variation of SiC and Si in their mixtures can be expressed as a function of Si composition (Fig. 2). This change of the zpc in the suspensions of SiC-Si mixture can be correlated with mixture composition in accordance with the theoretical predictions (4), already observed as concerns the alumina-silica system (5). Surface area equality of the solids in the mixture is a relevant factor present in the system already mentioned and delimiting the predictions regarding the zpc. In our case the BET specific surface (N_2 adsorption) was found to be 1.1 and 0.8 m^2/g for SiC and Si, respectively.

As noticeable in Fig. 2 the relationship obtained between the zpc of SiC-Si mixture and its composition is showing that a low surface interaction would exist between the two solids. This means that the forces between the active sites on the surface of each solid are neglectable. Thus the overall surface charge can be attributed to the relative number of each type of active site in the mixture. The increasing presence of Si on SiC contributes to modify surface charge of SiC, as reflected by the linear variation of the mixture zpc with the corresponding molar fraction when considering the experimental nature of the values obtained. Other experimental values obtained cannot be included because they concern mixtures with another type of electrolyte (KCl-HCl) in which instance a similar behaviour is being observed, but with deviations associated with the electrolyte change carried out in the system (5). A linear relation between the zpc of each solid and of the mixture would be:

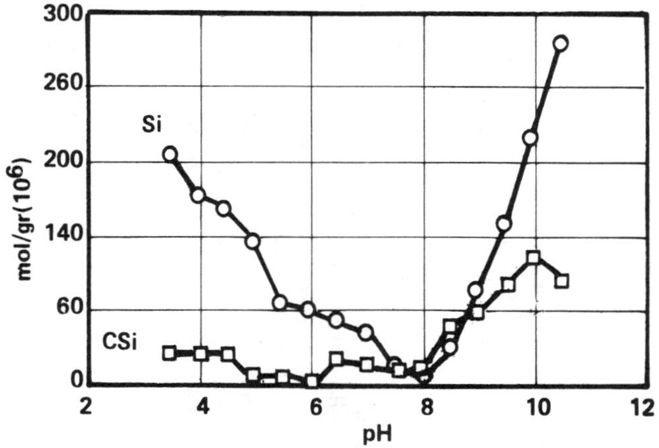

FIGURE 1.

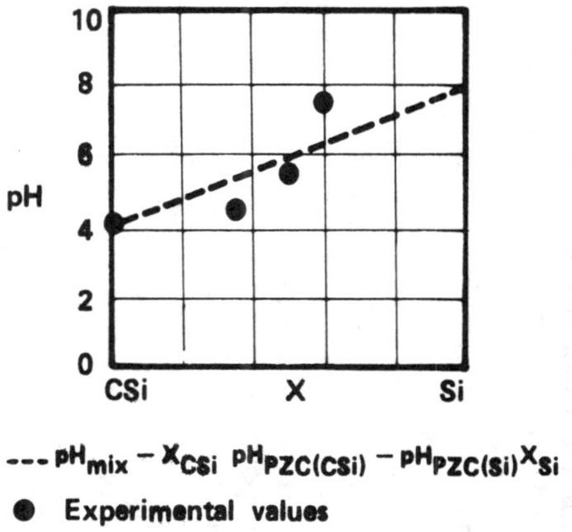

FIGURE 2.

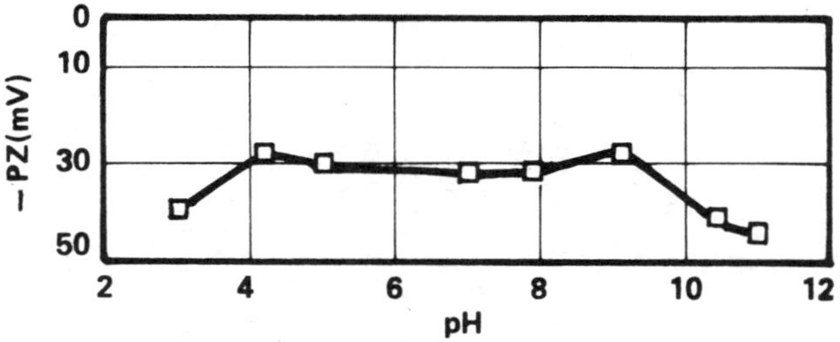

FIGURE 3.

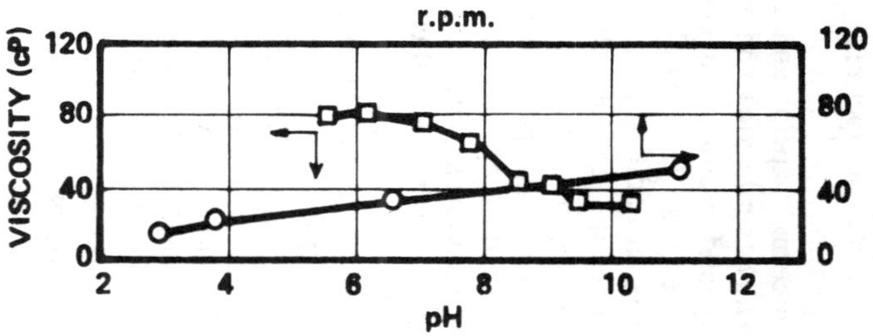

FIGURE 4.

$$zpc(mix) = zpc(a) \times (a) + zpc(b) \times (b) \qquad (1)$$

where zpc(mix), zpc(a) and zpc(b) are the zpc of the mixture and of each solid a, b, respectively, while x(a) and x(b) are the molar fraction of each of these two solids. Is this way and assuming a correspondence between the experimental points and those predictable (Fig. 2), it would be possible to conclude that the zpc of the potentiometric titration is an additive property, as shown by above relation (1).
In this case the zpc of solids mixture is showing that the variation in suspension pH would correspond to surface charge change on SiC or on SiC-Si mixture, that is to say, it would correspond to pH change on SiC surface. This apraisal constitues another condition delimiting the theoretical predictions regarding made of solids mixtures (4).
Similar behaviours have been observed with aqueous suspensions of mixtures of solids such as alumina and silica, in which case it is also possible to obtain a good prediction or the zpc in keeping with mixture composition. It is worth noting that said correlation has been obtained considering the variations of the charge/surface ratio of the potenciometric titration, and of the specific surface of each solid. According to likely appraisals of the zpc in systems with mixtures of solids, the application of a linear function, or simple graphical representation, of the different zero point of charge of every solid present in the mixture, would be feasible considering the variation, with the pH, of surface charge density (charge/mass) or of charge/surface ratio.
However, in the SiC-Si mixture studied herein, a linear correspondence has been obtained between the zero points of charge and concentration variation of charged species (mol/mass). If maintaining the limits of experimental measurements, this similar behaviour of mixture zpc would allow to establish that in the system discussed herein the surface charge variation on SiC is significant in suspensions that contain Si and hence can be detected through the concentration change of species charged, with the pH change the aqueous suspension.

Rheological behaviour
Viscosity variations in SiC suspension with 60% of solid content are showing that viscosity in an acid medium is larger than that in the alkaline range. The effect, though more pronounced and of behaviour varying with the pH, has been already obtained with other solids such as α-Al_2O_3 (6) and it reflecting the chemical affinity of the solid and its species for water.
Taking into account particle-size analysis and the high content of fine particles (90% under 20μ and 35% of these under 5 μ, a higher viscosity could be expected if accepting the behaviour exhibited by other solids in keeping with their grain size distribution, or the influence of fines on some solid suspension (7). However, in the case of SiC suspension, a low viscosity for a high solid content as well as a significant speed of flocculation are determined, which would show the weak chemical affinity of the solid in the aqueous suspension. This fact is noticed owing to a low variation of viscosity in the whole pH range studied, along with a zeta potential almost constant and negative (Fig. 3). An electrical potential produced between SiC surface and its aqueous medium an that does not vary with pH change, shows the reduced interaction between the solid and its hydrolysed species according to the potentiometric analysis already discussed.
Viscosity variation for a constant pH basic shear rate, i.e. rpm, shows a pseudoplastic behaviour of SiC suspensions (Fig. 4). The viscosity in the suspensions containing 60% SiC allows to observe a slight difference between the flocculated system (low pH) and the deflocculated system (high pH).
Si-particles presence in the suspension produces a shifting of all viscosity values plotted against pH towards higher pH values without altering viscosity range.

CONCLUSIONS

SiC suspensions are unstable within the pH range studied herein. Sole pH variation does not allow to obtain a completely deflocculated suspension.
The rheological behaviour of the aqueous suspensions of SiC exhibiting the grain size distribution used, shows a low chemical affinity for the medium. Si presence does not modify the system substantially.
The zero point of charge of a SiC-Si mixture is a linear combination of the individual zpc values.

Acknowledgements
This work has been partially financed by the Technical Department of Investigation of the University of Chile (Project Nr. I 2863-8812). In addition the authors gratefully acknowledge the collaboration of the students María Carolina Mancilla F. and Silvia Mónica Alvarez Q., and of Raymond Toledo and Oscar Morales for their help with the readying and printing of the manuscript, respectively.

REFERENCES

1. Hideaki Kita, Mariaki Henmi, Katsuski Shimazu, Hideshi Hattori and Hozo Tanabe. Meassurement of acid-base properties on metal oxide surfaces in aqueous solution. J. Chem. Soc. Faraday Trans. I, 1981, 77, 2451-2463.
2. J.A. Schawarz, C.T. Discoll and A.K. Bhanot. The zero point of charge of silica-alumina oxide suspensions.
J. Colloid Interface Sci., 1984, 97 (1), 55-61.
3. Karl A. Schwetz. Borcarbid, Siliciumcarbid und Ihre Werkstoffvarianten. II Deutsch-Franzoesische Tagung ueber Technische Keramik. Aachen. Maerz 1987, p. 155-179.
4. Jih-Fen Kuo and Teh Fu Yen. Some aspects in predicting the point of zero charge of a composite oxide system. J. Colloid Interface Sci., 1988, 121 (1), 220-225.
5. G. Piderit and P. Toro. To be published.
6. G. Piderit and P. Toro. Rheological behaviour of aqueous α-Al_2O_3 suspensions. 33 Congresso Brasileiro de Ceramica. Sao Paulo, Maio 1989.
7. Chang-Cheng Chou and Mamoru Senna. Correlation between rheological behaviour of aqueous suspensions of Al_2O_3 and properties of cast bodies: effect of dispersant and ultrafine powders. Am. Ceram. Soc. Bull. 1987, 66 (7), 1129-33.

PROCESSING OF ZRO_2 CERAMICS FROM SPHERICAL POWDERS OBTAINED BY SOL-GEL-PROCESS

W. Distler, G. Tomandl, R. Kohl, A. Stiegelschmitt
Universität Erlangen-Nürnberg, Inst. für Werkstoffwissenschaften III
Martensstr.5, 8520 Erlangen , W.-Germany
G. Rinn, Fraunhoferinstitut für Silicatforschung, Würzburg, W-Germany

ABSTRACT

Monodisperse ZrO_2 powders were synthesized by a controlled hydrolysis of dilute alcoholic solutions of zirconium alkoxides. Y-stabilized ZrO_2 powders were prepared by an emulsion-technique. The powders were characterized by using SEM, thermal analysis, infrared-spectroscopy and X-ray diffracion. Various methods to stabilize the Sol-Gel-derived powders were investigated. The state of powder aggregation in dispersions and powder packing in green bodies strongly affected the sintering behaviour. Uniformly packed powder compacts could not be achieved with these powders. Sintering of non regularly packed green bodies with an average density of 55 % th. d. resulted in finegrained microstructures with more than 97 % th. d. at a temperature of 1550 °C, which is lower than required to sinter conventional ZrO_2 powders.

INTRODUCTION

Improvements in ceramic fabrication processes are necessary in order to satisfy the stringent demands placed on modern engineering ceramics. The problems of property reliability stem from a failure to control precisely the microstructure of the ceramic product.

Our aim was to produce a ZrO_2-ceramic starting from spherical powders obtained by a sol-gel-process on the one hand and by an emulsion-technique on the other.

EXPERIMENTAL

Monodisperse ZrO_2-powders were synthesized by a controlled hydrolysis of dilute alcoholic solution of zirconiumalkoxide (G. Rinn, H. Schmitt and assistens). Y-stabilized ZrO_2-powders with a broad sphere-size distribution were prepared by an emulsion - technique (G. Rinn, H. Schmitt and assistents).

The powders were characterized by using SEM, thermal analysis, infrared-spectroscopy and X-ray diffraction.

Sol-Gel-Process

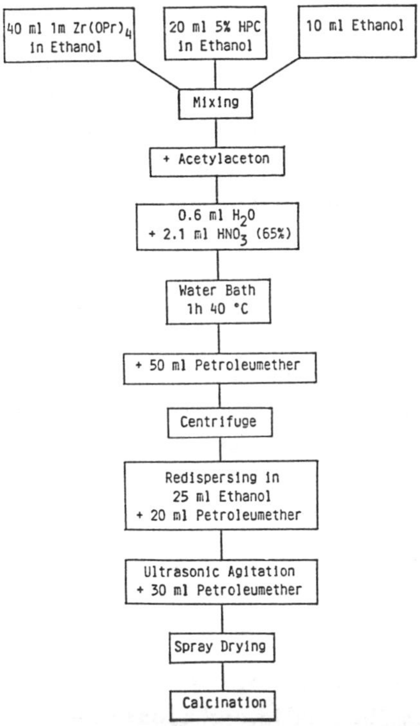

Emulsion-Technique

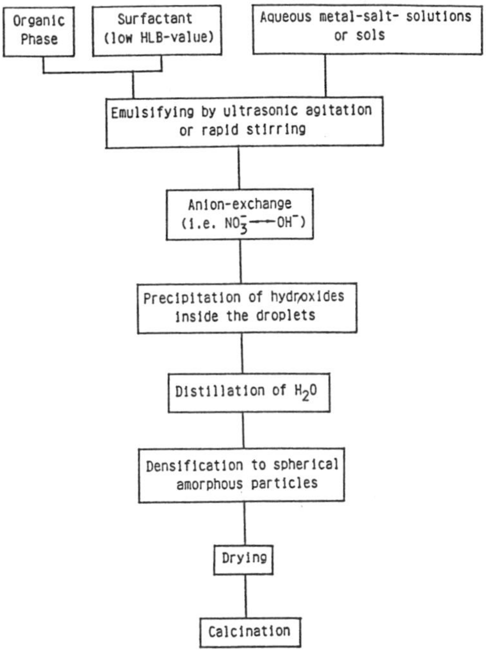

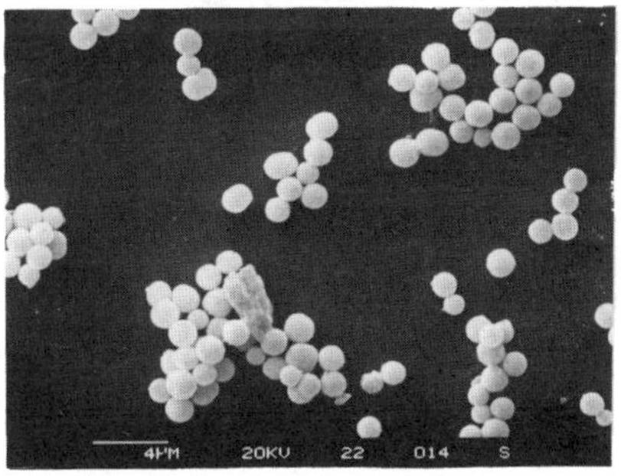

sol-gel derived powder

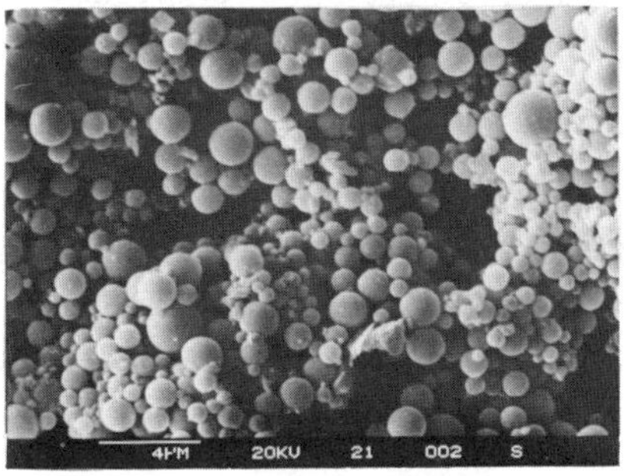

emulsion-derived powder (Y-stabilized)

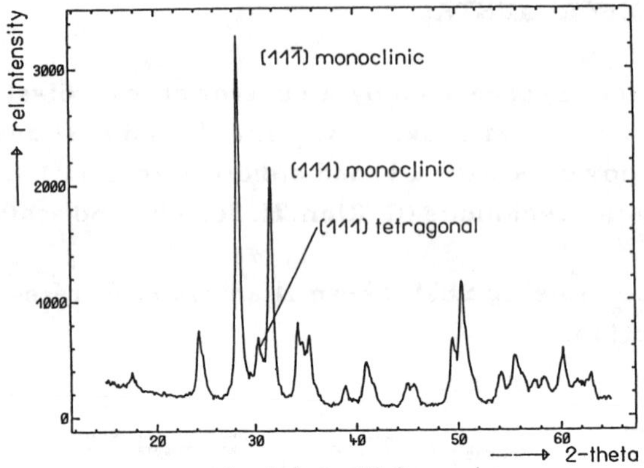

sol-gel-derived ZrO_2-powder

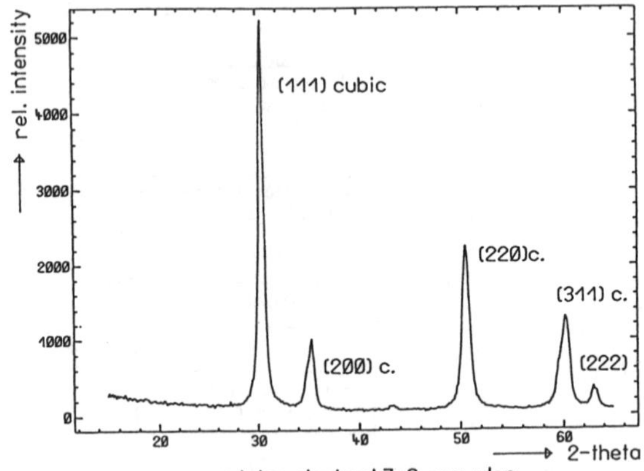

emulsion-derived ZrO_2-powder (Y-stabilized)

X-RAY CHARACTERIZATION

X-Ray-Diagramms of the sol-gel-derived powders showed amorphous powders up to 400 °C. First peaks of metastable tetragonal zirconia occured between 400 °C and 500 °C. The transformation to the stable monoclinic phase started at about 700 °C. Higher calcination temperature caused increasing intensities of the peaks of the monoclinic phase and a further decrease of tetragonal zirconia.

The emulsion-prepared powder heated up to 800°C gave a sharp diffraction pattern of single-phase cubic yttrium-zirconium-oxide.

THERMAL ANALYSIS

Thermogravimetry (TG) showed a rather similar behaviour of the two powders, calcinated at a temperature of 700 °C. The small weight loss of about 6 % indicated that a calcination temperature of 700°C is high enough for the decomposition of organic compounds and the following burnout of carbon.

The differential thermal analysis pattern (DTA) of the emulsion-derived powder showed no endothermic or exothermic peak. In the case of the sol-gel- powder the endothermic peak at about 300 °C could be related with the evaporation of surface-absorbed water.

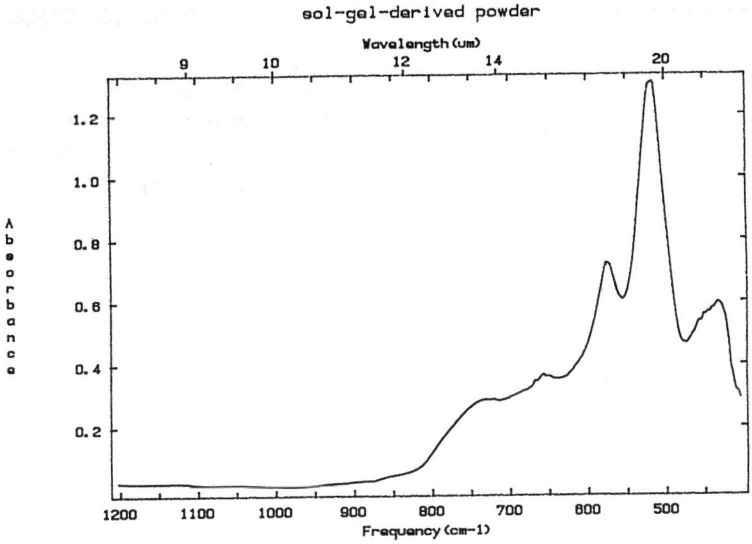

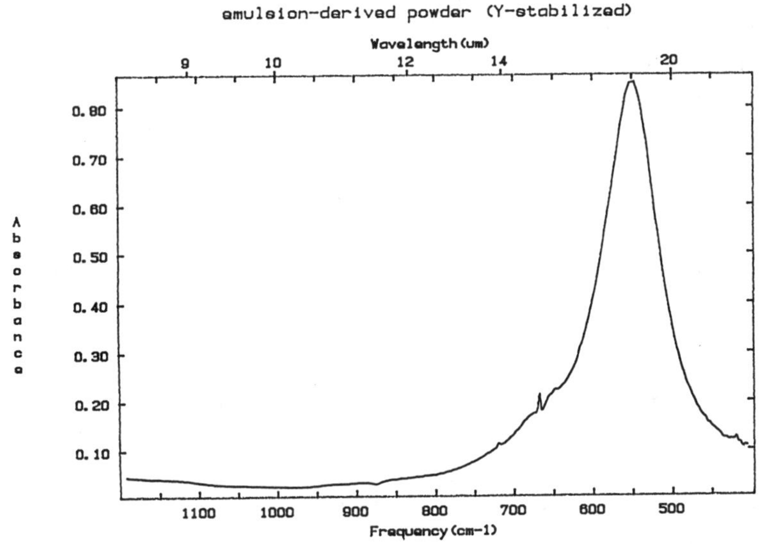

IR-SPECTROSCOPY

IR-spectra of calcinated sol-gel-powder showed the main absorption bands in the range of 400 - 800 cm^{-1} equal to absorption spectra of a commercially available, strain-free monoclinic ZrO_2-powder.
The broad absorption band of the emulsion-derived powder in the range from 450 cm^{-1} up to 650 cm^{-1} seemed to be related to structures similar to Y-stabilized cubic zirconia.

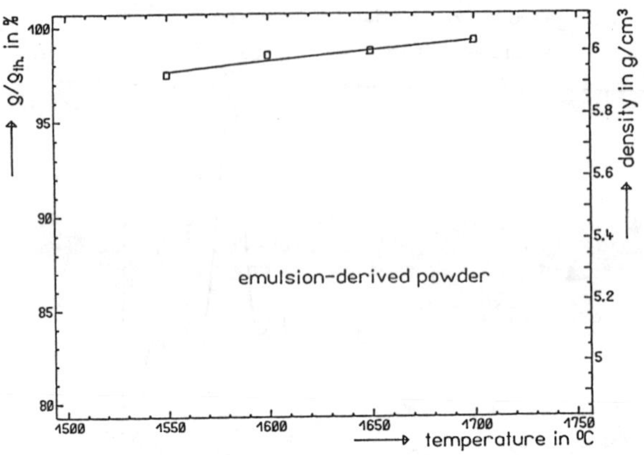

SINTERING BEHAVIOUR

Sintering of pellets (from non regularly packed emulsion - derived powder) with a green density of 55 % led to samples of more than 97 % th. d. at 1550 °C.

CONCLUSIONS

-The state of powder aggregation in dispersions and powder packing in green body strongly affected the sintering behaviour.

-Y-doped monosized sol-gel-powders could not be sintered to high density. The reason was the impossibility to obtain a perfectly uniform, close packed green body.

-The sintering of non regularly packed emulsion-derived compacts resulted in finegrained microstructures with more than 97 % th. d. at a temperature of 1550°C.

-Therefore a dense random packed green body, which could be achieved by the emulsion-derived powder with a broad sphere-size distribution has to be prefered in order to reach high density at lower temperature.

REFERENCES

1. Fegley, B.; White, P.; Bowen, H.K.; Processing and Charaterization of ZrO_2 and Y-Doped ZrO_2-Powders Am. Ceram. Soc. Bull. **64** (8) (1985) 1115 - 20
2. Barringer, E.A.; Bowen, H.K.; Formation, Packing and Sintering of Monodisperse TiO_2-Powder J. Am. Ceram. Soc.; **65** (12) (1982) C199 - C201
3. Ogihara, T.; Nobuyasu, M.; Masanori, K.; Processing of Monodispersed ZrO_2 - Powders, Ceramics Internat.,**13**, 35-40, Elsevier Applied Science Publishers Ltd.; England (1987)
4. Exner, H.E., Grundlagen von Sintervorgängen Hrsg. G. Petzow, Geb. Bornträger, Berlin-Stuttgart (1978)
5. Rinn, G.; Schmidt, H.; Prepar. of Monodispersed Zirkonia Powders from SolutionCeramic PowderProcessing Conference,Orlando(1987),i.Druck
6. Ingel, R.P.; Lewis III, D.; Lattice Parameters and Density for Y_2O_3-Stabilized ZrO_2, J.Am. Ceram. Soc. **69** (4) (1986) 325-32.

EFFECT OF ANIONIC POLYELECTROLYTE ON THE AGGLOMERATION PROPERTIES OF FUMED γ/δ ALUMINA AQUEOUS DISPERSIONS

B.C. BONEKAMP and W.H. VAN 'T VEEN
Netherlands Energy Research Foundation, ECN
P.O. Box 1, 1755 ZG PETTEN, The Netherlands

ABSTRACT

Stable sols of fumed γ/δ alumina contain hard agglomerates (clusters) of primary particles (20 nm) as the kinetic units. It is shown in this report that polyanions can induce flocculation (cluster-cluster agglomeration) of these clusters above a well defined polyelectrolyte content in dilute sols. No restabilization is observed at high polyelectrolyte adsorption values. The rheological behaviour of concentrated sols (gels) is markedly different below and above the mentioned cluster agglomeration transition.

INTRODUCTION

Polyelectrolytes are often used as processing aids in ceramic suspensions [1] in order to control the degree of agglomeration and in turn to control also the rheology and microstructure in the suspension and particle packing after compaction.
Dried and calcined gels of fumed alumina have a potential applicability as a ceramic membrane for ultrafiltration provided the membrane homogeneity and pore structure can be controlled. This notion led us to the question of how adsorbing polyelectrolytes affect agglomeration and gel properties in fumed γ/δ alumina sols, on which we report in this preliminary paper.
We found evidence in accord with others [2,3] that colloidally stable sols of the fumed oxide contain hard agglomerates (irreversible clusters) of primary particles. We show that anionic polyelectrolyte induces a cluster agglomeration transition and that the transition concentration is well defined. Further the difference in rheological behavior observed by us for concentrated sols (gels) below and above the transition concentration polyelectrolyte is briefly discussed.

EXPERIMENTAL

Fumed alumina (Alumina C) from the Degussa Compagny was used as received in the experiments. This pyrogenic γ/δ alumina has a primary particle diameter of about 20 nm as stated by the manufacturer and checked by us with TEM. The specific surface area of the powder was 99.4 m^2/g as determined with N_2 adsorption.

The polyelectrolyte used was a polycarboxylic ammonium salt (Dolapix CE 64) obtained from Zschimmer & Schwartz (BRD). the molecular mass of the polysalt is estimated by us to be about 5000 on the basis of capillary viscosimetry. The other chemicals used were of pro analyse quality and high quality dust free conductivity water was used.

Stock sols (10% w/w) of Alumina C were prepared by slowly dispersing the powder in water under stirring. Dilute sols (0.05% w/w) were made bij diluting the stock sol with water or polyelectrolyte solution. Concentrated sols for rheological measurements were prepared by evaporation of the stock sol (with polyelectrolyte) in a rotation evaporator.

Adsorbed amounts Γ were as usual determined by a depletion technique [4]. Polyelectrolyte concentrations were determined by acid/base titration and calibration bij means of known polyelectrolyte (DPE) solutions.

Electrophoretic mobilities were measured with a Zetasizer II C (Malvern) The average of 5 measurements at both stationary levels in a cylindrical cell was taken.

Particle size/agglomerate size estimates were obtained by dynamic light scattering using a Malvern Zeta- or autosizer.

Turbidity (optical density) measurements were performed in 1 cm cuvettes using a Beckman DU 40 spectrophotometer. Agglomeration was detected by measuring the size indicating light scattering parameter [5], $m=-d\ln \tau$(turbidity)/$d\ln \lambda$(wavelength) on dilute sols with increasing polyelectrolyte content ρ (mg/g).

Rheological measurements were carried out with a controlled stress instrument ('Deer Rheometer' van Bremen) with a cylindrical cell (gap 1.5 mm). Yield stresses from a quiescent state of the sol, i.e. 30 minutes after gently filling the cell, were taken as the stress level where the shear rate was 0.01 s^{-1}. The stress was increased with 12 mPa/10 s.

RESULTS AND DISCUSSION

The mean primary particle size of Alumina C is about 20 nm. The kinetic units observed by dynamic light scattering are in the range 50 - 200 nm. Hence clusters exist in the dilute sols. Ultrasonication (up to 20 minutes) or lowering the pH to 3 did not decrease the cluster size significantly. Break down of the clusters to the primary particle size does not occur. The highly positive electrophoretic mobility of Alumina C in water (pH is about 5): $3.7*10^{-4}$ $cm^2.V^{-1}.s^{-1}$ indicates that electrostatic repulsion between the irreversible clusters warrants the stability of the dilute sols in the absence of DPE polyelectrolyte.

The electrophoretic mobility u_e of Alumina C clusters decreases with increasing DPE content ρ as shown in figure 1. The zeta potential is zero at $\rho = \rho_0 \approx 35$ mg/g, which is in the plateau ($\Gamma_{max} = 0.18$ mg/m^2) of the high affinity adsorption isotherm (figure 2). It is more commonly observed that ρ_0 is about midway the steep part of the adsorption isotherm [6]. Probably the weak acid character of DPE plays a role here. DPE induces strong flocculation at a well defined polyelectrolyte content ρ as shown in figure 3. The cluster-cluster agglomeration midpoint lies at $\rho = \rho_m = 13$ mg/g DPE. This transition point (at $\rho_m = 12.7 \pm 0.7$ mg/g) was also observed bij means of batchwise and continuous turbidimetric titrations in a shear field (stirred vessel) and by means of dynamic light scattering. Hence the value of ρ_m is independent of the precise conditions of shear rate and sol concentration. ρ_m is about 0.7 times the plateau adsorption Γ_{max} and lies well below ρ_0. Because of the high electrical double layer thickness and the low molecular mass of DPE, polymer bridging cannot be the cause of the agglomeration transition. The fact that $\rho_m < \rho_0$ suggests that oppositely charged patches [6] on the Alumina C clusters due to adsorption of DPE plays a role

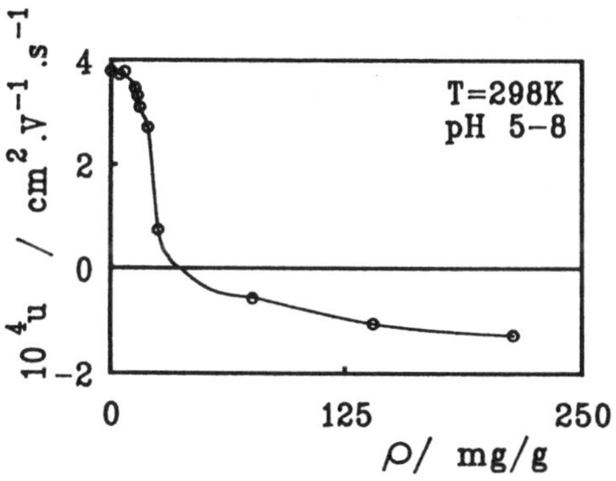

Figure 1.
Electrophoretic mobility of Alu-C in water (no added salt) as a function of the polyanion DPE content ρ in mg per gram Alu-C.

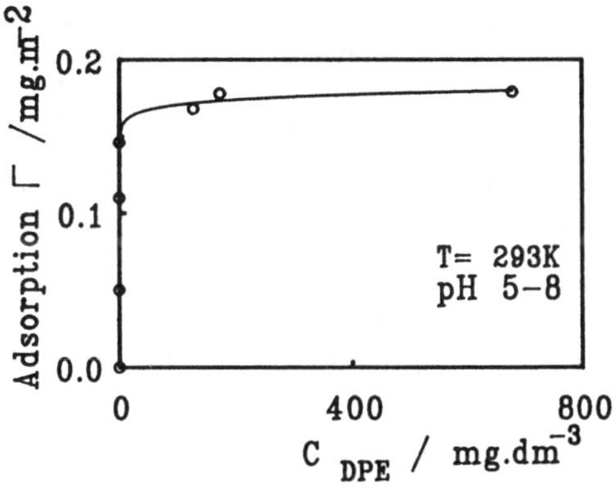

Figure 2.
Adsorbed amount DPE Γ as a function of the equilibrium concentration polysalt.

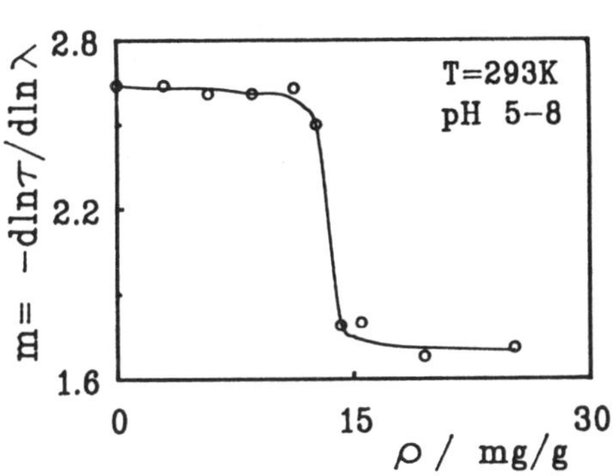

Figure 3.
The light scattering parameter m of Alu-C sols as a function of the DPE content ρ per gram Alu-C, which shows the agglomeration transition induced by the polyanions.

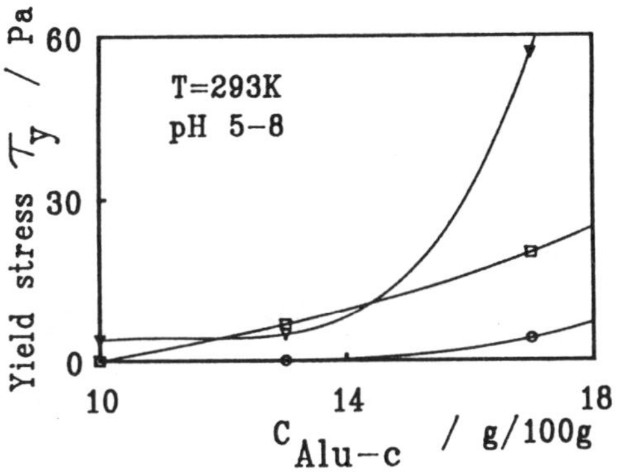

Figure 4.
Yield stress of concentrated Alu-C sols in the absence (o-o) and presence of DPE (-) 6 mg/ g; (□-□) 19 mg/g Alu-C.

in the mechanism of cluster-cluster agglomeration. This will be discussed in more detail elsewhere. Restabilization of Alumina C sols at high values is not observed, which may be caused by the low ionicity [6] of the DPE about pH 5. Also the low electrophoretic mobilities after charge reversal (Figure 1) point in that direction.

On increasing the volume fraction Alumina C, coagulation of the clusters occurs resulting in a gel state. This alumina volume fraction induced sol-gel transition occurs at a volume fraction solids $\phi_s \approx 0.05$ as evidenced by the yield value measurements shown in figure 4. The transition is probably reversible just as in the case of pyrogenic silica [7]. In the presence of DPE the occurrence of percolating clusters (gel forming) occurs at an even lower ϕ_s. Besides, the gel strength increases as evidenced by the greater yield values(figure 4). At low DPE surface coverage (0.06 mg/m^2) the yield values are the highest. At high surface coverage, above the flocculation transition in dilute sols, again lower values are observed. These observations are in line with the electrostatic patch model mentioned before.

ACKNOWLEDGEMENT

Financial support from the national program on Thechnical ceramics IOP-TK is gratifully acknowledged.

REFERENCES

1. Cesarano III, J. and Aksay, I.A., Processing of highly concentrated aqueous α-alumina suspensions stabilized with polyelectrolytes. J. Am. Ceram. Soc., 1988, 71, 1062-1067.

2. Strenge, K. and Pilgrim, H., The desaggregation of solid particle aggregates by ultrasonication. Colloid & Polym. Sci., 1983, 261, 855-857.

3. Ramsay, J.D.F., Characterization of colloids and gels. Mat. Res. Soc. Symp. Proc., 1988, 121, 239-304.

4. Cohen Stuart, M.A., Cosgrove, T. and Vincent, B., Experimental aspects of polymer adsorption at solid/solution interfaces. Adv. Colloid Interface Sci., 1986, 24, 143-239.

5. Sonntag, H. and Strenge, K., Coagulation kinetics and structure formation. Plenum Press, New York, 1987, pp 114.

6. Mabire, F., Audebert, R. and Quivoron, C., Flocculation properties of some water-soluble cationic copolymers toward silica suspensions. A semiquantitive interpretation of the role of molecular weight and cationicity trough a "patchwork" model. J. colloid interface Sci., 1984, 97, 120 - 136.

7. as 2 pp 129.

DISPERSION OF SiC WHISKERS IN WATER

K.J. Konsztowicz,
Atlantic Research Laboratory, NRC Halifax, N.S., B3H 3Z1 Canada

ABSTRACT

In this study the behaviour of as-received SiC whiskers in aqueous suspensions was compared to the behaviour of the same whiskers with washed surfaces. Viscosity and sedimentation measurements were performed. With proper surface treatment, stable aqueous suspensions of at least 15 vol % could be prepared without deflocculants at pH 3. Values of the electrophoretic mobility of SiC suspensions are related to the spatial arrangement of the whiskers in suspensions, as assessed by SEM examination.

KEYWORDS

SiC whiskers, dense suspensions, homogeneity, surface impurities, viscosity, sedimentation, electrophoretic mobility, agglomeration.

INTRODUCTION

The dispersability of SiC whiskers in aqueous suspensions can be enhanced greatly by removal of surface impurities. This type of improvement can be achieved either by oxidation (Adair et al, 1988) or by leaching (Sacks et al, 1988).

Because the suspension of SiC whiskers in an acid aqueous medium has many advantages, e.g. compatability with Al_2O_3 water slurries, the purpose of this work was to study in detail the conditions of dispersion of SiC whiskers with different surface treatments in H_2O of acid pH. Such suspension then can be added to suspensions of fine powders for the preparation of composites.

MATERIALS AND METHODS

As-received SiC whiskers[1] were examined for sedimentation and viscosity[2] in water suspensions with 10 weight % and 20 weight % solids content. Then similar tests were performed on SiC whiskers washed for 12h in acid (0.18N HNO_3) and base (0.15 N NH_4OH). Finally, another batch of whiskers was processed by washing in acid and base, then additionally rinsing in deionized H_2O several times until a constant pH was reached. Samples containing 15 w/o whiskers were tested for sedimentation and viscosity at a wide pH range, from 3 to 11. Out of this latter batch samples with solids content = 31 w/o were prepared by additions of small amounts of 0,1 N HNO_3 to bring the pH values to 2.5, 3, 3.5, 3.9 and 4.4.

---

[1] Tateho, SCW 1
[2] Brookfield, model LDTV

These samples then were centrifuged, and small amounts re-diluted (with pH adjustment) to the level of ∿300 p.p.m. in their respective supernatants for the measurement of electrophoretic mobility. Tenfold decrease of dilution, to ∿30 p.p.m. and subsequent vacuum filtering on a microporous membrane enabled the preparation of samples for SEM observations of dispersion of SiC whiskers at different pH values.

RESULTS AND DISCUSSION

Sedimentation

Measurements of sedimentation heights on as-received and washed-only (rinsed once after washing) whiskers with solids contents 3 and 7 vol %, and sedimentation times typical for ceramic powders (up to 300 min), did not give consistent results, irrespective of pH value applied.

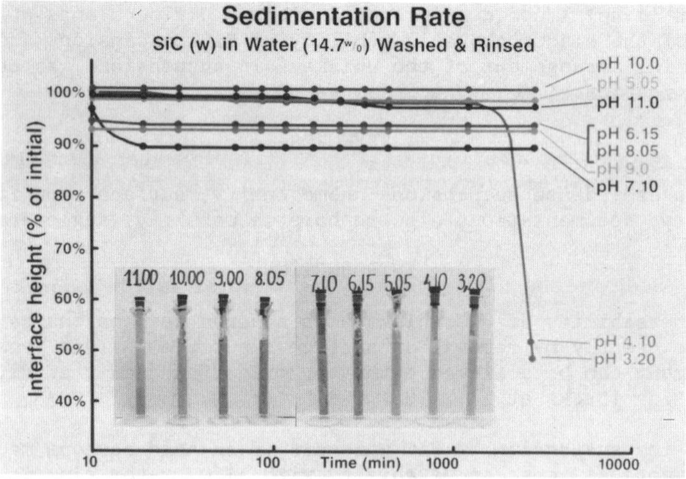

Figure 1 Sedimentation test of SiC whiskers washed in HNO_3 and NH_4OH, and rinsed in deionized water to constant pH. The values of pH adjusted by small additions of 0.1N HNO_3 or 0.1N NH_4OH only, in their respective ranges.

Fig. 1 shows that in the case of whiskers washed in HNO_3 and in NH_4OH, then properly rinsed in deionized water to constant pH value, after a relatively long time approaching ∿1000 min the interfaces can be seen in samples at pH 3 and 4. It is remarkable that only these suspensions with pH 3 and 4 (also pH 11 to some extent) are fluid and homogenous enough to be easily handled.

Electrophoretic Mobility and Viscosity

The results of the measurements of electrophoretic mobility[3] of SiC whiskers in water suspensions indicate that better dispersion can be achieved at low pH values (Fig. 2). It should be noted, however, that in the case of dense casting suspensions the increase of ionic strength attendant upon larger additions of acid would lead to overlapping and deformation of electrostatic double layers on particle surfaces and loss of stability towards flocculation.

---

[3] Rank Brothers, Mark III. Monoresearch Labs.

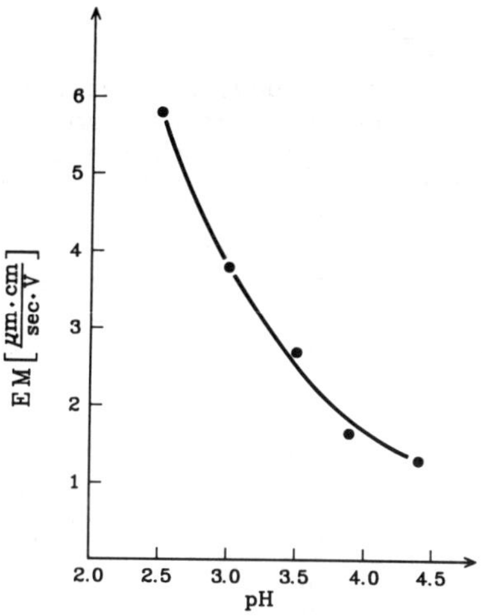

Figure 2 Electrophoretic mobility of diluted water suspensions (300 ppm) of SiC whiskers vs pH in acid range.

This tendency is confirmed by the values given in the "Viscosity vs pH" plot (Fig. 3) for dense (31.4% of solids) suspensions in the acid range. Despite the highest measured value of electrophoretic mobility at pH 2.5 (see Fig. 2), dense whisker suspensions show viscosities higher than that at pH 3. Further increase of viscosity follows the decrease of electrophoretic mobility, leaving a narrow "window" of practically useful pH values between 3 and 4 for the acceptable dispersion of SiC whiskers in water suspension.

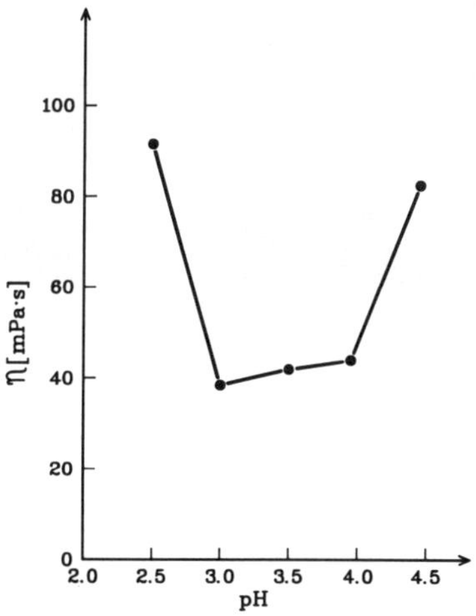

Figure 3 Viscosity of dense water suspensions of SiC whiskers (31.3v/o) in acid range of pH.

Microscopic Observations

The SEM[4] study of dispersed powders confirms the previously described results. Fig. 4a is the image of the sample dispersed at pH=2.5. Despite the good overall dispersion, the tendency to re-flocculation due to the increased ionic strength is obvious, with whisker agglomerates of 10-20 μm of diameter. Fig. 4b shows that the lowest viscosity combined with still good values of electrophoretic mobility results in the best dispersion of SiC whiskers at pH=3. With increase of pH the SiC whiskers show a tendency to agglomerate, in particular the smallest ones. Most of them form "balls" of remarkable size of dozens of microns (Fig. 4c). The increase of pH of water suspension to the value of 4.4 brings about almost complete agglomeration of the whiskers (Fig. 4d).

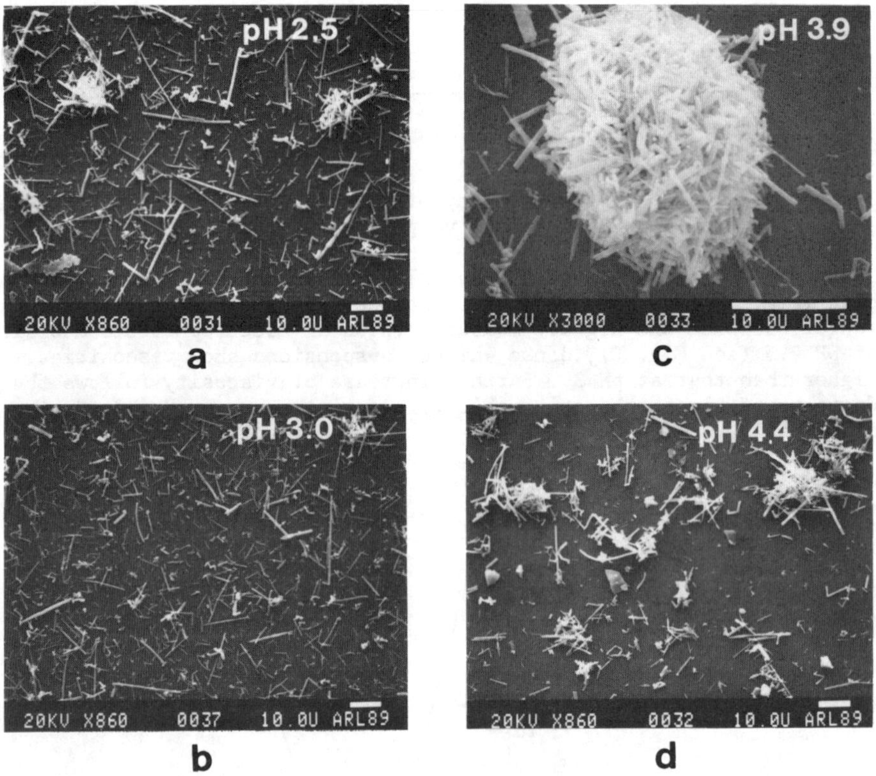

Figure 4 SEM micrographs of SiC whiskers strongly dispersed in water (30 ppm) a) pH 2.5; b) pH 3.0; c) pH 3.9 "ball"-like agglomerate of small whiskers; d) pH 4.4, strong agglomeration.

The dispersion of SiC whiskers with surfaces washed in HNO_3 and NH_4OH, then rinsed in H_2O, can be optimized in aqueous suspension at pH 3 giving a high solid content of 17 v/o (40 w/o) without the use of organic deflocculants. The slurry has good viscosity (Fig. 5) and is ready to mix with a dense suspension of the matrix powder.

---

[4] Jeol 2XA 35C

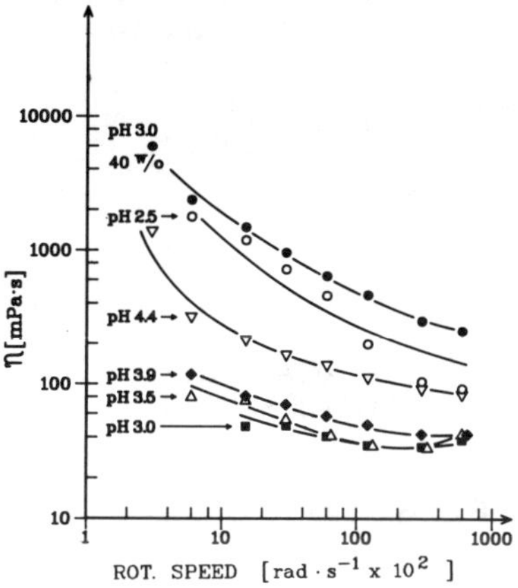

<u>Figure 5</u> Viscosity of dense water suspension of SiC whiskers (40w/o solids) at pH = 3 compared to less dense slurries (31.3%).

SUMMARY AND CONCLUSIONS

- Dispersion properties of SiC whiskers in aqueous suspensions have been examined over a wide range of pH, from 3-11.
- Washing in HNO_3 with following neutralization in NH_4OH significantly enhances dispersability of SiC whiskers in water suspensions. Subsequent intensive rinsing to constant pH, removing products of chemical reactions from whisker surfaces and a final suspension, additionally increases their dispersability.
- The lowest viscosities and highest solid loadings have been achieved in the acid range of pH of aqueous suspension.
- The values of measured electrophoretic mobility combined with those of viscosity, and related to SEM observation prove that the best dispersion of SiC water suspensions appears at a very narrow pH range, from 3 to 3.5.

REFERENCES

1. J.H. Adair, B.C. Mutsuddy, J. Drauglis, (1988) "Stabilization of Silicon Carbide Whisker Suspensions: I, Influence of Surface Oxydation in Aqueous Suspensions". <u>Advanced Ceramic Materials</u>, 3 [3] 231-234.
2. J.V. Milewski, (1986) "Efficient Use of Whiskers in the Reinforcement of Ceramics", <u>Advanced Ceramic Materials</u>, 1 [1] 36-41.
3. A. Roosen, H. Hausner, (1988) "Techniques for Agglomeration Control During Wet-Chemical Powder Synthesis". <u>Advanced Ceramic Materials</u>, 3 [2] 131-137.
4. M.D. Sacks, Hae-Weon Lee, O.E. Rojas, (1988) "Suspension Processing of Al_2O_3/SiC Whisker Composites". <u>J. Am. Ceram. Soc.</u>, 71 [5] 370-379.
5. P.K. Whitman, D.L. Feke, (1986) "Colloidal Characterization of Ultrafine Silicon Carbide and Silicon Nitride Powders". <u>Advanced Ceramic Materials</u>, 1 [4] 366-370.

FRACTIONATION OF A SILICA SUSPENSION BY SIZE DURING SEDIMENTATION

C F EDSER and S D LUBETKIN*

Department of Physical Chemistry,
University of Bristol,
Bristol BS8 1TS, U.K.

It has recently been shown (1) that sedimentation of a paucidisperse silica suspension produces an ordered sediment in which the interplanar spacing decreases in a monotonic fashion with increasing depth in the sediment. At some relatively small distance above the base of the cell in which the sediment is accumulating, this decrease in spacing is reversed, and the spacing increases. The decrease in spacing can be attributed to the increasing pressure with depth exerted by the weight of sediment gradually overcoming the electrostatic repulsion between the particles. The reversal of the trend near the base of the cell appears to be due to particle fractionation by size, and particle size analysis as a function of depth confirms this hypothesis.

INTRODUCTION

Much of the potential of ceramics as advanced engineering materials remains unrealised, chiefly because of the unpredictable nature of their fracture and failure behaviour. This unpredictability is linked to the presence of inhomogeneities in the particle packing at the green body stage, which may lead to porosity and internal crack formation, and thus finally to a defective ceramic product. A high and uniform packing density of the precursor particles is therefore important for the production of high strength, reliable ceramics for advanced engineering applications.

Theoretically, the use of colloidal particles of narrow particle size distribution provides a route to such high packing densities, and there is much interest in such particles as model systems, as well as the possibility that they may provide a practical route to the production of ceramics. Non-destructive methods for assessing the packing density and degree of order of compact beds of such particles are needed, and we here report two such methods.

First, when the particle bed is well-ordered, and the interparticle spacing is close to the wavelength of visible light, then Bragg diffraction of the light results, and the angular position of the Bragg peak allows a calculation of the interplanar spacing.

Second, the time of propagation of an ultrasound pulse through a known thickness of sample can be related to the particle volume fraction of the packed bed, if the appropriate parameters (such as the density and compressibility) are known.

In this paper, we report on studies of a model dispersion of silica particles in ethanol, using both these methods, together with electron microscopy. It will be shown that the packing of the particles is non-uniform with depth in the sediment, and reasons for this non-uniformity are put forward.

EXPERIMENTAL

(1) The dispersion. The suspension consisted of spherical particles of silica, prepared by the method of Stober *et. al.* (2), and suspended in pure absolute ethanol. The particle size was determined by electron microscopy, and photon correlation spectroscopy (PCS), and the mean diameter was found to be 450 nm, with a coefficient of variation of about 7%. The particle size analysis showed that the range of particle diameters was approximately 320 - 530 nm

(2) Visible light diffraction. The apparatus was based on an X-ray powder diffractometer (Hilger and Watts) with a θ - 2θ movement, and the X-ray source was replaced by a low-power helium-neon laser, (λ = 632.8 nm), which gave a well collimated beam of approximately 0.5 mm diameter. The sample was contained in an optical cell with dimensions 1 X 1 X 5 cm. This cell was mounted in such a way as to allow it to be moved in the vertical direction by known amounts, thus allowing Bragg angles to be measured at all heights in the sediment. The cell was surrounded by a cylindrical glass bath filled with liquid in order to reduce the effects of reflection and refraction at the cell walls.

(3) Acoustic measurements. The apparatus has been fully described elsewhere (1), and essentially consisted of a pair of piezo-electric transducers, mounted one either side of a glass (square section, 32 X 32 mm) cell. The first transducer was driven by single pulses (of approximately the resonant frequency of 6 MHz) and the other detected the arrival of the pulse after transmission through the sediment. Appropriate circuitry measured the time-of-flight of the pulse, and this time could be related to the volume fraction of particles in the sediment.

(4) Electron microscopy. Samples were carefully withdrawn from various heights in the sediment using an "Agla" syringe. Replicas were made of the particles, which were then photographed in a Jeol 100CX transmission electron microscope. The enlarged photomicrographs were analysed using a Carl Zeiss TGZ3 particle size analyser, calibrated with a photomicrograph of a diffraction grating of known line spacing.

THEORY

(1) Visible light diffraction. For a colloidal sediment, the diffraction angle, θ_d, observed for a given incident wavelength, λ, is related to the interplanar spacing in the sample, d, by a modified form of the Bragg equation:

$$(m\lambda)^2 = 4d^2 [n_s^2 - n_m^2 \cos^2(\theta_d)] \quad [1]$$

where m is the order of the diffraction peak, n_m is the refractive index of the bath medium, and n_s is the refractive index of the sediment. Thus, the diffraction angle is inversely related to the spacing between the particles, and directly related to the particle packing density.

The refractive index of the sediment, n_s, is related to the particle volume fraction, Φ, in the sediment, and to the refractive indices of the continuous phase and of the particles (n_{cont} and n_p respectively) by the equation:

$$n_S = (1 - \Phi)n_{cont} + \Phi n_p \quad [2]$$

The refractive indices are those at the appropriate wavelength, and these can be calculated from the Cauchy relationships for the materials involved.

(2) Acoustic measurements. The velocity of a transmitted ultrasound pulse depends on the velocity through the individual phases, and the composition of the dispersion, provided that the particle size is much smaller than the wavelength of the ultrasound. The observed velocity can be related to the composition of the dispersion by means of the Urick (3) equation:

$$v_o = [\{(1 - \Phi)\mu_1 + \Phi\mu_2\}\{(1 - \Phi)\rho_1 + \Phi\rho_2\}]^{-1/2} \quad [3]$$

where Φ is the volume fraction of particles, μ is the compressibility, and ρ is the density. In each case, the subscripts 1 and 2 refer to the dispersion medium and the particles respectively.

RESULTS

(1) Visible light diffraction. 15 days after the onset of sedimentation, the position of the diffraction peak as a function of depth (at 1 mm intervals) in the sediment was determined, and the results are shown in Figure (1).

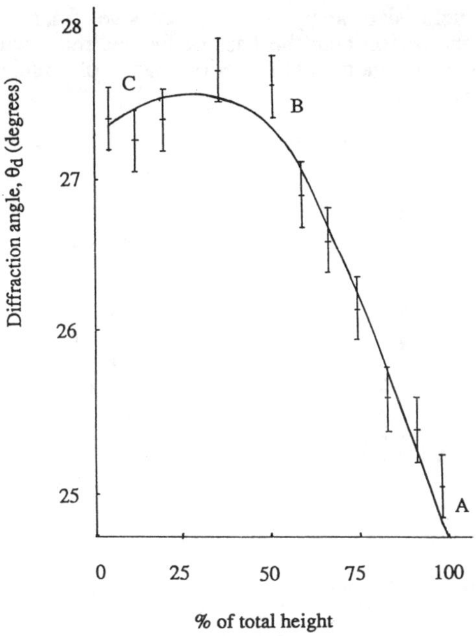

Fig. (1). *Bragg diffraction of monochromatic light of λ = 632.8 nm, plotted as the diffraction angle vs depth in the sediment. Since the Bragg angle increases as the interplanar spacing decreases, the results show a gradual decrease in interplanar spacing going down the sediment (A-B), which then levels out and finally increases again (B-C).*

(2) Acoustic measurements. It was not possible to measure the time-of-flight of the ultrasound pulse through the sediment until 50 days after the onset of sedimentation. At this time, the volume fraction profile was determined at 1 mm intervals throughout the depth of the sediment, and the results are displayed in Figure (2). Since we have not independently determined the adiabatic compressibility, nor the density of the particles, the absolute values of the volume fractions are in some doubt, but the general shape of the volume fraction profile agrees with that obtained by optical diffraction (Figure (1)).

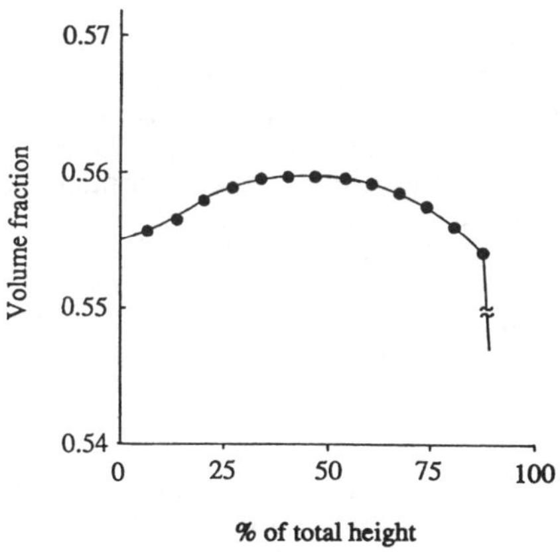

Fig. (2). *The acoustic measurements expressed as the volume fraction vs the depth in the sediment. The same general shape as shown in Fig. (1) is evident. It should be pointed out that the absolute values for the volume fractions are in some doubt, but that the general shape of the curve is correct.*

(3) Electron microscopy. Particle size analysis of successive fractions withdrawn from the sediment showed that the region near the base of the sediment was enriched in larger particles, and that near the top, there was a predominance of smaller particles. The results are shown in Figure (3).

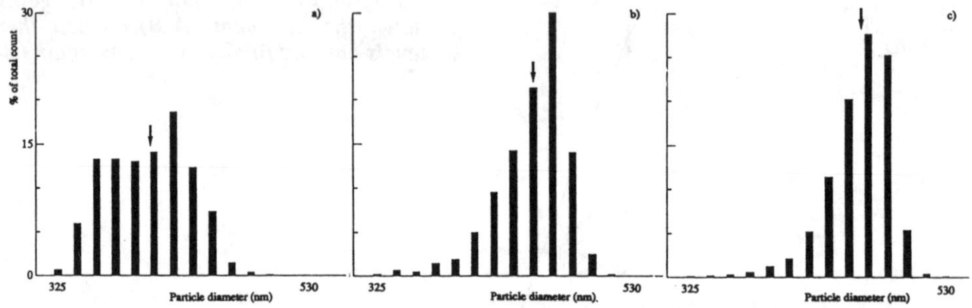

Fig. (3). The particle size analysis as determined by electron microscopy. (a) shows the distribution at the top of the sediment, (b) the overall distribution before sedimentation commenced, and (c) the distribution at the base of the cell. In each case, the mean value is marked by an arrow, and it is clear by comparing the initial distribution (b), with that shown in (a), that at the top of the sediment, the smaller particles predominate. Similarly, larger particles dominate at the base of the sediment, as shown by (c).

DISCUSSION

In general, the packing density was observed to increase with depth in the sediment, as shown by Figures (1) and (2). This behaviour is consistent with progressive compression of the sediment by the accumulated weight of the sediment above the height at which the measurements are made gradually overcoming the "soft" electrostatic repulsion between the charged silica particles, and this problem is discussed more fully elsewhere (1). However, both the optical measurements of Fig. (1) and the acoustic measurements, Fig. (2) show that the respectives curves pass through maxima near the base of the cell, and then decrease. For Fig. (1), this maximum corresponds to a minimum in the interplanar spacing, whilst for Fig. (2), it corresponds to a maximum in the volume fraction. Note that with a variable particle size, these two quantities (d and Φ) are not equivalent.

The particle size analysis shown in Fig. (3) reveals that there is an enrichment of the larger particles at the base of the cell, and an enrichment of the smaller particles at the top. During the sedimentation process, it is to be expected that some fractionation of the particles by size will take place, since the rate of settling of the particles, V_S, is (roughly) proportional to the square of the radius, R, as predicted by Stokes' law:

$$V_S = 2R^2\Delta\rho g/9\eta. \qquad [4]$$

It is not expected that Stokes' law will be accurately followed in our system, since the initial volume fraction of about 4% is large enough for deviations to be evident both from hydrodynamic and electrostatic interparticle interactions.

Clearly, the fractionation process would not be expected to produce a complete separation of particles by size, and the particle size distribution is simply expected to become increasingly skewed towards the larger diameters at the base, and smaller diameters at the top of the sediment. The diffraction technique effectively measures the interplanar distance in the ordered sediment, whilst the acoustic time-of-flight method is insensitive to particle size or degree of ordering, and measures the total solids content. Thus, the similar shapes of the curves produced by these two methods is in some measure surprising. A *monodisperse* sediment which was close-packed would be expected to give a volume fraction of about 72-74%, depending on the structure adopted, regardless of the particle size, but the interplanar spacing would increase linearly with the size.

It is clear, therefore that the explanation of the decrease in volume fraction towards the base of the sediment observed by the acoustic method does not lie *directly* in the increased mean particle size in this region.

The repulsive force between two charged spherical particles, F_r, separated by a distance, h, is proportional to the particle radius, R, as described by the Derjaguin approximation (4):

$$F_r(h) = 2\pi R\, W(h)_{plate} \qquad [5]$$

where $W(h)_{plate}$ is the energy of interaction per unit area of flat plates with the same electrostatic characteristics as the particles. It would thus be expected for the volume fraction to decrease where the larger particles predominate, since the repulsive force will be proportionately larger on average, and this will counteract to some extent the hydrostatic forces due to the sediment depth (5), and lead to a relative decrease in the volume fraction.

CONCLUSIONS

The volume fraction in a sediment of paucidisperse silica particles is non-uniform with depth in the sediment. Two regimes can be distinguished: First, for most of the distance down the sediment, the volume fraction increases, and this increase is related to the increasing hydrostatic pressure overcoming the soft electrostatic repulsion between the particles . Second, near the base of the sediment, a reversal of this trend occurs, and there is a decrease in volume fraction. This reversal is attributed to the increase in the force of repulsion between the particles as a result of an increase in the average radius, and this increase in average radius is itself due to partial fractionation by size during the sedimentation process. The importance of these effects will be greater in more polydisperse samples, such as those found in industrially significant ceramic precursor suspensions.

ACKNOWLEDGEMENT

We thank the SERC and BP Research Ltd for a studentship for CFE, and Shell Research Ltd., and in particular, Dr. D J Wedlock, for the use of the acoustic apparatus.

REFERENCES

[1] S D Lubetkin, D J Wedlock and C F Edser. **Colloids Surfaces** In press; see also S D Lubetkin. **J. Chem. Soc. Chem. Commun.** 335 (1989)

[2] W Stober, A Fink and E Bohn. **J. Coll. Int. Sci.** 26 62 (1968)

[3] R J Urick. **J. Appl. Phys.** 18 983 (1947)

[4] B V Derjaguin. **Kolloid-Z.** 69 155 (1934)

[5] We are grateful to R A Croot (personal communication) for pointing out this possible explanation to us.

EVALUATION OF GREEN BODIES RELATED TO THE FORMING TECHNIQUE.

C.GALASSI, E.RONCARI, C.CAPIANI, A.PIANCASTELLI
Research Institute for Ceramics Technology - CNR
Via Granarolo, 64 - 48018,Faenza,Italy

ABSTRACT

The dependence of the green body characteristics on the starting powders properties and on the processing parameters were evaluated for a ZTA (zirconia toughened alumina) material. Different forming techniques produce green texture, density and defect state which is dependent on the omogenization route, additives and technique of compaction. Green bodies obtained by slip casting, injection molding and cold isostatic pressing were compared; green density, porosity, pore size distribution and ZrO_2 dispersion were evaluated.
Correlations are drawn among green body properties and solid phase amount, type and quantity of additives and processing parameters.

INTRODUCTION

In a composite material the dispersion of the reinforcing additive must be optimized to improve the mechanical properties. In a previous work (1) a ZTA material was developed starting from the mechanical dispersion of partially stabilized ZrO_2 in the Al_2O_3 matrix (composite A) and the slip cast samples showed quite good mechanical properties but the improvement of the dispersion of ZrO_2 particles and the removal of forming defects were not optimized. In this study, a flocculation - deflocculation route (2) was chosen to obtain the dispersion of ZrO_2 in the Al_2O_3 matrix (composite B) and the effectiveness in the green body produced with different techniques was investigated.

MATERIALS AND METHODS

Commercial powders, Al_2O_3 Alcoa A 16 SG (10.2 m^2/g) and ZrO_2

TOYOSODA TZ-3Y (3 mol% Y_2O_3, 18.5 m^2/g) were mixed in the ratio 70/30 wt% (theoretical density 4.44 g/cm^3) with the flo-deflocculation procedure shown in Fig.1.

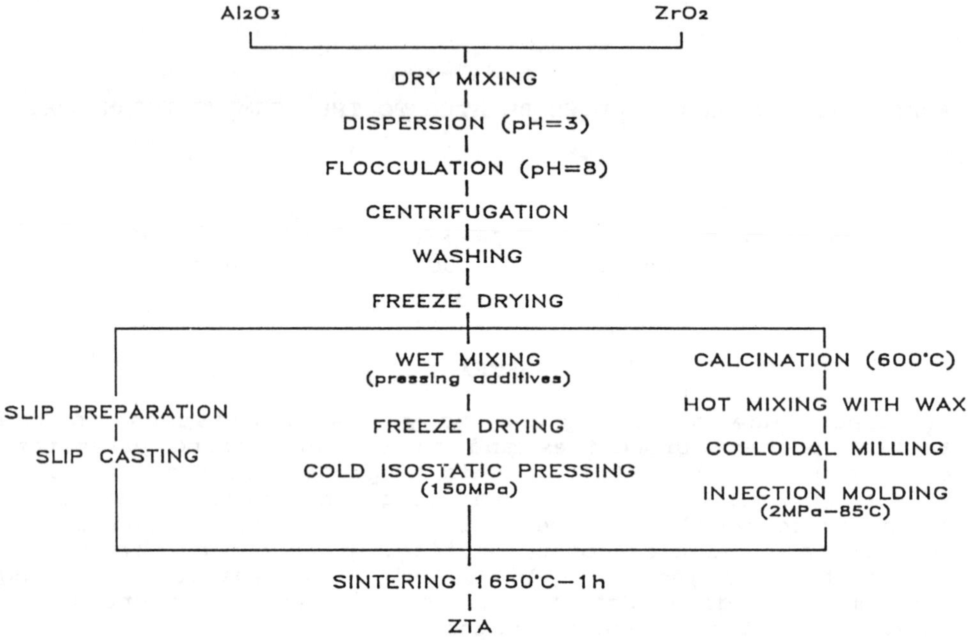

Figure 1. Processing route.

The freeze dried powder was treated and mixed with different forming additives. On the mixed powder granulometric distribution and ZrO_2 dispersion were evaluated. The water based slip was stabilized with Na-ligninsulphonate and different pH and solid phase concentrations were tested; the rheological properties of the slip were measured and cast samples were produced. The cold isostatic pressing was performed on the as received powder, and with PVA/PEG 60/40 (P) and XUS/PEG 95/5 (X) additives and green machinability was tested.
In order to prepare the moldable mix the powder was firstly calcined (600°C-1h).It was then hot mixed with the binder system, composed of paraffin wax (melting point 52°C) and oleic acid, in the ratio 80/20; the colloidal milled mix was injected at 2 MPa, 85°C, using a low pressure injection molding machine (Fig.1). Thermal extraction of binder was performed in air, with samples imbedded in Al_2O_3 powder (T max = 500°C). The processing conditions are reported in Tab.1.
The green samples were presintered at 1100°C (soaking time 1h) and sintered at 1650°C (soaking time 1h).

TABLE 1
Processing parameters, viscosity, green and sintered density, total shrinkage of the bodies (S.C.=slip casting, I.M.=injection molding, C.I.P.=cold isostatic pressing)

	Sample	Solid phase		Defloc.	pH	$\eta_{D=10\ s^{-1}}$	Density %		Total shrinkage
		wt%	vol%	wt%		Pas	green	sint.	%
S.C.	A-1	70	34	0.6	9.2	-	52.0	99.1	19.8
	A-2	75	40	0.6	9.2	-	55.1	99.6	19.1
	B-1	70	34	0.6	9.2	0.67	51.6	99.3	19.2
	B-2	70	34	0.6	10.0	0.85	51.1	99.2	18.9
	B-2bis	75	40	0.6	9.2	1.50	-	-	-
	B-3	75	40	0.6	10.0	1.36	53.6	99.6	18.7
	B-4	75	40	0.6	10.6	1.30	54.0	99.1	18.1
	B-5	77	43	0.6	10.0	4.38	56.3	99.4	17.7
	Sample	Solid phase				$\eta_{D=100\ s^{-1}}$	Density %		Total shrinkage
		wt%	vol%			Pas	green	sint.	%
I.M.	A1	79	43			13.9	54.6	-	-
	B1	77	41			11.5	49.3	97.8	24.8
	Sample	Additive				Machina-bility	Density %		Total shrinkage
		type	wt%				green	sint.	%
C.I.P.	A	-	-			low	55.2	99.3	18.1
	AP	PVA/PEG	3			medium	55.6	97.3	18.1
	B	-	-			low	55.6	98.9	17.7
	BP	PVA/PEG	3			medium	56.5	95.7	17.5
	BX	XUS/PEG	3			high	57.4	98.9	17.9

Green density, porosity and pore size distribution were measured with the mercury intrusion technique. On the sintered samples density and shrinkage were measured (Tab.1); the microstructure was evaluated by scanning electron microscopy (SEM).

RESULTS AND DISCUSSION

Powder preparation

The powder produced with the flo-deflocculation route results in a good dispersion of the ZrO_2 particles and absence of agglomerates; it is very finely divided, (d_{50}=0,3µm, 90 mass% <2µm).

Slip casting

The rheological measurement evidences that pwder B produces slip less viscous than powder A, with a plastic type behaviour.

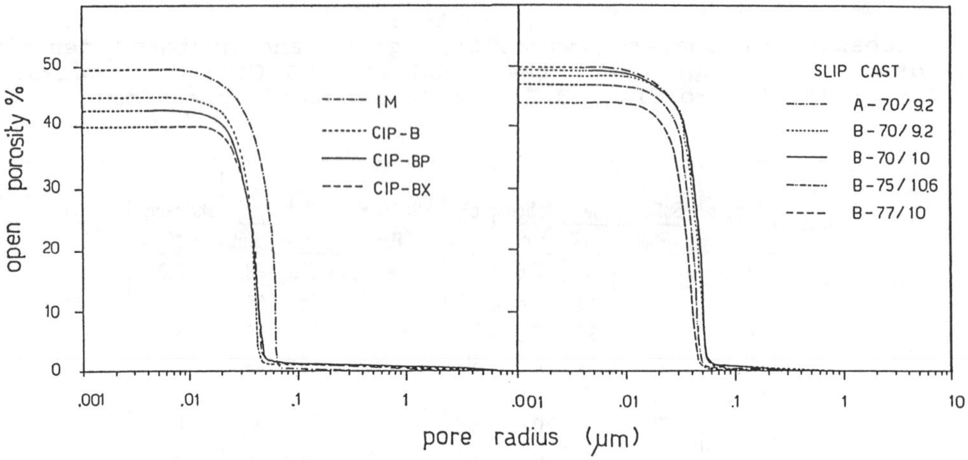

Figure 2. Pore size distribution in the green bodies (IM=injection molding; CIP=cold isostatic pressing).

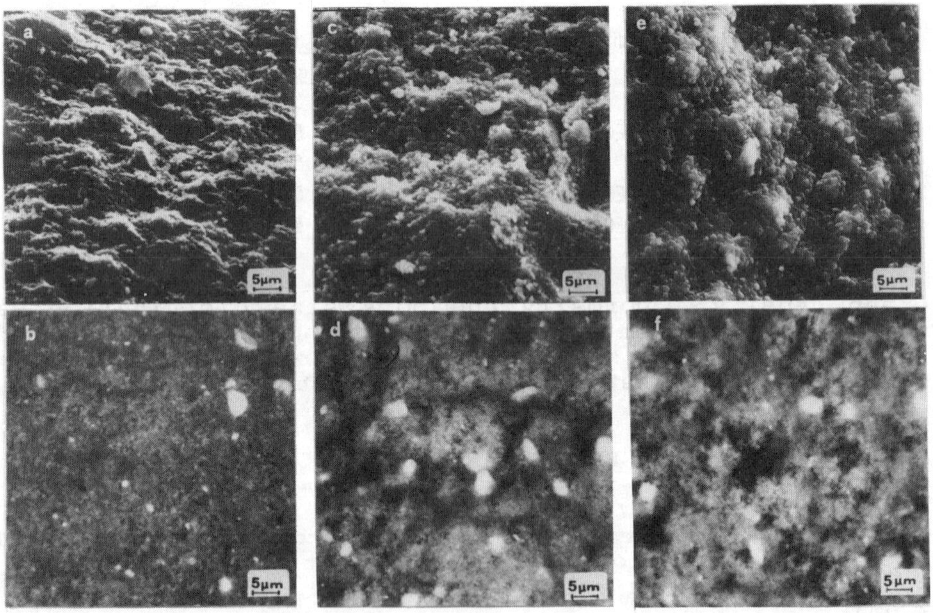

Figure 3. SEM morphology and BSE image of slip cast (a-b), injection molded (c-d), cold isostatically pressed samples (e-f) respectively.

A variation of the viscosity is detected also changing from pH 9.2 (slip as prepared) to higher values; pH 10 was chosen as the optimal because at this value 43 vol% solids is allowed, therefore best results are obtained with 40 vol%. The green bodies were free of macroscopic defects, the relative density ranged between 51 and 56% and the pore size ranged between 10 and 60 nm (Fig.2). The dispersion of ZrO_2 particles was better than in the samples produced starting from the mechanical mixing of the powder but few agglomerates of ZrO_2 still remain (Fig.3).

Cold isostatic pressing

Both powders A and B could be dry pressed without additives but the green body results very weak and the addition of the polyvinylalchol/poliethilenglycole (PVA/PEG) or cellulosic binder/poliethilenglycole (XUS/PEG) binders results in an increase of the green density and machinability (Tab.1).
The relative green density ranges between 55 and 60% and the pore size distribution ranges between 20 and 50 nm (Fig.2). ZrO_2 dispersion is good but with presence of few agglomerates (Fig.3).

Injection molding

For the preparation of the mix high volume (59%) of binder was necessary in order to reach the mouldability of the system. The green bodies resulted free of macroscopic defects, the relative green density was poor (49%) and the pore size ranges between 10 and 70 nm.

CONCLUSION

The forming techniques chosen result in different characteristics of the green bodies, mainly density, agglomerates and pores morphology. Slip cast body is more uniform and produces the highest final density; the content of additive for cold isostatic pressing should be optimized, but high green machinability was reached. Mixing of powder and binders for injection molding must be improved in order to reduce the quantity of organic phase and to produce satisfactory rheological performance.

REFERENCES

1. Bellosi,A., Galassi,C., Guicciardi,S., Slip casting of Al_2O_3 and Al_2O_3/ZrO_2 composite (in press).
2. Aksay,I.A., Lange,F.F. and Davis, B.I., Uniformity of Al_2O_3-ZrO_2 composites by colloidal filtration. J. Am. Ceram. Soc., 1983, 66, C 190-192.

A NEW TECHNIQUE TO STUDY POWDER COMPACTION DURING SLIP CASTING

L.A. CORREIA, R.J.G. BEENEN and H.J. VERINGA
Netherlands Energy Research Foundation, ECN,
P.O. Box 1, 1755 ZG PETTEN, the Netherlands

ABSTRACT

Powder compaction during slip casting is generally studied by measuring the development of the cake thickness during filtration. In this paper the feasibility of using ultrasonic interferometry is discussed. Using filtration models and phenomenological relations describing sound velocity in dispersions, the characteristics of the sediment are estimated. Preliminary studies show that manipulation of particle-particle interaction by increasing the pH increases the filtration rate caused by a decrease in the density of the compact formed.

INTRODUCTION

Ceramic processing involves several subsequent processes such as powder compaction, drying and densification. Each process has the potential of introducing heterogeneities, which will persist during further processing or even develop into new irregularities during densification. To increase the reliability of ceramics, microstructural control starting at the very first step is therefore essential. The colloidal approach shows to benefit microstructural control [1] because of the ability of manipulating particle-particle interaction.

During colloidal processing, e.g. slip casting, particle packing is mainly governed by the interaction between the particles. The density of the growing cake is reflected by the growth rate of the cake. Hence, the influence of particle-particle interaction on the cake structure can be studied by measuring the development of the cake thickness during slip casting.

The possibiliy for quantitative analysis of the cake density requires control of the development of all the subsystems involved, viz: the suspension, the cake and the filtrate in the mould. An acoustic technique, the ultrasonic interferometer, seems to offer a solution.

ULTRASONIC INTERFEROMETRY

The ultrasonic interferometer is used in different fields serving a large variety of goals. In materials, R&D, acoustic waves are used for flaw detection [2], while in the maritime field it is used for sounding the sea bottom. The propagation of sound waves through suspensions is used for determining the volume fractions and compressibilities of either the dispersed or the continuous phases e.g. probing of marine sediments [3] and monitoring of sedimentation of colloidal dispersions [4]. This technique is based on the phenomena that propagation of sound waves through a medium depends on the density and the compressibility of the material and while striking a boundary of materials of different densities the sound waves are partially reflected.
Propagation of waves through suspensions has been the subject of several studies. The simplest approach for calculating the velocity of ultrasound in suspensions is a phenomenological one in which the effective velocity depends on an effective density and an effective compressibility:

$$V = (\rho_{ef}\kappa_{ef})^{-\frac{1}{2}}. \tag{1}$$

If a suspension consists of a volume fraction ϕ of a phase with density ρ_2 and compressibility κ_2, dispersed in another phase of density ρ_1 and compressibility κ_1 both parameters can be evaluated using the law of mixtures [5]:

$$\rho_{eff} = \phi\rho_2 + (1 - \phi)\rho_1. \tag{2}$$

For suspensions of a high solid concentration the effective density is given by Johnson and Plona [6] as:

$$\rho_{eff} = \frac{\tau\rho_1(\phi\rho_2 + (1-1/\tau)(1-\phi)\rho_1)}{(1-\phi)\phi\rho_2 + (\tau - 2\phi + \phi^2)\rho_1} \tag{3}$$

where τ is the tortuosity.

ULTRASONIC SLIPCAST INTERFEROMETRY

A schematic view of the equipment is shown in Fig. 1. The acoustic waves propagating through the slipcast system will encounter three different boundaries, viz: the mould-cake, the cake-supsension and the suspension-air boundary, resulting in three different reflections (r) in succession as can be seen on the screen of the oscilloscope. By adjusting the reflection at the mould-surface to zero, the position of the reflections detected is a measure for the thickness of the different subsystems.

$$\begin{aligned} h_c(t) &= r_c(v_c/v_a) \\ l_s(t) &= (r_s - r_c)(v_s/v_a) \\ h_s(t) &= (r_s v_s - r_c v_s + r_c v_c)(1/v_a) \end{aligned} \tag{4}$$

where the subscripts c, s and a identify the cast, suspension and the adjusted sound velocity and t is the time.
The velocity of the sound waves through the cake cannot be estimated because of the unknown density of the cake. However, using the law of conservation of mass within the slipcast system in conjuction with the equation relating sound velocity and solid concentration in suspension (1)-(3), the

unknown sound velocity through the cake and the cake density can be evaluated. By measuring the reflection at the cake-suspension and the suspension-air boundary, the development of the cake thickness as well as the solid concentration in the cake can be calculated.

EXPERIMENTS

Slip casting is carried out on a porous ceramic mould with the suspension in a teflon cylinder with an inner diameter of 30 mm. The ceramic mould, developed in our laboratory, has a porosity of 47% and a mean pore size of 0.2 µm. The slurries, with a solid concentration of 30 vol % were prepared from Alcoa, A16SG alumina powder after upgrading the powder by classification through sedimentation. The mean particle size amounts 0.43 µm with 95 vol % of the particles smaller than 1.0 µm.
Particle-particle interaction was adjusted using nitric acid in the concentration range of 0.04 mmol up to 0.15 mmol per gram of powder. The suspensions were prepared by mixing in an attritor for three hours and aged until the pH was stabilized.

RESULTS AND DISCUSSION

For all the suspensions the evolution of the square of the thickness of the cast is given in Fig. 2. All samples exhibit the typical filtration behaviour, that is, a linear relationship between the square of thickness of the cake and time. Increasing the acid concentration results in an increasing growth rate caused by a decrease of the density of the cake as can be seen in Fig. 3. At an acid concentration between 0.039 and 0.075 mmol per gram of powder a maximum cake density of 69% is evaluated.

CONCLUSIONS

Ultrasonic interferometry showed to be an excellent technique to study cake formation during slipcasting.
The evaluated cake density decreases with increasing acid concentration in the suspension. A maximum density is evaluated between 0.039 and 0.075 mmol acid per gram powder.

ACKNOWLEDGEMENT

Financial support from the National Program on Technical Ceramics IOP-TK is gratefully acknowledged.

REFERENCES

1. Aksay, I.A., Schilling, C.H., in Ultrastructure processing of ceramics, glasses and composites, Eds. L.L. Hench, D.R. Ulrich, J. Wiley & Sons, New York, 1984, p. 439.

2. Vink, W.P.J., Versluis, N.H.R., Niet-Destructief Onderzoek, D.U.M. Delft, 1980.

3. Barrett-Gultepe, M.A., Gultepe, M.E. and Yeager, E.B., Coal Source Technol. 9 (1985) 441.

4. Howe, A.M., Mackie, A.R. and Robins, M.M., J. Dispersion Sci. Technol. 7 (1986) 231.

5. Ulrick, R.J., J. Appl. Phys. 18 (1947) 983.

6. Schwartz, L. and Plona, T.J., J. Appl. Phys. 55 (1984) 3971.

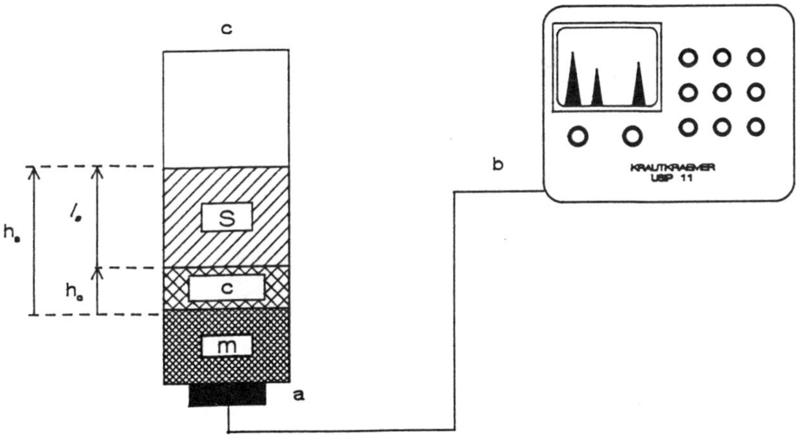

Fig. 1. Ultrasonic slipcast interferometer, consisting of a pulser receiver -4 MHz- (a), an oscilloscope (b) and a slipcast set up (c) with the subsystems; m: the mould, c: the cast and s: the suspension.

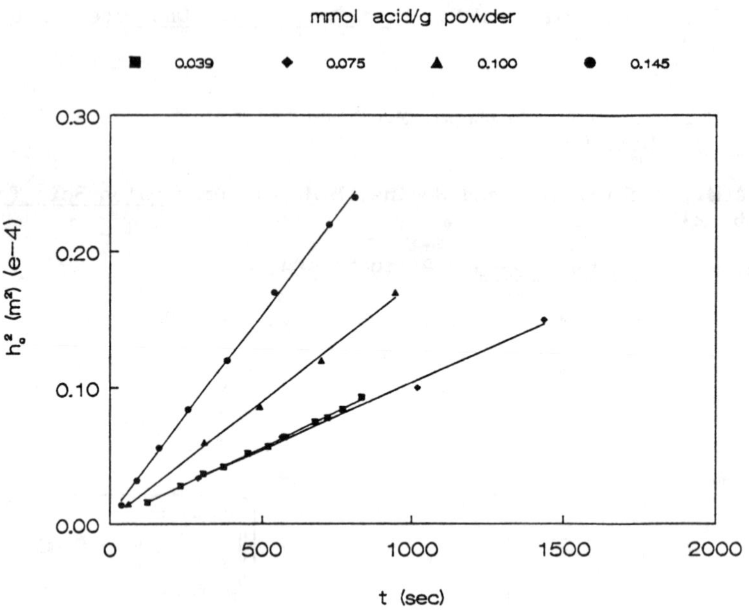

Fig. 2. Evaluation of the cake thickness during slipcasting for four different suspensions.

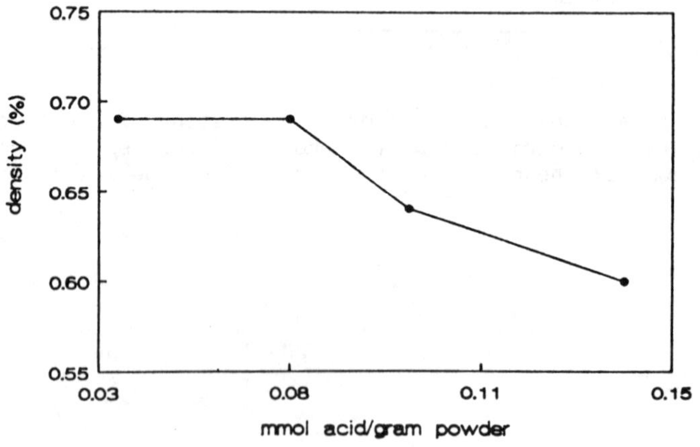

Fig. 3. The evaluated density of the cake formed during slip casting as a function of the acid concentration used for stabilizing the suspension.

VISCOSITIES AND COLLOIDAL FILTRATION PROPERTIES OF BIMODAL ALUMINA SUSPENSIONS

A.P. PHILIPSE and R.H.J. JONKER
Netherlands Energy Research Foundation, ECN,
P.O. Box 1, 1755 ZG Petten, the Netherlands

ABSTRACT

It is shown that the viscosity of bimodal alumina suspensions can be substantially lower than the viscosity of monomodal suspensions at the same concentration. This lowering is ascribed to an increase in packing fraction of the suspensions, which also explains qualitatively a trend in filter compact density as a function of suspension composition.

INTRODUCTION

The working hypothesis underlying this communication is that suspensions or pastes of ceramic particles with a multimodal size distribution are more interesting for ceramic shaping than their monomodal counter parts. Several arguments can be given to support this hypothesis.

Mixing particle size distributions leads to a multimodal system [1] which may have a substantially lower viscosity than the separate monomodal systems at the same overall solid concentration. This effect can be exploited, for example, in the processing of higly concentrated pastes which would have an unworkable high "monomodal viscosity". Further, multimodal systems can be employed to increase the density of green products. Ideally, voids in a particle packing are progressively filled up with increasingly smaller particles. In the third place, the presence of a fine powder component in the multimodal system may improve the sinteractivity of green products.

In this preliminary report some first results are discussed of ECN-investigations into the processing of bimodal suspensions, prepared from two types of commercial alumina powders. The results show that the viscosity of bimodal suspensions can indeed be significantly lower than viscosities of monomodal suspensions at the same solid concentration. A qualitative explanation of this viscosity lowering is given, which is supported by colloidal filtration experiments.

EXPERIMENTAL

Two types of alumina powder were used: DX-206 with an average particle size of about 6 μm and PN-202 with an average particle size of about 1.5 μm (specifications by the manufacturer, Martinswerke). Hereafter the powders will be referred to as D and P, respectively. For both powders aqueous stock suspension were prepared (ball milling for two hours) with solid volume fractions ϕ = 0.2, 0.3 and 0.4. Powder D was mixed with demi-water only, whereas for powder P 0.7 weight percent deflocculant (Dolapix CE-64) was added. The pH was adjusted to 9.7 for all suspensions with HNO_3. Bimodal suspensions where prepared by adding a P-suspension in various amounts to a D-suspension of the same concentration, until 50 (w/w)% P-particles, and the other way around for a higher P-content. Thus suspensions were obtained with a constant total volume fraction ϕ, but with different compositions ϕ_P/ϕ_D ($\phi = \phi_P + \phi_D$). After this addition suspensions were stirred for 30 sec. and viscosities were measured with a calibrated, rotating viscosimeter (Brookfield viscosimeter, spindel 4) operating at 50 rpm. Colloidal filtration experiments were performed by pouring suspensions, again after stirring, in a cylinder placed on a flat, porous alumina mold. The travelling distance of the upper suspension level was recorded as a function of filtration time. Solid volume fraction ϕ_c of final filter cakes were calculated from $\phi_c = V.\phi/V_c$, where V is the initial suspension volume and V_c is the final cake volume. Further information on this type of filtration can be found elsewhere [2].

RESULTS AND DISCUSSION

Fig. 1 shows representative examples of plots of suspension viscosity versus suspension composition ϕ_P/ϕ_D. It appears that the viscosity has a minimum, in the range ϕ_P/ϕ_D = 0.1 - 0.2, when the total volume fraction exceeds ϕ = 0.2. Note that for ϕ = 0.4 the minimum viscosity is more than an order of magnitude smaller than the viscosity of the separate D- and P-suspensions (Fig. 1). We observed the same minimum in duplo experiments, and a similar pronounced minimum for bimodal suspension of A16SG and a coarser CTB3SG powder (Alcoa). In ref. [3] a much less pronounced viscosity minimum is mentioned for a bimodal silica system, but no further explanation is given.

The occurrence of viscosity minima is a consequence of the concentration dependence of the viscosity at high concentration, as will become clear in the following consideration (see also ref. 1).

Consider the viscosity of a suspension of (polydisperse) particles at a given shear rate. Any plausible functional relation between the viscosity η and the concentration ϕ will have at least two features. 1) At low ϕ it reduces to a linear form:

$$\eta/\eta_o = 1 + \alpha\phi, \tag{1}$$

where α is a constant and η_o is the solvent viscosity. 2) The viscosity becomes infinite when ϕ approaches its maximum value $\phi = \phi_m$, which is attained when the particles in the suspension form a close-packed structure. Mooney's equation [4, 5]

$$\ln(\eta/\eta_o) = \alpha\phi/(1-\phi/\phi_m), \tag{2}$$

for example, has both features. The precise form of the concentration dependence, however, is not important here.

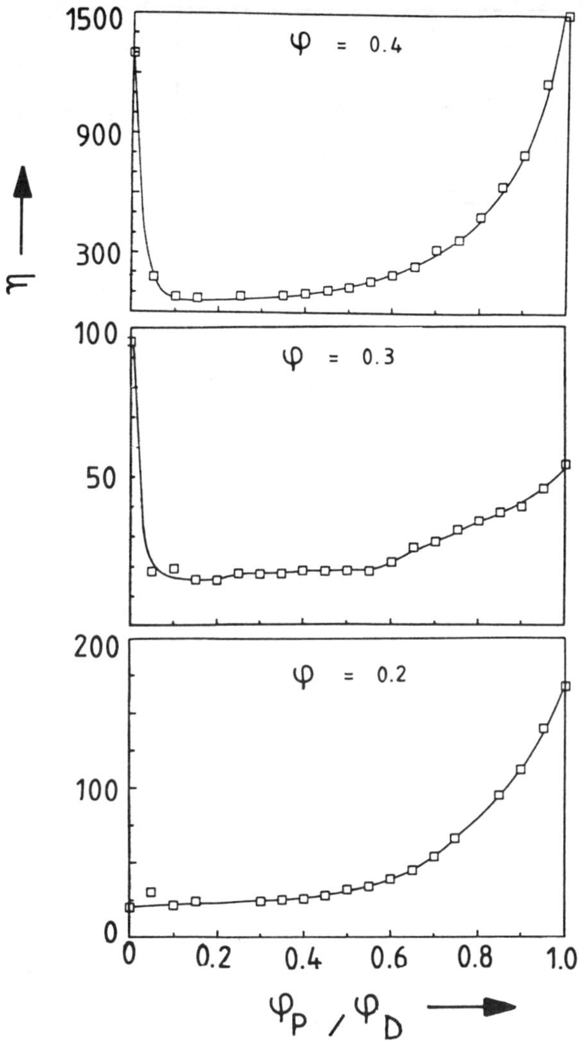

Fig. 1. Viscosity η in cp (at 50 rpm) versus volume ratio ϕ_P/ϕ_D of small alumina particles P and large alumina particles D at various total volume fractions ϕ = 0.1, 0.3 and 0.4.

It is clear from an equation like eq. (2) that an increase in the packing fraction ϕ_m at constant ϕ reduces the viscosity. But this is precisely what happens upon mixing small and large particles, because ϕ_m of the mixture is larger, and hence the viscosity is lower, than for the separate components. This lowering of the viscosity with respect to the "monomodal" values, leading to minima as in Fig. 1, will only be significant if ϕ is large enough. At small ϕ the viscosity is insensitive to changes in the packing fraction ϕ_m (eq. 1) and, hence, to changes in the suspension composition. This explains why there is no minimum in Fig. 1 for ϕ = 0.2. Nevertheless, for ϕ = 0.2 the viscosity is not constant, but gradually increases going from a pure D- to a pure P-suspension. The difference in viscosity of the monomodal suspensions must be due to factors such as shape and size of particles or aggregates, leading to different values of α in eq. (1).

To elucidate the role of the packing fraction in more detail, experiments are needed on mixtures of non-interacting, monodisperse spheres. For such

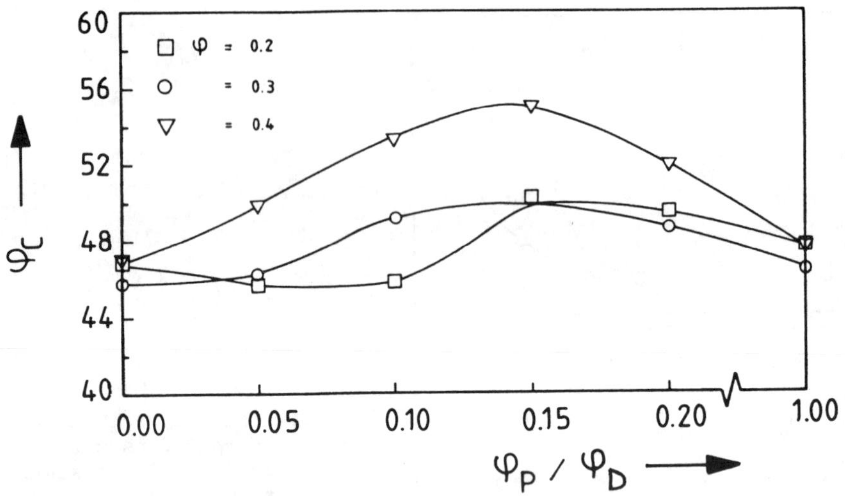

Fig. 2. Filter compact volume fractions ϕ_c versus suspension composition ϕ_P/ϕ_D.

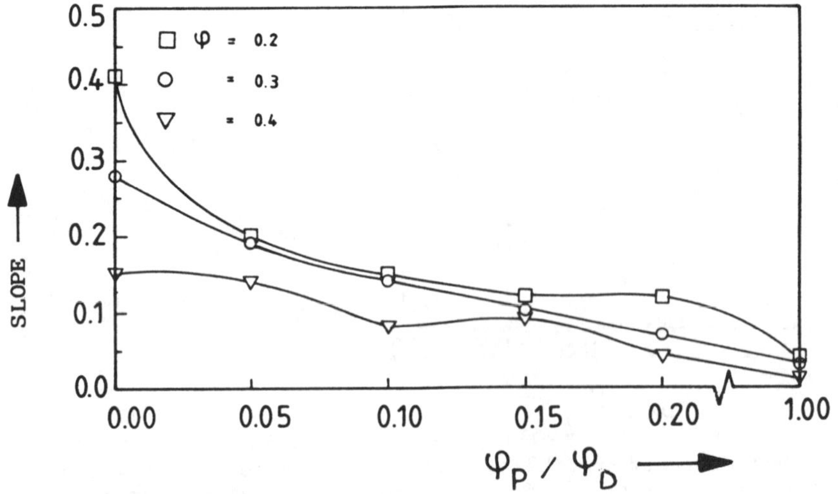

Fig. 3. Slope of h(t) versus $t^{\frac{1}{2}}$, in arbitrary units, as a measure of compact permeability (see eq. 3), versus suspension composition ϕ_P/ϕ_D.

mixtures plots as in Fig. 1 must be horizontal at low ϕ, whereas a minimum should be observed at higher ϕ at the suspension composition which would lead to the highest packing fraction ϕ_m.
For the powders in this study this composition apparently lies somewhere in the range ϕ_P/ϕ_D = 0.1 - 0.2 (Fig. 1). Therefore one expects that suspensions with a composition in this range form relatively dense filter compacts, as is indeed the case (Fig. 2). The absolute values of compact densities in Fig. 2 are not very high. However, the important point is that the increase in compact density is correlated to the decrease in suspension viscosity, both being a consequence of changes in ϕ_m.

The suspension composition also affects the compact permeability k and thus the filtration kinetics as described by [2]:

$$h^2(t) = \frac{k}{j}\frac{\Delta P}{\eta_o}\, t, \qquad (3)$$

where h(t) is the decrease in suspension level during filtration, j is a constant [2] and ΔP is the (constant) pressure drop across the compact. Fig. 3. shows the slope of h(t) versus $t^{\frac{1}{2}}$ plots (not shown here) as a function of the suspension composition. It is seen that the compact permeability decreases as the fraction of small particles increases, most likely because flow channels for liquid become smaller (see also ref. [3]). One would expect a local minimum in the permeability in the range ϕ_P/ϕ_D = 0.1 - 0.2, where the compact densities are highest. This only seems to be the case for ϕ = 0.4, but not for the other concentrations, for which we have no explanation at the moment.

CONCLUSIONS

The viscosity of a bimodal alumina suspension, with a concentration higher than $\phi \sim 0.2$, is lower than the viscosities of the separate monomodal suspensions. This is due to the relative large packing fraction ϕ_m of the bimodal systems, which also explains why filter compact densities are relative high at suspension compositions for which viscosities are low.
An interesting practical consequence is, that viscosity measurements can be used as a simple test to determine the suspension composition for a given set of powders which leads to the highest green product density.

ACKNOWLEDGEMENT

Trudy Zomerdijk is thanked for typing the manuscript.

REFERENCES

[1] Lee, D.I., "Packing of Spheres and its Effect on the Viscosity of Suspensions", J. Paint Technology 42, 579 (1970).

[2] Philipse, A.P., Bonekamp, B.C. and Veringa, H.J., "Colloidal Filtration and (Simultaneous) Sedimentation of Alumina and Silica Suspensions, Influence of Aggregates", submitted to J. Amer. Ceram. Soc.

[3] Gauthier, F.G.R. and Danforth S.C., "Packing of Bimodal Mixtures of Colloidal Silica", Ceramic Transactions, Vol. 1, Ceramic Powder Science II, Ed. G.L. Messing and E.R. Fuller, Amer. Ceram. Soc. (1988).

[4] Krieger, I.M. and Dougherty, T.J., "A mechanism for Non-Newtonian Flow in Suspensions of Rigid Spheres", Transactions Soc. Rheology III, 137 (1959).

[5] Mooney, M., "The Viscosity of a Concentrated Suspension of Spherical Particles", J. Coll. Sci. 6, 162 (1946).

OPTIMIZATION OF SLIP-CASTING TECHNIQUE FOR SHORT-FIBRE AND WHISKER-REINFORCED CERAMIC COMPOSITES

J. Lehmann, B. Müller and G. Ziegler
Deutsche Forschungsanstalt für Luft- und Raumfahrt (DLR)
Institut für Werkstoff-Forschung
D-5000 Köln 90, FRG

ABSTRACT

The effect of specially selected nonionic surfactants on the rheological characteristics and on the particle mobility is discussed. Based on this optimization the mechanical dispersion can be reduced or even eliminated. Furthermore, the advantage of pressure-assisted slip casting in the system Al_2O_3/SiC is demonstrated.

INTRODUCTION

Slip casting is an essential technique to produce complex shaped ceramic components. In the scope of developing whisker and fibre-reinforced ceramic composites this technique becomes more and more important in solving the problem of dispersion without damaging the reinforcing component.

With respect to the technological aspects, processing can be divided into two groups: whisker and chopped fibre-reinforced materials on the one side and continuous fibre-reinforced composites on the other side. In the case of whiskers and chopped fibres, the matrix material and the reinforcing components are the solid content in the slurry. To solve the problem of a simultaneous dispersion of matrix and reinforcing components, a specific combination of organic aids is necessary. In the case of continuous fibre-reinforced composites, fibre yarns or bundles have to be covered with slurry which contains the matrix material.
The goal of this work is to improve the slip-casting technique by using nonionic surfactants.

PHYSICO-CHEMICAL ASPECTS

In the present investigation a combination of nonionic surfactants in an aqueous medium has been developed. In addition, different waxes with binding effects were introduced into the slurry in the emulsified state. The emulsifying and dispersing properties of surfactants are shown in figure 1. If there is a solid particle or a wax or oil droplet, a quasi-hydrophilic colloidal particle is formed in any case.

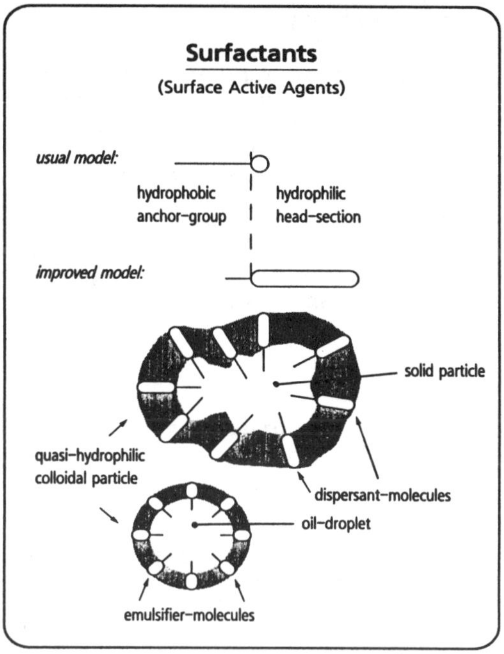

Figure 1. Emulsifying and dispersing properties of surfactants.

The overdosing of organic aids is a well-known problem in preparing slurries with a high solid content. One explanation of this behaviour is that surfactants can form micelle structures at a certain concentration. This value is called critical micelle concentration (CMC). The undesired intrinsically viscous behaviour of slurries is mostly due to the formation of mesophases with an aid concentration above CMC.
A promising method to investigate these surfactant and particle concentration effects and the dispersion stability is to determine the particle mobility using acoustophoresis (1).

The advantage in using nonionic surfactants is to obtain steric stabilization of various kinds of dispersions. The most important effect in this context is the formation of a disaggregation force by the spreading tension of the adsorbed surfactant layers on the particle surfaces (2). This surface chemistry could separate agglomerates or aggregates into single particles (3). Thereby an excellent dispersion and homogeneity can be achieved without any mechanical forces. Thus, in the case of coated whiskers or chopped fibres, their homogeneous introduction into the slurry is possible without any damage. To form complex shaped parts, however, low viscosity slurries with newtonian behaviour containing whiskers or chopped fibres are required. Therefore, an exact adjustment of the surfactant combination is needed to guarantee a non-thixotropic behaviour in the slip. This is easily possible with a controlled stress test under increasing and decreasing shear stresses as demonstrated by own experimental results with a Rheolab viscosimeter*.

TECHNOLOGICAL ASPECTS

Table 1 shows the improvement of slip-casting technique by applying nonionic surfactants. Homogeneously dispersed slurries formed due to repulsive interparticle forces by a steric stabilization give a high packing density in comparison with the flocced slurries as also proved by (4). In order to reduce the suction-time under atmospheric pressure (> 10 hrs) in the case of whisker and chopped fibre-containing slurries pressure assisted slip casting is advantageous. Pressure assisting in the range of 3 - 10 bars results in an improved drainage in the plaster-mould and thereby in the acceptable suction time of about 4 hrs.

TABLE 1
Optimization of Slip-Casting Technique (Al_2O_3)

	starting-level	after optimization
concentration of organic aids [w/o] (related to solid content)	10	3
solid content [w/o]	50	70-80
rheological behaviour	intrinsically viscous	newtonian
viscosity range [mPa s]	depending on shear stress	40-60 (related to raw material)
suction-time [h] (in plaster mould):		
monolithic/whisker-reinforced	≤ 15	1
short-fibre-reinforced	≤ 20	4
green density [%th.d.]	36-45	65
green strength	low, brittle	improved machinability

* courtesy Physica

MICROSTRUCTURE

An interesting point of view in forming whisker or short-fibre-reinforced parts is the orientation of the reinforcing components during feeding the mould. By optimizing the gate section of the plaster mould, composites with a certain fibre orientation can be achieved. As an example, Figures 2 and 3 show orientated and homogeneously distributed whiskers and chopped fibres in an Al_2O_3-matrix.

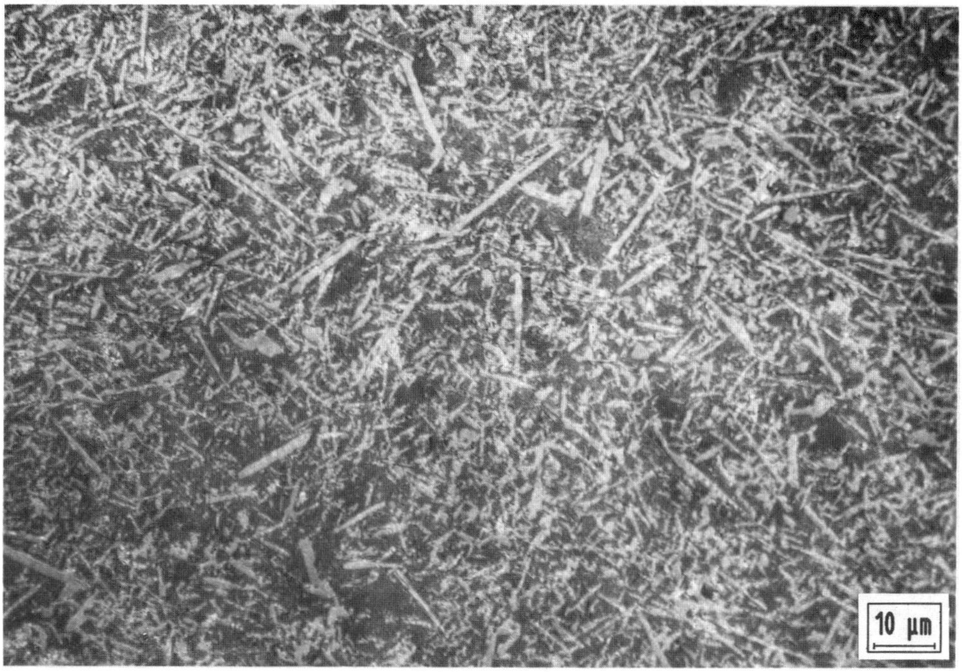

Figure 2. SiC-whisker-reinforced Al_2O_3 (hot-pressed, 40 vol.% SiC-whisker)

CONCLUSIONS

Covering the particle surfaces by nonionic surfactants leads to a masking of the original surface characteristics of the particles and to a steric stabilization of the slurries. Thereby disaggregation forces are formed to reduce or even substitute mechanical dispersion. In addition, tailoring of particle distributions by combining various raw materials can be used to improve the reproducibility and homogeneity of the microstructure of the green compact and the sintered component. The measurement of particle mobility seems to be a promising test to optimize the selection and amount of organic aids.

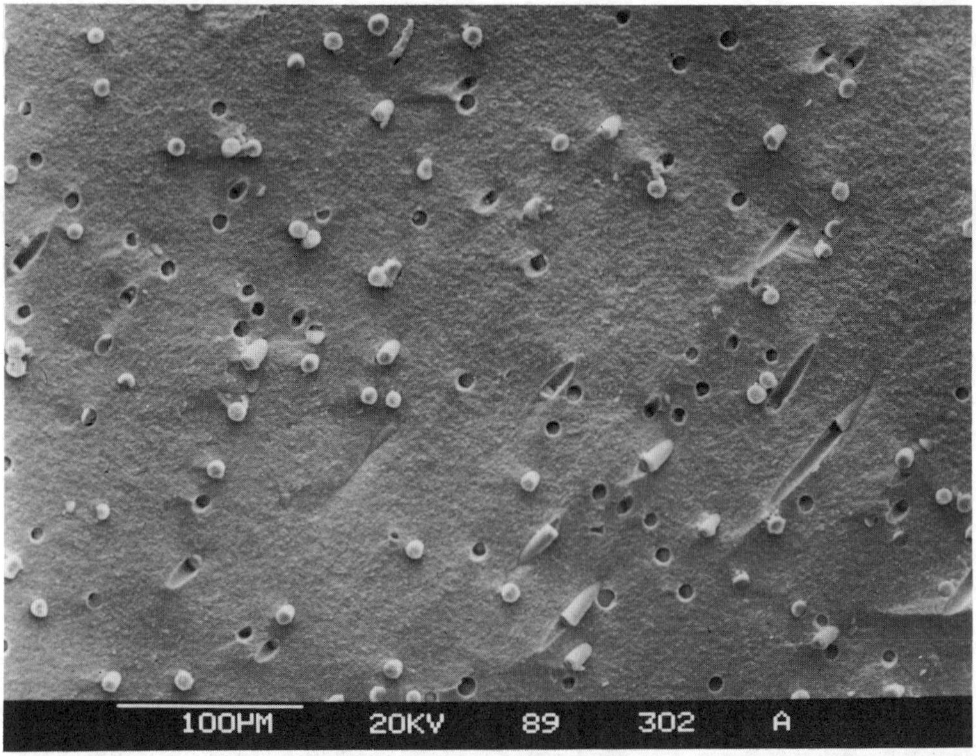

Figure 3. Short-fibre-reinforced Al_2O_3 (SiC-Fibres)

REFERENCES

1. Marlow, B.J. and Fairhurst, D., Colloid Vibration Potential and the Electro Kinetic Characterization of Concentrated Colloide. Langmuir, 4, 611, 1988.

2. Heusch, R., Grundlagen zur Emulgierwirkung von Tensiden. In Tensid-Taschenbuch, ed. H. Stacke, Carl-Hauser-Verlag, München, 1981, pp. 171-223.

3. Lange, H., Zur physikalischen Chemie der Tenside. In Waschmittelchemie ed. Henkel + Cie GmbH, Alfred Hüthig Verlag, Heidelberg, 1978, pp. 39-54.

4. Lange, F.F., Powder Processing Science and Technology for Increased Reliability. J.Am.Ceram.Soc., 1989, 72 [1], 3-15.

NON-AQUEOUS SLIP CASTING OF ALUMINUM NITRIDE

D.Bernhardt, A.Kranzmann, G.Petzow
Max-Planck-Institut für Metallforschung
Institut für Werkstoffwissenschaft
Pulvermetallurgisches Laboratorium
7000 Stuttgart 80, FRG

ABSTRACT

Non-aqueous AlN slurries were produced and rheologically examined. Green bodies were consolidated by colloidal filtration. Pressureless sintering of Y_2O_3 doped green bodies resulted in theoretical densities of more than 99 %. An AlN body with 2 wt.-% oxygen content achieved a thermal conductivity of 145 W/mK. This high value, compared with CIPed and sintered material, is due to an improved of microstructure by slip casting.

INTRODUCTION

Aluminum Nitride is well known as a material with good thermal conductivity. Complex shapes can be produced by use of slip casting. Furthermore, slip casting leads to an improved microstructure of the green bodies by avoiding microscopic inhomogeneities. This results in improved thermal conductivity of the sintered parts.

EXPERIMENTAL PROCEDURE AND RESULTS

Slip Composition and Rheology

The slurry (70 wt.% solids) was produced from:

Powder: AlN Grade C, (H.C. Starck Berlin)
CaO or Y_2O_3 as additives
Suspension vehicle: butylacetate or o'xylene
Deflocculants: Menhaden Fishoil, palmitic acid, palmitoleic acid, linolenic acid, linolic acid.

Relative contact angles were calculated for a variety of liquids by using the Washburn equation [1,2]. Butylacetate and o'xylene showed the best results.
Menhaden Fishoil and its three main components palmitic acid, palimtoleic acid, linolenic acid [3] and in addition linolic acid were tested in order to measure their effectiveness as flow modifiers. Palmitic acid showed no remarkable effect. The flow lines of the other substances are shown in fig.1.

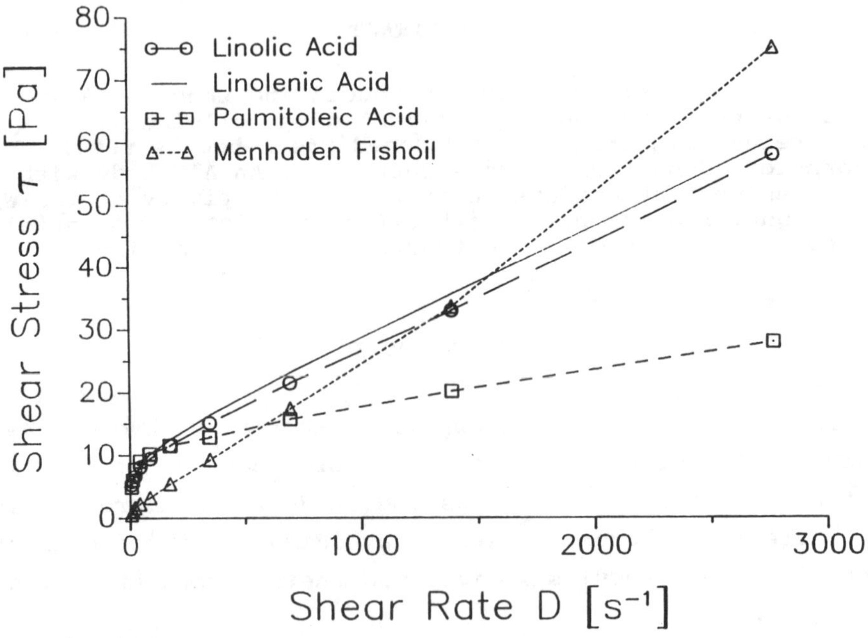

FIGURE 1. Flow properties of different deflocculants.

Only Menhaden Fishoil showed approximately Newtonian flow behaviour. Therefore it was choosen as deflocculant in the present experiments.
Compared with o'xylene as asuspension medium butylacetate demonstrated a significantly lower viscosity.

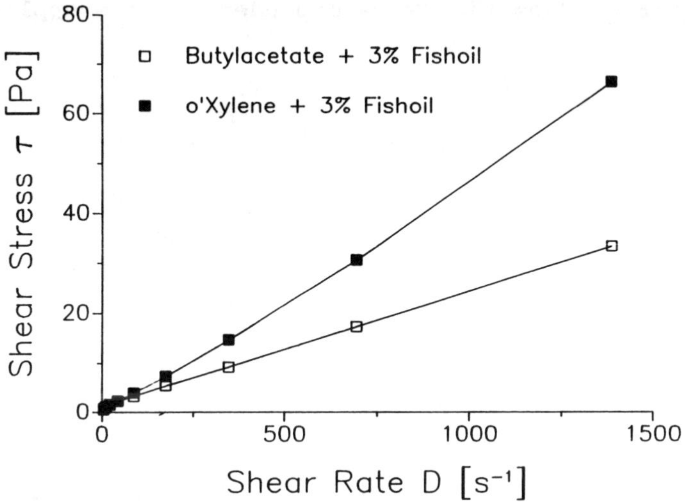

Figure 2. Flow properties of slurries with butylacetate and o'xylene as suspension vehicles.

Green Body Formation

The suspension was cast into a glass cylinder with a filter at the bottom. The vessel was pressurized with gas forcing the suspension vehicle through the filter [4]. The progress of green body formation was photographically documented.

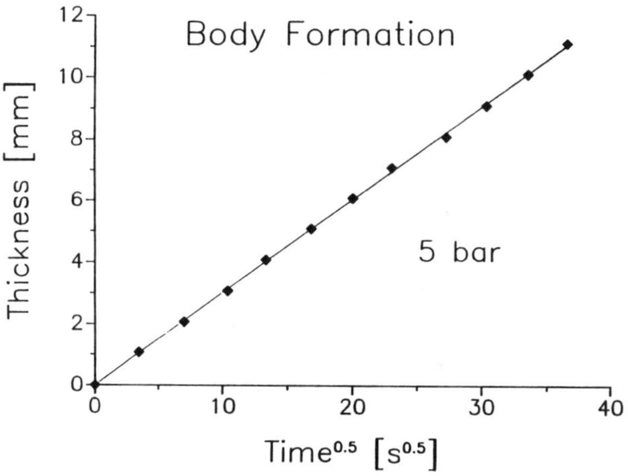

Figure 3. Green body formation at 5 bar load.

The achieved green densities were dependent on the applied gas pressure.

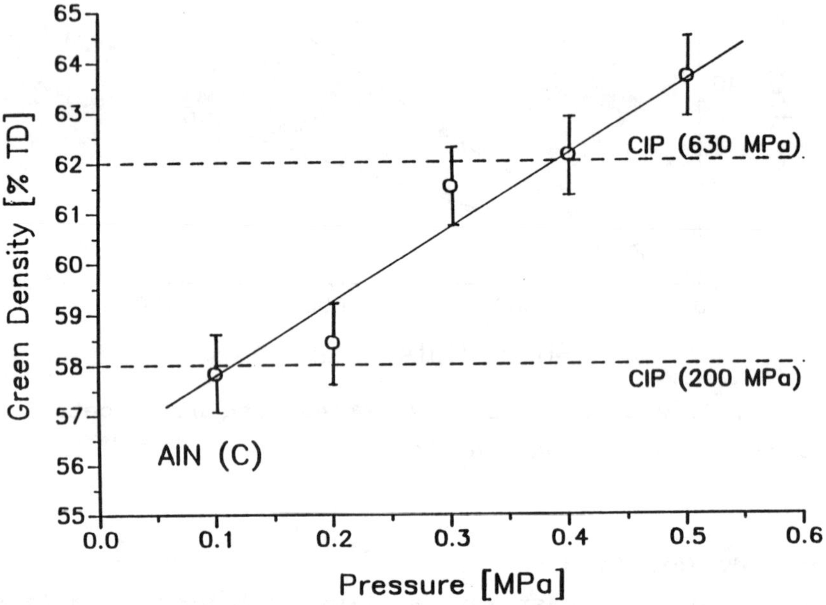

Figure 4. Comparison of green densities from cold isostatically pressed samples and pressure slip cast AlN specimens.

In CIPed samples there were local inhomogeneous areas.Slip cast specimens had a very homogeneous distribution of the powder particles.

Densification of Pressure Cast AlN with CaO or Y_2O_3 as Sintering Additives

The green bodies were pressureless sintered under nitrogen atmosphere within a temperature range of 1850 and 2000°C, debinding occured during a holding time at 500°C.

Y_2O_3 doped samples sintered to more than 99% TD. Use of CaO as sintering aid resulted in regularly dispersed 20 μm diameter pores. These pores can be attributed to a chemical reaction between CaO and fatty acids contained in the Menhaden Fishoil. This effect only occured in slip cast samples.

CONCLUSIONS

Pressure slip cast specimens doped with Y_2O_3 exhibited a remarkably high thermal conductivity. Its value of 145 W/mK with an oxygen content of 2 wt.-% is a result of the improved microstructure.

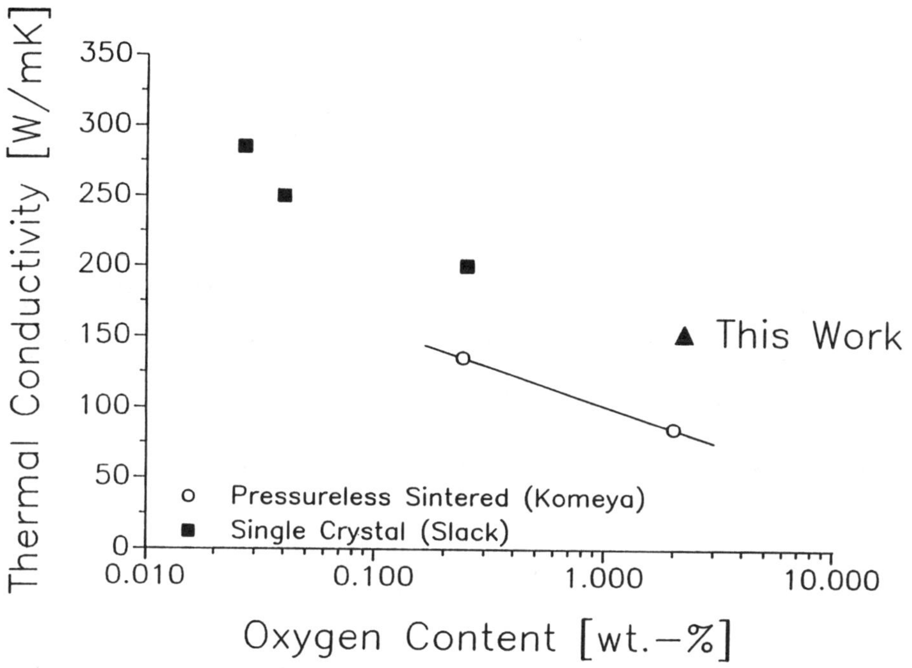

Figure 5. Thermal conductivity vs. oxygen content [5].

REFERENCES

1. Washburn, E.W., The Dynamics of Capillary Flow. Phys. Rev., **17** (3), pp. 374-375, (1921)

2. Bruil, H.G. and van Aartsen, J.J., The Determination of Contact Angles of Aqueous Surfactant Solutions on Powders. Colloid & Polymer Sci. **252**, pp. 32-38, (1974)

3. Handbook of Chemistry and Physics, ed. Weast, R.C..55th Edition, 1974-1975, p. D193

4. Velazquez, M. and Danforth, S.C., Casting of monodisperse Colloidal Silica. Adv. in Cer. IX, pp.105-114, (1984)

5. Data sheet from Tokuyama Soda, Japan

WET CONSOLIDATION OF FERRITES PREPARED BY PRECIPITATION.

A. Noordermeer and M.M.E. Vantilt
Philips Research Laboratories Eindhoven
P.O. Box 80.000, 5600 JA Eindhoven, The Netherlands

ABSTRACT

Powders of MnZn-ferrite were prepared with particle sizes between 0.03 μm and 0.33 μm and with a dispersity of 20-25%. These powders were consolidated by sedimentation/filtration of suspensions at different pH. The density of the compact strongly depends on the particle size, pH of the suspension and the counterion of the hydroxide. The density increases in the order NaOH, $(CH_3)_4NOH$ and $(C_4H_9)_4NOH$. Compacts with random packing and densities up to 63% have been realised.

INTRODUCTION

Ferrites are used for a wide range of magnetic applications. The increasing demands on the performance of these materials requires a better control of the properties and therefore a better control of the processing of the material. To achieve this, powders with a controlled particle size and size distribution as well as homogeneous and dense powder compacts are necessary. The powder preparation method we have used is a direct precipitation method [1]. Compaction by dry pressing of particles in the submicrometer range generally leads to low densities and inhomogeneities due to strong agglomeration. The use of suspensions has been proven to be a good method for preparation of homogeneous and dense compacts of these powders [2,3]. In suspension one enables the particles to move more independently of each other as single entities. The principle of the method is to achieve a maximal repulsion between the particles. In that case one expects to obtain maximum packing densities. If agglomeration occures during the consolidation random packings of particles are formed and usually lower densities are obtained. Agglomeration occurs in ferrite suspensions because the magnetic attraction between the particles often dominates the repulsive forces. The way in which agglomeration influences the properties of the compact depends, amongst others, on the relative agglomerate size and on the ability of the agglomerates to rearrange or, stated otherwise, on the ability of the particles to slide along each other. Generally the packing density of a sediment increases with increasing stability of the particles in the suspension. In this investigation we used a combination of sedimentation and filtration as consolidation method.

EXPERIMENTAL

Consolidation.
The chemicals used were of p.a. quality (>99.5 %) ($(CH_3)_4NOH$ (TMAH) and $(C_4H_9)_4NOH$ (TBAH): Fluka Chemie). Suspensions were made with a volume content of 1% ferrite. Part of the liquid was exchanged by a solution of the desired hydroxide of pH between 7 and 13. These suspensions were sonified for 3 minutes. The suspension was brought into a tube with at the bottom a cellulose filter (0.22 µm pore size, type GS, Millipore) on top of a coarse glass filter. After partial sedimentation vacuum was applied under the filter to filtrate the liquid. The cakes were dried slowly first at 40-50 °C and then in a vacuum furnace at 200 °C for at least 24 hours.

Measuring methods.
Particle sizes of the powders were determined from SEM-photographs using image analysis methods. Particles sizes of the agglomerates were determined with a Micromeritics Sedigraph 5000D. The agglomerate sizes measured with this method should be considered as relative quantities. The suspensions for these latter experiments contained 0.1 vol% of ferrite. Densities were measured by mercury porosimetry with a Micromeritics Poresizer P9310. Electrophoretic mobilities were measured with a Zetasizer II of Malvern. Zeta potentials were calculated from the electrophoretic mobilities according to the procedures described by O'Brien and White [4].

RESULTS AND DISCUSSIONS

Effect of the pH.
The electrostatic repulsion between the particles can be controlled by variation of the pH with e.g. NaOH. In fig. 1 the median agglomerate diameter is plotted as a function of the pH. An estimate of the electrostatic repulsion can be obtained by the zeta potential which is plotted in fig. 3. The curves can be split into two parts. Above pH=7 the particles are being negatively charged by the added hydroxide. As a consequence repulsion between the particles increases with pH and the degree of agglomeration decreases. At pH>9 the concentration of the counterions becomes the dominating factor. An increase in the electrolyte concentration leads to a decrease in the repulsion and agglomeration becomes more important again. As a consequence a minimum occurs at a pH of about 9.5. The minimum size of the agglomerates is about 4-5 µm which is considerably larger than the size of the primary particles which is about

0.2 µm. Apparently the magnetic interactions between the ferrite particles are larger than the repulsion causing agglomeration.

The question is then how the properties of the compact are related to the pH of the suspensions. In fig. 2 the density of a compact is plotted as a function of the pH. One observes a behaviour which is similar to the agglomeration. The optimum density and the minimum agglomeration degree are around the same pH of 9.5. Increasing agglomeration at either lower or higher pH leads to lower densities of the compacts.

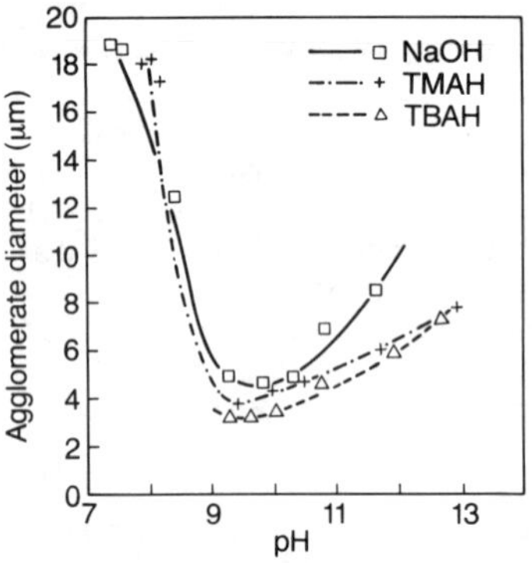

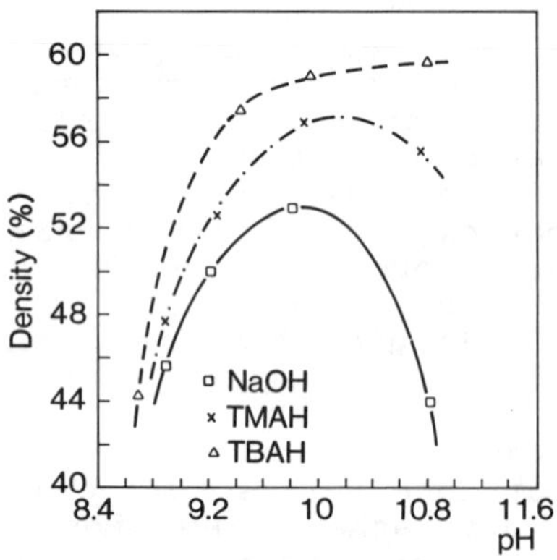

Fig. 1: Agglomerate diameter of particles in suspension vs pH for a primary particle size of 0.2 µm.
Fig. 2: Density of a compact vs pH of the suspension for NaOH, TMAH and TBAH, particle size 0.13 µm.

Effect of the counterions.

The effect of the counterion Na^+ at higher pH is the dominating factor in the obtained density. Therefore hydroxides with a larger counterion which have a lower polarizing capability e.g. TMAH and TBAH have been investigated. In fig. 2 for NaOH, TMAH and TBAH the density as a function of pH is plotted. First, the density increases considerably when using TMAH or TBAH. Second, for the different counterions the difference in the density at higher pH becomes more pronounced. The question is whether these results can be explained by electrostatic repulsion. If the zeta potential is plotted as a function of pH (fig. 3) we see that the curves of all three bases are at approximately the same level below pH=9. At higher pH the curve of TBAH is at a significant higher potential than TMAH. The potential one measures with the zeta potential is believed to be close to the potential in the Stern layer at a distance δ [5]. Two effects which act in the opposite direction, may play a role. Both are caused by the increase of size of the counterion in the order Na^+-TMA^+-TBA^+. First, the distance at which the counterions can approach the surface because of their finite size (assuming that specific adsorption does not occur), increases in the order NaOH-TMAH-TBAH leading to a larger potential drop over a distance δ . This leads to a lower zeta potential at a larger distance δ . Second if the size becomes larger a lower amount of counterions in the Stern layer may be adsorbed. This leads to less compensation of the negative surface charge in the Stern layer and therefore to a higher zeta potential at the distance δ . It depends on the relative size of these two effects how the zeta potential will behave. For pH<9 the zeta potential is of the same magnitude for all hydroxides but located at a larger distance δ in the order NaOH-TMAH-TBAH. Because also the attraction forces will be smaller at a larger distance the total potential will be less attractive. The decreased attraction in the order NaOH-TMAH-TBAH induces a decrease in agglomeration degree (fig. 1), which results in higher densities. At higher pH>9 the order in zeta potential is reversed for TMAH and TBAH. The order in agglomeration degree, however, remains the same as at lower pH indicating that also here the total potential is less attractive in the order NaOH-TMAH-TBAH.

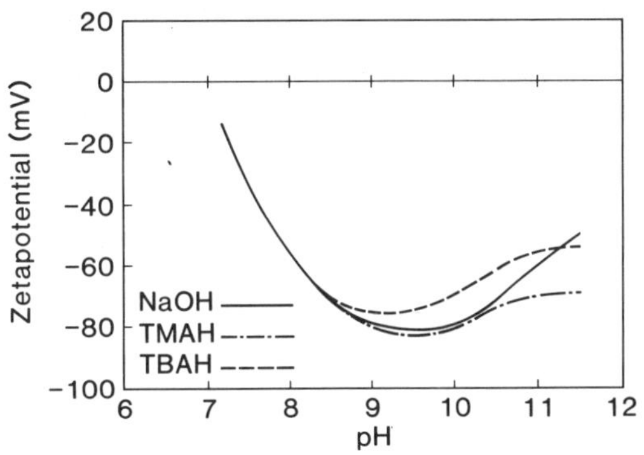

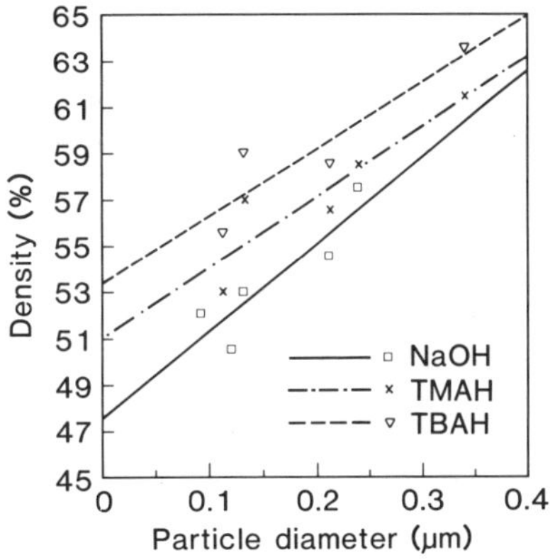

Fig. 3: Zeta potential vs pH of powder in fig. 2.
Fig. 4: Maximum densities from the density-pH curves as a function of particle diameter.

Effect of the particle size.
In fig. 4 we plotted the maximum densities from curves as in fig. 2 as a function of particle size. For every particle size the density increases in the order NaOH-TMAH-TBAH. Furthermore, the density increases with increasing particle sizes. This is probably due to a stronger agglomeration for the smaller particles. We have used the same amount of powder in the same volume and this means that with decreasing particle size the number of particles increases. This in turn results in a stronger possibility of agglomeration [6]. It should be noted that the maximum density for a random packing with equally sized spheres is theoretically about 64 percent [7] which for the largest particle size is almost reached.

Structure of the compacts.
If agglomeration occurs it might be possible to find the

structure of the agglomerates in the compacts. However, from SEM-photographs of compacts and from the pore size distributions as determined by mercury porosimetry any agglomeration structure cannot yet be detected. This indicates that the particles in these agglomerates are able to rearrange fairly easy during the final stage of sedimentation. Although (some) agglomeration is present, one is able to obtain densities with this method which are close to the maximum attainable.

CONCLUSIONS

The density of the compacts strongly depends on the pH and the type of cation of the hydroxide in particular the size of the cation which increases from Na^+ to TMA^+ to TBA^+. The effects can be explained largely by the behaviour of the electrostatic repulsion. A smaller particle size leads to an increasing agglomeration and therefore to lower densities. The maximum density for the largest particle size is close to the maximum attainable density for a random packing of equally sized spheres. The presence of agglomeration and the absence of the structure of agglomerates in the compacts indicate an easy rearrangement of the particles within these agglomerates.

Acknowledgement: We thank M. Severens for performing the electrophoretic mobility measurements and zeta potential calculations.

REFERENCES

1. Takada T., Kiyama M., "Preparation of Ferrites by Wet Method", Proc. Intern. Conf. on Ferrites, 69 (1970).

2. Pober R.L., Barringer E.A., Parish M.V., Levoy N. and Bowen H.K., "Dispersion and Packing of Narrow Size Distribution Ceramics Powders", Mater. Res. Sci., **17** 193 (1984).

3. Rhodes W.H., "Agglomerate and Particle Size Effects on Sintering Yttria-Stabilized Zirconia", J. Am. Ceram. Soc., **64**[1] 19-22 (1981).

4. O'Brien R.W. and White L.R., "Electrophoretic Mobility of a Spherical Colloidal Particle", J. Chem. Soc. Far. II, **74** 1607 (1978).

5. Overbeek J.Th.G., "Recent Developments in the Understanding of Colloid Stability", J. Colloid Interf. Sci., **58**[2] 408 (1977).

6. Sonntag H. and Strenge K., Coagulation Kinetics and Structure Formation, 1987, Plenum Press, New York/London.

7. Mason G.J., "A Model of the Pore Space in a Random Packing of Equal Spheres", J. Colloid Interf. Sci., **35** 279 (1971).

SLIP CASTING OF TiO_2

H. WOHLFROMM, P. PENA, J.S. MOYA, J. REQUENA
Instituto de Cerámica y Vidrio, C.S.I.C
28500 Arganda del Rey, Madrid, Spain

ABSTRACT

Rheology and casting behaviour of aqueous TiO_2-anatase slips were studied. Stable suspensions were obtained with Na-hexametaphosphate and Na-carboxymethylcellulose additions acting as deflocculant and binder, respectively. Drain and solid pieces with 51% green density were cast. Sintering at 1350 C yielded 97% dense rutile samples with 175$\pm$30 MPa bending strength.

INTRODUCTION

Uniaxial and isostatic pressing are commonly used for the shaping of advanced ceramics. Recently slip-casting techniques, normally associated with traditional ceramics, has been extended to the nonclay products (1,2).

Rutile ceramics are well known for its applications in electronics (3), but it is now being considered as a promising candidate for biomedical implants (4). Especially for the latter purpose, shaping of pieces with a complex geometry is an important requirement which justify studying the slip-casting route for TiO_2-ceramics. Nevertheless, only limited amount of work has been published dealing with stability and rheology of TiO_2-slips (5).

The aim of the present investigation is to obtain stable aqueous suspensions of an TiO_2-anatase powder, with low viscosity and adequate casting rate, and to cast pieces with elevated green density and strength with the final objective to obtain sintered rutile compacts with high density and mechanical strength.

EXPERIMENTAL PROCEDURE

A 99.9 pure TiO_2-anatase powder (Merck 808) with a specific surface area 9 m^2/g and average grain size of 0.4 μm was used as starting material. Na-hexametaphosphate (Nahmp) and Na-carboxymethylcellulose (Nacmc) were tested as additives being useful deflocculants for TiO_2-rutile powders (6). Aqueous slips prepared with solid/liquid ratios 70/30, 65/35, 60/40, 55/45, 50/50 by weight (37.7, 32.6, 28.1, 24.1, 20.7 vol.%) and Nahmp and Nacmc additions ranging from 0 to 0.4 wt.%. The corresponding viscosities were determined by a rotational viscosimeter (Haake Rotovisco RV20) at a constant temperature of 25 ^{0}C, providing a measure of both stability and castability of the suspensions. An adequate slip for the casting of both drain and solid pieces was selected based upon the criterion of low viscosity, high solid content, suitable casting rate and favorable drying behaviour.

The slips were then cast into plaster of Paris molds made with a water-to-plaster ratio of 70/100 by weight to form both drain cast pieces (crucibles, tubes, etc) and solid cast pieces (bars, cylinders, plates, etc). Wall thickness versus casting time was measured for drain casting. The cast bodies were carefully removed from the molds and dried at room temperature for 24h. Green densities were measured by the Archimedes method in mercury.

Suspensions with a 60/40 by weight solid-to-liquid ratio (28.1 vol% solids) were selected for casting because of their lower viscosity and high solids content, which results in a better-controlled process of wall thickness formation.

Cast specimens prepared from those suspensions were fired in air in the temperature interval 1250 to 1500^{0}C for times ranging from 2 to 8 hours.

The sintered pieces were characterized by density measurements (Archimedes method in distilled water) and its microstructure studied by reflected light optical microscopy (RLOM) on polished and thermally etched (1200 ^{0} C-18 min) specimens. The grain size was determined by the intercept line method. Bending strength were measured on rods of ~ 4.4 mm diameter using a 30 mm span and a crosshead speed of 0.005 cm/min.

RESULTS AND DISCUSION

SLIP CASTING

Fig. 1a shows the apparent viscosity of aqueous TiO_2-slips versus Nahmp concentration and solid content. Viscosity decreases with Nahmp addition up to a content of 0.2 wt.% but remains constant for Nahmp-content $\geq$ 0.2%. However, Nacmc, reported as deflocculant for rutile suspensions (6), shows a opposite effect on viscosity in the case of the anatase suspension examined in this study. Fig. 1b shows that viscosity is increasing with increasing Nacmc content.

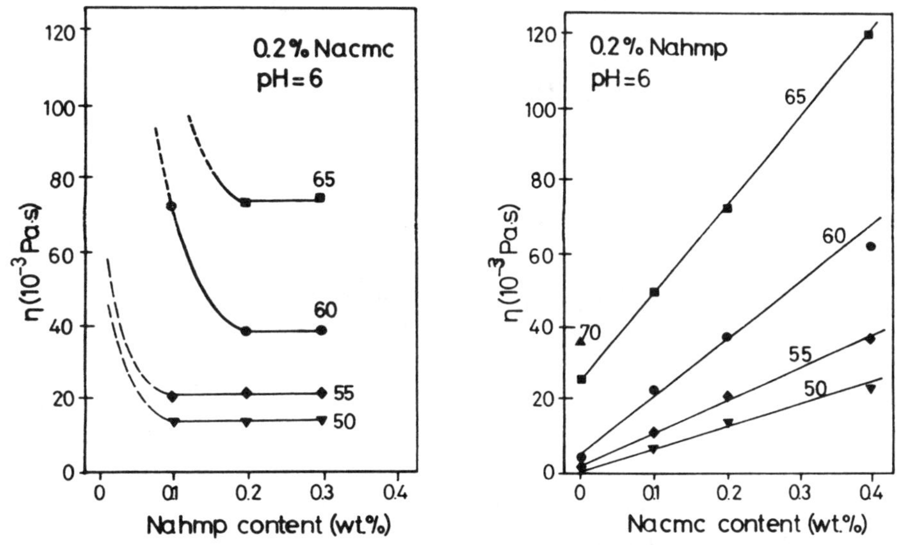

FIGURE 1. Effect of Nahmp (A) and Nacmc (B) additions on viscosity of aqueous TiO_2-slips for different solid contents (shear rate = 500 s^{-1})

Nevertheless, Nacmc was found to act as a binder enhancing the green strength of cast pieces (Table 1). In fact, only with Nacmc addition ≥ 0.2 wt.%, solid pieces could be cast.

TABLE 1.
Observations during drying of cast pieces

Nahmp (wt.%)	0.2	0.2	0.2
Nacmc (wt.%)	0	0.1	0.2
drain pieces	o.k.	o.k.	o.k.
solid pieces	cracked	cracked	o.k.

The flowcurves of TiO_2-slips containing 0.2% Nahmp and 0.2% Nacmc are showing low pronounced thixotropic cycles and pseudoplastic behaviour, Fig. 2. A suspension containing 60% by weight (28% by volume) of solids was choosen for slip casting since it represents an acceptable compromise between low viscosity and high solid content.

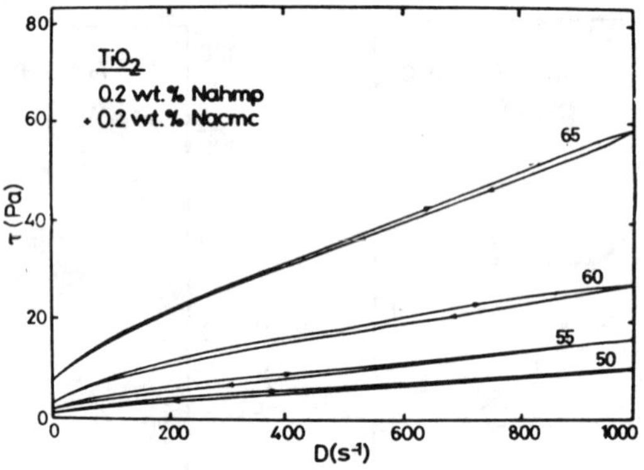

FIGURE 2. Flowcurves of aqueous TiO_2-slips containing 0.2wt.% of Nahmp + 0.2 wt.% of Nacmc at different solid contents.

The wall thickness versus casting time for this suspension is depicted in Fig 3, the casting rate of this suspension allows controlled casting of both drain and solid pieces.

Green pieces cast with this slip exhibit a green density of 51% and a bending strength of 1.2+0.3 MPa.

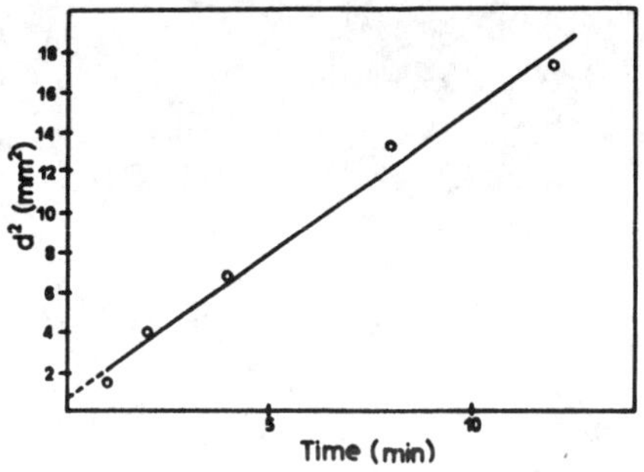

FIGURE 3. (Wall thickness)^2 versus casting time of an aqueous TiO_2-slip.

SINTERING

The sintering studies were performed with solid bars (60 mm length and 5 mm width) cast from the above described slip with 60% solid content and 0.2 Nahmp + 0.2 Nacmc addition. The relative densities of the specimens fired at temperatures ranging from 1250 to 1500^0 C are shown in Fig. 4. Fig. 5 displays the influence of soaking time on the obtained densities. A maximum 97% th density is achieved.

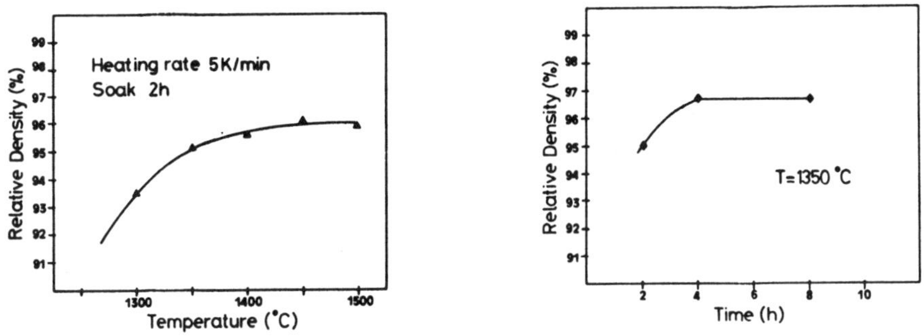

FIGURE 4. Relative density versus sintering temperature (A) and soaking time (B) of slip-cast bodies.

Specimens sintered at 1350^0 C for 4h are coarse grained (average grain size = 12 μm), displaying a bimodal grain size distribution and inter and intragranular pores (Fig. 5). The micrograph is suggesting that abnormal grain growth took place.

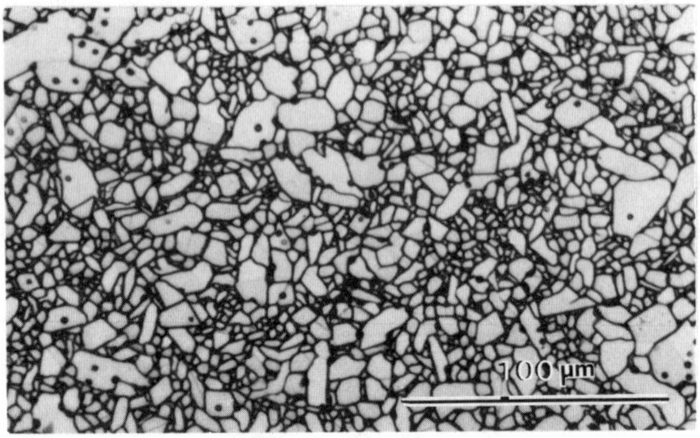

FIGURE 5. Optical micrograph of a specimen sintered at 1350 °C for 4h, after polishing and thermal etching.

A bending strength of 175±30 MPa was found for the as-sintered specimens. Weibull analysis of the strength data, Fig. 6, yielded a Weibull modulus of 9 which can be considered as quite acceptable.

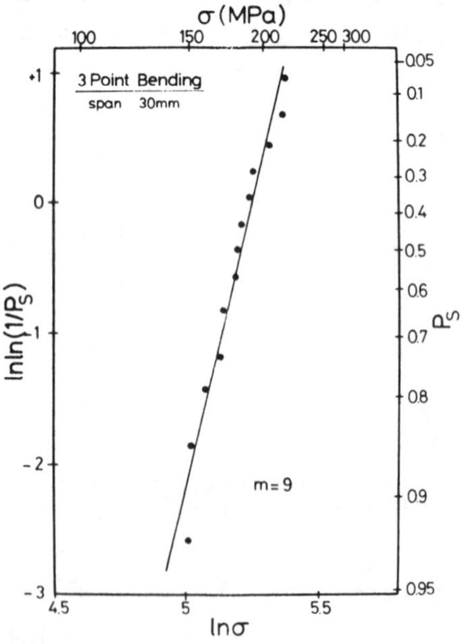

FIGURE 6. Weibull-plot of cast TiO_2-bars sintered at 1350 °C, 4h (P_S = survival propability). Mean strength = 175±30 MPa.

CONCLUSIONS

(1) Stable aqueous suspensions of TiO_2 can be prepared with a solid content of 60 wt.% (28 vol.%) using additions of 0.2 wt.% of Nahmp and 0.2 wt.% of Nacmc with viscosities as low as 37 10^{-3} Pas.

(2) Na-hexametaphosphate acts as an deflocculant and reduces the viscosity of the TiO_2-anatase suspensions whereas Na-carboxymethylcellulose improves the green strength of the cast pieces.

(3) Hollow and solid pieces can be cast from this suspension displaying a green density of 51% and a green strength of 1.2±0.3 MPa.

(4) Dense pieces (97% th) with 175 MPa. bending strength and reasonable good Weibull modulus m=9 were obtained after firing at 1350° C-4h.

REFERENCES

1. Adams, F., Slip-cast ceramics. In High Temperature Oxides, part IV, ed. Alper, A.N., , Academic Press, , New York, 1971, pp. 145-184.
2. Rado, P., Slip-casting of non-clay materials, Trans. J. Br. Cer. Soc. 1973, 72, 291-97.
3. Hill, D.C., Tuller, H.L., Ceramic Sensors: Theory and Practice, in Ceramic Materials for Electronics, ed. R.C. Buchanan, Marce Dekker, New York 1986.
4. Greim, J., Entwicklung einer Herzklappenprothese aus halbleitender Rutilkeramik, Ph.D. thesis, University Erlangen-Nürnberg 1985.
5. a) Rao, A.S., Effect of ph on the suspension stability of alumina, titania and their mixtures, Ceramics International, 1987, 13,233-241. b) Rao, S.A., Electrophoretic mobility of alumina, titania and their mixtures in aqueous dispersions, ibid. 19888, **14**, 71-76.
c) Rao, S.A., Effect of surface active agents on the electrokinetic and rheological behavior of alumina and titania slips, 1988, **14**, 141-146.
6. Efrenov, I.F., Kovylov, A.E., Safonova, V., Cumoxov, A.A., Colloidal and technological properties of TiO_2 hydrodispersions, Zh. Prikl. Khim. 1985, **58**, 1641-43.

ACKNOWLEDGEMENTS

Financial support from CICYT, Spain, under project number MAT88-0156 is acknowledged. One of the authors (H.W.) thanks the Commission of the European Communities for the award of a fellowship.

ON THE FORMING OF CERAMIC COATINGS BY CONCENTRATION POLARIZATION IN DEAD-END AND CROSS FLOW PRESSURE FILTRATION

B.C. BONEKAMP, R.A. TERPSTRA and H.J. VERINGA
Netherlands Energy Research Foundation, ECN
P.O. Box 1, 1755 ZG Petten, The Netherlands

ABSTRACT

In the filtration and membrane literature much work has been done to describe the phenomenon of concentration polarization, i.e. the forming of particle layers (cakes, compacts) on a filter surface due to a pressure drop across the filter. Both dead-end filtration and the practically more important cross flow pressure filtration have been dealt with. This note is an attempt to transfer some filtration knowledge to ceramic science and to discuss some aspects of the forming of particle cakes by static and tangential flow pressure filtration which are important for the application of colloidal filtration in ceramics.

INTRODUCTION

In membrane technology one of the goals is to prevent the build up of particle layers on the membrane (filter) surface in pressure filtration processes in order to enhance the permeate flux as much as possible [1]. In ceramic technology the phenomenon of cake build up (concentration polarization) can advantageously be used for the forming of ceramic coatings on a porous support [2]. After controlled drying and sintering such coatings may find application as a ceramic membrane or for example as a component in a solid oxide fuel cell composite material.
From the cross flow pressure filtration literature it appears that the cake properties such as thickness and particle composition depend on both the applied pressure and the velocity of the feed suspension.
Besides, the composition and properties of the casting suspension can be changed during the filtration process. This will have its repercussion on the properties of the coating (cake) formed. For example composite (multi) layer coatings could be prepared in this way. Cross flow pressure filtration (CFPF) in principle offers therefore extra possibilities for the control of cake formation and microstructure besides the normal (dead-end) colloidal pressure filtration. We think that the in ceramic processing new technique of CFPF can be exploited by ceramic process engineers in the future for the production of advanced coatings on porous substrates. In the

following some aspects of cake filtration static and dynamic (cross flow) of importance for the preparation of ceramic coatings will be briefly discussed.

STATIC COLLOIDAL PRESSURE FILTRATION

Static colloidal filtration (see fig. 1) is well known in ceramic science. See for example the treatments of Aksay [3] and Tiller [4] on slip casting and Lange [5] on pressure casting (filtration). Here we deal only with colloidal pressure filtration, which is called ultrafiltration for colloidal particles smaller than about 100 nm and microfiltration otherwise by membrane technologists.

For static filtration the flow J across the filter (membrane) can be written as:

$$J = \frac{\Delta P}{\eta(R_f + R_c)} \tag{1}$$

in which ΔP is the transmembrane pressure, η the solvent viscosity and R_f and R_c the hydraulic resistance of the filter and cake respectively.

When the pressure drop across the membrane is negligible compared to that across the cake and the cake has a uniform permeability it is straight forward to derive from (1) that the cake thickness δ varies with time t as:

$$\delta(t) \sim t^{\frac{1}{2}} \tag{2}$$

There are several cases in which deviations from (2) can occur, for example

1. in case of small colloidal particles, back diffusion takes place due to the concentration gradient between cake and bulk
2. in case of cross flow filtration, particles are removed from the cake by processes like erosion.

In all cases (1) changes to [8]:

$$J = \frac{\Delta P}{\eta\ (R_f + R_{c1} - R_{c2})} \tag{3}$$

where R_{c1} and R_{c2} are the cake resistances due to particle deposition and back transport respectively.

For the usual slip casting suspensions case 2 will be dominating when CFPF is used and this case is considered in more detail in the next section.

Consolidation of the particles in the cake ("trapping") occurs when the repulsion force F_E between the particles is smaller than the contact force F_c due to pressure difference across the cake. Sherwood [6] showed that for a cubic array of monodisperse spheres and a DLVO type electrical double layer interaction consolidation occurs at a depth $x = 2\ nd$, with n the particle layer number (n = 1 is the cake surface) and d the particle diameter, where

$$n = \frac{\varepsilon\ K\ \psi_0^2\ (1-\phi_c)^3}{6\ \pi\ \eta\ U \cdot \phi_c} \tag{4}$$

in which ε is the permittivity of the solvent, K the reciprocal Debye length, ψ_0 the surface potential, ϕ_c the volume fraction particles in the cake and U the fluid velocity normal to the cake surface.

Hence for filtration velocities normally encountered and a usual ψ_0 of

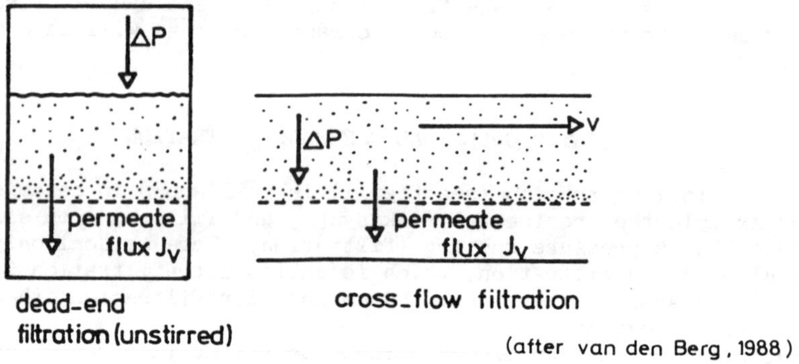

Figure 1. Schematical representation of dead-end (static) and cross-flow filtration.

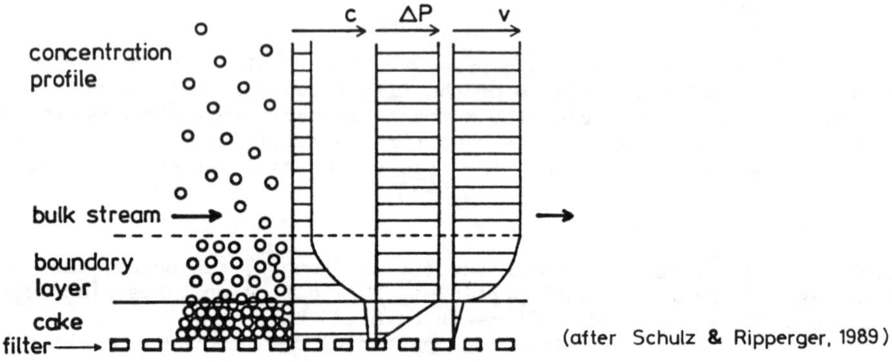

Figure 2. Polarization of particle concentration (c); pressure difference ΔP and fluid veclocity v at a membrane surface with cross flow pressure filtration (CFPF).

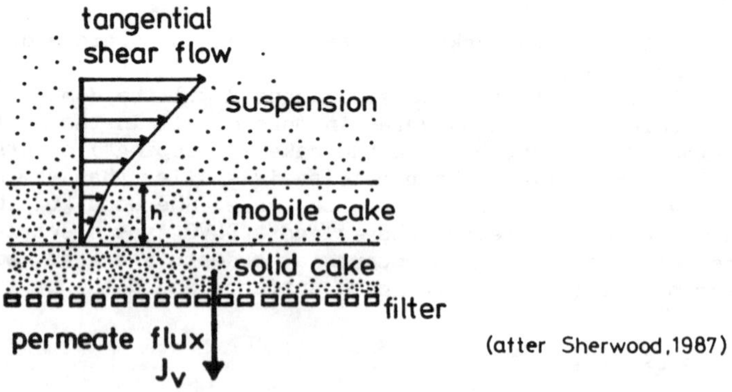

Figure 3. The filter cake can consist of two parts: a consolidated ("solidified") inner layer and a mobile zone adjacent to the suspension. Hydrodynamic instabilities at the interface cake-suspension may play a role in the back-transport of particles besides drag forces when a tangential flow is applied [7, 10].

about 50 mV liquid-like outer layers seem to be possible [7]. The consolidation front growths with $t^{\frac{1}{2}}$ just as the cake surface in the case of non-diffusive dead-end filtration. The forming of a solid like layer as described here is similar to the forming of a gel layer in ultrafiltration [8]. Whether the outer layer is liquid like or not is of course important for the possible ways of back transport.
In the case of (electro)steric stabilization a deep (primary) minimum in the particle interaction potential is absent and the wet coatings are probably liquid-like throughout. This may have (undesired) consequences for the handling of wet coatings formed from such suspensions.

TANGENTIAL COLLOIDAL PRESSURE FILTRATION

The filtration situation is depicted in fig. 1 and 2. Now back transport plays a role for all particle sizes and a balance is achieved between back transport and the convective transport towards the filter leading in principle (i.e. for monodisperse suspensions) to a steady state layer thickness and composition.
When diffusion is the main mechanism for back transport it can be shown [9] that the steady state flux J_s is given by:

$$J_s = \frac{D}{\delta} \ln \frac{c_c}{c_b} \tag{5}$$

where δ is the polarized particle layer thickness and c_c the particle concentration in the cake.
For larger particles (microfiltration) non-diffusive back transport mechanisms of a hydrodynamic origin are dominating of which scour (erosion) is the most important [7, 10].
Schulz and Ripperger [10] derived the following relation for the steady state flux for the case of back transport by particle erosion in the case of turbulent tube flow:

$$J_s = \sqrt{\frac{\xi \cdot \Delta P(\rho_c - c_b)\rho_l}{a_c \cdot c_b \cdot \rho_c}} \cdot \frac{v}{\eta} \tag{6}$$

where ξ is a back transport parameter, ρ_c is the density of the retained material, and ρ_l that of the solvent, a_c is the specific hydraulic resistance of the coating and v is the tangential velocity. This eq. predicts that the steady state solvent flux is proportional to the tangential velocity. The steady state layer thickness δ_c is given by:

$$\delta_c = \sqrt{\frac{\Delta P \cdot c_b \cdot \rho_c}{\xi \cdot a_c(\rho_c - c_b)\rho_l}} \cdot \frac{1}{v} \tag{7}$$

and it is seen that the layer thickness δ decreases with v^{-1} and that δ increases with $\sqrt{\Delta P}$. Hence the layer thickness can be controlled by variation of v. Furthermore it follows from $\delta \sim \sqrt{\Delta P}$ that control of δ by the transmembrane pressure becomes ineffective at high ΔP values. Baker et al. [11] found that for CFPF of TiO_2 suspension $J_s \sim v^{0.6}$ and $\delta \sim v^{-2}$. This deviation from the theoretical relations is probably due to the polydispersity of the TiO_2 suspensions.

The cause of the cake erosion is by Schulz and Ripperger assumed to be the drag force on a particle due to the cross flow velocity v. This force is larger for larger particles and hence in the case of a particle size distribution in the casting suspension the particle size distribution in the cake will shift to smaller particle sizes [10] with time the more the larger the cross flow velocity. This is observed indeed by Baker et al. [11]. Erosion of the cake can also be induced by hydrodynamic instabilities which can occur on the interface between a fluid like cake and the suspension (see fig. 3) [7]. It is out of the scope of this paper to discuss this back transport mechanism here further.
Finally we will remark that besides the application of a cross flow field for controlling the properties of a coating formed by pressure filtration also the application of an electric field for the introduction of a back transport mechanism belongs to the possiblities [12].

REFERENCES

1. Belfort, G., Fluid mechanics in membrane filtration: Recent developments, J. Membr. Sci. , 1989, 40, 123-147.

2. Hsieh, H.P. Bhave, R.R. and Fleming, H.L., Microporous alumina membranes, J. Membr. Sci., 1988, 39, 221-241.

3. Aksay, I.A. and Schilling, C.H., Mechanics of colloidal filtration, Adv. Ceram. 3, 1984, 9, 85-93.

4. Tiller, F.M. and Chun-Dar, T., Theory of filtration in ceramics: I,Slip Casting. J. Am. Ceram. Soc., 1986, 69, 882-887.

5. Lange, F.F. and Miller, K.T., Pressure filtration: consolidation kinetics and mechanics, Am. Ceram. Soc. Bull., 1987, 10, 1498-1504.

6. Fane, A.G., Ultrafiltration of suspensions, J. Membrane Sci., 1984, 20, (ref. 5), pp 250.

7. Sherwood, J.D., Erosion and instability of a colloidal filter cake, Europhysics. Lett., 1987, 4, 1273-1278.

8. Aimar, P., Taddei, C. Lafaille, J-P. and Sanchez, V., Mass transfer limitations during ultrafiltration of cheese whey with inorganic membranes, J. Membr. Sci., 1988, (ref 9, 10), pp 210.

9. Van den Berg, G., Concentration polarization in ultrafiltration, Thesis University of Twente, the Netherlands, 1988, pp 18.

10. Schulz, G. and Ripperger, S., Concentration polarization in cross flow microfiltration, J. Membr. Sci., 1989, 40, 173-187.

11. Baker, R.J., Fane, A.G., Fell, C.J.D. and Yoo, B.H., Factors affecting flux in cross flow filtration, Desalination, 1985, 53, 81-93.

12. Radovich, J.M., Behnam, B. and Mullon, C., Steady state modelling of electro-ultrafiltration at constant concentration, Separation Sci. Technol., 1985, 20, 315-329.

PORE PROPERTIES OF CERAMIC POROUS MEDIA PREPARED BY SLIP CASTING OF STARCH FILLED α-ALUMINA SUSPENSIONS

B.C. BONEKAMP, M.J. SCHOUTE and M.J.A.A. GORIS
Netherlands Energy Research Foundation, ECN
P.O. Box 1, 1755 ZG, Petten, The Netherlands

ABSTRACT

Porous alumina ceramics are formed by slip casting of starch filled α-alumina suspensions.
Pore properties of the macrovoid structure (in a dense alumina matrix) obtained are studied by means of bubble pressure porometry, Hg porosimetry, optical microscopy and SEM. The homogeneity of the macropore structure depends on the degree of flocculation of the alumina suspension. The total porosity of the porous bodies increases linearly with the volume % starch with respect to the total solids volume.
A percolation threshold for a connective void-throat network is observed between 15 and 20 % (v/v) starch depending on the suspension properties.

INTRODUCTION

Large macropores (10-100 μm) in a porous ceramic material can be obtained by the use of burnout fillers such as carbon black, graphite and woodflower [1,2,3] or the use of a sublimable compound such as melamine [2]. Usually extrusion or pressing are used as forming methods.
We prepared macroporous disks with large (2-100 μm) macrovoids by slip casting of alumina suspensions filled with potato starch particles. We studied some properties of the macrovoid structure in an otherwise dense alumina matrix as a function of the starch content originally present. It was found that the homogeneity (no porosity gradients) of the macro void structure can be controlled by adjustment of the colloidal stability of the suspensions. As expected a percolating pore structure of a characteristic void/throat type is formed above a threshold content of starch in the compact.
The porous ceramic structures described here may find application for example as a support material for ceramic membranes, in bioceramic applications and as an isolator material. Besides, the results indicate that the material is interesting for the experimental testing of theories concerning fluid transport and -imbibition/drainage in porous media.

EXPERIMENTAL

Macroporous ceramic disks (ø ≈ 37 mm, d ≈ 3 to 5 mm) were prepared by slip casting potato starch (AVEBE) filled α-alumina (A16SG Alcoa) suspensions on plaster moulds. The degree of flocculation of the suspensions (without starch), made bij ball milling, was controlled by the addition of poly-electrolyte deflocculant and the mixing/aging time and checked by measuring The Brookfield viscosity as a function of spindle (nr 3) rotation speed (rpm).
Starch powder was carefully added to the suspensions by means of a sieve (≈ 0.8 mm) under stirring. The stirring was continued for about 30 min. and then the slip casting of disks was started. The wet disks were carefully dried (to avoid cracking) at room temperature. Then they were slowly calcinated in air at 573-773 K in order to burn out the starch and finally the porous disks were sintered at 1873 K for 1 hour.
The overall porosity of the disks was determined from the dimensions and the weight. Further characterization was done with He pycnometry and Hg porosimetry using Micromeretics equipment, bubble point/gas flow-pressure porometry on samples initially completely saturated with a wetting fluid using a Coulter Porometer [4] and microscopy (optical and SEM) on polished transversal sections of the porous disks.
The particle size distribution of the starch was measured in dilute aqueous suspensions with Fraunhofer diffraction, using a Malvern instrument. The density of the starch was taken as 1.47 $g.cm^{-3}$ [7].

RESULTS AND DISCUSSION

The macroporous flat disks obtained were without major defects (cracks) as judged from optical (OM) and electron microscopy (EM). The distribution of voids in the slip cast direction is visible from optical micrographs (see fig 1.) The volume average void diameter distribution obtained from OM is very similar to the starch particle size distribution. In both cases the d50 ≈ 50 μm and hence it seems that d(void) ≈ d(filler particle) which is in contrast with results of Veselov et al [3] who found that for coke particles in a porous alumina matrix d(void) ≈ ½ d(filler particle). The dense matrix in our case is the origin of this difference. It was often observed that the voids were not completely empty, but contained a particle aggregate. This is possibly caused by alumina particles from the suspension which may be present in the starch particles. At high starch contents (> 30% (v/v)) a homogeneous void distribution was observed in all cases. At low starch content (< 15% (v/v)) a homogeneous distribution was only observed in the case of the strongly flocculated suspension a (see fig 2).
However, a strong porosity gradient was present with the deflocculated suspension c.
The cause of this phenomenon is the upward sedimentation (segregation) of the starch particles during slip casting which can take place in colloidally stable suspensions, but not in flocculated suspensions with a sufficiently high yield value. A characteristic example of the microstructure showing two neighbouring voids and the dense alumina matrix is shown in Fig. 1.3.
The total porosity of the porous disks increases linearly with the volume content starch originally present with a slope of 0.9 (see Fig. 3.) for all three suspensions. This indicates that only a small relative shrinkage of the voids with respect to the total volume solids occurs during sintering. Just as in the case of carbon black filled polymers, where a percolation

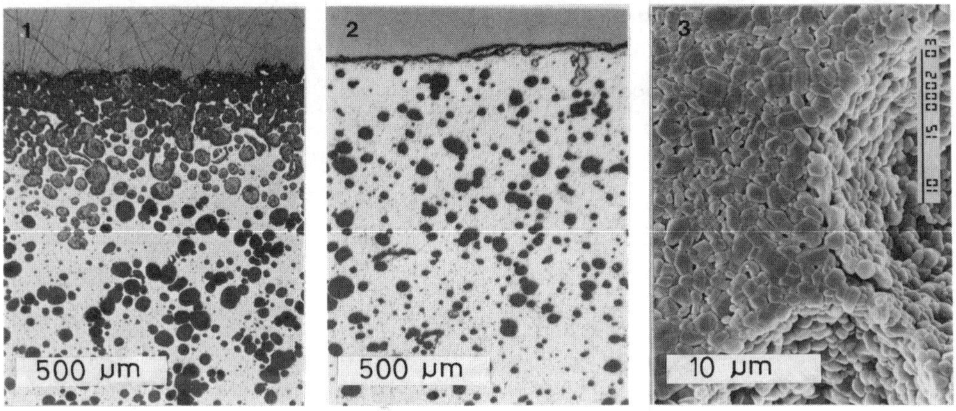

Figure 1. Optical micrographs of epoxy impregnated and polished sections (in the slip cast direction) of the starch derived porous ceramics at a volume fraction pores of 10% for suspension c (1) and suspension a (2) (see fig. 2). The suspension side surface of a disk is visible. (3) shows a SEM of an etched section. The microstructure of the dense matrix (with only closed pores) and two macrovoid are visible.

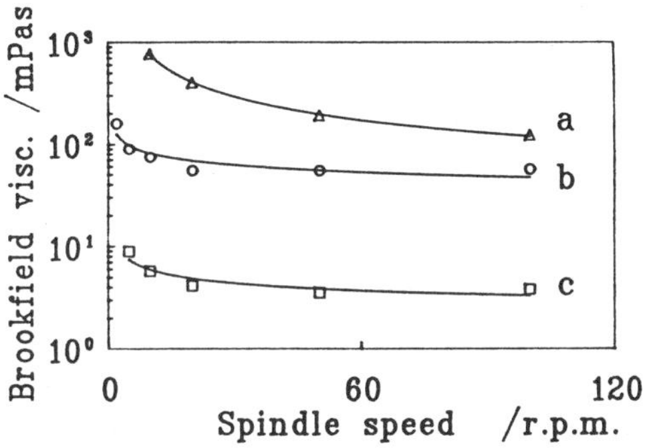

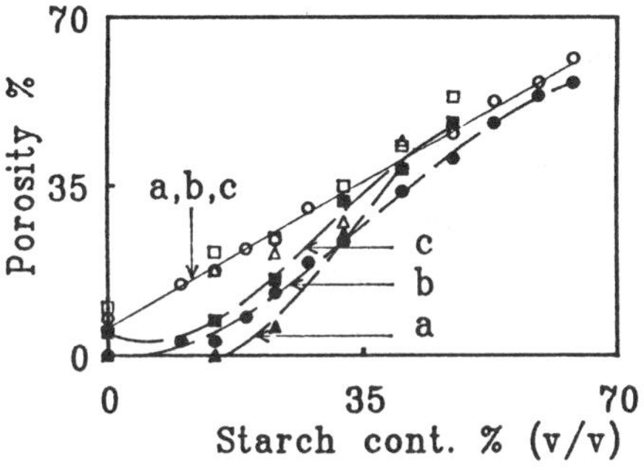

Figure 2. Brookfield viscosity of the alumina suspensions as a function of the shear rate for three suspensions used a,b,c with the same solids volume fraction of 0.3 but different degree of deflocculation.

Figure 3. Total porosity (—) and open porosity (--) (Hg porosimetry) of sintered disks as a function of the volume % starch with respect to the total volume solids in the casting suspensions.

threshold for conduction exist [5], a percolation threshold for fluid permeability (intrusion) is expected in our case.
The sudden appearance of open porosity as measured with Hg porosimetry (see fig 3.) and the appearance of a bubble point only above a starch content of about 15% (v/v) (see fig 5.) are indicative for this.

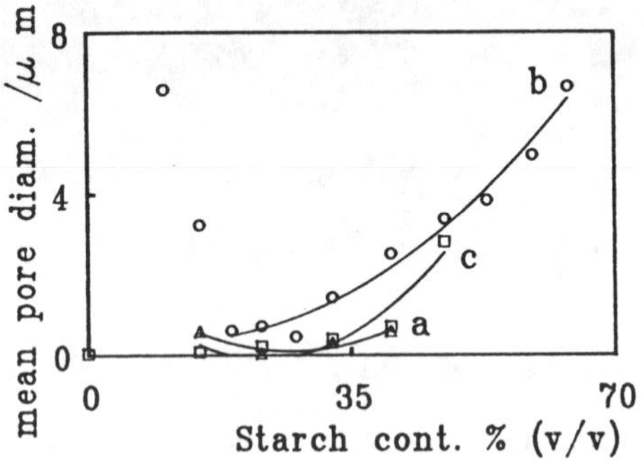

Figure 4. Mean pore diameter as measured with Hg porosimetry (cylindrical pores assumed) as a function of the starch content.

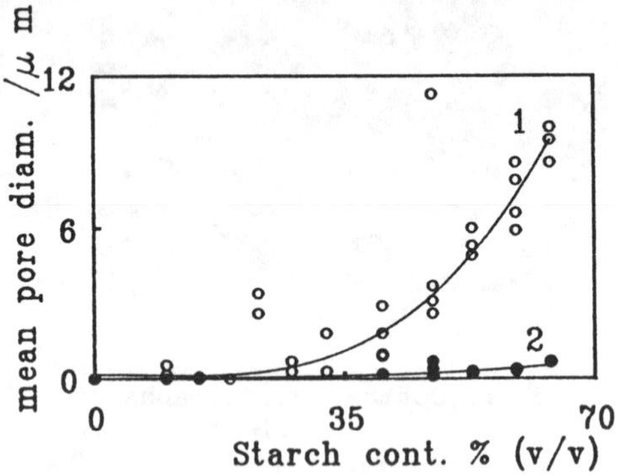

Figure 5. Bubble point pore diameter (1) and mean (flow) diameter (2) of the starch derived macro porous ceramic as a function of the starch content (susp b.)

With He pycnometry a density minimum also around 15% (v/v) starch was found. So below the threshold value merely closed non connected macrovoids and small void clusters exist and above the threshold the voids (clusters) become interconnected and macroscopically percolating. At a high volume fraction of voids the open porosity is within experimental accuracy the same as the total porosity and the fraction non-connected voids must be very small (see fig. 3). The threshold observed is more obvious (sharper) and at somewhat higher total porosity in the absence of void gradients as expected. (Fig. 3).
The void-network formed above the threshold value is of a typical void-throat type as evidenced by Hg porosimetry hysteresis (not shown) and by SEM. The latter showed throat diameters in the range 2-10 µm; an order of magnitude smaller than the void size.
Hg porosimetry data and bubble point (break through pressure) measurements analyzed with the usual cylindrical pore model give mean pore sizes in the range of a few µm too (see fig 4 and 5). The distributions obtained in this way are throat-size distributions [6] weighted in some (unknown) way according to the void sizes. Hence more advanced pore models are needed in order to obtain realistic distributions from the data.
The throat size seems to increase with the starch content (Fig. 4/5). An explanation for this must probably be sought in the changing throat/void network topology with increasing total porosity, but an increase in contact area between starch particles at higher starch concentration in the compacts is not excluded as an explanation. A quantitative model is clearly

needed here. Finally Fig. 5.2 shows also that the mean throat diameter as measured with the Coulter porometer is much smaller than the Hg throat diameter. This shows that on the average the narrowest constriction in the percolating pathways is rather small as expected.

ACKNOWLEDGEMENTS

AVEBE is thanked for their gift of a potato starch sample. Financial support from the National program on Technical Ceramics IOP-TK is gratefully acknowledged.

REFERENCES

1. Hammon, U. and Kotter, M., Herstellung von Formkörpern mit definierten porenstructur, Chem. -Ing. -Tech., 1984, 56, 455-463.

2. Trimm, D.L. and Stanislaus, A., The control of pore size in alumina catalyst support: A review., Appl. Catalysis, 1986, 21, 215-238.

3. Veselov, V.V., Levanyuk, A., Savinykh, V.V., Kushnarenko, T.I. and Morgunova, E.T., Burnout additive (petroleum coke) as a factor in regulating the porous structure of a catalyst support, Kinetika i Kataliz, 1973, 16, 1002-1008.

4. Wenman, R.A. and Miller, B.V., A novel automatic instrument for pore size distributions. In Particle Size Analysis 1985, ed. P.J. Lloyd, J. Wiley & Sons Ltd., 1987, pp 583-590.

5. Hsu, W.Y., Holtje, W.G. and Barkley, J.R., Percolation phenomena in polymer/carbon composites, J. Mater. Sci.Lett., 1988, 7, 459-462.

6. Dullien, F.A.L., Porous media fluid transport and pore structure, Academic Press New York etc., 1979, pp 98.

7. Van den Berg, C., Vapour sorption equilibria and other water-starch interactions; a physico-chemical approach. Thesis Agricultural University Wageningen, 1981, pp 77.

THE EFFECT OF THE GREEN MICROSTRUCTURE ON SINTERING OF ALUMINA

L.A. CORREIA, M.J.A.A. GORIS and H.J. VERINGA
Netherlands Energy Research Foundation, ECN,
P.O. Box 1, 1755 ZG PETTEN, the Netherlands

ABSTRACT

In this paper the influence of agglomerates, present in the powder as well as in the suspension, on the compaction behaviour of Al_2O_3, A16SG, suspensions and the densification of the slip casted compacts is discussed. It is shown that hard agglomerates present in the powder are mainly responsible for the rest porosity of the densified compacts. A minor contribution is from the intradomain structure developing during sintering by reorganization of the compact. The nature of the intradomain structure mainly depends on the microstructure of the compact formed during slip casting.

INTRODUCTION

In processing dense ceramic materials, submicron sized powders are not only used to reduce the sintering temperature, but also to obtain fine grained microstructures. The major problem in handling these powders is due to the spontaneous agglomeration of the powder particles. During compaction these agglomerates behave as large entities introducing a wide variation in pore size. The broad pore size distribution is caused by packing irregularities due to the large agglomerates. During subsequent drying and densification processes, due to capillarity forces and sintering stresses, rearrangement of the particles occur. In the compact dense domain structures develop with large intradomain pores [1-3]. Because of the difference in the rate of pore closure during densification, the larger pores will develop into a pore structure which can only be removed by prolonged sintering. In order to take full advantage of the submicron-sized powder agglomeration has to be prevented. Colloidal processing [4] has shown to facilitate deagglomeration. It also provides the possibility of eliminating, hard agglomerates present in the powder by sedimentation [5]. In this paper, the effect of agglomerates on the development of the microstructure of compacts, formed by slip casting, during processing will be discussed. For this purpose an alumina powder, A16SG (Alcoa), is used in two different grades viz., A16SG as received and A16SG classified by sedimentation. Deagglomeration is performed by using nitric acid.

EXPERIMENTAL

The suspensions

Alumina powder in two different grades are used, namely an Alcoa A16SG powder as received (AR) with a mean particle size of 0.8 µm containing particles as large as 80 µm. The second grade (FF) was obtained from the same powder by classification through sedimentation. The mean particle size amounts 0.4 µm with 95 vol. % below 1 µm. The suspensions, 30 vol. % Al_2O_3, are prepared by vigorously mixing the powder in water containing a certain amount of nitric acid. After deagglomeration the supsension is aged until a constant pH is obtained after which the suspension is used for slip casting.

Compaction

Compacts are formed by slip casting, using a mould having a mean pore diameter of 0.2 µm and a porosity of 47%. Samples with a thickness of up to 7 mm are slip casted and used for this study.

RESULTS AND DISCUSSION

Compaction behaviour

During compaction the development of the cake is studied with acoustic interferometry. This technique is discussed in a preceding paper [3]. The density of cake formed during slip casting of the FF suspension seems to be dependening on the acid concentration used for deagglomeration, Fig 1. Optimum cake density of 69% is obtained from suspensions with an acid concentration of up to 0.08 mmol acid per gram of powder. For structural analysis, using Hg-porosimetry and the SEM, compacts are pre-sintered at 1020°C in air for one hour. After this treatment the density of the compacts seems to be independent of the suspension used. This discrepancy between the density of the cast and the pre-sintered compacts could be due to the use of different techniques for analysis and/or sintering. However, an increase in the density of up to 10% is not expected at this low temperature. Therefore, the fragile green compacts were also characterized with Hg-porosimetry. All samples reveal the same density of 68%. Although pre-sintering induces a small decrease in the green density, it does not effect the relation between the density and the acid concentration of the precursor suspension.

During pre-sintering the mean pore diameter of the compacts increases from 50 nm to 64 nm.

Using suspensions of the AR powder the green density of pre-sintered compacts do show an acid concentration dependency, Fig. 2. An optimal density of 65% and mean pore diameter of 130 nm is obtained with a suspension of 0.075 mmol acid per gram of powder.

Sintering

After sintering of the samples at 1500°C in air for 2 hours, both the AR as well as the FF samples show a density between 3.85 and 3.90 $g.cm^{-3}$, using He-pycnometry. Here, again, the acid concentration of the precursor suspension is reflected in the density reached, Fig. 3. However, sintering for one hour at 1600°C do show a remarkable difference. Samples of the FF serie densify to 3.95 $g.cm^{-3}$. The samples of the AR serie are less dense with a maximum at 0.10 mmol acid per gram of powder of 3.93 $g.cm^{-3}$.

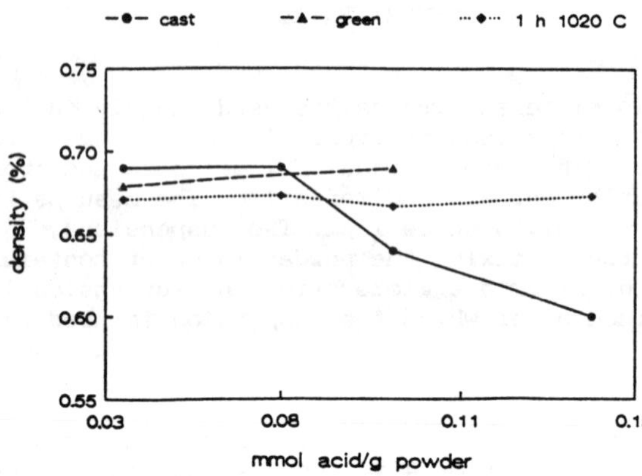

Figure 1. Density of the compacts formed by slip casting FF-suspensions with different acid concentration.

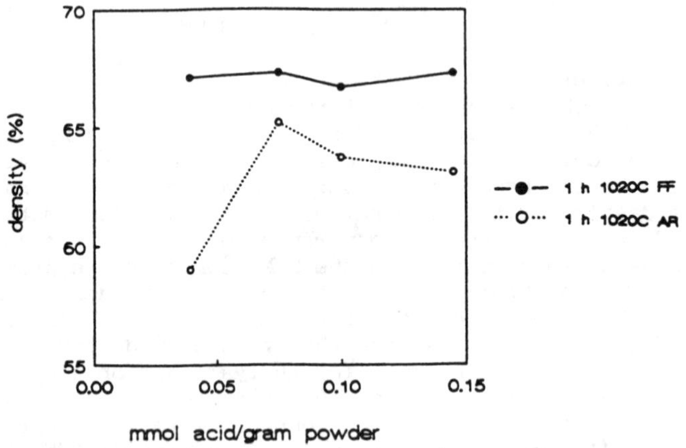

Figure 2. The denisty of the pre-sintered compacts formed by slip casting FF and AR suspensions with different acid concentration.

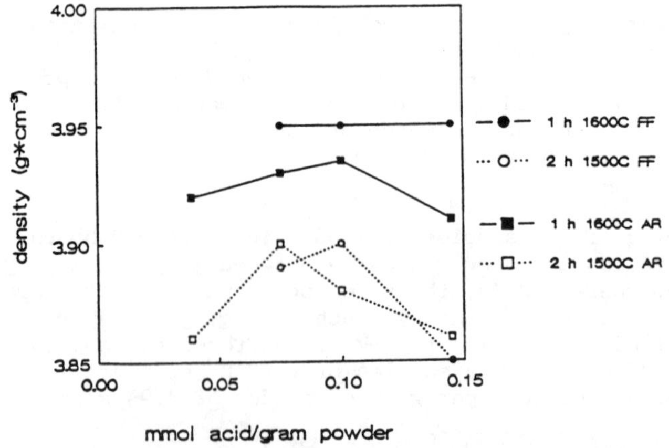

Figure 3. The density of the sintered AR and FF slip casted materials.

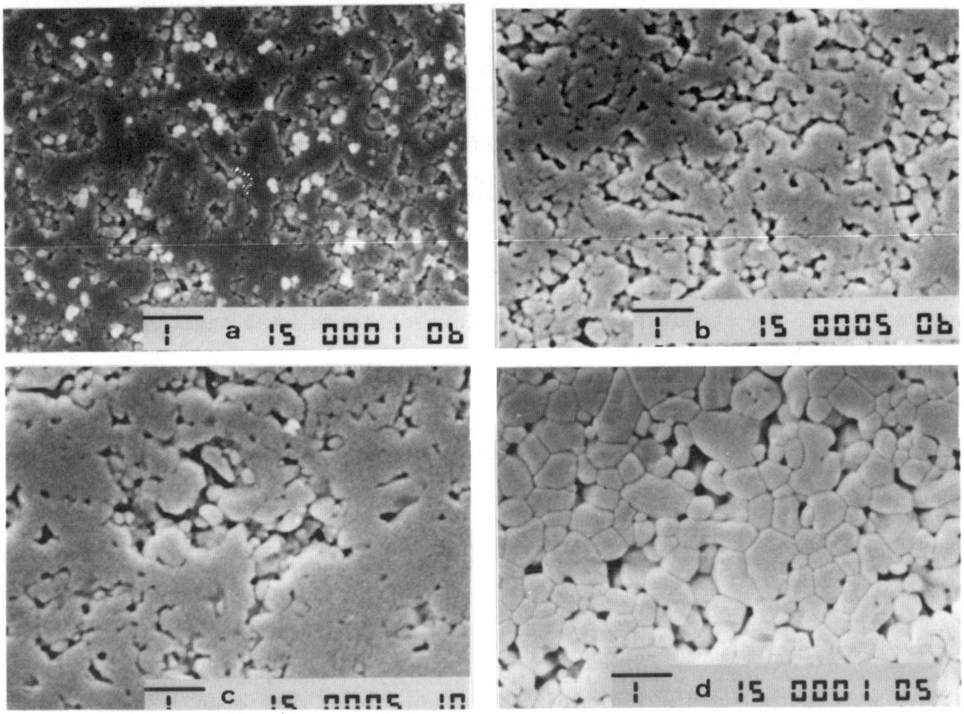

Figure 4. The development of the microstructure of an FF-sample (0.1 mmol acid per gram powder). All samples are sintered in air for 10 hours, at 900°C (a), 1200°C (b) and 1300°C (c and d). Fig. d shows the microstructure after etching the sample.

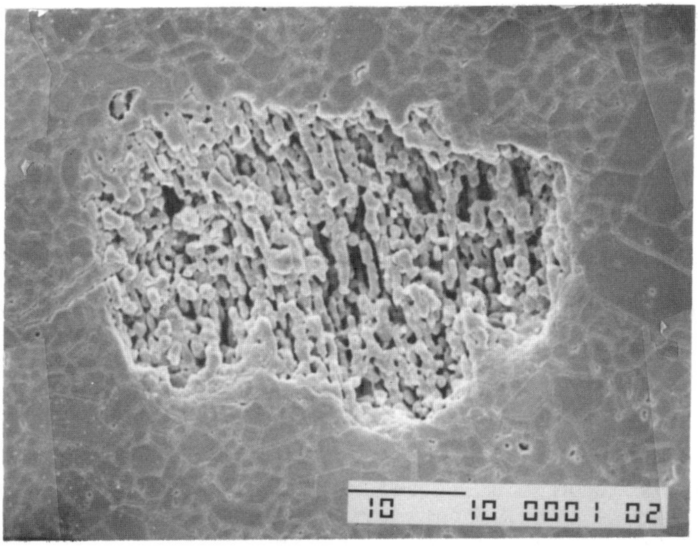

Figure 5. SEM-micrograph of an AR-sample with a large flaw (sample sintered at 1600°C).

Microstructural development
In Fig. 4, preliminary results on the development of the microstructure of an FF sample (0.1 mmol acid) is given. Increasing the sintering temperature from 900 to 1100°C results in a more or less opening of the structure while dense clusters (domains) are formed. At higher temperatures the dimension of the domains increases, which is accompanied by crystal growth within the domains and also increase of the porosity of the intradomain structure. At temperatures higher than 1300°C the intradomain structure densifies with simultaneous coarsening of the grains.
All samples sintered at 1600°C have a fine grained dense structure. SEM analysis of the AR samples reveals a high amount of large flaws, Fig. 5, consisting of parallel orientated chains of particles which apparently do not sinter. This structure resembles that of large hard agglomerates present in the powder.

CONCLUSIONS

Large hard agglomerates present in the powder reduces reorganization of the powder particles in slip cast compacts during drying. The same agglomerates, which show no densification, introduce rest porosity in the sintered compact.

Sintering of compacts of submicron powder particles is accompanied by reorganization, domain formation, of the particles introducing a intradomain structure with large pores. These pores densify at higher temperature than those within the domains formed. The precursor of this domain structure is the microstructure formed during slip casting the suspension.

ACKNOWLEDGEMENT

Financial support from the National Program on Technical Ceramics IOP-TK is gratefully acknowledged.

REFERENCES

1. Petzow, G., Exner, H.E., Z. Metallk., Bd. 6, H.9 (1976) 611.

2. Liniger, E.G., Raj, R., J. Am. Ceram. Soc., 79 (1988) C408.

3. Veringa, H.J., An alternative way of looking at sintering models and the development of instabilities, NKA-89-06, to be published.

4. Aksay, I.A., Schilling, C.H., in Ultrastructure processing of ceramics, glasses and composites, Eds. L.L. Hench, D.R. Ulrich, J. Wiley & Sons, New York, 1984, p. 439.

5. Parish, M., Bowen, H.K., Ceram. Intern. 10 (1984) 75.

6. Correia, L.A., Beenen, R.J.G., H.J. Veringa, A new technique to study powder compaction during slip casting, this volume.

CASTING OF SILICON NITRIDE

J.F.A.K. Kotte & J.A.M. Denissen
Centre for Technical Ceramics
P.O. Box 595, 5600 AN Eindhoven, The Netherlands

ABSTRACT

The wet consolidation techniques slip and pressure casting are investigated for a silicon nitride slip. As the slip ages its properties get better, resulting in less porous green products. Observed changes in Si_3N_4 slip properties are incompletely understood. Mould release, which is poor for silicon nitride, improves by treating the mould with alginate. Pressure casting experiments are performed at various pressure levels. Green product density depends on the degree of deflocculation and is independent of the applied pressure on the slip.

INTRODUCTION

Silicon nitride is a structural ceramic material, still gaining interest because of its good mechanical and chemical properties at low and particularly at high temperatures. Colloidal processing of ceramic powders is a promising preparation route, because these "wet" techniques result in more homogeneous packing of the particles compared to "dry" techniques. Two related wet-forming techniques are slip and pressure casting of a suspension with a high volume fraction of solids. Thus slip properties, e.g. viscosity and degree of deflocculation, are of great importance.
The highest density in slip casting can be obtained with completely deflocculated slips, whereas flocculated slips result in a low green density.
Pressure casting of a flocculated slip results in a more dense product when the applied pressure is higher [1]. A major point in this project was to investigate if a flocculated suspension will result in the same green density as is reached with a deflocculated slip by just applying a high pressure.

MATERIALS AND METHODS

The silicon nitride powder used in all experiments was LC-12 (Hermann C. Starck, Berlin, FRG) which has a B.E.T. specific surface of 19 - 22 m^2/g and an average particle size of 0.56 microns according to the manufacturer. The value of the specific surface was confirmed but the average particle size was lower according to our measurements (0.36 microns). Doubly distilled water was used as dispersing medium. Na_2SiO_3 (Fluka AG, Buchs, Switzerland) was found to be a good deflocculant. An accurate method of slip characterization is measuring the casting rate. This is performed in a glass measuring cylinder placed on a cube of Plaster of Paris (figure 1).

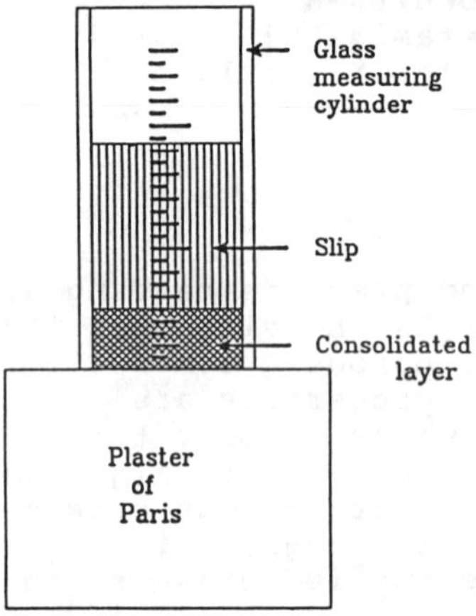

Figure 1.
Equipment for measuring the cake formation rate.

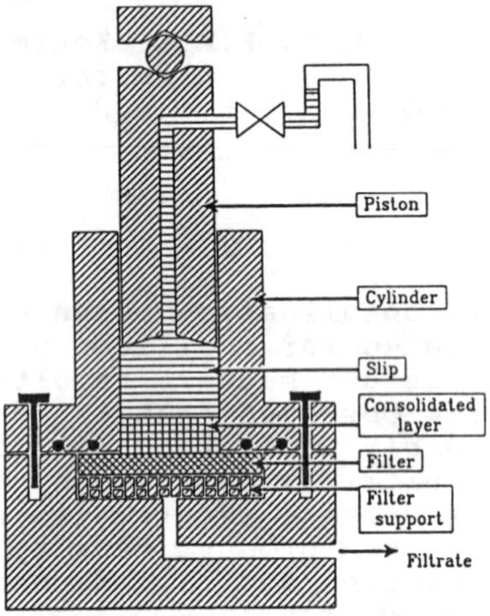

Figure 2.
Laboratory-scale pressure casting unit.

After filling the cylinder on t=0 s, time (t) is recorded at certain thicknesses of the consolidated layer (L). This results in the filtration constant (C) by the following relation, based on Darcy's law [2]:

$$L=C\cdot\sqrt{t} \qquad (1a) \qquad\qquad C=\sqrt{\frac{2\cdot P\cdot w}{(1-E-w)\cdot\eta\cdot\alpha}} \ . \qquad (1b)$$

In equation (1b) P is the pressure difference across the consolidated layer, w is the volume fraction of solids in the suspension, E is the wet porosity of the cake, η is the viscosity of the dispersion medium and α the specific cake resistance.
Measurements of the particle size distibution in the suspension (or more accurately, the floc size distribution) were performed on a Micromeritics Sedigraph 5100, with the slip diluted with distilled water.

Apart from slip casting some pressure casting experiments were performed on a laboratory scale pressure casting unit (figure 2) [4] which was placed between the clamps of a mechanical testing machine (ESH Testing Ltd.). Simultaneously the measured force on the piston (F_m) and the displacement of the piston (x) are recorded as a function of time (t). Equation (2) can be derived for this process:

$$t = \frac{\eta \cdot A \cdot (\alpha \cdot L_0 + R_p)}{F_m - F_f} \cdot x + \frac{\eta \cdot \alpha \cdot A \cdot w}{2 \cdot (1-w-E) \cdot (F_m - F_f)} \cdot x^2 \quad .(2)$$

In equation (2) A is the surface area of the cake, L_0 is the thickness of the cake at the arbitrarily chosen time t=0, R_p is the resistance of the filter and F_f is the force on the piston due to friction.
The density of the green bodies and their pore size distribution were measured with a mercury porosimeter (Micromeritics Pore Sizer 9310).

RESULTS AND DISCUSSION

With 0.6 w/o sodium metasilicate (SMS) (based on the ceramic content) a maximum volume fraction of 48 % of solids was reached while the slip was still pourable. The maximum volume fraction of solids in plain water is only 18 %.

Measurements of the filtration constant as a function of the age of the slip (figure 3A) show that the properties of the slip and the resulting green products change with time (figure 3B).

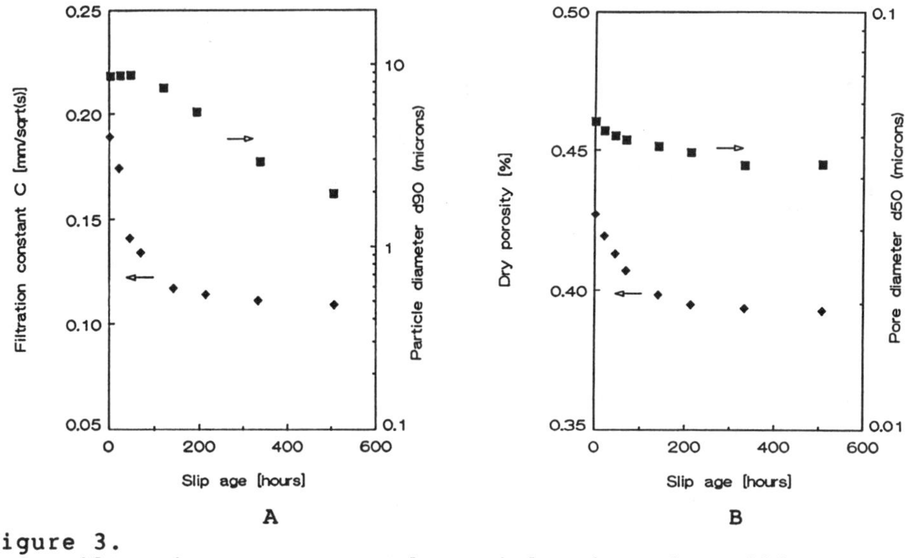

Figure 3.
A: Filtration constant and particle size of a silicon nitride slip (42 v/o) as a function of slip age.
B: Density and median pore size of green products from a silicon nitride slip (42 v/o) as a function of slip age.

The filtration process proceeds slower as the age of the slip increases. With the age of the slip the viscosity diminishes, the average pore size of the compact decreases, the green density of the product increases and the mould release improves. These latter changes are favourable but slip

properties which change in time are undesirable. The aging process could be explained either by an oxidation reaction of the Si_3N_4 with water, or the breaking up of agglomerates or a process that increases the slip stability.
To find out if oxidation causes the change in properties, a slip of a slightly oxidized Si_3N_4 powder was prepared. The filtration constant showed no difference in time dependence, only in absolute value compared to non-oxidized powder.
In figure 3A the particle size in the suspension is shown as a function of slip age. The particle size does not change when the filtration constant drops rapidly, at the time the particle size decreases, the filtration constant diminishes only slightly. The cause of slip aging is probably a combination of all three possibilities mentioned above, but the process is still not fully understood.
The mould release from Plaster of Paris is poor when a fresh slip is used and slightly better in case of an aged slip. A solution of this problem was found in treating the mould with ammonium alginate (Protamon S, Protan A/S, Norway) [3].

Treatment of a fresh slip with an ultrasonic horn [4] (Sonifier, Branson, USA) during a few minutes (power 175 W, 100 ml) resulted in properties which are comparable with or even better than those of a very old slip. The overall effect of ultrasonic treatment will be subject to further investigation.

Pressure casting experiments were carried out with two slips differing in degree of deflocculation. A deflocculated slip was prepared by ultrasonic treatment and with SMS as deflocculant (32 v/o Si_3N_4, 0.6 w/o SMS) and a flocculated suspension was prepared without deflocculant (17 v/o Si_3N_4). Results of these pressure casting experiments are presented in figures 4 and 5.

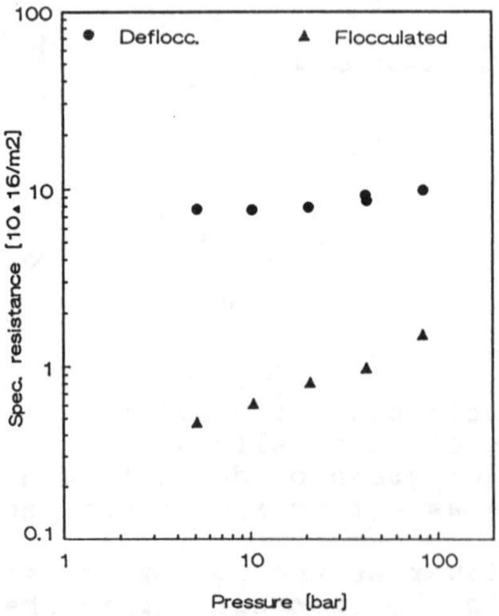

Figure 4.
Specific resistance α as a function of pressure.

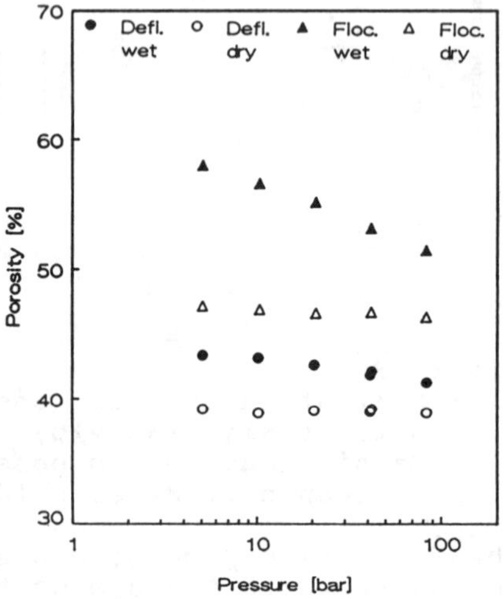

Figure 5.
Wet and dry porosity as a function of pressure.

Figure 4 shows that the specific resistance of a cake formed from a flocculated slip is more pressure sensitive than in case of a deflocculated slip. In agreement with this observation the wet porosity is lower when α is higher. A relation between the porosity and the specific resistance is given by Kozény-Carman [2]:

$$S_p^{\,2} = \frac{\alpha \cdot E^3}{5 \cdot (1-E)^2} \quad . \qquad (3)$$

S_p in relation (3) is the surface area of unit volume of solids.
Calculation of S_p in case of a deflocculated slip and recalculation to unit weight of solids, results in the value 19.5 ± 0.4 m^2/g. This is in agreement with the specific B.E.T. surface area of the raw material. In case of the flocculated slip this value is pressure dependent and ranges from 10 to 13 m^2/g when the pressure ranges from 5 to 83 bar.
Another phenomenon shown in figure 5 is that the dry porosity is almost constant irrespective of the wet porosity. This means that products cast at a higher pressure will have less drying shrinkage.

CONCLUSIONS

As the age of the slip increases, the filtration process proceeds slower, the average pore size of the compact decreases and the green density of the product increases. These desirable properties of an aged slip can be reached within a few minutes by treating the slip with an ultrasonic horn.
With pressure casting, the dry porosity depends only on the degree of deflocculation, irrespective of the wet porosity. This indicates that deflocculation (e.g. a change in zèta potential) might cause the, not well understood, ageing process of a Si_3N_4 slip.
The Kozény-Carman relation seems to hold for a deflocculated Si_3N_4 slip.

REFERENCES

1. Lange, F.F. and Miller, K.T.; Pressure filtration: consolidation kinetics and mechanics. Am. Ceram. Soc. Bull., 1987, **66**, 1498-1504.

2. Adcock, D.S. and McDowall, I.C.; The mechanism of filter pressing and slip casting. J. Am. Ceram. Soc., 1957, **40**, 355-362.

3. Rempes, P.E., Weber, B.C. and Schwartz, M.A.; Slip casting of metals, ceramics and cermets. Am. Ceram. Soc. Bull., 1958, **37**, 334-339.

4. Aoki, M., Ring,T.A. and Haggerty, J.S.; Analysis and modeling of the ultrasonic dispersion technique. Adv. Ceram. Mat., 1987, **2**, 209-212.

GRANULE STRENGTH AND COMPACTION BEHAVIOUR OF AGGLOMERATED MATERIALS

J. VAN DER ZWAN
TNO Institute of Applied Physics, Ceramics Dept
P.O. Box 595, 5600 AN Eindhoven, The Netherlands

ABSTRACT

Results are presented of an investigation into the relationship between granule properties and the compaction behaviour of pressing powders. Between granule diameter, granule weight and granule crushing force exists a clear relationship. The behaviour of the bulk granulate during compaction and the deformation and strength of each separate granule depend on powder preparation and binder distribution in the granule.

INTRODUCTION

In a long term research project on the compaction behaviour of agglomerated materials, the Ceramics Department of the TNO Institute of Applied Physics a great deal of effort was directed in clearing up the densification processes and compaction mechanisms of various types of pressing powders. From the observed phenomena during the compaction process, but also by SEM and light microscopy photographs, the authors concluded a distinctly different behaviour between the clay based powders and the ferrite powders. The former behaved more "brittle"; most of the ferrites showed a well defined "plastic" behaviour, resulting in different compaction mechanisms in the early compaction stages (1,2). This has raised the question what relation exists between granule properties (strength, deformation behaviour, density, structure) and compaction behaviour of the bulk granulate. More fundamental was the question, what granule properties control the whole densification process.

In 1978 the Ceramics Department of TPD-TNO started a one year study to determine the correlation between granule properties and densification behaviour of the granulate.

MATERIALS AND METHODS

Although in the whole project about 10 different pressing powders were investigated on their properties and behaviour, we focus in this publication on 3 granulates:
- a spray dried wall tile granulate, code SWT
- a spray dried ferrite granulate, code SF
- a pelletized ferrite granulate, code PF

Main properties of these industrial pressing powders are mentioned in table 1.

TABLE 1

Characterization of 3 industrial pressing powders

Granulate	Bulk density kg/dm³	tap density kg/dm³	d_{50} µm	BET surface m²/g
SWT	1,02	1,17	360	10,7
SF	1,26	1,47	140	2,7
PF	1,49	1,82	385	15,4

From the granulates narrow sieve fractions were produced by removing the granules which were stuck in the sieve openings after dry sieving with a soft brush.

The seperate granules were characterized by measuring the diameter of at least 100 granules of each fraction by light microscopy and by measuring the weight of a known number of at least 20 granules of the same fraction on a Mettler microbalance, type ME 22. From these data, the average granule density and porosity of each granule fraction can be calculated.

The strength, or better the crushing force, is measured by a special device, based on a standard electronic balance, type Sartorius 3716 (figure 1)

The crushing force of each of the 100 tested granules of each sieve fraction is registered on a Kipp & Zn recorder. The experimental set-up can easily be expanded with automatic data-aquisition and processing devices.

Figure 1: Granule strength apparatus

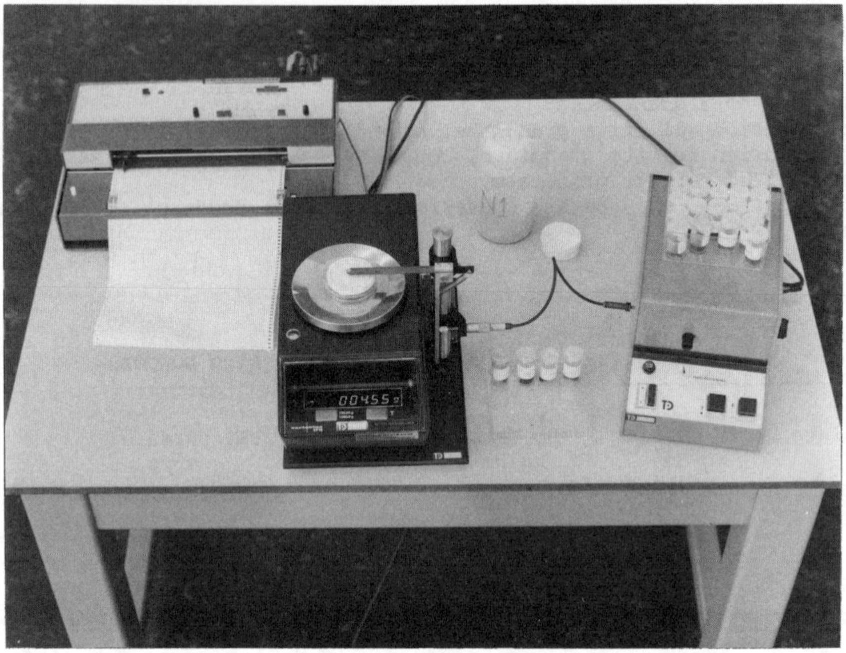

RESULTS AND DISCUSSION

The results of the crushing experiments are presented in figures 2 and 3. The average value and the standard deviation of the crushing force and granule diameter and weight are given. From the measuring points in the figures 2 and 3 a simple relation between force F and diameter D, or force F and weight or mass M can be derived. These relationships are conform the equations published by Capes (3) and Kapur (4):

$F = a.D^b$ [1], respectively $F = c.M^g$, [2]

where a, b, c and g are constants (F in gram, D in mm, M in μgram). From these two equations and the relation between granule diameter and granule weight a simple formula between the constants g en b can be derived:

$g \approx b/3$ [3]

The values of the constants mentioned above are given in table 2.

TABLE 2

Characterization of the separate granules

granulate	Capes: $F=a.D^b$ a	b	Kapur: $F=c.M^g$ c	g	behaviour of granulate under pressure
SWT	21,1	1,81	0,84	0,64	brittle
SF	20,0	1,76	0,35	0,58	plastic
PF	90,0	1,02	7,15	0,35	brittle

From the data in table 2 conclusions can be drawn about the internal distribution of the forces between the particles in the separate granules. When constant $b \cong 2$ the forces are uniformely devided in the granule; when the size index b (see Billam, 5) is close to 1, all internal forces are more concentrated near the skin of the granule. So the value of b (or g) give information about the uniformity of the granules and the way or quality of the powder preparation (spray drying vs pelletizing).

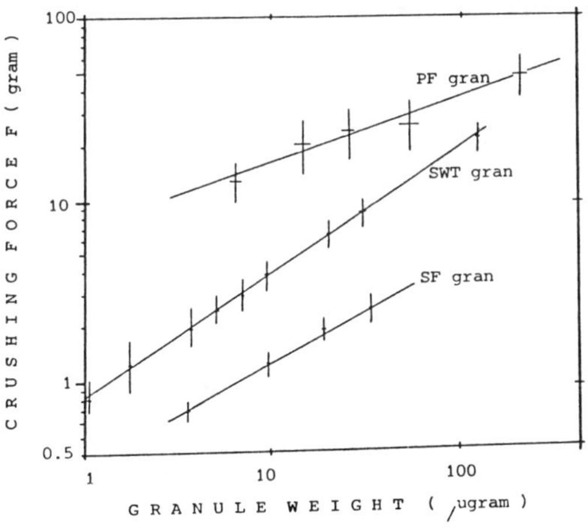

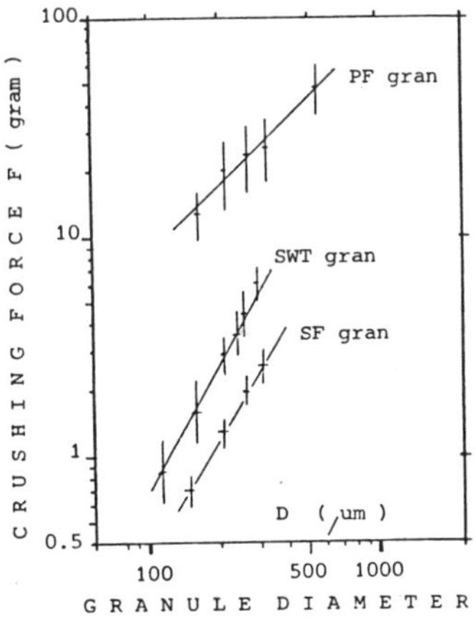

Figure 2: crushing force as function of granule weight
Figure 3: crushing force as function of granule diameter

This leads to the next stage of the research project: the correlation of granule properties and granulate behaviour during compaction.

In previous work it has been pointed out that the SWT-granulate, a clay based material with water as "binder", showed a brittle behaviour: the granules collapse and fragmentate, while the SF-granulate with an organic binder behaved "plastic". The pelletized ferrite powder, with acc. to figs. 2 and 3 the strongest granules (pellets) and an index b of appr. 1 shows a completely different behaviour compared with SF-granulate. PF can be compared with SWT, as compaction diagrams showed a typically brittle behaviour of the separate granules. Although not given here, the strength deformation diagrams registered during the granule crushing tests, give additional information on the brittle or more plastic behaviour of each granulate.

CONCLUSIONS

Between granule diameter, granule weight and granule crushing force exist a clear relationship. There is no explicit relationship between densification behaviour and granule properties. The behaviour of the bulk granulate during compaction and the deformation behaviour and strength of each separate granule depends on powder preparation, amount and distribution of the binder in the granule, binder recipe and water content. The "logarithmic or size index" according to Mott and Billam (5) gives valuable information about structure of the granule, but cannot predict the behaviour of the bulk granulate under pressure.

REFERENCES

(1) Zwan, J. van der, Siskens, C.A.M., Sci. Ceramics **10**, (1979), 159-168

(2) Zwan, J. van der, Siskens, C.A.M., Powder Technol.,(1982), **33**, 43-53

(3) Capes, C.E., Powder Technol.,1971,**4**,77-82

(4) Kapur, P.C., Fuerstenau, D.W., J. Am. Ceram. Soc.,1967,**50**,14-18

(5) Billam, J., Proc. Roscoe Memorial Symposium, Henley on Thames, England,1971,pp.69-80

COMPUTER SIMULATION OF AN ISOSTATICALLY PRESSED BODY AND ITS PACKING FRACTION AND COORDINATION NUMBER

YOSHIAKI HAMADA* AND KOZO ISHIZAKI
Department of Materials Science and Engineering
School of Mechanical Engineering
Nagaoka University of Technology
Nagaoka 940-21, Japan
*Present address: Nisshin Flour Milling Co., Ltd.

ABSTRACT

A process to obtain an Isostatically Pressed body has been computer simulated. There have been many reports on computer simulation of unidirectional packing, but none of them on isotropical packing to simulate Cold Isostatic Press (CIP) or a part of Hot Isostatic Press (HIP) process, so far to the knowledge of the authors. A relationship between the packing fraction and the mean coordination number of an isostatically pressed body was evaluated. Theoretical packing behavior, simulated uniaxial compaction models, and the isotropically densified bodies of the present work were compared to each other.

Relatively high mean coordination number of 6.5 and low packing fraction of 0.56 were obtained.

INTRODUCTION

Isostatic Pressing is one of the most effective compaction methods and is applied for CIPping as well as HIPping. The present work simulates isotropical packing behavior to help to understand these processes. First of all, the packing fraction and the mean coordination number of a simulated isostatically pressed body are calculated, and these values are compared with those of unidirectional and theoretical packing models.

The results also can be applied to help to figure out not only CIPped body, but also the initial state for HIPping process. This aspect is especially important to evaluate theoretical HIPping mechanisms (i.e., HIP maps).

SIMULATION PROCESS

In the present isotropical densification model all particles were considered rigid spheres of equal size. The densification process was assumed to progress by moving particles by an isostatically applied pressure. Particles were randomly generated in a given space. The process is explained schematically in Fig. 1. A particle A moved toward the center of the space as seen in Fig. 1(a), to make a contact with particle B (Fig. 1(b)). This particle A rolled over B until making the second contact with another particle C (Fig. 1(c)). Finally the particle A rotated over B and C until the third contact with particle D, maintaining the contacts with both particles B and C (Fig. 1(d)). This sequence was repeated with all particles. Let us nominate this kind of packing process as "initial triple contacts".

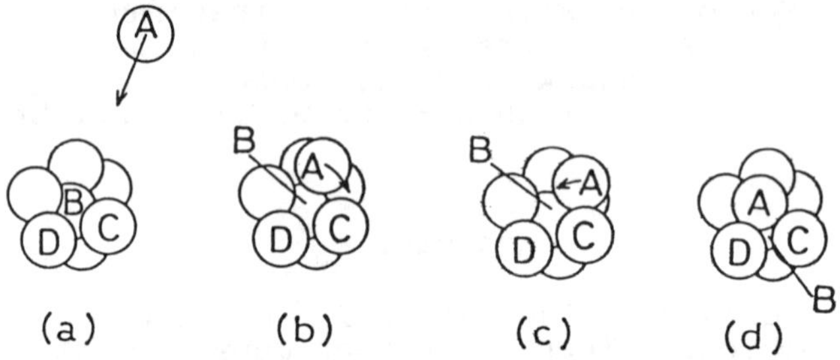

Fig. 1. Typical isotropical movement of a particle A to have contact with other particles B, C and D.

RESULTS AND DISCUSSION

The present rigid spheres model is well applied for engineering ceramics such as Si_3N_4 and SiC. Fig. 2 shows the results of other reports on a relationship of the packing fraction and the mean coordination number of uniform sized spheres. The solid and open triangles show experimental values of a loose and a close compaction models by ball bearing, packed into a cylindrical vessel [1-3]. A computer simulation model by uni-axial force is shown by the solid circle (Fig. 2) [4,5].

A relatively high value of coordination number of 6.5 with a low packing fraction of 0.56 was obtained by the present work as seen by the open square in Fig. 2. The mean coordination number of the isotropical compaction model was higher than the other models, even though the packing fraction of an isotropically packed body was lower than the other simulation works.

Relationships between the packing fraction as well as the mean coordination number and distance from the center of the simulated isostatically pressed body is shown in Fig. 3, using 5000 simulated particles. These values decrease toward the surface of the body. The packing fraction shows an almost constant value around the half way radius from the center to the surface of the body.

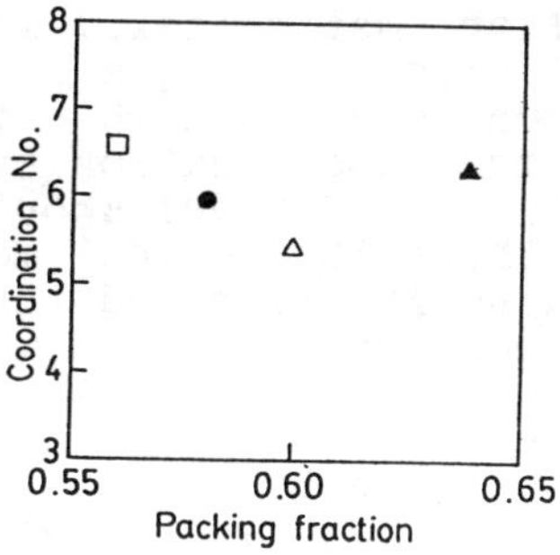

Fig. 2. A comparison of results on packing in various packing methods of spheres of equal size. □ : present work, ● : uniaxial simulation [4,5], △ ▲ : loose and close steel ball simulation [1,2].

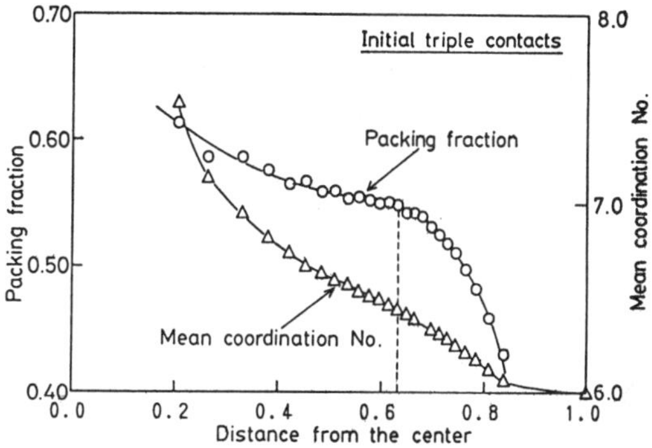

Fig. 3. Relationships between the packing fraction and the mean coordination number as a function of distance from the center of the simulated isostatically pressed body.

The computer simulated model of random packing has surface effects, therefore the packing fraction was decreased suddenly near the surface. The dashed line indicates the boundary line from which the surface effect is notably exhibited. The determination of this boundary location is explained later.

It may be possible to make a compaction model by placing particles at positions of regular structures. The basic packing systems which are suitable for this study are summarized in Fig. 4 [7].

Relationships between the mean coordination number and the packing fraction for the present work and the theoretical packing models is shown in Fig. 5. Open circles represent the present work, the circles to the right side correspond to the values near the center of the simulated body, the circles to the left side pertain the values toward the surface.

The mean coordination number of body centered cubic

(b.c.c.) and hexagonal prismatic (hex.) models are 8, and on the other hand that of simple cubic (s.c.) is 6. A randomly packed body may be considered a combination of these three models. Therefore the mean coordination number and the packing fraction of a randomly packed model should be located in a triangle formed by the three theoretical models as seen in Fig. 5. Most of the packing fraction and the mean coordination number of the present work were located within the triangle. Some of the simulated values of this study were placed outside of the triangle, and were affected by the surface. The dashed line in Fig. 3 was estimated by the intercept of the present work line and the triangle, and means the boundary line of the surface effect.

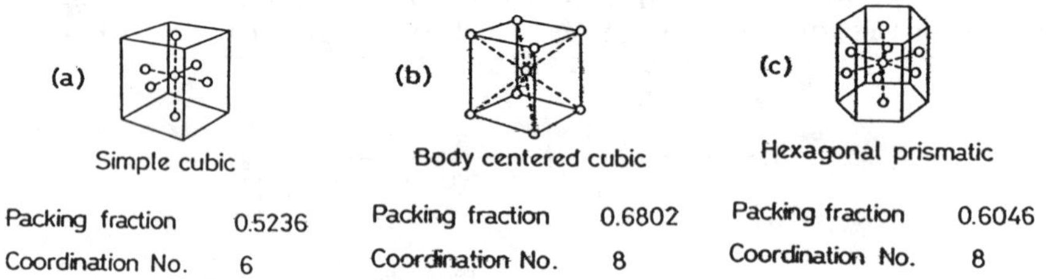

Fig. 4. A schematic representation of theoretical packing models.

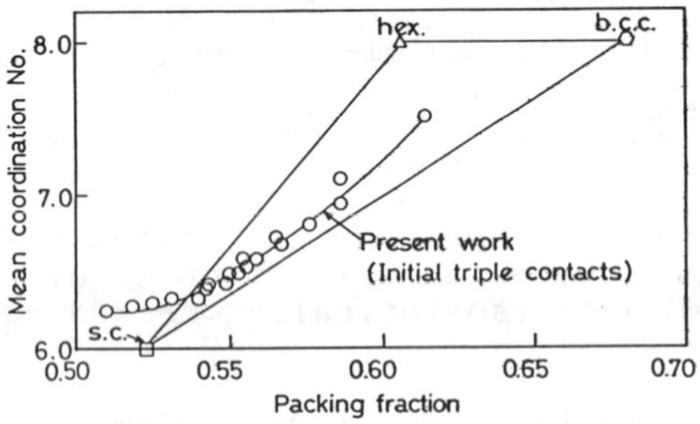

Fig. 5. A relationship between the mean coordination number and the packing fraction for the present work and the theoretical packing models.

Considering that the packed body is formed by a combination of three theoretical models, the fraction of each three models in the simulated model is calculated as seen in Fig. 6. Near the center of the simulated body, the fraction of hex.-like structured particles is higher than that of a s.c.-like structure, so that the packing fraction and the mean coordination number of the body are high. Toward the surface of that body, the fraction of b.c.c.-like and hex.-like particles decrease, and the former becomes even negative beyond the dashed line, which is located at the same position as in Fig. 3.

A simple combination of these three theoretical models does not allow to form a packed body beyond the dashed line due to the surface effect. The surface distance exerts its influence on four to five particle layers of the packed body. This surface thickness is a reasonable value considering inhibited surface diffusion.

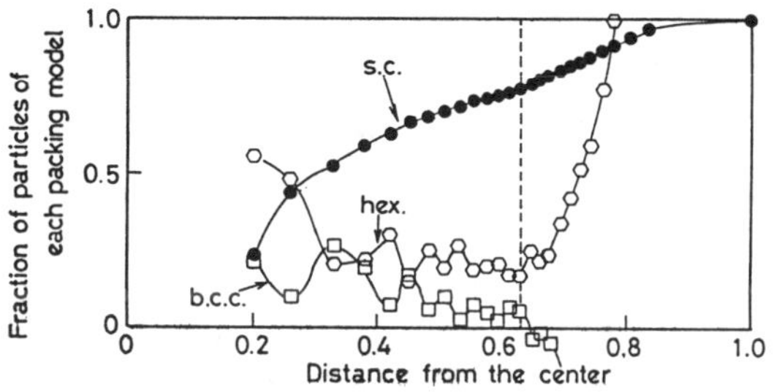

Fig. 6. Fraction of each theoretical packing model inside the simulated body.

CONCLUSION

An isostatically pressed body was made by computer simulation, and its packing fraction and mean coordination number were estimated. From the results, the following was concluded:
(1) The packing fraction and the mean coordination number of initial triple contact model are 0.56 and 6.5 respectively.
(2) The mean coordination number of an isotropically packed body is higher than that of an unidirectionally packed one.
(3) The surface distance exerts on four to five particle layers of a isostatically pressed body.

ACKNOWLEDGMENT

The present work was carried out by tele-communicating with a computer (HITAC M-682H) at the computer center of Tokyo University.

REFERENCES

1. G.D. Scott and D.M. Kilgour, Br. J. Phys. D, 1969, **6**, 863-6.
2. G.D. Scott, Nature, 1962, **194**, 956-7.
3. G. Mason, Nature, 1968, **217**, 733-5.
4. J. Rodriguez, C. H. Allibert and J.M. Chalx, Powder Technol., 1986, **47**, 25-33.
5. M.J. Powell, Powder Technol., 1980, **26**, 221-2.
6. A.S. Helle, K.E. Eastering and M.F. Ashby, Acta. Met., 1985, **33**, 2163-74.
7. A.K. Kakar and A.C.D. Chaklader, J. Appl. Phys., 1967, **38**, 3223-30.

FERRITE POWDER COMPACTION WITH ULTRASONIC ASSISTANCE

E. EMERUWA, J. JARRIGE, J. MEXMAIN
Laboratoire de Céramiques Nouvelles - CNRS U.A. 320
Université de Limoges - FRANCE

ABSTRACT

A newly developed method of powder compaction with ultrasounds is presented. The intrinsic curve of compaction with ultrasonic assistance is defined. It appears that the compaction behaviour of ceramic powders can be improved by ultrasounds (US), only when they are used at pressures lower than a critical value P_C. The effects of amplitude and time of application on the critical pressure P_C and on the compacts density, and the influence of ultrasounds on the microstructure of samples are discussed.

INTRODUCTION

Although much work has been done on the ultrasonic assistance of compaction, and has proved its efficiency on the increase of compact density (1,2,3,4,5,6,7), the best conditions for the application of ultrasounds have never been defined so far (8,9). In order to achieve the best results, a MnZn Ferrite granulated powder is compacted with ultrasonic assistance, and the intrinsic conditions for the application of ultrasounds are discussed.

MATERIALS AND METHODS

The compression experiments were performed with a Wolpert Machine. The compaction set includes an upper punch, a floating die in which 1 gram of the powder is put, and a lower piston brought into accord and connected to the ultrasonic circuit, which is composed of a generator (20 KHz), a transducer, and boosters for selecting the amplitudes of ultrasounds.

The ultrasonic pulse is transmitted to the lower punch through a device licenced (10) by LEGRAND company (Limoges, France). The samples density is calculated with the recorded sample thickness, and the diameter of the die (15 mm).

RESULTS AND DISCUSSIONS

Compaction without ultrasounds (US)

The ferrite powder was first compacted without ultrasonic assistance in a classical way.

The results are well described, as shown in Figure 1, by the equation proposed by Balshin (11) :

$$\text{Log } P = - C_1 \, V/V_o + C_2 \quad \text{or} \quad 1/d = C_3 - C_4 \text{ Log } P \quad [1]$$

where C_1, C_2, C_3 and C_4 represent constants, V_o and V the volume of the powdered sample before and after compaction respectively, d the density obtained under a pressure P.

Compaction with ultrasonic assistance

When an ultrasonic pulse is applied under a given load (frequency : 20 KHz, amplitude = 14 μm, time = 1 second), it leads to a suprising pressure drop equal to the compacting pressure (Figure 2). Above a critical pressure P_c, a reduction in the pressure drop is observed, then followed by a slight

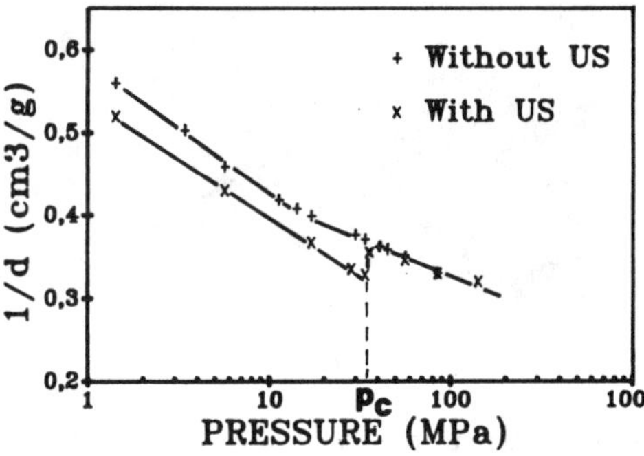

Figure 1. Inverse of compacts density versus pressure (x):intrinsic curve of ultrasonic assistance (A=14 μm, F=20 KHz, t=1s).

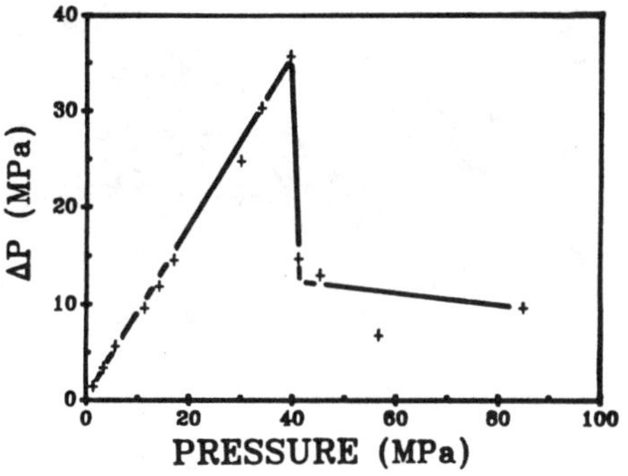

Figure 2. Pressure drop versus pressure of US application.

decrease. For pressures of ultrasonic assistance lower than P_C, a temperature rise of the samples up to 70°C occurs. This obviously results from the friction of the moving powder particles all along their rearrangment when applying ultrasounds.

Once pressure brought back to its initial compacting value, one can notice a reduction in the height of the compact which is equivalent to an increase in density. The inverse of the resulting density has been plotted versus pressure in Figure 1, where each spot, in coordinates (1/d, Log P) according to equation [1], represents a powdered sample compacted at a given pressure. This figure shows a break in the compaction behaviour of the powder on both sides of the critical pressure P_C = 34 MPa. For pressures lower than P_C, powder compaction is improved when an ultrasonic vibration is applied.

A previous knowledge of the intrinsic curve of ultrasonic assistance (Figure 1), and of the critical pressure P_C, is thus very important for all practical purposes.

Influence of ultrasounds on compacts micrsotructure

The microstructure of pellets compacted with and without ultrasounds are compared in figure 3.

a b

Figure 3. Sialon powder compacted under a load of 28,3 MPa : (a) without ultrasounds, (b) with ultrasounds.

Under a load of 28 MPa, without US (figure 3a), the rearrangement of granules is ended and just a few of them have been destroyed. But with ultrasounds, most of the granules are destroyed. The inter-granules pores are filled by the particules. The volume of pores and their distribution are considerably reduced by an ultrasonic impulse. Ultrasounds accelerate the process of densification during compaction (rearrangement and desaggregation of granules). The sintering behaviour of the compacts is thus improved.

Influence of the ultrasonic characteristics : Amplitude and time of application.

We now discuss in detail the influence of the vibration amplitude and the time of application.

Ultrasounds were applied at a given frequency (20 KHz) for the same time of application (1 s). For each amplitude, we studied the evolution of the resulting density as a function of pressure where ultrasounds are applied (cf. Figure 1) in order to detérmine the critical value P_C. The different values of P_C have been plotted versus the U.S. amplitude in Figure 4. At a given pressure (11,3 MPa), ultrasounds of different amplitudes have been applied, the calculated densities have also been plotted versus the U.S. amplitude (Figure 4). One can notice that P_C and density increase with amplitude.

The values of U.S. frequency (20 KHz) and amplitude (14 µm) are kept constant and the critical pressure P_C is determined for each application time by drawing the intrinsic graph (1/d, Log P) as before. The results, reported in Figure 5, show that P_C and density increase with time, and it is of no use applying ultrasounds longer than one second.

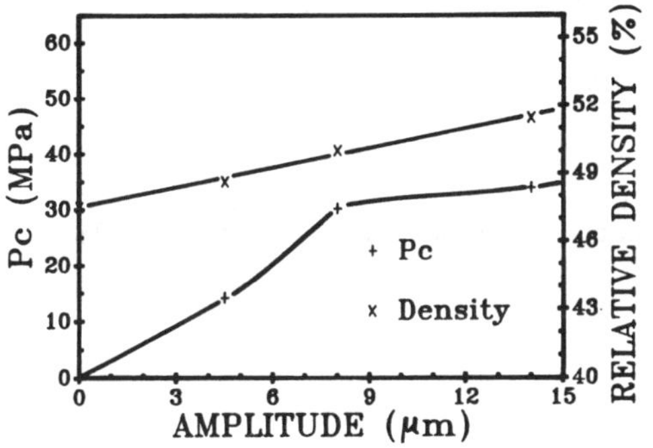

Figure 4. Critical pressure and relative density versus US amplitude (t=1s, F=20KHz)

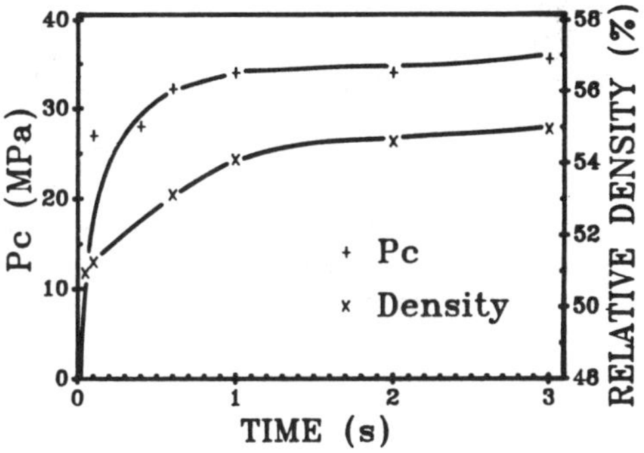

Figure 5. Critical pressure and relative density versus time of US application (A=14 µm, F=20 KHzl)

CONCLUSIONS

The experiments carried out enabled us to achieve the best conditions of ultrasounds application to improve the compaction behaviour of a powder. Therefore the following conclusions can be drawn :

- there is a critical compacting pressure P_C below which ultrasounds can only be applied with efficiency. For compacting pressures above P_C, ultrasounds have a mere effect on the green density.

- P_C and compact density increase with ultrasonic parameters such as amplitude and time of application.

- ultrasounds improve compacts porosity and microstructure for suitable sintering results.

The influence of the application of ultrasounds on the properties of samples after sintering will be the topic of a further paper.

REFERENCES

1. E. LEHFELD, Ultrasonics, 10 (1967) 219.
2. W.B. TARPLEY and al., in Aeroprojects AEC report, New-York, 1961, p. 9587 and 10005.
3. O. DRAGAN and A. PROTOPOPESCU, Proceeding of the 1st Symp. on high power ultrasonics, 1971, p. 109.
4. K. SIWKIEWICZ and S. STOLATZ, Powder Metall. Int., 18 (1986) 6.
5. B. ROGEAUX and P. BOCH, J. Mater. Sci. Letters, 4 (1985) 403.
6. J. JARRIGE, J. MEXMAIN and B. RIGONDEAU, Ind. Céramique, 803 (1986) 3.
7. J.G. THOMAS and al., in Aeroproject AEC report, New-York, 1958, p. 7921.
8. E. EMERUWA, J. JARRIGE, J. MEXMAIN, M. BILLY and K. BOUZOUITA J. Mater. Sci. (in press).
9. E. EMERUWA, Doctorat University of Limoges, France, 1989.
10. French Patent n° 82.07701.
11. M.Y. BALSHIN, Vestrik Metalloprom., 18 (1938) 124.

DRY-PRESSING OF CERAMIC POWDERS

ARTHUR P.S. REYMER
Ceramics Department
Institute of Applied Physics TNO
P.O.Box 595, 5600 AN Eindhoven, Netherlands

ABSTRACT

An overview is presented of past and present research by TPD-TNO Ceramics into the dry-pressing of ceramic powders. Relationships between pressure and density, the influence on compaction of moisture content of the powder, grain size, pressing technology and characterization of the green products are discussed. The R&D possibilities of new advanced experimental equipment for dry-pressing are highlighted. Dry-pressing is of great industrial importance for the floor- and wall-tile and electroceramic industry and will be increasingly used in the building and advanced ceramics industry.

INTRODUCTION

Dry-pressing as a production method is of significant economic importance for the ceramics industry. The production of wall tiles and floor tiles in the Netherlands in recent years was over 14×10^6 m^2, contributing about 25% of the total clay ceramic production in Dfl. Electroceramic products are also largely shaped by dry-pressing. Dry-pressing in the advanced ceramics field has started, in particular of spray-dried alumina powders. The brick and roofing tile industry still uses other well-known shaping technologies, but as the market pressure for fast, flexible and continuous processing further increases, structural clay ceramics as well as whitewares are likely to consider dry-pressing as well.

In view of this industrial importance the Ceramics Department of the Institute of Applied Physics TNO has undertaken a long term research effort into dry-pressing technology, which started in 1975. Recently, an advanced testing machine, specially developed in cooperation with industry for compaction experiments on ceramic powders has been commissioned. The goals of this effort are several. The main aim is to develop a thorough knowledge base on dry-pressing technology for use by industry. Several subfields of research have been outlined: pressing technology (pressing cycle etc.), influence of material properties (grain size, binder) on the compaction process, characterization of the green products (density,

density distribution, surface, defects, etc.). In turn, industry can ask to solve ad hoc pressing or powder problems. Existing technology (e.g. pressing cycles) can be optimized or new cycles can be developed for new materials, such as advanced ceramics.

A compacted product should have an optimal density, (ideally) a homogeneous density distribution, enough strength for further processing and a high quality surface structure. Furthermore, the pressing cycle should be as short as possible to maximize production per machine. Typical for dry-pressing are the all-important properties of the powder or granulate. Success in dry-pressing depends on the composition, grain-size distribution and moisture or binder content of the granulate. Many granulates are processed by the spray-drying method. While a proven technology, spray-drying is energy consumptive and not easy to control.

Since dry-pressing, as is implicit in the name, is applied to ceramic bodies with a low water or binder content (for clay granulates typically in the range of 6-8%), the method allows for fast production lines when combined with fast-firing furnaces. Moisture or binder content could be even further reduced before sintering using microwaves.

The following will illustrate some of the above mentioned topics in more detail.

DENSIFICATION UNDER PRESSURE

In previous reports [1, 2] the processes occurring in a granulate when compacted have been discussed. Electroceramic powders with relative plastic granules have been compared with the brittle properties of spray-dried clay granulates for wall tiles. The plastic or brittle behaviour of granules depends entirely on the powder processing method and the nature and distribution of the binder: there are also brittle ferrites, for example.

From the literature two general stages in the densification of massive grains emerge [3, 4, 5]: (1) rearrangement of grains and filling of voids with some elastic and/or plastic deformation and (2) filling of the small voids and spaces by grain-size reduction of the large granules or by plastic deformation.

With granulates these two stages can be expanded to four or five overlapping steps, taking into account that granulates are composed of porous grains (granules) and primary particles (the granule-forming minerals)[1].

The theoretical maximum density is that of the average density of the mineral components making up the ceramic body. In practice, optimal cold uniaxially pressed products have 50-70% of this theoretical maximum density.

The influence of moisture content and grain-size distribution

The moisture content of clay granulates or the binder content of advanced ceramic powders has a direct influence on the compaction behaviour. Granulates with a relatively high moisture content (>8%) can be easily compressed with low compaction pressures. At higher pressures, the water fills all interstitial spaces and no further densification can be obtained (unless under very low speeds the water is allowed to escape). Granulates which are too dry have little cohesion and insufficient lubrication to be adequately densified. Our research indicated optimum moisture contents of 6-8%, which are comparable with industrial practice.

The influence of the grain-size distribution on compaction behaviour has not been systematically investigated as yet. Normal granulates have a grain-size range of 100 - 500 μm, whereas the primary particles are in the micrometer range. The grain-size distribution is not only constrained by the compaction behaviour, but also by process technological aspects such as free-flowing properties and dust level. Flow properties are particularly important to guarantee a consistent filling of the die. In order to study this subject a computer-controlled filling device has been added to the testing facility.

Relationship between pressure and density

In previous reports [1,2] the relation between pressure and density of green products has been extensively discussed. As is well known, pressure and density follow a strongly non-linear relation.

In contrast to the older equipment, the new testing facility allows accurate and reproducible control of the pressing speed. This allows the influence of speed on the classic pressure-density curve to be investigated.

Preliminary results on ferrites show no large influence on density (fig. 1), but a clear maximum for product integrity. This latter effect is likely to be related to de-airing difficulties at high speeds of 75-100 mm/s.

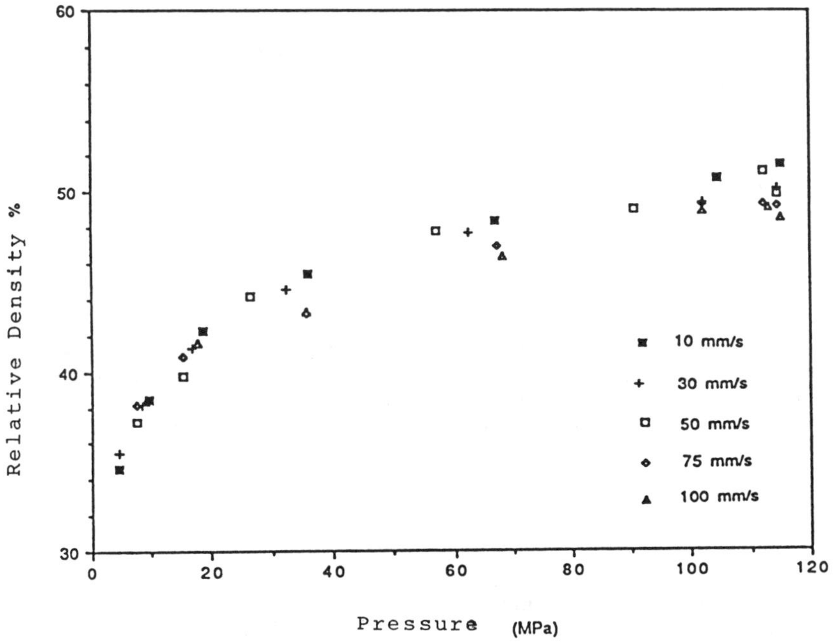

Figure 1: Pressure-relative density data for ferrite granulate at various pressure speeds.

In industrial practice, a compromise will be sought between economically favourable and technically possible pressures and the properties of resulting green products. With wall-tile production, these pressures are typically in the range of 20-40 MPa, resulting in relative densities of 70-75%.
The new experimental facility of the Department allows complex pressing cycles of both plunge and die to be programmed, stored and monitored. An example is presented in figure 2. In this example one pre-pressing stage was programmed. Commonly, displacement is programmed and pressure is the dependent variable. A simple plot between displacement and pressure for a particular batch of material allows quick pressure estimates to be made for further tests. The four experimental signals (displacement and pressure of plunge and die) are recorded and processed by computer with specially developed software. The net result are (colour) plots as in figure 2 and data sheets. After characterization of the green products, the characterizing parameters are related to the pressing variables by further computer processing.

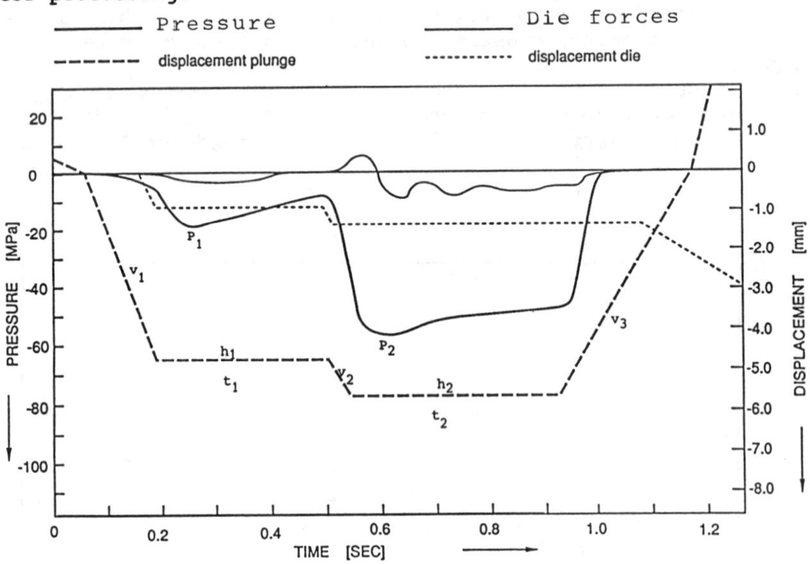

Figure 2: Output from experimental dry-pressing facility: pressure plunge, forces along die wall and displacements versus time. Note positive and negative forces along die wall and pressure drops during hold times t_1 and t_2.

CONCLUSIONS

There is little doubt that the future of ceramic processing lies in fast, flexible and continuous production lines, which are highly automatised and computer controlled. In general this requires higher expenditure on equipment, which is compensated by high production rates, minimal storage and the possibility of quick response to market demands. Energy and personnel reduction are further advantages. Dry-pressing, while not suitable to all ceramic products, will play an important role in modern ceramic processing.

REFERENCES

1. Zwan, J. van der and Siskens, C.A.M., The compaction and mechanical properties of some spray-dried ceramic materials, Science of Ceramics, 1980,**10**, 159-168.

2. Zwan, J. van der and Siskens, C.A.M., The compaction and mechanical properties of agglomerated materials. Powder Technology, 1982, **33**, 43-54.

3. Beddow, Initial stage of compacting metal powders, Int. J. Powder Metall., 1973, **9**, 127- 131.

4. Cooper, A.R. and Eaton, L.E., Compaction behavior of several ceramic powders, J. Am. Ceram. Soc., 1962, 45, 97-101.

5. James, P.J., Fundamental aspects of the consolidation of powders, Powder Metall. Int., 1972, 4, pp. 82,145,193.

WET PRESSING FOR FORMING ADVANCED CERAMICS

F. KOOLS AND O. FIQUET
Philips Components, Corporate Innovation Materials.
Bdg. BE-2, 5600 MD Eindhoven, The Netherlands.

ABSTRACT

First wet pressing for forming ceramics is discussed in general. Various aspects are highlighted briefly while constant rate filtration kinetics are treated in some detail.
Subsequently, wet pressing of hard ferrites is discussed, concentrating on recent results from laboratory experiments on filtration and compaction. A striking result is the absence of significant density gradients due to low frictional forces.

INTRODUCTION

Wet pressing is characterized by die pressing of a concentrated suspension (slurry) while liquid is removed by filtration. The use of a die and punches marks the difference from pressure filtration and (pressure) slip casting [1]. Compared with dry pressing the most striking advantage of wet pressing is the resulting homogeneous density. This is due to the well dispersed state of the slurry and to the low frictional forces, the liquid acting as lubricant. The application of wet pressing for forming ceramics has remained unsignificant for a long period. An important exception has been its application for the production of hard ferrites which is being done for more than thirty years [2]. However, since the advent of high performance ceramics, where homogeneous density is crucial, the interest in wet pressing has grown considerably [3].
This paper continues the discussion, started before [2], on wet pressing in general and on wet pressing of hard ferrites in particular. Various aspects are discussed only briefly, referring to [2]. Extensions concentrate on constant rate filtration kinetics and on the characterization of the filtration and compaction behaviour of hard ferrites.

GENERAL ASPECTS OF WET PRESSING FOR FORMING CERAMICS

Pressing cycle and pressing systems
The pressing cycle essentially consists of five steps [2] (Fig 1).
It can be realized according to different wet pressing systems:
filling with or without pressure, filtration at one or both punches,

pressing by one or two punches, ejection by moving punch or moving die.

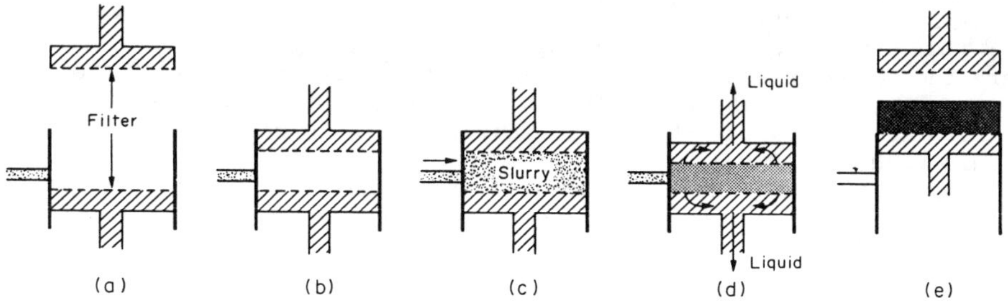

Figure 1: Stages in wet pressing : a) Starting position ; b) Upper punch in filling position ; c) after injection of the slurry ; d) filtration and compression ; e) after pushing out the compact.

Punch pressure filtration; constant rate filtration kinetics (fig.2)
Two stages are distinguished; cake formation (stage 1) and cake compression (stage 2). In the first stage, the slurry is gradually transformed to a filtercake while liquid is removed by filtration. At the end the cavity is completely filled with cake. In the second stage the cake is compressed by direct contact with the moving punch [2].
Constant rate filtration results in a gradual increase of the pressure during cake formation and a steep increase of the pressure during cake compression. Referring to fig. 2, the slope for both stages calculated, starting from the next two basic equations:

(1) $P = q.\eta.r_c.h_c$ (Darcy)
(2) $r_c = k\, d_p^{-2}(1-e_c)^2\, e_c^{-3}$ (Carman Kozeny)

For stage 1 it is valid that:
(3) $e_c = e_1$ (taken to be constant)
(4) $h_c = (1-e_s)(e_s-e_1)^{-1}\, \Delta S$ (mass balance)
(5) $q = v$ (by definition)
(6) $P = \eta\, k\, d_p^{-2}(1-e_1)^2\, e_1^{-3}(1-e_s)(e_s-e_1)^{-1}\, v^2\, t$ (eqs 1-5 combined)

This means that P increases linearly with time until $t = t_1$.

For stage 2 it is valid that:
(7) $e_2 = 1-(1-e_1)h_1 h_2^{-1}$ (mass balance)
(8) $h_2 = h_1 - \Delta S_2$ (moving punch in contact with cake)
(9) $q = \frac{1}{2}(1 + e_2).v$ (average of q near punch and near filter)

(10) $P = \frac{1}{2}\eta\, k\, d_p^{-2}(1-e_1)^2 h_1(1+e_1-2a)(1-a)(e_1-a)^{-3}\, v$ (eqs 1,2,7-9)
The parameter $a = \Delta S_2 h_1^{-1} = v(t-t_1)t_1^{-1}$ accounts for the time dependency.
We find as slope ratio:

(11) $dP/dt(2)/dP/dt(1) = \frac{1}{2}(e_1-a)^{-4}(3-e_1^2+2e_1a-6a+2a^2)e_1^3(1-e_s)^{-1}(e_s-e_1)$

Taking $e_1 = 0.6$ and $e_s = 0.7$, we find for the slope ratio as a function of a (displacement on time): 1.3 for a = 0.1 and 2.5 for a = 0.2. So , slope (2) increases with increasing punch displacement. Nevertheless, we often find a constant slope due to the fact that the press no longer can afford constant punch rate at high counter pressure.

Compaction characteristics

For stage 1, the overall density (= solid fraction) can be approximated by the empirical relation [2]:

(12) $1-e_c = \beta + \alpha.\log P_c = \beta + \alpha.\log x.P = \beta^1 + \alpha.\log P$

where the actual pressure in the cake (P_c) is a fraction (x) of the pressure exerted by the punch (P), x depending on the pressure distribution throughout the cake. Density gradients may be expected due to the distribution of P_c.

For stage 1, P_c, and hence the density, is higher near the filter, while for stage 2, P_c and resulting density is higher near the moving punch [1,2]

WET PRESSING OF HARD FERRITES

Hard ferrites ($MeFe_{12}O_{19}$, Me=Ba,Sr) used as ceramic permanent magnets, mainly as segments for DC motors or rings for loudspeakers. Alignment can be achieved by pressing in a magnetic field, provided that the particles are monocrystalline and separately mobile. More details about the material and wet pressing in production are found in [2]. We concentrate here on some laboratory pressing experiments aiming at the characterization of filtration and compaction.

Equipment

A 60 KN hydraulic press with moving underpunch is used, having a single cavity (Φ40mm) which is filled manually. Punch displacement and oil pressure are monitored. The cake formation stage is mostly characterized by constant rate filtration.

Filtration behaviour

An expression for the cake resistivity for standard porosity (e=0.55) has been derived based on eqs 1 and 2, called the normalized cake resistivity (r_{cn}).

(13) $$r_{cn} = 1.217\ P_1\ \eta^{-1}\ h_s^{-1}(1-e_s)^{-2}(e_s-\Delta S_1/h_s)^3(1-\Delta S_1/h_s)^{-2}(dS/dt)_1^{-1}$$

The parameters $(dS/dt)_1$, ΔS_1, t_1 and P_1 are derived from the experimental pressing curve (Fig 2,3). The parameters e_s, η and e_2 are determined separately while h_s is obtained from the relation:

(14) $h_s = h_2(1-e_2)(1-e_s)^{-1}$

The effect of particle size is shown in Fig 3. The resulting r_{cn} values (Table I) are proportional to the square of the specific surface area (s), in agreement with eq. 2. The slurry water content has influence on the slope of the P-t curve but not on the r_{cn} value (Table 1).

The alignment has a considerable influence on the rcn value, the isotropic cake resistivity being about 1.6 times higher than the anisotropic one. This results from the different structure of the cake giving rise to different k-values in eq. 2 [4].

Compaction behaviour

The overall density versus pressure has been studied for different cases (Fig 4). Density increases linearly with log pressure, having the same slope (α=0.08) for all cases. The difference between stage 1 and stage 2 (factor 1.6 in pressure) is attributed to a different pressure distribution and, hence, to a different x value (see before). We have e.g. for stage 1, x=0.5 (gradual decrease of x downstream [2] and for stage 2 x=0.8, which is conceivable when frictional forces are low. The difference between iso- and anisotropic is attributed to the higher resistance against

compression of the anisotropic cake due to the aligned pillars of platelet shaped particles [2].
The required pressure is low,(1/3 of the corresponding pressure for dry pressing) , which is another indication for the low frictional forces.
The density gradient in the pressing direction has been studied for two extreme cases: one sided filtration at the moving punch and one sided filtration at the still punch. In fact, no significant differences in density, were found stressing once more the absence of significant frictional forces.

Wall friction
Wall friction has been studied by measuring the pressure during ejection of the final compact. The ejection pressure is found to be lower than 4 % of the final compaction pressure. This is, indeed, very low as compared with dry pressing where values of 15-30 % are quite normal.

CONCLUSIONS

- The pressure during constant rate cake compression rises exponentially with time. Often a linear dependency is found due to the limitation of the press.
- The filtration behaviour can easily be characterized by the normalized cake resistivity (r_{cn}) derived from pressing curves.
- Specific observations for hard ferrite wet pressing are:
* the effect of particle size is according to Carman Kozeny relation.
* the isotropic resistivity is 1.6 times larger than the aniso. one.
* density increases linearly with log pressure
* the required pressure is relatively low
* density gradients are not significant
* wall friction is not significant

REFERENCES

1. Tiller F.M. and Tsai Ch., Theory of filtration of ceramics I slip casting, J. Am. Cer. Soc. 69 [12] (1986) 882-87
2. Kools F, Pressure filtration for forming advanced ceramics, Enc. Mat. Sc. and Eng. Ed. M.B. Bever (1986) 3916-20
3. Lange F.F., Powder processing Science and Technology for increased reliability, J. Am. Ceram. Soc. 72 [1] (1989) 3-15
4. Strijbos S, Pressure filtration of permanent magnetic powders. Proc. Conf. Hard magn. materials, Amsterdam 1974, 102-105

TABLE 1
Filtration behaviour for different slurries and/or conditions

Code	align-ment	e_s -	h_2 mm	e_2 -	η mPa.S	t_1 sec	P_1 mPa	ΔS_1 mm	$\Delta S_1/hs$ -	dS/dt_1 mm/sec	rcn $10^{14}m^2$
C1	iso	0.717	14.6	0.417	0.91	13.3	2.84	13.4	0.446	1.01	1.01
M1	iso	0.717	14.4	0.441	0.91	12.6	5.67	12.1	0.426	0.96	2.58
F1	iso	0.717	14.3	0.408	0.91	15.1	10.33	14.0	0.468	0.74	4.26
M2	iso	0.631	16.0	0.419	0.91	7.9	6.48	7.1	0.282	0.90	2.32
M3	aniso	0.717	15.1	0.445	0.91	12.1	3.24	12.0	0.405	0.99	1.58

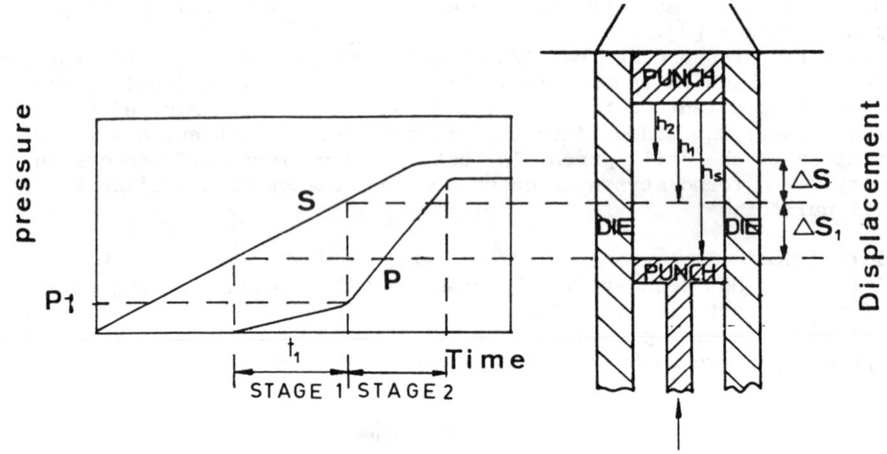

A = cross sectional area punch (m^2)
V = removed volume liquid (m^3)
* h = height slurry/cake/compact (m)
* S = punch position (m)
* ΔS = punch displacement (m)
* e = liquid vol. frac. (-)
* P = pressure exerted by punch (Pa)
* t = punch pressure filtr. time (sec)
q = A^{-1} dv/dt (liquid velocity) (ms^{-1})
v = dS/dt (punch velocity) (ms^{-1})
r_c = specific cake resistivity (m^{-2})
r_{cn} = normalized r_c (e=0.55) (m^{-2})
η = dynamic liquid viscosity ($Nm^{-2}s$)
d_p = particle size (m)
s = specific surface area (m^2/g)

*) suffix s,c refer to <u>s</u>lurry and <u>c</u>ake/compact respectively
suffix 1,2 refers to end stage <u>1</u> and end stage <u>2</u> respectively

Fig. 2. Reference pressing system and definition of symbols and units.

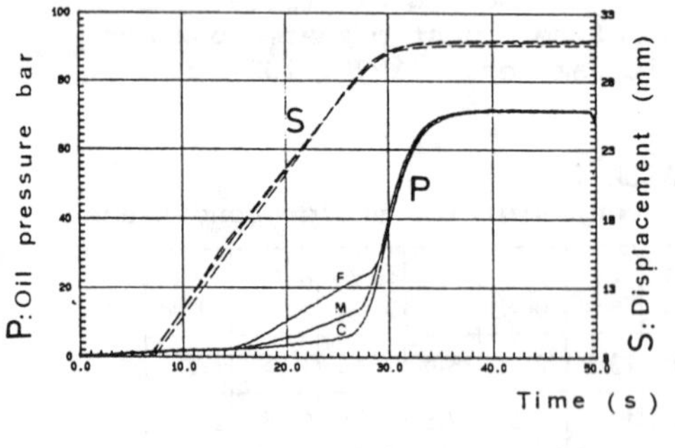

Fig. 3. Example of pressing curve: Effect of particle size (see Table 1) F,M,C = Fine,Medium,Coarse

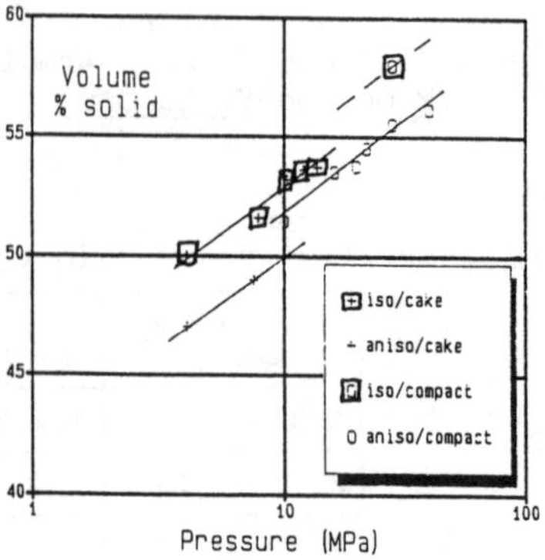

Fig. 4. Density vs pressure

DISPERSION OF CERAMIC POWDERS IN WAX : APPLICATION TO INJECTION MOLDING

P. VIVIER - B. CALES
CERAMIQUES TECHNIQUES DESMARQUEST
2, av. A. Einstein - 78190 Trappes - France

ABSTRACT

Injection molding is one of the most suitable technique for manufacturing complex ceramic parts. It requires the use of well controlled steps and especially a good dispersion of powder in the wax medium. However, the dispersion of ceramic powders in wax has not been extensively studied up to date. The objective of this study was to develop a procedure to evaluate the dispersion of various oxide and nitride powders in different waxes and to determine the influence of inorganic surfactants on the dispersion of these powders. The wetting of powder/wax systems has been studied as a function of temperature with a specific apparatus. The role of mineral additives in the powder or of organic surfactants in the wax on the wetting of the powder/wax systems has been analysed.

INTRODUCTION

The fabrication of ceramic components by injection molding presents many advantages when compared to shaping processes. Complex shapes can be obtained through a highly automated route with minimal additional operations. CERAMIQUES TECHNIQUES DESMARQUEST have been using this technique for a long time to produce ALBUZ ® spray nozzles. The manufacturing of more complex parts such as sintered silicon nitride turbine rotors requires an optimization of this process. It consists, for instance, in finding the best compromise between a good molding of the powder-wax mix that requires a high wax content and an easy dewaxing of the molded part that requires a low wax content. This difficult problem may be solved by optimizing the dispersion of the ceramic powders in wax medium. This was the objective of the present study.

EXPERIMENTAL PROCEDURE

Experimental Device

The experimental device used to investigate the dispersion of ceramic powders in wax is reported in figure 1. Already used to investigate penetration of pitches in coke (1), the present apparatus is slightly different from other similar devices described in the literature (2,3). It allows to measure the penetration of a wax sample, having a rod shape, in a ceramic powder bed, as a function of temperature. The driving of the wax rod into the powder bed is followed by a two dimensional photo-diode network. The lighting of the diodes is recorded as a function of temperature to precisely determine the geometrical variations of the wax rod over the whole temperature range.

The experiments are conducted as follows : a small wax rod (5 mm in diameter, 10 mm high) is put on the powder bed at room temperature. The ceramic powder has previously been shaped into granulates of 80 to 250 microns. The temperature of the powder bed is linearly increased up to 150°C at a rate of 20°C/mn, that corresponds to the heating rate currently used for the preparation of injection moldedg parts.

Wax Penetration Curves

The scheme of a typical experiment is presented on figure 2. Different domains are observed depending on the temperature.

In the first domain (A to B), the wax rod remains tense (i.e. it gives a rectangular shadow on the photo-diodes system) up to the softening point (B) of the wax where it abruptly takes a barrel shape. This point directly depends on the rheological characteristics of the wax (surface tension, viscosity...). For higher temperatures (B to C), the spheroidization of the wax samples increases and a slight penetration of wax in the powder bed may occur. This leads to an increase of the diodes lighting.

When the temperature reaches the wetting point (C) of the powder-wax system, the contact angle between wax and powder bed changes and the

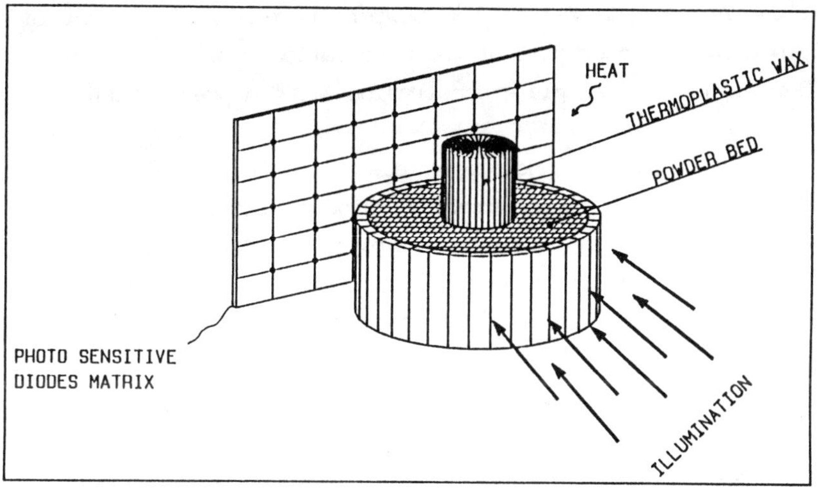

Figure 1. Schematic presentation of the experimental device used for the analysis of powder-wax systems.

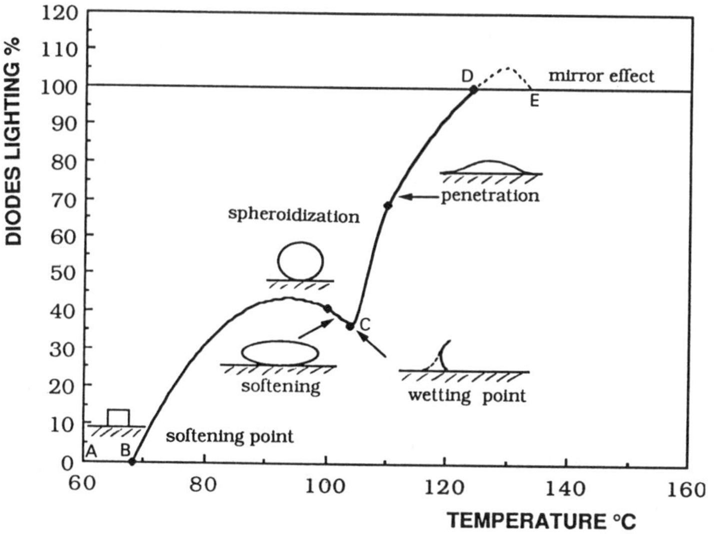

Figure 2. Typical penetration curve.

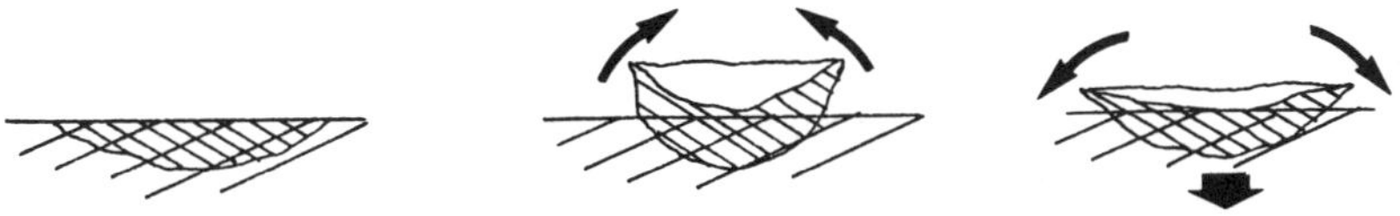

Figure 5. Schematic explanation of the "mirror effect" for the Si_3N_4-1 B wax system.

penetration rate of the wax in the powder increases. This part of the curve C to D is characteristic of the penetration kinetics of the wax in the powder bed up to a complete penetration (100% diodes lighting).

After point D a spurious phenomenon due to a "mirror effect" may be observed. It gives rise to an excessive diodes lighting and is due to a reflection of the incident light beam onto the surface of the wax. In the experimental procedure, it is necessary to wait the end of this transient effect to determine the real 100% diodes lighting.

The behaviour of wax-powder systems is directly dependent on the characteristics of the wax and the powder, such as the surface tension and the viscosity of the wax, the contact angle and the radius of the inter-granulates porosity in the powder bed.

In the present study, in order to compare the influence of the different components of the mixes (waxes, ceramic powders, mineral or organic additives), ceramic powders with the same granulate size have been used.

Studied Powders and Waxes

The studied waxes A and B are mainly made of paraffin and have different molecular weights. The Mettler softening point of the A wax is 96.5°C and 60.5°C for the B wax which has a lower molecular weight.

Dispersion of silicon nitride (Si_3N_4-1 and -2), alumine nitride (AlN), zirconia (ZrO_2) and alumina (Al_2O_3) powders have been analyzed. All these powders exhibit an average primary grains size lower than 1 micron. The two Si_3N_4 powders have been selected to determine the influence of the synthesis route on the wetting behaviour. Si_3N_4-1 powder has been obtained from imide precursors while Si_3N_4-2 powder has been synthetized by nitridation of silicon.

The influence of mineral additives, such as Cr_2O_3 and CO_3O_4, in ZrO_2 and in Al_2O_3 powders or organic surfactants, noted D1 and D2, in the wax has also been investigated.

RESULTS AND DISCUSSION

Influence of Powder Type

The wetting of the three nitride ceramic powders by the same wax is reported on figure 3. It may be observed that the softening point for a given wax is independent of the nature of the powder. However, different behaviours are noted between silicon nitride powders (1 and 2) and aluminium nitride powder. AlN powder is more easily penetrated by the wax and the wetting point corresponds to a lower temperature than for Si_3N_4 powders. No difference is observed between the two silicon nitride powders for temperatures higher than the temperature of the wetting point. Only a slight "mirror effect" is noted for Si_3N_4-1 powder. It thus may be concluded that the wetting of nitride powders by waxes appears to be more affected by the chemical composition of the powder than by its synthesis route.

Influence of Wax

The influence of the wax composition on the wetting of Si_3N_4-1 powder clearly appears on figure 4. The combination of Si_3N_4-1 and B wax (lower molecular weight) leads to the best wetting properties though it presents an important "mirror effect" (up to 114% lighting). Due to a "surface shrinkage" of the powder-wax mixture, the edges of the wetted zone raise up before a complete penetration of the wax in the powder bed, as schematically shown on figure 5. To avoid this phenomenon, the powder-wax parts for injection molding must be prepared at a temperature lower than that of the first 100% lighting (point D on figure 4).

Influence of Mineral Additives in the Ceramic Powder

The wetting properties of chromium oxide doped zirconia and cobalt oxide doped alumina are reported on figures 6a and 6b. On these two figures, we have compared the behaviour of the oxide powder with and without mineral additive. When mineral additives are introduced, the softening temperature (beginning of diodes lighting) is lowered and a rapid penetration of the wax in the bed is firstly observed for temperatures just higher than the softening temperature. When the temperature increases again, the penetration rate is lowered, when compared to

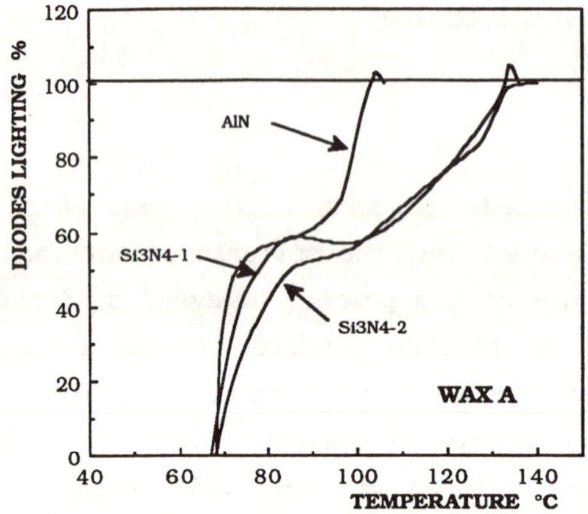

Figure 3. Influence of the nature of the powder on the wetting behaviour.

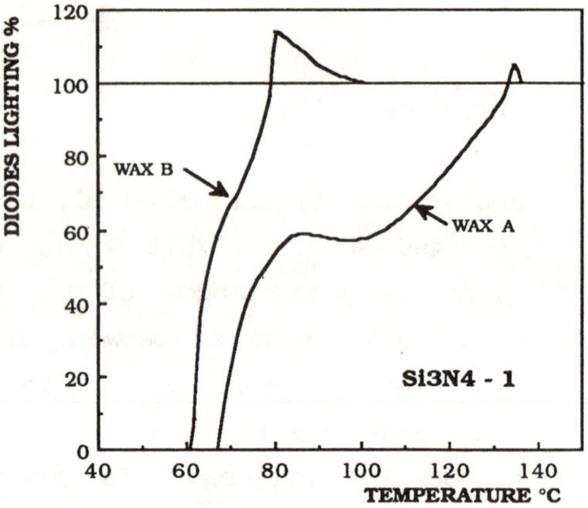

Figure 4. Influence of wax on the wetting of the system.

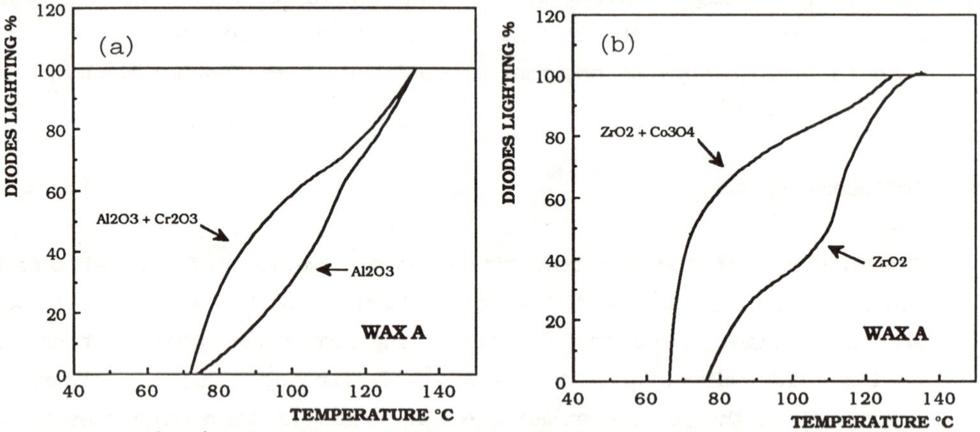

Figure 6 (a,b). Influence of mineral additives in the powders on the wetting behaviour.

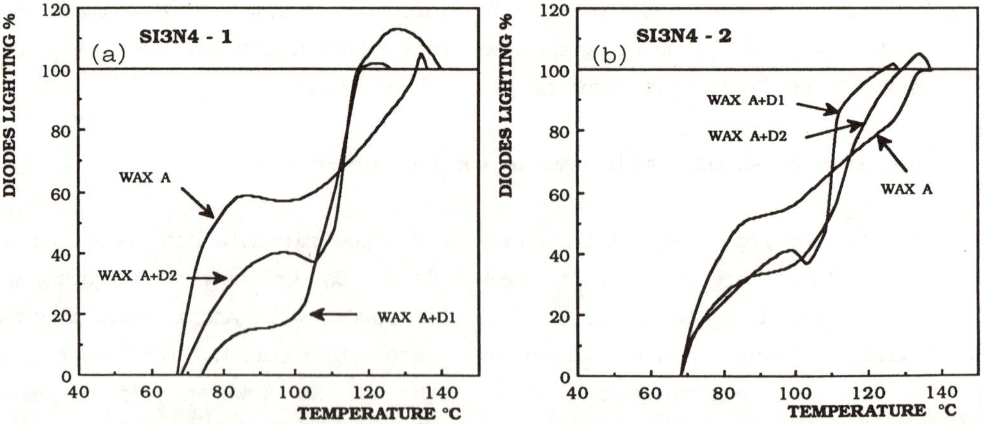

Figure 7 (a,b). Influence of organic surfactants on the wetting property of A wax on Si_3N_4 powders.

undoped powders and very poor wetting properties result. This behaviour could be at the origin of the higher difficulties encountered during the injection molding of such doped oxide powders.

Influence of Surfactants in the Wax

The influence of surfactants in the wax is shown on figures 7a and 7b for the two Si_3N_4 powders. As previously shown on figure 3, the two Si_3N_4 powders exhibit the same wetting characteristics in pure A wax. The addition of surfactants D1 and D2, even with a low content (4%) strongly modify the behaviour of the two Si_3N_4 powders. The wetting temperature is increased by the addition of surfactants but for temperature higher than the wetting point, the penetration rate of the doped wax in the powder bed is much higher than for pure wax.

CONCLUSIONS

The experimental procedure defined in this study allows to select powder-wax systems that can give rise to good wetting properties and consequently to good dispersion of powder in the wax. This method can also be used to optimize the powder composition, especially when a mineral additive is introduced, and the wax composition, by addition of organic surfactants.

It would then be interesting to verify the consequences of such optimization on the injection of powder-wax parts and on the dewaxing of injected parts. This is presently under investigation.

Acknowledgements

The authors thank Mr Hervieux and Mr Cantin from PECHINEY Casting Research laboratory (St Jean de Maurienne) for wetting experiments and discussions.

REFERENCES

1. Saint-Romain J.L., Cottinet D., Couderc P., Proceedings of the AIME Light Metals Meeting (1987).
2. European Patent Application n° 55637.
3. Couderc P. and Dhamelincourt P., Proceedings of the Int. Tar Conf. Biarritz, France (1983).

RELATIONSHIPS BETWEEN FABRICATION, MICROSTRUCTURE AND MECHANICAL PROPERTIES OF INJECTION MOULDED SI_3N_4

P. Lemaitre*, R. Gilissen*, A. Smolders*, J. Koenen**

* Studiecentrum voor Kernenergie, Materials Development Department, Boeretang 200, B-2400 Mol, Belgium

** DSM Research BV, Postbus 18, 6160 MG Geleen, The Netherlands

ABSTRACT

The effect of fabrication parameters on the mechanical properties of Si_3N_4 bars obtained by injection moulding and pressureless sintering were determined. The maximum four-point bending strength at room temperature was obtained for a sintering time of 2 hours and a powder bed thickness of 6 cm. The fracture surfaces were examined in order to correlate microstructure, mechanical strength and defects caused by processing. As compared to similar materials, reported in the literature, the developed material has satisfactory mechanical properties.

INTRODUCTION

This study forms part of a project which was undertaken to optimize the fabrication route of an in-house (SCK and DSM) developed injection moulded pressureless sintered Si_3N_4.

The article will focus on the influence of a number of fabrication parameters on mechanical properties. The sintering temperature was held constant at 1800°C. Three sintering times (t_s) were studied : 1, 2 and 4 hours. In order to avoid decomposition of silicon nitride during sintering the powder bed technique was chosen because of its technological simplicity [1]. One of the drawbacks of this technique for production of big batches is, however, the thermal inertia of the powder bed. Therefore the smallest powder bed thickness at which no exaggerated decomposition of Si_3N_4 occurs had to be determined in order to minimize this problem. Three powder bed thicknesses were investigated : 2, 4 and 6 cm. The fracture surfaces of test specimens were examined in order to determine the failure origins related to processing.

MATERIAL AND TESTING

The composition of the starting powder is as follows : 93 % Si_3N_4 (Starck LC 12), 5 % Y_2O_3 (Starck Fine Grade) and 2 % Al_2O_3 (Martinswerk ZPS502). The choice of sintering additives is a compromise between the good high temperature behaviour obtained with Y_2O_3 only and the aim to sinter to full density at 1800°C. Test bars were fabricated by injection moulding. After removal of the binder material the bars were pressureless sintered in a C-crucible at 1800°C. The powder bed used had the same composition as the test bars but 50 wt% BN was added to prevent sintering.

The four-point strength was measured at room temperature. The specimens were tested in the as fired condition with no additional surface preparation. The outer and inner rollers were spaced 30 and 10 mm apart, respectively. Test bars were 45 mm long bij $\pm$ 4.5 mm wide by $\pm$ 3.5 mm thick. All tests were carried out using a crosshead speed of 0.5 mm/min. Young's modulus was measured making use of the resonance technique. Density was obtained by weighing in water and applying the law of Archimedes. The fracture toughness was determined by the Vickers indentation method, following the relationship proposed by Anstis et al. [2].

RESULTS

Influence of Different Fabrication Parameters

<u>Sintering time</u> : Figure 1 shows the four-point bend strength as a

function of sintering time for different powder bed thicknesses. Each point represents the average of three measurements. The standard deviation is represented by vertical bars. The E-modulus and the density as a function of sintering time are represented in figure 2. The standard deviations of the density measurements are too small to be represented.

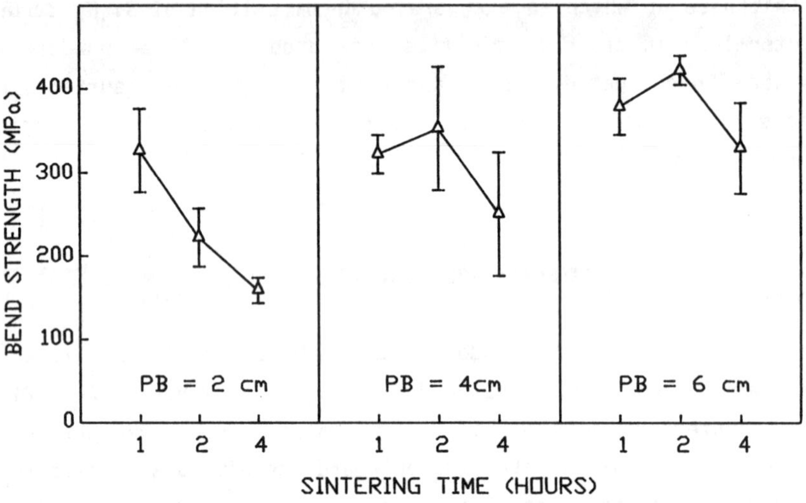

Figure 1. Four-point bend strength versus sintering time for different powder bed thicknesses (PB).

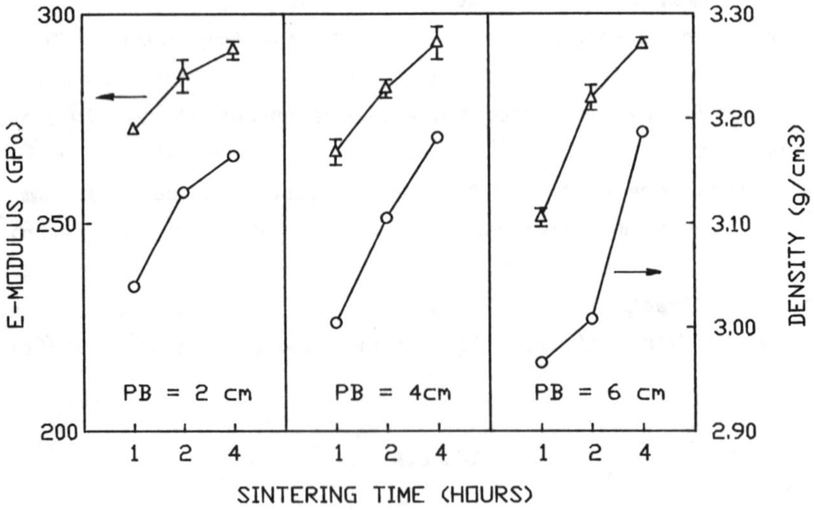

Figure 2. E-modulus (Δ) and density (o) versus sintering time for different powder bed thicknesses (PB).

In spite of the high scatter and the low number of specimens used it is clear that the highest strength occurs for a sintering time of 1 or 2 hours. Sintering during 4 hours gives in all represented cases the lowest strength for a given powder bed thickness. The E-modulus and the density increase significantly as the sintering time rises. Furthermore by comparing figures 1 and 2 it becomes clear that a higher density does not necessarily correspond with a higher strength.

Powder bed thickness : From figure 1 one can also deduce that the highest strength is obtained for a powder bed thickness of 6 cm, irrespective of sintering or injection moulding parameters. The E-modulus and the density (see figure 2) decrease with increasing powder bed thickness for sintering times of 1 and 2 hours whereas for a sintering time of 4 hours the reverse occurs. It is clear that the optimal powder bed thickness is 6 cm.

Fracture surface examination : The fracture surfaces of nineteen bars were examined. In nine of the examined bars porous zones and/or capriciously formed porosities were identified as failure initiation sites. These bars were sintered either during 4 hours or in a powder bed with a thickness of 2 or 4 cm. Eight bars ruptured at edge flaws whilst the fracture origin of the two remaining bars could not be identified.

Batch Sintered in Optimal Conditions

A batch of 108 bars was fabricated making use of the optimal processing parameters as determined before. Table 1 summarizes the mechanical properties.

TABLE 1

Mechanical properties of injection moulded pressureless sintered Si_3N_4 bars (T_s = 1800°C, t_s = 2 hours, PB = 6 cm).

number of bars tested	108
density (g/cm^3)	3.062 ± 0.005
E-modulus (GPa)	279 ± 2
four-point bend strength (MPa)	466 ± 106
Weibull modulus	5.3 ± 0.4
fracture toughness K_{IC} (MPa $m^{1/2}$)	4.5 ± 0.2

± x = ± standard deviation.

Fracture surface analysis was performed on 10 bars. In five specimens a porosity containing a small metallic particle could be identified as the failure initiation site. Three specimens ruptured due to an edge flaw whilst the fracture origin of the two remaining bars could not be identified.

DISCUSSION

Sintering during 4 hours or sintering in a powder bed with a thickness of only 2 or 4 cm lowers the mechanical strength significantly. Fracture surface analysis revealed that defects such as porous zones and capriciously formed porosities caused failure of bars sintered in these less optimal conditions. This can be accounted for by decomposition of Si_3N_4 and evaporation of the liquid phase. The highest four-point bend strength was obtained for a sintering time of 2 hours and a powder bed thickness of 6 cm. The E-modulus and the density decrease with decreasing powder bed thickness for sintering times of 1 or 2 hours. The reverse occurs for a sintering time of 4 hours. This can be attributed to the thermal inertia of the powder bed resulting in a somewhat lower effective sintering temperature as the powder thickness increases for sintering times of 1 and 2 hours.

ACKNOWLEDGEMENTS

The authors wish to thank DSM for financial support and permission to publish this work. They also want to express their gratitude to L. Knaepen for performing the SEM-work and to the secretarial staff for typing this text.

REFERENCES

1. G. Ziegler, J. Heinrich and G. Wötting, Review : Relationships between processing, microstructure and properties of dense and reaction-bonded silicon nitride, J. Mater. Sci., 1987, **22**, 3041-86.
2. G.R. Anstis, P. Chantikul, B.R. Lawn and D.B. Marshall, A critical evaluation of indentation techniques for measuring fracture toughness, J. Am. Cer. Soc., 1981, **64** (9), 533-38.

THE USE OF PLASMA SYNTHESIS AND PIGMENT COATING TECHNOLOGY TO PRODUCE AN YTTRIA STABILIZED ZIRCONIA HAVING SUPERIOR PROPERTIES

G. P. DRANSFIELD, K. A. FOTHERGILL AND T. A. EGERTON

Tioxide UK-Ltd., Central Laboratories, Portrack Lane, Stockton-on-Tees, Cleveland, TS18 2NQ, Great Britain

ABSTRACT

ZrO_2 has been synthesized from $ZrCl_4$, using a DC plasma. The powder is ultrafine (~16 m^2/g, median size <0.1 μm). Si, Fe and Na impurities are all less than 100ppm. Pigment coating technology has been used to add yttria stabilizer.

The coated powder has excellent sinterability. It may be sintered at 1350°C and above to >97% density. Peak strengths of ~1.4 GPa, have been recorded for 2 mole% yttria addition. The material also displays superior ageing under water at 180°C. The properties are compared with test pieces from commercially available yttria stabilized zirconias.

INTRODUCTION

Gas phase oxidation provides an attractive route for the preparation of fine non-aggregated powders of high chemical purity, but major practical difficulties must be overcome if such processes are to be operated at a commercial scale. Not the least of these, are the problems of heating reactants to the high temperatures required. In this work, a DC plasma has been used as a heat-source in the oxidation of zirconium tetrachloride. The zirconia crystals are then coated with stabilizing oxides and in this way an intimate mixture of zirconia and stabilizer is achieved.

MATERIALS AND METHODS

Zirconium tetrachloride is vaporized and fed into a plasma reactor where it reacts with oxygen at temperatures above 1200°C. The zirconia, which is formed is slurried in water and coated with a stabilizer, such as yttria, according to a proprietary method [1,2]. Typically, a coated powder contains 2-2.5 mole % yttria.

For densification studies, die-pressed samples were heated, at 10°C/min, to 1570°C in a sintering dilatometer. For strength tests 30 mm diameter discs were die-pressed and sintered at 1400 - 1500°C, with a soak time of 1-2 hrs.

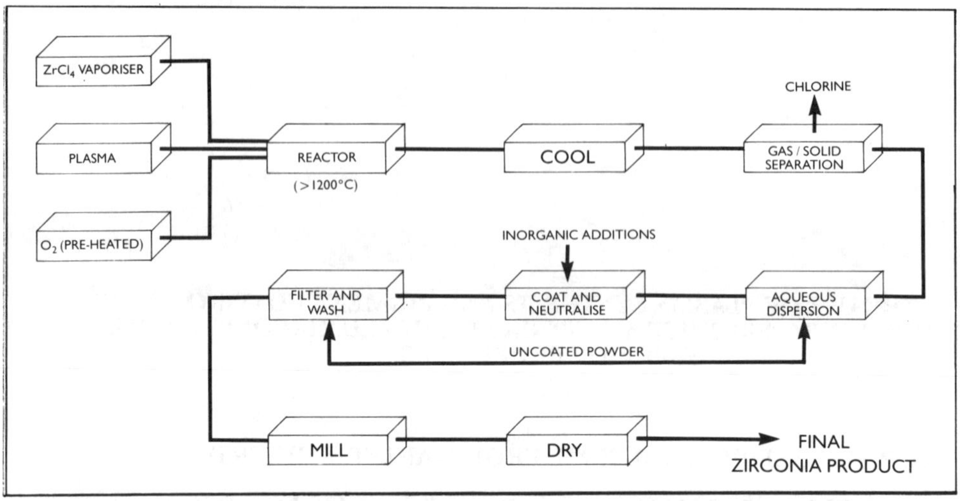

Figure 1: Process Flow Chart for yttria coated Zirconia powder.

The sintered bodies were strength tested by the biaxial disc flexure technique, described by Sivill.[3] For ageing studies : sintered discs were dipped in distilled water, in an autoclave at 180°C, following the method of Nakajima et al.[4] Transmission electron microscopy (TEM) was used to examine the starting powders and scanning electron microscopy (SEM) was used to examine the sintered microstructures.

RESULTS AND DISCUSSION

Plasma synthesis yielded a fine zirconia powder (crystal size 0.06 μm) with a narrow size distribution, as shown in Fig 2. The powder is low in Na and Si (<10 ppm), Fe, Ni (<100 ppm). The Al content is typically 800 ppm.

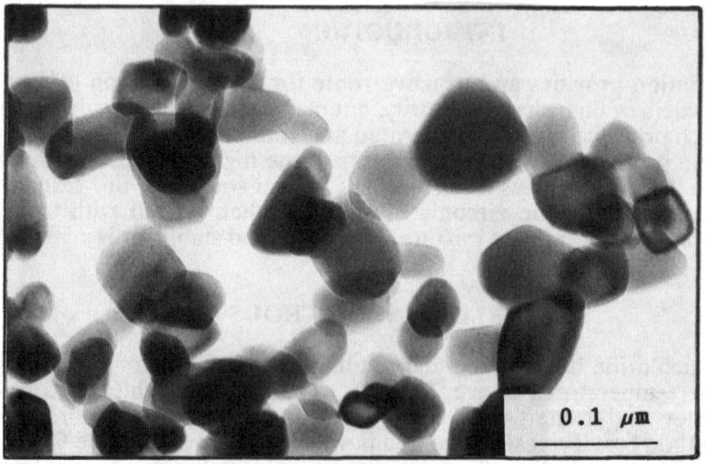

Figure 2: TEM Micrograph illustrating the morphology and size distribution of zirconia powder, produced by plasma oxidation

On coating with 2 - 2.5 mole % yttria, the surface area increases typically from 16 m^2/g to 22 m^2/g. Figure 3 shows that coating provides an intimate way of distributing the stabilizing oxide.

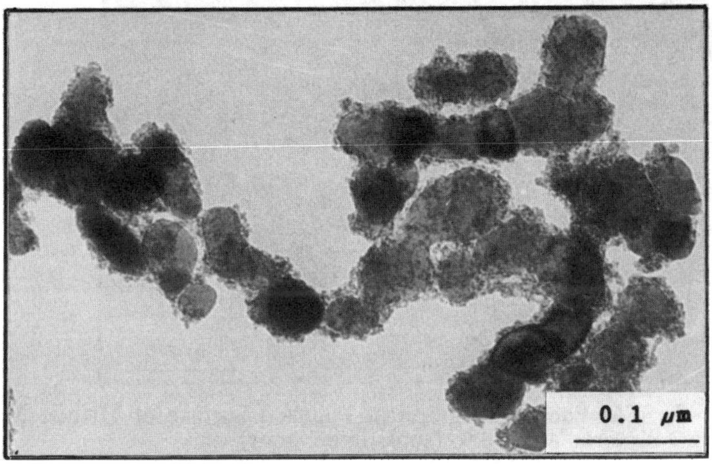

Figure 3: TEM Micrograph of coated zirconia powder

The coated powder processes to display excellent sinterablity, giving a final density exceeding 6 Mg/m^3 at temperatures below 1500^oC. A comparison of the sinterability of Tioxide powder with commercially available powders (denoted I, II and III) displays the relatively low temperature at which it densifies (see Fig 4).

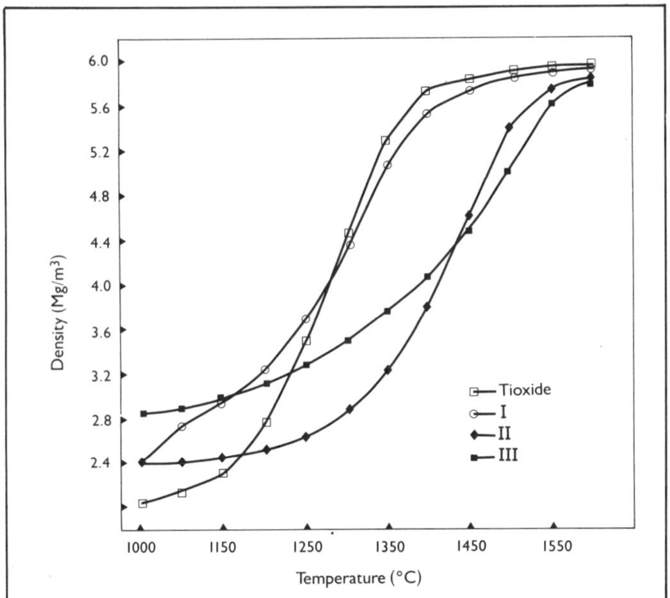

Figure 4: Densification Curves for Tioxide and Competitor Powders.

The sintered product consists of fine (~0.4 μm) tetragonal grains, as illustrated in Figure 5.

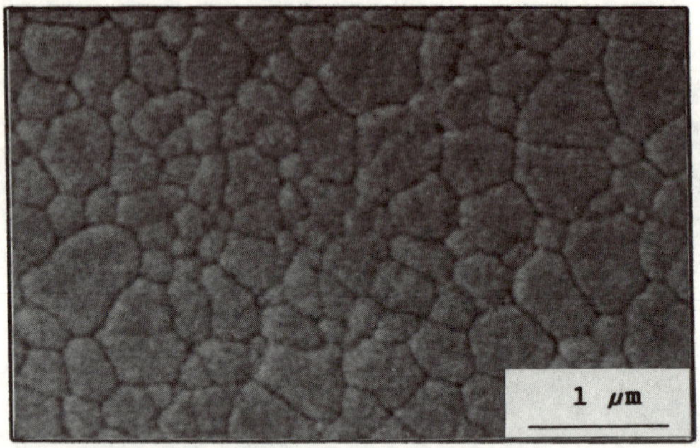

Figure 5: SEM Photomicrograph of Polished Surface of Tioxide Material, sintered at 1450°C (soak time 1 hour)

The resultant material is of high strength - up to 1.4 GPa for 2 mole% yttria addition. This low stabilizer requirement, relative to commercially available powders, indicates that a more homogenous product has been achieved.

The material displays superior resistance to ageing, as illustrated in Figure 6.

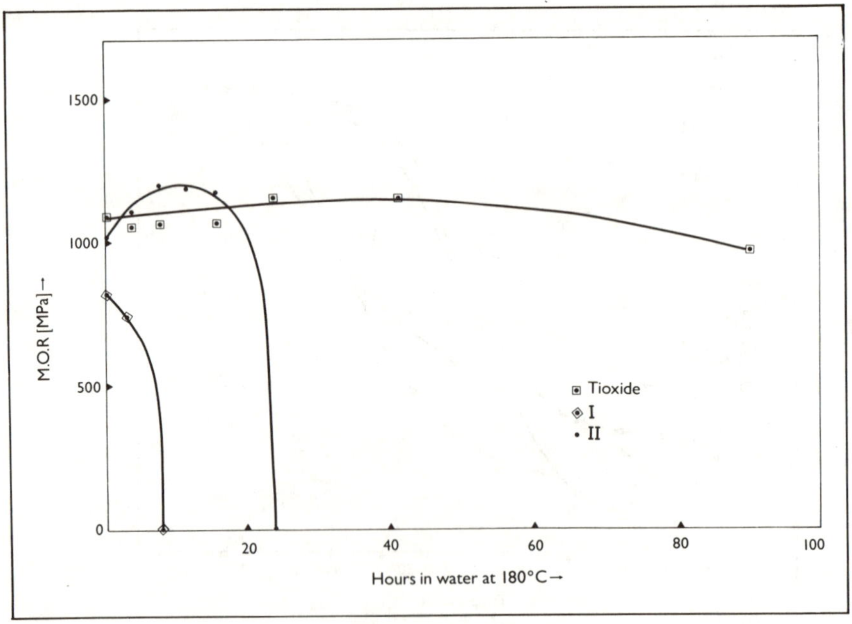

Figure 6: Comparative strengths of Tioxide and commercially available materials, upon ageing in water at 180°C

This superiority is believed to be due to the Tioxide material's high purity, small grain size and homogenous stabilizer distribution.

CONCLUSIONS

The work has shown that an ultra-fine powder, with narrow size distribution and high purity can be made at significant scale, using Tioxide's proprietary plasma technology. Furthermore, it has been demonstrated that pigment coating technique can be used to give a yttria stabilized powder with excellent sintering activity. The densified material has a very fine grain structure and superior mechanical properties at relatively low yttria loadings.

The material's resistance to ageing under water at elevated temperatures is especially worthy of note.

ACKNOWLEDGEMENT

The authors gratefully acknowledge Tioxide UK Ltd. for their kind permission to publish this paper.

REFERENCES

1. Egerton, T. A., Lawson, J. and Frost, P.W., Stabilized metallic oxides. British Application, 2, 181, 723A, 1986.

2. Egerton, T.A., Fothergill, K. A. and Dransfield, G.P., Stabilized metallic oxides. British Application 2,204,030A, 1988.

3. Sivill, A.D., Ph.D.Thesis, The development and application of a statistical basis for the design of brittle materials, Department of Mechanical Engineering, Nottingham University, 1974.

4. Nakajima, K., Kobayashi, K. and Murata, Y., Phase stability of Y-PSZ in aqueous solutions. In Advances in Ceramics Volume 12, Science and Technology of Zirconia II. ed. N. Claussen, M. Ruehle and A. H. Heuer, The American Ceramic Society, Columbus, Ohio, 1984, pp 399-403.

ZIRCONIA TOUGHENED ALUMINA COMPOUNDS FOR INJECTION MOULDING MADE USING A TWIN-SCREW COMPOUNDING EXTRUDER.

NASEEM THEILGAARD
Department of Plastics Technology, the Technological Institute, Gregersensvej, DK-2630 Taastrup, Denmark.

ABSTRACT

Zirconia based ceramics have now been developed to the stage where design of microstructure is possible by control of composition, fabrication route and subsequent processing to the final product.

This paper will describe a method whereby it is possible to produce a zirconia toughened alumina ceramic compound suitable for injection moulding using a twin-screw compounding extruder in conjunction with a high speed fluidising mixer. A comparative evaluation of the resulting dispersion will be made with a dispersion produced by a chemical precipitation method using SEM analysis.

The results of the 4-point bend strengths, Weibul moduli, K_{1C} values, density measurements and X-ray diffraction analysis tests carried out on the injection moulded samples produced by this method will be presented.

INTRODUCTION

The potential of zirconia for increasing both the strength and toughness of ceramics by utilizing the tetragonal -> monoclinic phase transformation of metastable tetragonal particles induced by the presence of the stress field ahead of a crack was first published by Garvie, Hannink and Pascoe in their seminal article "Ceramic Steel" [1].

A microstructure which uses a polycrystalline agglomerate of metastable zirconia in a matrix of fine grained alumina is described by Stevens and Evans [2]. Significant processing advantages are achieved by the use of alumina as a matrix material whereby raw material costs are lowered and at the same time there is an improvement in mechanical properties of the resulting ceramic. For the successful fabrication of such microstructures, it is essential that a high density is attained and the agglomerated zirconia polycrystalline is "tuned" to remain metastable, but that when the constraint is removed the maximum amount is transformed [1].

MATERIALS AND METHOD

The ceramic materials studied were made from ALCOA alumina powder CT 3000 SG and Dynamit Nobel Yttria stabilized zirconia powder, DYNAZIRCON F-5Y.

Two compounds were made, one which comprised entirely of alumina powder incorporated into a previously optimized binder system, TP IV, and the other which comprised of a ceramic powder consisting of approx. 14 vol% zirconia and 84 vol% alumina incorporated into the same binder system as for the alumina compound, TZP IV.

Figure 1 shows the process flow diagram and the equipment used at each stage.

Stage I	COMPOUNDING Ceramic Powder + Binder system	I. High Speed Fluidis-hing mixer II. Laboratory Twin Screw Compounding Extruder
Stage II	INJECTION MOULDING Granulated compound	Reciprocal Screw Injection moulding machine
Stage III	BINDER REMOVAL Injection moulded parts Green Bodies	Atmosphere/Vacuum Furnace - max Temp. 1100°C
Stage IV	SINTERING Presintered parts NET SHAPED PARTS (NEAR NET SHAPED PARTS)	High Temp furnace min. 1600°C

Figure 1. Process Flow Diagram and Equipment.

During the compounding stage all the components of the organic binder and the ceramic powder were premixed in a high speed fluidising mixer. The premix was then fed into a twin-screw compounding extruder. The compound was extruded through a single strand die, air cooled and pelletized.

Both compounds were subsequently injection moulded using a STÜBBE SKM 51 machine. The mould geometry used was suitable for producing samples which could be tested for 4-point bend strengths.

The resulting moulded parts made from the two compounds were then simultaneously heated in the vacuum furnace to remove the binder by thermal decomposition after which they were sintered at 1550 deg. C for 5 hours.

RESULTS AND DISCUSSION

Scanning electron microscopic analysis of the sintered samples was carried out to evaluate the homogeneity and microstructure of the sintered samples. Fig. 2 shows the micrograph of the sample made by the method described in this paper, and Fig. 3 shows the micrograph of a zirconia toughened alumina made by a chemical precipitation method*.

Figure 2. Micrograph of the TZP IV Sample. Magnification X4.00K.

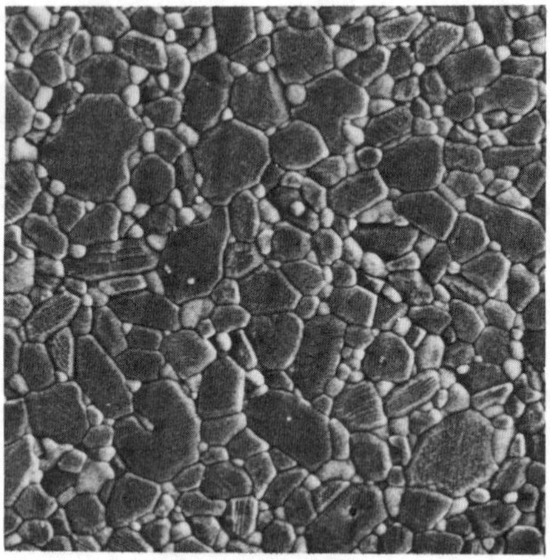

Figure 3. Micrograph of the zirconia toughened alumina made by chemical precipitation. Mag. X5.00K. * Courtesy of Risø National Laboratory.

The mechanical properties of the sintered samples were determined using 4-point bend strength tests and Weibul modulus evaluation. The K_{1C}, hardness and densities were also measured. The results are shown in Table I.

TABLE 1
Mechanical and Physical Properties of TP IV and TZP IV Samples after sintering.

Compound No.	TP IV	TZP IV
Volume fraction alumina in sintered body	100%	86%
Volume fraction zirconia in sintered body	0%	14%
App. sintered density g/cc	3.93	4.23
Open porosity	0.14%	0.49%
Bend strength σ_{50} MPa	282	457
Weibul modulus	7.3	17.8
No. of samples	20	13
K_{1C} MPa $m^{1/2}$	3.84+/-0.39	4.90+/-0.40
Hardness GPa	17.01+/-0.21	16.57+/-0.21

X-Ray diffraction analysis was also carried out on the strained surfaces of some of the samples used in the 4-point bend tests. The results are shown in Figure 4 and Figure 5.

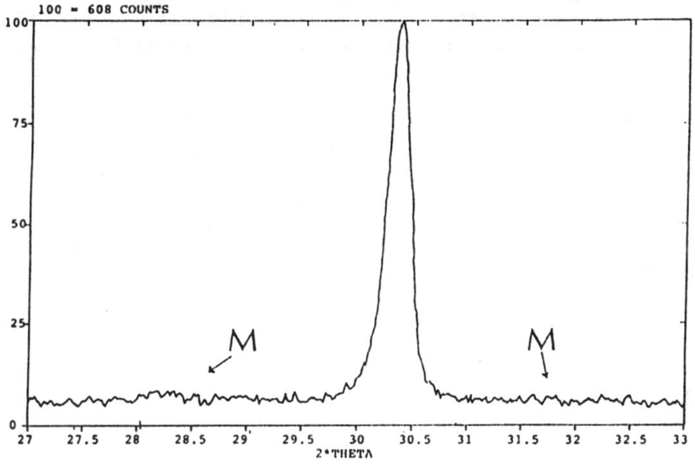

Figure 4. X-Ray Diffraction Results of a TZP Sample with Bend Strength = 271 MPa.

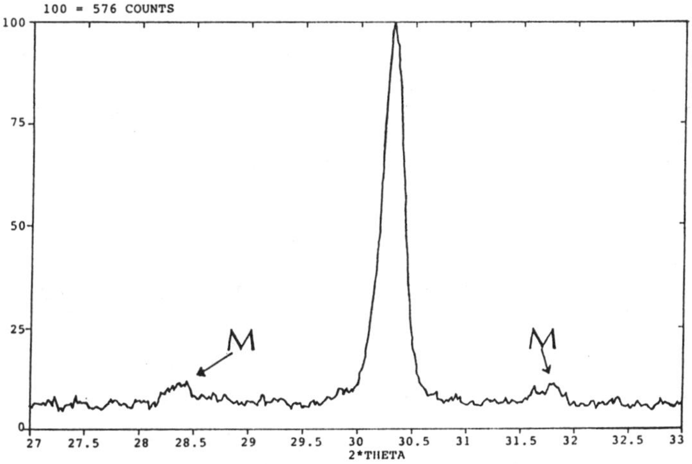

Figure 5. X-Ray Diffraction Results of a TZP Sample with Bend Strength = 453 MPa.

From the scanning electron micrographs it is seen that the homogeneity and dispersion of the zirconia toughened alumina produced by the method described here is comparable to that obtained by the chemical precipitation method.

The results for the increased 4-point bend strengths, i.e. from a maximum value of 381 MPa for alumina to 504 for the zirconia toughened alumina, as well as the increased K_{1C} values from 3.84 $MPa.m^{1/2}$ to 4.90 $MPa.m^{1/2}$ indicate that the presence of the zirconia incorporated into the alumina matrix by the method described results in a tougher ceramic. These values compare very favourably with the results obtained from the tests carried out on the samples made by the chemical precipitation method [3], where the max. 4-point bend strength measured was 346 MPa and the K_{1C} value was 5.16 $MPa.m^{1/2}$.

The toughening mechanism of phase transformation from tetragonal to monocline is further confirmed by the X-Ray diffraction results where the low strength failure does not give rise to a monoclinic phase whereas in the case of the higher strengths there is a corresponding increase in the monoclinic phase. The reasons for the low bend strengths can in many cases be traced back to defects introduced during the injection moulding process caused by the non-ideal flow behaviour of the material as well as insufficient venting in the mould resulting in premature failure.

CONCLUSION

The above results indicate that the method described in this paper can seriously be considered as an alternative method for producing zirconia toughened aluminas. The principle can possibly also be applied to other particulate ceramic composite systems and, by virtue of the relative ease with which the composites can be made, it could be commercially attractive.

ACKNOWLEDGEMENTS

The work reported in this paper has been financed entirely by the Danish Council of Technology and has been carried out in close co-operation with the Metallurgy Department at RISØ National Laboratory, Denmark, the Technical University of Denmark, the Chemical Laboratory and the Industrial Metallurgy Departments at the Technological Institute, Denmark.

REFERENCES

1. Stevens, R., An Introduction to Zirconia, Magnesium Electron Publication No. 113, July 1986. Ref. 6, p. 17.

2. Stevens, R. and Evans P.A., Transformation Toughening by Dispersed Polycrystalline Zirconia, Br. Ceram. Trans. J., 1984, **83**.

3. Maschio, S. and Toft Sørensen, O, Mechanical Properties and Microstructure of Zirconia Toughened Alumina. Risø-M2675, Risø National Laboratory, Denmark, 1988.

DEBINDING INJECTION MOULDED CERAMICS

MOHAN J. EDIRISINGHE
Department of Materials Technology,
Brunel University, Uxbridge, UB8 3PH, U.K.

ABSTRACT

Removal of polymeric binders used to process engineering ceramic powders by injection moulding depends on factors such as the geometry of the components, material properties and process control parameters. Identification of these factors and the mechanisms involved is essential to develop guidelines that will ensure low scrap rates which at present eludes the industry. Experimental evidence obtained from several investigations where polypropylene based formulations were subjected to binder removal by pyrolysis under different conditions have been used for this purpose.

INTRODUCTION

Injection moulding ceramic powder-polymer suspensions offers a very attractive route for automated production of complex shaped ceramic artefacts at high rates with good dimensional accuracy (1-3). The process consists of four stages. Firstly, a homogeneous blend is prepared by compounding the ceramic with polymeric binders in order to impart flow properties to the powder. Secondly, the ceramic-polymer formulation is injection moulded to the required shape. Thirdly, the binders are removed and finally the ceramic is sintered.

During the past 4.5 years the Ceramics Processing Group at Brunel University have investigated this fabrication route, stage by stage, with the intention of identifying suitable material properties and process control parameters. This has been achieved for the first two stages (4). Formulations containing polypropylene have shown superior flow properties (5, 6) and have been used to produce a variety of macro defect free components including a silicon nitride rotor shape which contains a central thick section of approximately 40 mm (7-9).

Debinding moulded bodies made from several polypropylene based formulations is being systematically investigated at present. Binder removal is performed by pyrolysis which is the most widely used technique (10). This involves controlled heating of mouldings over a period of about a week. Several factors such as the section size of the components,ceramic powder and polymer properties and process variables like heating rate and atmosphere in which debinding takes place affects the operation. These, together with binder removal mechanisms, have been identified. This paper brings together these factors and the guidelines the investigations have produced.

BINDER REMOVAL MECHANISMS

Pyrolytic debinding can take place in three different ways (11, 12). When the binders used are low molecular weight waxes chain scission does not take place and the organic species diffuse to the surface and evaporate. The use of high molecular weight organic vehicles like thermoplastics cause degradation to occur. Thermal degradation takes place in the absence of oxygen and the rate of production of degradation products is the same at all points in the polymer phase. The degradation products diffuse to the surface and evaporate. The presence of oxygen causes oxidative degradation to occur at the surface, the products of which evaporate. This additional surface reaction is superimposed on thermal degradation and therefore the diffusion distance for the thermal degradation products are reduced.

EFFECT OF SECTION THICKNESS

Whatever the mechanism it is clear that diffusion of organic species to the surface affects the debinding process. This effect becomes more important with increase in section thickness. Thin sections up to about 10 mm can be subjected to binder removal with very low scrap rates but as the section thickness increases the difficulties encountered are considerable, especially with fine powders as shown in Figure 1.

8 mm　　18 mm　　25 mm

Figure 1. Radiographs of cylindrical specimens made from a 61% vol. A16 alumina formulation* subjected to debinding using ramp 1** in flowing, oxygen-free nitrogen.

* Composition (wt.%): Alumina 87.27, isotactic polypropylene 5.66, atactic polypropylene 5.66, stearic acid 1.41. Sources and densities of materials have been given elsewhere (5).

** Ramp 1: 60°C/hour to 150°C, held at 150°C for 24 hours, 2°C/hour to 350°C, held at 350°C for 24 hours, 2°C/hour to 450°C and furnace cooled.

EFFECT OF CERAMIC POWDER

Modern developments in ceramic powders have favoured uniform, unagglomerated, fine particles (13), mainly to enhance sintering characteristics. However, the packing densities of fine powders could also be high and this reduces the size of the channels available for the outward diffusion of organic species during binder removal(14). Furthermore a proportion of the organic vehicle used is adsorped onto the powder particles (14). The remainder exists in a fluid state during debinding and allows diffusion. The high surface area of fine powders increase the proportion of adsorped species which causes a further reduction in the passages for diffusion.

The particle size distribution of A16 alumina[o],MA2LS alumina[o] and pilamec silicon[o] show that the mean particle size of the last two powders are approximately an order of magnitude greater than the A16 alumina. The measured specific surface area of A16 alumina is 9500 m^2/kg compared to 223 m^2/kg for MA2LS alumina and 5700 m^2/kg for silicon. Experiments have revealed (14, 15) that debinding components containing thick-sections made from the coarser powders was more successful. Figure 2 shows radiographs of 18 and 25 mm cylindrical specimens made using MA2LS alumina after debinding. Cracks and voids are absent.The reasons for these observations are those described above.

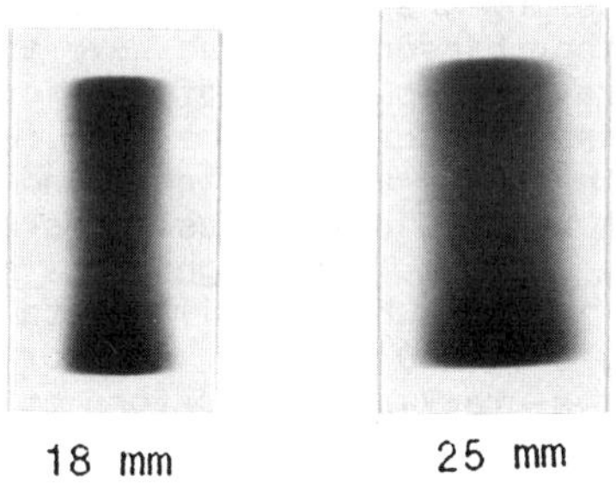

Figure 2. Radiographs of 18 and 25 mm diameter 61% vol. MA2LS***cylinders after debinding in flowing, oxygen-free nitrogen using ramp 1.

Similar components(Figure 1) made using fine A16 alumina powder severely bloated during debinding, under identical conditions, indicating that the degradation products were unable to diffuse out. Therefore, characteristics such as average particle size, particle size distribution, specific surface area and particle shape of fine powders used for ceramic injection moulding play a vital role at the debinding stage and should be carefully controlled during manufacture to suit the process.

[o] Source and details of powders are published elsewhere(5,14,15).

*** Composition (wt.%): Alumina 87.36, isotactic polypropylene 5.62, atactic polypropylene 5.62, stearic acid 1.40

It is also important to understand the behaviour of the powder during debinding (16). The relative viscosity (η) versus powder volume fraction relationship shows that η approaches infinity at some volume fraction V_{max}. Obviously, the ceramic volume fraction (V) of a formulation used for injection moulding must be less than V_{max} in order to permit flow into the mould. During debinding V approaches V_{max} and the ceramic particles move and should touch as $V \approx V_{max}$. When this occurs the formulation loses the properties of a viscous suspension and acquires those of a quasi-solid and therefore defects forming before V_{max} is achieved should take the form of bloating. Cracking should occur thereafter.It may be that in practice V_{max} is not achieved and the volume loading increases up to a limiting value V*, where $V < V^* < V_{max}$. Therefore both V_{max} (and V*) are important properties of the ceramic powder.

EFFECT OF THE BINDER SYSTEM

Binder removal studies (11,15) have revealed that it could be advantageous to use a blend of polypropylenes as the major binder rather than only isotactic polypropylene+ (IPP) . This modifies the molecular weight distribution of the IPP (17) which affects the rate of generation of volatalisable species. Atactic polypropylene+ (APP) makes a useful contribution in this respect (11, 15) and has the added advantage that it reduces the moulding shrinkage (7, 18) because of its lower crystallinity without significantly altering the flow properties (19).

The presence of a minor binder can also have an appreciable effect. Figure 3 shows thermogravimetric traces of two formulations SB4A° and SB3A°. The former contains microcrystalline wax+ (MCW) while the latter contains dibutyl phthalate+ (DBP) as the minor binder. This change gives rise to widely different weight loss versus temperature characteristics (Figure 3). SB4A shows a much smaller weight loss at low temperature (7% by 350°C) when diffusion is poor while similar data for SB3A shows that a weight loss of 17% occurs by 350°C. The cylindrical components made using SB3A showed extensive cracking. Therefore at low temperatures the weight loss (and rate of weight loss) of a binder system should be small.

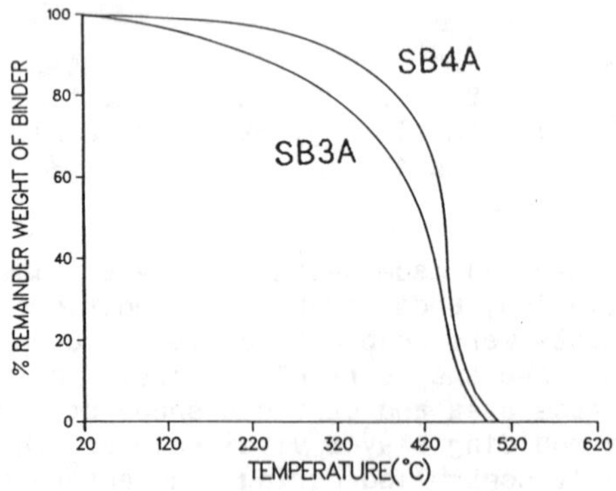

Figure 3. Thermograms for SB4A and SB3A

+Source and properties given elsewhere(5). °Details published earlier(15).

PROCESS CONTROL PARAMETERS

Heating Rate

A fairly fast heating rate up to the softening point of the ceramic-polymer formulation should be used in order to avoid defects during the initial stages (20). A 60°C/hour heating rate has been found to be suitable for IPP-APP blends (14, 20). Slow heating rates ($\leq$ 10°C/hour) tend to introduce cracks as a result of evolution of degradation products under conditions which are not helpful for diffusion. Fast heating rates ($\approx$ 280°C/hour) cause thermal shock damage. The post softening point heating rate should be very slow ($\approx$ 2 - 5°C/hour) up to about 450°C when most of the binder is removed (14, 15, 21).

Atmosphere

Evaporation processes are largely independent of the debinding atmosphere (12). Therefore the use of low molecular weight waxes as the organic vehicle will make the debinding process virtually independent of the atmosphere in which the operation is carried out. The weight loss-temperature characteristics of the binder system during thermal degradation is largely independent of the section thickness as shown in Figure 4. The presence of oxygen changes the situation significantly (Figure 4), with fine powders debinding in oxygen very quickly at rapid rates. The effect of the presence of oxygen diminishes as the component section thickness increases (Figure 4) and this agrees with the binder removal mechanisms described earlier. The presence of oxygen during binder removal has been preferred by some researchers (22, 23).

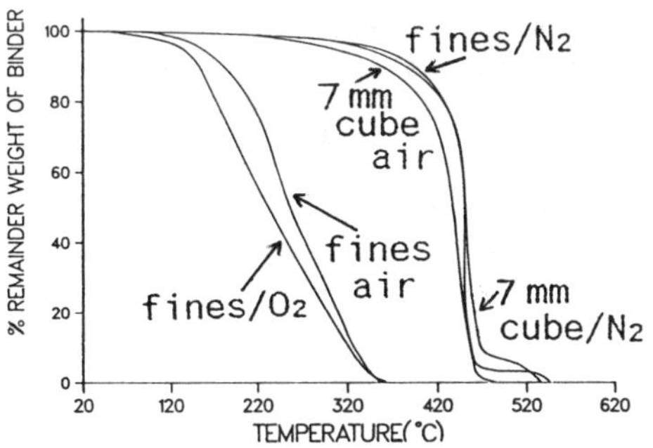

Figure 4. Thermograms of a silicon-IPP-stearic acid formulation. Details of this formulation have been published elsewhere (11).

Use of a Powder Bed

The use of a fine powder bed such as alumina to embed articles during debinding is common practice (10). However, although the powder bed has proved to be beneficial in the extraction of low molecular weight waxes that have a low viscosity (24), other binder systems have not responded favourably. Components made using composition SB4A showed bloating and cracking when subjected to binder removal in a powder bed. When the powder bed was not used identical components showed a substantial reduction in surface defects. The powder bed could be considered as an impediment (14) to the outward diffusion/evaporation mechanism described ealier. This can cause bloating and cracking.The detrimental effect of the powder bed was absent when coarse pilamec silicon powder replaced A16 alumina in formulation SB4A. Therefore this effect is dependent on the ceramic particle size of the components.

CLASSIFICATION OF DEFECTS

The extraction of organic binders from injection moulded components can be divided into three main stages and each could present characteristic defects. A catalogue of these defects and their causes have been published elsewhere (18). The first stage which includes heating to the softening point mostly gives rise to cracking due to poor diffusivity and residual stresses caused by injection moulding . Stage 2 (softening point to V*) could cause bloating due to poor diffusivity or a high rate of binder removal, and slumping if the components are poorly supported. During the third stage the material behaves as a brittle solid and obviously cracking is the main form of defect present due to a variety of reasons (18). Surface delamination and residue contamination could also occur at this stage.

CONCLUSIONS

This paper illustrates that debinding ceramic injection mouldings is a complex process which involves a variety of factors affecting the creation of defects. Unfortunately, any defects created are unforgiving and results in high scrap rates. Attempts to solve these problems using trial and error experiments have been shown to be inadequate. All these factors should be included in a numerical model that can predict suitable materials and other parameters. Thus a systematic method can be developed to overcome the problems. The Ceramics Fabrication Group of this laboratory hopes to take on this task shortly.

ACKNOWLEDGEMENTS

Support of S.E.R.C. (U.K.) for the injection moulding programme is gratefully acknowledged.

REFERENCES

1. Edirisinghe, M.J. and Evans, J.R.G., Review: Fabrication of Engineering Ceramics by Injection Moulding I. Materials Selection. Int. J. High Tech. Ceram., 1986, 2, 1-31.
2. Carlsson, R., The Shaping of Engineering Ceramics. Materials and Design, 1989, 10, 10-14.
3. Hendry, A., Processing of Engineering Ceramics. Powder Metallurgy, 1989, 31, 20-22.
4. Edirisinghe, M.J. and Evans, J.R.G., Systematic Development of the Ceramic Injection Moulding Process. Materials Science and Engineering, 1989, A(109), 17-26.
5. Idem, The Rheology of Ceramic Injection Moulding Blends. Trans. J. Br. Ceram. Soc., 1987, 86, 18-22.
6. Idem, Properties of Ceramic Injection Moulding Formulations I. Melt Rheology. J. Mater. Sc., 1987, 22, 269-277.

7. Idem, Properties of Ceramic Injection Moulding Formulations II. Integrity of Mouldings. ibid, 1987, 22, 2267-2273.
8. Zhang, J.G., Edirisinghe, M.J. and Evans, J.R.G., The Use of Modulated Pressure in Ceramic Injection Moulding. J. European Ceram. Soc., in press.
9. Idem, The Control of Sprue Solidification Time in Ceramic Injection Moulding. J. Mater. Sc., 1989, 24, 840-848.
10. Edirisinghe, M.J. and Evans, J.R.G., Review: Fabrication of Engineering Ceramics by Injection Moulding II. Techniques. Int. J. High Tech. Ceram., 1986, 2, 249-278.
11. Woodthorpe, J., Edirisinghe, M.J. and Evans, J.R.G., Properties of Ceramic Injection Moulding Formulations III. Polymer Removal. J. Mater. Sc., 1989, 24, 1038-1048.
12. Wright, J.K., Evans, J.R.G. and Edirisinghe, M.J., Degradation of Polyolefin Blends Used in Ceramic Injection Moulding. J. Amer. Ceram. Soc., in press.
13. Barringer, E.A. and Bowen, H.K., Formation, Packing and Sintering of Monodisperse TiO_2 Powders. J. Amer. Ceram. Soc., 1988, 65, C199-201.
14. Evans, J.R.G. and Edirisinghe, M.J., Pore Size and Molecular Size Effects in Ceramic Injection Moulding, to be published.
15. Edirisinghe, M.J., Removal of the Organic Binder from Moulded Ceramic Bodies, Proc. Br. Ceram. Soc., in press.
16. Wright, J.K., Edirisinghe, M.J., Zhang, J.G. and Evans, J.R.G., Particle mobility in Ceramic Injection Moulding, to be published.
17. Saito, K., Tanaka, T. and Hibino, T., Injection Moulding Composition, UK Patent 1426317, February 1976.
18. Zhang, J.G., Edirisinghe, M.J. and Evans, J.R.G., A Catalogue of Ceramic Injection Moulding Defects and Their Causes. Industrial Ceramics, in press.
19. Idem, to be published.
20. Idem, Initial Heating Rate for Binder Removal from Ceramic Mouldings. Materials Letters, 1988, 7, 15-18.
21. Quackenbush, C.L., French, K. and Neil, J.T., Fabrication of Sinterable Silicon Nitride by Injection Moulding. Ceram. Eng. Sci. Proc., 1982, 3, 20-34.
22. Mutsuddy, B.C., Oxidative Removal of Organic Binder from Injection Moulded Ceramics, presented at the Int. Conf. on Non-Oxide Tech. & Eng. Ceram., N.I.H.E., Limerick, Eire, July 1985.
23. Gilissen, R. and Smolders, A., Binder Removal from Injection Moulded Ceramics, poster presentation, 6th CIMTEC, Milan, Italy, 1986.
24. Peltzman, I. and Peltzman, M., Low Pressure Moulding of Ceramic Materials. Interceram., 1984, 4, 56.

STRUCTURE CONTROLLING OF CERAMICS BY PLASTIC FORMING

J.HAVRDA, J.TRÁVNÍČKOVÁ, F.OUJIŘÍ, J.MATOUŠEK
Department of Silicates,Prague Institute of Chemical Technology, 166 28 Prague 6, Czechoslovakia

ABSTRACT

The constitutive equation of composite paste was determined by the capillary tube viscometer method. The paste was composed of Al_2O_3-ZrO_2 matrix and t-ZrO_2 fibres. The constitutive equation is given in the form applícable to a general viscoplastic liquid. The velocity profiles were calculated from a mathematical model. It was observed that the fibre orientation in the matrix depends on the velocity gradient during extrusion and corresponds to the velocity profiles.

INTRODUCTION

The principle of preparing high quality structurally reinforced ceramics composed of Al_2O_3 and ZrO_2 issues from mastering the regulation of the products final structure. This can be regulated by the properties of the initial powder mixture and by the conditions under which the processing is conducted.

Al_2O_3 - ZrO_2 ceramics prepared from a homogeneous powder mixture of oxides having submicron grain sizes, high purity and reaction activity, exhibits a very high mechanical strength. Preparation of a body having a density close to the theoretical value and uniform in the body volume is a necessary precondition of proper technological processing of the mixture.

Composites with Al_2O_3 - ZrO_2 matrix and fibres, for example PSZ, show an increased mechanical strength. Uniform fibre distribution in the matrix volume and optimal fibre orientation are necessary for the fibres to have a positive effect. The orientation of the fibres can be influenced by the character of a velocity field during the flow of plasticized mixture /paste/. The velocity field established during paste flow can be described by a mathematical model of non-newtonian liquid provided its constitutive equation is known.

The objective of this work is to construct a mathematical model of the flow of the composite mixture through an extruder mouth and to use the model to forecast fibre orientation in the process of forming the body from the composite mixture composed of Al_2O_3 - ZrO_2 matrix and ZrO_2-Y_2O_3-Al_2O_3 fibres.

THEORY

If a tensor dynamics stress is unambiguously defined by a deformation tensor, a flow of isotropic incompressible paste with variable viscosity is described by the balance [1] :

$$\rho D_t \vec{v} = \rho \vec{g} - \text{grad } p + 2\eta/S/ \text{ div grad } \vec{v} + 2\overleftrightarrow{d} \text{ grad}\eta/S/ \qquad (1)$$

where $D_t \vec{v}$ is the substantional derivative of momentum, ρ is the paste density, $\vec{g}$ is gravity, p is pressure, $\overleftrightarrow{d}$ is the deformation tensor and $\eta/S/$ is viscosity $/S = 2\overleftrightarrow{d} : \overleftrightarrow{d}/$.

The constitutive equation of the paste is of the form:

$$\overleftrightarrow{\tau} = 2\ \eta/S/\ \overleftrightarrow{d} \qquad (2)$$

where $\overleftrightarrow{\tau}$ is the tensor of dynamics stress.

Solution of the equation (1) for the boundary conditions

$$t > 0 \qquad r = 0 \qquad \tau_{rx} = 0$$
$$r = R \qquad \tau_{rx} = \tau_s \qquad (3)$$

for isothermal stabilized paste flow through a circular extruder mouth of radius R and length L in the direction x produces an equation for the calculation of the velocity profile:

$$v_x /r/ = \int_0^R f /\tau_{rx}/ \, dr \qquad (4)$$

where $f/\tau_{rx}/$ is the inverse function to the equation (2) . The equation $\tau_s = R\Delta p/2L$ holds for shearing stress on extruder mouth wall and the equation $\tau_{rx} = r\Delta p/2L$ holds for a component of dynamic stress.

The final form of the constitutive equation of the paste for the calculation of the velocity profile can be determined by a capillary viscometer method.

PREPARATION OF THE COMPOSITE MIXTURE PASTE

Powders for Al_2O_3 matrix with 18% ZrO_2 and ZrO_2 fibres with 3% Y_2O_3 and 10% Al_2O_3 were prepared by a common precipitation of 0,4 mol solutions of respective nitrates by ammonia water in the reactor with intensive mixing at the temperature of 30°C. After washing, the precipitates were dried at the temperature of 120°C. The mixture of hydrous oxides ZrO_2-Y_2O_3-Al_2O_3 were plasticized by organic additives. The fibres were prepared by extrusion of the mixture through extruder mouth. After drying and calcining the fibres of 200 microns in diameter were treated to 3 mm length. The mixture of hydrous oxides Al_2O_3-ZrO_2 for the preparation of the matrix was calcined at the temperature of 650°C. The paste of composite mixture for the forming was prepared by plastifying the mixture of matrix and fibres.

DETERMINATION OF CONSTITUTIVE EQUATION OF THE PASTE

The constitutive equation of the paste was determined by the capillary viscometer method, which was based on a measurement of the mean velocity of paste flow through the capillary [2]. The constitutive equation of a general viscoplastic liquid

$$\tau_{rx} = \tau_0 + K \, /dv_x/\, dr/^n \qquad (5)$$

is illustrated in Fig.1 for the evaluated dependence $\tau_{rx} = f\, /dv_x/dr/$. The values of the constants are:

$\tau_0 = 154{,}7$ kPa, $K = 82{,}4$ kPa s^n, $n = 0{,}236$.

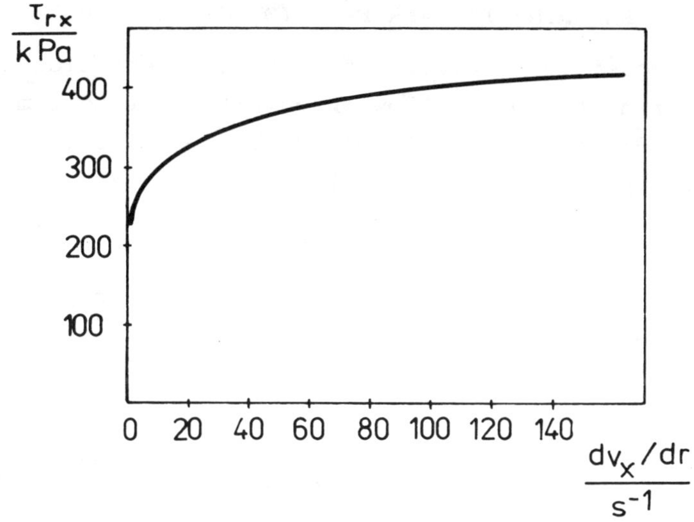

Figure 1. Plot of constitutive equation for plastified composite mixture paste.

MATHEMATICAL MODEL OF PASTE FLOW THROUGH A CIRCULAR EXTRUDER MOUTH

An equation for the calculation of the velocity profile was formulated by solving the equation (4) for the constitutive equation (5). The relation for the calculation of the velocity profile is valid within the intervals [3] :

a/ $r_0 \leq r \leq R$; $\tau_0 \leq \tau_{rx} \leq \tau_s$

$$v_x/r/ = 2L\,[K^{1/n}\,(1+1/n)\Delta p]^{-1}\,[(\tau_s-\tau_0)^{1+1/n} - (r\Delta p/2L-\tau_0)^{1+1/n}] \tag{6}$$

b/ $0 \leq r \leq r_0$; $0 \leq \tau_{rx} \leq \tau_s$

$$v_x/r/ = 2L[K^{1/n}(1+1/n)\Delta p]^{-1}\,(\tau_s-\tau_0)^{1+1/n} \tag{7}$$

where $r_0 = \tau_0 R/\tau_s$ is the radius equivalent to the end of piston flow.

EXPERIMENTAL VERIFICATION OF THE MODEL

If it is possible to control the fibre orientation in the matrix during the paste flow through the extruder mouth then the fibre direction is determined by the velocity profile. During the paste flow through the circular extruder mouth of a radius R = 5 mm the velocity profile in the Figure 2a was calculated from the model for a mean velocity flow $\bar{v}_1 = 2{,}85 \cdot 10^{-3}\ ms^{-1}$. The velocity profile in the Figure 2b was also calculated from the model for a mean velocity flow $\bar{v}_2 = 5{,}73 \cdot 10^{-2}\ m.s^{-1}$.

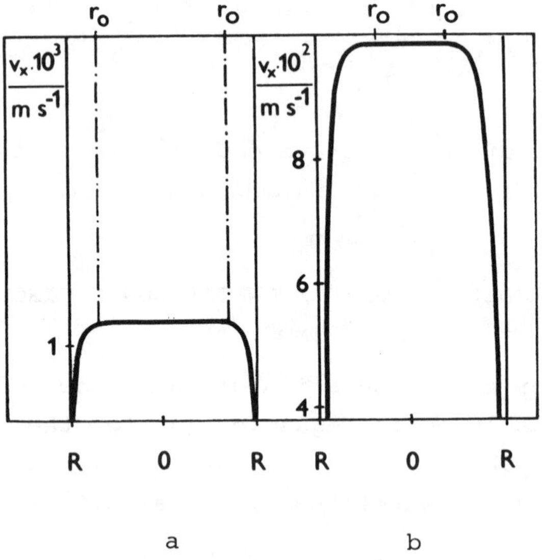

Figure 2. Velocity profile of paste flow through a circular extruder mouth.
a - $\bar{v}_1 = 2{,}85 \cdot 10^{-3} m.s^{-1}$; b - $\bar{v}_2 = 5{,}73 \cdot 10^{-2} m.s^{-1}$

It follows from the velocity profiles that the piston flow with zero gradient of velocity sets in with $\bar{v}_1$ in the interval $r \in \langle 0$ mm, $r_o = 2{,}67$ mm$\rangle$ and with $\bar{v}_2$ in the interval $r \in \langle 0$ mm, $r_o = 1{,}8$ mm$\rangle$.

The paste is transported during the flow as a whole in these intervals of radius and therefore the initial disordered fibre orientation must be maintained. The velocity gradient influencing an orientation of the fibres in the matrix during the paste flow originates in the interval $r \in \langle r_0; R \rangle$.

The results calculated from the model were experimentally verified by the shaping of the bodies from the prepared paste within two mean velocities $\overline{v}_1$ and $\overline{v}_2$. The orientation of the fibres was evaluated after cutting the bodies; the results are shown in the Figure 3.

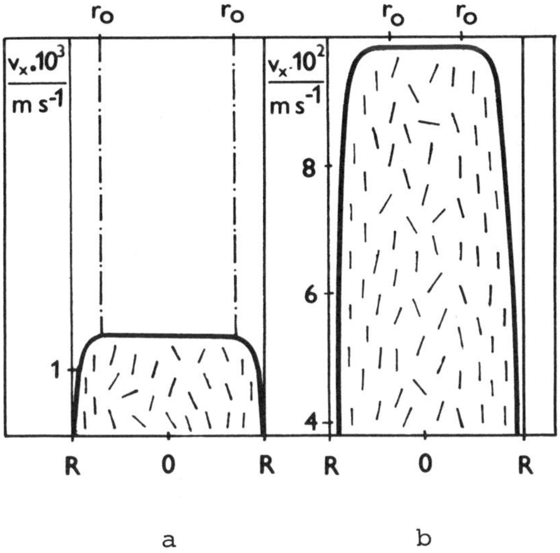

Figure 3. Orientation of fibres in the body shaped at the mean velocity a - $\overline{v}_1$ $2{,}85.10^{-3} m.s^{-1}$
b - $\overline{v}_2$ $5{,}73.10^{-2} m.s^{-1}$

CONCLUSION

The assumption about the influence of the velocity field on the fibre orientation furing paste flow was confirmed by a comparison of the evaluated fibre distribution with the corresponding velocity profiles. A mathematical model of the paste flow allows to control the conditions for reaching various required orientations of fibres ranging from unidirectionally oriented structure within the entire bulk up to structures having oriented fibres only at the surface layer of the body. The model approach allows controlled preparation of composite materials to suit specific applications. Thus the various orientations of fibres and their combinations provide a means of control over the directional mechanical strength of ceramics.

REFERENCES

1. Astarita, G., Marruci, G., Principles of Non-Newtonian Fluid Mechanics, McGraw-Hill, London, 1974
2. Havrda, J., Trávníčková, J., Oujiří, F., Determination of the Constitutive Equation for a Ceramic Paste. Silikáty, 1988, 32, pp. 305-311.
3. Havrda, J., Trávníčková, J., Oujiří, F., The Forming of Ceramic Paste. Part I - Mathematical Model. Silikáty, 1989, 33, pp. 1-5.

INFLUENCE OF ZrO_2 ON THE PREPARATION AND PROPERTIES OF TAPE-CAST, LAMINATED, ZrO_2-MULLITE COMPOSITES

MARIA HOLMSTRÖM
Swedish Ceramic Institute, Göteborg, Sweden, Box 5403, 40229
THIERRY CHARTIER, JEAN-LOUIS BESSON and PHILIPPE BOCH
UA-CNRS 320, ENSCI, 47 av. A. Thomas, 87065 Limoges, France

ABSTRACT

This study concerns the influence of ZrO_2 on the preparation and properties of laminated ZrO_2-mullite composites, processed by tape casting and reaction sintering. Al_2O_3, α-quartz and unstabilized ZrO_2 powders were used as starting powders. The preparation process of samples by tape casting is described. Microstructural features and the main characteristics of sintered materials are given and discussed, in comparison with materials prepared by dry pressing. Addition of ZrO_2 promotes densification and delays grain growth of the mullite phase, which improves the sintering behaviour and increases mechanical properties. Nearly dense ZrO_2-mullite ceramics can be sintered at temperatures which do not exceed 1560°C.

INTRODUCTION

There are several ways to increase the strength and fracture toughness of mullite ceramics. For instance, it is possible to use non-stoichiometric composition, such as SiO_2-rich mullite which shows a better high-temperature strength than 3-2 mullite. This is due to microplastic mechanisms, related to viscous SiO_2 residues (1). A more general method is to add ZrO_2, to increase toughness by stress-induced tetragonal-to-monoclinic transformation or by microcracking mechanisms (2). As far as homogeneity of green materials is concerned, tape casting generally leads to better homogeneity and smaller microcracks than other forming methods, in particular dry pressing (3). Finally, residual stresses in laminated composites produced by stacking thin layers of different compositions have a beneficial effect on the mechanical properties (4).

The aim of this study was to achieve high mechanical

properties for ZrO_2-mullite laminated composites. This requires dense, fully reacted materials, with uniform, finely grained microstructures. Therefore, this work involved three stages, namely i) the optimization of tape-casting slurries for mullite and ZrO_2-mullite systems, ii) the determination of suitable compositions (in particular the optimum ZrO_2 content), and iii) the choice of the sintering conditions.

EXPERIMENTAL

Starting materials
The powders used were alumina (A16SG, Alcoa, USA), α-quartz (C800 Sifraco, France), and unstabilized ZrO_2 (UPH12, Criceram, France). The mean particle sizes were 0.5 µm, 3 µm, and 0.3 µm, respectively. Before use, α-quartz was attrition milled (5h at 450 rpm in ethanol with ZrO_2 milling media) to reduce its mean particle size to less than 1 µm.

Three compositions were studied : one without ZrO_2, and the two others with ZrO_2 (5 and 15 vol%). The alumina-to-silica ratio was 68 wt% to 32 wt%, i.e. just below the 3-2 stoichiometric composition (71.8 wt% to 28.2%). This was expected to lead to improved mechanical properties (1). All compositions were mixed by attrition milling for 3h.

Preparation process
The flow chart of the preparation process is shown in figure 1.

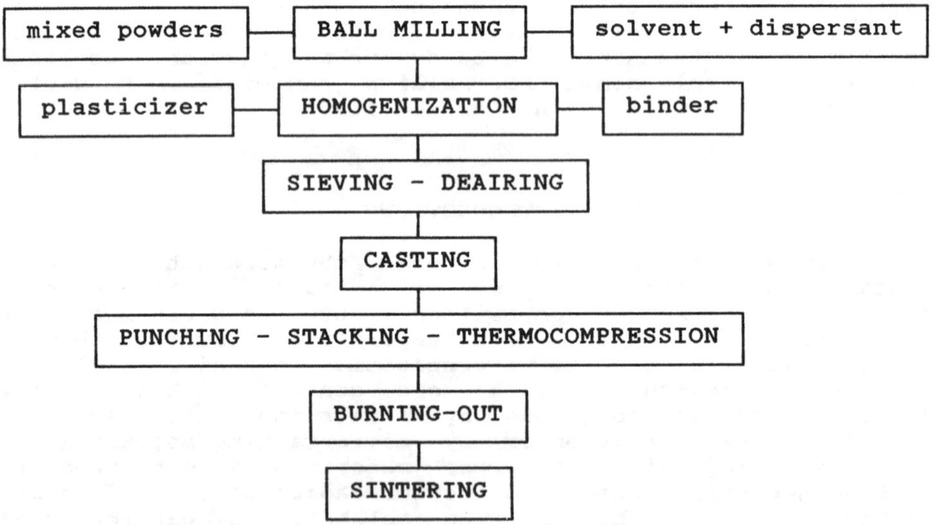

Figure 1. Preparation of laminated composites.

Organic components were burned out at a slow heating rate (0.2°C/min) from R.T. to 400°C with a plateau of 6h at this temperature. Slow heating is indeed required to avoid the cracking of samples. A two-stage sintering schedule was subsequently used, with a heating rate of 30°C/min from 400°C to 1350°C and then a heating rate of 60°C/min to the sintering temperature. It was observed that the first stage involves densification mechanisms only, whereas the second stage involves both densification and mullitization.

Characterization
Apparent density was measured by Archimedes' method in water. Semi-quantitative characterization of crystalline phases was performed using X-ray diffraction. Microstructural features were observed by scanning electronic microscopy on polished, thermally etched samples (etching treatment at a temperature of 100°C below the sintering temperature, for 10 min).

RESULTS AND DISCUSSION

Tape-casting slurries
The ZrO_2-(alumina + quartz) slurries were found to behave in a different manner than the ZrO_2-alumina slurries which were previously studied (5). This may be due to differences in the chemistry and the physical properties of the surface of powders, which have a great influence on the adsorption of polymeric dispersants (steric stabilization) and binders. The main requirements for tape-casting slurries are i) a high ratio between ceramic and non-ceramic components, ii) a low viscosity (less than 1200 mPa.s), and iii) homogeneity.

TABLE 1
ZrO_2-free mullite tape-casting slurry

material	function	wt%	vol%
$Al_2O_3+SiO_2$	ceramic powder	64.1	30.9
methyl ethyl keton	solvent	16.0	34.4
ethanol	solvent	8.3	17.7
phosphate ester	dispersant	0.6	1.0
polyvinyl butyral	binder	4.2	5.5
polyethylene glycol	plasticizer	4.9	7.4
dibutyl phtalate	plasticizer	1.9	3.1

Table 1 gives the formulation of a slurry which fulfils those requirements. Its ceramic to organic ratio is 30.9 vol% and its viscosity is about 1000 mPa.s. Its homogeneity is rather good. Furthermore, the green tape does not stick on the glass slab of the casting bench. This non-sticking behaviour is due to the fact the plasticizer is not free, but dissolved into the polymeric matrix.

Sintered materials

A previous study (6) on 3-2 mullite and ZrO_2-mullite ceramics prepared by dry pressing demonstrated that it is possible to obtain fully-reacted mullite materials, with a density better than 98%, for a sintering treatment of 2h at 1620°C. Shorter times do not allow the reaction to be completed and some Al_2O_3 residues were then detected. Figure 2 shows that the addition of ZrO_2 increases densification and accelerates mullitization, which is in agreement with results of Prochazka et al. (7).

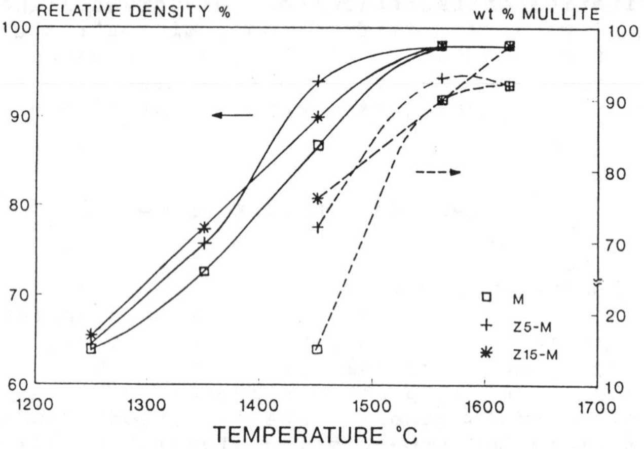

Figure 2. Relative density and mullitization of dry-pressed samples of ZrO_2-free mullite and ZrO_2-mullite (5 and 15 vol% of ZrO_2) sintered for 30 min at various temperatures.

In the present work on tape-cast, silica-rich mullite, a first improvement which was found was a better packing of the green state (65% instead of 60% for pressed samples). Nearly full densification (>98.5%) was achieved for a treatment of 2h at 1560°C only. As in the case of 3-2 mullite, the microstructure is a mixture of equiaxed (around 2 µm) grains and of elongated (around 1x4 µm) ones. Longer sintering times lead to grain growth and the development of elongated grains. For instance, a soaking time of 6h leads to a duplex microstructure with 3 µm, equiaxed grains and 2x8 µm, elongated needles.

ZrO_2-mullite materials exhibit smaller grains than their ZrO_2-free counterparts. For soaking treatments of 6h at 1560°C, the mean sizes are 0.8 µm, 1x2.5 µm, 0.8 µm, and 0.3 µm for equiaxed mullite grains, elongated mullite grains, intergranular zirconia inclusions, and intragranular zirconia inclusions, respectively. It must be pointed out that the beneficial influence of ZrO_2 for slowing down grain growth occurs only when the zirconia content is large enough. When the ZrO_2 content is too low, the formation of a solid solution between zirconia and mullite can favour grain growth, as indicated by I.M. Low and R. McPherson (8).

The stress-induced-transformation toughening depends on the ratio of tetragonal to non-tetragonal zirconia, and this ratio depends on the size of the ZrO_2 inclusions. They are larger in the 15 vol% ZrO_2 materials than in the 5 vol% ZrO_2 materials, due to coarsening during sintering. For a sintering of 6h at 1560°C, ZrO_2 remains 100% tetragonal in 5% ZrO_2 materials whereas it is 66% monoclinic in 15% ZrO_2 materials.

CONCLUSION

ZrO_2-free mullite and ZrO_2-mullite composites can be prepared by tape casting of mixtures of alumina, quartz, and zirconia, and subsequent reaction-sintering. The compositions of the tape-casting slurries were optimized to produce homogeneous green tapes, with a high packing density (65%), and to prevent adhesion on the casting bench.

The addition of ZrO_2 enhances both densification and mullitization. ZrO_2 also acts as a grain-growth inhibitor.

ZrO_2-mullite ceramics with a densification higher than 98.5% and a mean grain size of about 1.5 µm can be obtained by a reaction-sintering treatment of 2h at 1560°C.

REFERENCES

1. Boch, P., Chartier, T. and Rodrigo P.D.D., High-purity mullite ceramics by reaction-sintering. To be published in Advances in Ceramics, Am. Ceram. Soc., 1989.

2. Claussen, N. and Jahn, J., Mechanical properties of sintered, in situ-reacted mullite-zirconia composites. J. Am. Ceram. Soc., 1980, 63, 228-29.

3. Chartier, T., Besson, J.L. and Boch. P., Mechanical properties of ZrO_2-Al_2O_3 laminated composites. Advances in Ceramics, Science and Technology of Zirconia III, 1988, 24, pp. 1131-38.

4. Chartier, T. and Besson, J.L., Behaviour of ZrO_2-Al_2O_3 laminated composites loaded by various mechanical sollicitations. In Science of Ceramics 14, ed. D. Taylor, The Institute of Ceramics, 1988 , pp. 639-44.

5. Besson, J.L., Boch, P. and Chartier, T., Al_2O_3-ZrO_2 layer composites. In High Tech Ceramics, Materials Science Monograh, 38, ed. P. Vincenzini, Elsevier Publishers, 1987.

6. Holmström, M., Chartier, T. and Boch, P., Reaction-sintered ZrO_2-mullite composites. In Ceramic Materials Research, ed. R.J. Brook, Elsevier Publishers, 1989, pp. 105-9.

7. Prochazka, S., Wallace, J.S. and Claussen, N., Micro-structure of sintered mullite-zirconia composites. J. Am. Ceram. Soc., 1983, 66, C125-27.

8. Low, I.M. and McPherson, R., Crystallization of gel-derived mullite-zirconia composites. J. Mater. Sci., 1989, 24, 951-58.

HIPing GLASS-ENCAPSULATED SILICON NITRIDE TO FULL DENSITY

N. D. BUTLER*, M. A. HEPWORTH
T&N Technology Ltd, Cawston House, Cawston,
Rugby, Warwickshire, CV22 7SA, England.
I. ITURRIZA, F. CASTRO
Centro de Estudios e Investigaciones Tecnicas de Guipuzcoa (CEIT),
Barrio de Ibaeta, s/n, 20009 San Sebastian, Spain.

ABSTRACT

Cold isostatically pressed silicon nitride green compacts have been successfully glass encapsulated and HIPped to full density. The densification behaviour of several powder mixtures, using yttria as a sintering aid, has been investigated and compared to that of silicon nitride with no additions. The resulting microstructures have been characterised by TEM. The influence of composition, on both the relative amounts of α and β Si_3N_4 and the characteristics of the intergranular phases formed after HIPing, was investigated by XRD. The crystallization behaviour and the room temperature hardness of several specimens was also determined and represented as a function of the characteristic parameters of different HIPing cycles.

INTRODUCTION

Silicon nitride is a candidate material for high temperature applications in engineering components, having the desirable intrinsic properties of high decomposition temperature, good oxidation resistance, good thermal shock resistance and high strength to weight ratio. Due to the low self-diffusivity, Si_3N_4 ceramics are usually densified by the addition of oxides such as Al_2O_3, MgO or Y_2O_3 [1]. Densification occurs via a liquid phase sintering mechanism in which the $\alpha \longrightarrow \beta$ transformation of Si_3N_4 occurs [2]. The amount of oxides, which cause a degradation of high temperature strength, can be reduced if Hot Isostatic Pressing (HIP) [3] is used. This process involves encapsulation of the Si_3N_4 in a gas-impermeable membrane, and pressurization typically using Argon up to 300 MPa. Y_2O_3 is a favoured sinter aid [4], giving the most refractory grain boundary phases. In this work the densification behaviour of Si_3N_4 with up to 5 w/o Y_2O_3 has been studied using a high temperature HIP apparatus. The possibility of HIPing pure Si_3N_4 without additives [5] was also investigated.

EXPERIMENTAL PROCEDURE

The experimental powder mixes were prepared by adding controlled amounts up to 5 w/o of high purity Y_2O_3 (> 99.99%) to UBE E-10 Si_3N_4 powder of chemical composition (w/o): 1.9 O_2, 0.01 Fe, 0.2 C; sub-micron particle size and initial α/β of 95/5. After ball milling using silicon nitride milling media green compacts were cold isostatically pressed and glass encapsulated. HIPing was performed within a temperature range from 1675 to 1900°C and pressure range 10 to 150 MPa for 1 hour using an ASEA Q1H-6 hot isostatic press.

Density was measured using the Archimedes principle, and characterisation of the $\alpha \longrightarrow \beta$ transformation and secondary phases before and after crystallization treatment was made by X-ray diffractometry (XRD). The α/β Si_3N_4 ratios were calculated from the (210) and (101) reflections. Microstructural analysis in the transmission electron microscope (TEM) was made by ion-milling of mechanically ground slices to electron transparency, and coating with a carbon film to avoid charging. Analysis was carried out using a Philips CM12 with a LaB_6 filament.

RESULTS AND DISCUSSION

Figure 1 shows the recorded density as a percentage of theoretical density for Si_3N_4 with 2%Y_2O_3 and without additives as a function of HIP temperature for HIP pressure 150MPa. Compositions with 2-5%Y_2O_3 were all fully densified at HIP temperatures of 1675°C and over, indicating that sufficient liquid is generated. The maximum density achieved for Si_3N_4 without additives was 85% of the theoretical density at 1950°C and 150MPa. The effect of variation of HIP pressure at 1725°C showed that full densification of samples containing 2% and 4%Y_2O_3 was obtained at 50MPa or higher. At 1725°C and 10MPa these two compositions were sintered to 90% and 94% respectively of their theoretical densities, at which point a state of closed porosity has been reached. The measurement of α-Si_3N_4 content of Si_3N_4 with 2%Y_2O_3 as a function of temperature is shown in Figure 2. At HIP temperatures below 1800° full densification is achieved without complete $\alpha \longrightarrow \beta$ transformation. It was observed that transformation was greatest for compositions containing 3% and 4%Y_2O_3 at a given temperature, indicating the effect of the Y_2O_3/SiO_2 ratio on glass viscosity.

Crystallization treatments showed the formation of the y polymorph of $Y_2Si_2O_7$ [6] in amounts increasing with Y_2O_3 level. Compositions containing 5%Y_2O_3 also contained the apatite $Y_5(SiO_4)_3N$ and 1%Y_2O_3 compositions contained traces of Si_2N_2O. The appearance of these phases relates very closely to the position of the starting compositions in the Si_3N_4-SiO_2-Y_2O_3 phase diagram [7] taking into account the oxygen content of the Si_3N_4 powder. Dense ceramics in the Si_3N_4-Si_2N_2O-$Y_2Si_2O_7$ compatability triangle [7] have an excellent oxidation resistance, and this has been achieved for compositions containing 1-4%Y_2O_3.

Hardness measurements by Vickers indentation revealed an increase in Hv from around 2000 to 2500 in dense samples with the highest α-Si_3N_4 content, which is in agreement with previous reports [8].

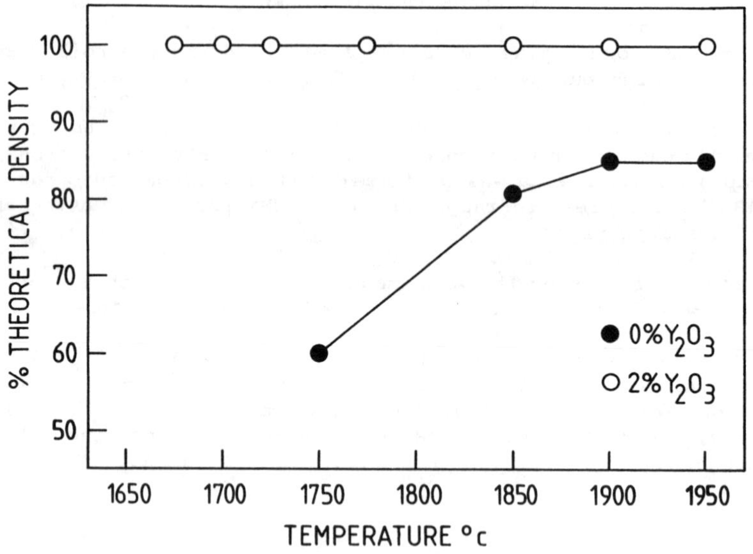

Figure 1. Percentage T.D. v HIP Temperature for HIP at 150 MPa 1 hour

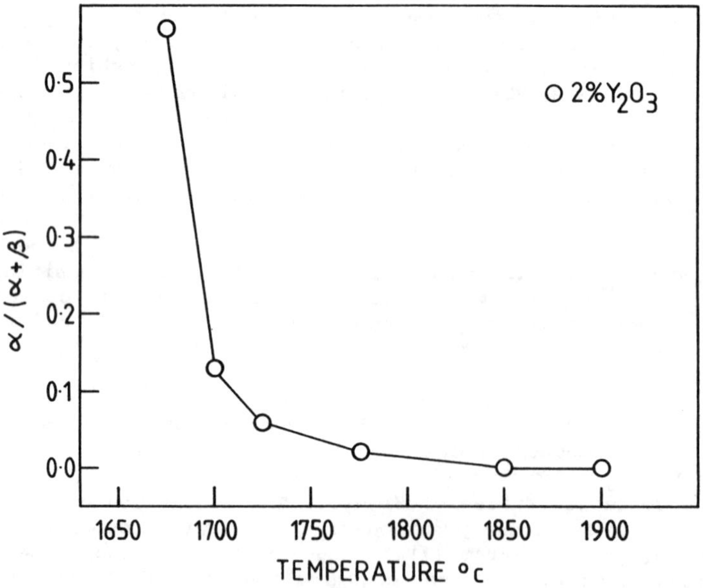

Figure 2. Transformation v HIP Temperature for HIP of Si_3N_4 + 2% Y_2O_3 at 150 MPa 1 hour

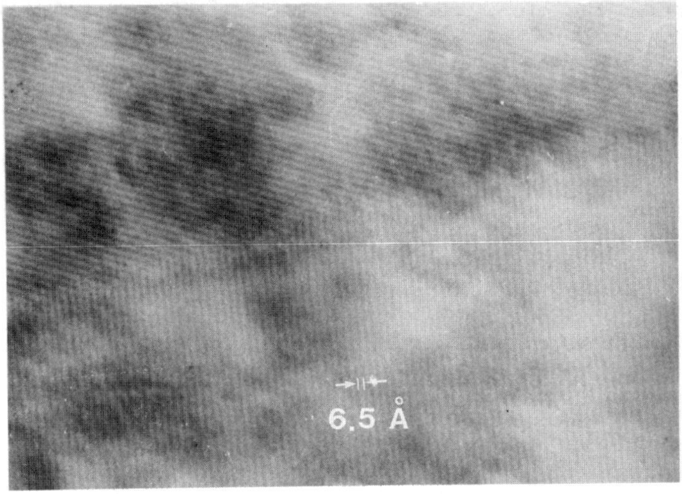

Figure 3. Glass-free Grain Boundary shown by HREM

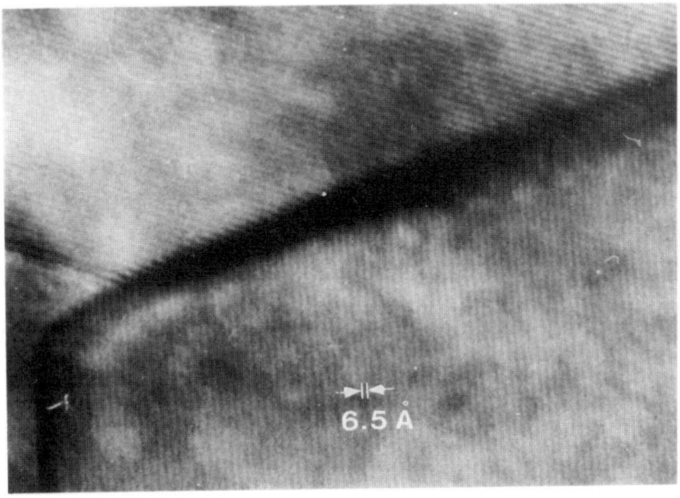

Figure 4. Direct Lattice Image showing Y_2O_3-rich Glass in the Boundary

Observations of the microstructure by TEM are shown in Figures 3 and 4 for Si_3N_4 containing 2% Y_2O_3. The majority of the grain boundaries are free of any second phase, but where there is an intergranular phase it is usually concentrated as small liquid pockets in triple points. Due to the high glass viscosity, there is a slower diffusion of α-Si_3N_4 and reprecipitation of β-Si_3N_4 occurs at a greater rate, causing high nucleation density and the observed small grain size of approximately 0.2 µm.

CONCLUSIONS

This study has shown that densification of Si_3N_4 with low Y_2O_3 levels can be achieved by glass encapsulation and HIPping at relatively low temperatures and pressures, without the need for complete $\alpha \longrightarrow \beta$ transformation. Successful densification of additive-free Si_3N_4 is dependent on powder characteristics and HIP method. Crystallization leads to oxidation resistant ceramics containing $Y_2Si_2O_7$. TEM observation shows the fine grain size and low amount of intergranular phase.

REFERENCES

1. G. Ziegler, J. Heinrich & G. Wotting, J. Mater. Sci., 1987, **22**, 3041-86.

2. S. Hampshire & K.H. Jack, Special Ceramics, ed. D. Taylor and P. Popper, 1981, pp. 37-49.

3. H.T. Larker, AGARD CP-276, 1979, pp. 18-1-4.

4. G.E. Gazza, Am. Ceram. Soc. Bull., 1975, **54**, 778-81.

5. K. Homma, Yogyo Kyokai-Shi, 1987, **95**, 91-96

6. K. Liddell & D.P. Thompson, Br. Ceram. Trans. J., 1986, **85**, 17-22.

7. F.F. Lange, S.C. Singhal & R.C. Kuznicki, J. Am. Ceram. Soc., 1977, **60**, 249-52

8. O. Yeheskel, Y. Gefen & M. Talianker, Mat. Sci. Eng., 1986, **78**, 209-16

EFFECT OF O_2-HIP FOR OXIDE CERAMICS

Y.MANABE, T.FUJIKAWA, M.UEDA, AND Y.INOUE
IP Center, Machinery Division, Kobe Steel, Ltd.
3-18, Wakinohama-cho 1-chome, Chuo-ku, Kobe 651, JAPAN

ABSTRACT

Sintered alumina and PSZ samples were HIPed under several different atmospheres and the differences in colour and mechanical properties before and after HIPing were investigated, focussing on the influence of the HIPing atmosphere, such as a heating element materials used and the pressure medium gases. The properties after an aging test in air (1000°C, 1200 hrs) were compared with the before aging values. Large differences in colour and slight differences in density and bending strength were observed. Microstructural differences were examined using X-ray diffraction, ESCA and TEM. The results obtained in this study suggest that HIPing oxide ceramics under an oxidizing atmosphere will lead to great changes in the manufacturing processes of the oxide ceramic industry.

INTRODUCTION

Oxide ceramics have a long history of development and therefore are the most familiar to us of the many existing ceramic materials. Their manufacturing process, however, is still being improved in order to meet new requirements in properties and reduced processing costs. About ten years ago, the HIP (Hot Isostatic Pressing) was introduced to manufacture alumina cutting tool inserts with exellent cutting properties at a reasonable cost. Since that, the application of HIP to oxide ceramics has eventually spread to other oxides such as Mn-Zn-ferrite, PZT, and PSZ. In these cases, HIP has been performed using a completely inert argon as a pressure medium gas. Through the experience gained from HIPing these oxide ceramics, it was recognized that the HIPing atmosphere affects the colour and other mechanical or electrical properties. Recently HIPing using an argon-oxygen mixuture gas as a pressure medium has been attracting attention as it avoids undesirable property changes and keeps the natural colour of the oxide ceramic.

In this study, the influence of HIPing atmosphere, such as the heating element materials used and the pressure medium gas, has been investigated using alumina and Y-PSZ ceramics.

MATERIALS AND METHODS

Alumina sintered samples were prepared in the following manner. Alumina powder (mean particle size: 0.6 m, purity: higher than 99.99%) were mixed with 0.25% MgO to inhibit abnormal growth. After adding an organic binder for forming, the powder was granulated using a spray-dryer. The granulated powder was then pressed at 500 kg/cm^2 in a metal die into green compacts with dimmensions of 6 x 6 x 50 mm. These green compacts were cold isostatically pressed at a pressure of 1,000 kg/cm^2 to improve the density uniformity. After dewaxing, the green compacts were sintered at 1530°C for 2 hrs in air.

Y-PSZ sintered samples were prepared from commercially available ZnO_2-3mol%Y_2O_3 co-procipitation method powder with a mean particle size of 0.3 m (Tosoh Corporation TZ-3YB). The preparation procedure was the same as the alumina except that it was sintered at 1500°C.

Chemical analysis data for the alumina powder are given in Table 1.

TABLE 1

Chemical analysis of Alumina sample

(%)

MgO	Fe_2O_3	SiO_2	CuO	Na_2O	CaO	B_2O_3	Cr_2O_3
0.23	0.0013	0.017	0.0003	0.004	0.004	0.003	0.0003

These sintered samples were HIPed at 1400°C, 1,500 kg/cm^2 for 2 hrs without encapsulation using a HIP unit (System 20) manufactured by Kobe Steel, Ltd. HIPing atmospheres investigated in this study are summerized in Table 2.

After HIPing some of the samples were subjected to an aging test at 1000°C, for 1200 hrs in air.

TABLE 2

HIPing atmospheres

Pressure medium gas	Mat. of Heating element	Mat. of inner surface of furnace	Mat. of sample container
Argon	Graphite	Graphite	Graphite
Argon	Molybdenum	Molybdenum	Alumina
Argon + 1% Oxygen	Pt-Rh alloy	Alumina	Alumina
Argon +10% Oxygen	Pt-Rh alloy	Alumina	Alumina
Argon +20% Oxygen	Pt-Rh alloy	Alumina	Alumina

For each sample, density was measured by the Arch. method and the measured for bending strength. In the bending test, the test piece size was 4 x 3 x 36 mm and the three-point bending method was employed with a 30 mm long span and 0.5 mm/min. cross-head speed.

In the Y-PSZ samples, the Charges in carbon concentration, microstructure and phase transformation before and after aging test were closely examined, using ESCA, SIMS, TEM and X-ray diffraction.

RESULTS

Table 3 gives the results of the density measurements for the samples before and after the aging test as well as before HIPing. A slight decrease in density was observed for the Y-PSZ processed using molybdenum and Pt-Rh heaters. While the Y-PSZ sample as HIPed using graphite heater was black, the other Y-PSZ samples exhibited a yellowish white colour. The colour of the black Y-PSZ sample changed to yellowish white after the aging test in air. For alumina, the color charges were more complicated. The colour before HIPing was pinkish white. After HIPing in an argon-oxygen gas mixture, it turned bright orange with a little translusence. Samples HIPed in pure argon with the molybdenum heater changed from a slightly transluscent white just after HIPing to orange after the aging test. The colour of the sample HIPed with the graphite heater after HIPing was white with dark portion.

TABLE 3

Results of density measurement

(g/cm³)

Heater Pres. mod. gas		Before HIP	Gr Ar	Mo Ar	Pt-Rh Ar+20%O_2	Pt-Rh Ar+10%O_2	Pt-Rh Ar+1%O_2
Alumina	Before aging	3.903	3.971	3.984	3.975	3.977	3.978
	After aging	3.907	3.977	3.978	3.978	3.976	-
Y-PSZ	Before aging	6.085	6.090	6.104	6.100	6.103	6.105
	After aging	6.086	6.091	6.089	6.083	6.086	-

The bending strength of alumina was improved by HIPing from about 471 MPa to near 686 MPa. The difference in HIPing atmosphere seems not to cause significant differences in bending strength. Figure 1 shows the results for Y-PSZ. The bending strength was enhanced from about 1,000 MPa to about 1,500 MPa or higher. Samples HIPed in argon-20% oxygen showed a slightly higher value than the other samples, but in general the strength after HIP seems nearly the same for all samples. After aging, only the sample HIPed using the graphite heater exhibited any remarkable decrease in the bending strength.

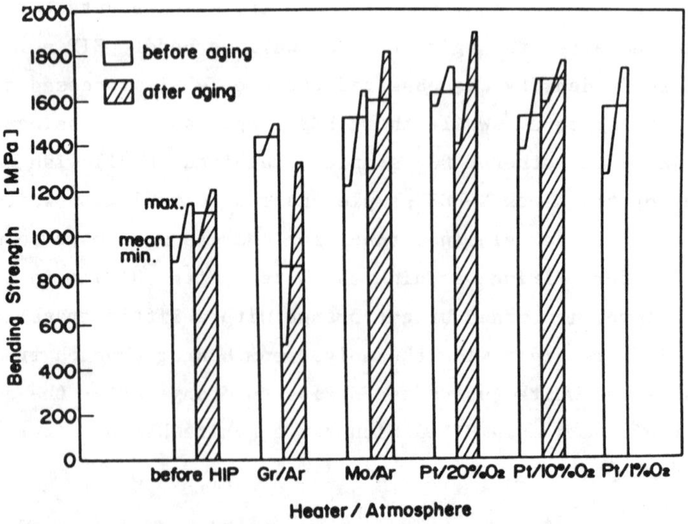

Figure 1. Results of 3-point bending strength measurements for Y-PSZ

DISCUSSION

In the case of alumina, the effect of HIPing on densification and improvement of the bending strength are clear, but the influence of the different atmospheres was not found for these properties. However the difference in colour was remarkable, especially between samples processed in argon and in argon-oxygen mixture gases. The cause of this color difference was not clarified but seems due to undetectable impurities contained in the samples.

For Y-PSZ, the improvement of the bending strength by HIPing is distinct, never the less the changes in density before and after HIPing were not large. This phenomenon is quite interesting, because this strength enhancement seems due to more than the elimination of the internal pores. The sample HIPed using the graphite heater was black and after aging its bending strength decreased about 40%. The cause seemed to lie in the presence of carbon. Carbon analysis using ESCA and SIMS supported this conjecture. By aging in air, this carbon seems to disappear and at the same time small pores regenerated along the grain boundary. The size of the pores were proved to be about 0.1 m from the microstructural observation using TEM. This pore regeneration will be one of the causes for the degradation in the bending strength. The mechanisms for strength improvement by HIP and also for this degradation are still unknown.

It is noticeable that by HIPing, high strength Y-PSZ either in white or black can be manufactured.

CONCLUSION

Alumina and Y-PSZ ceramics were HIPed under several different atmospheres and it was proven that in both cases the colour of the processed sample was dependent on the HIPing atmosphere, and that Y-PSZ HIPed using a graphite heater degrades in bending strength after aging in air.

The results of this study suggest that the O_2-HIPing of oxide ceramics is very effective in controlling the colour of the product, which is critical to its value and also in keeping a high level of mechanical properties such as bending strength. In the commercial production of some kinds of oxide ceramics, HIPing in an oxidizing atmosphere will be introduced in the future.

OXYGEN PARTIAL PRESSURE AND HIPPING OF CERAMICS - PROPOSED HIP PHASE DIAGRAMS

KOZO ISHIZAKI
Dept. of Materials Sci. and Engineering
School of Mechanical Engineering
Nagaoka University of Technology
Niigata 940-21, Japan

ABSTRACT

The oxygen partial pressure is an important factor on chemical reactions during ceramic sintering. The same aspect on HIP sintering must be even more critical due to the higher pressures applied. However this important aspect is not discussed commonly, because of difficulty to handle this problem in HIPping units. A HIP phase diagram proposed here is ΔG^0,i.e. partial oxygen temperature, as a function of total gas pressure and temperature. The HIP phase diagrams are conveniently applied HIP processes. Example cases on silicon nitride, aluminum nitride as well as super conducting ceramics are discussed.

INTRODUCTION

Thermodynamics can give an important criteria for many technical problems. The author proposed to construct a HIP Phase Diagram to improve HIP process.

Chemical stability is established by equating the Gibbs free energy of reactants and products. If there is high pressure gas component, e.g. argon gas, reactions will be influenced by the gas pressure. How can we consider this in the Gibbs free energy of reactions?

PRESSURE EFFECTS

If we consider a oxidation reaction for a metal M, as;

$$M + O_2 = MO_2 \,, \qquad (1)$$

the oxygen partial pressure is the dominant one in this oxidation reaction. If there are influences of other gases to the total pressure, normally (if an inert gas) is not considered. But if the pressure of the other gas is as high as thousands of atmospheres, such as in HIPping processes, are there any effects on the reaction? If any, how can we include the effects?

We have a phase diagram to see which is the most stable materials at a given temperature and oxygen partial pressure which is called an Ellingham (ΔG^0-T) diagram. We have to add a total pressure axis in the diagram to obtain chemical stability (i.e. phase) diagram as a function of temperature, total pressure and partial oxygen pressure. Let us call this as a "HIP Phase Diagram". The details of the method to obtain this diagram are discussed by the first time by the group of the author[1-3].

The pressure effects, say 200MPa have rather secondary effects for normal solids comparing to gases. At a given temperature the value of standard free energy difference ΔG^0 of Eq.(1) changes with pressure as,

$$\left(\frac{\partial \Delta G^0}{\partial P} \right)_T = \Delta V^0 . \qquad (2)$$

In Eq.(1) the change of volume is approximated by that of the gas, then

$$\Delta V^0 = V_{MO_2}{}^0 - (V_M{}^0 + V_{O_2}{}^0) \simeq -V_{O_2}{}^0 . \qquad (3)$$

The slope with respect to P at a reference temperature is expressed by Eqs.(2) and (3), consequently which is almost negative of one molar volume of a gas at a reference state. The reaction of carbon is very important because of common carbon heater in HIP equipments as well as silicon carbide formation.

The possible carbon oxidation reactions are;

$$C + O_2 = CO_2 , \qquad (4)$$
$$2CO + O_2 = 2CO_2 , \qquad (5)$$
and
$$2C + O_2 = 2CO. \qquad (6)$$

Using the ΔG^0 data at normal condition[4], we obtain a HIP Phase Diagram for carbon as in Figure 1. The three planes correspond to the above three reactions as seen in Figure 1. The ΔG-T-P space in the HIP Phase Diagram is divided in three different zones of stability for C, CO and CO_2 by these three planes. Different P_{O_2} (i.e. ΔG^0), T, and P give CO_2, C and CO as the most stable substance in the zones above the ABEF-BDE planes, below the ABEF-BGE planes, and between the BDE-BGE planes,respectively.

Similarly we may construct a HIP phase diagram for silicon as in Figure 2, based upon the following chemical reactions as,

$$Si + O_2 = SiO_2 , \qquad (7)$$
$$2SiO + O_2 = 2SiO_2 , \qquad (8)$$
and
$$2Si + O_2 = 2SiO. \qquad (9)$$

Now let us discuss some real applications using these HIP Phase Diagrams. Carbon impurity was introduced on silicon nitride powder to discuss the aforementioned effects as in the following chapter.

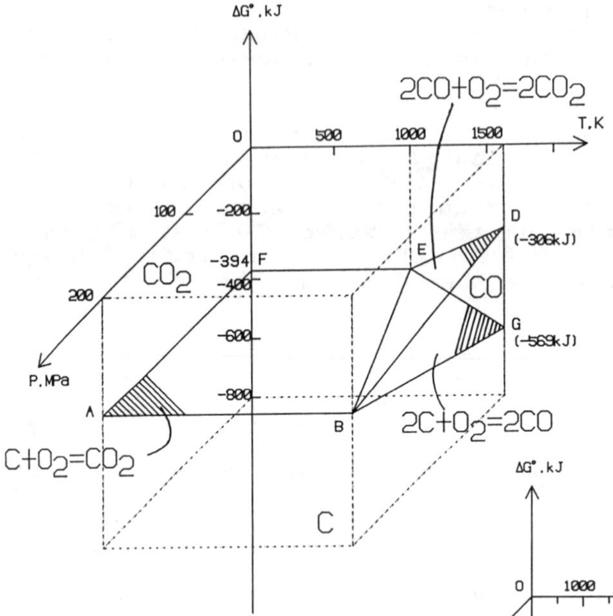

Figure 1. *HIP Phase Diagram for carbon to indicate stable zones for C, CO and CO_2. ΔG^0 of oxidation of carbon as a function of the total gas pressure P and temp. T. The three chemical reactions of Eqs. (4), (5) and (6) are indicated in the three planes.*

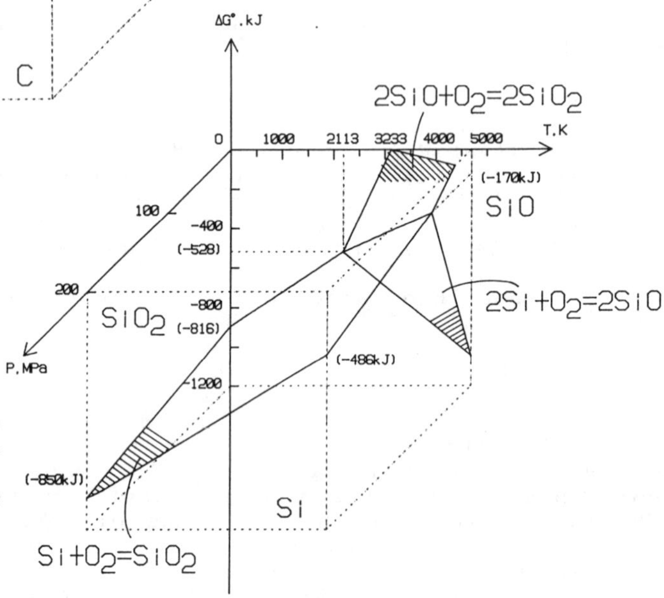

Figure 2. *HIP Phase Diagram for silicon to indicate stable zones for Si, SiO and SiO_2. ΔG^0 of oxidation reactions of silicon as a function of the total pressure P and temp. T. Eqs. (8), (9) and (10) are indicated. Three different zones have the stable form of Si, SiO and SiO_2 as indicated.*

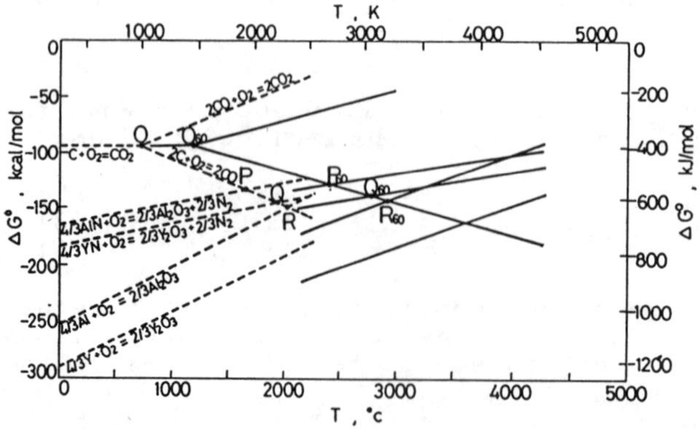

Figure 3. *A part of the HIP Phase Diagram of the related reactions at pressure 60MPa. The points P, Q and R shift toward the higher temperature to P_{60}, Q_{60} and R_{60} in accordance with the increase of the gas pressure. At points P_{60} and Q_{60}, alumina and yttria are no longer stable but metallic aluminium and yttrium.*

SILICON NITRIDE

When silicon nitride is sintered, there are many complicated problems. Some of them are oxide phase (probably SiO_2), carbon and so on as surface impurity on raw silicon nitride particles. Then the following reaction may proceed;

$$SiO_2 + 3C \rightarrow SiC + 2CO. \qquad (10)$$

The results of carbon contents after sintering is as follows: The original silicon nitride powder was contaminated up to 4000ppm by carbon. Carbon contents changed a little by HIP sintering, on the other hand carbon decreased to almost a constant value of 500ppm for hot pressed samples. We have to consider another reaction to examine the phenomena;

$$SiC + O_2 = SiO_2 + C. \qquad (11)$$

A HIP Phase Diagram was constructed for this reaction to show that SiC is stable at the sintering temperature if the total gas pressure is low, but under 200MPa SiO_2 is more stable under the oxygen partial pressure to equilibrate with graphite heater. It requires more about 2200K of temperature for SiC to be more stable(1-3).

This behaviour was expected by proceeding a chemical reaction as Eq.(10). If Eq.(10) is correct there must be silicon carbide formed in the grain boundary of hot pressed silicon nitride. On the other hand at the HIPping condition of 200MPa and carbon remains as graphite for HIPped silicon nitride.

Auger electron spectroscopy (AES) was employed to examine if the reminder carbon has SiC bonding or graphite bonding. To concentrate information on grain boundaries, samples were broken through grain boundaries in the AES ultra high vacuum chamber and the fractured surfaces were AES examined. SiC was found in hot pressed silicon nitride(1), and graphite-like AES peaks in HIPped samples(5). Hence as expected the HIP Phase Diagram proposed by the author worked correctly.

ALUMINUM NITRIDE

The oxygen contents, which influence the AlN thermal conductivity, are evaluated by using the HIP Phase Diagram for the oxidation reactions of carbon, the related metals and AlN as following.

Considering, again, a reaction $2C + O_2 = 2CO$ in a graphite furnace for normal-sintering, the oxygen partial pressure must be along the line corresponding to this reaction.

It is assumed that oxygen atoms exist as Al_2O_3 on the particle surface of the AlN powder. For normal- sintering with carbon heaters, the following reactions;

$$2/3Al_2O_3 + 2/3N_2 + 2C = 4/3AlN + 2CO, \qquad (12)$$
$$2/3Y_2O_3 + 2/3N_2 + 2C = 4/3YN + 2CO. \qquad (13)$$

may proceed.

These reactions occur above points P or Q in a part of HIP Phase Diagram as in Figure 3[6]. Considering both Eqs.(12) and (13) for the normal-sintered AlN, the oxygen contents could be decreased from the samples.

The cause of lower thermal conductivity for some of the high density HIPped samples than the low density normal-sintered AlN was explained well by the difference of oxygen contents which is explained by the proposed HIP Phase Diagram.

OXIDE SUPER CONDUCTORS

Superconducting ceramics ($BiSrCaCu_2O_x$ and $BiPb_ySrCaCu_2O_x$) were directly obtained by HIP densification without any other additional treatment for the first time by the group of the author. The details of this method are described elsewhere[7]. Another interesting aspect of this material is so called "partial melting." The high-Tc phase is usually obtained just below the "partial melting" point, at which temperature some of the constituent substances melt. There is strong influence of Po_2 on this partial melting[8].

The high Po_2 and the high total pressure were examined by the HIP Phase Diagram of higher Po_2. The conclusion is that the partial melting line also coincide with oxidation line at higher Po_2 and P_{total}. The details are discussed elsewhere by Seino, Ishizaki and Takata.

CONCLUSIONS

THE HIP PHASE DIAGRAMS. proposed in this work clarifies many HIP sintering problems which were not possible to be solved before. Using the proposed HIP Phase Diagram, silicon nitride with the silicon carbide reaction, aluminum nitride with abnormal thermal conductivity, and the partial melting phenomena of the oxide super conductor were explained.

REFERENCES

(1)Watari, K.,Ishizaki, K. and Kawamoto, M.,J.Ceram. Soc. Jpn. Inter. Ed., 1988,**96**, 128-134.
(2)Watari, K. and Ishizaki, K.,ibid., 1988,**96**, 535-540.
(3)Watari, K.,Ishizaki, K. and Kawamoto, M.,ibid., 741-748.
(4)For example, Wicks, C.E. and Block, F.E., Thermodynamic Properties of 65 Elements, U.S.Gov. Print. Office, 1963.
(5)Watari, K.,Ishizaki, K. and Kawamoto, M.,J.Ceram. Soc. Japan, 1988,**96**,127.
(6)Watari, K.,Ishizaki, K. and Hamasaki T.,J. Chem. & Phys. Solids to be published.
(7)Seino, H.,Ishizaki, K. and Takata, M.,Jpn. J. of Appl. Phys., 1989,**28**,L78-L81.
(8)ibid. to be published

ACKNOWLEDGEMENTS

This work was partially supported by Mon-busho (Ministry of Education and Culture) under cooperative programs (Kyo-do Kenkyu) with Nippon Kokan (NKK) on silicon nitride and Kobe Steel, Ltd.(KOBELCO) on O_2-HIPping. The author wish to express his gratitude to these organizations.

SINTER-HIP DILATOMETER FOR EXPERIMENTS AT PRESSURES UP TO 200 MPA

A. Kühne, R. Oberacker, F.Thümmler
Universität Karlsruhe
Institut für Werkstoffkunde II
Institut für Keramik im Maschinenbau (IKM)
Haid- und Neustr. 7, D-7500 Karlsruhe

ABSTRACT

A dilatometer for use in a commercial HIP-facility was developed at the IKM. Several tests showed that the dilatometer is capable to work correctly under any condition from RT to 2000°C/ vacuum and 2200°C/ 200MPa gas pressure. Sinter-, Post-HIP- and Sinter-HIP treatments with several ceramic materials were performed, showing the advantages achieved by this dilatometer.

INTRODUCTION

Regarding process development of sintered materials, dilatometers are an important help for cycle optimization. For normal sintering treatment there are several dilatometers commercially available. Due to their construction these dilatometers work under flowing atmosphere or even slight overpressures.
Build-on dilatometers for application in gas pressure furnaces up to 10 MPa recently have also reached the state of commercial availability [1].

Information on dilatometers for pressures up to 400 MPa as usual in hot isostatic presses (HIP) is rare. A few years ago a HIP-dilatometer was build at Batelle [2], but no further literature since then is known to the authors. At KfA Jülich a prototype HIP-dilatometer is in test runs [3]. The only commercial instrument for use in HIP [4] exhibits a measuring range of 1.2mm, allowing only very small specimens to be tested.
As a summary of the present situation it has to be stated, that unfortunately no dilatometer with a suitable measuring range is available for use in HIP-facilities. To study temperature- and pressure depending length changings during sinter-, Post-HIP- and combined sinter- and HIP-processes (Sinter-HIP) on specimens with reasonable dimensions a dilatometer had therefore to be developed.

DESIGN, TECHNICAL DATA AND FIRST TESTS

The dilatometer was designed in such a way, that it is able to work over the whole range of processing parameters possible with the instituts HIP (ASEA QIH6). The maximum condition parameters are 2000°C/ vacuum and 2200°C/ 200MPa Argon and/ or Nitrogen.
To minimize the measuring error the specimen should not be to small. Regarding the surface to volume ratio a small size also gives poor relevance for practical applications. With suitable specimen dimensions a measuring range of about 5mm should be available, allowing green sample lengths of up to more than 30mm, depending on their green density.

The IKM HIP-dilatometer is build up as a complete, separate unit with its own furnace (Fig. 1). For quick exchange with the standard furnace it can be easily plugged in the HIP-vessel. From this, no changes at the HIP-unit had to be done and many original parts could be used. For special tests the possibility to work with a powder bed is given. The choosen measuring system, having a linear range of 6mm at a resolution below 0.25µm, can withstand the conditions in the HIP vessel without any problem.

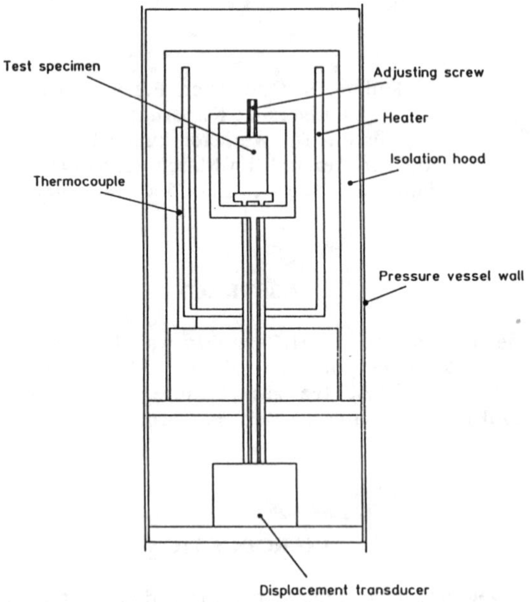

Fig. 1: Scheme of the IKM Sinter-HIP dilatometer

As to expect and with every dilatometer, there is a certain temperature and pressure influence. For example temperature gradients occour over the length of the dilatometers parts influencing the measuring signal. Extensive tests with dense ceramic specimens showed that the temperature/pressure-influence is nearly negligible compared with the length changes of samples during Sinter-HIP and even Post-HIP-treatments. Furthermore the influence is completely reproducible, allowing a relatively easy calibration and data correction by computer.

To test the dilatometer's accuracy comparative measurements with a precision dilatometer (NETZSCH) were performed. The transformation behaviour of zirconia thoughened alumina (ZTA) during heating up and cooling down under identical conditions was recorded with both dilatometers. Nearly the same dilatometer curves were detected, where the small differences may come from the accuracy in temperature-measurement and irreversible changings of the specimen.

Sinter-HIP cycles were carried out with SiC (Fig. 2) to test the dilatometer's behaviour at highest temperatures. The influence of nitrogen atmosphere, which is always somewhat critical for graphite parts, was tested in Sinter-HIP of Si_3N_4 (Fig. 3). Up to now more than 40 sinter-, Post-HIP- and Sinter-HIP-experiments where carried out successfully and without any problems. First results on Post-HIP and Sinter-HIP treatment studies on oxide ceramics were published in [5].

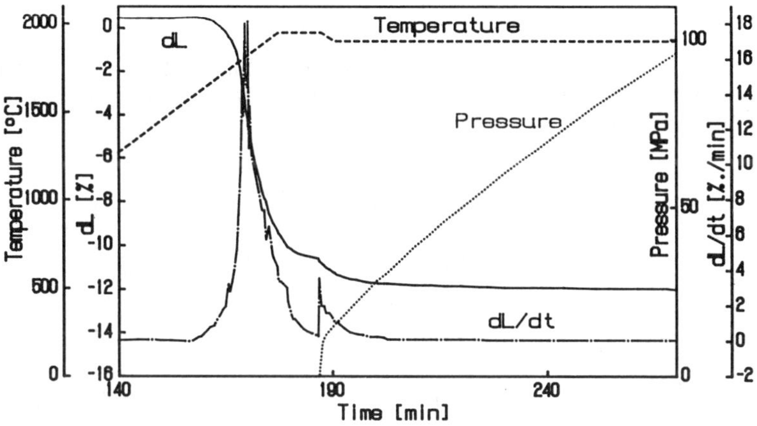

Fig. 2 : Part of SiC Sinter-HIP cycle (1950°C/150MPa Ar)

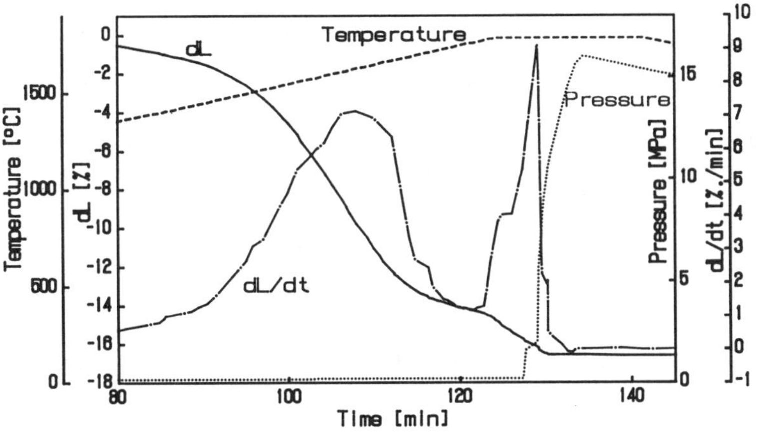

Fig. 3: Part of Si_3N_4 Sinter-HIP cycle (1800°C/16 MPa N_2)

PRACTICAL APPLICATION

Besides of monitoring the densification during sintering or HIPing the dilatometer gives the chance to pressurize the system at the rigth moment, which is important especially for Sinter-HIP cycles. This moment is reached when the specimen has sintered so far, that there are just no more open pores, allowing gas pressure to lead to maximum further densification. Pore closure normally is reached at a density of 93 to 95 % of the theoretical. If the exact value is known from sintering experiments, pore closure can be detected with help of the dilatometer by calculating the required linear shrinkage from the initial density and the density at the point of pore closure.

With the example of a dilatometer-curve resulting from Sinter-HIP of ZTA (Fig. 4) some advantages when using a HIP-Dilatometer can be shown. The processing parameters for first iteration of the cycle where taken after literature data.

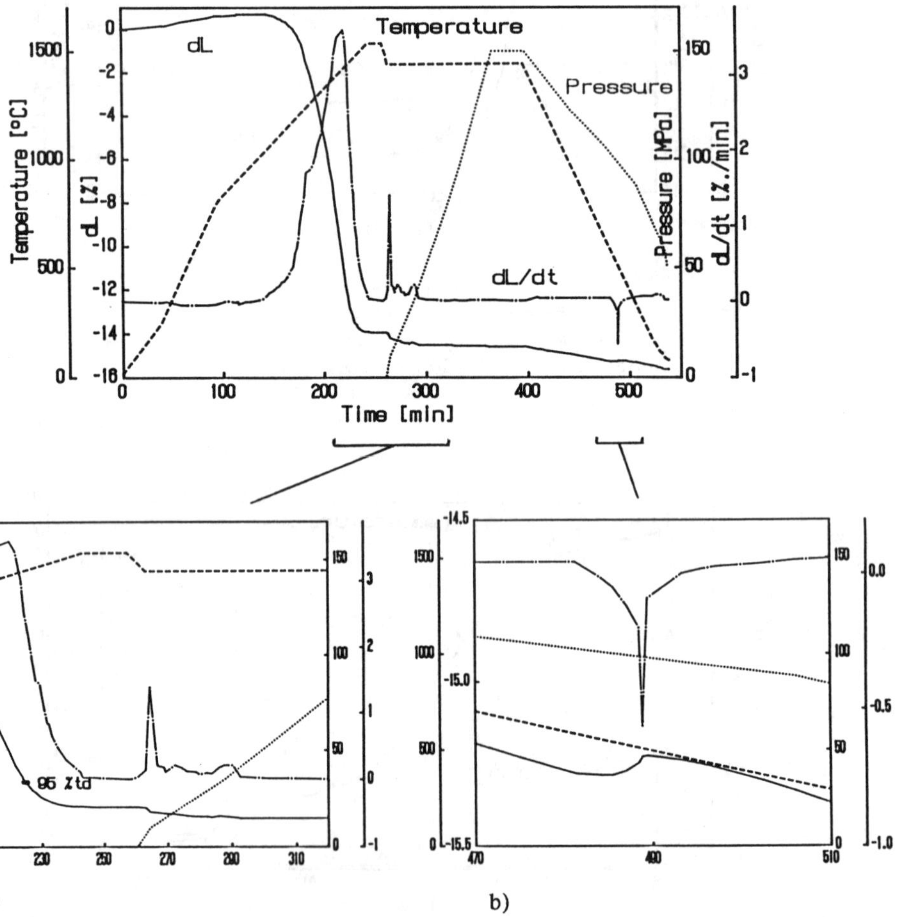

Fig. 4: Complete Sinter-HIP cycle of ZTA
a) pore closure and pressure assisted densification
b) zirconia tetragonal/ monoclinic transformation during cooling

As can be seen in the diagram, the point of expected pore closure (95%TD) had been reached far before pressure was applied. Therefore undesired grain growth could have taken place, without significant further densification. As a conclusion the sintering temperature could have been lower and/ or former pressurization could have been choosen. Looking at the pressure influence, it can be seen that no further densification occured at pressures beyond 50 MPa. With the knowledge of these facts, the treatment could have been more than 90 minutes shorter, also saving a lot of costs for the gas and the use of the HIP-unit.

Looking at the transformation during cooling down, no conclusions can be drawn by now. The use of the dilatometer however gives the possibility to study the pressure influence on the transformation temperature and volume effect.

Further fields of application are e.g. determining the influence of pressurization/ depressurization on densification, studying the pressure/ temperature influence on phase transformation or even development of a cycle with controlled densification. The verification of HIP-maps seems possible, too.

SUMMARY AND OUTLOOK

With the HIP-dilatometer developed at the IKM a powerful tool for a variety of densification studies at specimens of reasonable size is available.

The next step using the dilatometer system will be the realisation of a closed loop control system, allowing for temperature-, pressure- or even shrinkage-controlled treatments. Therefore online computer performed data aquisition and correction have to be realised.

REFERENCES

1. H.U. Kessel, H. Kolaska, K. Dreyer, Powder Met. Int., 20 (1988) 35-39
2. J.K. McCoy, R.R. Wills: Ceram. Eng. Sci. Proc., 6 (1985) 1146-1150
3. H.P. Buchkremer, IRE, KfA Jülich, private communications
4. ASEA Metallurgy AB in: Sinter/HIP-Technologie, Pulvermetallurgy in Wissenschaft und Praxis, Band 3, Verlag Schmidt, Freiburg, 1987
5. R. Oberacker, A. Kühne, F. Thümmler: Comparative Studies on Post-HIP and Sinter-HIP Treatments on Transformation Toughened Ceramics, 2nd Int. Conf. on Hot Isostatic Pressing, June 1989, Gainthersburg, USA, Proceedings in press

HOT ISOSTATIC PRESSING OF PURE BORON CARBIDE

JAN E. ADLERBORN, LENNART BJÖRK, HANS T. LARKER
ABB Cerama AB
S-915 00 Robertsfors, Sweden

ABSTRACT

The influence of maximum particle size and degassing temperature on density, content of free carbon, hardness, microstructure and strength on hot isostatically pressed (HIPed) pure boron carbide has been investigated. To reach optimal properties both separation of over-sized particles and agglomerates from the starting powder and a high degassing temperature of the preformed powder bodies are necessary. With these precautions a boron carbide material with homogeneous microstructure and high density, hardness and strength can be obtained.

INTRODUCTION

Boron Carbide (B_4C) exhibits high hardness and low modulus, low density and excellent wear resistance which makes it to an interesting candidate for bearings, nozzles, armor etc. However it is very difficult, if not impossible, to densify via pressureless sintering (1). Today Hot Pressing (HP) is commersially utilized for processing boron carbide. However, material with very low porosity is not easily obtained and this technique has furthermore very limited possibilities to produce components with complicated shapes.

Hot Isostatic Pressing (HIP) is an alternative which can combine flexibility in composition with flexibility in geometry and volume. Using the glass powder encapsulation a HIP technology has been developed for making dense ceramic parts to net shape with no or very limited finish machining. Internal cracks and voids in the green powder bodies can be healed and this technology allows production of high quality ceramic parts with complex shape (2, 3).

All commercially available boron carbide powders contain more or less boron oxide and free carbon as contaminants. A possible way to reduce these may be a heat treatment of the powder or the preformed powder bodies at such a high temperature that the free carbon reacts with the boron oxide under formation of carbon oxides and boron carbide. During conventional sintering or hot pressing the consolidation process is performed in such a way that any liberated gaseous species are free to disappear. However, when using the glass encapsulated HIP-process the material to be densified is enclosed in a gas-tight casing. Therefore such reactions must be carried out before HIPing.

Earlier investigations at ABB Cerama have shown that it is possible to HIP boron carbide powder to full density and with high strength at comparatively low temperature if over-sized grains and big agglomerates are separated from the starting powder before the consolidation (4).

In the present study the influence of powder particle size and degassing temperature on density, hardness, microstructure and strength of HIPed boron carbide has been investigated.

EXPERIMENTALS

Two boron carbide powders were used in this investigation. Powder properties and source are listed in table 1. For the separation of over-sized particles and hard agglomerates a wet sieving technique was utilized.

TABLE 1
Boron Carbide Powders

Powder code	Manufacturer	Grade	Particle size	Chemical analysis (wt%) B	C	O	N	Fe
A	H.C. Starck	F1500	1-3 μm	77.8	21.7	0.25	0.15	0.05
B	-"-	HCS-HP	<10 μm	75.9	22.5	0.8	0.5	0.03

The formed powder bodies, cold isopressed, were dewaxed and degassed in a vacuum furnace, encapsulated in glass powder and HIPed using the proprietary technique of ABB Cerama. Microstructure was studied in a light microscope and hardness was determined. The flexural strength at room temperature was determined on test bars, 3.0 by 3.0 mm, in a 3-point test rig with the span 20 mm. Crystalline phases present in the densified materials were identified by X-ray diffractometry (XRD) using monochromated Cu K_α radiation.

RESULTS AND DISCUSSIONS

The different treatments of powders and powder bodies together with the HIP-parameters used for the consolidation are listed in table 2.

TABLE 2
Powder Separation, Degassing Temperature and HIP-parameters

Sample code	Temporary binder	Wet sieving (μm)	Degassing temperature (°C)	HIP-parameters (°C, MPa, h)
A1	---	---	1400	1700, 200, 1
A2	---	- 3	750	1750, 200, 1
A3	---	- 3	1000	1750, 200, 1
A4	---	- 3	1400	1750, 200, 1

TABLE 2 cont.

A5	---	- 5	750	1750, 160, 1
B1	---	---	750	1750, 200, 1
B2	---	- 5	1200	1800, 160, 1
B3	---	- 5	1400	1800, 160, 1

A1 - A5: B_4C-powder A; B1 - B3: B_4C-powder B

Phase Analysis

The chemical analysis of the two powders used in this investigation show that both have an excess of carbon and the existence of free carbon was verified by XRD. This free carbon more or less remains in the degassed and HIPed materials. The amount is depending on original content, separation of over-sized particles as well as on degassing temperature. To determine the relative content of free carbon in the materials XRD was used. For the calculations the intensity of the reflex at d ≈ 2.803 Å was used for the B_4C and at d ≈ 3.35 Å for free carbon. The relative intensities for the free carbon are listed in table 3.

TABLE 3
Density, Relative Intensity for free Carbon, Hardness and Flexural Strength

Sample code	Density (g/cm^3)	$\frac{I_C}{I_C + I_{B_4C}} \cdot 100$	Hardness HK_{5N} (GPa)	Flexural Strength Average (MPa)	Flexural Strength Weibull modulus
A1	2.51	2	24.6 ± 0.2	523	6
A2	2.49	4	23.6 ± 0.3	712	17
A3	2.50	3	23.2 ± 0.2	645	18
A4	2.51	< 1	25.6 ± 0.1	653	9
A5	2.50	19	22.1 ± 0.2	588	8
B1	2.50	44	19.8 ± 0.2	439	11
B2	2.49	22	---	593	10
B3	2.51	19	---	562	12

The content of free carbon decreases both with increasing degassing temperature and the degree of particle separation. However, to get a very low level of free carbon it is necessary to use a combination of particle separation to ≤ 3 μm, high degassing temperature and a powder with low free carbon, sample A4. A degassing temperature ≤ 1000 °C and/or a coarser fraction of the powder A will give residues of free carbon in the consolidated material, A5.

In all the samples B1 - B3 relatively high carbon contents are observed. The degassing temperature has an influence, but it is not possible to reach the same level as with the A-powder. Obviously the starting level of free carbon in powder B is too high to be effectively reduced even at a high degassing temperature.

Density

The densities of the samples after HIP are at least 98.9 % of TD. (table 3). Any clear relation between the residue of free carbon detected and the density is not visible. However, a combination of separation of big particles and agglomerates and a high degassing temperature gives a density quite near the theoretical, 2.52 g/cm^3.

Hardness

Both Vickers- and Knoop-indentations with loads from 2N to 20N were tested. By using the Knoop-diamond with the load 5N it was possible to get indents of a reasonable size without cracking. The measured values are listed in table 3. It is evident that in order to reach a high hardness, HK_{5N} > 24 GPa, a low level of free carbon residue in the material is necessary.

Microstructure

The microstructure of the HIPed B_4C is strongly dependent on the powder preparation as well as on the degassing temperature. A low carbon powder, separated to ≤ 3 μm and degassed at 1400 °C, has after HIP a homogenous appearance with very small inclusions and no voids, fig. 1. The same powder, degassed at high temperature but not separated, also has a homogeneous appearance. However, some voids are found, probably originating from big, hard agglomerates in the raw material. With a low temperature degassing and separation to ≤ 5 μm a great number of small inclusions are found. If starting with a more coarse-grained powder, B, and degassing at a low temperature an inhomogeneous structure is obtained, fig. 2. Both voids and big inclusions are found. Obviously the big agglomerates existing in the starting powders are too hard to be densified at the HIP-parameters used in this study.

Flexural strength

The strength of the HIPed samples are listed in table 3. A high scatter characterizes the measured strength values except for A2 and A3. Most likely the scatter originates from the preparation of the test bars. The tests were performed on as-sliced bars and as pure B_4C normally has a very low fracture toughness even a small surface defect will seriously reduce the strength. However, it appears that in order to reach a high strength level, > 600 MPa, powder separation to at least ≤ 5 μm is necessary.

CONCLUSIONS

- Both separation of big particles and hard agglomerates and a high degassing temperature have a marked positive effect on the materials properties.

- It is possible to reduce the content of free carbon in the material by a heat treatment of the preformed powder bodies before HIP.

- To achive optimal hardness and strength both particle separation to -3 μm and a heat treatment at high temperature is necessary.

ACKNOWLEDGEMENT

The partial financial support by the National Swedish Board for Technical Development is gratefully acknowledged.

REFERENCES

1. Doyle, S.L. and Prochazka; "Densification and Microstructure Development in Boron Carbide." Cer. Eng. and Science Proc., **16**, No 7-8, 1151-60 (1985).

2. Larker, H.T.; "Recent Advances in Hot Isostatic Pressing Processes for High Performance Ceramics." Mater. Sci. and Eng. **71** (1985) p. 329-332.

3. Larker, H.T.; "On Hot Isostatic Pressing of Shaped Parts." In S. Somiya, E. Kanai and K-I. Ando (eds), Proc. 1st Int. Symp. on Ceramic Components for Engines, Hakone, October 17-19, 1983, KTK Scientific Publishers, Tokyo; Reidel, Dordrecht, 1984, p. 307-310.

4. Larker, H.T.; Hermansson, L. and Adlerborn, J.; "Hot Isostatic Pressing and its Applicability to Silicon Carbide and Boron Carbide." Ind. Cer. **8** (1) 17-20 (1988).

Fig. 1. Sample A 4. Boron Carbide powder A, separated to ≤ 3 μm. Degassed at 1400 °C. Bar = 10 μm

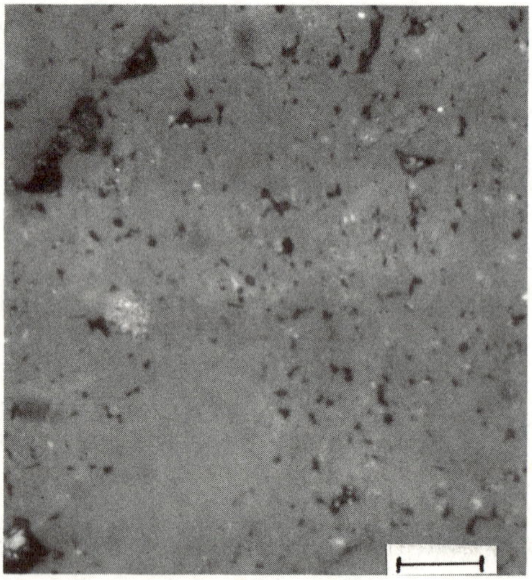

Fig. 2. Sample B 1. Boron Carbide powder B, not separated. Degassed at 750 °C. Bar = 10 μm

GAS-PRESSURE-SINTERING OF SILICON NITRIDE

D. Lange, H. Thaler, K.A. Schwetz
ELEKTROSCHMELZWERK KEMPTEN GmbH, West-Germany

INTRODUCTION

Silicon carbide (SiC), boron carbide (B_4C), aluminium nitride (AlN), boron nitride (BN) and silicon nitride (Si_3N_4) are non-oxide ceramic materials, which are now fabricated in sintered parts to a considerable extent. The future of silicon nitride in particular looks promising due to its favourable combination of hardness, toughness, strength and chemical durability. There are different methods to produce non-porous silicon nitride components; a common method is to densify pressed powders by liquid phase sintering. Oxide additions and the inevitable silicon dioxide on the silicon nitride powder form the liquid phase, which promotes the densification and results in a glassy phase after cooling down from the sintering temperature. A lot of literature deals with theoretical and practical aspects of manufacturing sintered silicon nitride parts (1,2,3,4,5). The increasing applications of silicon nitride have been described elsewhere (6). The main problem when sintering Si_3N_4 is its decomposition, which occurs at temperatures necessary for densification. Greskovich (7) has shown the partial pressure of nitrogen to reach 1 bar at 1900°C, nevertheless slow decomposition is observed at temperatures well below 1900°C. To avoid this undesired behavior, sintering is done in a "powder bed" or in an autoclave furnace. Because sintering in a powder bed is not economical for mass-produced articles i.e. indexable inserts, it is preferred to run the process in an autoclave furnace. At the beginning of commercial production of Si_3N_4, hot isostatic presses ($P_{gas} \leq 2000$ bar), which supplied an adequate gas pressure to prevent decomposition where used for sintering. Today it has been recognized that a gas pressure of 100 bar is sufficient for sintering and dedicated equipment is now available.

The present paper deals with the sintering behavior of different Si_3N_4-powders in nitrogen and argon gas atmosphere, respectively. The experiments show that carbon originating from the graphite-heating-element may be detrimental for the densification of silicon nitride.

MATERIALS AND METHODS

The powders used for the experiments are shown in table 1.
To promote sintering 8 wt% of metal oxides based on Al_2O_3-La_2O_3-MgO were added. The theoretical density of the sintered Si_3N_4-metal oxide mixture is 3.24 gcm^{-3}. For dispersing the sintering aids the powder mixture was grinded in silicon nitride jars with silicon nitride balls to avoid any contamination. Green bodies were pressed cold isostatically at 2000 bar.

The green density obtained by this moulding method is listed in table 1.

TABLE 1
Characterization of silicon nitride starting powder

Manufacturer	Type	Production route	O (wt%)	BET specific surface (m^2g^{-1})	α-phase content (vol%)	Green density* (%TD)
Denka	SN9Fw	Nitridation of Silicon	0.9	12	91	59.3
H.C.Starck	LC12		1.8	23	94	61.2
TOSOH	TSK	Imid-process	1.5	16.8	95	50.4

* cold isostatically pressed with 2000 bar

Sintering was carried out in a 100 bar laboratory gas-pressure-sintering furnace equiped with a dilatometer (FPW100, KCE-Sondermaschinen, Rödental, West-Germany). The size of samples suited for the dilatometer is d 8x15 mm. The sintered bodies (30x30x50 mm^3) were machined with diamond tools into flexure specimens (3.5x4.5x45 mm^3). Flexural strength was measured using 4-point loading (outer span 40 mm, inner span 20 mm); fracture toughness (K_{Ic}) using bridge indentation (12).

RESULTS

Fig.1 shows the optimized sintering parameters for powder SN9Fw. Using this processing cycle results in nearly theoretical density, i.e. 3.23 gcm^{-3}.

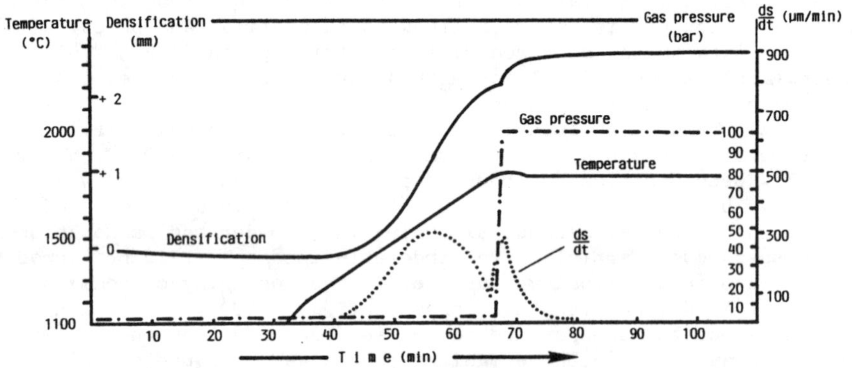

Figure 1. Cycle for Si_3N_4-gas pressure sintering

Preliminary experiments have suggested that it is favourable to keep the gas pressure reasonable low until the porosity of the material is closed. Up to 1780°C a pressure of 2.5 bar is sufficient to avoid considerable decomposition. The porosity is thought to be closed when the specimen's density approaches 92%TD. Since the shrinkage is monitored by the dilatometer the best moment for increasing the gas pressure is easily determined. After rising the gas pressure, the residual pores are compressed and the gas in the shrinking pores (N_2gas, Si_{gas}) condenses partially and is finally disolved in the liquid phase. The dilatometer shows a rapid densification (shrinkage rate up to 300µm/min) which is believed not to be diffusion-controlled. The rapid densification slows down after a few minutes and during subsequent sintering only little densification is observed. During this period the α/β-Si_3N_4 phase transformation completes. According to our observations 50 minutes are necessary for complete transformation. By cutting the holding time short, the α-contents can be increased up to 30vol% in a pore-free material.

Similar experiments with LC12 and TSK powders were not successful, i.e. densities >96% TD could not be obtained (see table 2).

TABLE 2
Density of sintered specimen using different materials for kiln furniture and the dilatometer (sintering program fig.1)

Si_3N_4-powder type	graphite gcm^{-3} / %TD	boron nitride gcm^{-3} / %TD	aluminium nitride gcm^{-3} / %TD
SN9Fw	3.23 / 99.7	3.23 / 99.7	3.23 / 99.7
TSK	3.11 / 96.0	3.22 / 99.4	3.23 / 99.7
LC12	3.08 / 95.0	3,13 / 96.5	3,23 / 99.7

Changing the temperature-pressure-program did not yield a density exceeding 96%TD.
Fig.2 shows the sintering behavior of LC12 when keeping the gas pressure (nitrogen) constant during sintering. Densification starts at higher temperatures when raising the gas pressure. Consequently the densification

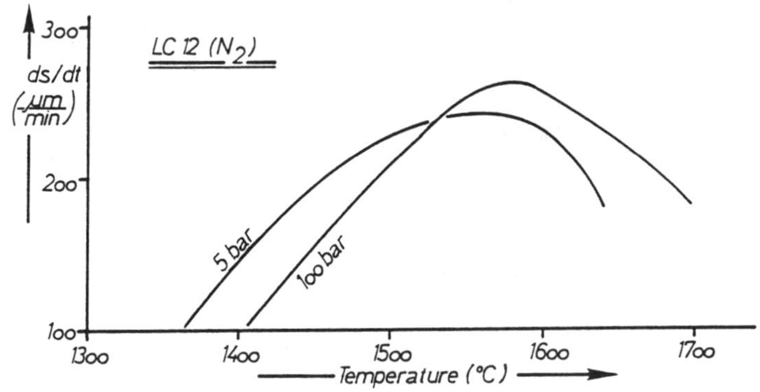

Figure 2. Sintering behavior of LC12 using constant gas pressure (nitrogen)

slows down at higher temperatures. The maximum densification rate increases only marginally upon increasing the gas pressure (5bar N_2:250µm min^{-1}; 100bar N_2:280µm min^{-1}).
Preceeding experiments have shown and literature data confirm (8) that graphite-heating-elements can have detrimental influence on the densification of non-oxide ceramic materials. For that reason the samples were shielded from the heating element by AlN- or BN-hats. The sample holder of the dilatometer was also made from AlN or BN, respectively. The results obtained by those modifications are shown in table 2. All samples reached nearly the theoretical density, when the sintering parameters according Fig.1 were used. Moreover it should be noted that the densification behavior of the three different powders is nearly identical when using AlN-shieldings.
The experiments confirm the known influence of the graphite heating-element. Another indicator for the influence of graphite is the formation of α- and/or β-Si_3N_4-whiskers, at the surface of the samples which is frequently

observed when sintering in nitrogen without any shielding to protect the sample. Presumably the whiskers grow via the reaction

$$3\ SiC + 2\ N_2 \longrightarrow Si_3N_4 + 3\ C \qquad (9).$$

The silicon carbide is formed by the reaction of carbon originating from the heating element and silicon originating from the decomposition products (e.g. silicon, silicon monoxide) of silicon nitride. The formation of Si_3N_4-whiskers in the furnace is shown in Fig. 3.

Fig. 3
Formation of Si_3N_4-whiskers in the graphite tube furnace during Si_3N_4 sintering under nitrogen gas pressure

Additional experiments using SN9Fw powder have shown, that sintering of silicon nitride is possible not only in nitrogen, but also in argon. Table 3 indicates that the density and the strength of sintered Si_3N_4 is independent of the kind of gas preventing Si_3N_4 decomposition.

TABLE 3
Properties of silicon nitride sintered under nitrogen and argon pressure, respectively

	nitrogen	argon*
density (gcm-3)	3.22	3.21
flexural strength (MPa)	820 ± 32	791 + 37
hardness (HK o.1)	1642	1630
max.temperature (°C)	1780	1760
max.gas pressure (bar)	100	100

* Sample was protected by an AlN-shielding

Nevertheless the microstructure of samples sintered in argon looks different from those sintered in nitrogen (see Fig. 4).

Samples sintered in argon exhibit a somewhat smaller particle size and the aspect ratio of the Si_3N_4-grains appears to decrease compared to N_2-atmosphere. The reason for this is still unknown. Formation of SiC on the specimen surface is promoted in argon. The depth of decomposition and reaction is of the order of 500µm. However such a layer should be avoided when near net shaping is desired. Again, the source of carbon is the heating element, but there is no nitrogen available for further reaction. For that reason formation of Si_3N_4-whiskers has not been observed when sintering in argon. Separating the sample from the heating element by an inert shielding prevents not only the formation of SiC, but also any decomposition at the Si_3N_4-surface. From an economical point of view the application of nitrogen would be prefered, but there is evidence that sintering in argon facilitates the crystallisation of the glassy bonding phase.

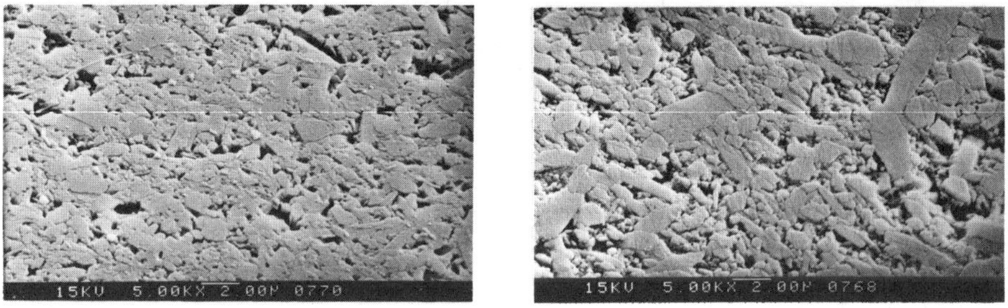

argon nitrogen

Figure 4. Microstructure of SN9Fw sintered in argon and nitrogen, resp. (Etched with boiling $HF/HNO_3/H_2O_2$ mixture).

MECHANICAL PROPERTIES

The mechanical properties of Si_3N_4 sintered in nitrogen are listed in table 4. There are only small variations in strength; best values were obtained with SN9Fw powder. Comparison of SN9Fw and TSK indicates, that the strength of SSN is nearly independent of the method by which the powder is produced. Hence other aspects are important for choosing the best powder for fabrication of components, e.g. compressibility, green density, green strength and sinterability.
As already mentioned, the $\alpha \rightarrow \beta$ transformation is incomplete immediately after the sample has reached full density. Therefore it is possible to retain a considerable amount of the metastable α-phase in the material. According to Greskovich (10) and Ingelström (11) increasing hardness may be expected with growing α-phase content. Hardness measured on samples having varying amounts of α-phase is listed in table 5. Only a marginal increase in hardness is observed. This behavior is attributed to the prominent role of the soft glassy phase in determining the overall hardness. As expected, strength has a maximum after the phase transformation is completed.

TABLE 4
Mechanical properties (σ_B, 4pt and K_{Ic}) of silicon nitride sintered under nitrogen pressure (process: see fig. 1)

Si_3N_4-powder	material of kiln furniture and shielding		
	graphite	boron nitride	aluminium nitride
SN9Fw	820 + 32 MPa 6.1 $\overline{M}$Nm-3/2	831 + 43 MPa 6.1 $\overline{M}$Nm-3/2	815 + 39 MPa 6.6 $\overline{M}$Nm-3/2
TSK	n.d.	760 + 57 MPa 5.9 $\overline{M}$Nm-3/2	797 + 51 MPa 6.1 $\overline{M}$Nm-3/2
LC12	n.d.	n.d.	709 + 63 MPa 5.8 $\overline{M}$Nm-3/2

TABLE 5

Flexural strength and Knoop hardness (100p load) as a function of isothermal sintering time (SN9Fw)

isotherm. sintering time	α/ß (vol%)	flexural strength (MPa)	Knoop hardness (HK o.1)	sintered density (gcm-3)
15'	30/70	870	1680	3.22
35'	20/80	909	1660	3.23
50'	0/100	910	1650	3.24
90'	0/100	840	1650	3.23

CONCLUSIONS

- liquid phase sintering of Si_3N_4 is possible not only in nitrogen, but also in argon.
- carbon originating from the heating element may have detrimental influence on the sintering behavior of Si_3N_4.
  Protecting the samples with hats of inert material (e.g. BN, AlN) provides an economical method to eliminate undesired reactions.
- formation of Si_3N_4-whiskers at the surface of specimens occurs only when sintering in nitrogen.
- varying the α-phase contents has only a marginal effect upon the hardness of sintered α/ß Si_3N_4.
- strength reaches a optimum directly after the α→ß-transformation is completed.

ACKNOWLEDGEMENT

The autors are very obliged to L. Sigl for careful review of manuscript and to M. Barry for graphically finalizing the paper.

REFERENCES

(1) Sorrell, C.C.: Silicon Nitride and Related Nitrogen Ceramics, Part 1, Journal of the Australian Ceram. Soc., Vol. 18, No. 2, Page 22-34
(2) Riley, F.L.: Production, Properties and Applications of Silicon Nitride Ceramics, World Ceramics, Vol. 2 (1985), Page 84-90
(3) Jack, K.H.: Nitrogen Ceramics, Endeavour, Vol. 11, No. 2 (1987), Page 80-87
(4) Ziegler, G., Heinrich, J., Wötting, G.: Relationship between processing, microstructure and properties of dense and reaction-bonded silicon nitride, J. Mat. Sci. 1987, 22 (9), Page 3041-86
(5) Knoch, H., Schwetz, K.A., Lipp A.: Microstructure Development in Silicon Nitride, Riley, F.L. (ed.) Progress in Nitrogen Ceramics 1983, Martinus Nijhoff Publishers, Page 381-392
(6) Steinmann, D.: Herstellung, Eigenschaften und Anwendung keramischer Werkstoffe auf der Basis von Siliziumnitrid
Teil 1: Der Zuliefermarkt März 1987
Teil 2: Der Zuliefermarkt Juli 1987
(7) Greskovich, C., Palm, J.A.: Development of High Performance Sintered Si_3N_4, AMMRC TR 80-46, September 1980
(8) Prochazka, S.: Das Sintern kovalenter Stoffe, Proceedings of the Int. Congress for Technical Ceramics 1.-3. March 1989, Wiesbaden/ West-Germany, Paper Mo3, Page 71-96; GE-Report 89CRD025, Feb.1989

(9) Nickel, K.G., Hoffmann, M.J., Greil, P.: Thermodynamic Calculations of SiC-Whisker Reinforced Si_3N_4. Paper presented at the 89th Annual Meeting ACerS; April 1987, Pittsburgh (39-C-87)

(10) Greskovich, C., Gazza, G.E.: Hardness of dense α- and ß-Si_3N_4 ceramics, J. Mat. Sci. Letters 4 (1985), Page 195-196

(11) Ingelström, N., Ekström, T.: Relation between Composition, Microstructure and Cutting Tool Performance of Alpha-Beta-Sialons. Journal des Physique, Suppl. 2, Tome 47, Feb. 86, Page C1-347

(12) Kölker, H., Henze, P., Schwetz, K.A., Lipp, A.: X-ray Microfocus and and Dye penetrant techniques for Crack Detection in Ceramics, Proc. 3rd Intern. Symp. "Ceramic Materials and Components for Engine, Las Vegas (Nov. 1988), in press.

HOT-PRESSING KINETICS OF AN AGGLOMERATED POWDER

INFLUENCE OF MICROSTRUCTURAL PARAMETERS

GERARD GAUTHIER, DIDIER BERNACHE-ASSOLLANT
Laboratoire de Céramiques Nouvelles (CNRS UA 320)
Faculté des Sciences, 123, Avenue Albert Thomas,
87060 LIMOGES Cédex (FRANCE)

ABSTRACT

An aluminium nitride powder, composed of large agglomerates (50 µm), was hot-pressed under different experimental conditions, in order to determine the influence of the sample microstructure on the kinetic law : sintering rate versus applied pressure. The evolution of the sintering rate versus pressure is drawn either from the shrinkage curves under isothermal conditions (continuous), or by modifying the applied pressure during the experiment (discontinuous). For a given pressure and a relative density the sintering rate varies strongly according to the chosen method. However the linearity of the pressure law suggests a diffusion process for the limiting in both cases step. Similar experiments were performed on a deagglomerated powder. The continuous and discontinuous methods led to similar values of the shrinkage rate.

INTRODUCTION

Recourse to the influence of the applied pressure on the deformation rate to discriminate hot-pressing mechanisms is now widely held (1,2,3). Practically a set of isothermal shrinkage curves under different pressures is constructed, then the dependence of the sintering rate upon pressure is deduced at a constant relative density. This approach relies on the hypothesis that the relative density is sufficient to describe the sample microstructure. Another way consists in sintering different samples up to a fixed density, using the same experimental conditions. The influence of the applied stress on the sintering rate is followed by increasing pressure at different levels. Both methods are compared on aluminium nitride samples in the following paragraphs. The discrepancy between the two methods arises from the morphology of the starting powder, particularly its agglomeration state. It is shown that only the second method must be used in the case of an agglomerated powder.

EXPERIMENTAL PROCEDURE

Materials

The commercial aluminium nitride powder was provided by Stark. The raw powder was crushed to pass a 350 mesh sieve (50 µm) . Surface specific area by nitrogen adsorption was determined as 3.9 $m^2\ g^{-1}$, which indicates an average particle diameter about 0.5 µm. The as-sieved powder contained hard agglomerates of sizes about 50 µm , and soft agglomerates smaller than 20 µm. Their relatives densities, estimated by mercury penetration were 0.75 and 0.45 respectively.

HOT-PRESSING KINETICS. CONTINUOUS AND DISCONTINUOUS METHODS

Continuous method. Isothermal shrinkage curves

Unloaded samples were heated up to 1700°C, pressure was applied for 40 minutes as soon as the temperature level was reached. An isothermal set of shrinkage curves was determined using different pressures. Densification rates were calculated from the density times curves using a fitting technique. In fig. 1 the relative densification rate is plotted versus pressure at a relative density ρ= 0.65.

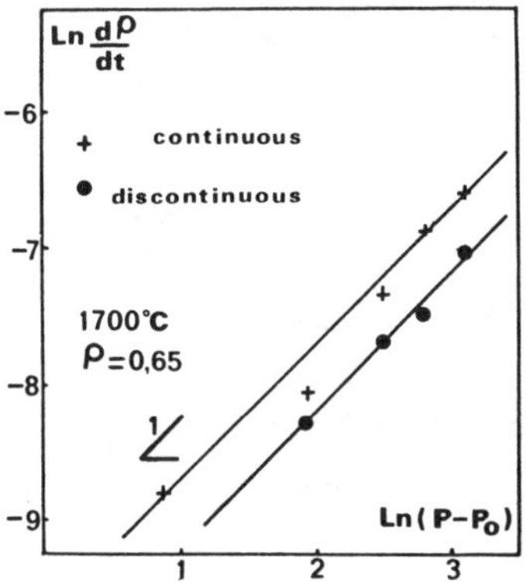

Figure 1. Sintering rate versus applied pressure for the continuous and discontinuous methods.

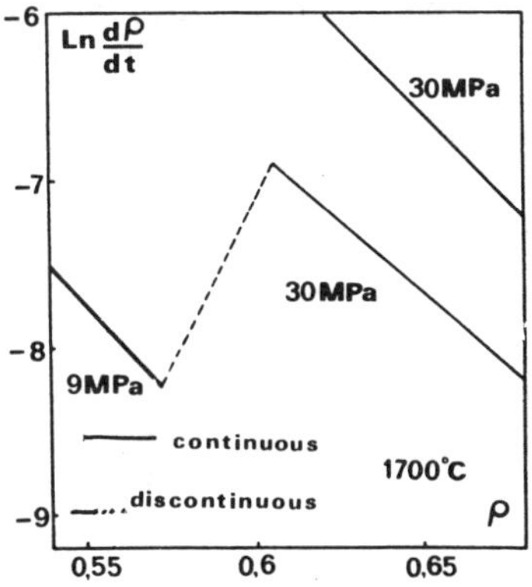

Figure 2. Sintering rate versus relative density for the densification of the raw powder.

The linear relationship between sintering rate and pressure is characteristic of a diffusion limiting process. Such a comparison between different rates for various pressures implies that samples with the same densities are identical. This means that the sample history should have any influence on the sample microstructure.

Discontinuous methods- Pressure increments

Different samples were heated up to 1700°C under atmospheric pressure. The same pressure (9 MPa) was then used to sinter the samples up to $\rho = 0.65$. Therefore all these samples had the same features, due to the same processing conditions. These compacts were hot-pressed at different pressures by suddenly increasing the external stress, in the range 10-30 MPa. Pressure dependence of the sintering rate for specimens tested in these conditions is presented in figure 1. From this last figure two main conclusions can be drawn :

sample history affects the sintering rate, but does not change the rate-pressure curve shape.

kinetical rates deduced from the discontinuous method are systematically smaller than those obtained from the continuous method.

The linearity of the rate-pressure relationship, in both cases, confirms that the rate limiting step consists of grain boundary diffusion.

The relationship between the densification rate and relative density is depicted in figure 2 for both the continuous and discontinuous methods.

Again a gap between the two procedures is experienced.

A provisional conclusion is that the relative density is not able to describe at all the sample microstructure. The sample history is an additional parameter which must be controlled also.

Deagglomerated powder

The previous powder batch was ground in order to reduce the agglomerate size. The ground powder, roughly deagglomerated, was then sintered with the continuous and discontinuous methods : 1700°C and 30 MPa for 40 minutes in the first case; 1700°C under 9 MPa up to $\rho = 0.65$, then 30 MPa for 40 minutes in the second case. Figure 3 shows the evolution of the sintering rate with density.

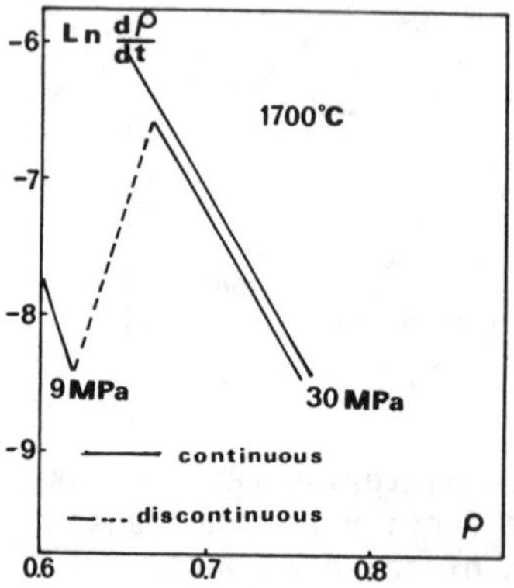

Figure 3. Sintering rate versus relative density for the densification of the ground powder.

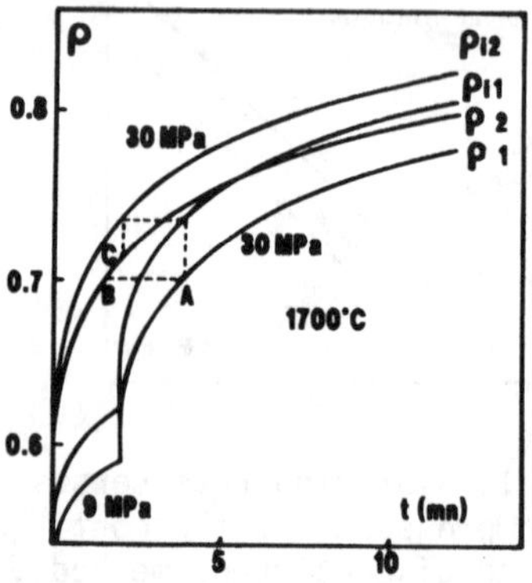

Figure 4. Isothermal shrinkage curves for the ground and the raw powders, using the continuous and discontinuous methods.

It is obvious in this case that both methods lead to similar behaviours ; this means that powder morphology is responsible of the great differences experienced with the agglomerated powder (fig.3).

DISCUSSION

Description of an agglomerated powder

An agglomerated powder contains two kinds of pores, small intraagglomerate pores, and coarser interagglomerate ones. This leads to distinguish two types of density :

the relative density inside the agglomerates : $\rho_i = V_{th} / V_{ag}$

V_{ag} is the agglomerate volume, V_{th} the theoretical material volume

The interagglomerate relative density : $\rho_e = V_{ag} / V$

V is the sample volume

The mean relative density ρ, given by geometrical considerations, is equal to $\rho_i \rho_e$.

Densification behaviours of an agglomerated powder and a ground powder

The ball milled and the raw powders were hot-pressed at 1700°C and 17 MPa. The fired densification ratios were 0.86 and 0.78 respectively. Pore size distribution of each sample was determined using a Carlo Erba porosimeter. The bimodal repartition for the agglomerated powder consists of two peaks, around 4 µm and 0.4 µm. The first one is due to the interagglomerate porosity, the second to the intragglomerate one.

Pore size distribution of the ground powder reveals that coarser pores have vanished. Moreover the average density of the deagglomerated powder and the intraagglomerate density of the agglomerated one are exactly the same, let 0.86. The main conclusion is that grains within the agglomerates have the same behaviour than those whithin the ground powder.Thus relative density time curves of the ground samples will be used later on to describe the evolution of the intraagglomerate density.

Interpretation of the discrepancy between the continuous and discontinuous methods

In figure 4, the symbols ρ_1 and ρ_{i1} refer to the agglomerated and deagglomerated powders sintered at 1700°C under 9 MPa up to 0.65 then under 30 MPa. The symbols ρ_2 and ρ_{i2} represent the relative density time curves at 1700°C under 30 MPa for both powders.

Let A the point $\rho = 0.7$ for the deagglomerated powder, the corresponding sintering time is about 4 minutes. The microstructural parameters for such state are :

$\rho_1 = 0.7 \qquad \rho_{i1} = 0.75$ thus $\rho_{e1} = 0.95$

The relation between ρ_i and ρ_e can be rewritten as followed : $1/\rho \; d\rho_e/dt = 1/\rho_i \; d\rho_i/dt + 1/\rho_e \; d\rho_e/dt$

$1/\rho_i \; d\rho_i/dt$ is the sintering rate within the agglomerates, $1/\rho_e \; d\rho_e/dt$ is the sintering rate between the agglomerates

From ρ_1 and ρ_{i1} it is easy to deduce the values of $d\rho_1 / dt$ at 0.7 and $d\rho_{i1} / dt$ at 0.735

$1/\rho_1\ d\rho_1/dt = 5.9\ 10^{-4}\ s^{-1}$ $\quad 1/\rho_{i1}\ d\rho_{i1}/dt = 5.1\ 10^{-4}\ s^{-1}$

Therefore : $1/\rho_e\ d\rho_e/dt = 0.8\ 10^{-4}\ s^{-1}$

Let B the point corresponding to the agglomerated powder sintered at 1700°C and 30 MPa up to the relative density 0.7 : $\rho_2 = 0.7$. For this sample the intraagglomerate and interagglomerate densities deduced from ρ_{i2} are : $\rho_2 = 0.7$ $\quad \rho_{i2} = 0.72$ $\quad \rho_{e2} = 0.96$

From ρ_{i2} and ρ_2 datas an estimation of $d\rho_2/dt$ ($\rho_2 = 0.7$) and $d\rho_{i2}/dt$ ($\rho_{i2} = 0.72$) is possible :

$1/\rho_2\ d\rho_2/dt = 0.72\ 10^{-4}\ s^{-1}$ $\quad 1/\rho_{i2}\ d\rho_{i2}/dt = 6.4\ 10^{-4}\ s^{-1}$

$1/\rho_{e2}\ d\rho_{e2}/dt = 0.8\ 10^{-4}\ s^{-1}$

Since for A and B the interagglomerate shrinkage rates are exactly the same, the discrepancy between the rates arises from the gap between the intraagglomerate sintering rates. Then to compare sintering rates the same intraaglomerate density must be chosen (points A and C) instead of the same apparent density (points A and B). Sintering rate at the point C is $6\ 10^{-4}\ s^{-1}$ close to that of point A. Obviously no difference is observed between both methods for a deagglomerated powder (ground powder) : ρ_{i1} and ρ_{i2} lead to the same rates for identical densities.

CONCLUSION

Determination of the sintering rate dependence on the applied pressure (resp. temperature) involves that samples to be tested possess identical microstructures. In the case of an agglomerated powder the relative density is not sufficient to characterize this microstructure. Thus, the available method consists of applying pressure increments at a constant relative density.

REFERENCES

1. Rahaman, M.N., Riley, F.L. and Brook, R.J., Mechanism of densification during reaction hot-pressing in the system SiAlON. J. Am. Ceram. Soc., 1980, 63, 648-653.

2. Bernache-Assollant, D. and Zhang, Y., Hot-pressing sintering, New ways to understand sintering mechanisms. J. of Inorg. Mater., 1988, 3, 289-304.

3. Pampuch, R., Mechanisms of hot-pressing of oxide powder compact. Mater. Sci. Monograph., 1981, 8, 216-247.

HOT PRESSING OF SILICON CARBIDE

EMILIJA KOSTIĆ, OLGA NEŠIĆ, SLAVICA ZEC
Boris Kidrič Institute of Nuclear Sciences, Laboratory 170
P.O.B. 522, 11001 Beograd, Yugoslavia

ABSTRACT

The hot pressing of silicon carbide powders in the temperature range from 1800° to 1950^oC in the presence of additives based on Al_2O_3-Y_2O_3 as well as Al_2O_3-Dy_2O_3 systems is applied to densify this material up to 95%TD. The additive amount controls reached density increasing at lower temperature.It is demonstrated that only the alpha SiC modification exists after hot pressing at 1950^oC. Fine grained materials are produced by this procedure.

INTRODUCTION

Our attempts to enhance the densification of α-SiC and β-SiC pointed to an Al_2O_3 and Y_2O_3 oxide mixture, as well as an Al_2O_3 and Dy_2O_3 one, as effective sintering aids. Both α-SiC and β-SiC can be sintered at 1950^oC up to 95%TD in the presence of 10 mass% of the mentioned additives (1).

The purpose of this paper is to produce ceramics at 1850°-1950^oC and decrease the amount of the secondary phase in dense SiC material as much as possible. The hot pressing technique with the above mentioned additives was used. The amount of additives was reduced to 0.75 mass%. The small amounts of additive present in SiC, consolidated at 1900^oC and 1950^oC, caused a fine grained microstructure of dense SiC ceramics. The obtained density and resulting mechanical properties can be strongly influenced by the microstructural characteristics. Keeping in mind the importance of microstructure in ceramic materials the special attention was paid to the grain growth process in the presence of the liquid phase of the produced SiC ceramics. This was discussed in terms of experimental observations and possible mechanisms. The additives were found to induce the β-SiC to α-SiC transformation at temperatures between 1900°-1950^oC.

EXPERIMENTAL

The starting materials were α-SiC and β-SiC powders produced by Herman Starck, Berlin (1). Powder mixtures of Al_2O_3 and Y_2O_3 (additive Y) as well as Al_2O_3 and Dy_2O_3 (additive D) of the molar ratio 0.77:0.23 were used as densification aids. The additive amount ranged from 0.5 up to 5 mass%. Hot pressing was carried out at 1800°, 1850°, 1900° and 1950^oC under 35 MPa pressure in protective gas. The isothermal heating time was 20 min. The heating and cooling rates were 15^oC/min. The obtained materials were characterized by density measurement, microstructural and X-ray analyses.

Hot pressed α-SiC powder at 1800°C in the presence of 5 mass% of additive exhibited a density of 2.74 Mg/m^3. High dense SiC can be obtained in the presence of 3 mass% additive at 1850°C. The attained densities at higher temperatures are demonstrated in Tables 1 and 2.

TABLE 1

Densities of hot pressed SiC powders at 1900°C (Mg/m^3)

Powder	Amount of additive (mass%)											
	0.5		0.75		1		1.5		3		5	
	Y	D	Y	D	Y	D	Y	D	Y	D	Y	D
α-SiC	3.03	-	3.18	3.08	3.25	3.17	3.23	3.23	3.22	3.24	3.22	3.20
β-SiC	3.05	-	3.16	3.15	3.25	3.26	3.22	-	3.23	-	3.15	-

TABLE 2

Densities of hot pressed SiC powders at 1950°C (Mg/m^3)

Powder	Amount of additive (mass%)				
	0	0.5		0.75	
		Y	D	Y	D
α-SiC	2.27	3.04	-	3.26	3.23
β-SiC	2.25	3.07	-	3.26	3.25

X-ray analyses of the starting powders, as well as of hot pressed samples at 1900°C and 1950°C with the additives were performed.

Figs.1 and 2 illustrate the resulting microstructures of hot pressed and sintered fully dense SiC materials.

DISCUSSION OF RESULTS

Densification

As is obvious from Tables 1 and 2, the density of hot pressed materials depends both on the temperature and additive amount. The results obtained on the additive amount needed for attaining 95%TD at given temperatures indicate that densification proceeds in the presence of the liquid phase (2). On the basis of the phase diagram Al_2O_3-Y_2O_3 (3) additive Y forms a liquid phase at about 1750°C. We assumed that additive D would behave in a similar way. In the presence of SiO_2 the appearence of the liquid phase is expected at lower temperatures (3), in small amounts. The presence of the liquid phase by which powder particles are wetted brings about their rearrangement causing density increases during hot consolidation, as well as other processes developing during liquid phase sintering. A relatively low density was obtained at 1800°C with 5 mass% additives meaning that the liquid phase did not encompass most of the grain boundaries between the SiC particles. In this case only 84%TD was obtained. At 1850°C to reach 95%TD one needs only

Figure 1. Fracture surface of sintered β-SiC at 1950°C with 10 mass% add.Y

a. 1900°C with 5 mass% add.Y b. 1950°C with 0.75 mass% add.Y

Figure 2. Fracture surfaces of hot pressed β-SiC

3 mass% of additives, while at 1900°C to attain the same density only 1-1.5 mass% of additives is necessary. The optimum additive amount for densification by hot pressing at 1950°C was 0.75 mass%. It is interesting to note for comparison that pure SiC under the same hot pressing conditions achieved only 70%TD.

The liquid phase amount is obviously proportional to the additive amount used to enable densification. With increasing hot pressing temperature the optimum additive amount for achieving full densification decreases which means that the properties of the liquid phase change with temperature. At higher temperatures the conditions for SiC particle wetting are better enabling the faster rearrangement of the particles.

X-Ray Data

A secondary crystalline phase with garnet structure was not detected with hot pressed samples by X-ray analysis which, however, was the case with sintered samples having higher additive content (1). The reason probably is in low amount of additives (less than 3 mass%) and also the thermal treatment conditions.

As shown by X-ray analysis, in the starting α-SiC powder the 6H polytype prevailed besides the 15R polytype which existed in smaller amounts. The starting β-SiC is mainly in 3C form with lower amounts of the 6H polytype. After hot pressing at 1800°-1900°C the obtained materials are composed of the same crystalline modifications, in the same amounts as they are in the starting powders. The α-SiC starting powder remains as such at 1950°C while the β-SiC starting powder transforms into the 6H and 4H polytypes along with traces of 3C.

The presence of Al_2O_3 in the used additives did not bring about the appearance of the 4H polytype in α-SiC which is mentioned in the literature. After the hot pressing β-SiC at 1950°C the ceramic material consisted of the 4H and 6H polytypes the last one being the major crystalline form.

Microstructural analysis

On the basis of the microstructural analysis of the fracture surfaces of samples sintered with 10 mass% of additives (Fig.1), the presence of the amorphous phase at the triple points and grain boundaries can be observed. This is direct proof for liquid phase existence during the hot consolidation of SiC powders in the presence of the Y additive. Its presence causes the rounding of the grain boundaries. Bearing this in mind it can be explained why the grain boundaries in samples hot pressed with 5 mass% of additive are more rounded in comparison with samples having 1-1.5 mass% of additives. Grain growth takes place during hot pressing in different ways depending on the starting powder characteristics, additive amount and temperature.

In samples obtained by hot pressing of the starting α-SiC at 1900°C the grain size ranges from 1-5 μm. The number of finest grains decreases with increasing additive amount up to 5 mass%. In the presence of a lower amount of additives, about 1 mass%, the number of bigger irregular grains is higher. A similar microstructure was observed both in the presence of additive Y and additive D.

In samples obtained from the starting β-SiC powder at 1900°C in the presence of 5 mass% of additives small grains are not very frequent, and the maximum grain size is about 4 μm. With lower additive amount the number of small grains (about 0.75 μm) increases within the hot pressed material. Large grains,even larger than the average value can also be observed. There is no difference in the microstructure of samples hot pressed with Y and D additives (Fig.2).

β-SiC powder hot pressed at 1950°C in the presence of 0.75 mass% of additive is composed of a great number of small grains (about 0.75 μm) which is not the case with the α-SiC starting powder hot pressed under the same conditions. In this case the grain size is about 1.5 μm. Larger grains can also be observed. It seem that grain growth was more intense in some areas (particular grains were up to 6 μm) (Fig.2).

The grain form is almost equiaxed independent of the starting powder modification and hot pressing temperature.

With ceramic material containing a lower amount of additive intergranular fracture is observed, with elements of transgranular fracture only through large grains. However, with samples having 5 mass% of additives, transgranular fracture is dominant.

The results of microstructural analysis indicate that the grain growth process proceeded by the action of two mechanisms, solution-reprecipitation and coalescence. The first one was more pronounced in samples having a higher amount of additives. It is also more pronounced in samples obtained from β-SiC. With material containing a lower amount of additives, coalescence is often obvious due to the low amount of the liquid phase which is not enough to encompass all the grains. Grain growth is more intense in some areas, giving rise to the formation of inhomogeneous microstructure. Starting with α-SiC the grain growth is more uniform. They get dissolved much slower than the grains of β-SiC, because the starting crystallites are larger than in β-SiC powder. As a consequence of the physical properties of the starting β-SiC, discontinuous grain growth is more pronounced.

Hot pressed β-SiC at 1950°C in the presence of 0.75 mass% of additives exhibits equiaxed grains contrary to sintered β-SiC at the same temperature with 10 mass% additives, which contains grains of elongated form (1). In both cases the β-SiC transformation into the 6H and 4H polytypes appeared.

CONCLUSIONS

By using hot pressing 95% TD of SiC ceramics was achieved at 1950°C in the presence of 0.75 mass% of additives from Al_2O_3-Y_2O_3 and Al_2O_3-Dy_2O_3 systems. The same density was obtained at 1900°C in the presence of 1-1.5 mass% of additives and with 3 mass% of additives at 1850°C. During hot pressing in the temperature range 1850°-1950°C a fine grained microstructure of high dense SiC was obtained. The mean grain size ranged from 1-3 μm. The starting powder which was in alpha crystalline form during hot pressing up to 1950°C did not change its polytype structure, i.e. it remained in the 6H and 15R forms while β-SiC transformed to the 6H and 4H polytypes above 1900°C.

REFERENCES

1. Kostić, E., Sintering of Silicon Carbide in the Presence of Oxide Additives, Powder Metall. Int. 1988, 20, 6, 28-29

2. Kostić, E., Kiš,Š., Bošković, S., Liquid Phase Sintering of Alumina, Powder Metall. Int., 1987, 19, 4, 41-43

3. Toropov, N.A., Barzakovskii, V.P., Lapin, V.V., Kurceva, N.N., Diagrammi sostoyaniya silikatnikh sistem-vipusk pervii, Izd. Nauka, Leningrad, 1969, 228 and 324.

GAS PRESSURE SINTERING OF α′-SIALON CERAMICS

G.Z. Cao, R. Metselaar & G. Ziegler
Centre for Technical Ceramics, P.O. Box 513
5600 MB Eindhoven, the Netherlands

ABSTRACT

α′-sialon ceramic densified via a gas pressure sintering process is an attractive subject at present. A gas pressure sintering process with three steps specially for the formation and densification of α′-sialon ceramics has been optimized. This investigation indicates the improvement of the characteristics of α′-sialon (also including α′+β′-sialon composite) ceramics by use of a gas pressure sintering process involving the formation of a transient liquid phase.

INTRODUCTION

While silicon nitride ceramics have been extensively developed for high temperature engineering applications, one of the open problems is the grain boundary glass phase degrading the mechanical properties at elevated temperatures. In an attempt to reduce this effect, α′-sialon ceramics have been studied, since they offer possibilities for developing single phase ceramics. In this case the metal oxides present in the starting mixtures as either sintering additives or impurities could be incorporated into the structure of the final ceramics.

In practice, it is very difficult to complete the formation and densification of α′-sialons, because the high nitrogen content results in a small amount and a high viscosity of oxynitride liquid [1]. Furthermore, as α′-sialon forms during the heating cycle, the amount of liquid gradually decreases until thermodynamically all liquid is incorporated into the α′-sialon lattice. Gas pressure sintering has been considered to overcome this problem. High pressure nitrogen hinders thermal decomposition [2], and therefore, higher sintering temperatures can be used, which allow a reduction of the amount of liquid [3]. Additionally, a high nitrogen pressure supports an extra sintering driving force for densification [4].

EXPERIMENTAL PROCEDURES

Mixtures were made of Si_3N_4 (LC-12, Starck), AlN (grade C, Starck), Al_2O_3 and Y_2O_3 powders by ball milling. Specimens (size: ø 12x7 mm) were pressed uniaxially followed by isostatic pressing. Sintering was carried out in a gas pressure furnace capable of 2200°C under up to 100 bar N_2. The material was characterized by density measurements, XRD and SEM investigations.

RESULTS AND DISCUSSION

A gas pressure sintering process with three sintering steps was adopted for the formation and densification of α′-sialon as well as composite α′+β′-sialon ceramics (Fig.1).

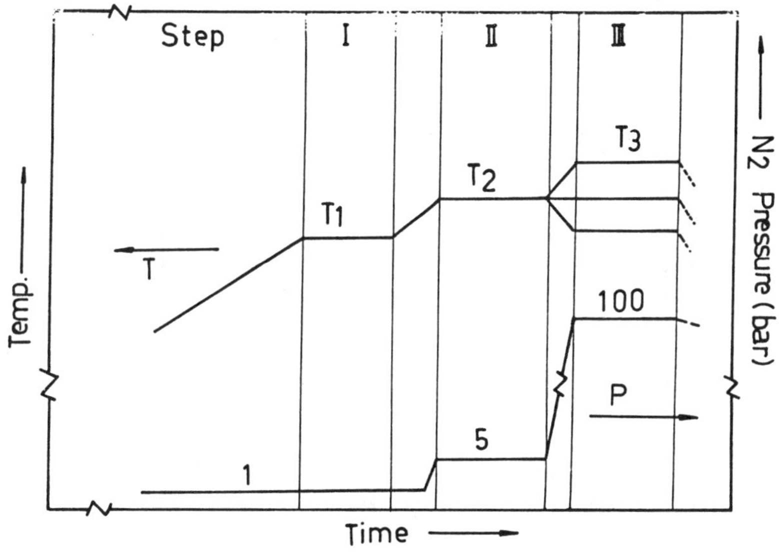

Fig.1 Schematic diagram of gas pressure sintering.

As in the case of silicon nitride ceramics [5] at least two sintering steps are necessary to obtain full densification:

a) sintering at low pressure to close all open pores (Fig.1, region II);

b) final densification at high pressure (Fig.1, region III).

In this investigation, the optimum sintering conditions were found to be about 1850°C for 10-60 min under low pressure and then increasing temperature up to about 1925°C and holding for 10-60 min at high pressure. In the low pressure step (region II) closed porosity and relative high density were achieved. However, the temperature sensitively influences weight loss. Fig.2 indicates that weight loss sharply increases with temperature above 1850°C because of thermal decomposition.

The high pressure step (region III, 1925°C/100 bar N_2) effectively improves the densification for all compositions. Fig.3 shows a comparison of the density values obtained via gas pressure sintering and via pressureless sintering. This figure also suggests that 2.5 mol% additives give the best densification behaviour.

Regarding the formation of sialon it should be pointed out that during heating with rising temperature above 1300°C Y_2O_3 will form an oxide eutectic liquid with SiO_2 on the surface of Si_3N_4 particles and Al_2O_3 on AlN [6]. As can be seen from Fig.4 at about 1380°C the dissolution of silicon nitride into liquid to form oxynitride liquid occurs. Concurrently the precipitation of α'-sialon as well as an

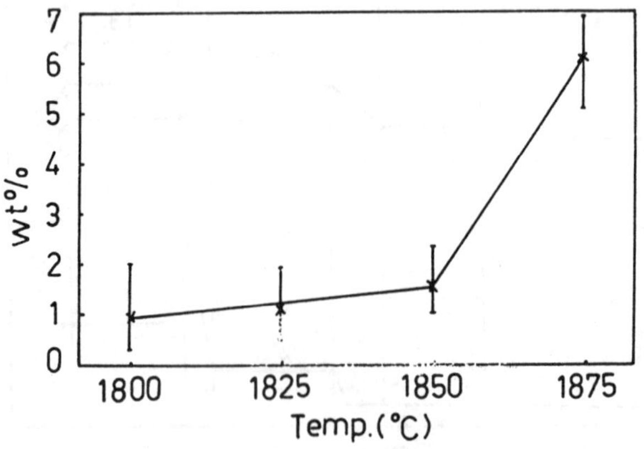

Fig. 2 The temperature influence on weight loss. heating rate: 10°C/min, holding time: 10 min

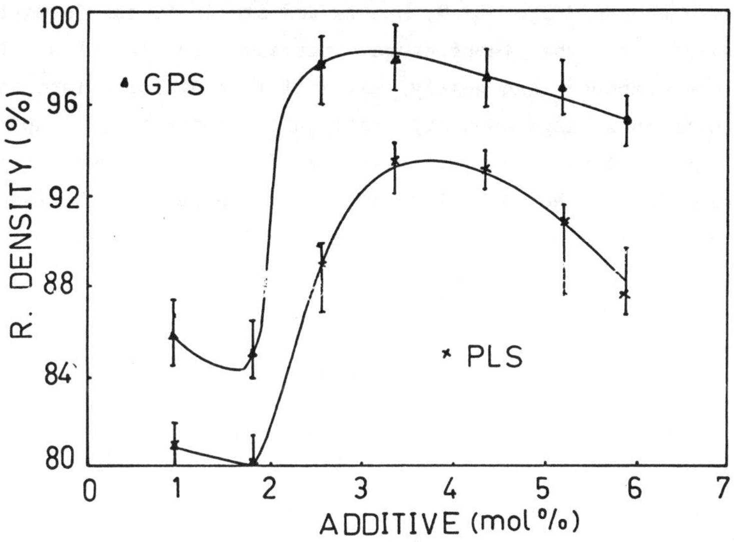

Fig. 3 A comparison of the densities obtained via gas pressure sintering and via pressureless sintering.

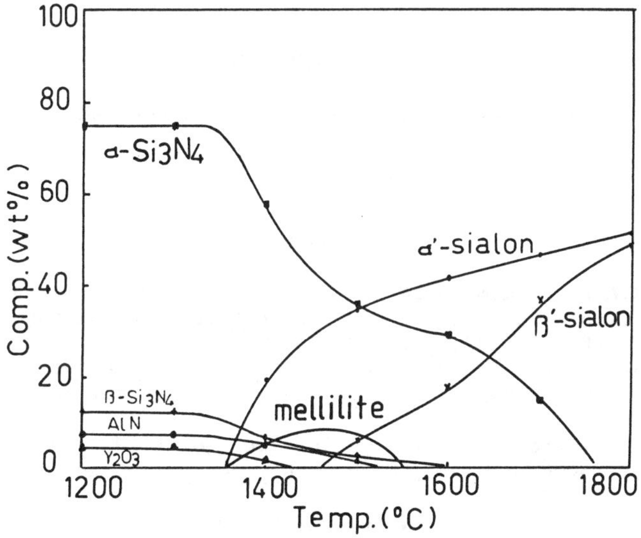

Fig. 4 The reaction sequences of sialon formation during heating cycle

intermediate compound $Y_2O_3 \cdot Si_3N_4$ starts and at 1480°C the formation of β'-sialon begins. As the temperature increases up to about 1550°C the intermediate compound and nearly all additives dissolve into the liquid, which results in a large quantity of liquid. Above this temperature the large amount of liquid promotes rapid grain growth with rising temperature. Thus, a holding time at low temperature is favourable to reduce the liquid content by incorporation into the sialon lattice, and, therefore, to get fine grained microstructures.

CONCLUSIONS

1. Gas pressure sintering effectively improves the densification of α'-sialon ceramics.
2. To achieve full densification of α'-sialon ceramics by gas pressure sintering, two sintering steps are required.
3. For α'-sialon ceramics, an extra presintering step is favourable to get fine grained microstructures.
4. With the three step sintering process, the best densification behaviour was obtained with 2.5 mol% additives.

ACKNOWLEDGMENT

The authors gratefully acknowledge Ir. F. Blömer and Ing. J.P.G.M. van Eijk for their help.

REFERENCES

1. Drew,R.A.L., Hampshire,S. and Jack,K.H., The preparation and properties of oxynitride glasses, Progress in Nitrogen Ceramics, ed. by Riley,F.L., 1983, p323-330.
2. Hirosaki,N. and Okada,A., Effects of N_2 gas pressure sintering on the surface and strength of Si_3N_4, Advanced Ceram. Mater. 3[5], 515-6, 1988.
3. Wötting,G. and Ziegler,G., Sintering of Si_3N_4 powder compacts under high nitrogen pressure, High-techn. Ceramics, ed. by Vincenzini,P., 1987, p983-991.
4. Coble,R.L., Reaction Sintering, Sintering. Theory and Practice, ed. by Kolar,D. et al, 1982, p145-151.
5. Greskovich,C. and Palm,J.A., Development of a sintering process for high performance Si_3N_4, AMMRC TR, 78-32, July 1978. SDR-77-178.
6. Levin,E.M. and McMurdie,F.H., Phase Diagrams for Ceramists, Amer. Ceram. Soc. Columbus, Ohio, 1975. Fig. 2586.

HOT PRESSING OF SIALON

E. KOKMEIJER AND R. METSELAAR
Centre for Technical Ceramics, P.O. Box 595, 5600 AN Eindhoven,
The Netherlands

ABSTRACT

Sialon samples, $Si_{6-z}Al_zO_zN_{8-z}$ with z=3, are prepared by uniaxial reaction hot pressing in a graphite die. Al_2O_3, AlN and Si_3N_4 powders are used as starting materials. The influence of the hot pressing procedure (time, temperature and pressure) on the reaction and densification process is determined and discussed. Addition of CeO_2 strongly decreases the temperature needed to obtain high density. High density samples without additives are prepared by hot pressing during 1800 s at 1800°C with a pressure of 21.6 MPa. X-ray diffraction shows complete conversion to β'-sialon. Micro vickers hardness of the sample is 14.1 GPa (5 N). Scanning electron microscopy shows an average grain size of 2 μm.

INTRODUCTION

$Si_{6-z}Al_zO_zN_{8-z}$ with z=3 can be prepared by carbothermal reduction of kaolin [1]. The advantage of this process is the low cost of the raw materials. The disadvantage is the impurity of the raw materials which might degrade the properties in comparison with pure sialons.
Previous to a study of the influence of impurities and sinteradditives on sialons the reaction hot pressing process of pure sialon has been studied [2,3,4]. Pure $Si_{6-z}Al_zO_zN_{8-z}$ with z=3 can be prepared according to:

$$Si_3N_4 + AlN + Al_2O_3 \longrightarrow Si_3Al_3O_3N_5 \qquad (1)$$

During hot pressing samples of the powder mixture, the reaction to β'-sialon takes place as well as a densification process. To develop a process to produce a high density β'-sialon sample the influence of time, temperature and pressure was studied.

EXPERIMENTAL PROCEDURE

A mixture of Si_3N_4, Al_2O_3 and AlN corresponding to a β'-sialon with z=3 was prepared by milling in isopropanol with silicon nitride grinding media.The specifications of the powders used are given in table 1.

TABLE 1
Specifications of the starting material. Given are the code and supplier, the median grain size (d_{50}),specific surface (s) and the purity.

material	code	supplier	d_{50} (μm)	s (m^2/g)	purity (%)	
Si_3N_4	LC12	HC Starck	0.5	20.2	96.3	3.68% SiO_2
AlN	grade C	HC Starck	1.2	3.7	96.4	3.61% Al_2O_3
Al_2O_3	CT3000	Alcoa	0.5	3	99.6	

The composition was corrected for the surface oxide on the powders. Five percent of polyethyleneglycol was added to the powder as a pressing agent. After evaporation of the isopropanol in a rotavapor and drying at 100°C, the powder was passed through a 63 μm sieve. Tablets of 20 g (ø34 mm) were uniaxially pressed under 2.5 MPa and afterwards isostatically pressed at 250 MPa. This resulted in a green density of $2.0\text{–}2.1 \cdot 10^3$ kg/m^3. Finally the tablets were slowly (0.008°C/s) heated up to 500°C to remove the pressing agent. The tablets were hot pressed in a graphite die (ø34 mm). To prevent reactions between the tablet and the die, the tablets were packed in graphite paper coated with boron nitride. The hot press conditions are given in table 2. In the experiments 30HP15, 30HP16 and 30HP18 full load was applied at room temperature. In the experiment 34HP58 full load was applied after reaching the maximum temperature. The other samples were preloaded with 2.5–5.4 MPa whereas full load of 21.6 MPa was applied between 1400–1550°C. The heating rate from 500°C up to the maximum temperature was 0.4°C/s. The final density of the tablets was determined by the archimedes technique. The compositions of the crystalline phase were analysed by X-ray diffraction. Semi-quantitative analyses were performed according to the adiabatic principle of XRD analysis of mixtures [5–7]. Macro vickers hardness measurements were performed with a load of 156 N. Micro vickers hardness was measured at a load of 5 N.

TABLE 2
Hot press conditions (temperature (T), time (t) and pressure(p)), macrovickers hardness (HV) and β'-sialon and 15R content of the crystalline phase.

code	T (°C)	t (s)	p (MPa)	HV (GPa)	β'-sialon (%)	15R (%)
30HP15	1750	1800	10.8	8.8	97	3
30HP16	1750	1800	21.6	10.3	98	2
30HP18	1750	1800	43.2	8.5	96	4
34HP37	1450	300	21.6	---	--	3
34HP38	1450	3600	21.6	1.4	5	7
34HP39	1550	300	21.6	0.8	20	5
34HP40	1650	300	21.6	2.5	45	6
34HP42	1700	300	21.6	5.2	58	5
34HP41	1750	300	21.6	9.5	70	5
34HP43	1800	300	21.6	9.5	79	4
34HP44	1850	300	21.6	11.8	98	2
34HP58	1800	1800	21.6	10.5	97	3
34HP60	1800	1800	21.6	13.3	98	2

RESULTS AND DISCUSSION.

Results of hardness measurements and X-ray analyses are listed in table 2. To illustrate the influence of temperature on the conversion and densification the relative density and β'-sialon content of 34HP37 to 34HP44 are given in figure 1. This figure shows that the reaction starts at 1450°C while the onset of the densification takes place at 1650°C. When pressure was applied after reaching the maximum temperature the density and hardness of the sample (34HP58) were lower than a comparable sample (34HP60) at which the pressure was applied earlier. Even at longer hot press times it was not possible to achieve full density at 1800°C if the pressure was applied after heating up. This indicates that the sinterability is improved by the reaction to β'-sialon [8]. At 1800°C the conversion to β'-sialon is nearly complete on account of which the sinterability is decreases and apparently is too low to produce a high density material.
In figure 2 the displacement of the punch of the hot press die is given as function of the temperature (during heating up) and time (at 1750°C) for three indicated pressures. The figure clearly shows that the part of densification taking place during heating up is strongly dependent on the applied pressure. This is in agreement with the fact that no high density is achieved if the pressure is applied after reaching the maximum temperature. A higher pressure results in a higher densification. At a pressure of 10.8 MPa it is not possible to achieve the same density as at 21.6 MPa at the same temperature. The difference between the hot press curves with a pressure of 21.6 and 43.2 MPa is much smaller. After 600 s at 1750°C the densification is the same in both experiments. Apparently above a certain pressure a further increase of pressure does not lead to a significant higher density. The effect of increasing pressure on the densification is similar to the effect of an increasing amount of liquid phase [8]. This is shown in figure 3. In this figure the hot press curves are given for powder mixtures with 0, 0.5, 1.0 and 3.0 wt% CeO_2 at a pressure of 21.6 MPa. With an increasing amount of CeO_2 the same final density can be achieved in the same time, with decreasing maximum temperature. The reduction of the maximum temperature per weight percent CeO_2 decreases at increasing CeO_2 content.

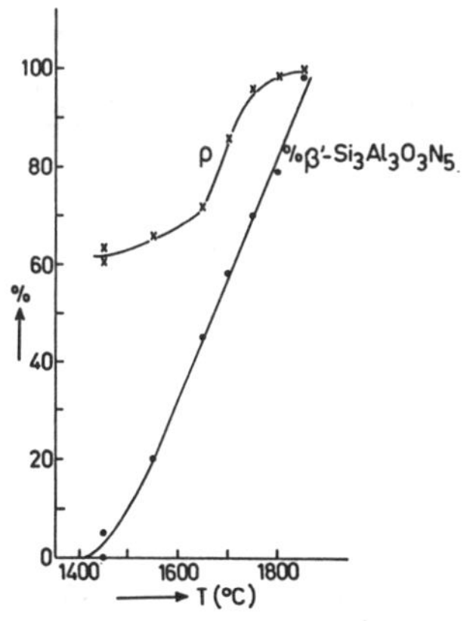

Figure 1. The relative density and the β'-sialon content as function of the maximum temperature (T) at a pressure of 21.6 MPa.

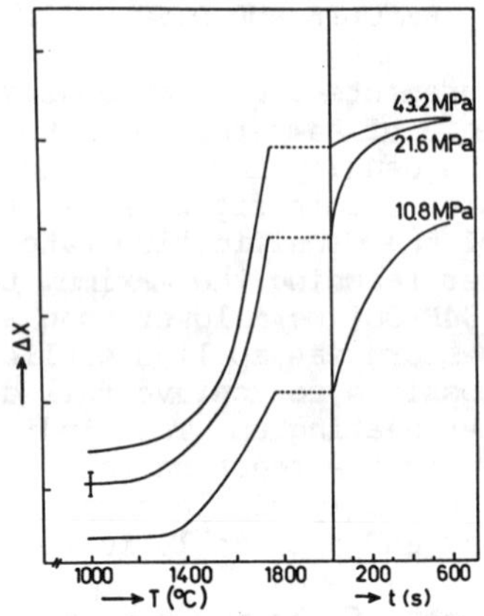

Figure 2. The hot press curves at different pressures. Given is the displacement of the punch Δx as function of the temperature T during heating up and the time t at 1750°C.

The reaction to β'-sialon is complete after 300 s at 1850°C; no Al_2O_3, AlN or Si_3N_4 is detected by X-ray diffraction. Most samples contain a few percent 15R as a second phase which is commonly ascribed to decomposition of β'-sialon at long sintering times. In the experiments considered here, this can not be the case since 15R is already detected in samples prepared at the lowest temperature. Possibly part of the SiO_2 (originated from the Si_3N_4) is present in the amorphous intercrystalline phase. This decrease in total amount of Si promotes the formation of 15R which is more Al rich than β'-$Si_3Al_3O_3N_5$. The hardness increases with the density (see table 2). The low hardness of sample 30HP18 can not be explained at this stage. At this moment the best result is obtained by hot pressing for 1800 s at 1800°C and a pressure of 21.6 MPa applied at 1500°C. The micro hardness of this sample is 14.1 GPa. The K_{1c} value, calculated from the cracks at the macrohardness measurement, is 5.4 $MPa/m^{-1/2}$. Scanning electron microscopy on an etched sample shows an average grain size of 2 μm.

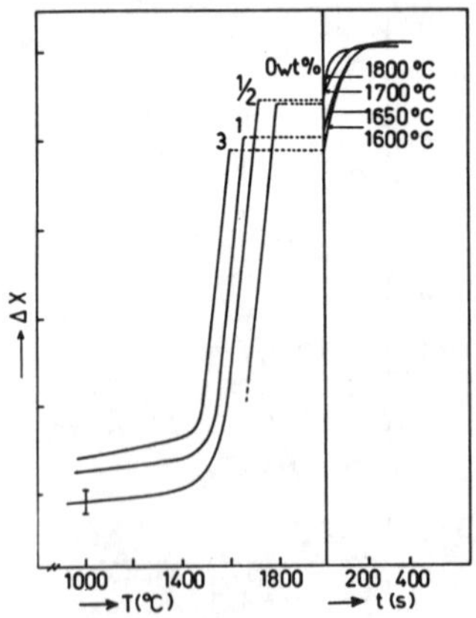

Figure 3. The influence of the amount of CeO_2 on the hot press curve. Given is the displacement of the punch Δx as function of the temperature T during heating up and time t at the indicated temperature.

CONCLUSIONS

It is possible to produce a high density sialon from a mixture of AlN, Al_2O_3 and Si_3N_4 without additives. Hot pressing for 1800 s at 1800°C with a pressure of 21.6 MPa results in samples with a porosity of less than 1% and a micro vickers hardness of 14.1 GPa (5 N). X-ray diffraction shows complete conversion to β'-sialon with a few percent 15R phase.
The highest density is obtained when pressure is applied at the beginning of the reaction to sialon. Higher pressure, higher temperature and longer times result in a higher density. The reaction to β'-sialon starts at 1450°C. If no sintering aid is added the densification starts at approximately 1650°C. Addition of CeO_2 lowers this temperature strongly. The maximum temperature for mixtures with 0, 0.5, 1.0 and 3.0 wt% CeO_2 is 1800, 1700, 1650 and 1600°C respectively.

ACKNOWLEDGEMENT

The investigations were supported in part by the Innovative Research Program Technical Ceramics (IOP) with financial aid from the Netherlands Ministry of Economic Affairs.

REFERENCES

1. Van Dijen, F.K., The carbothermal production of $Si_3Al_3O_3N_5$ from kaolin, its sintering and its properties. Ph.D. Thesis, Eindhoven University of Technology, 1986.

2. Mitomo, M., The formation of single phase β-sialon. In Proc. of International Symposium on Factors in Densification and Sintering of Oxide and Non-oxide Ceramics, ed. S. Somiya and S. Saito, Gakujutsu Bunken Fukyu-Kai, Tokyo, 1979, pp. 463-467.

3. Kuwabara, M., Benn, M., Riley, F.L., The reaction hot-pressing of compositions in the system Al-Si-N-O corresponding to β'-sialon. J. Mater. Sci., 1980, 15, 1407-1416.

4. Rahaman, M.N.A. and Riley, F.L., Densification behaviour during reaction hot-pressing of compositions in the β'-sialon system. Proc. Br. Cer. Soc., 1981, 31, 63-70.

5. Chung, F.H., Quantitative interpretation of X-ray diffraction patterns of mixtures. I Matrix-flushing method for quantitative multicomponent analysis. J. Appl. Cryst., 1974, 7, 519-525.

6. Chung, F.H., Quantitative interpretation of X-ray diffraction patterns of mixtures. II Adiabatic principle of x-ray diffraction analysis of mixtures. J. Appl. Cryst., 1974, 7, 526-530.

7. Chung, F.H., Quantitative interpretation of X-ray diffraction patterns of mixtures. III Simultaneous determination of a set of reference intensities. J. Appl. Cryst., 1974, 8, 17-19.

8. German, R.M., Liquid Phase Sintering, Plenum Press, New York, 1985, pp. 172-175.

THE PRODUCTION OF CERAMICS AND THE GASIFICATION OF CARBON

F.K. VAN DIJEN, DSM Research, Ceramics Department, P.O. Box 18, 6160 MD, Geleen, The Netherlands

ABSTRACT

The removal of carbon or carbon residue from ceramic powders or greenware by a gasification reaction is described. An inventory of the literature on the gasification and combustion of coal and carbon is given, paying attention also to reactions of gases and carbon with ceramic material. The gases studied were O_2, CO_2, H_2O and H_2. With some examples it is shown that the results of research on the gasification of coal/carbon can be applied for the removal of carbon or carbon residue in the production of ceramics.

INTRODUCTION

Ceramic powders and greenware often contain carbon or carbon forming materials. In most cases the carbon or carbon forming materials have to be removed before sintering. This is usually done by slowly heating up the greenware in air to the sintering temperature or by heating the greenware, for instance, at 600 °C in air for 1 hour prior to sintering. Under these conditions carbon or carbon forming material is burnt out of the greenware. A similar method is used for the removal of carbon or carbon forming materials from ceramic powders. However, often there is a need for oxygen free firing conditions, for instance in the case of non-oxide ceramics or metallized substrates (1).

In this paper we will discuss the removal of carbon or carbon residue in both an oxygen containing and oxygen free atmosphere. In general the carbon is removed by a chemical reaction with a gas. The carbon is transformed to a carbon containing gas: C_{solid} + gas = C_{gas} (+ gas).
Many gases can be used for the gasification of carbon. In this paper the use of O_2, CO_2, H_2O and H_2 as reactants, is discussed. The reactions of these gases with carbon are well discribed in literature concerning coal gasification or combustion. The aim of this paper is the definition of temperature-time prophiles for removing carbon or carbon residue from ceramic powders or greenware, using these gases, without negative effects for the product.

THERMODYNAMICS OF GASIFICATION REACTIONS

Four reactions are considered for the gasification of carbon

$C + O_2 = CO_2$ 1

$C + CO_2 = 2CO$ 2

$C + H_2O = CO + H_2$ 3

$C + 2H_2 = CH_4$ 4

Important features of the reactions are the change of the Gibbs free energy as a function of temperature, endothermic or exothermic character of the reactions, the kinetics and reactions of gases or carbon with the ceramic product. To avoid sintering, grain growth and neck growth of the ceramic the reaction temperature should be low enough. Table 1 gives some thermodynamic and kinetic data for reactions 1 to 4. From table 1 it can be seen that significant differences in chemical reaction rate are observed using different gases. It can also be deduced that a high temperature shifts the equilibrium of reactions 1, 2 and 3 in the direction of gasification: ΔG decreases with increasing temperature. For reaction 4 it can be seen that the gasification reaction is not likely to occur at high temperatures: ΔG increases with increasing temperature and is positive at temperatures higher than 820 K. An increase in temperature results in an increase of the chemical reaction rate for all four reactions. So, reaction 1, 2 and 3 are preferably carried out at high temperatures. Reaction 4 is preferably carried out at an optimum temperature considering both the thermodynamics and the kinetics of the reaction. According to thermodynamics the influence of the pressure on the equilibrium of reaction 1 is neutral, the influence of the pressure on the equilibrium of reactions 2 and 3 is a shift towards carbon formation and the influence of the pressure on the equilibrium of reaction 4 is towards gasification (3). Reaction 1 is fast and exothermic. Consequently, large differences in temperature in the greenware may occur and consequently, cracks can develop. Undesirable side reactions are oxidation of non-oxidic or metallic products by O_2, CO_2 and H_2O or reduction of oxides by H_2, CO or C. Usually, the chemical reaction rate of oxidation or reduction increases with increasing temperature.

TABLE I

The change of the Gibbs free energy, the relative reaction rate and the usual reaction temperature of the four reactions (2, 3)

reaction	change of the Gibbs free energy (J/mol)	relative chemical reaction rate at 1073 K and 0.1 atm	usual* reaction temperature (°C)
1 O_2	-395000-0.54T	10^5	> 500
2 CO_2	167000-171T	1	> 800
2 H_2O	133000-142T	3	> 800
4 H_2	-91000+111T	3.10^{-3}	< 800

*: 1 atm. total pressure

PHYSICAL ASPECTS

Important physical aspects of the reactions are: transport of reactants from the main gas stream to the powder, granulate or greenware, transport of reactants within pores, transport of gaseous reaction products within pores and transport of gaseous products from the powder, granulate or greenware to the main gas stream. Heat transfer from or to the powder, granulate or greenware is also an important physical aspect. In general it can be said that physically limited reaction rates are proportional to a driving force (3), for instance a temperature difference or a difference in concentration of the reactant in the gas flow and in the greenware or granule. Granulating the powder permits more freedom in the selection of a reactor for the removal of carbon from the powder. A specific physical aspect in this paper is the 'dilution' of carbon with the matrix material. In this paper no attempts are made to elaborate in the physical aspects mentioned.

CHEMICAL ASPECTS

Different types of carbon show large differences in absolute reactivity. The reactivity of the carbon, among others, depends on the pyrolysis conditions, the specific surface area, the presence of catalytic agents or impurities and catalytic interaction with the matrix material.

Pyrolysis is the process of forming carbon from carbon containing compounds, by heating. The reactivity of the carbon depends on the pyrolysis temperature (and time) (3). The higher the pyrolysis temperature the lower the reactivity of the carbon. Carbon black is formed at 1250 °C and is therefore not very reactive. A degrading polymer heated to a gasification temperature of 800 °C is more reactive than carbon black.
In general the reactivity of carbon can be described by equation 5 (3):

$$A = B \exp(-E_a/(R\,T)) \qquad \underline{5}$$

A is the reactivity or the total number of active sites present per mol of carbon (mol/mol), E_a the activation energy in J/mol of the deactivation of carbon, T the absolute temperature of the pyrolysis, R the gas constant and B a proportionality constant. The higher the specific surface area of the carbon the higher the total number of active sites per mol of carbon and the higher the total number of active sites per mol of carbon and the higher the reactivity. It is assumed that the number of active sites per m^2 is constant for a specific type of carbon during the gasification. Carbon or carbon forming materials contain impurities which are usually denoted as 'ash content'. Additives used in the production of ceramics and the matrix material also contain impurities. Some of these impurities, in the ppm range and higher, are able to increase the reaction rate of the gasification reaction. Wellknown elements which may catalyse gasification reactions and which are often found in carbon and ceramic materials are Na, K, Ca and Fe (3,4). Elements which catalyse gasification rate can be added to the powder or the greenware to enhance gasification rate. These elements are most active when they are well dispersed on the carbon surface. Deactivation of the catalytic agent can be due to volatilisation, reaction (with aluminosilicates for example) or sintering (3). The matrix material itself can also act as a catalyst. For the gasification of carbon with air it is reported that some oxides catalyse the gasification (5). Catalytic agents increase the number of active sites per mole of carbon and lower the activation energy of gasification. Activation energies for catalysed and uncatalysed gasification reactions are given in table 2 (3).

TABLE II
Activation energies for catalysed and uncatalysed gasification reactions in kJ per mol (3, 4)

reaction	catalysed	uncatalysed
1 O_2	170	250
2 CO_2	190-230	220-260
3 H_2O	140-180	140-200
4 H_2	150-200	150-200

The reaction mechanism of gasification generally consists of adsorption of gases on the carbon surface, surface reaction, and desorption of gaseous reaction products from the carbon surface. The adsorption and desorption reactions of the gases for the four reactions discussed in this paper are given in table 3. The rate determining step of the chemical reaction rate is the desorption of the reaction products. This is explained by the fact that very stable C-C bonds have to be broken in the desorption of carbon containing gas.

TABLE III
Reaction steps of gasification reactions (3, 4)

reaction	adsorption	desorption
1	$C^* + O_2 = CO_2^*$	$CO^* = C^* + CO$ and $CO_2^* = C^* + CO_2$
2	$C^* + CO_2 = CO + CO^*$	$CO^* = C^* + CO$
3	$C^* + H_2O = CO^* + H_2$	$CO^* = C^* + CO$
4	$C^* + H_2 = CH_2^*$	$2CH_2^* = 2C^* + CH_4$ $CH_2^* + H_2 = C^* + CH_4$

*: active site at carbon surfaces

Next, the influence of parameters on the chemical reaction rate according to measurements published in literature is discussed. The reaction rate is first order in the carbon concentration (3, 4). The reaction rate is first order in the reacting gas (O_2, CO_2, H_2O or H_2) at 1 atm. pressure (3, 4, 6, 7). A pressure of the reacting gas higher than 1 atm. only slightly increases the gasification rate in the case of CO_2 (8). The gasification rate with H_2 increases lineair with the pressure (4). The gasification rate with O_2 probably increases linear with the pressure (7). The gasification rate with H_2O is constant or decreases with increasing pressure for pressures higher than 10 atm. (9, 10). The rate of gasification can be reduced by reaction products (CO_2, CO, CH_4 and H_2). The gasification rate with O_2 is not reduced by CO_2, the gasification rate with CO_2 is reduced by CO and the gasification rate with steam is reduced by H_2 and slightly by CO and the gasification rate with H_2 is reduced by CH_4. Equations which describe the kinetics of the gasification for the four reactions are given in table 4. These results of published experiments will be discussed in relation to the theoretical features of the reactions. It is observed that the reaction is first order in the reactants. This is explained by the adsorption reaction for CO_2, H_2O and H_2, more active sites are occupied with increasing partial pressure. For O_2 is was expected that the order of the reaction rate would be 0 in the concentration of O_2, because all active sites are occupied by oxygen already at a low partial pressure of oxygen. The first order behaviour is explained by limitation of the reaction rate by diffusion in micropores (3, 4). The observation that the pressure is neutral on the rate of gasification with CO_2 and H_2O is explained by assuming that almost all active sites are occupied at 1 atm. Thermodynamics predict a decrease of the reaction rate with increasing pressure. An increase of the pressure in the gasification with H_2 is explained by assuming that an increasing number of active sites is occupied with increasing pressure and the chemical equilibrium shift to CH_4 with increasing pressure. The reduction of the gasification rate with CO_2, H_2O and H_2 by respectively CO, H_2 and CH_4 is explained by the desorption reaction. The observation in the literature that CO_2 does not reduce the gasification rate with O_2 is explained in the literature by assuming that there is desorption of CO instead of CO_2 and that CO reacts fast with oxygen (4).

TABLE IV
The equations which describe the reaction kinetics for the gasification of carbon

reaction	equation	literature
1 O_2	$r = k_1(O_2)$	(4)
2 CO_2	$r = k_1(CO_2)/((CO_2)+k_2(CO))$	(3)
3 H_2O	$r = k_1(H_2O)/((H_2O)+k_2(H_2))$	(3)
4 H_2	$r = k_1((H_2)^2-(CH_4)/K)/(1+k_2)(H_2))^2$	(3)
4*	$r = k_1((H_2)-(CH_4)/(K(H_2)))$	(11)

*: simplified. The values for k_1 and k_2 are different in each equation. K is the equilibrium constant. r in mol/sec.

The theoretical aspects of the gasification reactions can explain very well the results obtained by Giacobbe et. al. who studied the gasification of carbon in gas mixtures of H_2, N_2 and H_2O (12). Their study of the gasification of carbon originates from the removal of carbon or carbon residue from greenware and ceramic powders. However, they neglect to refer to literature on the gasification or combustion of coal or carbon.

EXPERIMENTAL, RESULTS AND DISCUSSION

Experiments were performed with carbon black, purified carbon black, SiC and Si_3N_4 powders and with mixtures of SiC or Si_3N_4 powder with 3 wt.%

carbon black. The carbon black was wet mixed with the ceramic powder using a colloid mill. The siliconnitride powder was Starch LC-12 powder. The siliconcarbide powder was Superior Graphite HSC powder or an experimental powder. Some experiments were done with a mixture of SiC and 9 wt.% free carbon after a carbothermal reaction. The carbon black was either Elftex 495 or Printer U. Elftex 495 could be purified by leaching with nitric acid followed by washing. Tablets containing ceramic powders, carbon black and powder mixtures, with a diameter and a height of 5 mm, were pressed. The density of the tablets made from ceramic powder or powder mixtures was 1.6 g/ml. The density of the tablets made from carbon black was 1.0 g/ml. Experiments were carried out in a Netzch STA 429 thermobalance. The tablets were placed on an alumina plate. The heating rate was 10 K per minute. The gas flow was 100 ml/minute at room temperature. The gas pressure was 1 atm. Experiments were also carried out in a microreactor coupled with a gas chromatograph at the University of Amsterdam. The volume of the packed bed in the microreactor was 1 ml. In these case broken tablets with a diameter between 0.1 and 0.5 mm were used in a packed bed at a reaction temperature of 975 °C. A total gas pressure of 1.5 atm. and partial CO_2 or H_2O pressures of 0.5 atm. were used. Oxygen analysis was performed with Strohlein equipment. Figure 1, 2 and 3 show the TGA of respectively Elftex 495 carbon black, Si_3N_4 and SiC in a gas flow containing 90 % CO_2 and 10 % He. From figure 1 it can be seen that the reaction starts at 900 °C. Figure 2 shows that the oxidation of SiC starts at 800 °C. Figure 4 shows the TGA of Si_3N_4 with 3 wt.% Printex U in a gas flow containing 90 % CO_2 and 10 % He. The isotherm part of the TGA is 3 hours at 1050 °C. From figure 4 it can be seen that carbon is removed without oxidation of the siliconnitride. This result was confirmed by an oxygen analysis of the product after removal of carbon. Figure 5 shows the TGA of SiC with 3 wt.% Printex U in a gas flow containing 90% CO_2 and 10 % He. The isotherm part of the TGA is 3 hours at 1050 °C. From figure 5 it can be seen that the weight loss is followed by a weight gain which suggests that SiC is oxidized, at least after removal of carbon. Oxygen analysis of samples during and after carbon removal showed that indeed there is oxidation of SiC during and after removal of carbon. Figure 6 shows the TGA of SiC, an experimental powder made by a carbothermal reaction with 9 wt.% free carbon, in 100 % H_2. The isotherm parts are 5 hours at respectively 800, 900, 1000 and 1100 °C. From figure 6 it can be seen that the reaction rate is very low indeed. Figure 7 shows the gasification of carbon in CO_2 and H_2O in the packed bed microreactor at the University of Amsterdam. From figure 7 it can be seen that the gasification in H_2O is about a factor 10 faster than the gasification in CO_2. The results obtained with the reaction of purified carbon black mixed with Si_3N_4 powder in the packed bed microreactor at the University of Amsterdam are given in figure 8. From this figure it can be seen that the reaction rate of the carbon in the powder mixture is equal to the reaction rate of the pure carbon. Again, the gasification rate with steam is about a factor 10 faster compared to the gasification with CO_2. In figure 9 the result of the gasification or purified carbon black with CO_2 in the microreactor at 2 different temperatures is given. From the figure the value of the activation energy was calculated. This value appeared to be 310 kJ/mol.

Suprisingly a difference was observed in oxidation behaviour of SiC and Si_3N_4 powder. It was expected that both powders would behave similar because the rate of oxidation would be controlled by a layer of silica around the particles. This layer should be equal for both powders. The difference in oxidation behaviour may be attributed to a layer of Si_2N_2O in the case of the oxidation of siliconnitride. This layer of siliconoxynitride is a better diffusion barrier than silica. The oxidation of carbon black by CO_2 starts at a relative high temperature of 900 °C compared to gasification of other carbon sources in the literature. The value of the activation energy of the gasification of carbon black with CO_2 is very high compared to values for (uncatalysed) gasinfication in the literature. Both observations are very well explained by the high purity of the carbon. The fact that the gasification starts at relative high temperatures is also explained by the high temperature at which carbon black is produced: 1250 °C. The observation that the gasification rate with H_2O is faster than the gasification rate with CO_2 is in good agreement with the literature. However, the difference in gasification rate is larger than predicted in the literature. The fact that the gasification rate with H_2 of (unreactive) carbon is very low indeed, is in good agreement with the literature.

CONCLUSIONS

Knowledge of the literature on the chemistry of coal gasification is very useful for the research on the removal of carbon or carbon forming materials from greenware and ceramic powders. It can be used to explain and predict qualitatively the reactivity of the carbon or carbon forming material. For instance: pure carbon after a carbothermal reaction at high temperature has a low reactivity and carbon residue after binder removal has a high reactivity. It can also be used to predict quantitatively the kinetics of the reaction. For instance: the reaction rate is first order in the carbon concentration and the value of the activation energy for gasification of pure carbon with CO_2 is about 250 kJ/mol. A lot of information is available on the reaction of carbon with O_2, H_2O, CO_2 or H_2.

For the removal of carbon from greenware or ceramic powder with a gasification reaction, the use of oxygen/air is the natural choice. Oxygen/air is very reactive and the process is economic. If the gasification with oxygen/air is not possible H_2O or CO_2 can be chosen, H_2O being more reactive than CO_2. If none of these methods are possible H_2 can be tried, especially H_2 at high pressures if the reactivity of the carbon is low. Finally, gases other than O_2, H_2O, CO_2 or H_2 can be tried, for instance $NH_3 + C = HCN + H_2$. However, reactions between carbon and other gases are relatively uncommon and require much more research.

Acknowledgement: J.A. Moulijn, F. Kapteijn, University of Amsterdam.

LITERATURE

1. Higgins R.J. The chemistry of carbon formation during binder burnout in ceramics, Report No 88, Ceramics Processing Research Laboratory, Massachusetts Institute of Technology (1987).
2. Turkdogan E.T., Physical chemistry of high temperature technology, Academic Press, New York (1980).

3. Kapteijn F., Moulijn J.A., Kinetics of catalysed and uncatalysed coal gasification, carbon and coal gasification, Martinium Nijhoff Publishers, Dordrecht (1986) 291-360.
4. Gmelins Handbuch der Anorganischen Chemie, C teil B lieferung 3, Verlag Chemie, Weinhem (1968).
5. Masia S., Calvert P.D., Rhine W.E., Bowen K.H., Effect of oxides on binder burnout during ceramics processing, Report No 94, Ceramics Processing Research Laboratory, Massachusetts Institute of Technology (1988).
6. Gmeling Handbuch der Anorganischen Chemie, C teil C2, Verlag Chemie GMBH, Weinheim (1972).
7. Tseng H.P., Edgar T.F. Combustion behaviour of bituminous and anthracite coal char between 425 and 900 °C, Fuel, 64 (1985) 373-379.
8. Kapteijn F., Peer O., Moulijn J.A., Kinetics of the alkali carbonate catalysed gasificátion of carbon, Fuel, 65 (1986) 1371-1376.
9. Suzuki, T., Mishima M., Takahashi T., Watanabe Y., Catalytic steam gasification of Yallourn coal using sodium hydridotetracarbonyl ferrate, Fuel, 64 (1985) 661-665.
10. Xian-quam R., You-ging W., Zuo-liang L., Shu-fen L., Effects of catalysis on gasification of Datong coal char, Fuel, 66 (1987) 568-571.
11. Hedden K., The formation of methane from hydrogen and carbon at high temperatures and pressures, fifth Biennial Carbon Conference, Pittsburgh, (1961) 125-131.
12. Giacobbe F.W., Franke B.A., Friedrich J.F., Carbon gasification at 900 °C in nitrogen and in varying mixtures of nitrogen, hydrogen and water vapor, Sprechsaal, 121 (1988) 776-780.

APPENDIX

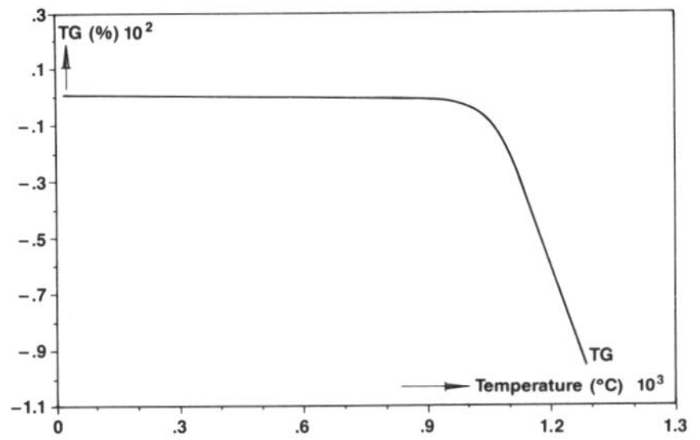

Figure 1.
TGA of carbon black (Efltex 495) in 90 % CO_2 and 10 % He. The heating rate is 10 K per minute. The oxidation starts at 900 °C.

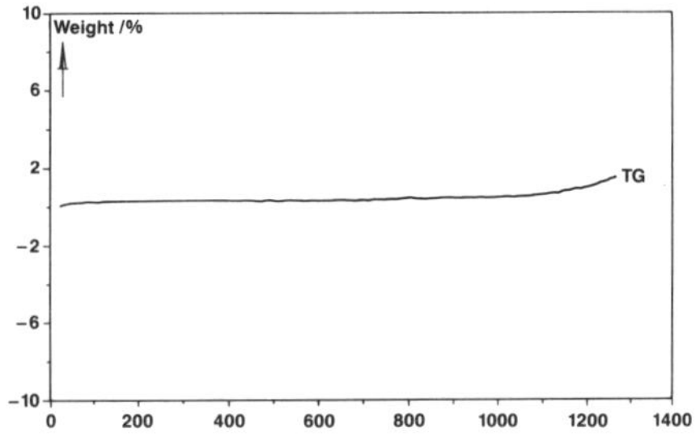

Figure 2.
TGA of siliconnitride (Starck (LC-12) in 90 % CO_2 and 10 % He. The heating rate is 10 K per minute. The oxidation starts at 1100 °C.

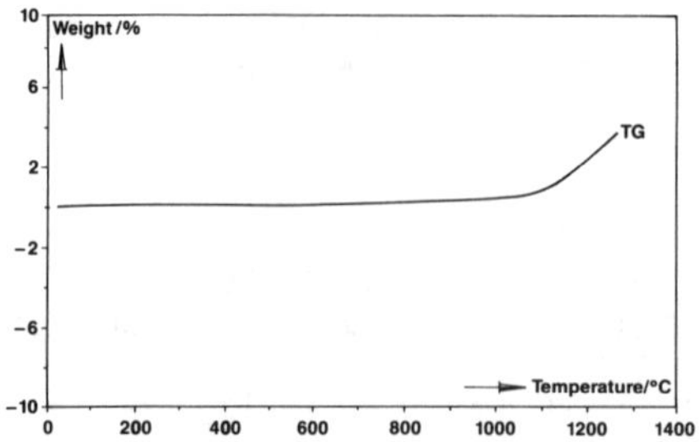

Figure 3.
TGA of siliconcarbide (HSC-SiC) in
90 % CO_2 and 10 % He.
The heating rate is 10 K per minute.
The oxidation starts at 800 °C.

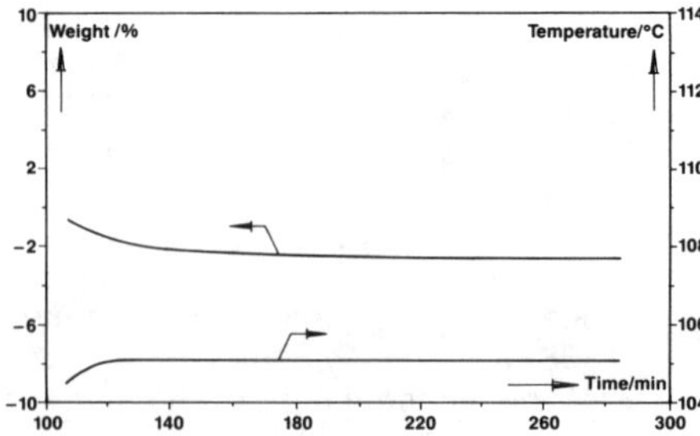

Figure 4.
TGA of siliconnitride with about
3 wt.% carbon black (Printex U) in
90 % CO_2 and 10 % He at 1050 °C.
The reaction is completed within
2 hours.

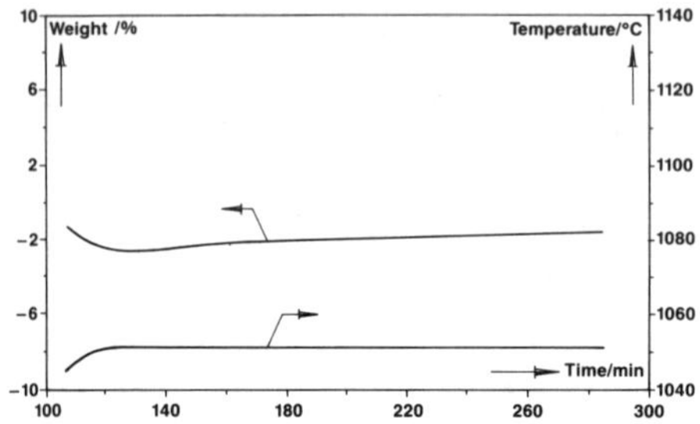

Figure 5.
TGA of siliconcarbide with about
3 wt.% carbon black (Printex U)
in 90 % CO_2 and 10 % He at 1050 °C.
First there is a weight loss, then
there is a weight gain.

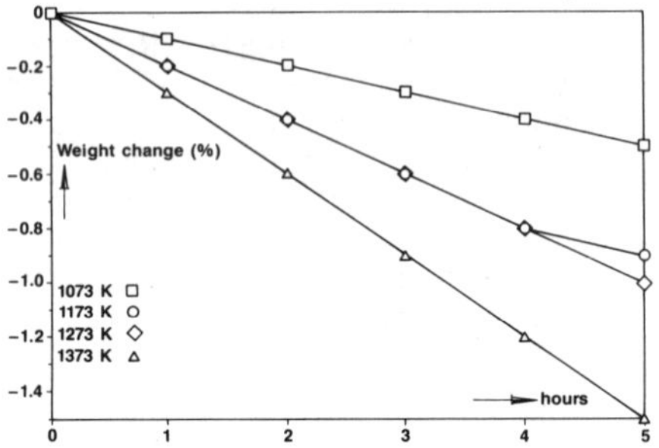

Figure 6.
TGA of siliconcarbide with 9 wt.%
carbon in 100 % H_2 at 800, 900,
1000 and 1100 °C. The powder mixture
is made at DSM Research.

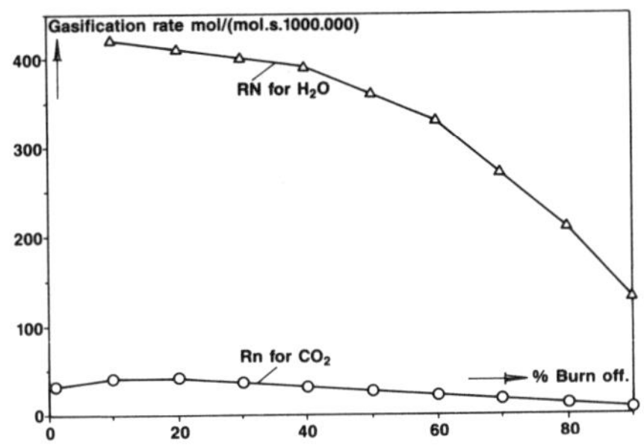

Figure 7.
The gasification rate of purified carbon black (Elftex 495) in CO_2 and H_2O in mol/(mol.s) at 975 °C at 1.5 atm.
The concentration of CO_2 or H_2O is 33 %. The reaction rate Rn is calculated as: Rn = x.F/C, with C the initial amount of carbon, F the flow of CO_2 or H_2O and x the conversion of CO_2 or H_2O.

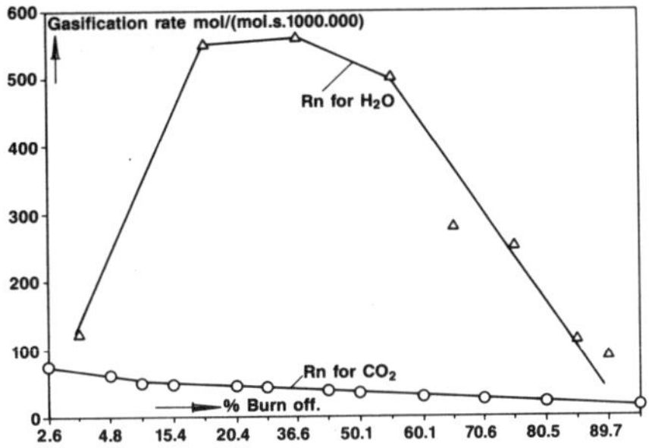

Figure 8.
The gasification rate of purified carbon black (Elftex 495) in CO_2 and H_2O in mol/(mol.s) at 975 °C at 1.5 atm. The concentration of the reactants is 33 %. The carbon black is mixed with siliconnitride powder. The mixture contains 3 wt.% carbon black.

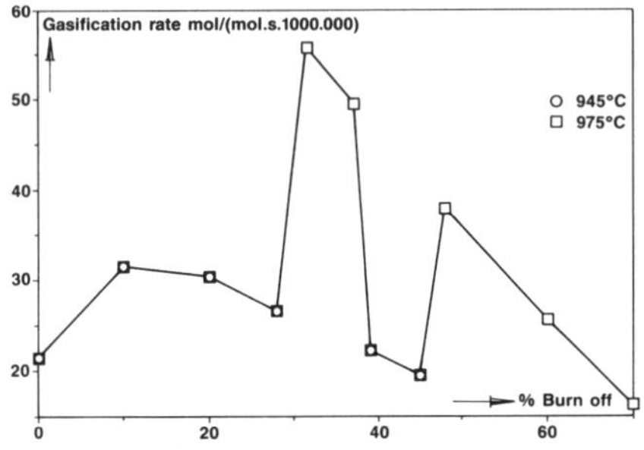

Figure 9.
The gasification rate of carbon black (Elftex 495) in CO_2 in mol/(mol.s) at 975 and 945 °C at 1.5 atm. The concentration of CO_2 is 33 %.

GRAIN BOUNDARIES DYNAMICS IN THE CERIA-ZIRCONIA CERAMICS

S. Meriani, E. Bischoff* and V. Sergo
Appl. and Ind. Chem. Inst.- University of Trieste-ITALY
* Max Planck Inst. Metallforsh.-Stuttgart-FRG

ABSTRACT

Grain boundaries in SiO_2 doped ceria-zirconia alloys migrate under the driving force due to the coherency lattice strain introduced into the coexisting phases by the asymmetric concentrations of the (Ce^{+3}) species in the surface layers facing the interphase material.

INTRODUCTION

Grain boundaries (GB's) control and design become increasingly important challenges in materials science and engineering. They influence sintering processes and they can enhance or reduce electrical and mechanical properties. It has been reported that within the grain boundaries of most ceramic materials there is often an intergranular second phase rich in silica, an ubiquitous contaminant which is often a solvent for other impurities or doping elements. This is also the case in ceria-zirconia ceramics.

For these materials we have previously reported that in the presence of a silica rich film, which we named "zip"interface, GBs exhibit features which recalled phenomena as diffusion and chemically induced grain boundaries migration (DIGM or CIGM) (1,2). Wavy grain boundaries in ceramic materials have been observed only in a few other cases (3,6) whereas in metals there are various alloys which show grain boundaries migration either through liquid film (LFM) or via vapour or solid solute sources (CIGM-DIGM). Both mechanism rely for their driving force on the coherency lattice strain of grains layers facing the interface (either liquid or solid) which acts as a source or sink for solute atoms (7).

As regards the ceria-zirconia ceramics, from the previous investigations we have gatherd the following correlations: "zip" boundaries were found if a minimum amount of silica was present; after firing above1450°C, increasing with temperature. So far we were of the opinion that "zip" features should be confined, or at least strongly enhanced, to the surface or in large pores where a surface atmosphere exchange is possible. In fact, we postulated the following set of redox reactions, partially based on the experimental evidence of the thermoanalitycal behaviour of these materials (8)

(a) $2(CeO_2\text{-}ZrO_2)$ (b) = $Ce_2O_3\text{-}2ZrO_2$ (b) + $1/2O_2$

(b) $Ce_2O_3\text{-}2ZrO_2$ (b)+ SiO_2 (gb) = 2($CeO_2\text{-}ZrO_2$) g + SiO (g). (??).

where (b) is bulk, (gb) grain boundary and (g) gas. However, the mechanism implying the SiO contribution could not be unanbiguosly proved.

This investigation has been undertaken to identify the nature of the GB's and the mechanism through which they assume the wave-like configuration.

EXPERIMENTALS

Samples were made with 40 mol % CeO_2 and 60 mol % ZrO_2 of commercial origin, 99.99% purity, either as pure or doped mixtures. Doping consisted in the intentional addition of 0.1 weight % of fine SiO_2 to the ball milling media made of zirconia balls and isopropanol in a plastic jar.

According to the phase diagram (9) at 1600°C this compostion should give two phases, one tetragonal (Ce-TZP) containing 18 mol% ceria (CZ18) and one cubic of about 50 mol % ceria (CZ50). This latter on cooling could eventually undergo a transformation into a Ce-TZP' tetragonal phase of the same composition (10). Our sample exhibited an XRD pattern showing both these tetragonal phases.

TEM investigations have been carried out on samples taken inside the pellet whereas SEM microstructures revealed the surface features after thermal etching.

RESULTS AND DISCUSSION

The GB of both the clean and doped materials can be seen in Fig. 1(a) and (b). This latter shows and example of the waved "zip" interface. The small particles in Fig. 2(a) gathered in clusters are made of the CZ18 phase surrounded by a thick layer, in relief, and by the CZ50 matrix. The Ce/Zr ratio and the Si concentration obtained by SEM microanalyses show that Si is present both at the CZ5O/CZ18 and the

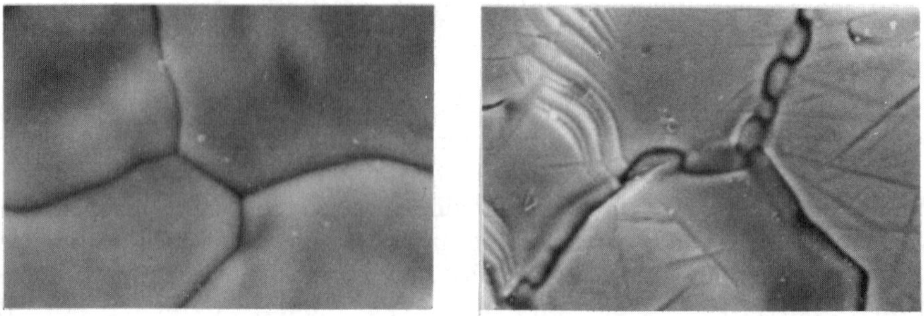

Fig.1. (a) normal GB in pure CZ50; (b) SiO_2 doped CZ50 with "zip" GBs

CZ18/CZ18 interfaces, Fig. 2(b). The rim can be seen only between CZ18 and CZ50. The Ce/Zr ratio of the rim suggests a CZ60 phase composition, richer in ceria if compared with the matrix composition.close to CZ50. The origin and the location of this rim of different composition should be related with the nature of the GB within the bulk of the samples.

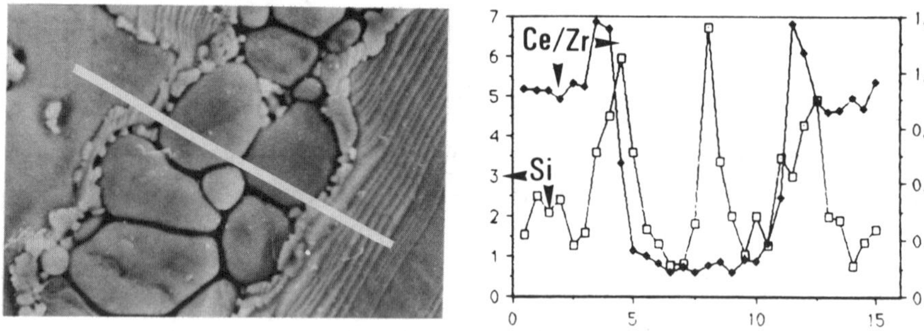

Fig. 2 (a) SEM image of CZ18 cluster in CZ50 matrix; (b) Ce/Zr ratio and Si (at%) across the particles. White line is15 microns.

In fact the investigation undertaken via analytical TEM revealed three different cases: (a) isolated pockets and glass filled triple points made of 96% SiO_2; (b) some clean GB's a few nm thick and (c) a thick crystalline layer surrounding some grains or clusters of them. Fig. 3 shows a detail of this interlayer found mainly where CZ18 and CZ50 grains face each other.

The Zr, Ce and Si concentrations and the Ce/Zr ratio are reported in Fig. 4(a) and (b) respectively. It is worth pointing out that Si (SiO_2) concentration in this layer is in practice reduced to an impurity level. This may suggest that dewetting of the GB has occurred with the gathering of the silica glass at the triple points and in the pockets mentioned before.

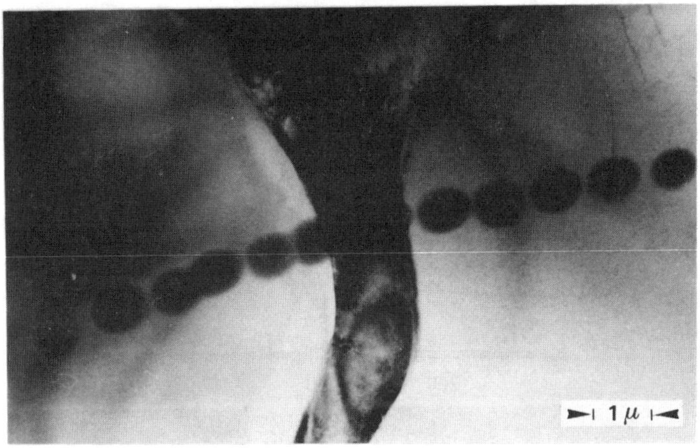

Fig. 3 - TEM analysis of GB between CZ18 and CZ50 along dark spots. -

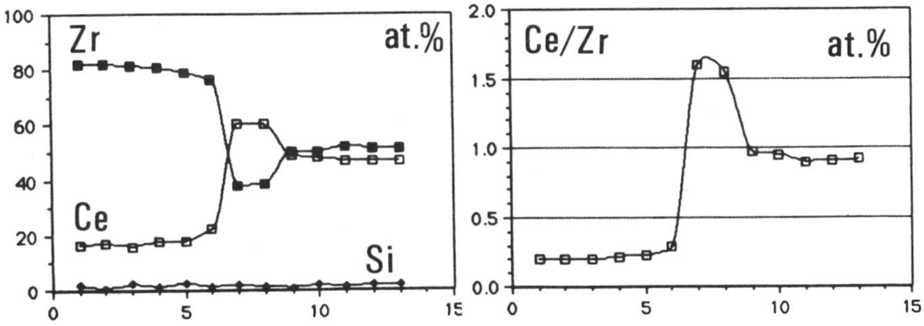

Fig. 4 -(a) Concentrations profiles (b) - Ce/Zr concentration ratio.

The interphase is made of a material with a CZ60 composition which cannot be explained on the basis of the binary diagram that requires a CZ18/CZ50 equilibrium pair. Furthermore, some bulges protruding from the CZ18 grains have been observed facing an opposite concavity within the CZ50 grain. Fig. 5 shows one of these protruding structures and the related TEM analysis. In practice it was found that the newly grown material has a composition of the CZ18 type. Questions rise how to explain the reason why CZ18 grows at the expenses of the CZ50, both originally at equilibrium.

The phenomenon can be interpreted via a LFM or CIGM if there are appropriate driving forces to overcome the excess surface energy increased by the extension of the CZ18 and the regression of the CZ50 phases. In fact both surfaces are moving away from their centers. of curvature.

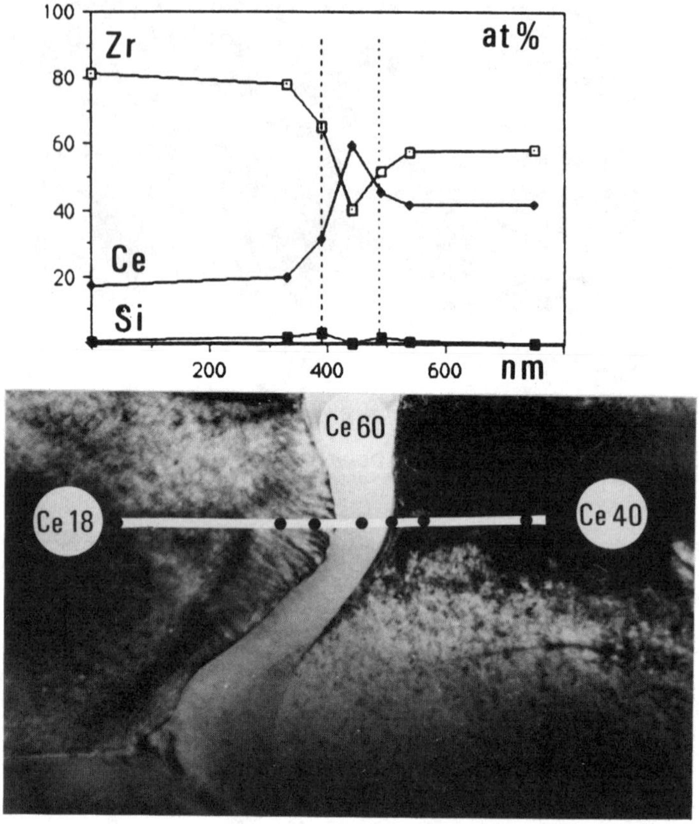

Fig. 5 -TEM analysis of the GBs between phases. Numbers are at.%. Ce

It has been previously mentioned that LFM and CIGM phenomena find their driving forces in the coherency lattice strains (7). In the CZ ceramics based on the fluorite structure, or on its distorted versions (10), the Ce(+4) ions may become reduced to a Ce(+3) state with the contemporary formation of oxygen vacancies We have measured the unit cell volumes of the reduced and oxidized forms of CZ50: (0.1473 nm^3).and (0.1504 nm^3) respectively. The reduced CZ5O exhibits a cubic XRD pattern of the fluorite type with a volume increase in excess of 2%. which is due to the combined ionic radius increase (Ce^{+4}=0.101 nm; Ce^{+3}=0.114 nm) and oxygen vacancies formation. If we consider that the reduction occurs at high temperature and that it must start at the GB surface to be controlled by the oxygen diffusivity from the bulk throughout the reduced layer, some parts of the fluorite lattice becomes coherently strained by the 2% volumetric expansion of the reduced material.

On the other side the CZ18 phase undergoes the same type of redox - vacancies formation and its lattice will exhibit a coherency strain proportional to the Ce(+3) concentration. Although at this stage of our investigation we cannot yet give the quantitative unit cell expansion of the CZ18 after reduction, we can safely predict that the coherency strain should be ultimately proportional to the amount of the Ce present in the structure. Since CZ18 contains 2.75 times less Ce than CZ50, it will be proportionally less strained. If we consider that the Gibbs free energy increases proportionally to the strain, the two phases CZ18, CZ50 originally in thermodynamic equilibrium, move to different free energy levels. Being CZ50 the most displaced to higher values, there will be a transfer of matter from the CZ50 to the CZ18 phase to reduce the overall energy content, Fig. 6.

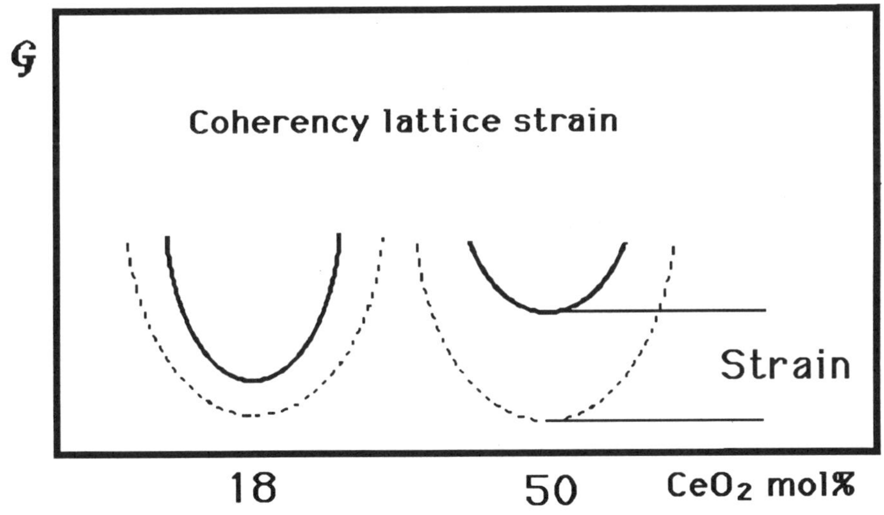

Fig. 6 Gibbs free energy vs mol % CeO_2 of phases CZ18 and CZ50 in equilibrium (lower curves) and under coherency lattice strain (upper curves), arbitrary energy units

To accomplish this transfer the GB plays the role of sink and source of both Ce and Zr ions. Furthermore the higher Ce concentration in the GB layer (CZ60) can be easily explained if we think that from the CZ50 phase the ions dissolve in a 1:1 ratio but they precipitate at the lower free energy phase CZ18 in which the Ce/Zr ratio is 1:4, in this way an excess of Ce ions are left behind in the GB material, where the composition moves from from CZ50 to CZ60 .

CONCLUSION

The GB migrations observed in ceria - zirconia ceramics are most probably caused by the coherency lattice strain introduced by the increased Ce^{+3} dimensions through the liquid intergranular siliceous film which acts as sink and source of matter. The relative quantity of the reduced species is a function of the original cerium content and of the combined effects of temperature and oxygen partial pressure. Further work is in progress to investigate the GBs behavior in single phase materials as for instance CZ60.

REFERENCES

1. Meriani S., Journ. de Phys., 1986, 47 Suppl. au n.2, 485-89 Science of Ceramics 13.
2. Chiodelli, G., Magistris, A., Lucchini, E., and Meriani, S., In Science of Ceramics 14, Ed. Taylor, D., The Institute of Ceramics Publ. Stoke on Trent, U.K., 1988, pp. 903-906.
3. Butler, E.P. and Heuer, A.H., J. Amer. Ceram. Soc. 1985, 68, 197 - 202.
4. Chaim, R., Heuer, A.H., and Brandon, D.G., J. Amer. Ceram. Soc 1986, 69, 243 -248.
5. Kim, J.J., Song, B.M., Kim, D.Y. and Yoon, D.N., Amer. Ceram. Soc. Bull. 1986, 65, 1390-1392..
6. Chang, E.K., Wang, Z.Y., Smyth, D.M., and Harmer, M.P., 88th Annual Meeting Am. Ceram Soc. Ceramographic exhibition.
7. Won-Hyuk Rhee, and Yoon, D.N., Acta Metall., 1989, 37, 221-28.
8. Meriani,S., Schmid C. and Sergo.V.; "Zirconia '88 - Proceeding Ed. Meriani S. and Palmonari C., Elsevier Publ. U.K. (in print).
9. Tani, E., Yoshimura, M. and Somiya, S., J. Amer. Ceram. Soc., 1983, 66(7), 506-510.
10. Meriani, S.and Spinolo, G., Powder Diffraction, 1987, 2 (4), 254-56

ACKNOWLEDGEMENT

Authors gratefully acknowledge the Ministry of Public Education financial support through the University of Trieste.

MICROSTRUCTURAL DEVELOPMENT DURING PROCESSING OF A Si_3N_4/ZrO_2 MATERIAL

L.K.L. FALK
Department of Physics,
Chalmers University of Technology and University of Gothenburg,
S-412 96 Gothenburg, Sweden

M. HOLMSTRÖM
Swedish Ceramic Institute,
S-402 29 Gothenburg, Sweden

ABSTRACT

The development of microstructure in an Si_3N_4/ZrO_2(+3 mol% Y_2O_3) composite ceramic material has been characterized by analytical electron microscopy in combination with x-ray diffractometry. The composites were formed with 30 wt% partially stabilized ZrO_2 powder, and densification was carried out by pressureless sintering at 1650 °C. Partially sintered compacts showed that a near fully dense body was obtained during heating to the sintering temperature. This was achieved by formation of viscous phases and was associated with only a limited conversion of α to β Si_3N_4. During heating to the sintering temperature, the ZrO_2 exhibited grain growth which was followed by a solution/reprecipitation process at 1650 °C. This resulted in tetragonal ZrO_2 grains, and TEM indicated that a significant part of the ZrO_2 had the non-transformable t' structure. From oxidation experiments carried out at 1200 °C, it is suggested that N entered the ZrO_2 lattice during densification.

INTRODUCTION

Composite ceramic materials based on Si_3N_4 and partially stabilized ZrO_2 are recognized as having a high potential for Si_3N_4-based ceramics with high strength and improved fracture toughness [1,2]. However, a major problem associated with the formation of this type of composite ceramic is that oxidation prone phases such as zirconium oxynitrides and ZrN may form during processing [2,3].

Si_3N_4/ZrO_2 composites formed using unstabilized ZrO_2 have been found to contain $ZrO_{2-3x/2}N_x$ which, upon oxidation at temperatures above approximately 500 °C, transforms to monoclinic ZrO_2 [2,3]. This oxidation is accompanied by a molar volume increase which introduces internal stresses in the material, and a prolonged exposure to an oxidising environment may cause disintegration of the ceramic [2,3]. It has been suggested that the use of

Y_2O_3 stabilized ZrO_2 starting powder reduces the N uptake of the ZrO_2 during densification and suppresses a formation of zirconium oxynitrides [2].

Previous work has indicated that the atmosphere during densification also affects the formation of N containing phases in the $Si_3N_4/ZrO_2(+Y_2O_3)$ composite ceramic [4]. Particularly, the O_2 partial pressure plays an important role in determining relevant phase equilibria. It was shown that it is possible to avoid formation of ZrN on the surface of the ceramic by controlling gas environment and temperature, and it was concluded that this is probably also applicable to the formation of zirconium oxynitrides [4].

The present paper discusses the development of microstructure in an $Si_3N_4/ZrO_2(+Y_2O_3)$ composite ceramic material which is formed by pressureless sintering.

MATERIALS AND METHODS

Formation Process

Powder mixtures containing Si_3N_4 (UBE E10), 30 wt% ZrO_2 (Toyo Soda TZ3Y, prereacted with 3.0 mol% Y_2O_3) and 5 wt% Al_2O_3 (Sumitomo AKP 50) added as sintering aid were prepared by ball milling in water using Si_3N_4 milling media. After a milling time of 86 h, 3.75 wt% polyvinylalcohol (PVA) and 1.25 wt% glycerol were added as pressing aids, and the mixtures were subsequently milled an additional 2 h. Powder compacts were prepressed

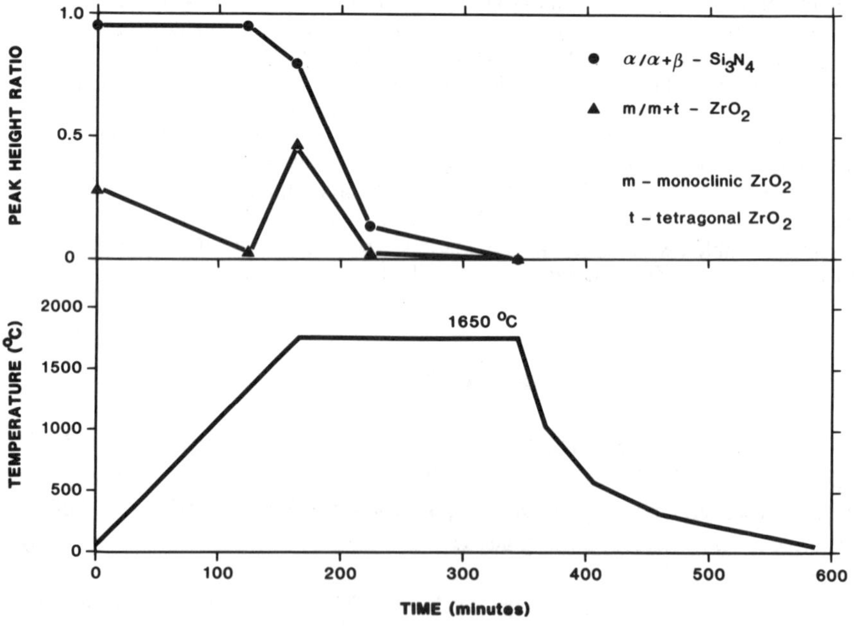

Figure 1. Temperature/time programme used for pressureless sintering. The Si_3N_4 and ZrO_2 phase contents of the starting powder as well as partially and fully sintered compacts are shown.

uniaxially, which was followed by cold isostatic pressing at 280 MPa. The green bodies had a density of 1.85 g/cm^3 which corresponds to 50 % of the calculated theoretical density.

Pressureless sintering was carried out in a N_2-atmosphere according to the temperature/time programme in Figure 1. The compacts were embedded in a Si_3N_4 protective powder bed and the vacuum furnace was heated by Kanthal elements. The resulting oxygen partial pressure made it possible to avoid the formation of ZrN [4]. In order to investigate the sintering process in more detail, partially sintered compacts were analysed.

Post-sintering heat treatments of fully sintered material were carried out in static air at 1200 °C for 2 h.

Analytical Techniques

Phase compositions of partially and fully sintered compacts were determined by x-ray diffraction from polished sections. Analytical transmission electron microscopy was carried out using a JEOL 2000-FX TEM/STEM/SEM instrument with an attached Link EDX system for quantitative analysis. The microstructure of the sub-scalar region in oxidised samples was studied on tranverse sections through the oxide scale into the bulk material.

RESULTS AND DISCUSSION

Partially Sintered Compacts

The powder compacts reached near theoretical density during heating to the sintering temperature 1650 °C. As shown in Figure 1, this was associated with only a limited conversion of α- to β-Si_3N_4. TEM showed that the first step in the densification process was the formation of viscous phases which caused consolidation of the starting powder particles, Figure 2. This was followed by extensive grain growth of the ZrO_2 and the onset of the α- to β-Si_3N_4 transformation.

After heating to 1350 °C, the amount of monoclininc ZrO_2 in the compacts was reduced and the majority of the ZrO_2 appeared tetragonal in x-ray diffractograms, Figure 1. This was most likely caused by mechanical constraints of the starting powder particles imposed by the surrounding matrix in the partially sintered compact [5]. The subsequent increase in monoclinic content of compacts which had been heated to the sintering temperature 1650 °C, can be explained by the presence of larger ZrO_2 grains which would not allow retention of the high temperature tetragonal structure to room temperature [5]. EDX in the STEM showed that the Y_2O_3 contents of these large grains varied significantly, from below detectability up to around 3 mol%. At 1650 °C the grain size of the ZrO_2 was reduced considerably. The ZrO_2 grains had an irregular shape filling out space between adjacent grains, which suggests that the ZrO_2 participated in the liquid phase formation and that the observed decrease in grain size was achieved by a solution/reprecipitation process via the liquid phase. Only the tetragonal structure could be identified by x-ray diffractometry of compacts held 3 h at this temperature.

Fully Sintered Compacts

Towards the end of the sintering process the amount of liquid phase was reduced considerably, which resulted in a microstructure with an extremely small volume fraction of residual glass, Figure 2. Composites held at 1650 °C contained Si_2N_2O in the form of large rod-shaped

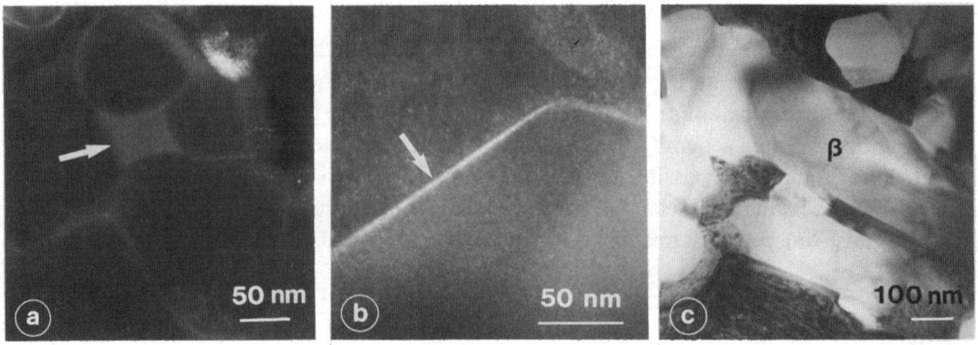

Figure 2 (a)-(b). TEM centered dark field images formed using diffuse scattered electrons showing the presence of amorphous phases (arrowed). (a) compact heated to 1650 °C; (b) fully sintered material; (c) ZrO_2 grains in a matrix of β-Si_3N_4 in fully sintered material.

grains, and during holding at this temperature a Si_3N_4 matrix consisting of high aspect ratio β grains developed, Figure 2.

Centered dark field imaging in the TEM using tetragonal reflections that are forbidden in the cubic fluorite structure, showed that the ZrO_2 grains consisted of tetragonal regions often in the form of laths, Figure 3. Anti-phase domain boundaries were frequently observed, which suggests that a significant part of the ZrO_2 had the non-transformable t' structure [6]. The Y_2O_3 contents of the ZrO_2 varied both within and between grains. Y_2O_3 concentrations between 1.7 and 3.7 mol% were measured by EDX in the STEM.

Oxidised Material

After oxidation at 1200 °C for 2 h, a major part of the ZrO_2 grains in an approximately 200 μm

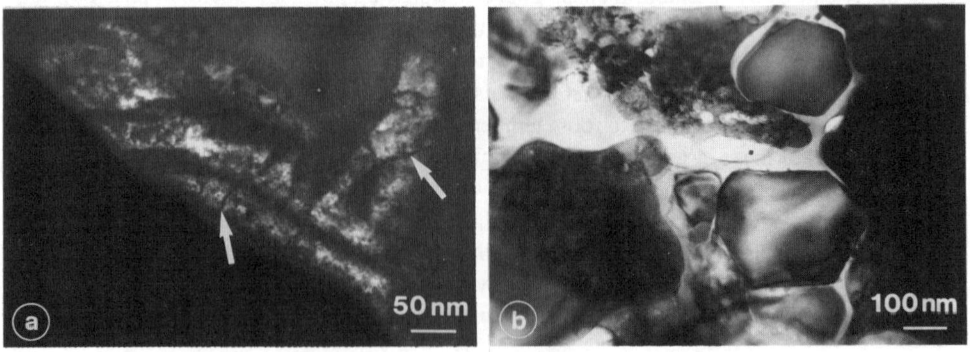

Figure 3. Internal structure of ZrO_2 grains. (a) TEM centered dark field image formed using a (112) reflex showing tetragonal regions containing anti-phase domain boundaries (arrowed) in a ZrO_2 grain in fully sinteted material; (b) subscalar region after oxidation at 1200 °C for 2 h.

wide subscalar region had decomposed into porous microcrystalline areas penetrated by Si-rich amorphous films. In this part of the thin foil grain boundary cracking and cavitation was observed, which indicates that the decomposition of the ZrO_2 introduced internal stresses in the material. This implies that the ZrO_2 lattice had incorporated some N during densification, and it is suggested that the decomposition is due to internal oxidation of a N and Y_2O_3 containing tetragonal ZrO_2. Close to the oxide scale, the volume fraction of glass increased, and the Si_3N_4 grains developed an irregular shape, Figure 3. The increased amount of amorphous phase in this region is the product of oxidation of the Si_3N_4.

CONCLUDING REMARKS

A Si_3N_4-based ceramic material with an extremely small volume fraction of residual glass can be formed by pressureless sintering of Si_3N_4/ZrO_2(+ 3 mol% Y_2O_3) compacts at 1650 °C. Formation of viscous phases and particle rearrangement during heating to the sintering temperature resulted in a nearly fully dense body, but the major part of the α- to β-Si_3N_4 transformation took place at 1650 °C. At the sintering temperature, the ZrO_2 undergoes a solution/reprecipitation process via the liquid phase which results in formation of a N and Y_2O_3 containing oxidation prone tetragonal ZrO_2 structure.

ACKNOWLEDGEMENT

This work was supported by the Swedish National Board for Technical Development (STU).

REFERENCES

1. Carpenter, H.W., Schnittgrund, G.D. and Brockmeyer, J.W., Transformation-toughened silicon nitride. In Proceedings of the Twenty-Fifth Automotive Technology Development Contractors Coordination Meeting, Society of Automotive Engineers, Inc., Warrendale, 1988, pp. 161-173
2. Lange, F.F., Falk, L.K.L. and Davis, B.I., Structural ceramic composites based on $Si_3N_4/ZrO_2(+Y_2O_3)$ compositions. J. Mater. Res., 1987, **2**, 66-76
3. Lange, F.F., Compressive surface stresses developed in ceramics by an oxidation-induced phase change. J. Am. Ceram. Soc., 1980, **63**, 38-40
4. Tjernlund, A.K., Pompe, R., Holmström, M. and Carlsson, R., On the influence of atmosphere on the phase reactions involving the system Si_3N_4-ZrO_2. In Special Ceramics 8, ed. S.P. Howlett and D. Taylor, Institute of Ceramics, Stoke-on-Trent, 1986, pp. 29-34
5. Lange, F.F., Transformation toughening. Part 3: Experimental observations in the ZrO_2-Y_2O_3 system. J. Mater. Sci., 1982, **17**, 240-6
6. Lanteri, V., Chaim, R. and Heuer, A.H., On the microstructures resulting from the diffusionless cubic -> tetragonal transformation in ZrO_2-Y_2O_3 alloys. J. Am. Ceram. Soc. 1986, **69**, C-258-61

CRYSTALLIZATION OF THE INTERGRANULAR PHASE IN SILICON NITRIDE CERAMICS WITH RARE EARTH OXIDE SINTERING ADDITIVES

C. BOBERSKI, M. PEUCKERT, P. SELGERT
Hoechst AG, Postfach 800320,
D-6230 Frankfurt/M. 80, F.R.G.

ABSTRACT

The pressureless sintering of silicon nitride powder with rare earth oxides: Y_2O_3, La_2O_3, CeO_2, Pr_6O_{11}, Nd_2O_3, Sm_2O_3, Gd_2O_3, Tb_4O_7, Dy_2O_3, Ho_2O_3, Er_2O_3, Yb_2O_3 as sintering aids in addition to Al_2O_3 is reported. It was found that the crystallization of an intergranular apatite phase increases both room temperature strength and strength at 1200° C. Also the nature of the glassy phase effects the bending strength of the ceramics: with decreasing ionic radius of the rare earth element the strength was found to increase.

INTRODUCTION

In the past, numerous research groups have intensively studied the sintering of silicon nitride ceramics with a vast variety of additives, namely the oxides of Al, alkaline earths, rare earths, and some transition metals. Despite all these efforts the quest for the optimum Si_3N_4 sintering aid continues [1-10]. Beside the choice of the proper additive, also the subsequent crystallization of the glassy intergranular phase has been studied as a means to further increase strength. Especially, systems containing Y_2O_3 and Al_2O_3 can be crystallized by heat treatment between 1100° C and 1400° C either to Y-silicates [4,6,11] or Y-garnet $Y_3Al_5O_{12}$ [7]. Systematic studies on the effect of the various rare earth oxides with Al_2O_3 as sintering additives for silicon nitride ceramics on glass phase crystallization and mechanical strength is still lacking.

MATERIALS AND METHODS

Silicon nitride powder (LC-12 from H. C. Starck) was intensively mixed with oxide sintering additives. The RE-oxide RE_2O_3 and Al_2O_3 were prepared in a molar ratio of 2:1. The total amount of additives was chosen such that the concentration of glass forming, sinteractive oxide was kept constant, i.e. the concentration of oxygen was always 0.2 mol per 100 g of SSN ceramic. (All oxides from Ventron, Karlsruhe, REO 99.99%; Al_2O_3 type A-16 from Alcoa). The premixed starting materials were wet milled in an attritor mill; the dried powders were isostatically pressed. These samples were pressureless sintered at 1800° C for 1h in N_2. After sintering the specimens were annealed in 0.1 MPa of N_2 for 24h at 1350° C for crystallization of the amorphous intergranular phase. The density was determined by the buoyancy method. Grain size and aspect ratio was measured on SEM pictures of overetched specimens as described in ref.[12]. Strength measurements have been performed by 4-point-bending tests.

RESULTS AND DISCUSSION

The samples were analysed with regard to their sintered density, to the grain size and aspect ratio of the β-Si_3N_4 grains. Unfortunately, no simple correlation between these structural properties or melting points of the oxides eutectics with Young modulus E, stress intensity factor K_{IC} or bending strength could be observed [13]. Nevertheless, a distinct increase of bending strength at room temperature as well as at 1200° C with decreasing radius of the RE^{3+} ions was found, see figure 1. A possible explanation might be a difference in the nitrogen content of the intergranular Si-RE-Al-O-N glass [14].

Crystallization of the intergranular glass can, in principle, lead to a variety of phases which for most rare earths exist as shown in figure 2 [15]. Basically, different aluminates like $RE_3Al_5O_{12}$ or $RE_4Al_2O_9$, silicates like RE_2SiO_5 and $RE_2Si_2O_7$, and mixed oxynitrides have been observed.In the present study only apatite has been detected after crystallization of the glassy phase. From comparison of the d-values with the known substances $Y_4(SiO_4)_3$ and $Y_5(SiO_4)_3N$ it can be concluded that in all cases the nitrogen containing apatite has been formed. In figure 3 measured d-values are plotted over the radii of the RE^{3+} ions. In our experiments the apatite $RE_5(SiO_4)_3N$ has always been found except for ceramics with Yb, Er and Y which formed silicates or garnet. Non of the ceramics has formed YAM ($RE_4Al_2O_9$), the phase with the same RE_2O_3: Al_2O_3 (2:1) ratio like the starting composition of the additives.

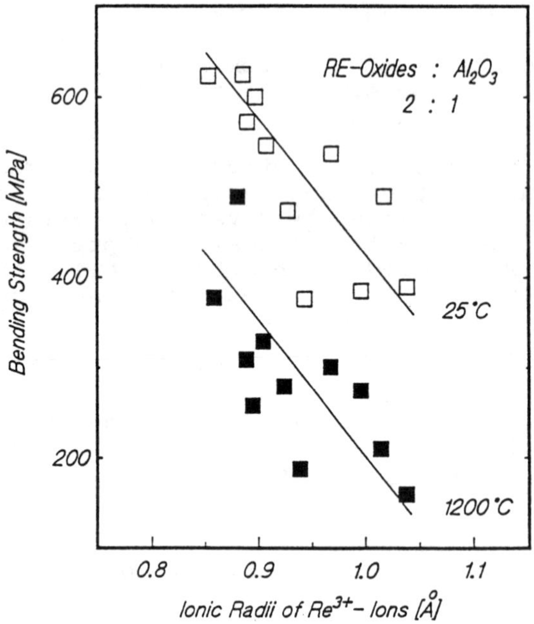

Figure 1: Bending strength as a function of the radius of the RE^{3+}ion in ceramics with an amorphous intergranular phase

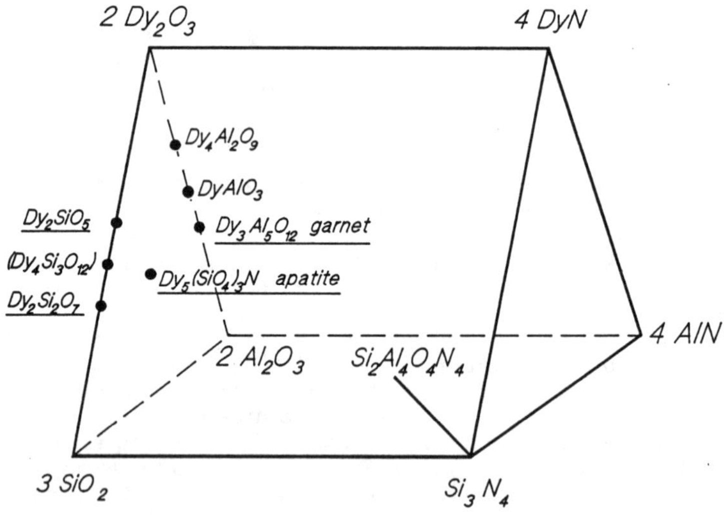

Figure 2: Phases in the system Si_3N_4-Al_2O_3-Dy_2O_3-DyN-AlN-SiO_2 [15]

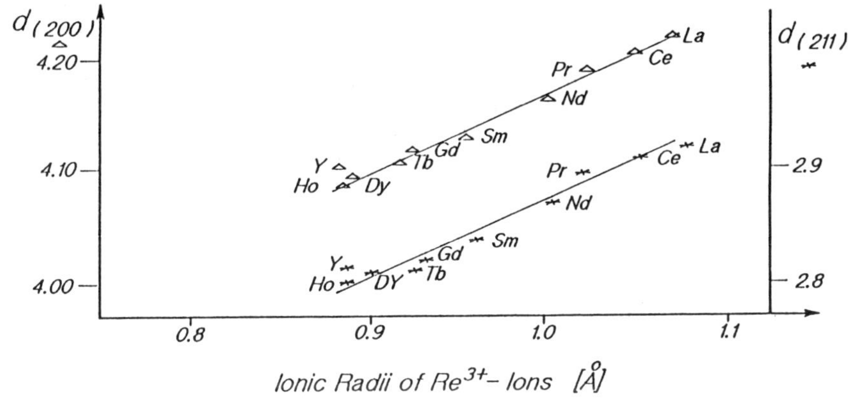

Figure 3: d-values from apatite $RE_5(SiO_4)_3N$ with different RE^{3+} ions

The crystallization of the glassy phase yield with the oxides Y and Er a decrease in high temperature strength compared with the untreated samples. In all other ceramic systems crystallization led to an increase of room temperature strength as well as high temperature strength, see Figure 4.

This may be explained by a good fit of the thermal expansion coefficients of apatite and β-Si_3N_4, although a detailed study of the microstructures still has to be done.

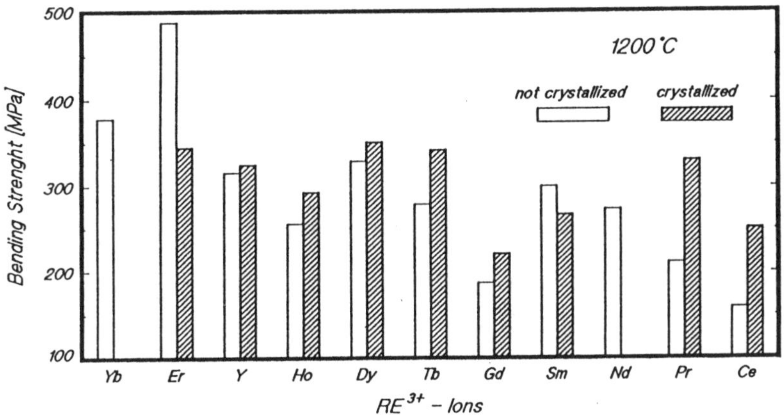

Figure 4: Influence of crystallization of the intergranular phase on the bending strength at 1200° C

REFERENCES

[1] Sanders, W.A. and Mieskowski, D.M., Am. Cer. Soc. Bull., 1985, **64**, 304-309.

[2] Tani, E., Nishijima, M., Ichinose, H.; Kishi. K. and Umebayashi, S., Yogyo-Kyokai-Shi, 1986, **94**, 300-305.

[3] Tani, E., Umebayashi, S., Kishi, K., Kobayashi, K. and Nishijima, M., Am. Ceram. Soc. Bull., 1986, **65**, 1311-1315.

[4] Tsuge, A., Nishida, K. and Komatsu, M., Am. Ceram. Soc., 1975, **58**, 323-326.

[5] Falk, L.K.L. and Dunlop, G.L., J. Mater. Sci., 1987, **22**, 4369-4376.

[6] Smith, J.T. and Quackenbush, C.L., Am. Ceram. Soc. Bull., 1980, **59**, 529-537.

[7] Bressiani, J.C., Thesis, Univ. Stuttgart (FRG), 1984.

[8] Mukerji, J., Das, P.K., Greil, P. and Petzow, G., Ceram. Int., 1987, **13**, 215-221.

[9] Ueno, K., Yogyo-Kyokai-Shi, 1986, **94**, [7], 702-705.

[10] Xu, Y.R., Fu, X.R. and Yan, D.S., Physica, 1988, **B150**, 276-282.

[11] Lee, W.E., Drummond III, C.H., Hilmas, G.E., Kiser, J.D. and Sanders, W.A., Ceramic. Eng. Sci. Proc., 1988, **9** [9-10], 1355-1366.

[12] Wötting, G. and Ziegler, G., Science of Ceramics, 1983, **12**, 361-370.

[13] Peuckert, M., Boberski, C. and Selgert, P., Proc. 3rd. Int. Symp. Ceramic Materials and components for Engins. Las Vegas, 1988., in press.

[14] Rocherulle, J., Verdier, P. and Laurent, Y., Materials Science and Engineering, 1989, **B2**, 265-268.

[15] Reser, M.K. (ed.), In Phase Diagrams for Ceramists, Am. Ceram. Soc.; Columbus, OH, Vol. I-III.

GRAIN BOUNDARY PHASE ELIMINATION AND MICROSTRUCTURE OF ALUMINIUM NITRIDE

FUMIO UENO, AKIHIRO HORIGUCHI
TOSHIBA Research and Development Center,
Komukai Toshiba-cho, Saiwai-ku Kawasaki J-210, JAPAN

ABSTRACT

Rare earth oxide is a good sintering additive for aluminium nitride (AlN). During sintering, yttrium oxide reacts with surface oxide layer of AlN powder, and forms yttrium-aluminium-oxide grain boundary phase. This grain boundary phase was eliminated from AlN ceramic by heating with nitrogen gas under carbo-reductive atmosphere at above ca. 1800°C. During this process, the yttrium-aluminium-oxide phase was changing into yttrium oxide and finally no trace of oxide nor pore were found in the ceramic after several hours. The surface of this pure AlN ceramic was coverd by yttrium nitride (YN) in some cases, but this YN was easily removed from AlN ceramic by hydrolysis. This phenomenon was interpreated as follows. The grain boundary oxide phase is carbo-reduced at the surface of the ceramic to form AlN and YN. As a result, oxide concentration gradient is developed in the ceramic, which is relaxed by the grain boundary diffusion of oxygen and yttrium. However, this was not diffusion controlled reaction.

INTRODUCTION

Although AlN has no melting point, it sinters with additive such as alkali earth or rare earth oxide. AlN powder containing oxygen can also sinter with no additives, so oxygen impurity seems to be essential for the sintering of AlN. However, thermal conductivity of AlN is largely affected by the oxygen contamination. Thermal conductivity of the ceramics can be increased by sintering with rare earth or alkali earth compounds. This is interpreted by a oxygen gettering by Y_2O_3 to form aluminium-yttrium-oxide grain boundary phase [1]. Oxygen impurity in AlN powder is nessary for sintering and additive such as Y_2O_3 is useful for both sintering and high thermalconductivity, but these oxygen or oxide are needless for the high thermal conductivity AlN product. This grain boundary oxide had found to be eliminated from the ceramic even after sintering in a certain condition [2,3]. The method and the mechanism of elimination was examined and discussed in this paper.

MATERIALS AND METHODS

AlN powder from Tokuyama soda with 0.8 wt% of oxygen and of 0.6μm diameter and a 99.99% pure yttrium oxide from Shin-Etsh Chemical were used. Density of a ceramics was measured by buoyancy in water. Thermal conductivity was measured by Laser flush technique at 20°C. Quantitative analysis of yttrium by titration and of oxygen by radio activation analysis was carried out.

Experiment 1

AlN powder with 3 or 5 weight percent of yttrium oxide was mixed with organic binder, pressed and sintered at 1800 or 1900°C for 2 to 96 hours in a electric furnace with carbon heater. A carbon or a AlN sagger was used for the sintering. AlN powder was also hot-pressed at 1800°C, 400kg/cm^2 for an hour. This was again heated at 1800°C for 24 hours in the carbon sagger. Sintering and hot-pressing were carried out under atmospheric pressure of nitrogen gas.

Experiment 2

Many pieces of four kinds of AlN ceramics were prepared by a sintering (1800°C, 2 h) with 5, 10, 20 wt% of Y_2O_3 or with both 5wt% of Y_2O_3 and 2 wt% of Al_2O_3. All of these ceramics were fully densified and were named 5Y, 10Y, 20Y and 5Y2A, respectively. These samples were finished to 4.5x24.0x24.0 mm^3. Then, these were heated at 1830, 1880, 1930 and 1980°C for 10 minutes to 96 hours in a carbon or a AlN sagger with atmospheric pressure of nitrogen or argon gas. A electric furnace with a carbon or tungsten heater was used for this treatment. After the heating, YN was removed, if necessary, from the sample surface in boiling water for several hours.

RESULTS AND DISCUSSION

Thermal conductivity and density of the samples in experiment 1 are listed in Tables 1 and 2. Prolonged sintering time increased thermal conductivity and decreased density of AlN ceramics with Y_2O_3 additive. This effect was more obvious when a carbon sagger was used instead of a AlN sagger. The sample sintered for 24 hours in the carbon sagger had no pores nor grain boundary phases from SEM observation, and was transparent and colorless. On the contrary, density and thermal conductivity of the hot pressed AlN with no additive was changed only a little after it was heated in the carbon sagger.

TABLE 1

Balk density (d) and thermal conductivity (κ) of sintered AlN.

additive	sintering conditions	d / g cm^{-1}	κ / W m^{-1} K^{-1}
3wt%Y_2O_3	1800°C, 2h, in AlN sagger	3.31	170
	1800°C, 24h, in AlN sagger	3.28	193
	1800°C, 24h, in carbon sagger	3.26	258
non	hot-press:1800°C,400kg/cm^2,1h	3.26	93
	↓	↓	↓
	heated at 1800°C,24h in carbon	3.26	98

As is shown inTable 2, oxygen and yttrium content of the ceramics had been decreased as the sintering time prolonged, and the composition of the grain boundary phase became changed to yttrium rich phase.

TABLE 2
Sintering time dependency of some properties of AlN sintered with 5 wt% of Y_2O_3 at 1900°C in a carbon sagger under 1 atm. of nitrogen.

time/hour	$d/g\ cm^{-3}$	O/wt%	Y/wt%	$\kappa/\ W\ m^{-1}K^{-1}$	grain size/ μm	grain boudary phase
2	3.312	1.00	2.75	210	7.4	$Al_2Y_4O_9 + Y_2O_3$
6	3.287	0.41	1.95	239	9.3	$Y_2O_3 \gg Al_2Y_4O_9$
12	3.264	0.05	0.08	247	10.5	Y_2O_3 (trace)
24	3.263	0.04	0.03	254	13.2	non
96	3.263	0.04	0.02	266	15.0	non

At a temperature higher than 1720°C, reaction (1) occurs from a thermodynamical calculation. In fact, $Al_5Y_3O_{12}$ and Y_2O_3 kept in the carbon sagger at 1900°C changed to AlN and YN. Therefore, reaction mechanism of the grain boundary phase elimination under carbo-reductive atmosphere with nitrogen gas can be expressed as equation (2). Then, oxide concentration gradient is developed in the ceramics because of the surface reaction (2), and is relaxed by the grain boundary diffusion of the oxide (figure 1)

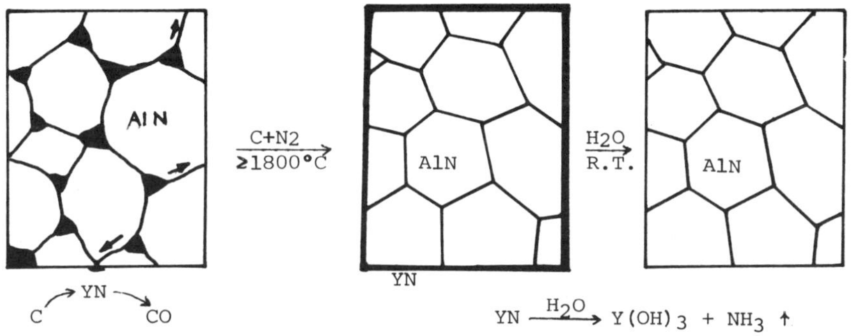

$$Al_2O_3 + 3C + N_2 \rightarrow 2AlN + 3CO \qquad (1)$$

$$xAl_2O_3 \cdot yY_2O_3 + 3(x+y)\,C + (x+y)\,N_2 \rightarrow 2xAlN + 2yYN + 3(x+y)\,CO \qquad (2)$$

Figure 1.Reaction sheme of grain boudary phase elimination from AlN ceramic

Experiment 2
Figure 2 shows weight loss of AlN (5Y) in both AlN and carbon sagger under atmospheric pressure of nitrogen at 1930°C. Surface of the samples after heating in the carbon sagger was covered with black YN film. This film was removed before weighting the sample. No YN was found on the surfaceof itin AlN sagger,but YN deposited on inner surface of the AlN sagger was found. Contribution from evaporation of AlN to this weight loss was negligible. For example, rate of weight loss was $(3.8 \pm 0.4) \times 10^{-6}\ g\ cm^{-2}\ s^{-1}$ at 1930°C in the carbon sagger and evaporation rate of pure AlN under the same condition was $1.3 \times 10^{-7} g\ cm^{-2} s^{-1}$, which was determined from the weight loss of pure AlN ceramics with no grain boundary phase, which was prepared by the method in this paper. In the case of both saggers, whole of the grain boundary phase was eliminatied from the ceramics after a certain time. After this,weight of the ceramic became almost constant. Time dependency of weight loss (ΔW) obeyed eqation (3) with a slope n of 1.0 ± 0.1.

$$\log \Delta W = n \log t + const. \qquad (3)$$

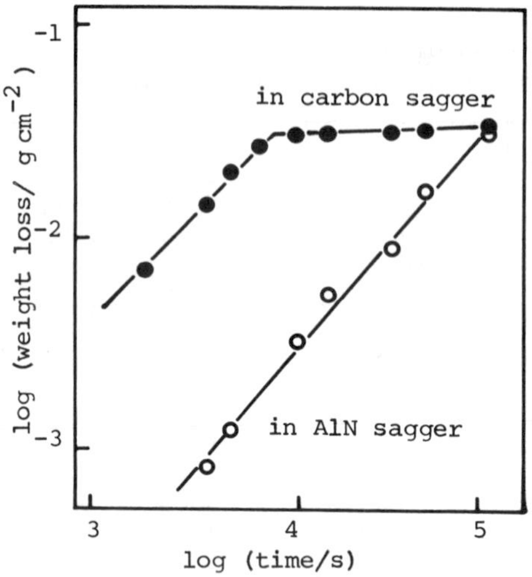

Figure 2. Weight loss of AlN ceramics (5Y) in different saggers at 1930°C under 1 atm. of N_2.

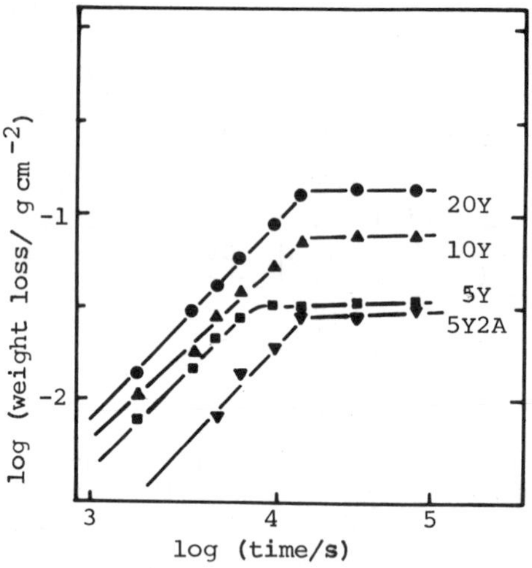

Figure 3. Time dependency of AlN weight loss at 1930°C in carbon sagger.

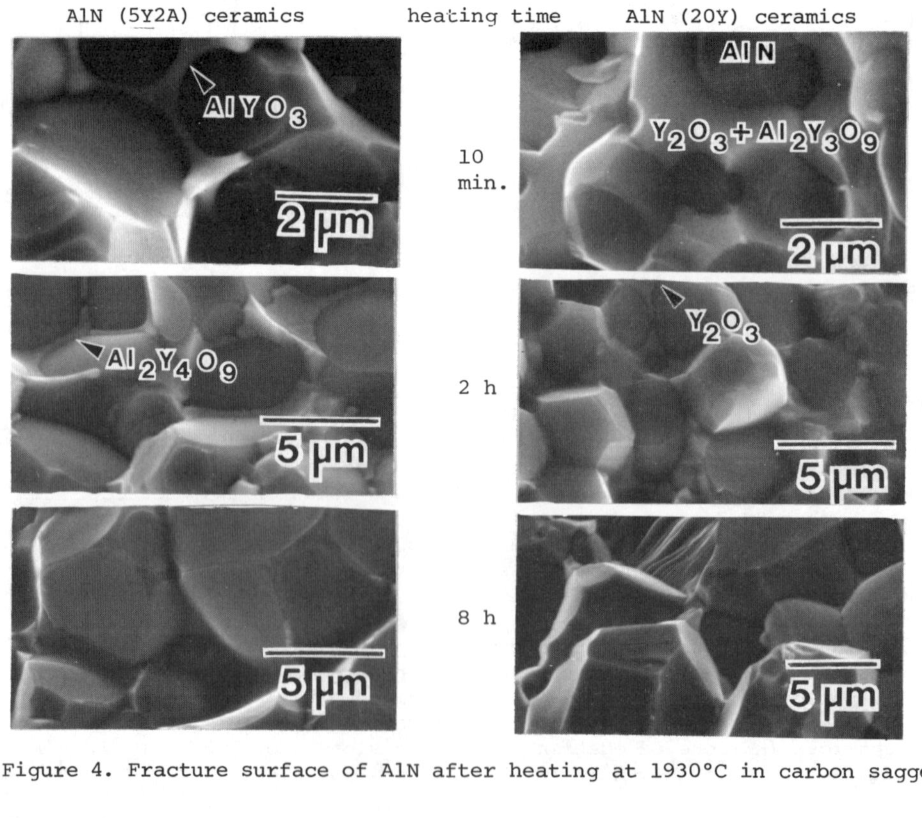

Figure 4. Fracture surface of AlN after heating at 1930°C in carbon sagger.

However, rate of weight loss was differnt in two housing conditions, and what was differnt in the two cases seems to be the carbon vapor supply. Therefore, weight loss is not controlled by a mass diffusion in the ceramic. Furthermore,weight loss is not controlled by the evaporation of YN, because no deposition of YN and slower rate of weight loss was found in the case of AlN sagger. AlN (5Y) was heated in the carbon and the AlN sagger under atmospheric pressure of argon gas 1930°C for 6 hours. In this case, evaporation of pure AlN in the carbon sagger, 5.3% i.e. $2.67x10^{-2}$ g cm^{-2}, was remarkable.Neverthless, AlN (5Y) heated in the carbon sagger was covered by yellow carbide film which also decomposed in water, density was decreased from 3.33 to 3.02 g cm^{-3}. AlN (5Y) was also heated in a metal furnace with tungsten heater without any sagger under one atmosphere of nitrogen. In this case, only negligible amount of weight loss was found. These results show grain boundary oxide phase were reduced by carbon gas and reacted with nitrogen to form AlN and YN. Figure 3 shows the weight loss courvé of AlN ceramics with different amount of additives. From the results of XRD, grain boundary phase was changing from alumina rich phase to yttria rich phase during heating (Figure 4). After the weight loss became constant, it was nearly equal to the amount of added Y_2O_3 in each case. For example, AlN (5Y) contained about 6.7% of grain boundary oxide. However, weightloss of(5Y)was not 6.7% but 5.5% And that of AlN (5Y2A) was not 8.7% but also 5.6%. This means YN was removed from the ceramics but the AlN produced from oxide phase was remained in the ceramics. For the elimination of yttrium oxide, both yttrium and oxygen must be transported to the surface of the ceramics, but in the case of aluminium oxide in AlN, aluminium may not necessary to move to the surface and diffusion rate of aluminium oxide elimination may be determined by the oxygen diffusion. Faster disappear of aluminium oxide than yttrium oxide may because of different diffusionrateof oxygen and yttrium. Nevertheless, rate of weight chang was independent to the chemical species of grain boundary phase. This result agrees with the fact that the grain boundary phase elimination was not diffusion controlled. The elimination process of grain boundary phase seems to be controlled by surface reaction or carbon gas supply. However, which controls the process is not clear from this experiment.

CONCLUSION

Grain boundary phase in AlN ceramics sintered with yttrium oxide was eliminated by carbo-reduction and nitridation, followed by diffusion of oxide phase, which relax concentration gradient developed in the ceramics. This process was not controlled by mass diffusion in the AlN ceramics nor surface evaporation of yttrium nitride.

REFERENCE

1. K.Shinozaki, K.Anzai, T.Takano, A.Tsuge and K.Komeya, Proceedings of 22nd Symposium on Basic Science of Ceramics, Ceramic Soc. Jpn., Jan, 1984, p.43.

2. A.Horiguchi, M.Kasori, F.Ueno, Y.Sato, A.Tsuge, M.Hayashi, H.Endo and K.Sinozaki, Japanese patent disclosure (Kokai), 63-303863, 1988.

3. A.Horiguchi, F.Ueno, M.Kasori, K.Shinozaki and A.Tsuge, Proceedings of 25th Symposium on Basic Science of Ceramics, Ceramic Soc. Jpn., Jan., 1987, p.155.

TWIN-ASSISTED GRAIN GROWTH DURING LIQUID-PHASE SINTERING OF $BaTiO_3$ CERAMICS

G. KÄSTNER, R. WAGNER, V. HILARIUS*
Institute of Solid State Physics and Electron Microscopy,
Academy of Sciences of the German Democratic Republic,
PF 250 Halle DDR 4010
*Kombinat Keramische Werke Hermsdorf, ZFT
Hermsdorf 6530

ABSTRACT

Based on crystallographic findings at liquid-phase sintered $BaTiO_3$ ceramics, models are derived for the formation of twinned grains during shrinking (densification) and for the preferential (anomalous) growth of doubly twinned grains.

INTRODUCTION

Among the forces and processes that control liquid-phase sintering (1), consequences of crystallographic anisotropies have been considered very occasionally until now. They can affect the grain shape, interaction of grains, and their growth including twin-assisted anomalous growth. The governing principle means that surface areas of low crystallographic index appear as facetes if as usual their (normal) growth speed is low enough, depending on various conditions.

MATERIALS AND METHODS

Specimens were prepared by the mixed oxide technique with few percent Ti excess or Si in order to obtain a liquid phase. Besides, Ca, Mn, and Li were added for obtaining semiconductivity with positive temperature coefficient (PTC ceramics). Pellets were air-sintered either 50 h at 1250°C (below or close to the eutectic) or 0.5 h at 1340°C well within the liquid-phase range. The grain morphology of thinned samples was studied by optical and electron microscopy (SEM and 1MV-TEM) combined with diffraction for determining orientations.

RESULTS

Grain growth in $BaTiO_3$ ceramics tends as known (2...6) to behave discontinuous: On the expense of normal growing grains, few abnormal grains grow preferentially until they stop each other. In the extreme case no normal grains remain.

Normal Grain Growth

During 1340°C sintering, the grains tend to form facetes of crystallographic (100)-type. Particular well defined (100)-facetes were observed in contact with amorphous pocket phases as shown in Fig. 1. This minimum growth speed of the (100)-

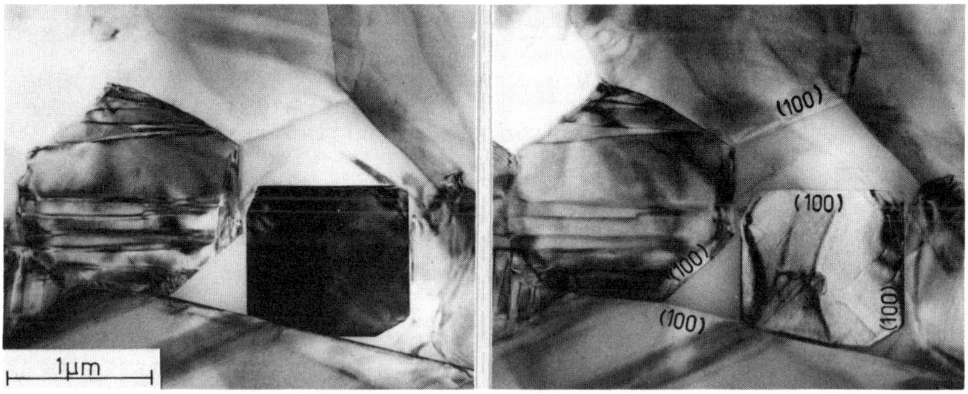

Figure 1. TEM images (differing by minor specimen tilt) of normal grains of $BaTiO_3$ exposing (100)-facetes against amorphous material. Sintering 0.5 h 1340°C.

face can be reasoned by the atomistic bond theory where (100) is a F-face according to the PBC theory of crystal growth (7).

On the other hand, 1250°C sintering tends to favour (111)-type facetes. The reason is not yet understood. The (111)-lattice planes consist of alternating BaO_3^{4-} and Ti^{4+} layers, thus exposing polarity, in contrast to the (100)-planes where layers of BaO and TiO_2 alternate. - An extreme case of (111)-facetes are the (111) twin boundaries of twinned grains. Their occasional occurence within the size fraction of normal grains (8) could be confirmed for both sintering conditions. More rarely, doubly twinned grains were found as shown in Fig. 2.

Orign of the Twins

It is yet an open question how the twinned grains orign. Eibl et al. (8) observed narrow ($\lesssim$ 30 nm) twin lamellae after calcining. They were considered to arise from some solid-state chemical reaction but the question has been put how such narrow lamellae transform into the thicker ones usually observed (2...6) in well-sintered $BaTiO_3$. A further work (9) leads to

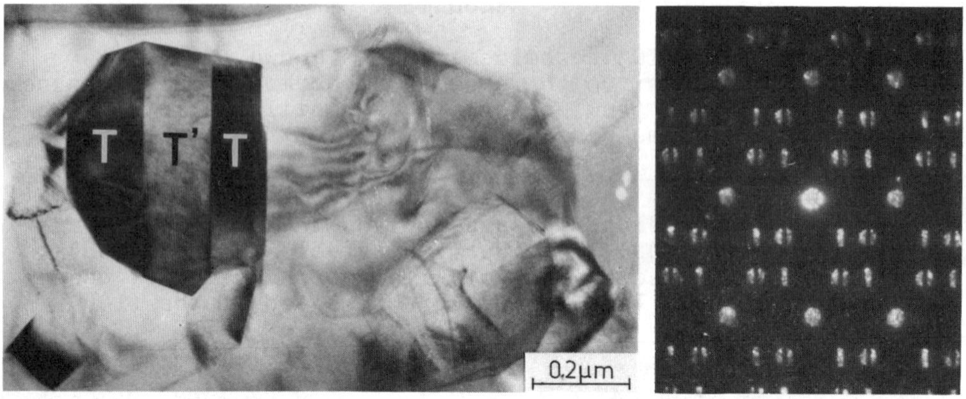

Figure 2. TEM of a grain composed of twins T, T'. Its [110]-zone diffraction pattern is slightly defocused for revealing the orign of the spots from the twins.

the question whether nuclei of the hexagonal, high-temperature phase of $BaTiO_3$ may act as twin nuclei.

According to our findings the thickness of the twin lamella is comparable to the grain size prior to sintering, and twinning is favoured by a fast speed of shrinking (5). Hence we assume that twins are formed during shrinking by particle rearrangement of two (111)-faceted grains, favoured by the low twin interfacial energy and by sufficient mobility due to the liquid phase (6). A double twin (twin lamella) may be formed by rearrangement with a third grain. If this happens at every 10^3th or 10^4th grain, this low probability would be sufficient to explain the low density of nuclei for anomalous grains that contain a twin lamella.

Anomalous Growth of Doubly Twinned Grains

Preferentially growing (anomalous) grains of $BaTiO_3$ ceramics were often found to contain a central twin lamella some 0.1 to 1 μm thick (2...6). Depending on the sintering conditions, two different grain shapes can be distinguished as schematically drawn in Fig. 3. The shape of platelets a) is well known from sintering near 1250°C (2...6) whereas at $\gtrsim$ 1340°C, the grains tend to a "globular" (2,3,4) or "cube" shape (5,6) to be related to Fig. 3b. - The reason for preferential growth of the platelets has been discussed by (3,4) in terms of theoretical re-entrant facete angles at the twin as encircled in Fig. 3a. Experimentally, such facete angles could be found neither at the platelets a) nor at the cubes b) (6). A recent example is shown in Fig. 4. Across the twin, the grain boundary does not alter generally its mean geometrical orientation. The latter is one of the lowest growing speed as supposed. Therefore the emerging surface of the twin deviates, in general, from this orientation. Hence it tends to grow faster, thus initiating growth on the whole surrounding boundary. This holds for

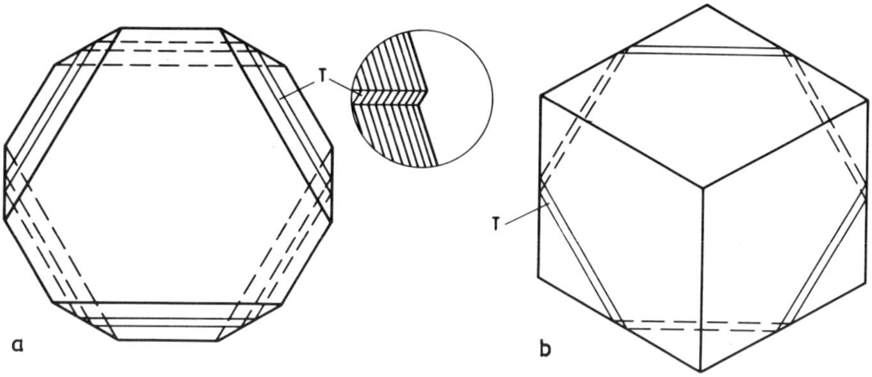

Figure 3. Polyhedral crystals containing a twin lamella T parallel to the drawing plane (located between the double lines). a) platelet bounded by (111)-type surfaces, b) cube bounded by (100)-type surfaces.

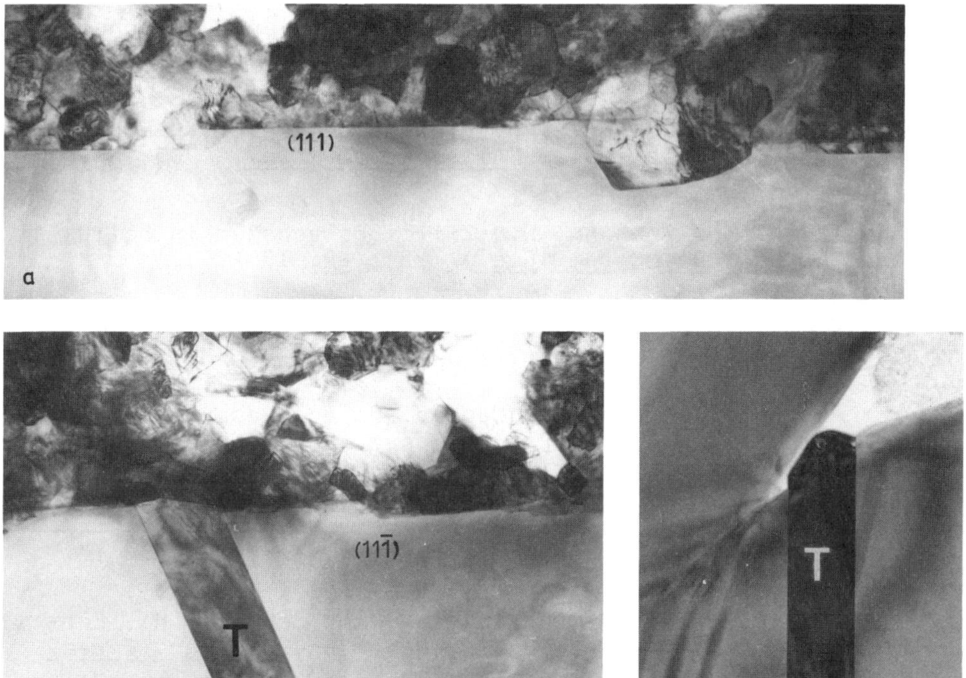

Figure 4. TEM of boundary regions of anomalous grains,
a) platelet, (111) top face against normal grains,
b) same platelet, side face with twin lamella T,
c) globular grain, twin T protruding to free pocket.

all 6 cube faces of Fig. 3b and for the 6 side faces of the platelet. Nucleation of steps (Fig. 4a) may control the thickness of the platelet as well as of the twin lamella (6). The question remains how growth inhibitors (10) affect anisotropy.

CONCLUSIONS

Crystallographic faceting of ceramic grains favours distinct grain-boundary orientations and, in particular, formation of twins during shrinking (densification). Doubly twinned grains (containing a twin lamella), e.g. platelets or cubes, can grow preferentially at the faces where the twin emerges. The governing principle is the crystallographic anisotropy of the growth speed. It can be affected by various sintering conditions and probably by growth inhibitors.

REFERENCES

1. German, R.M., Liquid Phase Sintering, Plenum, New York 1985.
2. Schmelz, H. and Meyer, A., Nachweis anomalen Kornwachstums in $BaTiO_3$-Keramik unterhalb der Eutektikalen. Ber. Dt. Keram. Ges., 1982, 59, 436-43.
3. Oppolzer, H. and Schmelz, H., Investigation of Twin Lamellae in $BaTiO_3$ Ceramics. J. Amer. Ceram. Soc., 1983, 66, 444-50.
4. Schmelz, H. and Thomann, H., Zwillingswachstum in $BaTiO_3$-Keramik. Ber. Dt. Keram. Ges., 1984, 61, 199-204.
5. Bühling, D., Hilarius, V. and Kästner, G., Microstructure Development in Semiconducting Barium Titanate Ceramics. Ber. Dt. Keram. Ges., 1989, in press.
6. Kästner, G., Hilarius, V., Wagner, R. and Lacayo, G., Twins and Grain Growth During Liquid-Phase Sintering of $BaTiO_3$ Ceramics. J. Mater. Sci. Lett., 1989, in press.
7. Hartman, P., Modern PBC Theory. In Morphology of Crystals, ed. I. Sunagawa, Terra Scient. Publ. Co., Tokyo, 1987, pp. 269-319
8. Eibl, O., Pongratz, P., Skalicky, P. and Schmelz, H., Formation of (111) Twins in $BaTiO_3$ Ceramics. J. Amer. Ceram. Soc., 1987, 70, 195-7.
9. Eibl, O., Pongratz, P. and Skalicky, P., Crystallography of (111) Twins in $BaTiO_3$. Philos. Mag. B, 1988, 57, 521-34.
10. Drofenik, M.H., Grain Growth During Sintering of Donor-Doped $BaTiO_3$. J. Amer. Ceram. Soc., 1986, 69, 8-9.

COMPACTION AND SINTERING BEHAVIOUR OF NANO-CRYSTALLINE Y-TZP

A.J.A. WINNUBST, G.S.A.M. THEUNISSEN and A.J. BURGGRAAF
University of Twente, Faculty of Chemical Technology,
Lab. for Inorganic Chemistry, Materials Science and Catalysis,
P.O. Box 217, 7500 AE, Enschede, The Netherlands.

ABSTRACT

A dense fine-grained ceramic (55 nm) of yttria doped tetragonal zirconia polycrystals (Y-TZP) is obtained using a nano-crystalline (8 nm) powder. The change in aggregate and agglomerate structure during compaction and sintering of these powder compacts is reviewed. The weak agglomerates are fractured during isostatic compaction. The aggregate morphology is a key parameter in microstructure development during sintering.

INTRODUCTION

Ceramic materials with grain sizes in the nanoscale range show special properties such as superplastic deformability (1). In order to prepare these ceramics attention has to be paid to the homogeneity of the ultra-fine grained starting powder (e.g. composition, crystallite size, crystal structure, pore morphology) (2). Furthermore grain growth has to be controlled and low sinter temperatures are necessary. Despite the use of fine grained powders literature does not report the succesfull synthesis of dense ceramics with a grain size less than 100 nm.

The nano-crystalline starting powders are often clustered to aggregates and/or agglomerates. In aggregates the primary crystallites are held together by necks. In agglomerates the aggregates or crystallites are held together by relatively weak attractive forces. In this paper the change in aggregate and agglomerate structure of nano-crystalline Y-TZP powders during compaction and sintering is summarized and critical steps for nanoscale ceramic preparation are identified.

EXPERIMENTAL PROCEDURE

Tetragonal 3 mol% yttria doped zirconia powders were prepared by two gel precipitation techniques, using metal chloride or metal alkoxide solutions as precursor (resp. "Chloride" and "Alkoxide" powders). The third powder was a commercial TZ3Y powder (Tosoh Co. Ltd, Tokyo). Crystallite sizes were determined by X-ray line broadening or TEM pictures. Specific surface areas were analysed by the BET method. The powders were isostatically pressed up to 500 MPa. Overall densities were measured by the Archimedes technique. After isostatically pressing at 400 MPa the powder compacts were sintered at several temperatures. The (open) pore size distribution and pore volume in the green and sintered compacts were calculated from a nitrogen adsorption/desorption isotherm at 72 K or by Hg intrusion porosimetry. More experimental details are given in (3,4,5).

RESULTS AND DISCUSSION

Agglomerate properties

Relative densities of powder compacts as a function of the logarithm of the applied pressure can be devided in two linear portions (Fig. 1). The model for this compaction behaviour is described in (3,4,6,7) and is summarized below. At pressures below the point of intersection the agglomerates only rearrange. The point of intersection is a measure for the agglomerate strength. The agglomerates are fragmented when the powder is applied to a pressure which lies in the high pressure linear region. The agglomerates of all 3 powders are fragmented prior to sintering.

The pore size distribution after compaction at 4 MPa is given in Fig. 2A. Clearly 2 types of pores are present. The pores smaller than 100 nm in radius do not change in size and volume if pressures less than the agglomerate strength are applied to the powder (3,4). From the volume of these intra-agglomerate pores the agglomerate density can be determined (4). The results are given in table 1. Clearly a difference in intra-agglomerate pore size and porosity is present between Chloride and Alkoxide powders, which however have the same crystallite size.

Aggregate properties

From the specific surface area (S_{BET}) the sizes can be calcultated of particles, which are accessible for the absorbing gas. These "equivalent spherical diameters" are 8, 10, 47 nm for Chloride, Alkoxide and Tosoh

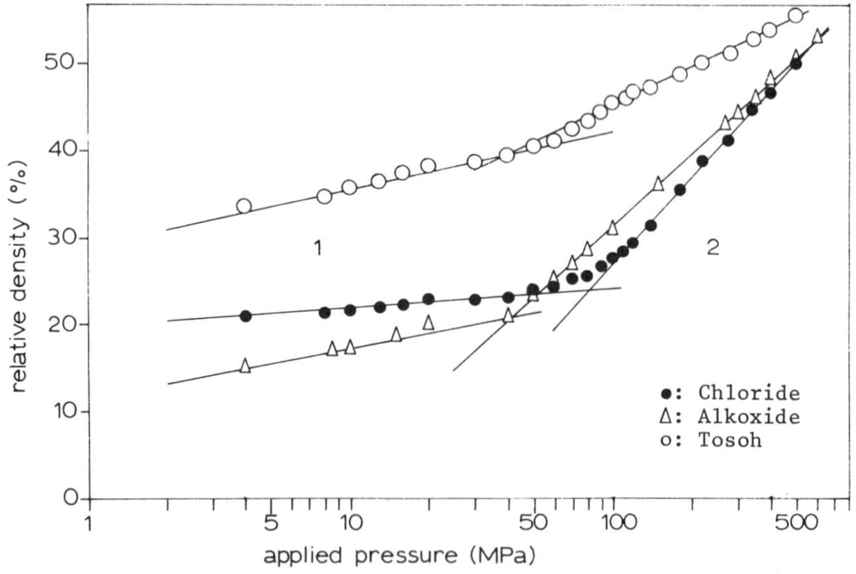

Figure 1. Densification during isostatic compaction.

TABLE 1
Powder properties; all powders have a composition of $(ZrO_2)_{0.97}-(Y_2O_3)_{0.03}$

Powder*	Crystallite size (nm)	S_{BET} (m^2/g)	Agglomerate properties strength (MPa)	density (%)	Aggregate density (%)	size (nm)
Chloride	8	124	80	30	-	-
Alkoxide	8	101	45	22	70	18
Tosoh	34	21	40	36	90	49

* Chloride and Alkoxide after calcination (820K 2 hrs.) Tosoh: as received

resp. The mismatch of these values with the crystallite sizes can be explained by clustering of crystallites to aggregates (the absorbing gas can not completely cover the surface of the aggregated crystallite) (3).

The aggregate density is determined from the pore size distribution data after compaction at 400 MPa (Fig. 2B). The Alkoxide compact shows a bimodal pore size distribution. From the pore volume of the shoulder around 2 nm (the intra-aggregate pores) an aggregate density of 70 % is calculated. The Alkoxide aggregate size is 18 nm, assuming closed packed stacking of 9 primary crystallites which are held together by necks. The Tosoh compact only shows one type of pores. The mismatch between equivalent sphere diameter and crystallite size can be explained by the presence of almost dense aggregates with a size of 49 nm. Simple cubic packing of these spherical Tosoh aggregates enclose pores with a radius of 17 nm, which fitts well with compaction data at 400 MPA (Figs. 1 and 3). According to this discussion the Chloride powder contains no aggregates. However in these powders crystallite "chains" are present which form small crystallite "clusters" (about 5 crystallites) at chain intersections (5,8). The aggregate structures as discussed above determine initial sintering of the compacts.

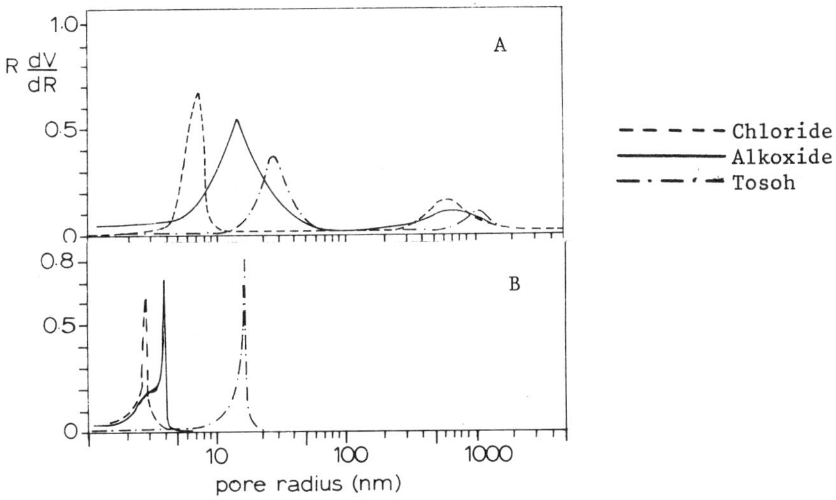

Figure 2. Poresize distribution after isostatic compaction at 4 MPa (A) and 400 MPa (B).

Sintering behaviour
Prior to sintering the powder compacts (400 MPa) have a relative density and pore size distribution as given in Figs. 1. and 2B. During initial sintering no overall densification occurs, but the internal microstructure changes. After sintering at 950 K the Alkoxide and Chloride compacts contain crystallites with sizes of resp. 13 and 16 nm. For the Tosoh compact a crystallite size of 47 nm is found after intitial sintering at 1250 K. The same values of crystallite sizes are found if a calculation is made for densification of the aggregates as mentioned in the previous paragraph. Simple cubic packing of these recrystallized aggregates enclose pores with radii of resp. 5, 6 and 17 nm, which fits with experimental data. This microstructure obtained after initial sintering determines further densification and grain growth. The Chloride compact (with the smallest recrystallized aggregate) is the best start for obtaining a nano-crystalline ceramic. In Fig. 3 the microstructure development during final sintering of a Chloride compact is shown. A grain size of 55 nm and a relative densify of 97 % are obtained after sintering for 6.5 hours at 1315 K. A dense Chloride compact is also obtained after sintering for 200 hours at 1270 K. For a dense Alkoxide and Tosoh ceramic minimum grain sizes of 200 nm are reported (3,5,8).

A detailed analysis of densification and grain growth of a Chloride compact is performed. Densification as a function of temperature can be devided in two different temperature regions. At temperatures below 1150 K densification occurs by grain boundary slip assisted by grain boundary diffusion. Densification by grain boundary slip is observed when crystallite sizes are less than 20 nm. Between 1150 and 1350 K densification occurs by grain boundary diffusion but probably also some grain boundary slip is present in this temperature region.

In zirconia-yttria ceramics a segregation layer of yttria is observed in the grain boundaries. This segregation behaviour decreases grain boundary mobility due to the yttrium gradient. The grain gowth during sintering is therefore retarded. Small aggregate size and grain growth retardation results in an ultra-fine grained Y-TZP ceramic. The densification and grain growth of these nano-crystalline ceramics will be described in more detail in (5).

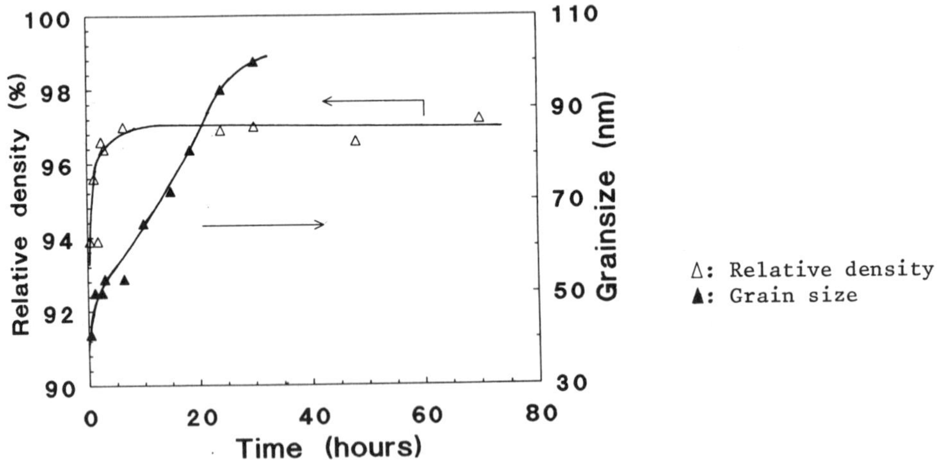

Figure 3. Grain size and density development of a Chloride compact after sintering at 1315 K.

CONCLUSIONS

- By means of gel-precipitation weakly agglomerated zirconia-yttria ceramic powders can be prepared with a primary crystallite size of 8 nm.
- All agglomerates are completely collapsed prior to sintering.
- The Chloride agglomerates consist of crystallite chains which form small cluster "aggregates" at the chain intersections of about 5 crystallites.
- The Alkoxide aggregate size is 18 nm (rel. density 70 %)
- The Tosoh aggregates are almost dense.
- During initial sintering recrystallization of the aggregates to "single" crystallites takes place, which is the starting grain size for further grain growth.
- It is possible to densify a Chloride compact to almost full density after 6.5 hours of sintering at 1315 K. The resulting grain size is 55 nm.

Acknowledgement
This research was partly supported by the Innovative Research Program on Technical Ceramics (IOP-TK) with financial aid of the Dutch Ministry of Economic Affairs.

REFERENCES

1. Burggraaf, A.J., Theunissen, G.S.A.M. and Winnubst, A.J.A., Synthesis and properties of nanophase ceramics and composites. Proc. 1st Eur. Ceram. Soc. Conf. (This issue).

2. Brook, R.J., Tuan, W.H. and Xue, L.A., Critical issues and future directions in sintering science. In Ceramic Transactions Vol. 1 ed. G.L. Messing, W.R. Fuller and H. Hausner, The Am. Ceram. Soc., Westerville (Ohio), 1988, pp. 811-23.

3. Groot Zevert, W.F.M., Winnubst, A.J.A., Theunissen, G.S.A.M. and Burggraaf, A.J., Powder preparation and compaction behaviour of fine grained Y-TZP. J. Mater. Sci., accepted for publication.

4. Van de Graaf, M.A.C.G., Ter Maat, J.H.H. and Burggraaf, A.J., Microstructure and sintering kinetics of highly reactive ZrO_2-Y_2O_3 ceranics. J. Mater. Sci., 1985, 20, 1407-18.

5. Theunissen, G.S.A.M., Winnubst, A.J.A. and Burggraaf, Sintering kinetics of nano-scale Y-TZP ceramics. to be published.

6. Niesz, D.E., Bennet, R.B. and Snyder, M.J., Strength characterization of powder aggregates. Am. Ceram. Bull., 1972, 51, 677-80.

7. Halloran, J.W., Role of powder agglomerates in ceramic processing. In Advances in Ceramics, vol. 9, ed. J.A. Mangels and G.L. Messing, Am. Ceram. Soc. Inc, Columbus (Ohio) 1988, pp. 67-75.

8. Theunissen, G.S.A.M., Winnubst, A.J.A., Groot Zevert, W.F.M. and Burggraaf, A.J., Microstructure developement during sintering of ultra-fine grained Y-TZP. Paper presented at 7th, SIMCER, Zirconia 88,Bologna, December 14 - 18, 1988.

MICROSTRUCTURE DEVELOPMENT DURING GAS PRESSURE SINTERING OF Si_3N_4

FRANC FRIEDRICH*, MILOŠ KOMAC, DRAGO KOLAR
"Jožef Stefan" Institute, University of Ljubljana, and *"Belinka",
Ljubljana, Yugoslavia

ABSTRACT

Powder mixtures consisting of Si_3N_4 and small amounts of sintering additives i.e. MgO, Al_2O_3 and metallic silicon, were used to study sintering and microstructure development during the heat treatment of Si_3N_4 ceramics. Samples were sintered under nitrogen gas overpressure to high density (98- 99% T.D.) and subsequently annealed at 1400°C for different periods of time. Generally, by increasing the sintering temperature and/or pressure the coarsening of β-Si_3N_4 grains was promoted, while during thermal treatment the aspect ratio of β-Si_3N_4 crystals was found to decrease.

An attempt is made to correlate the microstructural peculiarities and mechanical properties of Si_3N_4 ceramics.

INTRODUCTION

At present, silicon nitride is considered to be a highly useful engineering material. Its good mechanical properties arise mainly from microstructural peculiarities. Among various methods of densification of this typically unsinterable covalent material, gas-pressure sintering is one of the most promising technologies for the fabrication of dense Si_3N_4 ceramics for structural applications[1]. Hence, an understanding of the factors influencing microstructure evolution during pressure assisted sintering and post-sintering treatment is a prerequisite for the fabrication of material with superior mechanical behaviour.

The aim of the present work was to study the densification process and microstructure development by the application of different processing conditions. The interdependence of microstructural variables and mechanical properties was studied according to modified Griffith equation[2]:

$$\sigma F \sim a . \frac{1}{\sqrt{l}}$$

where σ F is the fracture strength, a is the average aspect ratio and l is the length of the b crystals. As can be seen from this equation, microstructural variables a and 1 have to be known as a function of processing conditions and starting powder properties. The experiments showed that the grain size distribution of annealed dense samples can have much more influence on mechanical strength than the aspect ratio of Si_3N_4 grains.

MATERIALS AND METHODS

The silicon nitride powder from H.C. Starck, LC-12, with surface area > 19 m^2/g, was mixed with 3,0 wt.% Al_2O_3, 1.2 wt.% MgO and up to 6 wt.% metallic silicon, homogenized and milled for six hours in an attritor plastic mill with Al_2O_3 balls in isopropyl alcohol. Test samples were CIP-ed to a green density of 62% T.D. and sintered in a high temperature autoclave according to the P, T, t diagram in Fig. 1.

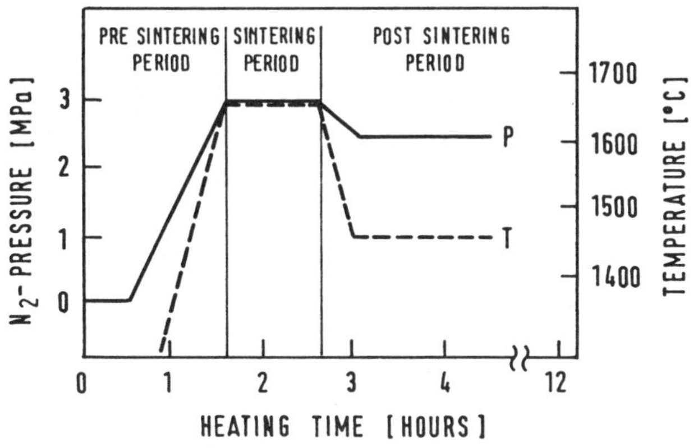

Figure 1. Pressure-Temperature-time regime during heating.

During the presintering period, the furnace was evacuated while the temperature was raised to 900°C. After half an hour's soaking time, the temperature and N_2-pressure were raised to the desired sintering conditions, i.e. T = 1650°C, P_{N2} = 3.0 - 9.0 MPa, t = 1h. In the post-sintering period the temperature was decreased to 1400°C and held at this value for different periods of time.

The characterization of samples comprised determination of the density and room temperature flexural strength, and quantitative evaluation of the microstructure according to [2].

RESULTS AND DISCUSSION

The sintering behaviour of the samples is illustrated in Fig. 2. There is a large increase in relative density at temperatures between 1500°C and 1650°C (A), at a N_2 pressure between 1.5 and 3.0 MPa (B), after one hour of sintering (C). No weight loss was observed during sintering except when the temperature exceeded 1700°C. In this case the weight loss rapidly increased (A).

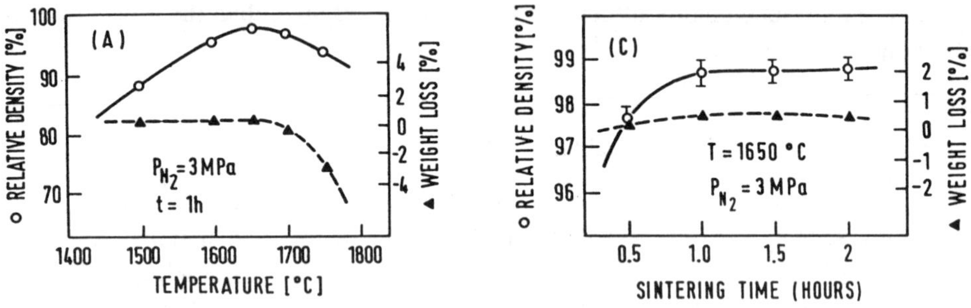

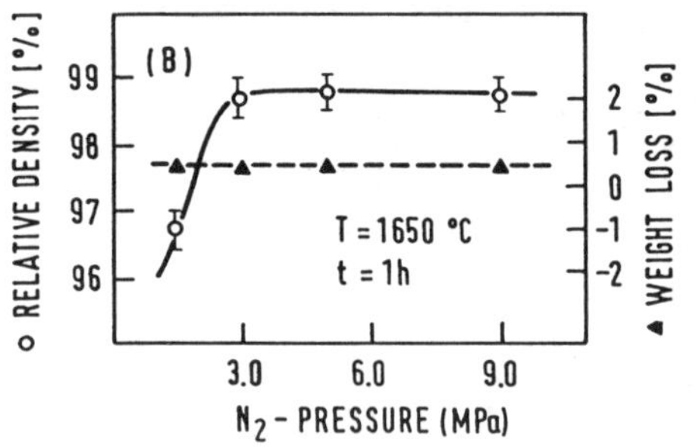

Figure 2. Relative density and weight loss as a function of sintering temperature (A), N_2-pressure (B) and time (C).

It is known that the densification and α/β phase transformation during liquid phase sintering of Si_3N_4 are stimulated by using fine starting powders, adding metallic silicon as a sintering aid[3] and by applying a sufficiently high N_2-overpressure. Simultaneously, under such conditions thermal decomposition of Si_3N_4 will be suppressed which, in addition, enhances densification[4]. During sintering and post-sintering treatment, metallic silicon can react with the N_2 atmosphere and crystallization of β-Si_3N_4 crystals from the Si melt can occur. As shown in Fig. 2 (A) and (B), a slight increase in weight can be observed after sintering, indicating that nitridation of Si has occurred.

About 11 wt.% sintering aids (the amount of SiO_2 was calculated from the oxygen present in the Si_3N_4 powder) form the silicate melt, whose viscosity and composition determines the size and morphology of the rod-like β-Si_3N_4 grains forming during the densification and α/β phase transformation. Due to the high amount of liquid phase, densification can readily take place; however, a high volume of glassy phase can result in inferior mechanical properties. Hence, the samples were subjected to post- sintering annealing at 1400°C. It was assumed that at this temperature the β-Si_3N_4 phase crystallizes from the Al, Mg silicate melt. Simultaneously, aluminium and oxygen ions can dissolve in β-Si_3N_4. The amount of glassy phase will thus be reduced.As shown in Fig. 3, during annealing the aspect ratio decreases with time in both cases, if 3 wt.% or 6 wt.% of metallic silicon were present. The decrease of the aspect ratio of rod-like β- Si_3N_4 grains results in a decrease of flexural strength of the samples, as also shown in Fig. 3.

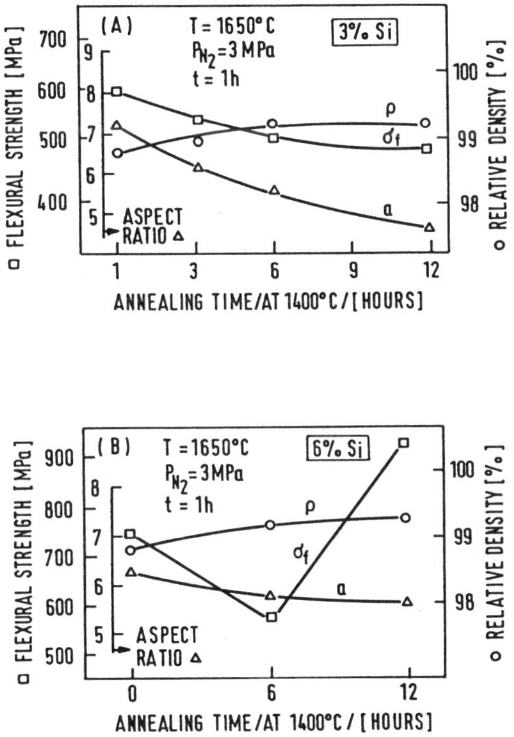

Figure 3. Flexural strength, relative density and aspect ratio versus annealing time

(A) 3 wt. % Si added
(B) 6 wt. % added

However, anomalous behaviour can be observed in the case of samples with 6% metallic silicon additive. After prolonged annealing, the room temperature flexural strength increases quite substantially, in spite of the decreasing aspect ratio. A tentative explanation can be given for this phenomenon on the basis of an inspection of the microstructure. It was found that the grain diameter distribution becomes increasingly uniform with annealing time (Fig. 4), which together with diminution of the amount of liquid phase due to secondary crystallization of β- Si_3N_4 phase could result in a strength increase. The importance of the amount of metallic Si additive is not clear at present.

CONCLUSIONS

Dense Si_3N_4 ceramics can be fabricated by using Al_2O_3, MgO and metallic Si as additives, the sintering temperature being around 1650°C and the N_2 gas pressure in the furnace in the range 1.5-3 MPa. During post-sintering annealing at 1400°C globularisation of $\beta-Si_3N_4$ grains occurs giving rise to a decrease of aspect ratio and consequently to strength decrease. Prolonged annealing results in the evolution of a rather homogeneous microstructure which could be one of the origins of the observed strength increase.

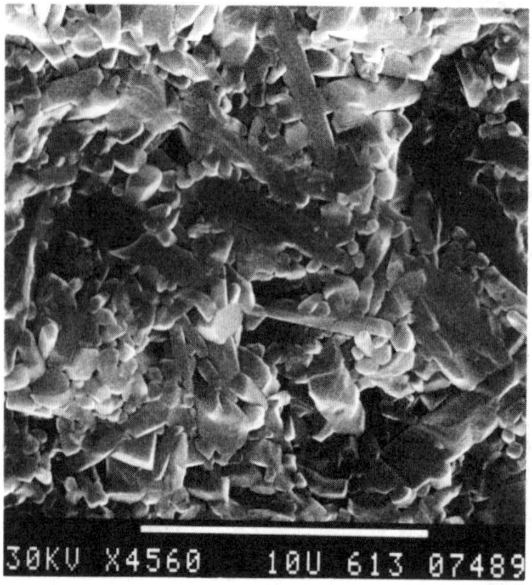

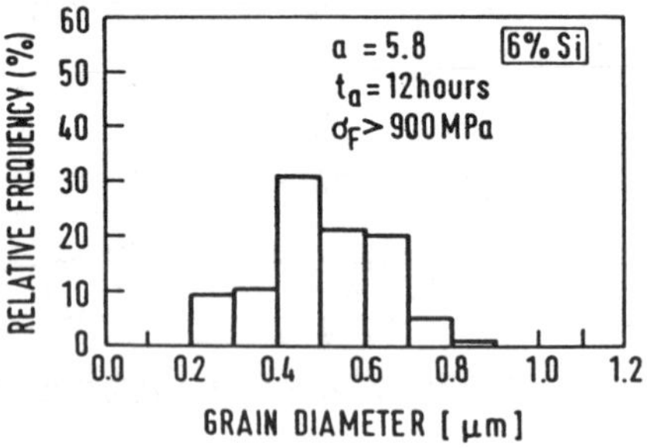

Figure 4. Well developed microstructure of an annealed Si_3N_4 sample and the corresponding grain diameter distribution diagram

REFERENCES

1. Hattori, Y., Tajima, Y., Yabuta, K., Matsuo, J., Kawamura, M. and Watanabe, T., Gas pressure sintered silicon nitride for turbocharger applications. In Ceramic Materials and Components for Engines, eds. V. Bunk and H. Hausner, Deutsche Keramische Gesellschaft, Bad Honnef, 1986, pp. 165-72.

2. Wötting, G., and Ziegler, G., Characterization of microstructural variables of sintered, hot-pressed and HIP Si_3N_4 and their correlation with mechanical properties. In Science of Ceramics 12, ed. P. Vincenzini, Ceramurgica, 1984, pp. 361-70.

3. Park, Y., Kim, R.J. and Kim, H.C., Effect of free silicon on the α to β phase transformation in silicon nitride. J. Amer. Ceram. Soc. Comm., 1987, 70, C240-C242.

4. Baik, S. and Raj, R., Effect of silicon activity on liquid- phase sintering of nitrogen ceramics. J. Amer. Ceram. Soc. Comm., 1985, 68, C124-C126.

SINTERING OF AlN POWDERS OBTAINED FROM CARBOTHERMAL REDUCTION

M.P. CORRAL, J.S. MOYA, R. MARTINEZ
Instituto de Cerámica y Vidrio (CSIC)
28500 Arganda del Rey, Madrid, SPAIN
A.P. TOMSIA
Materials and Chemical Sciences Division
Lawrence Berkeley Laboratory, Berkeley, CA 94720, USA.

ABSTRACT

The influence of Al_2O_3 particle size, specific surface area, impurities and carbon stoichiometry on the production of AlN powders by carbothermal reduction has been studied.
The effect of 2 wt. % addition of CaO, CeO_2 and Y_2O_3 in the sintering and microstructure of AlN compacts was also investigated.

INTRODUCTION

Aluminium nitride has been recognized as a suitable material for many high temperature applications and as a substrate for electronic packaging, due to its good thermal and mechanical properties and low thermal expansion (1,2,3). For those substrate applications where cooling is most important AlN materials should have high thermal conductivity (4,5). However, the inherent thermal properties of AlN bodies are critically affected by their purity and oxygen content as well as for their microstructure and grain boundary phases. The purity, particle size, oxygen content and specific surface area of AlN powders directly affect the sinterability and properties of the sintered compacts.

The present work describes the synthesis and sintering process of AlN powders obtained by carbothermal reduction of alumina. By using different starting alumina powders, the influence of several factors in the carbothermal process such as the nature of the alumina powders, impurities content, grain size and carbon content has been examined.

EXPERIMENTAL PROCEDURE AND RESULTS

Four different commercial alumina powders (Table 1) have been used as starting materials. Three of them (A2, A3 and A4) were ball milled in aqueous media with alumina balls in order to reach an average particle size of 0.5-2 μm. The fourth powder (A1) was used in as-received condition.

TABLE 1
Alumina powders used in the carbothermal reduction process

alumina	Specific surface area (m^2/g)	Aver. part. size (μm) as-received	Aver. part. size (μm) after mill.	Impurity content (wt. %)
A1[1]	8	0.5	--	MgO (0.1) Na_2O (0.1)
A2[2]	3	30*	1.2	MgO (0.001) Na_2O (0.1)
A3[3]	5	8*	0.9	< 0.01
A4[4]	0.8	3.6	2	SiO_2 (0.03) Fe_2O_3 (0.03)

[1]Alumina Alcoa CT3000 SG. [2]Alcoa CT3000. [3]Alumina Fluka.
[4]Alumina Alcan G-71.
*agglomerates

Synthesis of AlN powders.
Alumina powders and carbon black were mixed in isopropyl alcohol using a high-speed mixer. Al_2O_3/carbon molar ratio 1:3.6, that is 20 wt.% excess carbon over the stoichiometric amount, was used. In the case of sample A1 a mixture of stoichiometric alumina/carbon ratio, was also prepared. After drying, pellets were made by uniaxial pressing (≈ 20 MPa).

Carbothermal reduction and nitridation were carried out in a tubular furnace with a graphite camera inside, at temperatures ranging from 1480°C to 1560°C for 1 to 10 hours. Nitrogen flow during heat treatment was 0.3 l/min. After nitridation, the excess of carbon was removed from the samples by burning at 700°C in air. Phase composition in the reaction product was determined by quantitative X-ray analysis. SEM was also used to characterize the carbothermal reaction product. Fig. 1 shows micrographs of powders obtained from sample A1 after 1, 4 and 10 hours of carbothermal treatment at 1560°C.

The kinetic curves show the degree of reaction expressed as unreacted α-Al_2O_3 determined by quantitative XRD analysis. Fig. 2 shows the advance of the carbothermal reduction at 1560°C vs. reaction time. Similar curves were obtained for all temperatures studied. The only crystalline phases detected by XRD in the reaction products were α-Al_2O_3 and AlN for all samples and temperatures used.

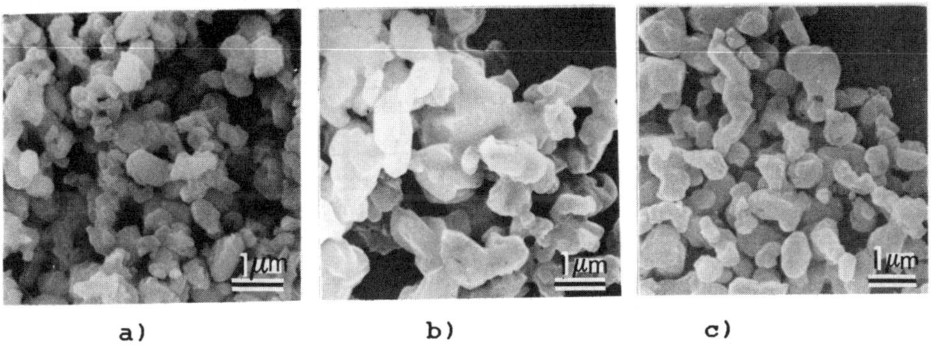

Figure 1. SEM micrographs of alumina powder A1 after carbothermal treatment at 1560°C for: a) 1 hour, b) 4 hours, and c) 10 hours.

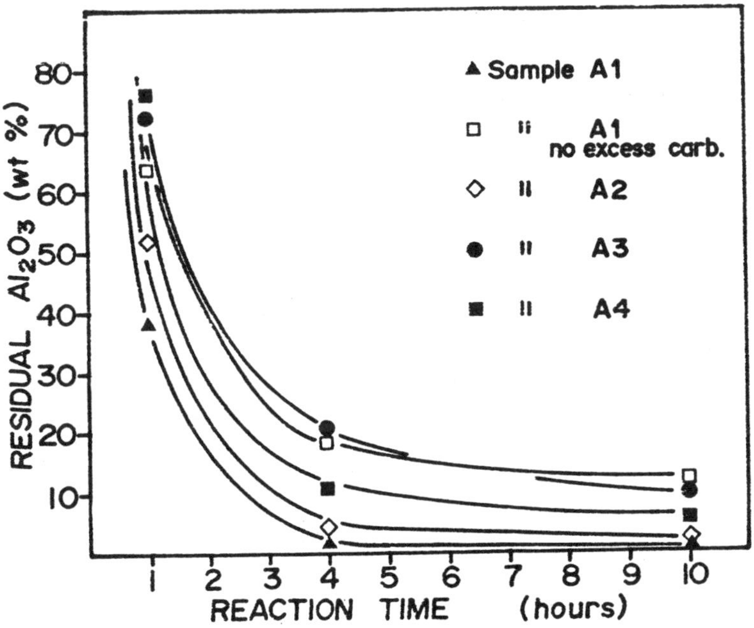

Figure 2. Advance of nitridation reaction expressed as residual alumina (wt %) vs. time at 1560°C.

The analysis of the kinetic curves and observation of SEM micrographs (Figs. 1 and 2) indicate the following steps in the carbothermal process:

Initially, nitridation of the finest single grains of the Al_2O_3 powders occurs in all tested samples. At the same time, neck formation between Al_2O_3 grains occurs, forming agglomerates as observed in Fig. 1. When the degree of nitridation of these agglomerates is high enough, breaking of the agglomerates in the neck areas takes place due to the volume change from alumina (d=3.96 g/cm^3) to aluminium nitride (d=3.26 g/cm^3).

Because of this fact, a fine AlN powder, nearly free of agglomerates is obtained.

From the results presented in Fig. 2 and Table 1 it is evident that the rate of nitridation of Al_2O_3 powders is dependent on: a) starting particle size of Al_2O_3; b) carbon stoichiometry and c) impurity content.

Aluminas with similar average grain size as A2 and A3 have different rate of nitridation (Fig. 2). The purest of them (A3) shows the lowest nitridation rate. This fact indicates that the impurity content (Table 1) plays an important role in the kinetics of the nitridation process.

Sintering

AlN powder obtained from alumina A1 after 10 hours of nitridation treatment at 1560°C was selected for sintering studies, due to the absence of Al_2O_3 (Fig. 2).

The characteristics of this powder are: 0.9 µm average particle size and 3.2 m^2/g specific surface area (BET). Oxygen content determined by neutron activation analysis (NAA) was found to be 1.7 wt. %.

Three additional AlN batches were prepared with 2 wt% of Y_2O_3, CeO_2 or CaO (as $CaCO_3$) as sintering aids. Aluminium nitride and additives were mixed by attrition milling with silicon nitride balls in isopropyl alcohol for 30 minutes. Cylindrical bars (1 cm diameter) of pure AlN and AlN with additives were isostatically pressed at 200 MPa.

Samples were sintered in a graphite furnace at 1850 °C in 1 atm of nitrogen for 2 hours.

The effect of additives on the microstructure and densification of sintered AlN samples was investigated by scanning electron microscope. SEM observations were carried out on fractured (Figs. 3a, 3b, 3c) and on polished surfaces (Fig. 3d). Oxygen content in sintered AlN specimens was determined by electron-probe microanalysis (EPMA) on polished surfaces. Table 2 shows the main characteristics of the AlN compacts obtained after sintering.

The SEM micrographs taken from the fractured surfaces of sintered compacts show uniform microstructure and high density. These data are in good agreement with the results presented in Table 2.

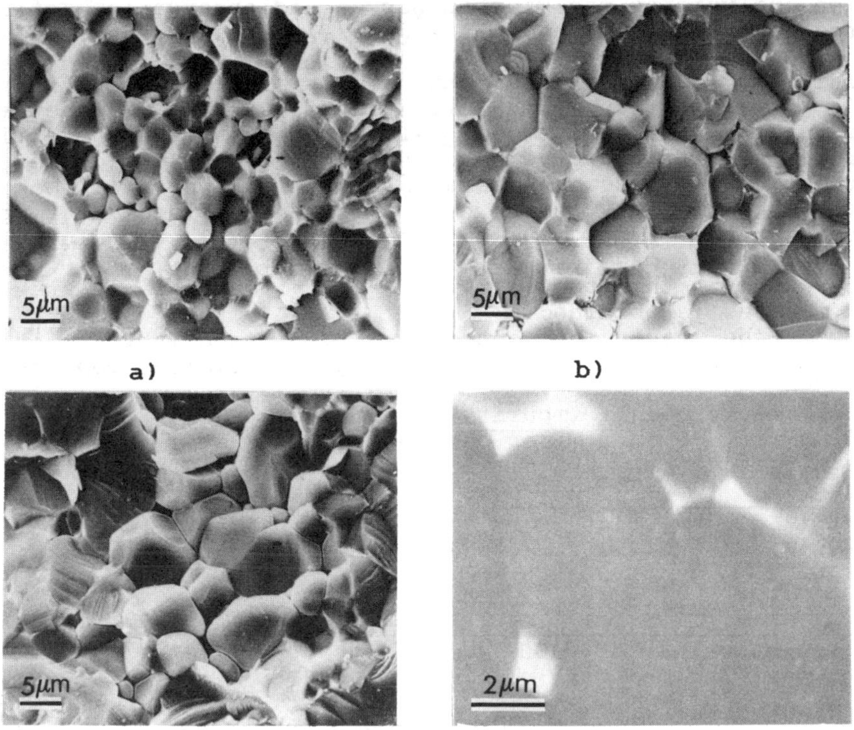

c) d)

Figure 3. SEM micrographs of fractured surfaces of sintered samples; a) AlN-CaO, b) AlN-CeO_2, c) AlN-Y_2O_3 and d) SEM micrograph on AlN-Y_2O_3 polished surface.

Figure 4. SEM micrograph of pure AlN sintered sample on fracture surface.

The high porosity observed in the case of CaO-containing specimen is probably due to the formation of a gaseous phase during the decomposition of calcium carbonate (Fig. 3a).

Almost theoretical density was obtained through a liquid phase sintering for yttria and ceria containing specimens. In both cases, glassy phase is homogeneously distributed mainly at the triple points (Figs. 3b, 3c, 3d).

Few trapped pores were detected in the additive-free AlN compact as observed in Fig. 4. This fact explains the smaller density obtained in this case (Table 2).

TABLE 2
Final density, phase analysis (XRD) and oxygen content (EPMA) of sintered AlN specimens

Sample	Density (g/cm^3)	Crystalline phases	Oxygen content (wt.%)*
AlN+2wt%CaO	3.10	AlN	1.40
AlN+2wt%CeO_2	3.29	AlN+$CeAlO_3$	1.21
AlN+2wt%Y_2O_3	3.30	AlN	1.25
pure AlN	3.19	AlN	0.94

*Average value of 400 points tested by EPMA on polished specimens. The accuracy of EPMA for oxygen was ± 100 ppm.

DISCUSSION

The results obtained in the present investigation show that it is possible to obtain a free of agglomerates micron-size AlN powder with a low oxygen content (1.7 wt%) and high sinterability by selecting appropriate starting alumina powder and following a low cost carbothermal reduction process. This powder has lower oxygen content than the one obtained from polymerized Al_2O_3 (2.65 wt%) (6) or by nitridation of alkoxide-derived oxides (3.1 wt%) (7). In addition, this powder shows excellent sinterability, reaching a high final density (Table 2).

The average values of oxygen content in sintered AlN compacts are in the same range than commercially available AlN. As previously reported in the literature (8), the oxygen content determines the thermal conductivity of AlN.

According to the results obtained in the present work, it is reasonably expected that these AlN materials have similar thermal conductivities as the AlN compacts having the same range of oxygen content (> 120 W/mK) (9).

Work is in progress to further optimize the processing conditions and properties of AlN compacts using the reported carbothermal reduction process.

REFERENCES

1. Kölker, H., Grellner, W., Aluminium nitride - a new substrate material. Ceramic Forum International /Ber.DKG. 1988, 3/4, 75-79.

2. Kuramoto, N., Taniguchi, H. and Aso, I., Development of Aluminium Nitride Ceramics. Am. Ceram. Bulletin., 1989, 68 (4), 883-887.

3. Reports., AlN - Many uses, little used?. Interceram, 1989, 2, pp. 64-65.

4. Richards, V.L., Tien, T.Y. and Pehlke, R.D., High-temperature electrical conductivity of aluminium nitride. Journal of Materials Science, 1987, 22, 3385-3390.

5. Zulfequar, M. and Kumar, A., Electrical conductivity and dielectrical behavior of hot-pressed AlN., Advanced Ceramic Materials, 1988, 3 (4), 332-336.

6. Silverman, D. L., Carbothermal Synthesis of Aluminium Nitride. Advanced Ceramic Materials, 1988, 3 (4), 418-419.

7. Mimoto, M. and Yoshioka, Y., Preparation of Si_3N_4 and AlN powders from alkoxide-derived oxides by carbothermal reduction and nitridation. Advanced Ceramic Materials, 1987, 2 (3A), 253-56.

8. Kuramoto, N., Taniguchi, H. and Aso, I., Translucent AlN Ceramic Substrate. IEEE Transactions on CHMT, Dec. 1986. 9 (4), 386-390.

9. Kurokawa, Y., Kazuaki, U. and Takamizawa, H., Development and microstructural characterization of high-thermal conductivity aluminium nitride ceramics. J. Am. Ceram. Soc., 1988, 71 (7), 588-94.

SINTERING OF Si_3N_4-TiN COMPOSITES FORMED BY SLIP CASTING*

H. MOSTAGHACI
Atlantic Research Laboratory, National Research Council
Halifax, Nova Scotia, Canada

ABSTRACT

The use of ultrafine silicon nitride powders for the pressureless sintering of parts with a wide variety of shapes has become a common practice. However, ultrafine powders generally exhibit a poor packing capability, and forming techniques such as pressing, which are based upon the application of mechanical force to the particles, are often responsible for the presence of defects in the final microstructure. In this study sintering behavior and microstructural evolution of colloidally consolidated ultrafine Si_3N_4 and Si_3N_4 composites containing up to 40 wt% TiN were investigated at temperatures up to 1850°C and pressures up to 6 MPaN_2 pressure. Alumina and yttria have been used as sintering aids.

INTRODUCTION

Indications from sintering experiments suggest that inhomogeneities introduced during powder processing step remain in the material even after a successful sintering process(1). The conventional ball milling and dry powder routes produce green compacts with poor homogeneity due to the prevalence of agglomeration in dry powders. It has been shown that inhomogeneities inhibit sintering(2), and this in turn, degrades the mechanical properties and reliability of silicon nitride-based materials. Therefore, it is desirable to find powder processing routes that produce homogeneous green compacts in order to produce materials with optimum properties. Slip casting has been reported to be a suitable method for manufacturing parts from ultrafine silicon nitride powders (3-5). However the properties of the slips are dependent on the properties of the starting powders, and more importantly, the deflocculation of the slips; the latter is particularly critical in the case of silicon nitride due to the addition of sintering aids to the starting powders.

*NRCC No. 30322

The aim of the present study was threefold; (i) rheological characterization of Si_3N_4-based compositions containing Y_2O_3 and Al_2O_3 as sintering aids and TiN as reinforcement material, (ii) slip casting of uniform green compacts from appropriate slips and (iii) gas-pressure sintering of these compacts to full density. Sintering experiments and microstructural studies were used to examine the effect of the process on the fabrication of silicon nitride materials.

EXPERIMENTAL PROCEDURE

The as-received Si_3N_4 powder (UBE, SN-E10) had oxygen and chlorine contents of 1.47 Wt% and 100 ppm respectively. The powder consists of equiaxed particles that are mainly submicron in diameter. This powder is ∿100% crystalline consisting of >95% α-Si_3N_4. The physical characteristics of the Si_3N_4 powder and the additives are given in Table I.

TABLE I - RAW MATERIALS

Powder	Oxygen Content (wt%)	Density(*) (Mgm^{-3})	Mean Particle size(+) (μm)	Specific Surface Area(#) (m^2g^{-1})
Si_3N_4(UBE, SN-E10)	1.39	3.162	0.66	10.11
TiN (ICD chemicals)	0.53	5.274	1.08	3.16
Al_2O_3 (Reynolds, RC-HP-DBM)	____	3.964	0.66	7.60
Y_2O_3 (Aldrich Chemicals)	____	4.808	2.09	8.48

(*) Measured using a He gas pycnometer (Micromeritics Multivolume Pycnometer - Model 1305)

(+) Measured using a particle size distribution analyzer (HORIBA, Capa-700)

(#) Measured using a Micromeritics Flowsorb II 2300

Suspensions with up to 60 wt% solids content consisting of Si_3N_4 powder plus 10 wt% of both Y_2O_3 (Aldrich Chemical Co.) and Al_2O_3 (Reynolds, RC-HP-DBM) and up to 40 wt% TiN (ICD Chemicals) were prepared by dispersion of the powder mixture. Viscosity as a function of pH for the suspensions was measured using a Brookfield EX-200 digital viscometer. For electrophoresis measurements, 0.04 wt% suspensions of powder in aqueous solutions of 10^{-3}M KNO_3, were used. A range of pH conditions was examined; NaOH, NH_4OH, HNO_3 and HCl were used to achieve a desired pH value. Suspensions were stirred for 2 hours by means of a magnetic stirrer prior to determining electrophoretic mobility using a Malvern Zeta Sizer II instrument. Suspesions were cast using plaster of paris molds.

Sintering was carried out in an electically heated graphite furnace capable of operating up to 2200°C in 10 MPa of inert gas pressure. Samples were packed in loose powders of the same composition as the pellet and placed in graphite crucibles. The temperature was measured using a W5Re/W26Re thermocouple installed close to the sample in the hot zone of the furnace.

RESULTS AND DISCUSSION

(a) <u>Viscosity Measurements</u> The suspensions of Si_3N_4 with or without additives, were prepared by dispersion of the solid phases in de-ionized water with vigorous stirring in a blender, followed by an ultrasonic treatment. The pH of the medium was adjusted over a wide range using NH_4OH and HNO_3 of various concentrations. The results of the viscosity measurements as a function of shear rates (Frequency) for Si_3N_4, Y_2O_3, Al_2O_3 and TiN are shown in Figure 1.

(b) <u>Mobility Measurements</u> The variation in zeta potential as a function of pH, for the conditions employed, is shown in Figure 2. Particles of Si_3N_4 and TiN exhibit similar zeta potentials over the pH range studied (pH=2-12), with an isoelectric point at pH value close to 4 for TiN and 5 for Si_3N_4. The sintering aids, Al_2O_3 and Y_2O_3, have much higher isolectric points, around pH=8 (Figure 2). These results indicate that aqueous suspensions of Si_3N_4 and TiN mixtures should be stable against flocculation in slightly alkaline media. However, if Al_2O_3 and Y_2O_3 are present the pH should be raised to a value close to 10 or above in order to achieve a sufficiently high Zeta-potential to promote repulsive electrostatic interparticle forces between all the components. Figure 3 shows the variation of viscosity as a function of shear rate (Frequency) for three Si_3N_4-based slips containing up to 40 wt% TiN and 10 wt% of

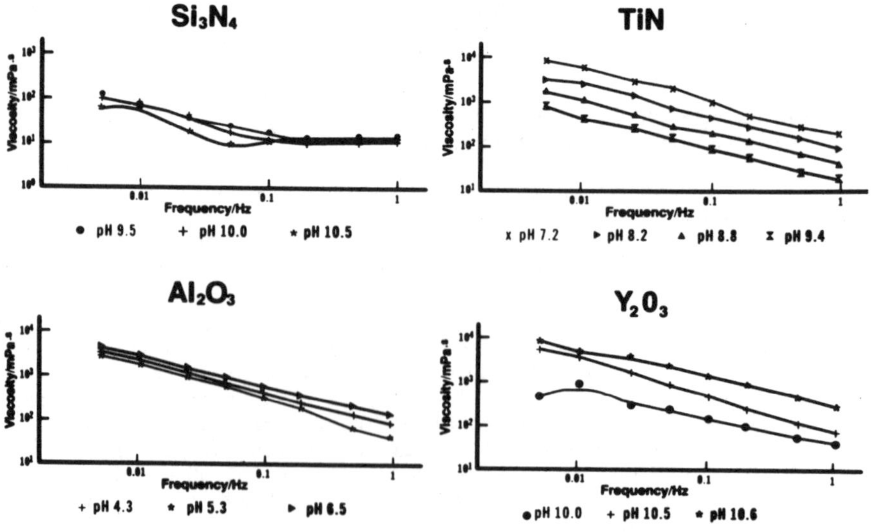

Figure 1 - Viscosity as a function of shear rate (Frequency) for the components involved in the fabrication of Si_3N_4-TiN composites.

both Y_2O_3 and Al_2O_3 as sintering aids. All the slips were prepared at pH 9 - 9.5, as was suggested from the zeta-potential results. As can be seen from Figure 3, the viscosity of a Si_3N_4 slip containing 10 wt% TiN as dispersoid was independent of shear rate and showed favourable viscosity (∿20 mPa•s) over a wide range of shear rates (Frequency). Samples in the form of a disk or ring were then cast successfully from these slips and a green density of about 55% of theoretical density was achieved.

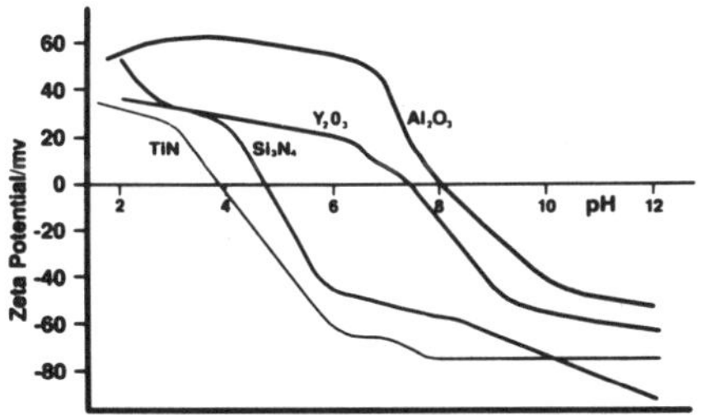

Figure 2 - The variation in Zeta potential as a function of pH.

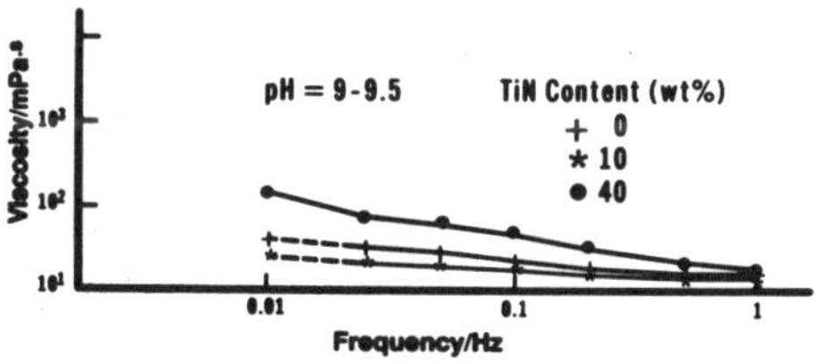

Figure 3 - Viscosity as a function of shear rate (Frequency) and TiN content for Si_3N_4/TiN composites.

(c) Sintering and Microstructure Development A two-step gas pressure sintering process was employed for sintering of the cast parts. The process described here is used primarily for its effectiveness during the final stages of the sintering as suggested by other workers(6). At this stage the pore phase transforms into discrete, disconnected pores that are filled with gas at the seal-off conditions. Experimentally, in this work, the slip cast parts from Si_3N_4-based suspensions were sintered to the closed-pore stage, by firing for 1/2 h at 1850°C under 0.2 MPa of N_2 pressure. After completion of this step and while the samples remained in the furnace, the second step consisted of increasing the N_2 pressure to typically 5-6 MPa, for rapid densification. The pressure-temperature-time cycle used during the gas-pressure sintering process is illustrated in Figure 4, and a summary of sintering results is given in Figure 5.

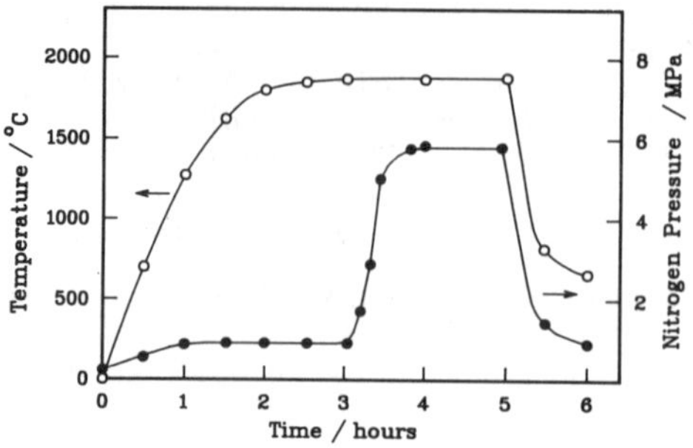

Figure 4 - A typical temperature-time-pressure profile of a two-step gas-pressure sintering cycle.

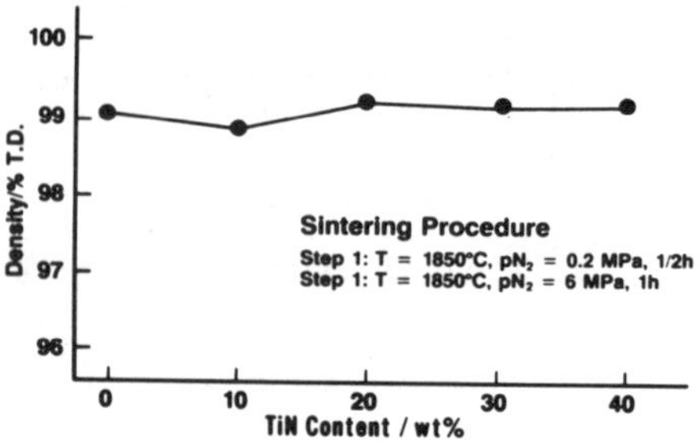

Figure 5 - Sintered densities as a function of TiN content in Si_3N_4 composites materials after a two-step sintering cycle.

As can be seen, samples of Si_3N_4 containing up to 40 wt% TiN were densified to >99% of theoretical density by the two-step gas pressure sintering. However, densification of Si_3N_4 composites during sintering was found to be independent of the TiN content of the composite. A typical microstructure of sintered Si_3N_4 containing 40 wt% TiN having a relative density of >99% is illustrated in Figure 6.

Figure 6 - SEM micrograph from the fracture surface of a Si_3N_4-based composite containing 40 wt% TiN.

The microstructure shows a uniform distribution of second phase within the Si_3N_4 matrix, which is a result of a uniform green structure obtained by the slip casting of Si_3N_4-based suspensions.

CONCLUSION

(1) Suspensions containing over 60 wt% solid content of the compositions Si_3N_4+Y_2O_3+Al_2O_3+TiN were successfully cast into parts with uniform green structures.

(2) Dense Si_3N_4-TiN composites containing up to 40 wt% TiN were fabricated to densities >99% by two-step gas-pressure sintering.

(3) Densification of Si_3N_4-TiN composites was found to be independent of the TiN contents of their compositions.

ACKNOWLEDGEMENTS

The author is grateful to Cynthia Gillis for invaluable experimental assistance, to Drs. M. Patel and S. J. Milne for Zeta-potential measurements and to Dr. S. G. Whiteway for reviewing the manuscript.

REFERENCES

1. Bowen, H.K., "Basic research needs on high temperature ceramics for energy applications", Mater. Sci. Eng., 44 (1980), 1-56.

2. Huppmann W.J. and Bauer, W. "Characterization of the degree of mixing in liquid-phase sintering experiments", Powder Metall., 18 (1975), 249-58.

3. Persson, M., Hermansson L. and Carlsson, R., "Some aspects of slip casting of silicon nitride and silicon carbide", In Ceramic Powders, ed. P. Vincenzini, Elsevier Scientific Publishing Company, Amsterdam, 1983, pp. 735-742.

4. Hartman, W.J.A.M., VanDijen, F.K., Metselaar R. and Siskens, C.A.M. "Slip casting parameters for commercial Si_4N_4 powders", in Science of Ceramics, 13, Eds.: P. Odier, F. Cabannes and B. Cales, les éditions de physique, Franace 1986, pp. 79-83.

5. Babini, G.N., Galassi, C., Lapasin, R. and Lucchini, E., "Rheological properties of aqueous suspensions of Si_3N_4-Si-Al_2O_3-ZrO_2 powders", ibid, pp. 73-78.

6. Greskovich, C.D. and Palm, J.A., "Development of high performance sintered Si_3N_4", AMMRC TR 80-46, September 1980, SRD-80-111.

MICROSTRUCTURES OF Si_3N_4 CERAMICS FORMED BY HIP

E. MARIA KNUTSON and LENA K.L. FALK
Department of Physics,
Chalmers University of Technology,
S-412 96 Gothenburg, Sweden

THOMMY EKSTRÖM
AB Sandvik Hard Materials,
Box 42056, S-126 12 Stockholm, Sweden

ABSTRACT

Analytical electron microscopy and x-ray diffractometry have been used to characterize the effect of process parameters on the microstructure of Si_3N_4-based ceramics formed by HIP using different additions of Y_2O_3, Al_2O_3 and ZrO_2. These results have been related to room temperature fracture toughness and hardness of the materials. The α- to β–Si_3N_4 transformation during densification was promoted by increased additions of sintering aids as well as a prolonged sintering time. A reduction in the α–Si_3N_4 content was accompanied by a decrease in hardness. High aspect ratio β–Si_3N_4 grains, which gave a fibrous microstructure and increased fracture toughness, were only formed in the presence of an Y_2O_3 rich liquid phase.

INTRODUCTION

Densification of pure Si_3N_4 powder compacts is inhibited by the strongly covalent character of the Si-N bond which results in an extremely low self diffusivity of the Si_3N_4. Additions of oxide sintering aids, e.g. Y_2O_3, Al_2O_3, MgO, ZrO_2, makes it however possible to obtain a dense Si_3N_4 body through liquid phase sintering. At elevated temperatures, the oxide additives react with the silica present on the Si_3N_4 powder particles and some of the Si_3N_4 to form a liquid phase which serves as a mass transport medium during densification [1]. Upon cooling from the sintering temperature, this liquid forms a residual intergranular glassy phase, and possibly also secondary crystalline phases, which often have a negative effect on the performance of the ceramic. It is therefore of considerable interest to minimize the amount of sintering aids added during processing of the Si_3N_4 material. This is made possible by the use of glass encapsulation and hot isostatic pressing (HIP) developed by ABB Cerama AB [2] which enables the formation of a dense Si_3N_4 body using only small amounts of oxide additives.

This paper presents results from a microstructural characterization of Si_3N_4 ceramics formed by HIP using small additions of different oxides. These results have been related to room temperature fracture toughness and hardness of the ceramics.

EXPERIMENTAL

Si_3N_4 ceramics with four different compositions were formed by glass encapsulation and HIP using 200 MPa at 1750 °C, see TABLE 1. The microstructures of the eight samples were characterized using analytical electron microscopy (TEM, SEM, STEM, EDX) and x-ray diffractometry. The relative amount of residual α-Si_3N_4, the Q-value, was determined from peak height ratios from the x-ray diffractograms.

Specimens for SEM were prepared by plasma etching of polished surfaces. This technique preferentially removes the Si_3N_4 and Si_2N_2O grains, resulting in a network containing the secondary phases. Thin foils for TEM/ STEM/EDX were prepared using standard techniques.

Hardness was determined by Vickers indentation using a 10 kg load, and fracture toughness was calculated according to the method of Anstis [3].

RESULTS AND DISCUSSION

Development of microstructure

The densification process is generally described by the three stages of Kingery's model for liquid phase sintering viz:particle rearrangement; solution of α- and precipitation of β–Si_3N_4; followed by coalescence [4]. During HIP, the applied pressure enhances densification, and full density may be achieved mainly by particle rearrangement and coalescence without attaining equilibrium in the solution/precipitation process.

TABLE 1
Powder compositions and phase compositions of the examined specimens.

Specimen	Oxide additives (wt%)	Phase composition	Q-value	Density % TD	Sintering time at 1750 °C
A1	2 Al_2O_3	α, β, SNO	0.08	99.8	1 h
Y1	2 Y_2O_3	α, β, SNO	0.08	99.8	1 h
AY1	2 Al_2O_3, 2 Y_2O_3	α, β, SNO	0.03	99.8	1 h
AZ1	2 Al_2O_3, 2 ZrO_2	α, β, SNO, t/c	0.03	99.8	1 h
A2	2 Al_2O_3	α, β, SNO	0.02	99.8	2 h
Y2	2 Y_2O_3	β, SNO	0	99.8	2 h
AY2	2 Al_2O_3, 2 Y_2O_3	β, SNO	0	99.8	2 h
AZ2	2 Al_2O_3, 2 ZrO_2	α, β, SNO, t/c	0.01	99.5	2 h

Legend: α = α–Si_3N_4, β = β–Si_3N_4, SNO = Si_2N_2O, t = tetragonal ZrO_2, c = cubic ZrO_2, Q = α/α+β Si_3N_4

The four different compositions all resulted in a dense Si_3N_4 body after 1 h at 1750 °C. Specimens AY1 and AZ1, which were formed with a larger addition of sintering aids, had a lower content of residual α-Si_3N_4, see TABLE 1. However, during a prolonged holding time, the presence of Y_2O_3 in the liquid phase appeared to promote the α to β tansformation. Only in material containing an addition of 2 wt% Y_2O_3 had a complete α to β-Si_3N_4 transformation taken place after 2 h at 1750 °C.

The morphology of the Si_3N_4 grains was dependent upon the type of additives. When Y_2O_3 was added (samples Y1, Y2, AY1 and AY2), a majority of the β-Si_3N_4 grains had a hexagonal cross section and a relatively high aspect ratio. These materials also contained a comparatively large volume fraction of residual glass, Figure 1. This shows that the Y_2O_3 addition resulted in the formation of a sufficient amount of a liquid phase that allowed growth of β in an isotropic liquid phase environment.

Although considerable transformation of α- to β-Si_3N_4 took place at an early stage in the sintering of samples A1, A2, AZ1 and AZ2, the Si_3N_4 was present as equiaxed grains and only thin grain boundary films of glass separated adjacent grains, Figure 2. This implies that the liquid present during densification of these compacts favoured rearrangement and coalescence rather than growth of β.

When Al_2O_3 was added to the ceramics, around 2 wt% Al, with respect to the Si and Al contents, was dissolved in the β-Si_3N_4 grains which formed during densification. These grains were thus a low substituted β SiAlON. Since these Al contents correspond to a significant proportion of the Al_2O_3 added, this reduced the volume fraction of liquid phase resulting in a lower amount of residual glass in samples A1 and A2. In addition, samples formed with both Al_2O_3 and ZrO_2, AZ1 and AZ2, contained crystalline ZrO_2. The ZrO_2 grains were present in irregular shaped pockets, Figure 2, suggesting that some of the ZrO_2 added participated in the liquid phase formation. Formation of this secondary crystalline phase would then also reduce the amount of liquid phase. The ZrO_2 appeared tetragonal, and possibly also cubic, in x-ray diffractograms which shows that the tetragonal structure was retained to room temperature by

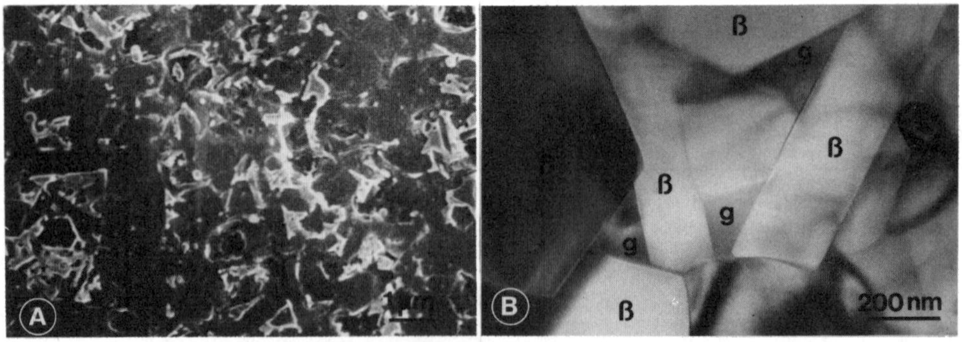

Figure 1. Microstructures of samples formed with additions of Y_2O_3. (a) Plasma etched surface of sample Y2, showing a wide distribution in grain size and the presence of a relatively large volume fraction of glass (bright areas). (b) TEM bright field image showing β-Si_3N_4 grains (β) separated by glass (g) in sample AY1.

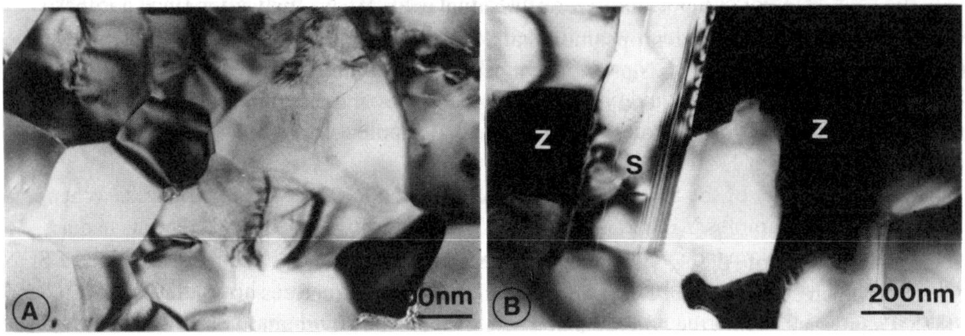

Figure 2. The microstructures of samples formed with additions of Al_2O_3 contained extremely small volume fractions of glass (TEM). (a) Equi-axed grains in sample A2. (b) Si_3N_4 and Si_2N_2O (S) grains separated by smaller grains of ZrO_2 (Z) in sample AZ2.

mechanical constraints from the surrounding matrix [5]. It is also possible that nitrogen entered the ZrO_2 structure during densification, affecting the crystal structure [6].

All samples contained approximately the same volume fraction of Si_2N_2O which was mainly present as larger, heavily faulted, elongated grains, Figure 2.

Mechanical properties

The room temperature fracture toughness showed a clear dependence upon composition and microstructure of the Si_3N_4 ceramics, Figure 3. Material formed with Y_2O_3 had a comparatively high toughness, which could be explained by the presence of high aspect ratio β-Si_3N_4 grains which may act as crack deflectors [7]. When Al_2O_3 was added separately or together with ZrO_2, the equiaxed morphology of the Si_3N_4 was reflected in the lower fracture toughness. The increase observed with the addition of ZrO_2 could be a result of transformation toughening [5].

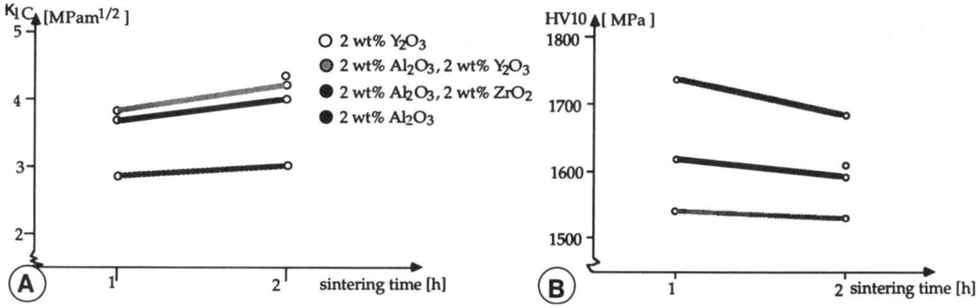

Figure 3. Room temperature mechanical properties of the examined samples. (a) Hardness. (b) Fracture toughness.

Specimens with the highest amount of residual α-Si_3N_4 also had the highest hardness. This is in agreement with the proposition that the α polymorph of Si_3N_4 has a higher hardness than the β polymorph [8]. It has previously been suggested that an increased amount of glass would lower the hardness [9], and this is also indicated by the diagram in Figure 3a.

CONCLUDING REMARKS

Different additions of Al_2O_3, Y_2O_3 and ZrO_2 as sintering aids, results in the formation of Si_3N_4 ceramics which show pronounced differences in their microstructures after HIP at 1750 °C and 200 MPa pressure. Only the additions of Y_2O_3 promoted the formation of a fibrous β-Si_3N_4 structure with a high room temperature fracture toughness. High hardness was obtained when the microstructure of the ceramic contained a higher proportion of residual α-Si_3N_4 or less intergranular glass.

ACKNOWLEDGEMENT

This project was supported by the Swedish Board for Technical Development (STU).

REFERENCES

1. Hampshire, S. and Jack, K.H., The kinetics of densification and phase transformation of nitrogen ceramics. In Special Ceramics 7, ed. D. Taylor and P. Popper, British Ceramic Research Association, Stoke-on-Trent, 1981, pp. 37-49.
2. Larker, H.T., HIP of ceramics; technology and applications today. In Proceedings of the International Conference on HIP, ed. T. Garvare, CENTEK publishers, Luleå, 1988, pp. 19-26.
3. Anstis, G.R., Chantikul, P., Lawn, B.R. and Marshall, D.B., A Critical Evaluation of Indentation Techniques for Measuring Fracture Toughness: II, Strength Method. J. Amer. Cer. Soc., 1981, **64**, 539-43.
4. Kingery, W.D., Densification during Sintering in the Presence of a Liquid Phase. J. Appl. Physics, 1959, vol. 30, **3**, 301-10.
5. Lange, F.F., Transformation toughening. Part 1-2. J. Mater. Sci., 1982, **17**, 225-239.
6. Lange, F.F., Falk, L.K.L. and Davis, B.I., Structural ceramics composites based on $Si_3N_4/ZrO_2(+Y_2O_3)$ compositions. J. Mater. Res., 1987, **2**, 66-76.
7. Ziegler, G., Heinrich, J. and Wötting G., Review, Relationships between processing, microstructure and properties of dense and reaction-bonded silicon nitride. J. Mater. Sci., 1987, **22**, 3041-86.
8. Greskovich, C. and Gazza, G.E., Hardness of dense α– and β–Si_3N_4 ceramics. J. Mater. Sci., 1985, **4**, 195-196.
9. Ingelström, N. and Ekström, T., Silicon nitride made by the encapsulated HIP technique. In Proceedings of the International Conference on Hot Isostatic Pressing, ed. T. Garvare, CENTEK publishers, Luleå, 1988, pp. 307-314.

Microstructural Investigation of Silicon Nitride with Ceria Zirconia Additions

J.M. Nel , A.W. Paterson and H.K. Schmid
Division of Materials Science and Technology, CSIR
Pretoria, South Africa

Abstract

Ceramics of nominal composition 85% Si_3N_4, 12.5% ceria-zirconia and 2.5% Al_2O_3 were fabricated by sintering under nitrogen at atmospheric pressure using a variety of sintering schedules. The microstructures of the materials produced were evaluated by a number of techniques including SEM and TEM. Phases present were identified using X-ray diffraction. It was found that the microstructural characteristics varied as a function of the distance from the specimen surface. Near surface effects include the formation of ZrN agglomerates. In the bulk the zirconia grains are not reduced to the nitride but the ceria is expelled from the zirconia lattice resulting in the transformation of the zirconia to the monoclinic phase. The ceria is redistributed in the liquid phase composed of silicon oxynitrides which assist densification of the material. The relevance of these findings to other zirconia-silicon nitride systems and the densification of sintered silicon nitride materials is discussed.

Introduction

There have been many investigations of the Si_3N_4-ZrO_2 ceramic systems using Y_2O_3-ZrO_2[1,2,3]. Lange et al[4] found that 5-30 mol% ZrO_2 in the Si_3N_4 system resulted in an increase in the fracture toughness. A significant amount of Zr-oxynitride, which oxidized to monoclinic ZrO_2 at <800°C, was identified by XRD. In a later study by Lange et al[3] the main phases identified were β-Si_3N_4, tetragonal and monoclinic ZrO_2 and Si_2ON_2. The presence of Si_2ON_2 is regarded as an indication of the presence of Zr-oxynitride from the following equation:

$$xSi_3N_4 + ZrO_2 \rightarrow ZrO_{2-3x/2}N_x + 3x/2Si_2ON_2$$

After a detailed microstructural investigation it was found that the Y_2O_3 content of the Zr grains after sintering was considerably less than in the original powder. This excess Y_2O_3 was considered to be redistributed in the liquid phases produced during sintering. A critical amount of Y_2O_3 was found to be necessary to preclude the formation of Si_2ON_2 and thus Zr-oxynitride. This amount is expected to be dependant on the composition of the Si_3N_4/ZrO_2 composite, the amount and type of densification aid and the sintering conditions.

Experimental Procedures

The starting powders used were Toya Soda TSK Si_3N_4 TS7 with 12.5 wt% CeO_2-ZrO_2 powder, (DMST), and 2.5 wt% Al_2O_3. The powders were attrition milled in isopropanol to a median particle size (d_{50}) of <0.5 μm. The dry powders were uniaxially compacted at 5 MPa and then cold isostatically pressed at 108 MPa. Sintering was carried out at 1700 °C and 1750 °C in a controlled atmosphere furnace with nitrogen.

Microstructural studies were carried out using the scanning and transmission electron microscopes. Polished, fracture, and etched surfaces were studied in the SEM operating at 25kV after gold coating. The etching was undertaken in molten KOH for a few minutes. Thin films for transmission electron microscopy (TEM) investigations were prepared by mechanical grinding and ion beam thinning. Quanititative micro-analysis with high spacial resolution was carried out by means of analytical TEM (STEM) using a scanning transmission electron microscope operating at 120 kV. X-ray diffraction peak data was obtained using Cu-Kα radiation.

Results and Discussion

The sintered surface and sections of sintered compacts were analysed using x-ray diffraction for the phases present. Table 1 gives the comparison of the phases present in the materials sintered at 1700 °C, 1750 °C and those obtained by Lange et al[3].

Table 1: Comparison of phases found in this study and by Lange[3]

GROUP	STUDY	SURFACE	INTERIOR
DMST	1700°C	β^a, m^b, t^c, ZrN	β, m, t
DMST	1750°C	β, m, t, ZrN, AlN	β,m, t, ZrN, AlN
Lange et al	1700°C	β,m,Si_2ON_2, t, c^d	-

a - β-Si_3N_4 c- tetragonal ZrO_2
b - monoclinic ZrO_2 d - cubic ZrO_2

No trace of Si_2ON_2 or Zr-Oxynitride were found in any of the materials studied. At 1700°C some of the ZrO_2 appears to have reduced completely to ZrN on the surface. At 1750°C significant amounts of AlN were also formed. The ZrN appears to be mainly a surface effect at lower temperatures and the peak intensities decrease and often are not present in the interior of the material.

The SEM backscattering imaging mode was used to study the zirconium distribution on the polished surfaces. Zr-rich grains in the bulk material were uniformly distributed, while in the near surface region there were distinct agglomerates of larger zirconium rich grains. Figure 1 shows very clearly the distinct difference between (a) the bulk material and (b) the near surface material. From this it would appear that the ZrO_2 is more soluble in the near surface region and

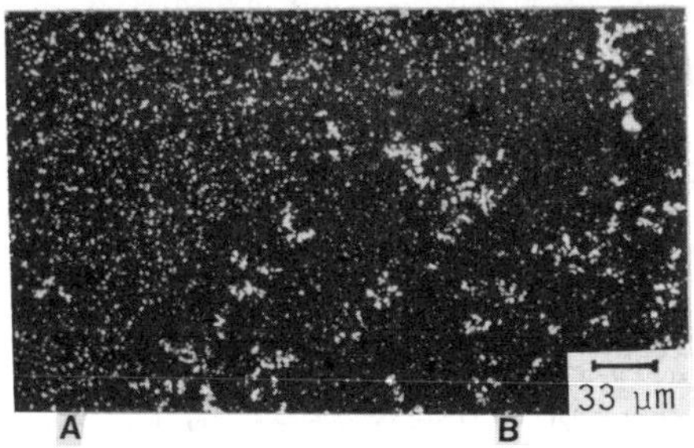

Figure 1: SEM micrograph of the bulk material (a) and the near surface material (b) taken in backscatter mode.

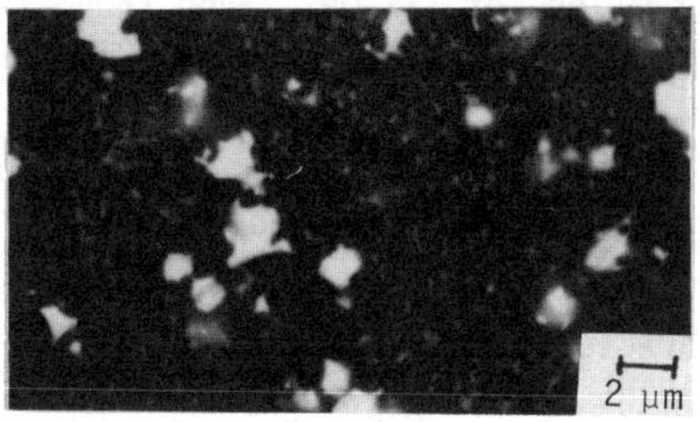

Figure 2: SEM micrograph showing the Zr rich agglomerates and smaller light regions which could be Ce-rich glassy phases.

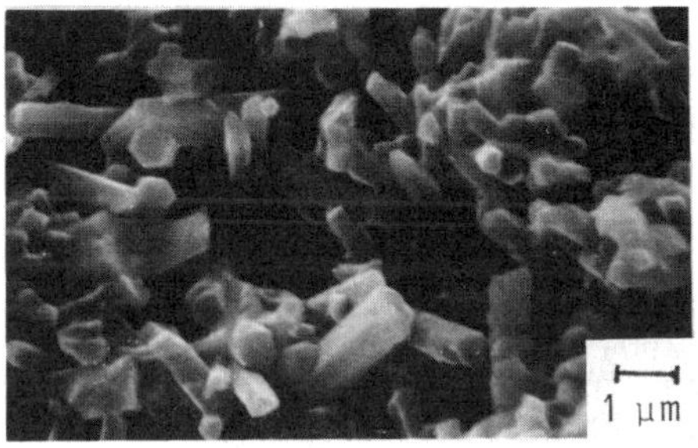

Figure 3: A SEM fracture surface micrograph with interlocking grains.

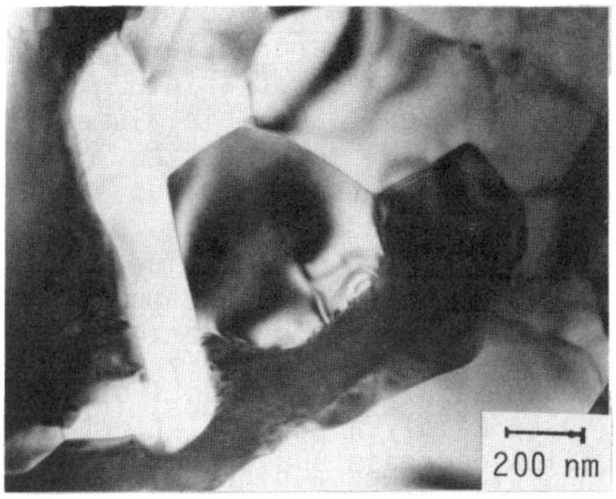

Figure 4: TEM bright field micrograph showing the grian morphology of Zone II.

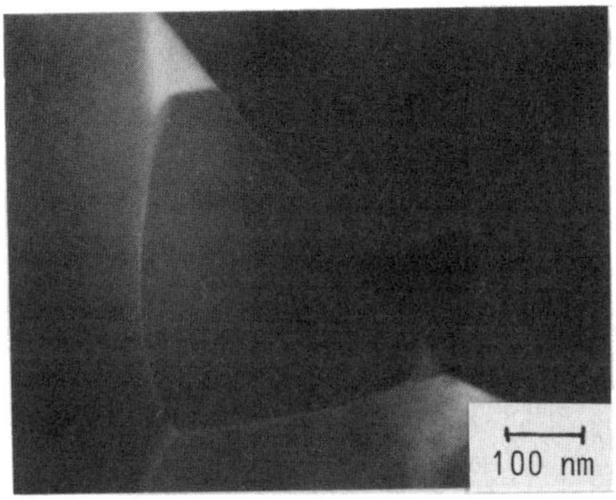

Figure 5: Zone II, intergranular glassy phases in pockets at triple grain junctions and along grain boundaries in dark phase.

therefore the solution-reprecipitation here is higher and Ostwald ripening is enhanced. There also appears to be a region where ZrO_2 depletion occurs. Figure 2 shows the same area at a higher magnification. Light regions between the Zr agglomerates are observed. It is thought that these areas could be the Ce-rich glassy phases which are observed in TEM studies.

When examining the etched surfaces in the secondary electron mode, the characteristic elongated, acicular grains of silicon nitride were evident. It was also noted that the zirconium content decreased considerably. This would imply the association of the zirconia with the glassy phases as they are removed together on etching. The β-silicon nitride grains varied between 1 and 3 μm in length.The grain morphology is seen more clearly on the fracture surfaces. The hexagonal, interlocking grains, Figure 3, are characteristic of the high strength β-Si_3N_4 materials.

Preliminary TEM investigation of the material indicates that it consists of three different regions, termed Zones I, II and III. It is thought that these three zones correspond to the bulk material, a Si_3N_4 rich ZrO_2 depleted area and the near surface region respectively. Zone I is composed essentially of pure β-Si_3N_4 and ZrO_2 grains with approximately 5wt% intergranular glassy phase. A micrograph of this region is shown in Figure 4. The β-Si_3N_4 grains are easily recognized by the characteristic hexagonal cross-sections. Some of the pure ZrO_2 grains showed characteristic twinning, which indicates that the grains have transformed from tetragonal to monoclinic symmetry. This is confirmed by the x-ray diffraction results. There were concentrations of Si in the ZrO_2 grains in the order of several wt%, while the Ce content is very low and in some cases below the detection limit of ~0.1 wt%.

The three phases noted above were also observed Zone II, however the fraction of glassy phase observed in this region was considerably higher, in the order of ~15 wt%. The pockets of intergranular glassy phase varied widely in size ranging from 5 nm pockets at triple grain junctions (TJ), as shown in Figure 5, up to areas 300 nm in diameter which are comparable with the average Si_3N_4 grain size. The analysis of the glassy phases situated in pockets of various sizes indicated an increase of Si concentration and decrease of Ce concentration with increasing size of pockets. TEM dark field images revealed the existence of a continuous intergranular phase completely wetting the Si_3N_4 grains (Figure 5). These grain boundaries are much thinner than those reported in the Si_3N_4-ZrO_2-Y_2O_3 system[5]. These films are assumed to be dependant on composition[2] and the structure of the grains next to them[6]. Beside pure Si_3N_4 and ZrO_2 grains, many grains in Zone II showed a high density of finely dispersed precipitations. Spectra acquired from a large sample area again indicated high Si concentrations well above the nominal value, whereas the Zr concentration showed a minimum, marginally above the detection limit. An EELS analysis was carried out on a pure Si_3N_4 grain and a large glassy body. The atom ratio N:O was roughly 8:1 in the Si_3N_4,indicating some oxidation of the grain surfaces (ZrO_2 content was 0.3wt%) In the glassy phase an atom ratio N:O of 1.2:1 was observed. This atom ratio would be consistent with a $CeO_2.Si_2ON_2$ state of the glassy phase if a ratio of cations Si:Ce = 1:1 is assumed. If the actual Si:Ce ratio in the glass phase is lower, the observed N:O ratio could be made up by assuming the presence of some SiO_2 in the glassy phase.

Zone III appears to be a conglomerate of ZrO_2 and Si_3N_4 grains mingled with a large fraction of obscure glassy phase interspersed with precipitations of various crystalline phases. The spectrum acquired from a large sample area indicates a rather low Si concentration in this zone well below the nominal composition, whereas Zr seems to be enriched by a factor of three over the nominal value. It is assumed that these regions are associated with the solution and reprecipitation of zirconia grains and the formation of glassy phases due to the reduction of the cerium in the zirconia sublattice.

Conclusion

When considering the x-ray results and the microscopic findings it can be concluded that the surface of the material is rich in ZrN. It would seem that the ionic size of the dopant in the ZrO_2 is too large to be accommodated in the lattice when introduced to the Si_3N_4 system. The CeIV might be reducing to CeIII and be too large to be accommodated in the ZrO_2 lattice. Reaction with the Si_3N_4 and SiO_2 surface films to form a cerium-silicate oxynitride glassy phase results. It then appears that the ZrO_2, which is no longer stabilized, is also reduced in the presence of the nitrogen glasses and atmosphere. In previous studies[3] the ZrO_2 was reduced to the oxynitride. However in this study the reduction seems to have been more complete and the pure ZrN is retained on cooling to room temperature. From the TEM analyses it would also seem that there is effective transfer of Zr ions in oxynitride liquids resulting in coarsening of the near surface zirconium nitride grains.

The association between the ZrO_2 and the glassy phases which is observed on etching and the presence of the ZrN may contribute to the low toughness values that were obtained for this material. The mechanism for this assumption is not yet known.

References

1. P. Vincenzini, A. Bellosi, and G.N. Babini, Thermal Instability of Si_3N_4/ZrO_2 Composites, Ceram. Int., 1986, **12** 133-145.

2. W.A. Sanders and D.M. Mieskowski, Strength and Microstructure of Si_3N_4 Sintered with ZrO_2 Additions, Adv. Ceram. Mat, 1986, 1 (2), 166-173.

3. F.F. Lange, L.K.L. Falk and B.I. Davis, Structural ceramics based on Si_3N_4-$ZrO_2(+Y_2O_3)$ compositions, J. Mat. Res., 1987, **2** (1), 66-76.

4. F.F. Lange, Compressive Surface Stresses Developed in Ceramics by an Oxidation-Induced Phase Change, J. Am. Ceram. Soc., 1980, **63** (1-2), 38-40.

5. D.R. Clarke, On the equilibrium thickness of intergranular glass phases in ceramic materials, J. Am. Ceram. Soc., 1987, **70** (1), 15-22.

DENSIFICATION MECHANISM OF Si_3N_4 WITH AZ-TYPE NON-TOXIC NON-OXIDE ADDITIVES *

GE CHANG-CHUN, XIA YUAN-LOU AND CHEN LI-MIN
Division of Powder Metallurgy
University of Science and Technology Beijing
Beijing 100083, China

ABSTRACT

Si_3N_4 ceramics with AZ-type non-toxic, non-oxide additives can be highly densified. In this work densification behavior and mechanism of Si_3N_4 with AZ-type additives were investigated.

INTRODUCTION

We have found that non-toxic, non-oxide additives AlN, ZrN or AlN-ZrN (abbreviated as AZ-type additives) can be used as sintering aid instead of generally used oxide additives and make Si_3N_4 highly densified during hot-pressing (HP)(1). Some of them show enhanced high temperature mechanical properties and superior oxidation resistance. The aim of this work is to investigate the densification behavior and mechanism of Si_3N_4 ceramics with AZ-type additives.

MATERIALS AND METHODS

Commercial Si_3N_4 powders (0%-2.25%, BET-Surface area Sg – 5.8 m^2/g) and self-made AlN powders (0%-2.00%, Sg – 6.75 m^2/g), ZrN powders (0%-2.90%, Sg – 4.1 m^2/g) were used as raw materials. Phase analysis of Si_3N_4 powders shows α-93.4%, β-5.9%, free Si – 0.7%. MgO and $MgO \cdot Al_2O_3$ powders of reagent-grade were used for making Si_3N_4 with typical oxide additives as reference samples. Si_3N_4 powders with different additive were mixed and ball-milled with ethanol as medium, vacuum dried, sieved through-200 mesh, and then hot-pressed in a small hot-press under N_2 atmosphere with 10 MPa at different temperatures. The shrinkage was measured with a transducer installed on the press. Porosity measurement and XRD analysis were made on the

* The Project Supported by National Natural Science Foundation of China and National Education Committee of China

samples. The symbols for various specimens are expressed as follows: AN-Si_3N_4 + AlN, Zn – Si_3N_4 + ZrN, ZAN – Si_3N_4 + ZrN + AlN, MAO – Si_3N_4 + $MgO \cdot Al_2O_3$, NC-132 – Si_3N_4 + MgO.

RESULTS

1. Densification Behavior

Various compositions in AlN-ZrN system are selected as non-oxide additives for HP of Si_3N_4. Dependences of the relative density on the HP temperature and the atom ratio of Al to Zr are shown in Fig.1,2. The highest density was obtained after HP at 1800°C. Densification of Si_3N_4 with additive composed of AlN + ZrN is obviously higher than single component additive AlN or ZrN, no matter whether 1750°C or 1800°C was taken as HP temperature.

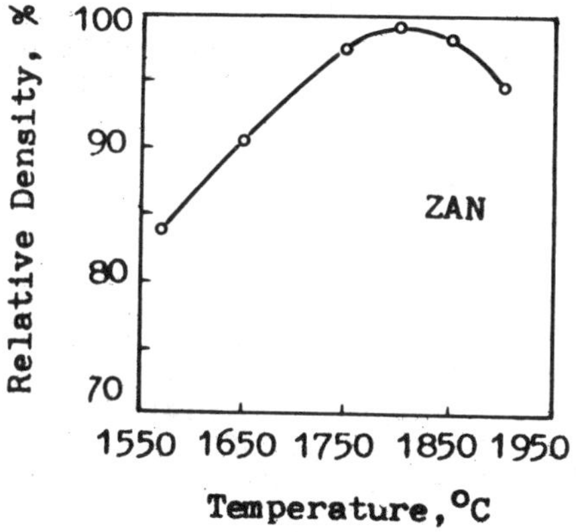

Fig.1 Dependence of relative density on HP temperature

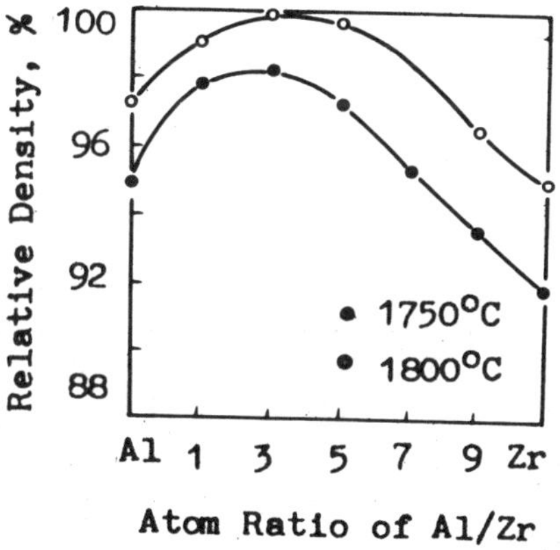

Fig.2 Dependence of relative density on atom ratio of Al to Zr

2. Shrinkage Kinetics in Hot Densification

As shown in Fig. 3, $K_1 = 3.33 \times 10^{-5}$ m/s, $K_2 = 1.86 \times 10^{-5}$ m/s, $K_3 \doteq 1.28 \times 10^{-5}$ m/s, $K_1' = 4.2 \times 10^{-7}$ m/s, $K_2' = 3.3 \times 10^{-6}$ m/s, $K_3' = 6.63 \times 10^{-6}$ m/s.

K_1, K_2, K_3 —— maximum shrinkage rate for ZAN, AN and ZN;

K_1', K_2', K_3' —— shrinkage rate for ZAN, AN and ZN at 1800°C

So $K_1:K_2:K_3 = 2.6:1.5:1$ and $K_1':K_2':K_3' = 1:7.9:15$

ZAN sample showed the highest maximum shrinkage rate. As HP temperature reached 1800°C, ZAN was nearly full densified and hence showed lowest shrinkage rate at 1800°C.

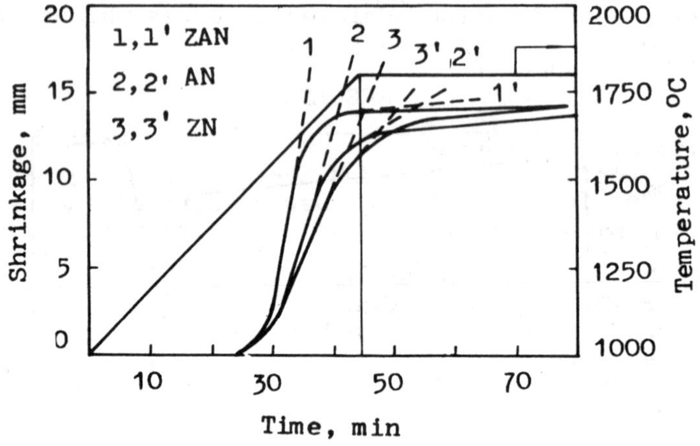

Fig.3 Shrinkage kinetics of ZA, AN and ZAN

Fig. 4. showed the relation between relative shrinkage and time at HP temperature in logarithm scale for ZAN, AN and ZN, compared with those of MAO and NC-132.

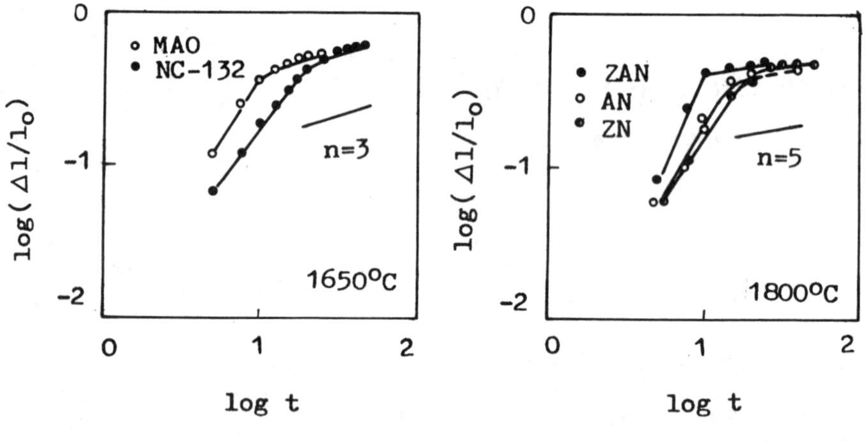

Fig.4 Log. shrinkage V. log. time (minutes) plots

3. $\alpha \rightarrow \beta$ Transformation During Densification

As shown in Fig.5, only a small amount of α transformed to β for samples hot-pressed at 1570°C. As HP temperature increased to 1650°C, $\alpha/\beta \approx 1$. $\alpha \rightarrow \beta$ transformation was nearly complete when HP temperature increased to 1750°C, though the densification process was still going on. Maximum relative density was obtained only when the HP temperature increased to 1800°C. The fact that α—β transformation had completed before maximum densification achieved is a resemblance to the result for Si_3N_4 with Y_2O_3 as additive, reported by S. Hampshire and K. H. Jack[2].

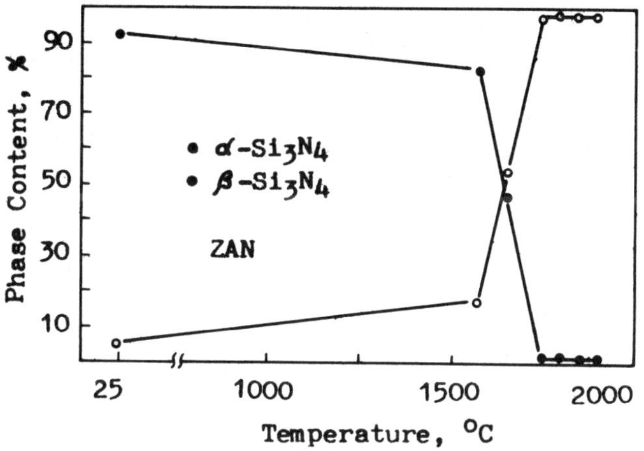

Fig.5 Dependence of relative contents of α and β on HP temperature

4. Phase Analysis of AN, ZAN and ZN after Cooling

XRD analysis was done on samples AN, ZAN and ZN, HP at 1750°C. For AN, only β'- Si_3N_4 appeared; for ZAN, apart from β'- Si_3N_4 ZrN was found. While β - Si_3N_4, ZrN, ZrO_2, Si_2ON_2 and a small amount of α- Si_3N_4 was detected in the XRD pattern of ZN.

DISCUSSION

1. Densification Achieved through Liquid Phase Sintering(LPS)

As above-mentioned, α-β transformation dramatically proceeded and completed at 1750°C, while high densification achieved just at the range of 1750—1800°C. This fact can be only explained with LPS mechanism. The AZ-type non-oxide additive with a small amount of oxides (Al_2O_3, ZrO_2) reacted with surface silica on the Si_3N_4 powders to give an oxynitride liquid that promoted shrinkage. α- Si_3N_4 dissolved in LP and reprecipitated in the form of β- Si_3N_4. The LP dramatically accelerated the transport rate of atoms, that made α→β transformation rapidly completed and directly promoted the densification process of Si_3N_4.

Phase analysis of AN can be realized according to the pseudo-ternary diagram Si_3N_4-AlN-Al_2O_3-SiO_2 proposed by I. K.

Naik, J. Gauckler and T. Y. Tien(3). At temperatures above 1700°C LP dissolved more AlN, $\alpha \rightarrow \beta$ transformation rapidly completed through the Al-rich liquid. Finally, only β' existed in AN after cooling. Phase analysis of ZN can be understood according to the behavior diagram of SiO_2-Al_2O_3-ZrO_2-Si_3N_4 at 1700°C(4), in which 3 phases are in equilibrium in the compatible angle Si_2ON_2-Si_3N_4-ZrO_2. All these phases were found in the XRD pattern of ZN.

More effective densification of ZAN than AN or ZN can be interpreted with the behavior diagram of Si_3N_4-SiO_2-AlN-Al_2O_3-ZrN-ZrO_2 system (5). ZrO_2 in the system can lower the solidus temperature and expand the composition range of liquid phase field (6). Our earlier work shows(7), ZAN can be highly densified even in gas-pressure sintering, while ZN or AN can only be densified with HP.

2. Application of Kingery's LPS Model

The kinetics of densification can be interpreted by kingery's model of LPS. There are 3 successive stages during LPS: (1) particle rearrangement; (2) solution-diffusion-precipitation; and (3) coalescence. In the (2) stage, $\frac{\Delta V}{V_0} \propto t^{\frac{1}{n}}$ ($\frac{\Delta V}{V}$ - volume shrinkage, t-time), where n=3 if the rate-controlling process is solution into or precipitation from the liquid and n=5 if diffusion through the liquid is rate-controlling.

For MAO and NC-132, rapid densification occurred through particle rearrangement at 1650°C and n=3 in the (2) stage as shown in Fig. 4. It suggests that solution or precipitation is rate-controlling. For Si_3N_4 with AZ-type additives, n=5 in the (2) stage of LPS. The difference is due to the smaller liquid volume and the higher liquid viscosity with AZ-type additive compared to MgO or $MgO \cdot Al_2O_3$. For ZAN, ZN or AN, diffusion through the more viscous liquid and smaller liquid volume is rate-controlling.

REFERENCES

1. Ge Chang-Chun, Xia Yuan-Luo and Chen Li-Min, Paper presented on 1988 Int. Conf. of P/M, Orlando.
2. S. Hampshire and K. H. Jack, Special ceramics 7, P.37.
3. I. K. Naik, L. J. Gauckler, and T. Y. Tien, J. Am. Ceram. Soc. 61 (1978) 332.
4. A. W. J. M. Rae, D. P. Thompson, and K. H. Jack, Ceramics for High Performance Applications, P.1039.
5. J. Weiss, L. J. Gauckler, H. L. Lukas, and G. Petzow, T. Y. Tien, Paper Submitted to J. Mat. Sci.
6. Y. Inomata, Y. Hasegewa, Y. Matsuyama, and Y. Yajima, Yogyo Kyokai Shi 84, 600.
7. Ge Chang-Chun, Xia Yuan-Luo and Chen Li-Min, Paper Presented on 1986 Int. Conf. of P/M, Düsseldorf.

PROCESSING AND SINTERING BEHAVIOUR OF 3Y-TZP PREPARED BY A SPRAY PYROLYSIS METHOD

B. DUBOIS, F. CABANNES, D. RUFFIER and P. ODIER
CNRS-CRPHT 1D avenue de la recherche scientifique
45071 ORLEANS, FRANCE.

ABSTRACT

3Y-TZP monodisperse particles in the range 0.2 µm to 1.8 µm with a narrow size distribution (standard deviation 30%) have been prepared by a spray pyrolysis method. An appropriate thermal treatment of the mist of an aqueous solution of mixed zirconyl chloride and yttrium nitrate produced by an ultrasonic nebulizer, allows to obtain slightly micro-porous particles with a well controlled size.
The dynamic sintering behaviour of these powders studied with a high temperature dilatometer shows that the HERRING scaling law is appropriate for predicting the potential benefit from particle size reduction. The sintering temperature can be decreased from 1750°C down to 1250°C by reducing the particle size from 1.8 µm to 0.2 µm. The sintering of the different powder compacts results in very uniform fine grained microstructures with a density >99% of theoretical density and no exaggerated grain growth. The analysis of shrinkage rate data emphasizes the fractal character (and its benefits) of the powder prepared by this route.

INTRODUCTION

Recently, RING (1) has rewieved the evolution of ideas that lead to present-day thinking about desired ceramic green bodies in terms of constituent particle size distribution and particle packing. Random packing is preferable. It can be obtained with a particle size distribution having a standard deviation larger than 10%. In order to get fast sintering the standard deviation should be less than 35%. In order to get low sintering temperature, the particle should have a fractal geometry.
One of the most promising preparation methods of such powders with respect to sol-gel processing is the spray pyrolysis method. This continuous process involves the atomization of the starting solution. The mist is dried and then decomposed in a hot furnace. Each droplet acts as a "container" and reactions can be conducted within the droplets to yield spherical particles.
The sintering behaviour of these powders has been studied with a high temperature dilatometer. We comment on the correlation between the temperature of the maximun shrinkage rate and the most probable pore radius.

RESULTS AND DISCUSSION

3Y-TZP particles in the range 0.2 µm to 1.8 µm have been prepared by the spray pyrolysis of aqueous solutions of zirconyl chloride and yttrium nitrate using an ultrasonic mist generator (2). This process yielded spherical particles with a narrow size distribution (Fig.1). The slightly porous spherical particles consisted of cristallites = 12 nm in diameter after calcination at 700°C. These particles have a fractal geometry which fractal dimension, D, is defined by :

$$D = 3.0 + \ln P / \ln S$$

where P is the packing fraction of each aggregate and S is the generation to generation size ratio. For particles of Fig.1, P = 0.70 and S = 67, therefore the fractal dimension of particles is 2.91.

A relative green density of = 55% is achieved with all powders (by dry isostatic pressing at 480MPa). During compaction, microspheres are not broken like the ones obtained by emulsion type sol-gel synthesis (3), (Fig.1). In all compacts the average initial particle radius is ≃ 6 times larger than the average initial interagglomerate pore radius.

The sintering behaviour of these powders has been studied with a high temperature dilatometer (SETARAM) at a constant heating rate a=250°C/h. The results presented on Fig.2 show that the HERRING scaling law (4) is appropriate for predicting the potential benefit from particle-size reduction, as already experimentally shown by RHODES (5). The sintering of different compacts results in very homogeneous microstructures (Fig.3). The final density in all samples is > 99% of theoretical density. After sintering, the grain size is close to the initial particle size.
As the particle have a fractal geometry, the sintering of such particles occurs in two stages : a first one corresponding to the intraagglomerate sintering and a second to the interagglomerate sintering. Both different stages of sintering development are clearly observed on shrinkage and shrinkage rate vs. temperature curves. Initial, intermediate and final sintering steps occur in each stage. In order to study the different steps of microstructural development during the first stage, pore size distributions were determined by Hg porosimetry in powders calcined from 700°C to 1000°C for 2h before compaction (480 MPa), (Fig.4). The intraagglomerate pore size increases with T as a result of grain growth. A part of the pore volume within the agglomerates is transfered towards the surface.
In order to determine the activation energies of the different transport mechanisms occuring in the intermediate sintering step, we used an adaptation of the COBLE's model for the non-isothermal sintering developed by GENUIST and al. (6). In that step, the system is modelized by uniform grains having faces in common and the pore phase is approximated by an interconnected continuous cylinder network along the edges of the grains assimiled to the KELVIN's tetrakaidecahedron.
The plots obtained with bulk or grain boundary diffusion relation, presented a similar shape and the magnitude of the calculated activation energies correspond preferably to a bulk diffusion mechanism (7). Therefore, we only consider bulk diffusion in the following :

$$\mathrm{Log}\left(T.G^3.\frac{1}{(R+1)^4}.\frac{dR}{dT}\right) = -\frac{Q}{RT} + C^{te}$$

where, T is the absolute temperature, G is the grain size, R is the shrinkage, dR/dT is the shrinkage rate and Q is the activation energy. The average activation energy is found near of 450 kJ/mol. This value certainly corresponds to the Y^{3+} lattice diffusion which is the slowest diffusing species.

By considering the work of ROOSEN and al. (8,9), a tight correlation appears between the temperature of the maximun shrinkage rate and the most probable pore radius in nearly monodispersed systems. Our data clearly show such tendency (Fig.2). It says that a pore of radius r is associated with a temperature characterizing its elimination. We assume this temperature to be the temperature of maximun shrinkage rate in a monodispersed system. The rate of change of pore volume for a cylindrical pore, dV/dT is for bulk diffusion :

$$\frac{dV}{dT} = \frac{56\,\pi\,\Omega\,\gamma\,D_0}{a\,k\,T}\exp\left(-\frac{Q}{RT}\right)$$

where, V is the volume of cylindrical pore, a is the constant heating rate, Ω is the molecular volume, γ is the surface energy, D_0 is the bulk diffusion coefficient, Q is the activation energy of the operating mechanism, T is the absolute temperature, k is the BOLTZMANN's constant. The volume of a pore is $\pi . r^2 . l$ with $l = (0.3561/6)*r$ (using the approximation of COBLE). We have :

$$\int_{r_0}^{0} r^2 dr = \int_{0}^{Te} \frac{K}{T} \exp\left(-\frac{Q}{RT}\right) dT \qquad K = C^{te}$$

where r_0 is the initial pore radius and Te its elimination temperature. By integration (using the approximation of WOOLFREY (10) :

$$\frac{Q}{R} \cdot \frac{1}{Te} - \text{Log } Te = -2 \text{ Log } r_0 + K' \qquad K' = C^{te}$$

Between 1200°C and 1600°C, Log(Te) can be considered constant compared with the variation of Q/RTe. Therefore the equation can be approximated by :

$$\frac{1}{Te} = -3 \frac{R}{Q} \text{ Log } r_0 + K'' \qquad K'' = C^{te}$$

which agrees very well with our experimental observations. Such a curve fits very well our own experimental results (fig.5) if one assimilates the temperature Te to the temperature of the maximun shrinkage rates and plots Te versus the logarithm of the average initial pore radius r_0. The activation energy determined with the slope is close to 340 kJ/mol. The magnitude of this value is lower than the previous one. One reason of this difference is perhaps due to increasing of pore size during the sintering process which is not taking into account in this model.

CONCLUSION

3Y-TZP were prepared by a mist decomposition method using an ultrasonic atomization technique. The sintering behaviour is governed by the fractal geometry of powders desirable to get low sintering temperature. The HERRING scaling law is appropriate for predicting the potential benefit in sintering from particle size reduction. We found a correlation between the temperature of maximun shrinkage rate and the pore radius in a monodispersed system.

ACKNOWLEDGMENTS

This work was supported by RHONE-POULENC and CNRS.

REFERENCES

1. T.A. RING, "Continuous Production of Narrow Size Distribution Sol-Gel Ceramic Powders", M.R.S. Bull. oct.-nov. 34-8 (1987).
2. B. DUBOIS, D. RUFFIER and P. ODIER, "Preparation of Fine, Sperical Y_2O_3-Stabilized ZrO_2 by the Spray Pyrolysis Method", J. Am. Ceram. Soc. 72 [4] 713-15 (1989).
3. M.A.C.G. van de GRAAF, J.H.H. ter MAAT and A.J. BURGGRAAF, "Microstructural Development During Pressing and Sintering of Ultrafine Zirconia Powders", 'Ceramics Powders' edited by P. VINCENCINI (Elsevier Scientific Publishing Cie, Amsterdam, 1983) pp.783-794.
4. C. HERRING, "Effect of Change of Scale on Sintering Phenomena", J. Appl. Phys. **21** [4] 301-303 (1950).
5. W.H. RHODES, "Agglomerate and Particle Size Effects on Sintering Yttria-Stabilized Zirconia", J. Am. Ceram. Soc. **64** [1] 19-22 (1981).
6. C. GENUIST and J.M. HAUSSONNE, "Sintering and Control of Microstructure : Theoretical Models and Definition of the Sintering Schedule", Sciences of Ceramics 14, 279-284 (1988)
7. M.A.C.G. van de GRAAF, J.H.H. ter MAAT, A.J. BURGGRAAF, "Microstructure and Sintering Kinetics of Highly Reactive ZrO_2-Y_2O_3 Ceramics", J. Mater. Sci. **20** 1407-1418 (1985).

8. A. ROOSEN and H. HAUSNER, "The Influence of Processing Conditions on the Sintering Behaviour of Coprecipitated Calcia-Stabilized Zirconia Powders", 'Ceramics Powders' edited by P. VINCENCINI (Elsevier Scientific Publishing Cie, Amsterdam, 1983) pp.773-782.
9. F. DOGAN, A; ROOSEN and H. HAUSNER, "Agglomerate Formation, Compact Characteristics and Sintering Behaviour of Ceramic Powders", J. Phys. 47 C1 231-35 (1986).
10. J.L. WOOLFREY and M.J. BANNISTER, "Nonisothermal Techniques for Studying Initial-Stage Sintering", J. Am. Ceram. Soc. 55 [8] 390-94 (1972).

Fig.1 SEM fractured surface of a compact (mean particle size of 0.8 µm). bar = 5 µm.

Fig.3 Microstructure by SEM, a sintered sample with a 0.3 µm powder. bar = 1 µm.

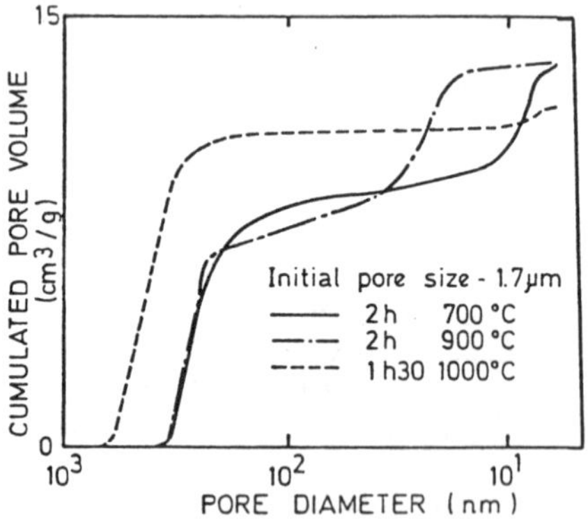

Fig.4 Pore size distribution in powders calcined from 700°C to 1000°C.

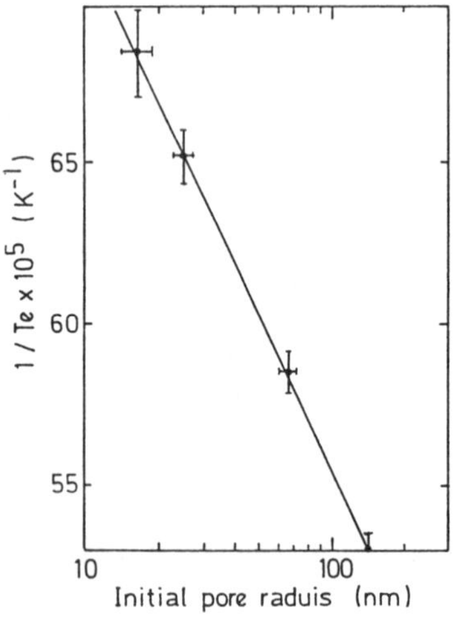

Fig.5 Relation between the temperature of the maximun of shrinkage rate and the most probable pore size.

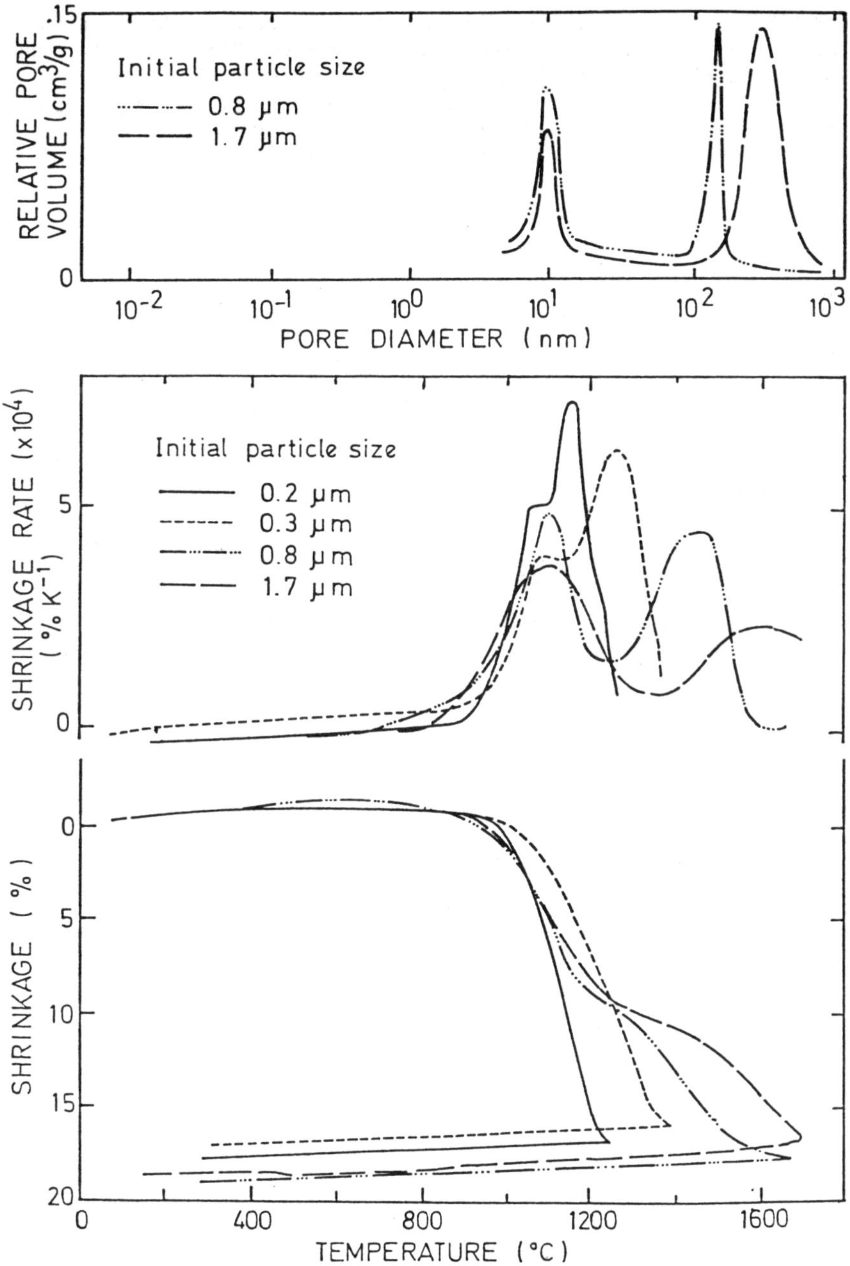

Fig.2 Pore size distribution, shrinkage rate and shrinkage for several powders.

SINTERING, MICROSTRUCTURAL DEVELOPMENT AND MECHANICAL PROPERTIES OF CeO_2-TZP CERAMICS

M. González, C. Moure, F. Capel, J.R. Jurado and P. Durán
Instituto de Cerámica y Vidrio (CSIC)
Electroceramics Department
Arganda del Rey, 28500 Madrid, Spain

ABSTRACT

Compositions ranging from 0 to 20 mol% CeO_2 were prepared by a wet-chemical route in the ZrO_2-CeO_2 system. The tetragonal phase stability region as a function of temperature and time of sintering was stablished, and the best sintering condictions for those compositions retaining tetragonal structure at room temperature were studied. Microstructural development on the sintered bodies and the influence of the grain size on the tetragonal phase stability have also been studied. Finally, the influence of the sintering conditions on the microhardness (H) and fracture toughness (K_{IC}) of the sintered bodies have been studied.

INTRODUCTION

Recent several results (1-4) have shown the importance of the ceramic materials based on solid solutions of the ZrO_2-CeO_2 system. Its particular electrical behaviour lead it to its application in some electrical systems.

ZrO_2-CeO_2 is also considered as a toughened material since a stress-induced phase transformation can be used to increase the fracture toughness. Such process is important because of the extensive tetragonal solid solution in this system which can be retained to room temperature depending critically on the size of the microstructure (5).

The present work was undertaken, in its first part, to study the influence of powder preparation on the microstructure and mechanical properties of some compositions in the ZrO_2-rich region of the ZrO_2-CeO_2 system.

EXPERIMENTAL PROCEDURE

Compositions ranging from 8 to 20 mol% CeO_2 were prepared by the partial precipitation using a CeO_2 fine-powder suspension (99.9%) and $Zr(Butox)_4$ as the starting materials.

Calcined powders were isostatically pressed at 200 MPa and then pellets of about 3 mm thick for 10 mm diameter were sintered in air at 1400º, 1500º and 1600ºC for various periods of time.

The Archimedes technique was applied to measure the density of sintered samples. Phase identification of calcined powders and sintered surfaces was carried out by XRD. Fracture toughness and microhardness were measure by the Vicker's indentation method with loads from 30 to 50 Kp, and data were calculated on the basis on Evans' equation (6). The average grain size was measured by the interception method (7) by using SEM.

RESULTS AND DISCUSSION

Phase Evolution

FIGURE 1 shows the partial phase diagram for ZrO_2-CeO_2 system recently proposed by Durán et al. (8). They indicate the probable existence of an eutectoide reaction for a 15 mol% CeO_2 composition at about 150º±50ºC. At 500ºC the tetragonal field is extended from - 12 to 17 mol% CeO_2 and above the monoclinic-tetragonal transformation it goes from 0 to 18 mol% CeO_2 .

Densification

Sintering process was carried out at 1400º, 1500º and 1600ºC for various periods of time, showing the higher density value at 1600ºC for 6 hours. FIGURE 2 shows the density as a function of the CeO_2 content at 1600ºC for 6 h, and as a function of the sintering time for the composition containing 16 mol% CeO_2 at 1600ºC.

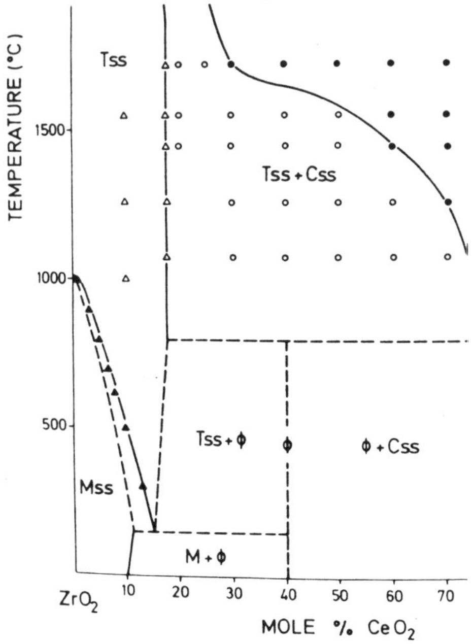

Figure 1. Partial ZrO_2-CeO_2 phase diagram proposed by Durán et al (8).

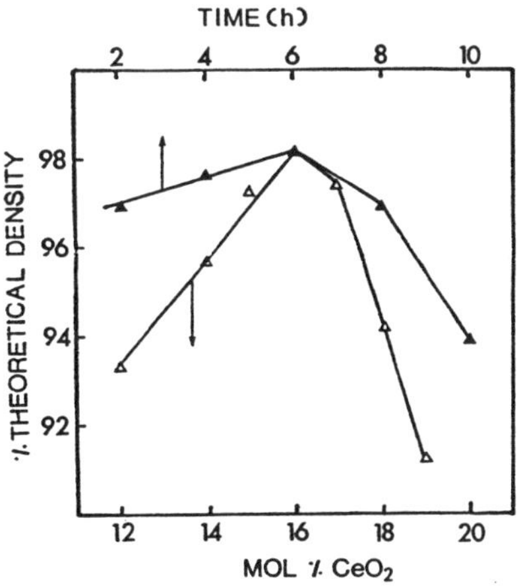

Figure 2. % Theoretical density evolution with CeO_2 content and sintering time.

Compositions with 8 and 10 mol% CeO_2 show cracking after sintering due to the big amount of monoclinic phase because of the tetragonal to monoclinic transformation suffered on cooling. Above 10 mol% CeO_2 the density increases up to a maximum located for 16 mol% CeO_2 . Higher CeO_2 contents produced a strong density decreasing. Such a behaviour could be explained by a combination of two factors, **a)** Below 16 mol% CeO_2 density and grain size could be playing an important role in retaining tetragonal phase, and **b)** Above 16 mol% CeO_2 a critical grain size could be controlling the tetragonal to monoclinic transformation and therefore the density decreasing.

Also, a density evolution versus sintering time is plotted in FIGURE 2, where at 6 h the higher density value is achieved. The behaviour here shown could be related to the above explanation where longer periods of time would increase the grain size over the critical value giving low densities.

Microstructural Development

FIGURE 3 shows the effect of CeO_2 addition and sintering time on the average grain size. The grain size remain almost constant with CeO_2 content in those samples sintered at 1600ºC for 6 h until the CeO_2 amount is enough to retain the tetragonal single phase. From this composition a grain size growing is observed along the monophasic field.

On the other hand, a progresive grain size increasing is also observed for a 16 mol% CeO_2 composition sintered at 1600ºC for various periods of time.

This behaviour can be clearly seen by SEM micrographs on FIGURE 5.

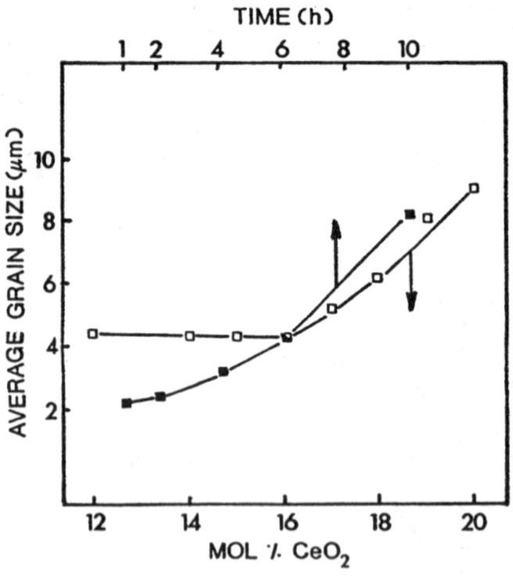

Figure 3. Average grain size evolution as a function of CeO_2 content and sintering time.

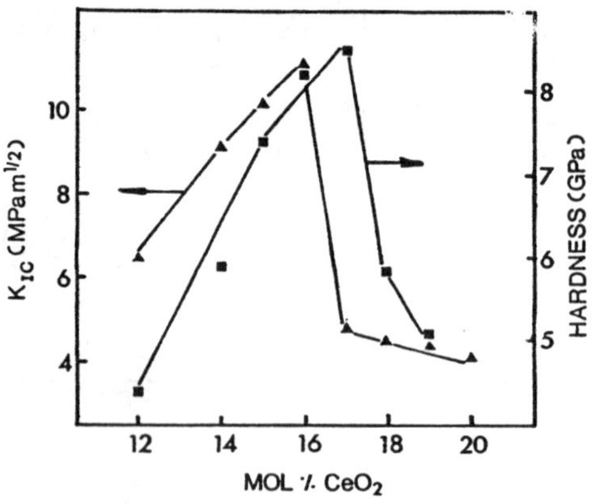

Figure 4. K_{IC} and hardness represented against CeO_2 content for samples treated at 1600ºC-6h.

Fracture Toughness

The critical stress intensity factor (K_{IC}) and hardness are represented in FIGURE 4 against CeO_2 content. The evolution shown can be explained on the basis of the tetragonal metastability evolution and grain sizes: Above 12 mol% CeO_2 the tetragonal metastability increases up to 16 mol% CeO_2 in which a K_{IC} maximum was measured. Beyond this point the tetragonality (c/a) diminished and the induced transformation toughening phenomenon plays a less important role and, in such a way, the K_{IC} sharply decreases. According to the FIGURE 4 it seems to exist a critical grain size about which the K_{IC} decreases. Such a critical grain size was of 4 μm in the particular case of the sintered bodies here fabricated.

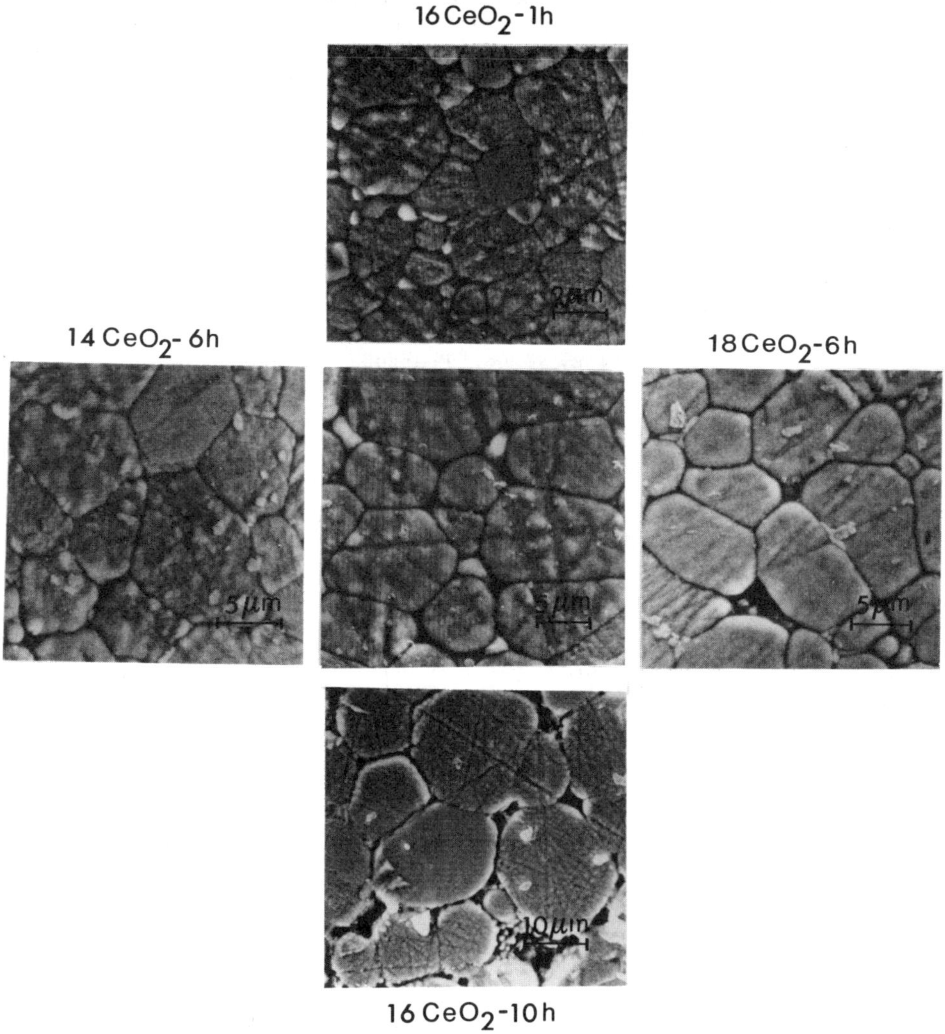

Figure 5. SEM micrographs obtained for several compositions and treatments.

Acknowledgment
This work was done under the auspices of the Commission of the European Communities, Contract EN3E/0167/E.

REFERENCES

1. Palguev, S.F. and Volchenkova, Z.S., Electrical Conductivity and Transport Numbers in the CeO_2-ZrO_2 System, Russ. J. Phys. Chem., 1960, **34**, 211-3.

2. Lange, F.F., Transformation Toughening, Part 5, Effect of Temperature and Alloy on Fracture Toughness, J. Mater. Sci., 1982, **17**, 255-62.

3. Tsukuma, K. and Shimada,M., Strength, Fracture Toughness,and Vickers Hardness of Ce-Stabilize Tetragonal Zirconia Polycrystals (Ce-TZP), J. Mater. Sci., 1985, **20**, 1178-84.

4. Tsukuma, K., Mechanical Properties and Thermal Stability of CeO2 Containing Tetragonal Zirconia Polycrystals, Am. Ceram. Soc. Bull., 1986, **65**, 1386-9.

5. Durán, P., Recio, P., Jurado, J.R., Pascual, C. and Moure, C., Y(Er)-Doped Tetragonal Zirconia Polycrystalline Solid Electrolyte. Part 2, J. Mater. Sci., 1989, **24**, 708-16

6. Evans, A.G. and Charles, E.A., Fracture Toughness Determination by indentation, J. Am. Ceram. Soc., 1976, **59**, 371-28.

7. Fullman, R.L., Measurements of Particle Sizes in Opaque Bodies, J. Metals. Trans AIME, 1953, **197**, 447-53

8. Duran, P., González, M., Moure, C., Jurado, J.R. and Pascual, C., Phase Equilibrium Diagram for the ZrO_2-CeO_2 System: A New Tentative, To be published.

MULTIPHASE α/ß SiAlON CERAMICS

S. BOSKOVIĆ and K.G. NICKEL*
Boris Kidrić Institute of Nuclear Sciences, IM- 170, Belgrade, Yugoslavia
*Max- Planck Institut für Metallforschung, PML, Stuttgart, F.R.G.

ABSTRACT

Phase relations in Si_3N_4- rich corner of Si_3N_4- $SiAl_2O_2N_2$- YAl_3N_4 plane were investigated. It was shown that within a defined compositional field α and ß SiAlONs coexist with minor amounts of a glass phase. In this field the α/ß ratio is controlled by Y content and is nearly temperature independent between 1600- 1850°C. However, density and microstructure of synthesized composites change with varying α/ß ratio, temperature and hot- pressing time.

INTRODUCTION

SiAlONs are widely used as cutting toos, primarily because of good hardness, toughness and wear resistence. There are many investigations concerning ß- SiAlONs. Recently (1,2) the research has been focused to α- SiAlONs, which exceed ß- SiAlONs in hardness and thermal shock resistence. It was shown that multiphase α/ß- SiAlONs can have higher hardness (3) with increasing α content. Thus a property tailoring seems possible.

Phase relations in the Y- Si- Al- O- N system were therefore the subject of a numbers of studies (4- 6) but it remained unclear whether there was strong temperature influence on α- SiAlON formation (6) or just a compositional dependence (4).

In this study we concentrated on Y- SiAlONs in the Si_3N_4- rich corner of Si_3N_4- $SiAl_2O_2N_2$- YAl_3N_9 plane of Y- Si- Al- O- N system. Changes in α/ß ratio, density and microstructure with varying hot- pressing temperature and composition, were investigated. Some results concerning the Nd- Si- Al- O- N system which is currently being studied, are included as well.

EXPERIMENTAL WORK

Materials were prepared from Si_3N_4 and AlN (H.C. Starck), Al_2O_3 (Alcoa) Y_2O_3 (Ventron) and Nd_2O_3 (Merck) powders. The powder mixtures (Tab.1) have been homogenized in isopropanol in attritors (Y- Si- Al- O- N) and in vibratory mills (Nd- Si- Al- O- N). Hot pressing was done at 1600- 1850°C for 2 hours in nitrogen atmosphere. Change in density and microstructure were monitored. The α/ß ratio was determined by XRD.

TABLE 1
Compositions of starting mixtures /wt%/

	Y- Si- Al- O- N system				Nd- Si- Al- O- N system			
sample	Si_3N_4	Y_2O_3	Al_2O_3	AlN	Si_3N_4	Nd_2O_3	Al_2O_3	AlN
A	82.96	2.5	5.8	8.6	82.38	4.58	5.38	7.65
B	90.62	3.5	0.0	5.9	89.06	5.13	0.10	5.60
C	85.80	3.5	3.5	7.2	84.50	6.14	3.27	7.01
D	82.40	3.5	5.5	8.8	81.62	5.09	5.40	7.84
E	78.90	5.7	4.3	11.1	77.10	7.85	4.20	10.84
F	79.90	7.1	1.6	11.3	76.19	10.33	1.56	11.92
G	73.53	7.1	5.2	14.2	71.67	10.19	5.01	13.12
H	75.70	7.4	3.4	13.5	72.51	10.74	3.25	13.49
I	75.56	9.2	0.8	14.4	72.10	13.07	1.00	13.70
J					72.50	8.50	6.86	12.12
K					67.55	15.33	1.04	16.08

RESULTS AND DISCUSSION

Compositions of mixtures listed in Tab.1 were chosen to enable synthesis of α/ß SiAlON composites from the two systems with varying α- phase amount at equilibrium. Identification of phases and relative α/ß ratio were determined for each composition in the Y- Si- Al- O- N system and the results are shown in Fig.1. Within the field II α and ß SiAlON composite materials can be obtained.

The effect of hot pressing temperature on phase relations between α- and ß- SiAlON is not considerably pronounced. At 1600°C the equilibrium composition has been achieved after 4 hours (Fig.2). It can be seen that a particular α/ß ratio can be obtained either by varying hot pressing time at low temperaturs or establishing equilibrium for a given composition.

Phase composition of α/ß SiAlON composites, however, depended strongly on the concentration of added modifying ion as shown in Fig.3, given as mol% $Y_3Al_9N_{12}$. Relationships in the Nd- Si- Al- O- N system will be discussed separately.

For practical purposes of manufacturing a particular composite from Y- Si- Al- O- N system it is of no consequence whether α/ß ratio is obtained as metastable or equilibrium value. However, the mechanical properties of the composites may vary because of differing bulk and phase compositions.

A strong dependence of density on Y and Nd concentration was found. Density increased between 3.20 - 3.30 and 3.26 - 3.48 (g/cm^3) for Y- SiAlONs and Nd- SiAlONs with increasing Y/Nd content.

Full densification is achieved during heating up at 1500°C while equilibration reactions ($\alpha \rightarrow$ ß) preceed. Intense grain growth takes place above 1600°C (Fig.4) Solution- reprecipitation as well as grain coalescence take place during grain growth. Coalescence was observed at 1800°C in samples containing Y (Fig.4). Nd- SiAlON

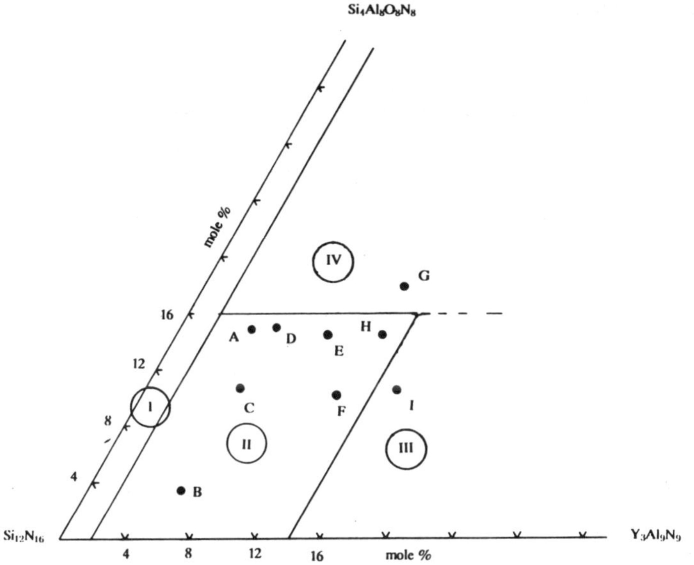

Figure 1. Si_3N_4 rich corner of the compositional plane SiN- $Y_3Al_9N_{12}$- $Si_4Al_8O_8N_8$. Letters mark the positions of the compositions in Tab.1. Lines separate phase fields at 1600- 1800°C: I: ß- SiAlON, II: α/ß- SiAlONs, III: α- SiAlON, IV: α/ß- SiAlON + phase x, (7).

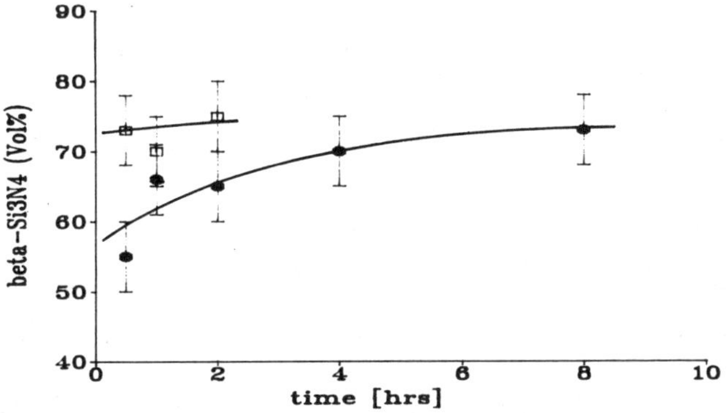

Figure 2. Time vs. α/ß- SiAlON ratio (Vol- %) for hot pressed starting material A at 1600°C (dots) and 1800°C (squares).

composites exhibit different grain sizes. Finer grains (Fig.5) in Nd- SiAlON are probably due to a different homogenization procedure.

Preliminary results indicate that in both Nd- and Y- SiAlON system the hard

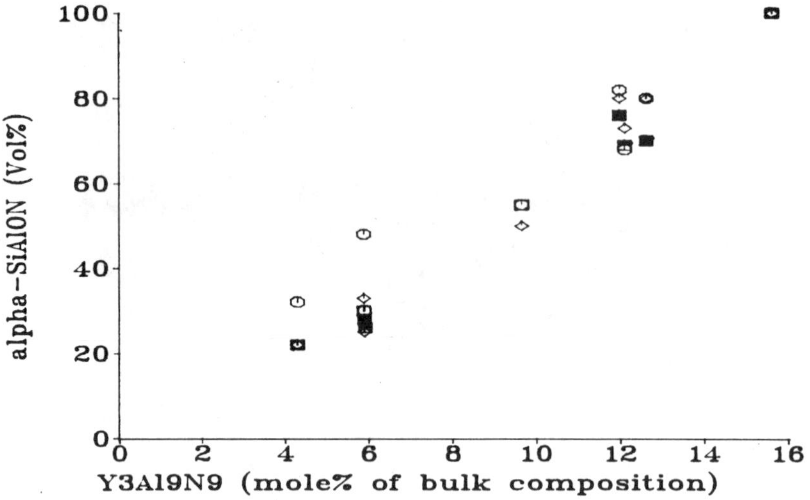

Figure 3. Vol % of phase as a function of Y content at 1600° C (circles), 1700°C (diamonds), 1800°C (squares).

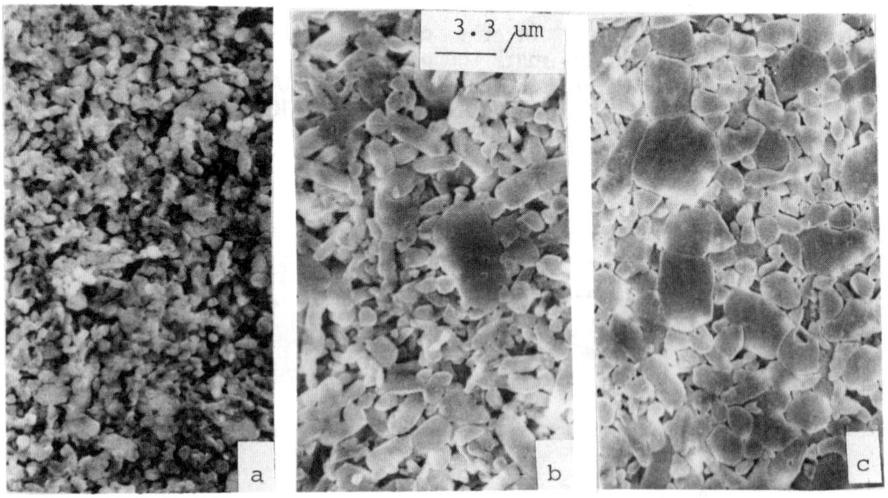

Figure 4. Etched surfaces of Y- SiAlON with 3,1 equiv.% hot pressed at a) 1600° C, b) 1700° C, and c) 1800° C, 2 hours.

ness increases with increasing α- content, but correlates inversly with the fracture toughness.

CONCLUSIONS

Coexistence of α and ß SiAlON was found in $Si_{12}N_{16}$- $Si_4Al_8O_8N_8$- $Y_3Al_9N_{12}$ plane in the range of Y concentration between one to 3.1 equiv.% (≈ 4 - 15 mole % YAl_3N_9)

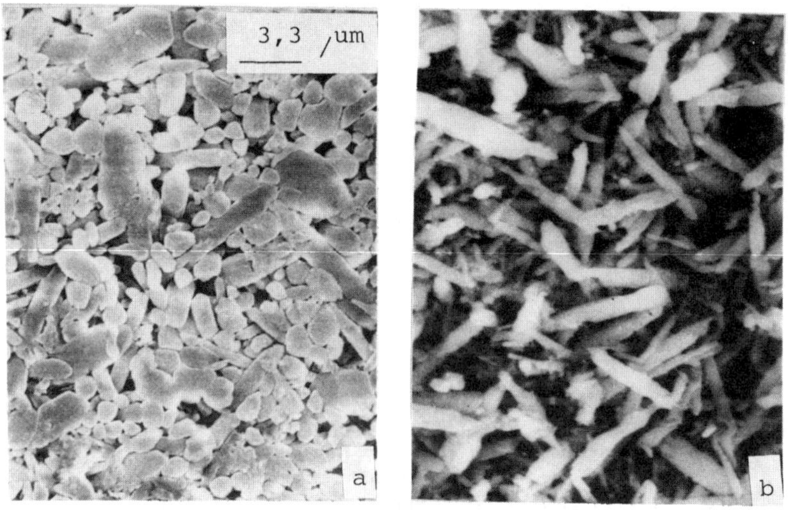

Figure 5. Microstructures of Y- and Nd- SiAlON containing the same amount of Me ions (1.8 equiv.%): a) Y- SiAlON 1800°C, 2 hours, b) Nd- SiAlON 1750°C, 2 hours.

and up to 5.5 equiv.% O (≈ 15 mole % $SiAl_2O_2N_2$). The α/ß- ratio within the composite material depends strongly on the Y- content and is nearly independent of temperatur in the range 1600 to 1850°C. The hot pressed density increases with increasing Y- and Nd- content.

REFERENCES

1. Jack, K.H. and Wilson, W.I, Ceramic Based on Si- Al- O- N and Related Systems, Nature Phys. Sci., 1972, **328**, 28- 29.

2. Huang, Z.K, Greil, P., and Petzow G, Formation of α- Si_3N_4 Solid Solutions in the System Si_3N_4- AlN- Y_2O_3, J. Am. Ceram. Soc., 1983, **66**, C 96- 97.

3. Greskovic, C. and Gaza, G.E., Hardness of Dense α and β Si_3N_4 Ceramics, J. Mat. Sci. Letters, 19855, 4, 195- 196.

4. Peuckert, D., Herstellung und Untersuchung von mehrphasigen Si_3N_4- Mischkristallkeramiken, Ph.D. Thesis, University of Stuttgart, 1987.

5. Sun, W.Y., Huang, Z.K., Cao, G.Z., and Yan, D.S., Phase Relationships Involving Mixed O'- ß'- SiAlONs in the Y- Si- Al- O- N System, Int. J. High Techn. Ceramics, 1987, **3**, 277- 284.

6. Sun, W.Y., Wu, F.Y., and Yan, D.S., Studies of Formation of α and ß SiAlON, Mat. Lett., 1987, **6**, 11- 15.

7. Bosković, S., Nickel, K.G., Petzow, G., Phase Relations in the Y- Si- Al- O- N System: Controlled Manufacturing of α/ß SiAlON Composites, Proc. German- Yugoslav Meeting on Mat. Sci., 1989, Hirsau, Germany (in press).

THERMITE REACTION ASSISTED SINTERING OF TiB_2 UNDER HIGH PRESSURE AND ITS SINTERED PROPERTIES

S. Takeda, T. Maruyama* and K. Ikeda**
Research Division, Material Lab.,
KOMATSU LTD.
1200 Manda, Hiratsuka, Kanagawa 254, Japan
***Research Laboratory for Nuclear Reactors,**
Tokyo Institute of Technology
Ookayama, Meguro-ku, Tokyo 152, Japan
****Depertment of Metallurgy,**
Iwate University
4-3-5 Ueda, Morioka, Iwate 020, Japan

ABSTRACT

A fine TiB_2 powder without additives was densified up to 99 % theoretical density by the instantaneous alumino-thermite heating (THP process) under 2 GPa. The effect of pre-heating for the powder compacts before sintering on resultant microstructures was examined. The differences in grain size and morphology are discussed regarding the compositions in the vicinity of grain-boundary facets. Several physical properties were measured : elastic moduli at room temperature, thermal conductivity, thermal expansion, and electrical resistivity. Thermal conductivity over the range room temperture to 1000°C was found to be independent on temperature.

INTRODUCTION

Rapid sintering of ceramics have been received much attentions for many years for enhancing densification and suppressing grain growth during heating in the field of monolithic materials. For achieving the above two objectives simultaneously, high pressure techniques which assist pore elimination, would offer the especially effective means in materials with poor sintering behavior if a high-temperature heating is carried out for a short time. One of the recent novel process is that applied shock-wave technologies. An representative study is a likely entitled : " Dynamic Compaction of Ceramic Powder using Exothermic Reaction Heat [1] ". SiC powder was densified to values close to theoretical by this technique.

As described in the previous paper[2], one of the author demonstrated the new process named : " Thermite Reaction Assisted Sintering & Synthesis under High Pressure (THP process) ". It is characterized as instantaneous heating process using exothermic reaction, such as

$2Al + Fe_2O_3 \longrightarrow Al_2O_3 + 2Fe + 204\ kcal.$

In this investigation, paticular attention is focused on clarifying the microstructures of TiB_2 fabricated by THP process and its sintered properties.

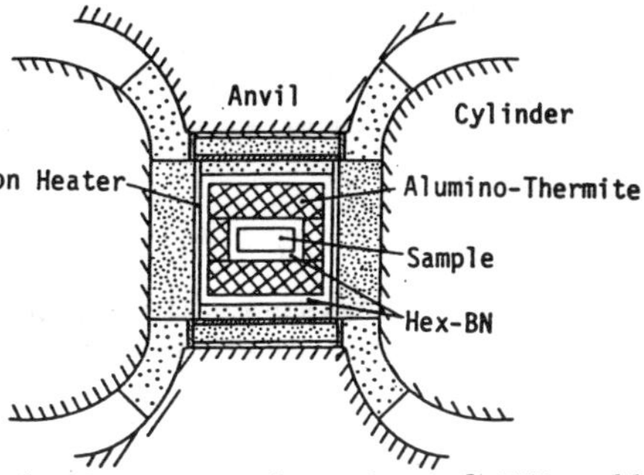

Fig. 1. Cross section view of THP cell assembly in belt apparatus.

EXPERIMENTAL PROCESURE

A fine TiB_2 powder* derived from the carbothermic process was used for this sintering work. The average particle size (FSSS) is ~1 um with a surface area of approximately 3.0 m^2/g, as determined by BET analysis. Principal impurities by wt % are : O_2 (0.75), N_2 (0.52), and Fe (0.02).

The powder was cold pressed at 100 MPa (12.8 mm in diameter and 5 mm long), and pre-heated at 1300 ~ 1700°C, 3hr under vacuum ($2x10^{-3} \sim 1x10^{-4}$ torr) before sintering. Pre-heating was conducted for the purpose of either decreasing the surface area or lowering the level of impurity.

The sample was embedded at the center of the THP cell (Fig. 1). The thermite mixture of 41.9 g (40 kcal equivalent was set in place in the THP cell in each runs. The pressure insides was raised to 2 GPa at room tenperature and electric power was applied to the carbon heater for igniting the alumino-thermite mixture. A typical THP process is shown in Fig. 2.

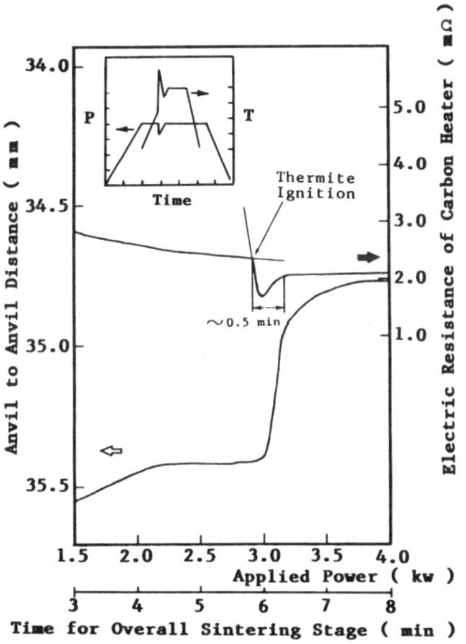

Fig. 2. Typical schedule for THP process.

The compositions in the vicinity of grain-boundary facets were analyzed by using micro-Auger electron spectroscopy** (MAES). Elastic moduli were estimated in the determination of sonic velocities by using the Scanning laser acoustic microscope[3]***(Acoustic Spectral Interferometry method). The electrical resistivity measurement (DC current method) on the specimens (2 by 2 by 10 mm) were made using a four point contacting arrangement.

RESULTS AND DISCUSSION

Table. 1 summarizes the characteristics of THP sintered TiB_2. Densities exceeding 99% of theoretical (theoretical value : 4.50 g/cm^3) can be achieved by vacuum pre-heating for powder compacts. The difference in sintered densities between the each specimens is not significant.

Although the significant difference in the level of impurity oxygen was not detected in each sintered bodies, the differences in grain size and morphology were observed. Particularly, anisotropic grain growth but uniform matrix grain in its size distribution is evident for the

* supplied by Hermann C. starck Berlin. ** JAMP-10S, JEOL.
*** Sonomicroscope System 130, Sonoscan.

Table. 1. Characteristics of THP sintered TiB_2 Compacts

Specimen	Density* (g/cm^3)	Grain size** (μm)	Significant impurities (wt%)*** Oxygen	Nitrogen	Carbon+
CF5	4.4_0	~ 1	0.73	0.32	(0.70)
CF4-1300	4.4_4	anisotropic grain growth	0.79	0.12	(0.21)
CF5-1500	4.4_6	1 - 3	0.64	0.041	(0.33)
CF6-1700	4.4_5	2 - 5	0.58	0.039	——

*Bulk water immersion. **SEM analysis. ***Infrared absorption, TCD, and ICP.
+Impurity level of powders before THP process.

specimen CF4-1300. With increasing the pre-heating temperature, moderate grain growth is observed, and grains are equiaxed. The SEM micrographs of fracture surfaces shown in Fig. 3 illustrate such differences depending on the pre-heating temperature.

The question is why pre-heating for powder compacts influences the resultant microstructures during THP process. To explore such effect mentioned above, a MAES analysis of fracture surfaces in TiB_2 was undertaken. The compositions, especially impurities, in the vicinity of grain-boudary facets were analyzed by using the depth profiling technique. The operation condition such as beam voltage, beam current, and spatial resolution were 10 keV, 10 nA, and less than ~1 μm respectively. Sputter-Etching condition of as fractured grain-boundary facets is described (Fig. 4). The depth profiles for two specimens (CF5 and CF6-1700) are shown in Fig. 4. For the specimen CF5, which is not pre-heated, the region of grain-boundary facets is enriched in impurity oxygen, and carbon is also superimposed with oxygen profile. B as well as Ti (not described in Fig. 4) depletion associated with the oxygen segregation mentioned above in the vicinity of grain-boundary facets were observed. On the other hand, the relative concentrations of oxygen & carbon remained relatively unchanged. It can be recognized that pre-heat treatment of powder compacts could improve the heat transmission efficiency of thermite heat into the powder for sintering. Under this circumstance, the dramatic change in microstructure might occur even in a short time, accompanied by the moderate grain growth.

Elastic moduli calculated from sonic velocities are approximately 530 GPa for Young's modulus and 220 to 240 GPa for Shear modulus at room

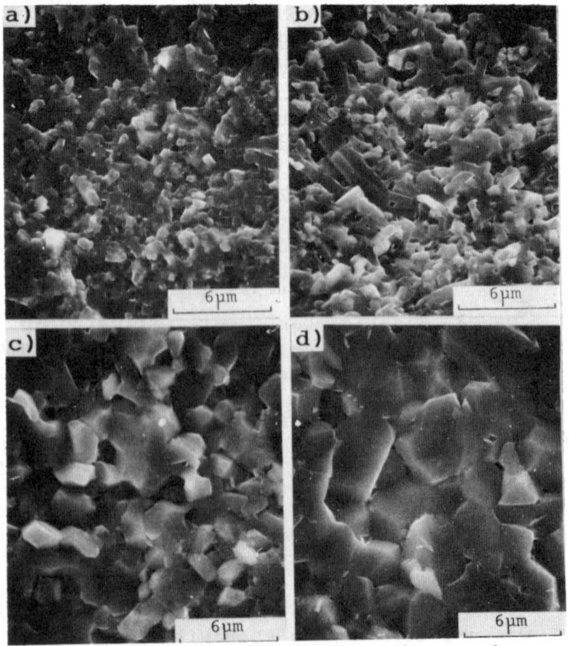

Fig. 3. Effect of pre-heating for powder compacts on resultant microstructures of TiB_2 in THP process ; a) non-heating, b) 1300°C x 3h, c) 1500°C x 3h, d) 1700°C x 3h.

temperature (R.T.) respectively. These values agree very well with hot-pressed TiB_2 with additives [4] and the previous reports [5-6] within the error of sonic velocity determination.

The thermal conductivity by the laser flash method are shown in Fig. 5. The highest value was achieved in the specimen CF6-1700 with 90 W/m·K at R.T.. The thermal conductivity over the range R.T. to 1000°C seems to be independent on temperature. It might be the reason why the specific heat of TiB_2 increases linearly with increasing the temperature until around 1000°C . The previous value [7] also indicate the same trend as the present results.

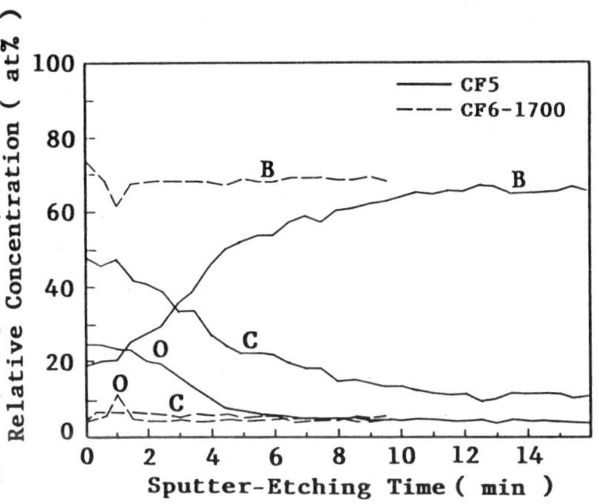

Fig. 4. Depth profiling measured from grain-boundary fracture surfaces of THP sintered TiB_2, because of uncertainties in sputter yield, relative concentrations are plotted against the sputter-etching time. Etching condition is as follows : argon ion incidence; 3 keV, ion current density;120~160 μA/cm^2. (under this condition, etching rate corresponds to 200 Å/min in SiO_2 film)

The thermal expansion data on heating between R.T. and 1300°C are given in Fig. 6. There is no significant difference in the thermal expansion vs temperature curves between the three specimens with different microstructures. The average coefficient of thermal expansion α_{AV}(R.T.-1000°C) for these specimens were 7.4~7.6 x 10^{-6} (°C^{-1}), and their values indicate the good agreement with the value of polycrstalline (7.6 x 10^{-6}(°C^{-1})) calculated from the provisional values for the axial thermal expansion [8].

The electrical resistivity of THP sintered TiB_2 are presented in Table. 2. For comparision purposes, recent data on 98.2% pure and 99% dense fine-grained (4 μm) polycrystalline TiB_2 and single-crystal [9] , in addition, hot-pressed TiB_2

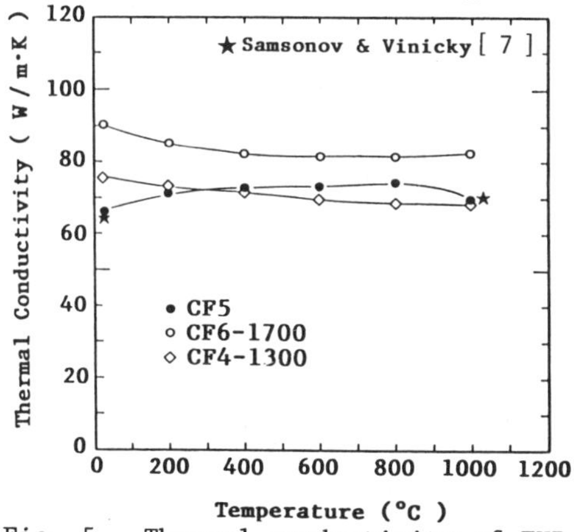

Fig. 5. Thermal conductivity of THP sintered TiB_2 in the temperature range 25°C to 1000°C.

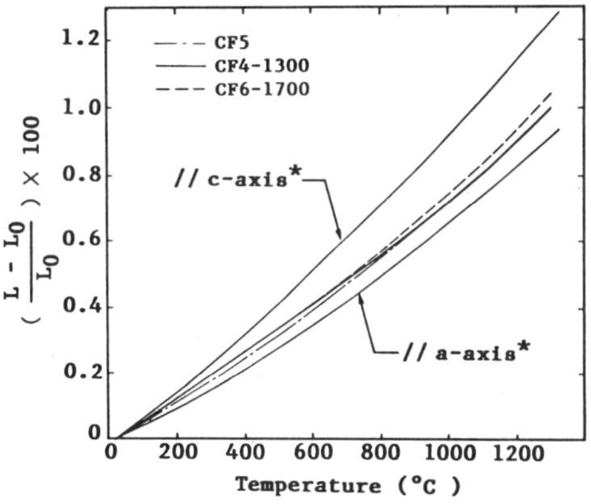

Fig. 6. Thermal expansion of THP sintered TiB_2 in the range 25°C to 1300°C. (*/ /a-axis, */ /c-axis : Provisional values for the axial thermal expansion for powdered sample [8].

Table. 2. Electrical resistivity of THP sintered TiB_2

Specimen		ρ_{77K} ($\mu\Omega$-cm)	ρ_{300K} ($\mu\Omega$-cm)	$d\rho_H / dT$ gradient in resistivity ($n\Omega \cdot cm$ / K)	α avrage coefficient of resistivity (mK^{-1})
CF5		4.4_9	11.2_5	39	3.5
CF4-1300		3.5_9	10.3_9	41	3.9
CF6-1700		3.3_8	10.4_5	44	4.2
K***		3.4_1	11.9_3	43	3.6
Ref. [9]	P+	—	7.4 ±0.2*	42.2 ±0.4	5.7 ±0.2**
	S+	—	6.4 ±0.2*	35.2 ±0.3	5.3 ±0.2**

*Interpolated resistivity as ρ_{298} at 298K   +P = polycrystalline
**Value as defined by $\alpha = d\rho_H / dT / \rho_{298}$ +S = single-crystal
***Hot-pressed TiB_2 with additives (supplied by Dr. T. Watanabe, Goverment Industrial Research Institute, Kyushu) ; TiB_2-5%TaB_2-1%CoB-2%TiC [4]

with additives are also included. It is evident that the resistivity at 77 K decreases with decreasing the impurities (oxygen plus nitrogen). The difference in gradient in resistivity, $d\rho_H/dT$, between CF5 and CF6-1700 is not significant. However, the value of average coefficient of resistivity, α, in CF6-1700 shows slightly higher than that of CF5 as well as K specimen. Mcleod, Haggerty, and Sadoway [9] have recently suggested that α is a measure of the intrinsic electrical behavior of TiB_2. Our values of THP sintered TiB_2 are lower than that of theirs. Polycrystalline TiB_2 they measured shows non-stoichiometric composition, B/Ti = 1.97 (boron poor). It might be especially pointed out that the influences of non-stoichiometry as well as impurities in THP sintered TiB_2 on α still remain to be clarified with respect to their microstructures.

ACKNOWLEDGEMENT

The authors express appreciation to Dr. G. Schwier, Hermann C. Starck Berlin, for supplying TiB_2 powder, Dr. T. Watanabe, Goverment Industrial Research Institute, Kyushu, for supplying TiB_2 sintered compacts with additives, and Mr. C. Kato, Rigaku Corporation Application Laboratory, for thermal conductivity measurements. The technical assistance of Mr. H. Shishiba for sintering work, micrography, and manuscript preparation should be noted.

REFERENCES

1. Akashi, T. and Sawaoka, A. B., Advanced Ceramic Materials, 1988, 3 [3] 288-90.
2. Takeda, S., Sintering '87 (Proc. 4th Int. Sympo. Sci. and Tech. of Sintering held in Tokyo), Elsevier Applied Science, 1987, 2 1076-1081.
3. Oravecz, M., and Less, S., Acoustical Imaging. (Proc. of the Thirteenth Int. Sympo.) Plenum Press, 1983, 13 397-408.
4. Watanabe, T. and Shoubu, K., J. Cerm. Soc. Jpn., 1988, 96 [7] 778-82.
5. Finch, C.B., Becher, P.F, Angelini, P., Baik, S., Bamberger, C.E., Brynestad, J., Advanced Ceramc Materials, 1986, 1 [1] 50-54.
6. Baumgartner, H.R., and Steiger, R.A., J. Am. Ceram Soc., 1984, 67 [3] 207-212.
7. Samsonov, G.V., and Vinicky, I.M., Handbook of Refractory Compounds, (Metallurgiya Press in Moscow 1976).
8. Touloukian, Y.S., Kirby, R.K., Taylor, R.E., and Lee, T.Y.R., Thermal Expansion-Nonmetallic Solids. (Thermophysical Properties of Matter 13), IFI/Plenum 1977, 778-782.
9. McLeod, A.D., Haggerty, J.S., and Sadoway, D.R., J. Am. Ceram. Soc., 1984, 67 [11] 705-708.

INFLUENCE OF ADDITIVES ON SURFACE AND STRUCTURE STABILIZATION OF GAMMA Al_2O_3

JELISAVETA KATANIĆ-POPOVIĆ AND SLAVICA ZEC
"Boris Kidrič" Institute for Nuclear Science, Beograd,
P.O.B. 522, Yugoslavia

ABSTRACT

Nitrate salts of magnesium, aluminum and lanthanum were used as precursors of oxide additives to gamma Al_2O_3. The effect of these dopes in pseudoboehmite or gamma Al_2O_3 powders, compacts as well as sintered samples was investigated. The samples with 0.06 M/Al ion ratio and equimolar ratio additive to Al_2O_3 were analysed. Specific surface area, porosity distribution and density changes were followed. The structure development was characterised by X ray. The specific surface area of sintered doped samples increases compared to pure Al_2O_3, even when porosity decreases. This effect is more pronounced in gamma Al_2O_3 treated with nitrate salts than in the case of doped calcined pseudoboehmite. MgO and Al_2O_3 addition causes some structural destabilization of gamma alumina. La_2O_3 has been confirmed as the beneficial additive concerning the surface and structure stabilization of gamma alumina.

INTRODUCTION

Stability of gamma alumina catalyst support is very important for the applications at relatively high temperatures (600-800^oC). This research deals with the influence of MgO, Al_2O_3 and La_2O_3 on surface and structure stabilization of gamma Al_2O_3, obtained from synthesized pseudoboehmite powder. These additives were choosen on the basis of different dimensions of metal ion radii as well as expected solid state reactions on the gamma alumina particle surface. Specific surface area, structure characteristics, porosity and pore size distribution were examined in order to explain complex effect of these dopes on gamma alumina thermal stability.

MATERIALS AND METHODS

The starting pseudoboehmite powder (P) was prepared by precipitation of aluminum chloride and ammonium hydroxide /1/. Gamma Al_2O_3 powder (G) was obtained by pseudoboehmite calcination at 500^oC for 2^h. The nitrates (pro analysis quality): $Mg(NO_3)_3.6H_2O$, $Al(NO_3)_3.9H_2O$ and $La(NO_3)_3$, $6H_2O$ were used as precursors for oxide dopes. The additive to aluminum (M/Al) ion ratio was 0,06. The mixtures with equimolar compositions of components were also investigated. These relatively great quantities of oxide dopes were used in order to faciliate phase identification by X ray diffraction method. Cylindrical compacts were made from dryed powders by cold pressing at 50 MPa. Sintering tests were carried out for 24^h at 800°C in air. Densities and open porosities of sintered samples were measured by xylene immersion method. Open pore size distributions were determined by mercurry porosimeter. Specific surface areas were measured by areameter with nitrogen adsorption at -196°C. X ray diffraction analysis was carried out with Cuk_α radiation.

RESULTS

The pseudoboehmite powder shows better compressibility than gamma alumina. The nitrate addition reduces green porosity (Table 1) the nitrates melt below 100°C and decompose up to 600°C, releasing relatively great amount of gases. The sintered samples are characterized by greater porosity comparing to green compacts.

TABLE 1
Characteristics of compacts and sintered samples obtained from pseudoboehmite (P) and gamma alumina (G)

Sample number	Dopant nitrate	Compacts porosity (%)	Sintered samples	
			porosity (%)	specific surface area (m^2g^{-1})
1P	-	61.40	78.41	118
2P	MAGNESIUM	52.33	67.55	129
3P	ALUMINUM	51.20	67.97	120
4P	LANTHANUM	52.01	66.14	131
1G	-	81.39	82.73	116
2G	MAGNESIUM	72.64	80.50	135
3G	ALUMINUM	70.85	81.07	124
4G	LANTHANUM	63.02	75.72	139

It was established that MgO and La_2O_3, being in equimolar ratio with gamma Al_2O_3, formed compounds at 800°C, 24^h. These reactions were more intense in doped pseudoboehmite, reffering to doped gamma alumina (Fig.1). Alpha Al_2O_3 was detected in calcined powder with equimolar ratio of aluminas, obtained from pseudoboehmite and aluminum nitrate. Poorly crystallized gamma Al_2O_3 with theta phase in traces was detected in sintered samples. The diffraction lines of $MgAlO_4$ and $LaAlO_3$ were not found. It is evident, from the porosity data, that oxide dopes, being on the particles surfaces, significantly influence pore size distribution, especially in the region from 10 to 100 nm (Fig.2).

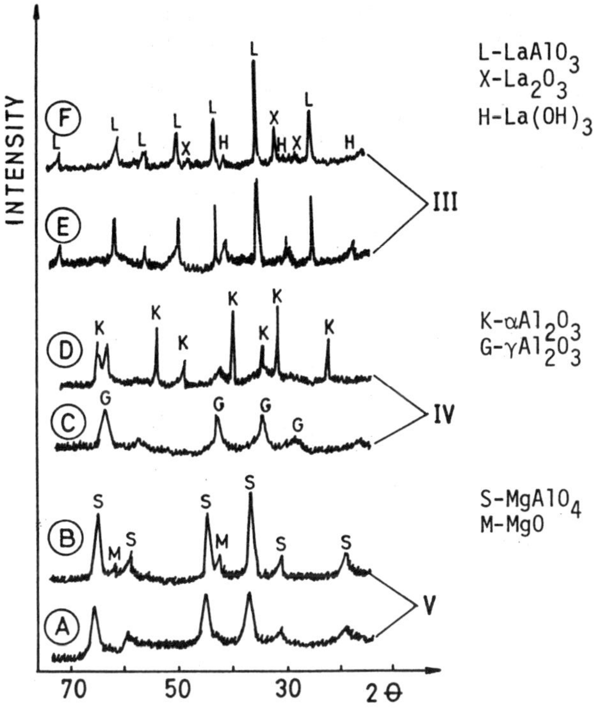

Figure 1. X-ray diffraction of calcined powders with equimolar ratio of of La_2O_3 (III), Al_2O_3 (IV) and MgO (V) with gamma Al_2O_3. Initial mixtures are corresponding nitrate and gamma alumina (A,C,E) or pseudoboehmite (B,D,F)

DISCUSSION

The results presented in Table 1 indicate that the addition of examined dopes increases specific surface area, compared to pure gamma alumina, in spite of smaller porosities. The additive deposited on the particles surface reduces surface diffusion, causing smaller changes of specific surface area. This effect is more pronounced in gamma alumina treated with nitrates, than in doped pseudoboehmite. This fact is in accordance with better crystallinity of the reaction products of calcined pseudobohemite powders with equimolar ratio of components (Fig.1). The appearance of theta and alpha phases, as well as better crystallinity of gamma Al_2O_3, decreases specific surface area. La^{3+} with the greatest ion radius can not enter the defect spinel structure of gamma alumina and shows a positive effect on structure stabilization. Al^{3+} and Mg^{2+} ions, with significantly smaller ion radii, could be placed in the vacant cation positions of the gamma alumina lattice, causing some structural destabilization. The additions of MgO and Al_2O_3 facilitate gamma to theta and gamma to alpha Al_2O_3 transformations, respectively.

The study of open size distributions for sintered samples confirmed the assumption /2/ that oxide additives fill up some pores and decrease

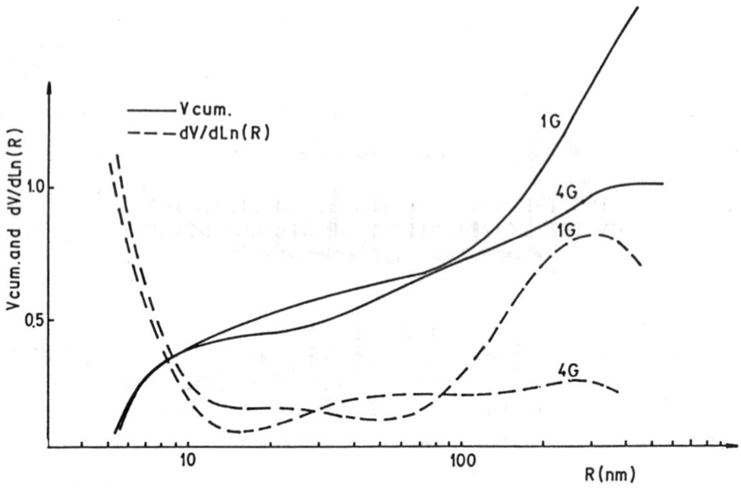

Figure 2. Pore size distribution foe sintered gamma alumina (1G) and La_2O_3 doped (4G) samples

initial specific surface area. The pore filling in some extent is more pronounced in sintered samples obtained from doped pseudoboehmite.

CONCLUSIONS

MgO, Al_2O_3 and La_2O_3 dopes, with 0,06 M/Al ion ratio, decrease porosity, change the pore size distribution but in spite of that increase specific surface area of gamma alumina at 800^oC. Better surface stabilization is obtained for doped gamma alumina as compared to doped pseudoboehmite samples.

The addition of MgO and Al_2O_3 causes some structural destabilization of gamma alumina. MgO facilitate gamma to theta and Al_2O_3 gamma to alpha phase transformations.

The La_2O_3 has been confirmed as the benefitial additive concerning the surface and structure stabilization of investigated gamma alumina.

REFERENCES

1. Katanić-Popović, J., Gašić, M. and Kostić-Gvozdenović, Lj., Formation and sintering of gamma Al_2O_3 with MgO addition from coprecipitate hydrates, J. de Physique, 1986, 47, pp.31-35.

2. Shaper, H. and Van Reijen, L.L. Influence of dopes on the stability of gamma alumina catalyst supports in Sintering-Theory and Practice ed. D.Kolar and S.Pejovnik, Elsevier Scientific Publishing Company, 1982, pp.173-176.

THE INFLUENCE OF POWDER CONDITIONING ON THE SINTERABILITY OF SILICON NITRIDE

ROBERT GILISSEN
Materials Development Department / Ceramic Technology
Studiecentrum voor Kernenergie (SCK/CEN)
Boeretang 200, B-2400 Mol, Belgium

ABSTRACT

When fabricating sintered Si_3N_4 from fine sinterable powders, it is common practice to admix the sintering additives by intense milling. In addition to intimate mixing, this powder conditioning step affects particle size distribution, specific surface area and oxygen content of the powder. The purpose of this study is to evaluate the specific influence of each of the afore-mentioned powder characteristics on the sinterability of the powder. The relative importance of the milling step is obtained from the comparison of a standard commercial powder with a commercial disagglomerated powder recently launched on the market.

INTRODUCTION

It is well known that the characteristics of the silicon nitride starting powder, the type and the amount of additives as well as the sintering conditions, temperature, time and pressure affect densification behaviour, microstructure and material properties. The influence of the powder characteristics on the sinterability is well documented (1,2,3). Effective sintering additives are known (4) and their role in liquid-phase sintering is well understood (5,6). It is common practice to mix the sintering additives to the silicon nitride by intense milling. The effect of milling on the specific surface area (7) and on the degree of dispersion of the sintering additives (8) have been studied. However, the overall effect of this powder conditioning step on the sintering behaviour and on the materials properties has not been reported. It is the aim of this paper to study the correlation between milling and sinterability of some commercial silicon nitride powders.

MATERIALS AND METHODS

Two commercial silicon nitride powders A and B (H.C. Starck, grade LC12 and LC12S) produced by the nitridation of silicon have been compared. Basically both only differ in the degree of agglomeration. The specific surface area, oxygen content, metallic impurities content, phase composition and particle morphology are practically identical. A modified powder C prepared from powder A has been included in the comparison. This powder has been milled to increase the specific surface area and the oxygen content. The powder characteristics are summarized in table 1. The mean particle size of powder B is 0.7 µm with less than 10 wt.% above 2 µm. The mean particle size of powder A is 2 µm with 10 wt.% higher than 10 µm, which shows the high degree of agglomeration of this powder.

The different silicon nitride powders were mixed with 5 wt.% Y_2O_3 and 2 wt.% Al_2O_3. The additive concentration is a compromise to obtain high densities by pressureless sintering with acceptable high temperature materials properties and acceptable oxidation behaviour. The yttria powder (H.C. Starck, finest) has been used as received. The alumina powder (Martinswerk ZPS502) has been milled to reduce its particle size. The powder characteristics of both sintering additives are included in table 1.

TABLE 1
Powder characteristics of the starting powders

code	powder grade	specific surface (m^2/g)	O (wt.%)	C (wt.%)	Fe	Al (ppm)	Ca	Mg	Phase Compos. $\alpha/(\alpha + \beta)$ (%)
A	Si_3N_4 LC12	19.9	1.5	0.17	50	590	40	30	92
B	Si_3N_4 LC12S	18.7	2.0	0.27	50	470	15	60	94
C	Si_3N_4 (*)	22.8	2.3	0.13		n.m.			92
-	Y_2O_3	13.7	-	-	15	-	<10	<3	-
-	Al_2O_3 (**)	20.3	-	-	150	-	570	100	-

(*) LC12 modified by 64 h planetary ball milling
(**) ZPS502 modified by 66 h planetary ball milling

The sintering additives are dispersed in the silicon nitride powders by wet planetary ball milling for different times using 0.5 l polyamide jars and 12 mm diameter silicon nitride milling balls (powder/liquid ratio approx. 1:2, powder/ball ratio 1:3).

Specific surface area was measured by the five-point BET method (Micromeritics), particle size distribution by sedimentation (Sedigraph Micromeritics), oxygen content by fusion gas extraction (Leco). The additive dispersion was followed by backscattered electron image (BEI) (Jeol). Sintering was performed on 7.5 mm diameter pellets compacted at 200 MPa at 1800°C for 2 h under 1 bar N_2 pressure. Densities after sintering were measured by immersion in water.

RESULTS AND DISCUSSION

The effect of the milling time on the density after sintering and on different powder characteristics is represented in figure 1.

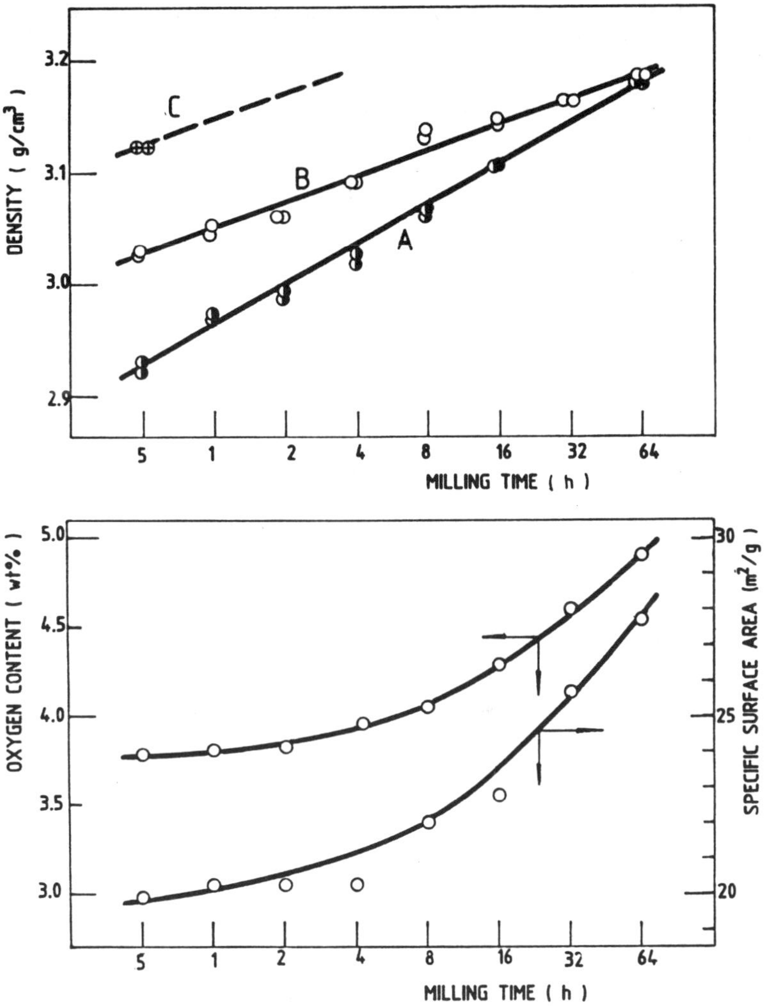

Figure 1. Influence of the milling time on the density after sintering, on the specific surface area and on the oxygen content.

The curve for the agglomerated powder A, which is representative for most commercial nitrided powders, shows that density increases with increasing milling time. The effect of milling on the powder characteristics is an increase in specific surface area and a proportional increase in oxygen content. At the same time the dispersion of the sintering additives is improved and hard agglomerates are broken down. The dispersion of sintering aids has been followed by BEI on sectioned green compacts (fig. 2). It shows that after 8 hours of planetary ball milling the agglomerated additives are broken down to their crystallite size. The effect of hard agglomerates in the silicon nitride starting powder, resulting from wet processing, is often overlooked. Micrographs of shortly milled material clearly show that those agglomerates give rise to heterogeneous sintering (fig. 3). The large porosities in these pictures are formed by the melting of the agglomerated additives.

The use of disagglomerated powder (curve B in fig. 1) gives a substantial rise in the density after sintering, but again density increases with increasing milling time. Oxygen content and specific surface area change in the same way for both powders. Powder C, made from high specific surface area silicon nitride scores much higher densities after half an hour of milling. The fact that the density is still lower than the results for powders A and B after intense milling is due to the bad dispersion of sintering additives after this very short milling time. Again, about 8 h of milling should be necessary to obtain a good dispersion. Whether oxygen content or specific surface area are responsible is difficult to estimate. From recent literature (3) it appears that under our sintering conditions both contribute to better sintering.

7 min 1 h 8 h 50 µm

Figure 2. Dispersion of Y_2O_3 in Si_3N_4 after different milling times (BEI).

5 min 30 min 20 h 100 µm

Figure 3. Effect of Si_3N_4 agglomerates on microstructure of sintered material (1 h at 1600°C).

CONCLUSIONS

The aim of this study was to evaluate the influence of powder processing on the sintering behaviour of commercial silicon nitride powders made by nitridation of silicon. It should be stressed that the conclusions of this study relate only to these powders (and probably to similar nitrided powders from other sources), to the applied amounts of sintering additives and to pressureless sintering. It has been shown (3) that the relative importance of powder characteristics may be totally inversed for other types of powders.

Agglomerated powders need a moderate milling time to break down the agglomerates of the silicon nitride and disperse the sintering additives throughout the mixture.

Better results are obtained with disagglomerated powders. Again, moderate milling times are necessary to disperse the sintering additives in the silicon nitride.

Prolonged milling is not needed to obtain homogeneous powder mixtures but to reduce the specific surface area of the silicon nitride. Only in this way high densities can be achieved.

If powders with a higher specific surface area should be available, the powder conditioning step could be drastically reduced to the level of homogeneous dispersion of the sintering additives.

REFERENCES

1. Braue, W., Wötting, G. and Ziegler, G., Influence of Impurities in Different Processed Si_3N_4-Powders on High-Temperature Properties of Sintered Materials. In British Ceramic Proceedings, 1986, 37, Special Ceramics 8.

2. Wötting, G. and Ziegler, G., Powder Characteristics and Sintering Behaviour of Si_3N_4-Powders. Powder Metallurgy Int., 1986, 18, 25-32.

3. Nietfeld, G., Wickel, U. and Wötting, G., Tailoring of Si_3N_4-Powder Properties concerning Sintering Behaviour and Mechanical Properties. 2nd Int. Conf. on Ceramic Powder Processing Science, 1988, Berchtesgaden, FRG (in press).

4. Negita, K., Effective sintering aids for Si_3N_4 ceramics. J. of Materials Sci. Letters, 1985, 4, 755-758.

5. Kingery, W.D., Densification in the Presence of a Liquid Phase. J. of Applied Physics, 1959, 30, 301-306.

6. Ziegler, G., Heinrich, J., Wötting, G., Relationships between processing, microstructure and properties of dense and reaction-bonded silicon nitride. J. of Materials Sci., 1987, 22, 3041-3086.

7. Herbell, T.P., Freedman, M.R. and Kiser J.D., Particle-Size Reduction of Si_3N_4-Powder with Si_3N_4 Milling Hardware. Ceram. Eng. Sci. Proceedings, 1986, 7, 817-827.

8. Zarnon, L.C. and Drew R.A.L., Dispersion Methods of Yttria in Silicon Nitride by Comminution. J. Am. Ceram. Soc., 1989, 72, 495-98.

EFFECT OF SiO_2 ON GRAIN GROWTH AND DENSIFICATION OF ALUMINA PREPARED BY SOL - GEL TECHNIQUE

Hai - Doo KIM and Hwan - Kyu KANG
Korea Institute of Machinery & Metals,
Changwon, Korea

1. Introduction

Alumina is one of the most important technical ceramics and has been intensively investigated because of its versatile applications. In alumina technology each property is strongly dependent on the microstructure. For example, mechanical and electrical properties will be improved when grain size and porosity are reduced to minimum; that means, grain growth should be suppressed and densification be enhanced during sintering process.

There are many ways to meet such requirements, one of which is the use of additives. If small amount of additive (normally 2+ or 4+) is dissolved in alumina lattice, Al^{3+} site is replaced by the additive cation, which increases the concentration of defects, thus, enhances the diffusion process.

The most popular additive for alumina sintering is MgO. The most widely accepted Coble theory (1-3) says oxygen ions transport faster along grain boundaries accompinied with the volume diffusion of aluminium ions which is the rate-determining, which leads to the idea that both 2+ and 4+ additives enhance the sintering of alumina due to the formation of aluminium interstitials and aluminium vacancies, respectively.

Recently, the enhanced grain boundary diffusion of oxygen ion has been argued by Reynen and Kim (4), based on the works of Kröger (5-7), in which the diffusing species on grain boundaries is not oxygen ions but oxygen atoms. Reynen and Readey theory (8-11) predicts that sintering process of alumina would be retarded by doping 4+ due to the decrease in oxygen vacancy concentration .

In this study, change in sintering behavior with a SiO_2 additive (4+) is discussed in relation with microstructural development.

In paticular, the general sintering temperature of alumina is over 1600°C, which is above the eutectic temperature in binary system of Al_2O_3 - SiO_2 and ,therefore, the use of SiO_2 as sintering additive for alumina of which sintering temperature is over 1600°C has not been considered in previous. However, recent research trend in alumina technology is towards the low temperature sintering of alumina; Yeh and Sacks (12) suceeded to sinter alumina near theoretical density (TD) at 1150 °C, which was difficult to imagine in previous days. Furthermore, some alumina powders produced in industries sinter to near TD at $<$ 1550 °C which is below the eutectic temprature of Al_2O_3 -SiO_2 system. Threfore, it's meaningful to assess the effect of SiO_2 as an additive for alumina powder which sinters at very low temperature.

2. Experiments

0, 100, 1000, 10000 ppm of SiO_2 were doped to alumina through sol - gel process in order to be sure that small amounts of SiO_2 additive are homogeneously mixed (Fig. 1) . Table 1 shows the impurities of the starting materials . Especially alumina seed (< 0.5 μm) has been prepared by sedimentation method . Gels have been dried at 70 ^{o} C for two weeks in drying oven and gel fragments of over 1 cm in size were collected for sintering experiments.

DTA-TG analysis was performed in order to determine the transformation temperature of boehmite to α - Al_2O_3.

A series of gel fragments was sintered at 1350 ^{o}C, 1450 ^{o}C, 1550°C for 3, 30, 300 mins. in air . Up to 900 ^{o}C heating rate was 3 ^{o}C/min in order to avoid the cracks followed by 10 ^{o}C/min to the desired temperatures.

Table 1. The impurities of the materials used

(wt.%)

	AlOOH*	SiO_2**	Al_2O_3***
Al_2O_3		0.05	
SiO_2	0.022		0.04
TiO_2		0.03	
Fe_2O_3	0.016	0.003	0.01
Na_2O	0.002		0.03
carbon	0.4		
HCl		0.005	
average particle size (mm)	11	7	300
BET(m²/gr)	180	380	5
LOI		< 2.5	0.1
H_2O		< 1.5	0.2

* AlOOH Disperal R Alumina, Condea Chemie, W. Germany

** SiO_2 Amorphous fumed silica, Degussa, W. Germany

*** Al_2O_3 Ceralox HPA alumina, Ceralox, U. S. A.

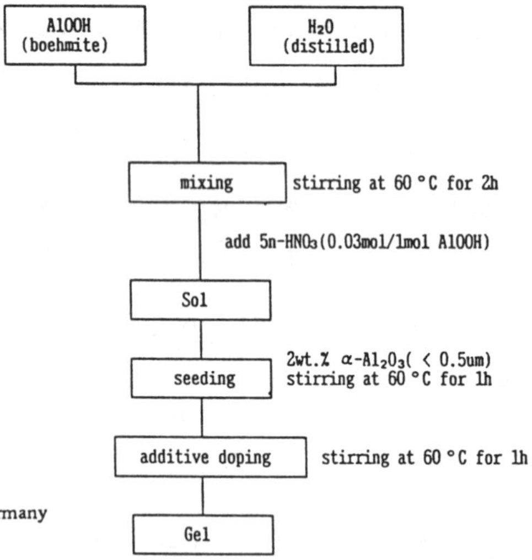

Fig. 1. Schematic diagram for the preparation of gel specimens

The bulk density after sintering was measured by Archimedes principle and, for the specimen with open pores, they were coated with lacquer before measuring density.
The microstructures were investigated by scanning electron microscope after polishing and thermal etching.

3. Results and Discussions

Fig.2 shows the DTA - TG results of each gels which has been dried for 2 weeks at 70 ^{o}C. Boehmite transforms to $\alpha - Al_2O_3$ at 1210 ^{o}C while seeded specimen as well as seeded specimen doped with SiO_2 transform to α - Al_2O_3 at 1100 ^{o}C. The transformation temperature of 1100 ^{o}C is relatively higher than the previous reports (13-15) which may reflect no special treatment for boehmite powder to eliminate the agglomerates.
The density of the sintering at various conditions were shown in Fig. 3. At low temperature and short sintering time, seeded specimens without SiO_2 additive show higher density than those with SiO_2 additive while, at higher temperature and prolonged sintering time, they show at least same density or reverse.
Without SiO_2 additive sintered density does not go above 99 % TD even with 1550 ^{o}C, 300 min sintering while , with SiO_2 additive, it goes above 99 % TD.
This possibly suggests that SiO_2 additive decreases the grain growth and densification during initial and intermediate stage sintering, however, also inhibits the onset of discontinuous grain growth, thus , leading to higher density.

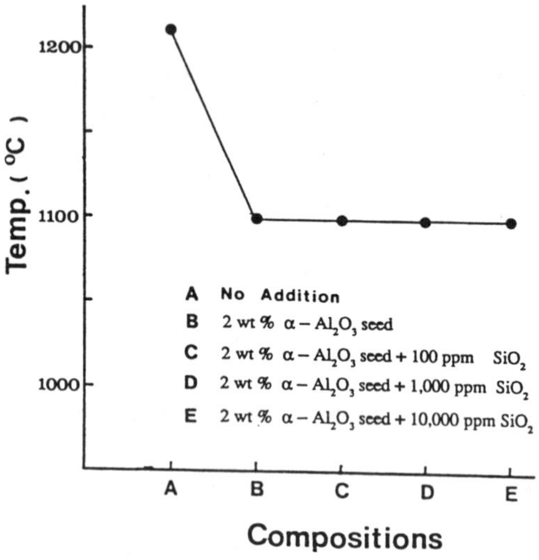

Fig. 2. Phase transformation temperatures to α - Al_2O_3

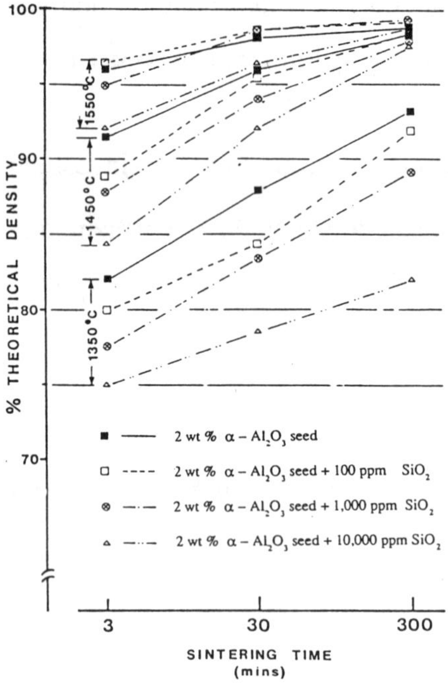

Fig. 3. Change in density at various conditions

This view is further supported by the plot of grain size against percent theoretical density (Fig. 4).

The seeded specimen without SiO_2 additive shows the increase in grain size with the increase in density followed by the rapid increase in grain size as the density approaches the theoretical density. The seeded specimens with SiO_2 additive show very little grain growth until ~ 93 - 95 % TD, beyond which relatively lesser grain growth occurs than the above case.

It seems clear that SiO_2 additive retards the grain growth of alumina until the end of intermediate stage, enables the pores remain attached to grain boundaries, thus , enables them to sinter to near theoretical density. This also means that grain growth rate is relatively slow when pores are interconnected while it becomes relatively rapid when pores are isolated. Without SiO_2 additive grain growth is not retarded and , thus , an isotropic grain growth and discontinuous grain growth are quite possible.

The microstructural evidence for above arguments is presented in Fig. 5.

Fig. 5 shows the difference in microstructures between seeded specimen and seeded + 1000 ppm SiO_2 specimen sintered at 1550 °C for 3 and 300 mins. Without SiO_2 additive, large anistropic and discontinuous grain growth occurred at 1550 °C, 300 mins sintering while, with SiO_2 additive, no anisotropic and discontinuous grain growth occurred and its microstructure is characterised by high density with small grain size.

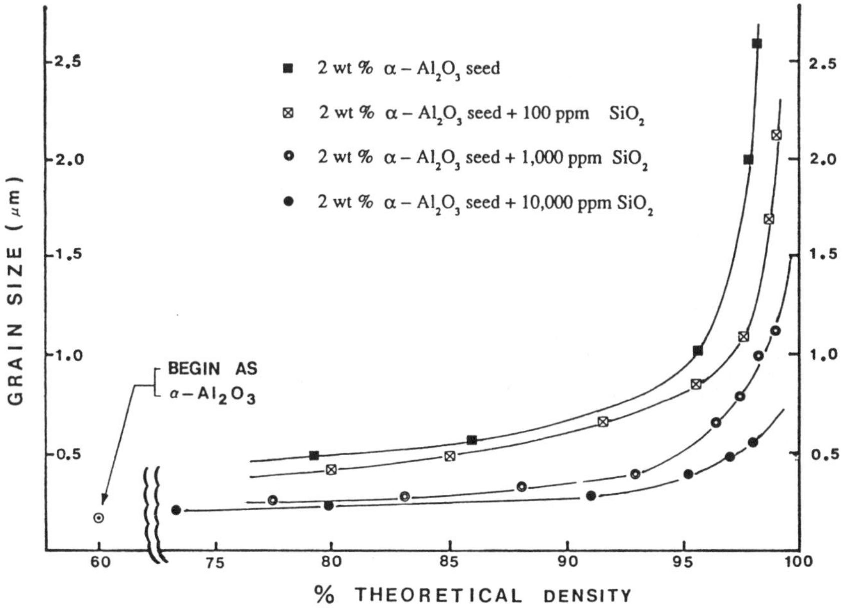

Fig. 4. The plot of grain size vs percent theoretical density. The sintered density just after transformation to α-alumina (sintered at 1120°C) was measured to 60 % TD and grain size was calculated from BET data.

Especially, the seeded specimen with 1000 ppm SiO_2 sintered at 1550 ^oC, 3 min shows ~ 95 % TD and 0.6 μm mean grain size, which seems to be just the end of intermediate stage. This specimen shows that all the pores remain at grain corners or grain boundaries, which is the favourable situation for densification to TD, thus, resulting in the near TD.

Fig. 5. Microstructural development of seeded specimen with and without SiO_2 additive sintered at 1550 ^oC in air

The solubility of SiO_2 in Al_2O_3 at high temperature is not known. If we assume that the solubility of SiO_2 in Al_2O_3 is very low, the remarkable retarding effect of SiO_2 on the grain growth is possibly due to the second phase formation, i.e. mullite. The formation of mullite through sol - gel process has been reported by Ismael et al. (16) in which mullite starts to be observed at 1300 ^oC and , at 1400 ^oC - 1 hour, all XRD peaks correspond to mullite. The sintering temperatures employed in this experiment are high enough to give mullite, which was, however, not yet detected because of its extreme small grain size and small quantity.

4. Conclusions

Without additive grain growth is not retarded and, thus, anisotropic and discontinuous grain growth are inevitable at high temperature.

SiO_2 additive retards the grain growth of alumina until the end of intermediate stage, enables the pores remain attached to grain boundaries, thus, enables them to sinter to near theoretical density.

The remarkable retarding effect of SiO_2 on grain growth of alumina is possibly due to the second phase formation, i.e. mullite.

References

1. Coble, R. L. " Sintering crystalline solids : I. Intermediate and final stage diffusion models", J. Appl. Phys. 32, 787 - 92, (1961)
2. Coble, R. L. " Sintering crystalline solids : II. Experimental test of diffsion models in powder compacts", J. Appl. Phys. 32, 793 - 99, (1961)
3. Paladino, A. E. and Coble, R. L. " Effect of grian boundaries on diffusion -controlled processes in aluminium oxide", J. Am. Ceram. Soc. 46, 133-36, (1963)

4. Reynen,P. and Kim,H.D. "The influence of Na_2O on sintering and creep of alumina doped with MgO and TiO_2", cfi/Ber. DKG 63, 272-79, (1986)
5. Hou,L.D., Tiku,S.K., Wang,H.A. and Kröger,F.A. "Conductivity and creep in acceptor-dominated polycrystalline Al_2O_3", J. Mater. Sci. 14, 1877-89, (1979)
6. Wang,H.A., and Kröger,F.A. "Chemical diffusion in polycrystalline Al_2O_3", J. Am. Ceram. Soc. 63, 613-19, (1980)
7. El-Aiat,M.M., Hou,L.D., Tiku,S.K., Wang,H.A. and Kröger,F.A. "High temperature conductivity and creep of polycrystalline Al_2O_3 doped with Fe and/or Ti", ibid, 64, 174-82, (1981)
8. Reijnen, P., " Non-stoichiometry and sintering of ionic solid", in Reactivity of Solids, Proc. 6th Int. Symp. Reactivity of Solids, p. 99-114, (1968)
9. Reijnen, P. " Sintering behavior and microstructure of aluminates and ferrites with spinel structure with regard to deviation from stoichiometry", Sci. Ceram., 4, 169-88, (1968)
10. Reijnen, P. "Nonstoichiometry and sintering in ionic solids", in Problems of Nonstoichimetry, Ed. Rabenau, A., North Holland Pub. Co., p, 219-38 (1970)
11. Readey, D.W., " Mass transport and sintering in impure ionic solids", J. Am. Ceram. Soc., 49, 366-69, (1966)
12. Yeh,T.S. and Sacks, M.D., " Low temperature sintering of aluminum oxide", ibid, 71, 841-44, (1988)
13. Kumagai, M. and Messing, G.L. "Enhanced densification of boehmite sol - gel by α - alumina seeding " , ibid, 67, c-230 - c-231, (1984)
14. Kumagai, M. and Messing G.L. " Controlled transformation and sintering of a boehmite sol - gel by α – alumina seeding ", ibid, 68, 500-05, (1985)
15. Kim, H.D. " Microstructure Control in Alumina", ph.D. thesis, Univ. Aachen, W. Germany, (1987)
16. Ismael,M.G.M.U., Nakai, Z., Minegishi, K. and Somiya. S. "Synthesis of mullite powder and its characteristics", Int. J.High Tech.Ceram. 2, 123-34, (1986)

THE EFFECT OF BARIUM OXIDE ON THE DENSIFICATION AND DECOMPOSITION OF ZIRCON

by

N S Carr and A W Paterson
CSIR, Pretoria

ABSTRACT
The effects of barium oxide, calcium oxide, temperature, time and combinations of these factors on the densification and decomposition of zircon were studied using factorial designed experiments. Measurements were made of density, apparent porosity, Young's modulus and relative zircon XRD intensities. A microstructural study was also carried out on an interfacial reaction zone prepared from zircon and barium oxide powder compacts. The results show evidence of strong capillary action and intergranular ionic diffusion resulting in a broad spectrum of phases. These were studied in the SEM using EDS. The phases observed and the factorial effects are interpreted with reference to the $BaO-SiO_2-ZrO_2$ compatibility diagram.

1. Introduction

This series of experiments was designed to investigate the densification and decomposition behaviour of zircon in the presence of barium oxide as a dopant. Two experimental techniques were used. The first was a quantitative study of the effects of BaO, CaO, sintering time and temperature on the bulk density, percent apparent porosity, Young's modulus and X-ray diffraction intensities of zircon. The second approach was a qualitative study of the effect of BaO on zircon using SEM and EDX techniques. The decomposition and densification of zircon is important in its own right and, in conjunction with alumina, in the processing of zirconia-mullite which has been widely studied [1-3].

2. Experimental

2.1 Factorial Design

Factorial experimental designs make possible the investigation of a number of variables of interest in a single composite experiment. This is organized in such a way that information derived from analysis is specific to each variable. The technique also allows for the quantification of interactions between variables. This separation of effects in complex systems facilitates a clearer understanding of the mechanisms at work. The technique is described more fully elsewhere by Carr and Paterson [4]. The design used in this case was a 2^4 factorial. The results are presented as normal probability plots. Any points falling off the straight lines are considered real non-random effects.

Sample Preparation and Testing for Factorial Studies
All the samples were prepared by milling 250 g superfine (0.7μm) zircon and analar $BaCO_3$ and $CaCO_3$ (where applicable) in 200 g demineralized water with 10 g polyethylene glycol (M.Wt.200) for 1 hour in an attritor. The powders were then dried at 75 C for 16 hours. Green compacts were uniaxially pressed at 5 MPa and CIPed at 200 MPa. After sintering the samples were tested for density, apparent porosity, Young's modulus and zircon XRD intensities on selected peaks.

Sample Preparation for SEM and EDX Study
Powder compacts were prepared as illustrated in Fig. 1. A loose compact (A) of zircon was formed in a 20 mm diameter die. A smaller compact of $BaCO_3$ (B) was formed in a 20 mm diameter die and placed in the centre of A. Finally zircon (C) was added on top of A and B and uniaxially pressed to a pressure of 6 MPa and CIPed at 245 MPa. The sample was fired for 4 hours at 1 450 C after which it was sectioned as shown in Fig. 2. After polishing the section was studied using electron microscopic techniques.

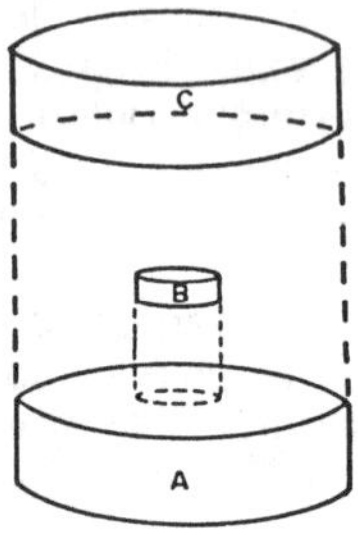

FIGURE 1: Compact formation for interfacial reaction study

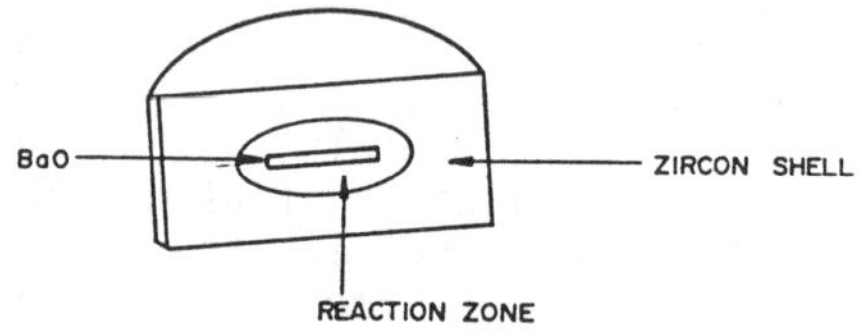

FIGURE 2: Cut-away section of compact after sintering

RESULTS
2^4 Factorial Design
The results of the 2^4 factorial experiment are summarized in Figs. 3a - d. Fig. 3a shows that only CaO and BaO had a significant effect on the zircon XRD intensities. A 2wt.% addition of CaO and BaO resulted in 17 and 7,3% reduction in intensities respectively. If the molar ratios of CaO and BaO are computed the extent of the effect expressed as a ratio is 1:1.3 Figure 3b summarizes the variable effects on density. In this case temperature, time and the BaO/CaO interaction all tend to increase density, with temperature having the greater effect. The other two are closely equivalent. Figure 3c summarizes the effects on porosity. Here only temperature and time are significant with temperature having the greater effect. Figure 3d summarizes the Young's modulus effects. All four variables have a significant effect including the dopant interaction. Temperature and time have a strongly enhancing effect which would be expected due to the observed effect on porosity and density. The dopant effects are both negative which correlates with the decomposing effect observed on the XRD intensities. An unusual feature is despite the dopants having a negative effect on Young's modulus, the interaction of both is positive. This can be explained by the positive effect of the interaction on density presumably due to eutectic formation assisting densification.

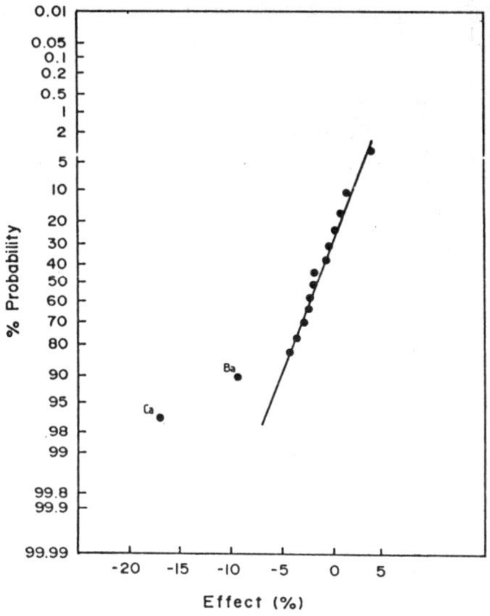

a) Effects on XRD zircon intensity

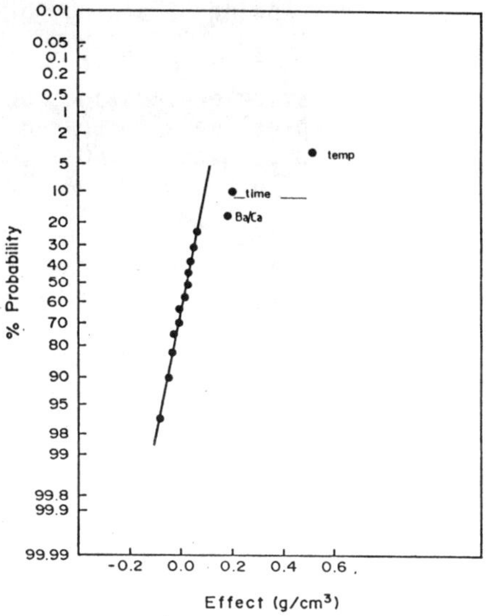

b) Effect on density of compacts

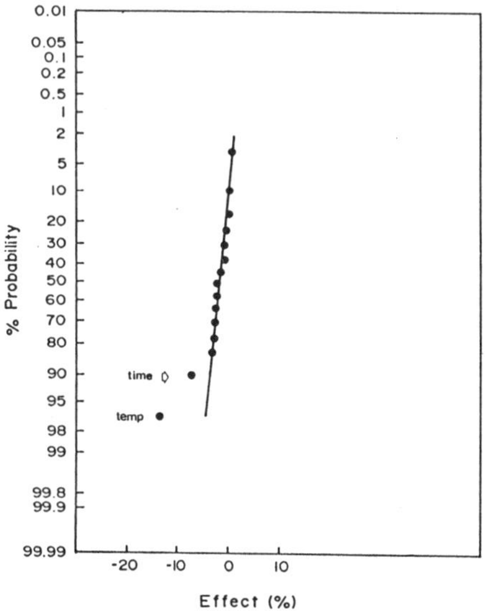

c) Effect on porosity of compacts

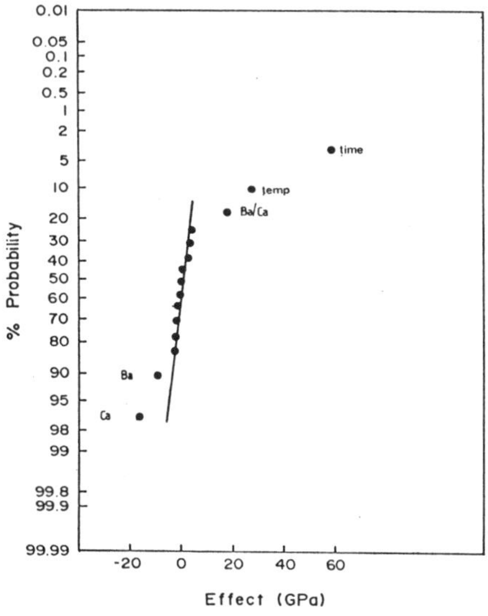

d) Effect on Young's modulus

FIGURE 3: Normal probability plots of effects shown in 2^4 factorial experiment.

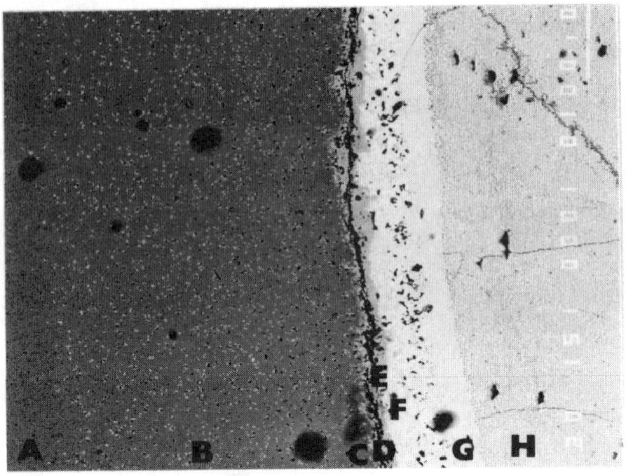

PLATE 1: General view across reaction zone

TABLE 1: Description of zones across reaction interface

Zone	Description
A	zircon - homogeneous porosity
B	zircon - inhomogeneous porosity - ZrO_2 precipitates
C	zircon - inhomogeneous porosity - few ZrO_2 precipitates
D	ZrO_2 precipitates, crack
E	ZrO_2 precipitates and barium zirconium silicate (BAS) matrix
F	Large barium zirconate crystals and BAS matrix
G	Fine barium zirconate crystals and BAS matrix
H	Barium oxide and barium zirconate

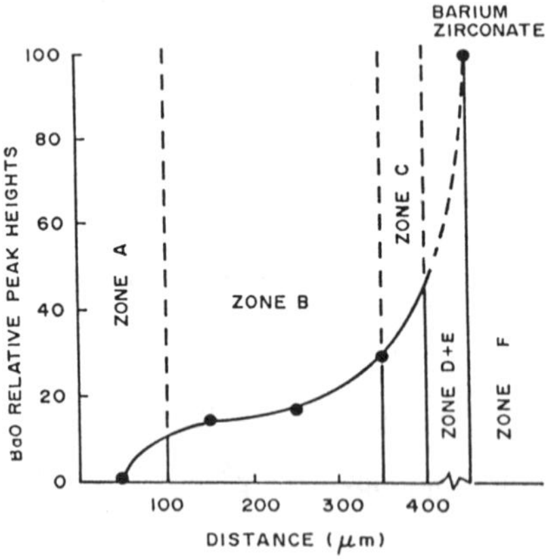

FIGURE 4: Plot of relative barium intensities across the $BaO-ZrSiO_4$ interface
$BaO-ZrSiO_4$ interface

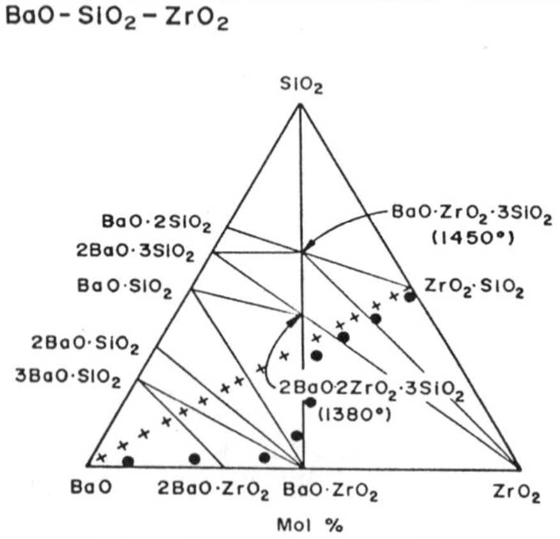

FIGURE 5: Compatibility diagram of the $BaO-SiO_2$ system after Chukhlantsev and Galkin [5]

SEM Study
The interfacial reaction zone between barium oxide and zircon was studied using scanning microscopic (SEM) and energy dispersive X-ray spectroscopy. Plate 1 gives a general view across the reaction zone. For descriptive purposes 8 zones were identified by morphology differences. These are labelled in plate 1 and described in Table 1. These were identified by EDS spectra. Figure 4 shows the relative barium concentration across the interface derived qualitatively from the EDS spectra. The points at which intensities were measured are marked on plate 1.

DISCUSSION

Barium has been shown to have a substantial effect on the decomposition and sintering of zircon at 1 450 C. Reaction appears to begin with the formation of an amorphous barium zirconium silicate with zirconia precipitates. The silicates provide a medium for couterflow of barium and zirconium ions driven by thermodynamic requirements. This has two major effects, namely capillary action which wets the grain boundaries of the crystals enhancing densification at the same time as providing a medium for rapid ionic mobility and grain growth. The silicate liquid allows the rapid coalescence of pores which are then trapped and cannot be sintered out. Zircon sintered in presence of barium oxide has a proportion of non-homogeneous porosity. Fig. 5 shows the compatibility diagram for the $BaO-SiO_2-ZrO_2$ system [9]. It has been reported that the two ternary compounds could be formed by heating the oxides at 1 300 C for 24-30 hours. The 2:2:3 compound melts incongruently at approximately 1 380 C with formation of the 1:1:3 compound and liquid. The 1:1:3 compound melts congruently near 1 450 C. The compositional variation expected in the system is shown by crosses from BaO through to zircon in Fig. 5. The EDS spectra however shows that the path followed is closer to that illustrated by the dotted line. There is very little SiO_2 present in the barium core despite the presence of zirconia. There are problems with clear interpretation since very little work has been reported on this system and any correlation with the $CaO-Al_2O_3-SiO_2$ system is invalid owing to many more phases being present.

REFERENCES

1. E. Claussen, J. Jahn, Mechanical properties of sintered in situ-reacted mullite-zirconia composites, J. Am. Ceram. Soc., 3-4, [64], (1980) 228-229.
2. E. Di Rupo, M.R. Anseau, R.J. Brook, Reaction sintering: Correlation between densification and reaction, J. Mat. Sci., [14], (1979), 2924-2928.
3. P. Pena, P. Miranzo, J.S. Moya, S. De Aza, Multicomponent toughened ceramic materials obtained by reaction sintering, Part 1, ZrO_2 - Al_2O_3 - SiO_2 - CaO system, J. Mat. Sci., [20], (1985), 2011-2011.
4. N. Carr and A.W. Paterson, Optimization of properties in zirconia-mullite, in preparation.
5. Phase diagram for Ceramicst, 1975 supplement, J. Am. Ceram. Soc. Inc., p. 221, Fig. 456.

ALUMINIUM TITANATE-MULLITE MATERIALS OBTAINED BY REACTION SINTERING OF TITANIA-BAUXITE MIXTURES

A. Caballero, P. Pena, A. L. Cavalieri*, and S. de Aza
Ceramics Department,
Instituto de Cerámica y Vidrio, C.S.I.C. Arganda del Rey
28500- Madrid, SPAIN

ABSTRACT

Aluminium titanate materials have been obtained by reaction-sintering of bauxite-titania mixtures. Dynamic sintering and phase evolution were analyzed as a function of firing temperature. The mechanical properties, microstructure and thermal stability of these aluminium titanate ceramic materials was also studied. Finally, the coefficient of thermal expansion was determined.

INTRODUCTION

Aluminium titanate is an attractive material for application in the manufacture of engines, kiln furniture, glass molds, pipes and launders due to his low thermal expansion characteristics (1). However this materials display low mechanical strength and turned out to be unstable in the temperature range from 1000 to 1200 C (2), this characteristics rendering it apparently uselless for industrial aplications.

A way of increasing both mechanical properties and thermal stability of Al_2TiO_5 may be suggested considering the isobaric pseudoquaternary system Al_2O_3-SiO_2-TiO_2-Iron oxide in air (3) as observed in Figure 1. In this system there is a two phase field where Al_2TiO_5 (ss) and mullite (ss) coexist. Consequently, if appropiate mixtures of raw bauxite, where iron oxide is always present (1-3 w%), and titania are used, aluminium titanate based materials could be obtained.

The aim of this work is to determine the factibility to obtain aluminium titanate materials by reaction sintering of bauxite-titania mixtures according to the reaction:

----

* Permanent adress in CETMIC-CONICET-CIC, La Plata ARGENTINA

$$(85\%\ Al_2O_3+TiO_2+Fe_2O_3)+TiO_2 \rightarrow Al_{2-x}Fe_xTiO_5 + Al_{6-y}Fe_ySi_2O_{13}$$
$$\text{bauxite} + \text{titania} \rightarrow \text{aluminium titanate} + \text{mullite}$$

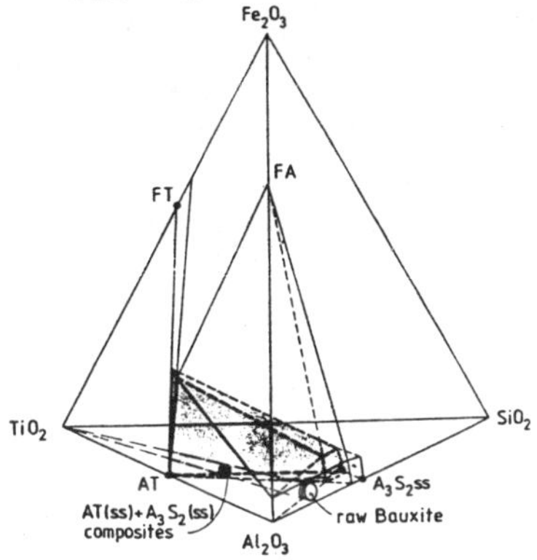

Figure 1.- The system Al_2O_3-SiO_2-TiO_2-Fe_2O_3 in air showing the Al_2O_3 solid state compatibilities in the temperature range 1308-1390 C.

EXPERIMENTAL PROCEDURE

Titanium oxide and Surinam bauxite were used as starting materials. Weighed powders were wet mixed by attrition milling with Al_2O_3 balls in water, dried, sieved and calcined at 920 C for 1h. Powders were then isostatically pressed to rods of approximately 4 mm diameter at 200 MPa. The chemical analysis of the as milled powders are 36.90 % Al_2O_3, 3.55 % SiO_2, 36.90 % TiO_2, 0.90 % Fe_2O_3, 0.10 % MgO, 0.18 % ZrO_2, 0.03% CaO, 0.05% Na_2O, 0.04% K_2O and 1.88 % L.O.I.

Green samples were fired in an electrical furnace in the temperature range 1200 to 1450 C in air.The phase content was determined by X-ray diffraction using the CuKα radiation of a Philips 1010 diffractometer with nickel filter. The rupture modulus was measured in an Inström machine using three-point load, a 20 mm span, and a cross-head speed of 0.5 mm/min. The microstructure was studied by scanning electron microscopy (SEM) on polished and thermally etched surfaces.The shrinkage behaviour during constant heating was established in a dilatometer Adamel Lhomargy and the thermal expansion measurements were made in the same equipment using a silica device and an alumina standard calibration.

RESULTS AND DISCUSSION

During constant heating rate sintering, Figure 2, the sample exhibits high densification rate between 1000 and 1250^0 C. The densification rate slows down abruptly at 1270 ^{0}C, indicating the expansive formation of aluminum titanate (90 vol%). The skrinkage curve shows another two expansive effects at ~980 and 1400^0 C. These effects are relative to primary and secondary mullite formation (4).

The phase contents of the specimens as a function of

sintering temperature are depicted in Figure 3. From this figure, it can be deduced that aluminum titanate formation takes place at temperatures higher than 1250 °C, which is in good agreement with dinamic sintering results (Figure 2). The mullite formation starts at 980° C but it must be raised up to 1450 °C to reach a complete reaction.

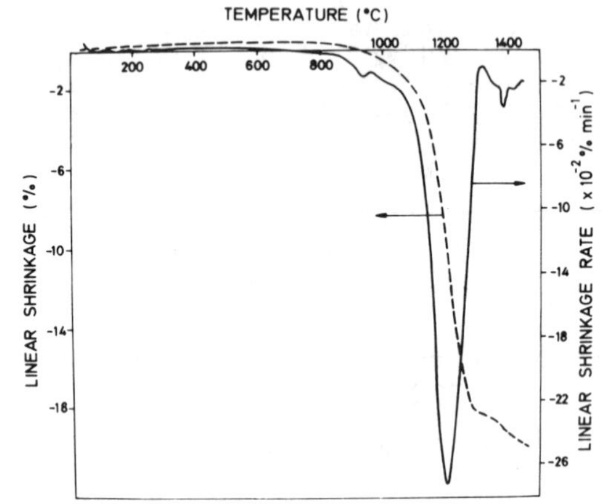

Figure 2.- Linear shrinkage and shrinkage rate of powder compact. (Heating rate, 2 °C/min.).

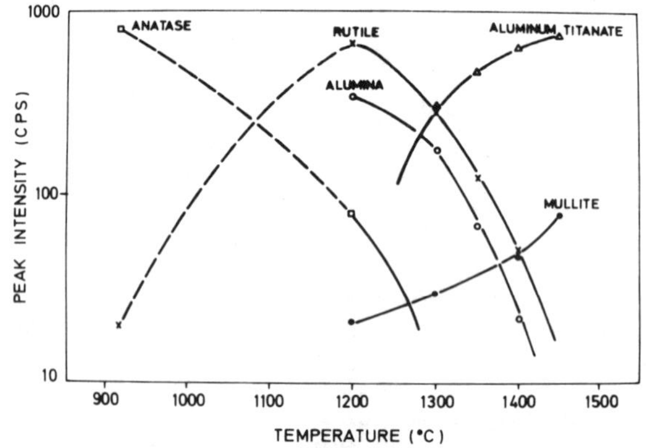

Figure 3.- Phase content of the sintered specimens as a function of sintering temperature from the following XRD-peaks: Alumina=Al_2O_3(113); Rutile=r-TiO_2(110); Anatasa=a-TiO_2(101); Mullite=$Al_6Si_2O_{13}$(331); Aluminium Titanate= β-Al_2TiO_5(023).

After sintering at 1450° C for 2 hours, only mullite and aluminium titanate were found. Iron oxide is present in solid solution in mullite and aluminium titanate crystals, as confirmed by EDX microanalysis.

The aluminium titanate formation takes place through corundum and rutile as can be seen in Figure 2, although all the starting powders contained TiO_2 in the anatase form and alumina as gibssite.

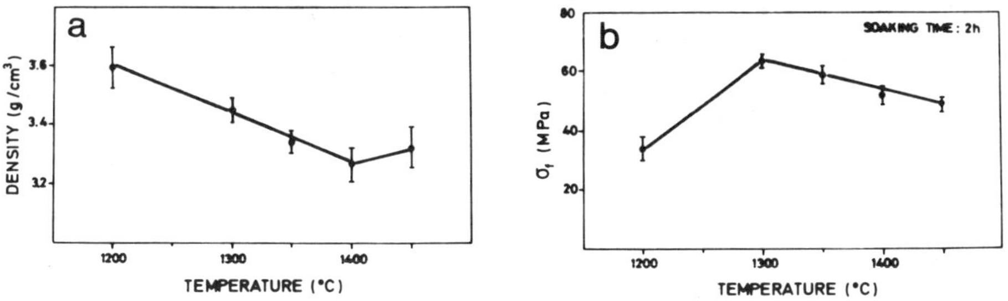

Figure 4.- a) Bulk density versus sintering temperature.
b) Bending strength versus sintering temperature.

The density and bending strength of specimens sintered for 2h at temperatures ranging from 1200° to 1450° C are shown in Figure 4a. The density value decreases when temperature increase up to 1400°C as a consequence of the aluminium titanate formation. At temperatures higher than 1400°C, small amounts of liquid phase, due to the presence of minor impurities, appears increasing the density (5).

The bending strength (σf) is plotted versus sintering temperatures in Figure 4b. The (σf) value at temperatures higher than 1300°decreases linearly with increasing sintering temperature because of the aluminium titanate formation. The sample heat treated at 1200°C shows a small value of σf and high density as a consequence of the partial sintering and lack of aluminium titanate formation at this stage (Fig.2).

The microstructure of the sample sintered at 1450°C for 2h (Figure 5) shows only two crystalline phases and a low porosity. Aluminium titanate is present as matrix and a few mullite grains are mainly located in grain boundaries and triple points. Aluminium titanate grains are equiaxials with a grain size about 3 μm while mullite crystals have columnar shapes with square sections and dimensions smaller than 1 μm. The porosity is mainly intergranular and only a few intragranular pores are present. Finally, only a small amount of microcracking is observed.

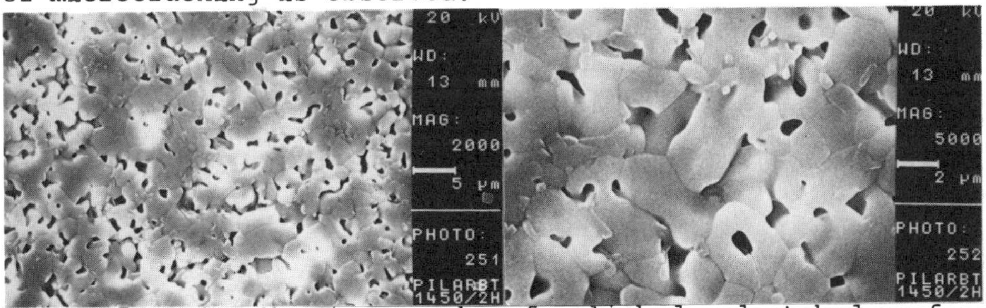

Figure 5.- S.E.M. micrographs of polished and etched surface of sample sintered at 1450°C/2h.

The sample sintered at 1450°C for 2h was treated at 1150°C during 70 hours. After heating, about 80 % aluminium titanate is even present. This value is close to that obtained in MgO stabilized aluminium titanate materials (6).

The thermal expansion behaviour of the specimen sintered at 1450 °C for 2h is shown in Figure 6. The sample shows a low thermal expansion ($\alpha_{25-900^{\circ}} \simeq 3 \cdot 10^{-6}\ K^{-1}$) and the heating-cooling hysteresis cycle is the typical one for aluminium titanate based materials (7). The relative high value of CTE is due to the low microcraking effect as can be seen in the sample microstructure showed in Figure 5.

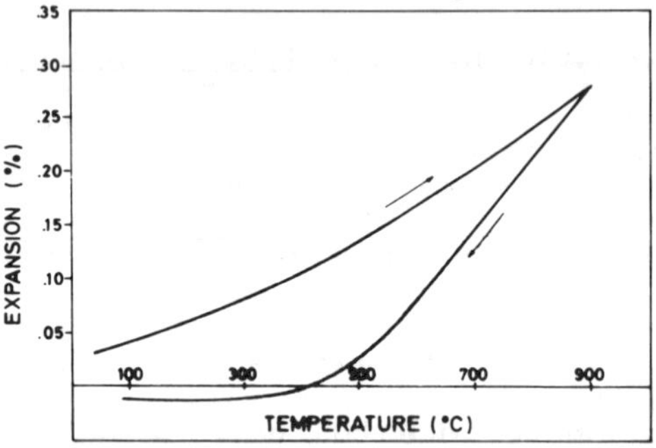

Figure 6.- Thermal expansion curve between room temperature and 900 °C. (Heating rate 2°C/min).

FINAL CONSIDERATIONS

Alumimium titanate materials have been obtained from bauxite and titania by reaction-sintering. These materials exhibit good mechanical and thermal properties similar to those obtained from pure oxides.

REFERENCES

1. Stingl, P.,Heinrich, J. and Huber, J., Properties and applications of aluminium titanate components. In Proc. of the 2nd Int. Symp.on Ceram. Mat. and Comp. for Engine, ed. W. Bunk and H. Hausner , DKG, Bad Honnef, 1986, pp.69-80
2. Kato, E., Daimon, K.and Takahashi, J., Decomposition temperature of -Al_2TiO_5. J. Am. Ceram. Soc., 1980, 63, 355-56
3. Caballero, A. and de Aza, S., Phase equilibria relations in the system Al_2O_3-SiO_2-TiO_2·iron oxide in air. In High Tech Ceramics, ed. Vicenzini, Elsevier Sci. Publishers, Amsterdam, 1987, pp. 229-36
4. Moya, J. S. and de Aza, S., Equilibrium diagrams. A tool for designing new ceramics. In Sci. of Ceramics 14 ,ed. D. Taylor. The Inst. of Ceramics, U.K., 1988, pp. 27-40
5. Caballero, A., Refractory bauxites: constitution and thermal behaviour. Interpretation in base of Al_2O_3-SiO_2-TiO_2--Fe_2O_3 system. Thesis, Univ. of Madrid, 1985.
6. Wohlfromm, H., Moya, J.S., Pena, P., Effect of $ZrSiO_4$ and MgO additions on reaction sintering and properties of Al_2TiO_5-based materials. To be published in J. Mat. Sci.
7. Buessem, W.R., Thielke, N.R. Sarakauskas, R.V., Thermal expansion hysteresis of aluminium titante. Ceramic Age, 1952, 60, 38-40

Financial support from CICYT,Spain (Project MAT88-0156)

THE SINTERING AND OXIDATION BEHAVIOR OF ALUMINIUM NITRIDE

P. Barthelme, F. Ansorge, M. Kulig, T. Hofmann*, C. Rüssel
Institut für Werkstoffwissenschaften III (Glas und Keramik)
Universität Erlangen-Nürnberg, D-8520 Erlangen, Martensstr.5

ABSTRACT

Commercially available ultrapure aluminium nitride powders, doped with yttria, were pressurelessly sintered under nitrogen atmosphere in a carbon tube furnace. At temperatures higher than 1800 °C, a theoretical density of >98% was reached. Depending on the sintering conditions, different yttria containing phases appeared in the microstructure. The aim of this treatment was to localize all oxygen contaminations of the aluminium nitride starting material at separated phases, at the grain boundaries or in the triple points.
AlN-powders and compact AlN-ceramics were oxidized in different gas atmospheres like dried and wet air. On the sintered ceramic, a dense layer, solely composed of alumina, was observed. Kinetics of the oxidation obey a parabolic rate-law. Aluminiumoxynitride phases could not be detected at any temperatures investigated. The influence of different gas atmospheres is discussed.

INTRODUCTION

Potentially, AlN ceramics present high thermal conductivity and high electrical insulating properties and are expected to serve for special applications in microelectronics. However, the theoretically excellent properties are influenced for example by purity and particle size of the raw material and by the sinterability and microstructure of the ceramics [1]. AlN can only be sintered economically, that is pressurelessly, by using appropriate additives for the formation of a liquid phase, e.g. with earth alkali oxides and rare earth oxides [2]. One aim was to examine the influence of sintering conditions on occuring phases and microstructure of Y-doped AlN-ceramics.
The possibility of using AlN as an engineering ceramic depends upon its ability to maintain its mechanical strength during long time use under high temperatures [3]. The conditions under which high temperature oxidation of AlN ceramics occurs are described to clarify this question. For a more complete explanation of the oxidation mechanism, additional oxidation tests with AlN powders took place.

EXPERIMENTAL

Commercially available AlN powders were doped with yttrium containing sintering aids (1-3% Y_2O_3 or YF_3), pressed isostatically into cylindrical shapes and sintered pressureless in a graphite tube furnace under nitrogen at temperatures

between 1700°C and 2000°C.
Microstructures and phases in the sintered AlN were analyzed (SEM and XRD); the relative densities and the oxygen content of the ceramics were measured.

Thermogravimetric- and Differential-Thermo-Analyses (TGA and DTA) of an ultra-pure industrially manufactured AlN powder were made. In addition, AlN powder specimens were heated in a muffle furnace under lab-atmosphere for one hour, at temperatures between 800°C and 1200°C.

Industrially manufactured AlN substrates were oxidized at different atmospheric humidities. The oxidation-time was varied between one and six hours at temperatures between 1100°C and 1400°C.

The resulting oxidic layers were examined by using measurements of the relative increase of weight, of the layer-thickness determined by light-microscopy, and of oxygen- and nitrogen-contents. Phases existing in the oxidized substrates and powders were determined with XRD. The boundary area between AlN and the oxidic layer was analyzed with SEM. A scan of the oxygen concentration in this region was recorded by wavelength dispersive X-ray analysis (WDX).

SINTERING BEHAVIOR

Ceramics with a theoretical density ≥ 98% could be produced from all commercially available AlN- powders, according to the sintering activity of respective powders at sintering temperatures between 1800°C and 1950°C. Very short soaking times of less than one hour are sufficient. Dense ceramics contained less oxygen (0.5-1%) than the starting material (0.7- 1.9%). Table 1 shows the relation between sintering conditions and resulting phases in AlN ceramics.

TABLE 1
Sintering parameters, -additives, and phases identified in AlN-ceramics (XRD)

sintering temp. / °C	soaking time / h	additives		phases identified (XRD)		
		% Y_2O_3	% YF_3	YAG	$Al_2Y_4O_9$	none
1700	0.5	1	-	x	-	-
1800	0.5	1	-	x	-	-
1900	0.5	1	-	-	x	-
2000	0.5	1	-	-	x	-
1750	0.5	-	3	-	-	x
1800	0.75	-	3	-	-	x
1800	3	-	3	-	-	x
1900	0.5	-	3	-	x	-
1800	0.5	-	2	-	-	x

When sintering Yttria-doped AlN at temperatures up to 1800°C, yttriumaluminiumgarnet ($Y_3Al_5O_{12}$="YAG") crystallizes during cooling. YAG can be localized at triple points and grain boundaries. By this way, oxygen can be fixed in a separate phase. But by contrast, during cooling from temperatures above 1800°C, monoclinic $Y_4Al_2O_9$ crystallizes.

Using Yttriumfluoride as a dopant at sintering temperatures below 1800°C, no additional phase can be identified in these ceramics. YF_3 is assumed to react with the oxygen contaminations of AlN to form Y_2O_3, developing AlF_3 is vaporized. During this reaction YAG should be formed, but Yttrium gets lost by vaporization: the higher sintering temperatures (for constantly Y-doped AlN) were applied, the lower Y-contents of ceramics were analyzed (1700°C: 0.8%Y; 1800°C: 0.6%Y; 1900°C: 0.3%Y). In many cases, only a small amount of less than 3wt.% YAG is present, which cannot be identified with XRD.

The fracture surface of YF_3-doped AlN ceramics is shown in the SEM-picture of fig.1. Liquid phase sintering affects the slight roundness of the hexagonal grains with a mean grain size of about 5-8 microns. At some points, the pulled off triple point- and grain boundary-phase is recognizable.

Fig.1: Fracture surface (SEM) of YF_3-doped AlN-ceramics (sintered 1h at 1850°C)

OXIDATION BEHAVIOR

At temperatures above 900°C, the AlN powder oxidizes considerably. Catastrophic oxidation takes place at temperatures starting at 1100°C .

Kinetics of the oxidation of AlN obey the parabolic rate-law. Fig.2 shows the increase of the growing oxidic layer according to this squareroot rate-law.

Comparison of the three curves from fig.3 shows the influence of atmospheric humidity on the oxidation behavior of AlN. More interesting than the absolute values of the relative increase of weight is the comparison of changing slopes. At long oxidation times in an atmosphere of high humidity, oxidation is slightly faster than in dry air. However, at short heating times, oxidation is considerably faster than in dry atmosphere.

Only two phases, AlN and α-Al_2O_3, are appearing at all XRD-examinations of oxidized substrates and powders. The resulting white oxidic layers are single phased, cubic α-Al_2O_3. The bulk of the material is scarcely oxidized.

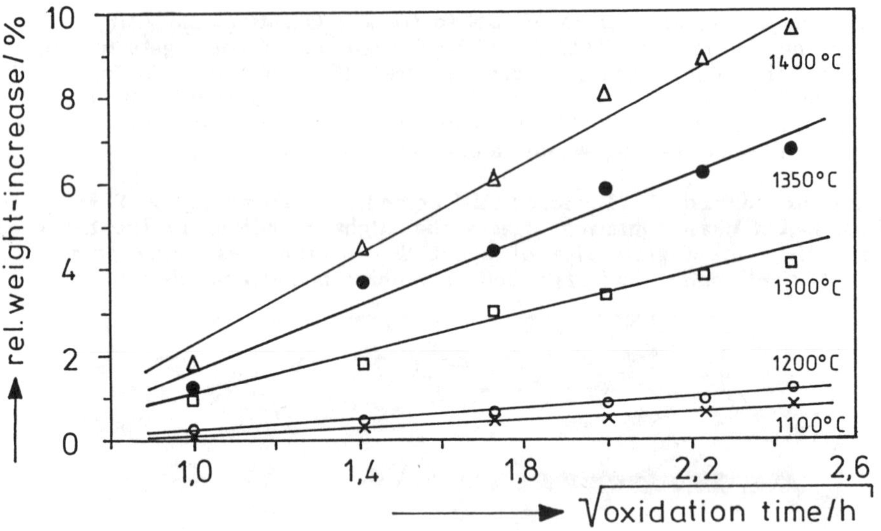

Fig.2: Rel. weight–increase vs squareroot of oxidation time (Oxidation in lab–atm.)

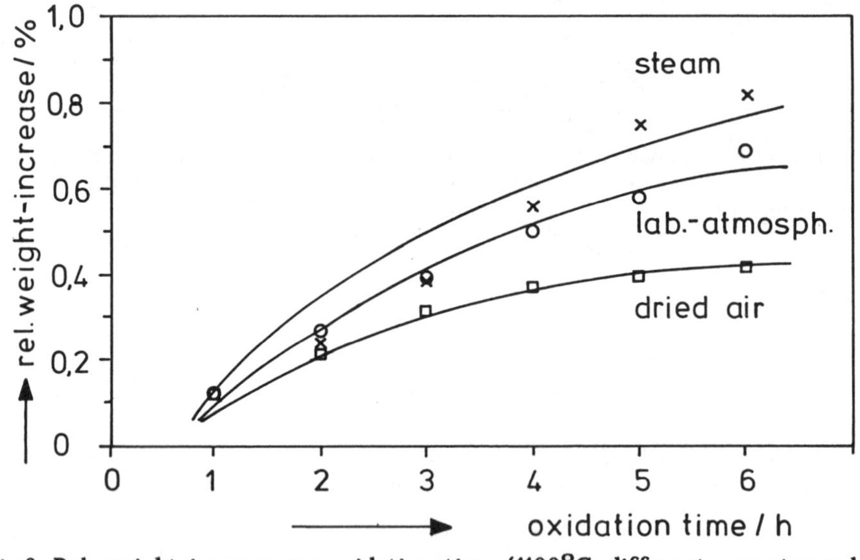

Fig.3: Rel. weight-increase vs oxidation time (1100°C, different gas atmospheres)

Oxynitrides were not found, neither in the oxidic layer nor at an interface between α-Al_2O_3 and AlN-phases. WDX-scans perpendicular to the borderline between the oxidic layer and the AlN ceramic show an oxygen content of two different levels but a constant value in each of both phases.

A cross-section of an oxidized AlN-substrate is shown in fig.4. The cracks in the bright alumina layer presumably are spreading during cooling of the samples because of the different thermal expansions of AlN and Al_2O_3.

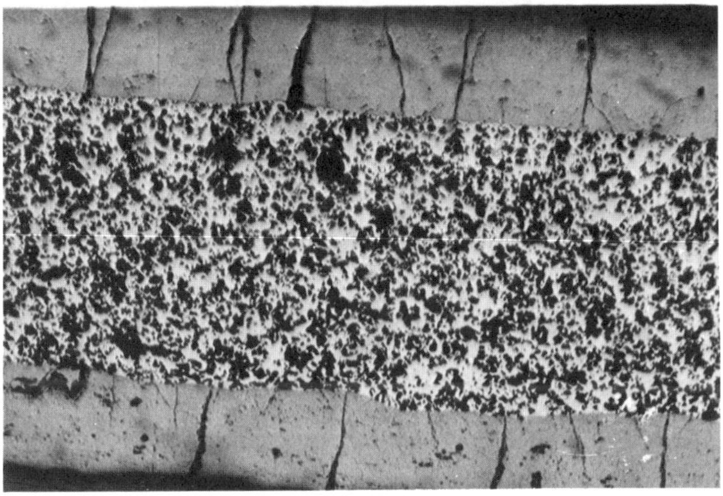

Fig.4: AlN, oxidized for 6 hours at 1400°C in lab-atmosphere (magnification 104 x)

CONCLUSIONS

Oxygen impurities in AlN ceramics partially can be removed by vaporization, or they can be fixed in separate phases. Appropriate sintering conditions effect decreasing contents of yttrium and oxygen in AlN ceramics.

At temperatures below 800°C, AlN-powders are slightly sensitive to oxidation in an oxygen containing atmosphere, while AlN ceramics are oxidized noticeably at temperatures above 1000°C .

A single phased cubic Al_2O_3-layer grows when oxidation occurs. Intermediate oxidation states (e.g. in appearance of aluminiumoxynitrides) were not identified.

Up to 1000°C, Al_2O_3 coatings protect the bulk material by impeding the diffusion of oxygen. At rising temperatures, cracks in the oxidic layer result in accelerated diffusion of oxygen. Consequently, the alumina layer is growing in squareroot of the oxidation time, so that losses in mechanical strength are to be suspected.

REFERENCES

1. Kuramoto,N., Taniguchi,H., Aso,I.: Development of translucent AlN ceramics Cer.Bull. **68** (1989) , 883-887
2. Schuster, J.C.: Phase diagramms relevant for sintering aluminium nitride based ceramics. Rev. Chim. min. (1987), 676-685
3. Billy,M.,Mexmain,J.: Processing and properties of AlN, a new candidate for high temperature applications. Sprechsaal **3** (1985), 245-249

PREPARATION OF SIALONS BY THE NPS TECHNIQUE.

I.W.M.BROWN
Chemistry Division, DSIR, Lower Hutt, New Zealand
R.POMPE AND R.CARLSSON
Swedish Ceramic Institute, Göteborg, Sweden.

ABSTRACT.

The Nitrided Pressureless Sintering (NPS) technique has been used to prepare sialon compositions in the O'-ß' field, along the Si_3N_4-Al_2O_3 join. Following pre-nitridation at 1250°C, samples were fully nitrided in less than 4 hours at 1370°C, conditional upon achieving sufficiently high surface areas during milling. TG and XRD analyses indicate that the nitridation mechanism involves the early formation of a yttrium silicate phase during the pre-nitridation step which is in turn consumed below the nitridation temperature, probably in favour of a partial Si_3N_4-SiO_2-Y_2O_3-Al_2O_3 liquid phase. Yttrium aluminium garnet (YAG), whose volume fraction increases with time at temperature and with increasing level of Al addition, is formed promptly under these nitridation conditions. Sintering of the fully nitrided materials in nitrogen was carried out in the temperature range 1500-1830°C, with specimens achieving maximum density by 1600°C. XRD analysis reveals that the sole product of sintering low alumina compositions was ß'-sialon. Specimens in which the residual glass phase was crystallised at 1200°C developed $Y_2Si_2O_7$ and mullite as secondary phases.

INTRODUCTION.

The Nitrided Pressureless Sintering (NPS) technique has been developed by Pompe et al. [1-3] to assist the densification of Si_3N_4-based materials. The principal characteristics of the technique are to use mixed Si:Si_3N_4 powders, typically of mass ratio 5:5 - 7:3, milled to achieve submicron particle size. Typically, Y_2O_3 (6wt.% with respect to the nitrided Si_3N_4) and Al_2O_3 (2wt.%) are added as sintering aids. The role of the added Si_3N_4 is two-fold: it acts as a dispersing agent to reduce agglomeration of fine Si particles and helps maintain reaction control by modifying the pore structure during nitridation and by moderating the effect of the exothermic reaction. Si_3N_4 bodies prepared by the NPS technique are characterised by fast nitridation times, relatively low shrinkage, high green densities and narrow pore size distributions when compared to other techniques of reaction bonding or pressureless sintering. A further advantage is the option of machining after the nitridation step, prior to sintering.

This study has been undertaken to demonstrate the earlier described [2] applicability of the NPS technique to sialon compositions, with particular emphasis on the low alumina end of the Si_3N_4-Al_2O_3 join. This region of the sialon behaviour diagram offers the opportunity to achieve mixed O'-ß' sialon microstructure with the inherent properties of improved oxidation resistance due to the O' phase (Si_2N_2O) and high strength due to the ß' phase [4]. The combination of high green density, narrow pore size distribution and fine particle size featured by this technique is anticipated to enhance the sinterability of sialons resulting in materials of fine-grained microstructure and improved mechanical properties.

MATERIALS AND METHODS.

Four sialon compositions were prepared having 15.7, 19.7, 29.7 and 36.2 wt.% Al_2O_3 respectively. Compositions 1 and 2 represent points on the behaviour diagram where the proposed O'-ß' tie-lines of Thompson [4] and Jack [5] intersect with the Al_2O_3-Si_3N_4 join. In all compositions 75% of the required Si_3N_4 was added as Si. Further, 6wt.% Y_2O_3 (with respect to the nitrided composition) was added as a sintering aid. The powders for NPS sialon preparation were as follows with measured BET surface areas as indicated :

Si	KemaNord Sicomill 2D	1.9 $m^2.g^{-1}$
Si_3N_4	HC Starck LC10	11.9 $m^2.g^{-1}$
Al_2O_3	Sumitomo AKP50	9.4 $m^2.g^{-1}$
Y_2O_3	HC Starck "Fine"	16.2 $m^2.g^{-1}$

The powders were prepared as 150g batches, transferred to polythene bottles and milled for up to 96 hours in butanol using Si_3N_4 milling media. 3wt.% of oleic acid was added as a pressing aid. After milling, the slurry was wet filtered through 10 micron cloth, followed by rotary evaporation. 2-4wt.% residual butanol was left in the powder to assist the oleic acid pressing aid. The powders were sieve granulated (500 micron), pre-pressed in an acrylic mould to form 10g test bars of approximate dimensions 70x15x8mm and subsequently CIPped to 50,000psi. Finally, the organics were removed using the Rate Controlled Extraction (RCE) technique [6] at temperatures up to 500°C. Two series of the four compositions were prepared, namely series A and series B, having mid-range and high surface areas, respectively. Measured BET surface areas for series A were in the range 9.5-12.4 $m^2.g^{-1}$ and for series B, 20.2-22.9 $m^2.g^{-1}$. The green densities for series A samples lie in the range 58-62% theoretical, whereas the series B samples lie in the range 54-56% theoretical.

NITRIDATION.

The initial nitridation conditions were defined with TGA techniques on test pieces of mass 1.2-1.5g using a Mettler TA1 thermobalance. The samples were placed in an alumina crucible and the balance system was evacuated and flushed twice with nitrogen (AGA "Plus" grade : <5ppm O_2, <5ppm H_2O). Samples were heated at 15°C.min^{-1} to 1250°C for 30 min. as a pre-nitridation step, then at 4°C.min^{-1} to the chosen nitridation temperature, followed by soaking at temperature until no further weight gain took place. Figure 1 shows the dependence of nitridation rate on surface area for the two series of four compositions nitrided at 1350°C. Clearly, a surface area above 13-14 $m^2.g^{-1}$ is required to reduce the nitridation time to below 4 hours.

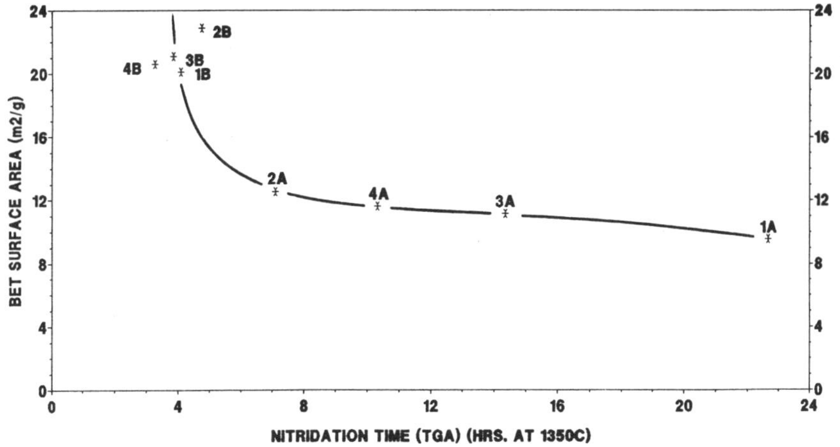

Figure 1. Dependence of nitridation rate on powder surface area for NPS sialon compositions.

The effect of temperature on nitridation rate for composition 1B has been measured at 20°C intervals between 1330 and 1410°C. There is a linear reduction in nitridation time between 1330°C (5.2 hrs.) and 1410°C (1.3 hrs.) showing that extremely fast nitridation is achievable at elevated temperatures. However, it should be noted that this temperature is very close to the melting point of the silicon powder, given the additional factor of the exothermic nature of the nitridation reaction. Further, the first partial liquid phase in the system Si_3N_4-SiO_2-Y_2O_3-Al_2O_3 is likely to have formed by this temperature [7]. Although this appears not to inhibit nitridation rate, it may have an undesirable impact upon grain growth of ß-Si_3N_4 present in the nitrided microstructure. In subsequent experiments a compromise temperature of 1370°C was used to maximise nitridation rate while retaining process control.

Test bars were nitrided in a Balzers graphite resistance furnace following a heating scheme similar to that described above. The test bars were placed on a silicon nitride powder bed in a graphite box (20x20x30 cm.). Specimens from series A were heated for 3, 6, 9 and 12 hours at 1370°C. Changes in density, linear shrinkage, mass and phase composition (by XRD) were used to monitor the progress of the nitridation reaction, including the effect of the pre-nitridation step.

The pre-nitridation step in NPS processing is accompanied by early reaction of all Y_2O_3. One of the principal functions of this sintering aid is to react with the small quantity of silicon oxide and oxynitride [8] associated with the Si and Si_3N_4, thus exposing fresh Si surfaces to nitrogen. XRD of pre-nitrided specimens showed no free Y_2O_3 but small quantities of poorly crystallised Y_2SiO_5 and $YSiO_2N$ (YN-wollastonite) were apparent. Such products are consistent with the following reactions :

$$Y_2O_3 + SiO_2 \longrightarrow Y_2SiO_5 \quad (1)$$

$$Y_2O_3 + Si_2N_2O \longrightarrow 2YSiO_2N \quad (2)$$

Usually 10-12% conversion of Si to Si_3N_4 accompanies the pre-nitridation step. Complete nitridation of series A specimens was relatively slow, consistent with the TGA results.
Figure 2 shows the phase development in series A samples nitrided at 1370°C for 6 and 12 hours. It is clear that $Y_3Al_5O_{12}$ (YAG) is an important phase in the nitridation chemistry of all sialon compositions. By extrapolation back to very low alumina levels, such as those used in NPS Si_3N_4 preparation, we may infer that YAG formation is possible in circumstances where Al_2O_3 is added only as a sintering aid. This may have a negative influence by competing for Y_2O_3, which would then have reduced availability for removal of the silicon oxide and oxynitride film on the particles. Figure 2 also shows YAG increasing with increasing Al_2O_3 content and this probably provides the only significant mechanism for Al_2O_3 reaction at these temperatures. Note that there is significant conversion of α-Si_3N_4 to ß-Si_3N_4 in these samples. This might be due to stabilising of the ß structure by minor amounts of Al, but Guinier XRD unit cell measurements show little deviation from regular ß-Si_3N_4 values, even for the highest Al_2O_3 compositions. The densities of fully nitrided samples lie in the range 70-73% of theoretical density. Linear shrinkage on nitridation ranged from 1% (composition 1A) to 3% (composition 4A).

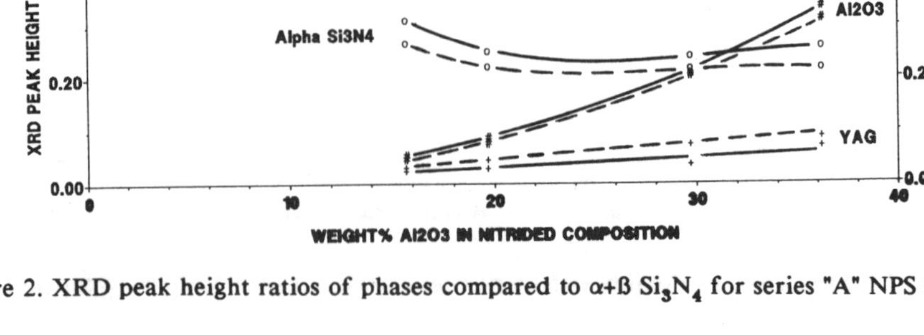

Figure 2. XRD peak height ratios of phases compared to α+ß Si_3N_4 for series "A" NPS sialons.

Nitridation of series B test bars (high surface areas) in the graphite resistance furnace showed expected consistency with the TGA results; all compositions could be fully nitrided in less than four hours at 1370°C. This compared favourably with previous results for NPS Si_3N_4 [1] in which nitridation times of less than 3 hours at 1350°C are normal. The reaction intermediates and products appear to be the same as those for series A ie. early reaction of Y_2O_3 and progressive formation of YAG. Nitrided densities lie in the range 67% (1B) to 74% (4B) theoretical. Linear shrinkage on nitridation ranged from 2.7% (1B) to 5.5% (4B), the higher shrinkage when compared to series A samples reflecting the lower green densities.

SINTERING.

Test bars for sintering were packed in an Si_3N_4 powder bed in a graphite box before firing in a Balzers graphite resistance furnace. The furnace was evacuated and refilled twice with nitrogen prior to heating at 20°C.min^{-1} to 800°C. At this temperature the furnace was fully evacuated to below 10^{-5} bar for 3 hours, refilled with nitrogen and the heating program resumed at 20°C.min^{-1} until the desired sintering temperature was achieved. Specimens of compositions 1A to 4A and 1B were sintered for 3 hours at a range of temperatures between 1500°C and 1830°C. The degree of sintering was monitored by measurements of density and porosity (by water intrusion), linear shrinkage, weight loss and phase composition changes (by XRD).

Figure 3 shows the densities obtained after sintering. Maximum density is achieved at relatively low temperatures (1575-1600°C) and for compositions 1 and 2 this density is maintained through the entire test sequence, declining slightly at 1830°C. Compositions 3 and 4, with higher alumina content, drop steadily in density immediately after achieving this maximum. In fact, these compositions show significant bloating when heated above 1600°C. Open porosities for all samples heated to 1575°C are less than 0.1%; for compositions 1 and 2 this is unchanged at 1675°C, but compositions 3 and 4 have bloated to 2% and 11% respectively. The weight loss on sintering at 1575°C is 0.5-0.7% for all samples; for compositions 1 and 2 this rises to 0.7-0.8% by 1675°C, but compositions 3 and 4 have risen to 1.5 and 2.8% loss, respectively.

XRD shows that for compositions 1 and 2 the only crystalline phase is ß'-sialon. All YAG and free Al_2O_3 has been consumed. For compositions 3 and 4 there is significant free Al_2O_3 at 1575°C, when the samples have achieved their maximum density. The implication is that the rate controlling step in the alumina take-up is the ability of the sialon structure to incorporate the Al at 1575°C, rather than any problems due to alumina particle size. Detailed examination by Guinier XRD of all specimens sintered at 1775°C showed that each composition had formed a ß'-sialon whose Al content was exactly that predicted by translation of the starting composition from the Si_3N_4-Al_2O_3 join down onto the ß'-sialon line on the sialon behaviour diagram.

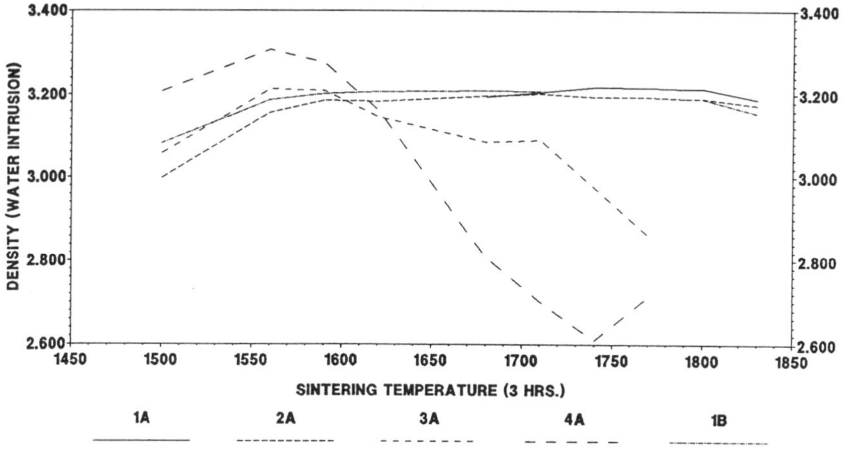

Figure 3. Density as a function of sintering temperature for NPS sialons.

CRYSTALLISATION.

Experiments designed to crystallise the residual glass phase in all compositions were carried out in nitrogen at 1100, 1200 and 1300°C for 3 hours. Crystallisation was apparent in the 1200°C samples only. Subsequently, samples were heat treated for 3, 12 and 48 hours at 1200°C. Maximum development of the crystalline phases was somewhat faster for series A samples (12 hours) than for series B samples (48 hours). However, XRD revealed all crystalline products other than ß'-sialon as very minor phases only. The principal secondary phases were $Y_2Si_2O_7$ and mullite. O'-sialon (Si_2N_2O) was noticeably absent. These crystallised specimens form an integral component of an on-going study of the mechanical and microstructural properties of NPS sialons.

CONCLUSIONS.

This paper has demonstrated the applicability of the Nitrided Pressureless Sintering technique to low alumina sialon compositions. The important features of NPS sialon fabrication are as follows :

1. The nitridation rate is critically dependent on control of the milling conditions.
2. Nitridation times under 3 hours are achievable.
3. Sintered bodies achieve maximum density at low temperatures, typically 1575-1600°C.
4. There is a significant difference in sintering behaviour between the compositions corresponding to the extended O'-ß'[5] and X-ß' phase regions, the latter compositions showing a narrow firing range.
5. Crystallisation of the residual glass phase gives ß'-sialon, yttrium disilicate and mullite.

ACKNOWLEDGEMENTS.

The authors wish to thank Ms. Maria Fanto and Mr. Martin Sjöstedt, SCI, and Mr. Jörgen Sjöberg, Department of Inorganic Chemistry, CTH, Göteborg for experimental assistance.

REFERENCES.

1. Pompe, R., Hermansson, L. and Carlsson, R., Development of commercially advantageous techniques for fabrication of low shrinkage Si_3N_4-based materials. Sprechsaal, 1982, 115, 1098-1101.

2. Pompe, R., SE 8103269-0 (1981), US 463,884 (1983), EP 82-00161 (1982).

3. Nyberg, B., Falk, L.K.L., Pompe, R. and Carlsson, R., Some features of nitrided pressureless sintered (NPS) silicon nitride materials made by modified preparation routes. In Ceramic Materials and Components for Engines, Proceedings of the Second International Symposium, Lubeck, 1986, ed. W.Bunk and H.Hausner. Verlag Deutsche Keramische Gesellschaft, Berlin 1986, pp.155-163.

4. Thompson, D.P., Sun, W.Y. and Walls, P.A., O'-ß' and α'-ß' sialon ceramics. *ibid*, pp. 643-650.

5. Jack, K.H., Sialons and related nitrogen ceramics. J. Mater. Sci., 1976, 11, 1135-1158.

6. Johnsson, A., Carlström, E., Hermansson, L., and Carlsson, R., Rate-controlled extraction unit for removal of organic binders from injection-moulded ceramics. In Ceramic Powders, ed. P.Vincenzini. Elsevier, Amsterdam, 1983, pp. 767-772.

7. O'Meara, C., Dunlop, G.L. and Pompe, R., Formation, crystallisation and oxidation of selected glasses in the Y-Si-Al-O-N system. J. Mater. Sci., 1989, (in press).

8. Bergström, L. and Pugh, R.J., Interfacial characterisation of silicon nitride powders. J. Am. Ceram. Soc., 1989, 72, 103-109.

ON THE REACTION MECHANISM AND MECHANICAL PROPERTIES OF NITRIDED PRESSURELESS SINTERED (NPS) Si_3N_4-ZrO_2 (+Y_2O_3) COMPOSITE CERAMICS

KENT RUNDGREN*, ROBERT POMPE, ROGER CARLSSON
Swedish Ceramic Institute, Box 5403, S-402 29 Göteborg, SWEDEN

KAZUSHI KISHI, SEIKI UMEBAYASHI
Government Industrial Research Institute, Kyushu, Tosu-shi, 841 Saga-ken, JAPAN

ABSTRACT

ZrO_2-toughened Si_3N_4 ceramics have been prepared by nitridation of Si:Si_3N_4/ZrO_2 submicron powder compacts (NPS-technique) followed by pressureless or gas-pressure sintering. Thermal analyses combined with X-ray diffraction analyses showed that monoclinic (*m*) and tetragonal (*t*) ZrO_2 reacts in argon atmosphere with Si forming $ZrSi_2$ and SiO_2 at 1150°C (Eq. 1). By the stabilization of cubic (*c*) ZrO_2 with 10 mol% Y_2O_3 prior to mixing with Si the formation of $ZrSi_2$ was precluded at 1150°C. Preliminary data for not fully densified samples made from Al_2O_3-coated Si:Si_3N_4 powder mixtures with addition of ZrO_2 and Y_2O_3 as colloidal sols showed a fracture toughness of 5.8 ± 0.6 MPa·m$^{1/2}$ and a room temperature flexural strength of 700 ± 60 MPa.

INTRODUCTION

A novel method for the fabrication of complex shapes of dense Si_3N_4-based ceramics, the Nitrided Pressureless Sintering technique (NPS technique), has been developed [1,2]. It involves the nitridation and subsequent pressureless sintering of submicron Si and Si_3N_4 powder compacts containing addition of sintering aids.

ZrO_2 has been shown to be an effective densifying and toughening aid for Si_3N_4 [3-5]. However, a main drawback exists with these types of composite materials. The reaction between Si_3N_4 and ZrO_2 produces less desirable by-products such as ZrN and Zr-oxynitride. In the system Si-ZrO_2 Rundgren and Pompe [6] showed that *m*-ZrO_2 reacts with Si in an argon atmosphere forming $ZrSi_2$ and SiO_2 according to

$$ZrO_2 + 3\,Si \rightarrow ZrSi_2 + SiO_2. \qquad (1)$$

The aim of the present study has been to investigate two ways to prevent the Si/ZrO_2 reaction: (1) by alloying ZrO_2 with Y_2O_3 and (2) by coating Si/Si_3N_4 mix with Al-isopropoxide. ZrO_2 and Y_2O_3 were added as (a) powder and (b) colloidal sols. The resulting microstructural and mechanical properties of the latter material have also been examined.

---

* Formerly at Dept. of Inorganic Chemistry, Chalmers University of Technology and University of Göteborg, S-412 96 Göteborg, SWEDEN.

EXPERIMENTAL

For the reaction studies between ZrO_2 and Si in an argon atmosphere three different ZrO_2/Si powder compacts were prepared. The monoclinic (TZ-0#), tetragonal (TZ-3Y#) and cubic* (10 mol% Y_2O_3) ZrO_2 powders were milled in a molar ratio ZrO_2/Si = 1:3 with Si powder A in an agate mortar and uniaxially pressed (10 MPa) to discs. The samples were heated to 1150° and 1250°C in Ar atmosphere in a thermobalance for one hour.

The Si*:Si_3N_4# mass ratio was 7:3 and the oxide additives used were ZrO_2 (+Y_2O_3) and Al-alkoxide in such amounts that the nitrided body contained an equivalent of 5 wt% Al_2O_3, and 23 wt% ZrO_2. The Y_2O_3 addition was 3.0 mol% with respect to the amount of ZrO_2. The powder processing route is illustrated in Fig. 1. The ZrO_2 was added as colloidal sols$ of ZrO_2 (+Y_2O_3) or *t*-ZrO_2# powder. Before adding ZrO_2, all Si:Si_3N_4 powder mixtures were prepared either by milling to sub-micron size according to the NPS proprietary process [2] or by mixing with submicron Si† powder. The specific surface areas of the powders were 11.9 and 33.0 m^2/g for powder A and B, respectively.

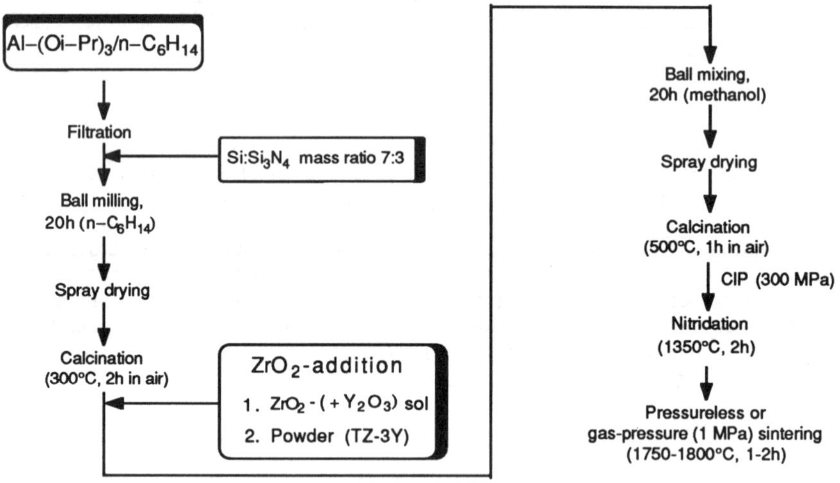

Figure 1. Flow chart for powder processing.

Rectangular discs were pre-pressed in a steel die under 10 MPa and then cold isostatically pressed under 300 MPa. They were then nitrided in a graphite resistance furnace in N_2 gas (5 l/min) with 2% H_2 at atmospheric pressure at 1100°C for one hour and 1300° or 1350°C for 2 hours. The nitrided specimens were packed in granulated Si_3N_4 powder in a boron nitride crucible and then pressureless sintered at 1750°C for two hours in a graphite resistance furnace in an atmosphere with a N_2:Ar ratio of 1:9. Gas-pressure sintering (GPS) was performed at 1 MPa total pressure at 1750° or 1800°C for one to two hours. The samples were cut and groundwith a 600# diamond wheel to test bars with the dimension of 3 by 3 by 30 mm.

The phase composition of as-sintered and ground surfaces was determined by X-ray diffraction (XRD) analysis. The bulk density and open porosity were measured by the water

* Zircar ZrO_2 (8.0 wt% Y_2O_3) was prereacted in air at 1400°C for 3 hours with the addition of submicron Y_2O_3 (H. C. Starck) to a total amount of 10 mol% Y_2O_3.
* SICOMILL, Kema Nord Industrikemi AB, Sweden.
# Si_3N_4 (TZ-7) and ZrO_2 (TZ-3Y), Toyo Soda Mfg. Co., Ltd., Tokyo, Japan.
$ Nyacol Products Inc., USA.
† Kurosaki Refractories Co., Ltd., Japan.

immersion method. The flexural strength measurements were performed using a three-point loading device with a span of 20 mm and a cross-head speed of 0.5 mm/min. The fracture toughness was measured by Vickers indentation method [7] using a 20 kg load. SEM-EDX was used to characterize fracture surfaces and the microstructures of polished surfaces.

RESULTS AND DISCUSSION

X-ray diffraction analysis of as-fired and bulk ZrO_2/3Si compacts heated to 1150° or 1250°C in an argon atmosphere for one hour showed that both monoclinic and tetragonal ZrO_2 (3.0 mol% Y_2O_3) reacts with Si forming zirconium silicide, $ZrSi_2$, and SiO_2 at 1150°C (Eq. 1). Cubic ZrO_2, stabilized with 10 mol% Y_2O_3, showed only SiO_2 on the surface of the sample at 1150°C, probably originating from the evaporation of SiO from the surface silica on the Si powder. However, $ZrSi_2$ was found at 1250°C.

These results show that the reaction could be avoided at 1150°C by alloying ZrO_2 with enough of a stabilizing agent such as Y_2O_3, in this way forming *c*-ZrO_2. As a consequence of the overall, very low driving force [6] for the reaction, the expected lowered activity of ZrO_2 may be enough to prevent this reaction at 1150°C. In nitrogen atmosphere at 1150°C surface nitridation of Si particles is usually achieved. This nitride layer may then protect Si from the reaction with ZrO_2.

Nitridation of Al_2O_3-coated Si:Si_3N_4 powder compacts without addition of ZrO_2 was found to give a lower rate for nitridation than previously observed [1]. This may suggest that a dense protective Al_2O_3-coating on the Si particles forms, which would prevent nitrogen from reacting with the silicon. However, in samples with ZrO_2 addition, the rate of nitridation was instead found to be higher, than observed earlier[1].

After one hour nitridation at 1100°C all ZrO_2 in sample B (without Y_2O_3) had reacted with Si forming $ZrSi_2$ in the bulk and ZrN on the surface. In samples with Y_2O_3 addition some ZrO_2 still remains after one hour at 1100°C. ZrN had formed both in the bulk and at the surface. The amount of $ZrSi_2$ was found to decrease somewhat with Y_2O_3 addition. The absence of $ZrSi_2$ in all samples at 1250° and 1300°C and the increasing amount of ZrN suggests that the following reaction occurs

$$3\ ZrSi_2 + 51/2\ N_2 \rightarrow 3\ ZrN + 2\ Si_3N_4. \qquad (2)$$

These results indicate that coating the Si:Si_3N_4 powder mixture with Al_2O_3 was not sufficient to prevent the reaction of Si and ZrO_2 to give $ZrSi_2$ and later ZrN on nitridation. Hence, somewhat surprisingly, the Al_2O_3 layer has not been a sufficient barrier to the diffusion within the ZrO_2/Si couple. At 1250°C, where 90 % of the nitridation had been completed, $ZrSi_2$ disappeared in all samples and ZrN was found in the bulk in all samples and on the surface of sample B. SiO_2 was also found at 1250° and 1300°C on the surface of all samples.

TABLE I.
Sintering data for Si_3N_4/ZrO_2 (+Y_2O_3) composites.

Sample	Sintering Temp/time (°C/h)	Atmosphere N_2/Ar (MPa)	Linear Shrinkage (%)	Weight-loss (%)	Bulk Density (g/cm^3)	Open Porosity (%)
A. 23 wt% TZ-3Y powder, Gas-pressure sintered.						
A1.	1600/1800/1	0.1/0.9	11	1.8	3.29	0.75
A2.	1750/2	-"-	9	0.6	3.25	1.4
B. 23 wt% ZrO_2 and 3 mol% Y_2O_3 as colloidal sols, Pressureless sintered.						
B1.	1750/2	0.01/0.09	14	3.8	3.31	0.4

Sintering data for the Si_3N_4/ZrO_2 composites are shown in Table I. The results show that these nitrided Si_3N_4/ZrO_2 compacts could be sintered to more than 90 % of theoretical density. Despite lower green density, the sintering shrinkage was higher for the composites fabricated from colloidal sols than for TZ-3Y powder. This could be due to a higher specific surface area for powder B and a homogeneous distribution of ZrO_2, and possibly a better pore size distribution for green bodies of powders prepared from colloidal sols.

The results from XRD confirm that ZrN forms at the surface of the samples when using pressureless sintering at 1700-1800°C. The formation of ZrN on the surface was suppressed when gas-pressure sintering at 1750° to 1800°C with the N_2:Ar ratio of 1:9 was applied. The ZrN formed during nitridation in the bulk was found to disappear on sintering of all samples, presumably due to a reaction with ZrO_2 to Zr-oxynitride in the presence of a sintering liquid.

A complete α- to β-Si_3N_4 conversion was found. In addition to β-Si_3N_4 and *c*-ZrO_2/Zr-oxynitride approximately 10 wt% Si_2N_2O was present in all samples, as determined on powdered specimens. The presence of Si_2N_2O could be an indication that the ZrO_2 reacts with Si_3N_4 to form Zr-oxynitride and Si_2N_2O. However, it can just as well be due to reaction of Si_3N_4 with the native silica layer on the silicon nitride particles, the amount of which could have increased further during the heat treatment of the sol-based powders in air (Fig. 1).

The scanning electron micrographs in Fig. 2 show the microstructure of sample A and B. The ZrO_2 particles in samples fabricated from colloidal sols (B) were small and well-dispersed in the Si_3N_4 matrix. The presence of large ZrO_2 particles and agglomerates in all samples fabricated from TZ-3Y powder (A), indicates some agglomeration and/or insufficient dispersion of the powder during the processing steps.

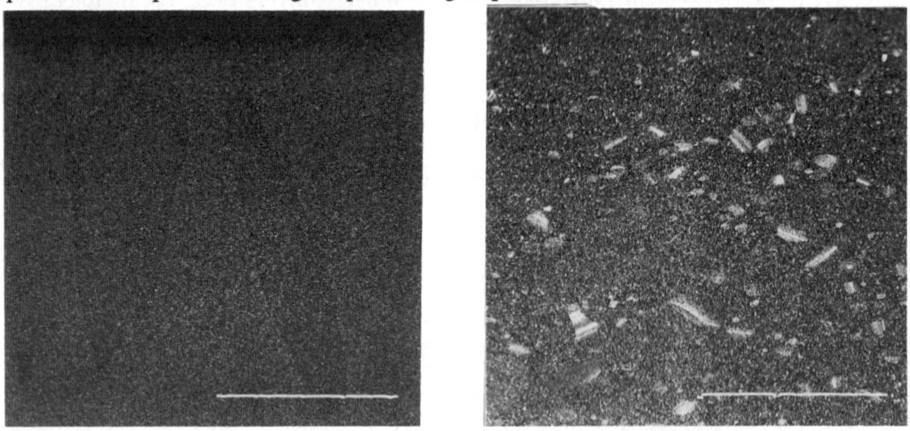

A B

Figure 2. Scanning electron micrograph (BEI) showing A) homogeneous distribution of ZrO_2 in sample B1, pressureless sintered at 1750°C, 2h, and B) ZrO_2-agglomerates in sample A2, gas-pressure sintered at 1750°C, 2h (bar = 1000 μm). ZrO_2 particles have the brightest contrast.

The mean room temperature flexural strength, 3-point bending, for sample series A, fabricated from TZ-3Y powder and gas-pressure sintered, was only about 250 MPa. Pores and large agglomerates of ZrO_2, formed during powder processing, were found as fracture origins in samples of series A. ZrO_2-agglomerates and pores were found as fracture origins in these composites. The mean room temperature flexural strength for sample B1, made from colloidal sols and pressureless sintering at 1750°C for 2 hours, was 700 MPa. The predominant flaw responsible for the failure in sample series B was pores in the bulk.

The fracture toughness, K_{IC}, was of the order of 6 MPa·$m^{1/2}$. These preliminary results suggest that the ZrO_2-toughened NPS materials can be made with mechanical properties as good as those of hot-pressed Si_3N_4/ZrO_2 composites [5].

CONCLUSIONS

Homogeneous, fine-grained Si_3N_4/ZrO_2 (Y_2O_3) composite ceramics were made by the NPS-technique using an Al-alkoxide coated Si:Si_3N_4 powder mixture with addition of ZrO_2 (+Y_2O_3) as colloidal sol. This coating could not sufficiently prevent the reaction of Si with ZrO_2 to give $ZrSi_2$ and later ZrN on nitridation. The ZrN has been shown to disappear on sintering, presumably due to reaction with ZrO_2 to form Zr-oxynitride in the presence of sintering liquid. Monoclinic (*m*) and tetragonal (*t*) ZrO_2 react in argon atmosphere with Si to form $ZrSi_2$ and SiO_2 at 1150°C. However, by stabilization of cubic (*c*) ZrO_2 with 10 mol% Y_2O_3 prior to mixing with Si, the formation of $ZrSi_2$ has been precluded at 1150°C.

Preliminary data for not fully densified samples (≈ 90 % of theoretical density) made from Al_2O_3 coated Si:Si_3N_4 powder mixtures with addition of ZrO_2 and Y_2O_3 as colloidal sols showed a fracture toughness of 5.8 ± 0.6 MPa·m$^{1/2}$ and a room temperature flexural strength of 700 ± 60 MPa.

Acknowledgements: Financial support from the Swedish Board for Technical Development (STU) and the Swedish Natural Science Council (NFR) is gratefully acknowledged. The authors thank Thomas F. Swank, Nyacol Products Inc., USA for providing colloidal sols. The authors also thank Lars Eklund and Camilla Petersson at the Swedish Ceramic Institute (SCI) for help with the SEM-study. Many thanks go to Richard Warren and Tomas Hansson , Dept. of Mechanical Engineering, Chalmers University of Technology, Sweden and Inga-Lill Ekberg at SCI for their help with the Vickers indentation measurements. Finally thanks to Dr. Ian Brown, Chemistry Division, DSIR, New Zealand, for proof-reading the text.

REFERENCES

1. R. Pompe, L. Hermansson, and R. Carlsson, "Fabrication of Low-Shrinkage Silicon Nitride Material by Pressureless Sintering", Engineering with Ceramics **11**, Proc. Brit. Ceram. Soc., edited by R.W. Davidge (British Ceramic Society, Stoke-on-Trent, 1982) pp. 65-74.

2. R. Pompe, SE 81 03 269-0 (1981)/US 463,884 (1983)/ EP 82-00161 (1982).

3. R. W. Rice and W. J. McDonough, "Hot-Pressed Si_3N_4 with Zr-Based Additions", J. Am. Ceram. Soc, **58** [5-6] 264 (1975).

4. F. F. Lange, "Compressive Surface Stresses Developed in Ceramics by an Oxidation-Induced Phase Change", ibid, **63** [1-2] 38-40 (1980).

5. K. Rundgren, P. Elfving, K. Kishi, S. Umebayashi, and R. Pompe, "Hot-Pressed Si_3N_4-ZrO_2 Composites: Effect of Different ZrO_2 Additions on the Mechanical Properties", to be published in J. Am. Ceram. Soc.

6. K. Rundgren and R. Pompe, "On the Mechanism of Silicon Nitridation in the Presence of ZrO_2", J. de Phys., **47** [2] C1-515-19 (1986).

7. B. Marshall and A. G. Evans, "Reply to “Comment on ‘Elastic/Plastic Indentation Damage in Ceramics: The Median/Radial Crack System, ’”" J. Am. Ceram. Soc., **64** [12] C182-C183 (1981)

CRYSTALLIZATION BEHAVIOUR OF $MgO-Al_2O_3-SiO_2$ GLASS CERAMICS DERIVED FROM ALKOXIDE GELS AND GLASSES

Dechun Fu and J. Engell
Institute of Mineral Industry,
Technical University of Denmark, DK-2800 Lyngby, Denmark

ABSTRACT

The crystallization behaviour of two enstatite rich compositions in the system $MgO-Al_2O_3-SiO_2$ has been studied as a function of TiO_2 addition and thermal treatment. The behaviour of both alkoxide derived gels and quenched glasses of the same compositions has been studied. Crystallinity curves and TTT-diagrams are presented.

INTRODUCTION

The formation of glass ceramics in the system $MgO-Al_2O_3-SiO_2$ has received much attention over the past years (1-4, 6-7, 9-10, 12, 17-20). The interest have centered around compositions close to the silica-cordierite join (Fig. 1). An extensive field of metastable liquid immiscibility exists in the silica rich part of the system $MgO-Al_2O_3-SiO_2$ (Fig. 1; 5). However, addition of a nucleating agent is necessary in order to obtain uniform crystallization of bulk samples (cf. 2, 3, 10), otherwise the crystallization proceeds from the surface of the sample (cf. 9, 10). The choice of nucleating agent and its concentration affects the crystallization behaviour markedly (1, 3, 10). In general the use of TiO_2 gives the best results (9, 1). For cordierite-type compositions 10-12 wt% TiO_2 appears to be optimal (2). Addition of TiO_2 to this level promotes metastable liquid immiscibility, and the subsequent formation of dendritic pseudo-brookite (MAT; cf. 1) as the first crystalline phase. The crystallization of $MgO-Al_2O_3-SiO_2$ glasses involves the formation of several metastable solid solution series before equilibrium eventually is reached. The most important are the high-quarts$_{ss}$ (14, 1), Mg-petalite$_{ss}$ (15, 16) and the enstatite-Mg-tschermak series (Fig. 1).

The present contribution concerns the crystallization behaviour of two enstatite rich compositions (Fig. 1) as a function of TiO_2 addition and thermal treatment. Both alkoxide derived gels and quenched glasses of the same compositions have been studied. The polymorphism of enstatite and the possibility of using these for transformation toughening of ceramics have recently been discussed by Lee & Heuer (8).

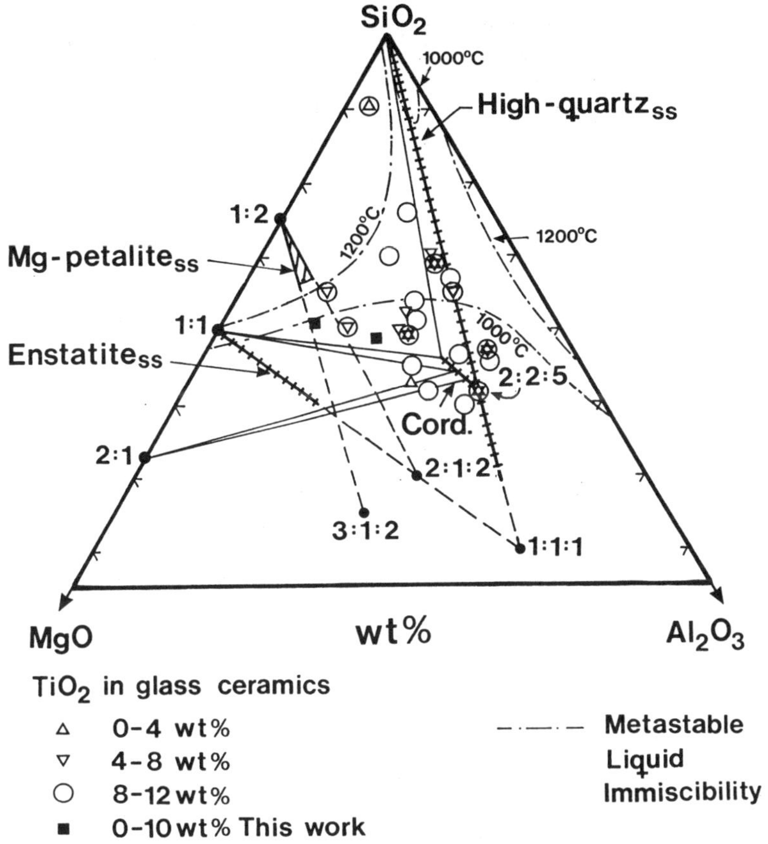

Fig. 1. Phase diagram for the system $MgO-Al_2O_3-SiO_2$ showing stable (13) and metastable phase relations (5, 14, 1, 16) and compositions of glass ceramics studied in the present work and by others (1-4, 6-7, 9-10, 12, 17-20).

EXPERIMENTAL

The materials studied include the base compositions MAS-1 (SiO_2 58.9wt%, Al_2O_3 18.9wt%, MgO 22.2wt%) and MAS-2 (SiO_2 61.wt%, Al_2O_3 10.6wt%, MgO 28.4 wt%) as well as the corresponding materials containing respectively 9.1wt% TiO_2 (MAST-1) and 9.5wt% TiO_2 (MAST-2).

The gels where prepared from $Si(OEt)_4$ diluted with Pr^nOH (50wt%) and partially hydrolyzed under acid conditions (H_2O:$Si(OEt)_4$:HNO_3= 1:1:1 mole), $Al(OBu^s)_3$ diluted with Bu^sOH (50wt%) was added while stirring, after 2h the desired amount of $Mg(NO_3)_3{*}6H_2O$ dissolved in Pr^nOH (30wt%) was added. Finally undiluted $Ti(OPr^i)_4$ was added to the MAST compositions. In order to facilitate gelation additional H_2O was used (10:1 mole). This results within a few hours in the formation of translucent di-phase gels containing $Mg(NO_3)_3{*}6H_2O$. The gels were dried in a vacuum oven (80°C, 600mbar plus 110°-150°C,<1mbar) and calcined in air at 600°C for 5h (heating ½°C/min). The resulting materials are coarse grained powders of XRD amorphous porous glasses with surface areas in the range 150-200 m^2/g.

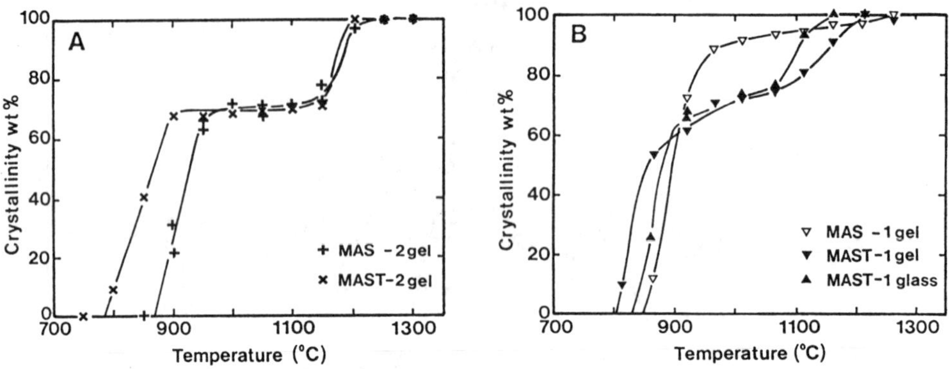

Fig. 2. Crystallinity as a function of temperature for gels and glasses isothermally heat treated for 2h. Heating rate 10°C/min for MAS-2 and MAST-2, and 150-270°C/min for MAS-1 and MAST-1.

The glasses where prepared by melting the corresponding alkoxide derived gels at 1470°C for 5h in Pt-crucibles, and quenching the poured melt in air to form ca. 10 mm thick slabs of clear colourless glasses.

For the construction of the TTT-diagrams the samples (ca. 0.25cm^3) were treated isothermally in a preheated electrical tube furnace at temperatures between 750°C and 1350°C (<±5°C) for periods between 1h and 45h. Assuming the heating up time to be 5 min this correspond to average heating rates in the range 150-270°C/min.

XRD-data for phase identification and the determination of crystallinity were collected by step scan (0.05°/step, 2Θ 5°-65°) on a fully computerized Philips diffractometer with automatic divergence slit using $Cu_{K\alpha}$ radiation.

The crystallinity of the heat treated samples was determined as described by Ohlberg & Strickler (11). Fine powders of respectively the quenched glasses or gels calcined at 600°C were used as reference materials of zero crystallinity. Similar materials heat treated at 1200°-1300°C for extended periods were used as reference materials of 100% crystallinity.

RESULTS AND DISCUSSIONS

The MAS-1 composition is on the boundary between the cordierite and the enstatite primary fields in the equilibrium phase diagram (13), whereas MAS-2 is in the primary field for enstatite.

Microscopic examinations show that the TiO_2-free glasses crystallize via a surface mechanism. The TiO_2-containing glasses crystallized uniformly. The change in crystallinity as a function of temperature is shown in Fig. 2 for samples heat treated isothermally for 2h. Crystallization begins near 800°C in the gels containing TiO_2 as compared with 850°C for the TiO_2-free gels. For gels and glasses of the same composition the gels begin to crystallize first. The step like shape of the curves for the MAS-2 and MAST-2 gels (Fig. 2A) reflects an early crystallization of aluminous orthoenstatite followed by the formation of high-quartz$_{ss}$ and then protoenstatite, cordierite and cristobalite at temperatures around 1200°C.

The difference between the crystallinity curves for the MAS-1 gel and the MAST-1 materials (Fig. 2B) reflect the evolution of different metastable phase assemblages as shown in the TTT-diagrams (Fig. 3). The sequence of

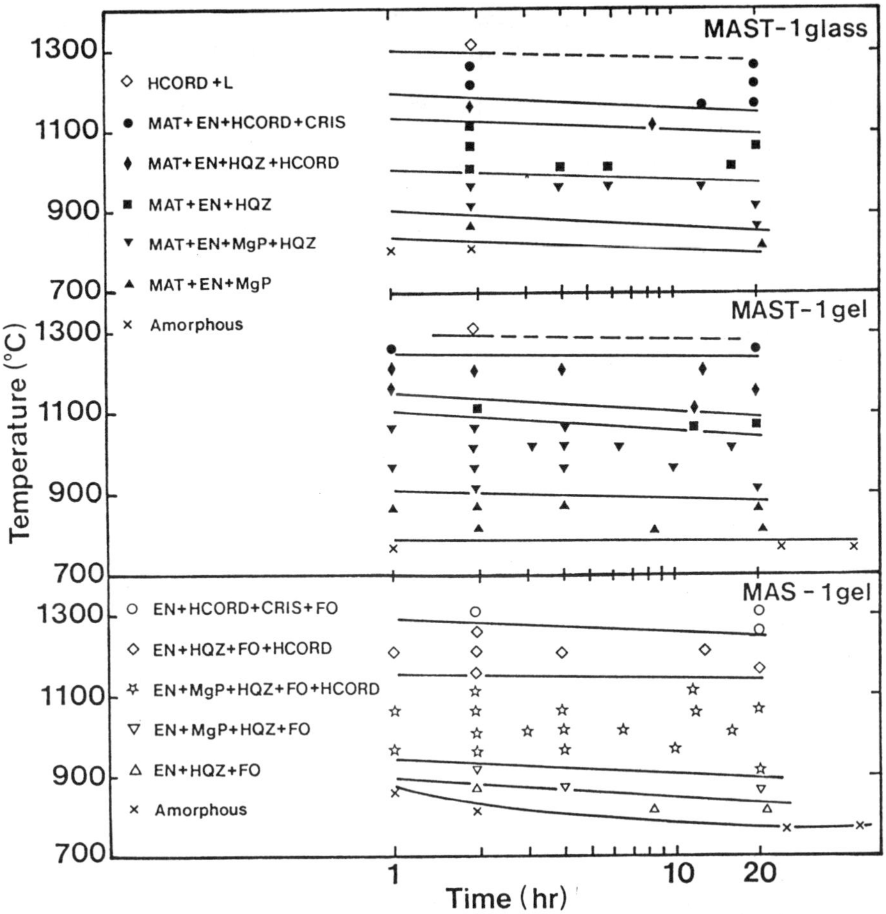

Fig. 3. TTT-Diagrams for the gels and glasses studied.

phase assemblages developed in the MAST-1 gel and glass are identical, but the temperature intervals in which they occur differ (Fig. 3). The MAS-1 and MAST-1 materials are similar in that enstatite (EN) is present throughout the crystallization interval. Unit cell calculations indicate a variation from an aluminous ortho-EN at low temperature, to proto-EN at high temperature where high cordierite (HCORD) and cristobalite (CRIS) are formed. Except for the formation of the pseudo-brookite phase (MAT) in the MAST compositions, the main differences between the MAS-1 and MAST-1 materials are: At low temperatures high-quartz$_{ss}$ (HQZ) is a major phase in MAS-1, while Mg-petalite (MgP) is a major phase in MAST-1. The unit cell dimensions of the HQZ (14, 1) indicate a SiO_2 content of respectively 62wt% and 71wt% in the MAS-1 and MAST-1 materials after heat treatment at 1010°C for 16h. Although the amount of Mg-petalite in the MAS-1 gel was barely detectable by XRD, this phase persisted to higher temperatures in the MAS-1 gel than in the MAST-1 materials (Fig. 3). A further difference is the small amounts of forsterite (FO) present throughout the crystallization interval for the MAS-1 gel. This phase is not observed in the MAST-1 materials.

ACKNOWLEDGEMENT

This work was supported by a Grant from Peoples Republic of China to Mr. Dechun Fu.

REFERENCES

1. Barry, T.I., Cox, J.M. & Morrell, R.: Cordierite glass-ceramics - effect of TiO_2 and ZrO_2 content on phase sequence during heat treatment. J. Mat. Sci., 1978, **13**, 594-610.
2. Corning Glass Works: Method of making Ceramics and product thereof. The Patent Office, London, 1957, **829,447**, 1-23
3. Doenitz, F.-D. & Vogel, W.: The Crystallization Mechanism in High Strength Glass-ceramics. Kristall und Technik, 1980, **15**, 891-894.
4. Duvigneaud, P., Höland, W., Plumat, E.R. & Vogel, W.: Thin films of glasses and glass ceramics from the SiO_2-Al_2O_3-MgO-system derived from gels. CFI Ber. DKG, 1986, **63**, 523-526.
5. Galakhov, F. Ya., Aver'yanov, V.I., Vavilonova, V.T. & Areshev, M.P.: The Metastable Liquid-Phase Separation Region in the MgO-Al_2O_3-SiO_2 System. I.V.Grebenshchikov Inst. of Silicate Chem., 1976, **2**, 412-416.
6. Helgesson, C.I.: Properties of Cordierite Glass Ceramics Produced by Sintering and Crystallization of Glass Powder. Sci. of Ceram., 1975, 347-361.
7. Kumar, S. & Nag, B.B.: Electrical Properties of Crystallized Glasses in the System MgO-Al_2O_3-SiO_2-TiO_2. J. Am. Ceram. Soc., 1966, **49**, 10-14.
8. Lee, W.E. & Heuer, A.H.: On the Polymorphism of Enstatite. J. Am. Ceram. Soc., 1987, **70**, 349-360.
9. Leger, L. & Bray, J.: An experimental study of the controlled crystallization of alkaline earth silico-aluminate glasses. Glass Technol., 1966, **7**, 134-142.
10. Maier, V. & Müller, G.: Nucleation and crystallization in Mg-Al-silicate-glass-ceramics. CFI Ber. DKG, 1988, **65**, 208-212.
11. Ohlberg, S.M. & Strickler, D.M.: Determination of Percent Crystallinity of Partly Devitrified Glass by X-ray Diffraction. J. Am. Ceram. Soc., 1962, **45**, 170-171.
12. Ohlberg, S.M., Golob, H.R. & Strickler, D.M.: Crystal Nucleation by Glass in Glass Separation. Symp. Nucl. & Cryst. in Glasses & Melts, Am. Ceram. Soc., 1962, 55-62.
13. Schreyer, W. and Schairer, J.F.: Compositions and structural states of anhydrous Mg-Cordierites: A Re-investigation of the Central Part of the System MgO-Al_2O_3-SiO_2. J. Petrology, 1961, **2**, 324-406.
14. Schreyer, W. and Schairer, J.F.: Metastable solid solutions with quartz type structures on the join SiO_2-$MgAl_2O_4$. Z.f.Krist., 1961, **116**, 60-82.
15. Schreyer, W. and Schairer, J.F.: Metastable Osumilite- and Petalite-Type Phases in the System MgO-Al_2O_3-SiO_2. Am. Min., 1962, **47**, 90-104.
16. Varshal, B.G., Baiburt, L.G., Gel'berger, A.M., Malakhovskaya, N.N. & Naumkin, A.P.: Synthesis of Metastable Petalite-like Phases Containing Magnesium and Zinc. Izv. Akad. Nauk SSSR, Neorgan. Mat., 1971, **7**, 712.
17. Varshal, B.G., Gel'berger, A.M., Knyazher, G.B. & Naumkin, A.P.: Crystallization of glasses in the system SiO_2-Al_2O_3-MgO with added TiO_2. Izv. Akad. Nauk SSSR, Neorgan. Mat., 1973, **9**, 2206-2212.
18. Vekey, R.C. de & Majumdar, A.J.: The role of TiO_2 in the formation of cordierite glass ceramics. Phys. Chem. Glasses, 1975, **16**, 36-43.
19. Vogel, W. & Höland, W.: Nucleation and crystallization kinetics of a MgO-Al_2O_3-SiO_2 base glass with various dopants. In Simmons, J.H., Uhlmann, D.R. & Beall, G.H. (Eds.) Adv. in Ceramics, **vol 4**, 1982, 125.
20. Zdaniewski, W.: Crystallization and structure of a MgO-Al_2O_3-SiO_2-TiO_2 glass-ceramic. J. Mat. Sci., 1973, **8**, 192-202.

EMISSION MIXES FOR HIGH-INTENSITY DISCHARGE (HID) LAMPS - AN APPLICATION REVIEW -

R. K. DATTA
Advanced Technology Department
GE Lighting
Nela Park
Cleveland, OH 44112, USA

ABSTRACT

An emission mix (polycrystalline, inorganic, fine-powder) coating is used in HID lamp electrodes to improve lamp maintenance and life. The exact composition of this mix depends on the type, chemistry and operating conditions of lamps. Alkaline-earth oxides, tungstates, thorates, rare-earth oxides, etc., are commponly used as emission mixes. During lamp-operating conditions interaction between the emission mix and tungsten electrode generates low work function activating species which migrate to the electrode surface lowering the work function of the overall electrode. Based on the known commercial emission mixes, a generic classification of the activating process is proposed. Certain guidelines for design and search of emission mixes are suggested.

INTRODUCTION

A typical emission mix is a refractory, polycrystalline (oxidic), fine powder material. The use of an emission mix in HID lamps to reduce cathode work function (Ø) and thereby improve lamp performance is a common and long-standing practice in lamp design. The improvement is reflected on and measured in terms of lamp life, maintenance (color, efficacy, etc.), and lamp startability.

Depending on the lamp type and wattage, the size and shape of the electrodes may vary. The operating temperatures may also vary widely. The electrodes of an HID lamp is, however, commonly made of tungsten. The general design of an electrode consists of a tungsten shank with tungsten coil wound around the top. Excellent reference material on lamp electrodes including emission mixes for different lamp types are available in the literature [1,2,3]. **Table 1** summarizes the major HID lamp types, electrode temperatures, arc tube and arc materials.

TABLE 1
High-Intensity Discharge Lamp Types [1,2,3]

Lamp Type	Electrode Temp. Range °C	Arc Material	Arc-Tube Material
Metal Vapor (MV)	900 - 2300	NaI, TlI, ScI_3, ThI_4, etc.	Fused Silica
High-Pressure Mercury Vapor (HPMV)	820 - 2200	Hg	Fused Silica
High Pressure Sodium (HPS)	1200 - 1600	Na, Hg	Alumina

Usually the emission mix is applied as a coating on the coiled, tungsten lamp-electrodes and also retained in the interstices of the coils. As the lamp life progresses, emission mix is continuously lost from the electrodes through (1) sputtering and (2) vaporization.

The chemical process of interaction between the emission mix and the electrodes--commonly described as the 'electrode-activation' process, lowers the work function of the electrodes and is the key to the improved lamp performance. The process depends on the lamp type and is usually defined by the emission mix composition, lamp chemistry (arc material) and lamp-operating conditions (temperature, pressure, etc.).

The exact nature of the electrode-activation process is a complex one and all the steps involved are not precisely understood, nor are they all experimentally verified. One generally accepted [1,2,3,4,5] activation mode of HID lamp electrodes is that during the activation process, low work-function* materials (neutral atoms, ions, oxides, etc.) generated by chemical interactions between the tungsten electrode, emission mix/arc materials, etc., migrate to the outer surface and the tips of the electrodes with the formation of a layer of the activating material on the tungsten surface. This layer is responsible for lowering the work function and significant improvement in the electron-emission properties of the W-electrode [6,7,8,9]. The task of a material scientist is to optimize the activation process by designing the composition and process of the mix matching the lamp type.

Although certain basic guidelines are emerging, most commercial emission mixes are designed phenomenologically. The activation processes involved in many lamp types, nevertheless, are well understood to the extent that a generic classification of the process is possible.

---

*Work function lower than that of tungsten metal (4.5 ev)

PRESENT SCOPE

The specific objectives of the present paper are to:
1. Outline the functions and requirements of an emission mix.
2. Describe and review compositions and activations in the most common HID lamps from the point of view of a ceramic scientist.
3. Present a generic classification of the activation processes.
4. Suggest certain guidelines for design and search of E-mixes.

Any description of the emission mix synthesis, characterization, electrode treatment, etc. are beyond the scope of the present paper.

FUNCTIONS AND REQUIREMENTS OF AN EMISSION MIX

Goal of an emission mix is to lower the electron-emission characteristics of the refractory electrode (W) in HID lamps. Lowered work-function (Ø) of the electrode results in:
1. Easier startability of the discharge (lower needed starting-voltage and temperature)
2. Lower electrode-operating temperature thus reducing electrode erosion and darkening of the arc-tube wall
3. Improved lamp life and efficacy.

The essential requirements of an emission mix are discussed below:
1. It is imperative that the emission mix contain activating species with work-function lower than that of the electrode metal (W). Table 2 lists common metal/oxides with lower Ø used as ingredients in designing emission mixes.
2. The designed composition must react with tungsten during lamp operation and the reaction products must include the activating atoms, ions or oxides.
3. The generated activating species must diffuse from the interface, through the layers of the mix and be transported to the electrode surface and the tip.
4. The activation process should have appropriate kinetics, so that the process lasts during the useful life of the lamp.
5. The vapor pressure of the emission mix and the generated activating species should be regulated, so that they are retained on the electrode surface and not depleted through rapid vaporization.

TABLE 2
Measured Workfunction (Ø) Values
of Metals and Oxides on Tungsten Substrate [19]

	Workfunction (Ø)ev
W	4.5
Na	1.6 - 2.1
Ca	2.24 - 2.40
Sr	2.20
Ba	1.6 - 2.8
La	2.2 - 2.7
Y	2.70
Th	2.6 - 3.4
CaO	2.1
SrO	1.1
BaO	1.1 - 2.0
ThO_2	2.2 - 2.5

COMPOSITION AND ACTIVATION OF COMMERCIAL EMISSION MIXES

Because of their lower work-function (Ø<4.5 ev), alkaline-earth and rare-earth oxidic compounds are mostly used in commercial HID lamps. Chemical and thermal compatibilities, however, restrict their uses in specific lamp types. Most common emission mixes and their activation processes in various HID lamps are described below.

Metal Vapor (MV) Lamps

These lamps contain various metal iodides (NaI, TlI, ScI_3, ThI_4, etc.)[2] as arc materials. Hence alkaline-earth based compositions are not used as emission mixes as they would, otherwise, form more stable iodides (BaI_2, CaI_2, etc.) and vaporize almost instantaneously from the electrodes. Two types of emission mixes are common in MV lamps:
(1) Low work-function elements, ions etc., (Na, Th, Sc, etc.) are adsorbed on the W-electrode surface from the dissociated, vapor-phase arc materials (NaI, ScI_3, ThI_4, etc.)[2]
(2) Low work-function (3.2 ev) of W-electrode is attributed to interaction between W-surface and a defect semiconductor (P-or n-type)emission mix, $Th_{1-x}Y_xO_{2-.5x}$[10,11].

High-Pressure Mercury Vapor (HPMV) and High-Pressure Sodium (HPS) Vapor Lamps

Emission mixes of these two lamp types contain BaO,CaO, etc., and the composition design is based on crystal chemistry, phase equilibria, reaction kinetics, rate of vaporization, etc. Electrodes in HPMV lamps operate at about 200°C higher temperature than those of HPS lamps. In addition, HPS lamps have sodium as arc material. Hence, activation processes in these two lamp types are apt to be different, requiring different emission mixes. Some of the emission mixes tried in HPMV and HPS lamps are listed below.

1. Because of its successful application in vacuum tubes [5,6], BaO-Oxide coated electrode was also used in HPMV and HPS lamps. BaO, however, has high vapor pressure and depletes rapidly from the electrode, causing early lamp failure.
2. Other oxides with low vapor pressure, such as ThO_2 or rare-earth oxides (Gd_2O_3, Y_2O_3, Dy_2O_3, etc.) are also used [12,13]. Through interaction with W-metal, these oxides release a mono-layer of Th or rare-earth metal on the W-surface [12], resulting in good emission properties for the electrode. ThO_2- and R_2O_3(R=Gd, Dy, Y)- coated electrodes, however, remained inferior to the BaO-coated cathodes [13].
3. After extensive out-of-lamp investigation (phase equilibria, vaporization, work-function, etc.) of the system BaO-CaO-WO_3, phase-pure Ba_2CaWO_6 was used in HPS lamps with excellent lamp performance [14,15, 16]. By and large, this is the most extensively used emission mix in commercial HPS lamps. The assumed activation process is:

$$2Ba_2CaWO_6+W = 3(Ba_{1/3}Ca_{2/3})\ WO_4 + 3Ba(g) \qquad \text{I.}$$

The reaction takes place in the interface. Later, high-temperature Knudsen cell (Sapphire-lined W cell) mass-spectrometry showed direct generation of the activating Ba atoms also by thermal dissociation of Ba_2CaWO_6 at 1300°C. [16].

4. Based on the phase equilibria and actual lamp tests (voltage drop across the electrodes), another viable family of emission mixes in the system Ba_2CaWO_6-Ba_2CaMoO_6 was characterized [12]. It was found that compositions on the Ba_2CaMoO_6-side of the series show better emission characteristics than Ba_2CaWO_6. Activation process and generation of Ba may be attributed to the reactions similar to reaction (III) or thermal decomposition of Ba_2CaMoO_6 at the lamp-operating temperature [14].
5. Another series of compositions in the system $Ba_3Y_2WO_9$-$Ba_3Y_2MoO_9$ was also characterized and lamp-tested with acceptable performance. Activation process was demonstrated to be generation of Ba and BaO species through thermal decomposition [12, 16].

CLASSIFICATION OF ACTIVATION PROCESS

As mentioned earlier, each step involved in the overall chemical process of activation is not precisely understood. The activation process may be initiated by special electrode treatment prior to lamp assembly, or during lamp start. However, activation continues and reaches its peak during lamp operation [1,2,3]. Based on the discussion in the previous sections, a simplified classification and correlation with the band configurations of metal and semiconductor can be attempted [18]. There are three distinct types of activation:
1. Metal on metal
2. Bulk semiconductor
3. Defect semiconductor

In addition, there may also be overlapping of these types.

<u>Metal on Metal</u>: According to the classical band model of metals, work-function (Ø) is the energy required to remove an electron from the Fermi level. Since the location of Fermi level in each metal is different, each metal has its own work function. However, it can be modified. Lowering of work function (emission enhancement) of tungsten can be achieved by introducing a layer of more electropositive atoms on its surface. In case of HID electrodes, such a layer can be created by:
1. Alloying with a lower work function metal (Th, Zn, etc.)[2].
2. Vapor dispersed adsorbate [2]. This requires certain conditions for heat of vaporization/adsorption of metal species [17,18].
3. Ion implantation.

<u>Bulk and Defect Semiconductor</u>: In this type, activation takes place through reduction of an oxidic semiconductor (emission mix) by tungsten at the emission mix/tungsten interface, followed by the release of an activating species on the electrode surface at elevated temperature. The following reactions illustrate the phenomenon [6]:

$$6BaO + W \rightarrow Ba_3WO_6 + 3Ba \qquad \text{II}$$

$$2Ba_3WO_6 + W \rightarrow 3BaWO_4 + 3Ba \qquad \text{III}$$

$$2Ba_2CaWO_6 + W \rightarrow 3(Ba_{.33}Ca_{,66})WO_4 + 3Ba \qquad \text{IV}$$

Ba atoms thus generated at the interface can migrate through the emission mix coat to the W-surface. This is a case of transition between semiconductor to metal-on-metal type activation. In the above example, during the reaction process (II-IV), Ba^{+2} ion generated can be incorporated in the BaO or tungstate lattice creating an oxygen-deficient, defect semiconductor with lower work function. Similarly, a defect semiconductor (extrinsic type) with lower Ø can be created by controlled valency doping prior to the lamp start. Example: Solid solutions between ThO_2 and Y_2O_3[10]; $BaMO_x$(M=Y,Sc,W) and ThO_2,Ta_2O_5, etc.[13].

GUIDELINES FOR SEARCH/DESIGN OF EMISSION MIXES

Based on the foregoing discussion, certain guidelines for search and chemical design of emission mixes can be outlined:

1. The emission mix should contain low work-function (<4.5 ev) materials (activating species).
2. The material should release Ba or other activating agents through interaction with tungsten or thermal decomposition at the lamp-operating temperature.
3. The activating species should be able to diffuse through the mix to the electrode surface.
4. The kinetics of the activation process should be such that a steady release of the activating species is maintained throughout the lamp life.
5. The mix should have a low vapor pressure and high, congruent melting point.
6. The mix and the vaporized species thereof should be chemically compatible with other lamp and arc materials (tungsten, Al_2O_3, SiO_2, Na, Hg, rare-earth iodides).
7. The material should have good adherence to tungsten surface at the lamp-operating temperature and should not shrink or fall off the electrode.
8. The emission mix stored in the interstices of the electrode coil should not sinter into coarse material during lamp life.
9. For easy handling and storage, the emission mix should be stable in H_2O and CO_2 atmosphere (air).

ACKNOWLEDGEMENTS

Many technical discussions with C. I. McVey and J. M. Strok of GE Lighting and J. Deluca (GE CR&D) were very helpful. K. A. Smith encouraged and supported my participation in the 1st European Ceramic Congress.

REFERENCES

1. Elenbaas, W., High Pressure Mercury Vapor Lamps and Their Applications, Philips Technical Library, Eindhoven, The Netherlands, 1965.
2. Waymouth, J.F., Electric Discharge Lamps, MIT Press, Cambridge, MA, USA, 1971.
3. deGroot, J.J. and vanVliet, J.A.J.M., The High-Pressure Sodium Lamp, Philips Technical Library, Deventer, The Netherlands, 1986.
4. Rittner, E.S., On the Mechanism of Operation of the Type B Impregnated Cathode, 1977, 48, 4344-4346.
5. Rittner, E.S., Ahlert, R.H., and Rutledge, W.C., Studies on the Mechanism of Operation of the L Cathode I, J.Appl.Phys., 1957, 28, 156-166.

6. Rutledge, W.C. and Rittner, E.A., Studies on the Mechanism of the L Cathode II, J.Appl.Phys., 1957, 28, 167-173.
7. Haas, G.A., Gray, H.F., and Thomas, R.E., Effects of S, Ba and C on Impregnated Cathode Surfaces, J.Appl.Phys., 1975, 45, 3293-3301.
8. Levitskii, V.A., Hekimov, Yu and Gerassimov, Ja.I., Thermodynamics of Double Oxides, II. Galvanic-cell Study of Barium Tungstates, J.Chem. Thermodynamics, 1979, 11, 1075-1087.
9. Forman, R., A Proposed Physical Model for the Impregnated Tungsten Cathode Based on Auger Surface Studies of the Ba-O-W System, Applications of Surface Science, 1979, 2, 258-274.
10. Datta, R.K., Thoria-Yttria Emission Mixture for Discharge Lamps, U.S. Patent No. 3,919,581. 1975.
11. Potter, R.M., 1984, Private communication.
12. Bhalla, R.S., Improved Electron Emission Materials for HID Lamps, J.Illum.Eng.Soc., 1979 6, 8, 174-178.
13. DeLuca, J., 1984, Private communication.
14. Smyser, W.E., and Speros, D.M., Discharge Lamp Thermionic Cathode Containing Emission Material, U.S. Patent No. 3,708,710. 1970.
15. Kreidler, E.R., Phase Equilibria in the System $CaO-BaO-WO_3$, J.Am.Cer.Soc., 1972, 10, 514-519.
16. Hirayama, C. and Bhalla, R.S., The Dissociative Vaporization of Alkaline Earth-Containing Mixed Oxide Thermionic Emission Materials, J.Illum.Eng.Soc., 1980, 9, 240-246.
17. Stull, D.R. and Prophet, H., JANAF Thermochemical Tables, 2nd ed. Nat.Bur.Standards, Washington, D.C., USA, 1971
18. Herrmann, G., The Oxide Cathode, Vol.2, Transl. S.Wagener, Chapman & Hall, London, U.K., 1951, pp. 160-207.
19. Fomenko, V.S., Emission Properties of Materials, Joint Publication Research Service (JPRS-56579), Dist. by Natl.Tech.Info.Service, U.S.Dept. of Commerce, Springfield, VA, USA, 1972.

PHASE COMPOSITION AND SINTERABILITY OF THE MIXTURES IN THE SYSTEM $CaO-Cr_2O_3-TiO_2$

JANUSZ ZBOROWSKI AND WOLFGANG KRÖNERT
Institut für Gesteinshüttenkunde der RWTH Aachen, Mauerstr. 5, D-5100 Aachen, FRG

ABSTRACT

Experiments were carried out to check the ability of TiO_2 to promote the sintering process of calcium chromite ($CaO \cdot Cr_2O_3$). Starting compositions lying within the triangle $CaO \cdot Cr_2O_3$-$CaO \cdot TiO_2$-Cr_2O_3 (CK-CT-K) were fired at 1300 and 1600 °C and cooled down naturally within a furnace. Apparent and relative densities were determined and SEM-observations of the microstructure were carried out. Phase composition was examined using X-ray diffraction and semi-quantitative EDX-analysis. Under applied experimental conditions CT-K and CK-CT solid solutions were proven to be coexistent in compositions lying on the respective tie lines. No ternary compounds were found within the examined area which turned out to be mainly a three phase field. Only some compositions lying on the K-CT tie line revealed a good sinterability and reached relative densities over 90% after firing at 1600 °C.

INTRODUCTION

The early studies (1-5) concerning the course of synthesis and basic properties of calcium chromite allowed the formulation of some concepts (6,7) concerning its technical application as a refractory.

A dense sintering of calcium chromite, however, still remains an unsolved problem. In spite of prolonged heat treatment at 1600 °C the relative density of pure calcium chromite does not exceed 70% (7). The sintering process of this compound is believed to be inhibited due to characteristic needle-shaped morphology of its crystals.

Taking into account the influence of TiO_2 on sintering behaviour of Cr_2O_3 (8,9) its addition could be considered as a possible way to affect advantageously the sintering process of calcium chromite also.

Since no literature data concerning ternary area of the system

$CaO-Cr_2O_3-TiO_2$ (C-K-T) were available the experiments were carried out to determine the compositions revealing both good sinterability and acceptable refractoriness.

EXPERIMENTAL

Starting materials were precipitated calcium carbonate*, chromium oxide ** and titanium oxide*. After homogenization the mixtures were briquetted and fired at 1300°C (4h soaking period, N_2 atmosphere). Then the pellets were ground (< 63um) and the samples (15mm diameter and 5mm height) were pressed under 100 MPa, fired at 1600 °C (8h soaking period, N_2 atmosphere) and cooled down naturally with the furnace. Apparent and relative densities were determined to compare a sintering ability of the examined mixtures. Phase composition of the samples was determined using X-ray diffraction analysis after both steps of firing. After firing at 1600°C their polished sections were examined in SEM and phase composition was additionally assessed by mean of the semi-quantitative EDX determination of the elements' concentrations. Some starting mixtures were examined using high-temperature X-ray analysis. In this case the diffractograms were registered first at room temperature, then after heating up to the maximal temperature of 1300°C (heating rate 10K/min, 10 min. soaking period) and finally again at room temperature after the rapid cooling down (1 min.). Data concerning $CaO-Cr_2O_3$ (1-3), $Cr_2O_3-TiO_2$ (10,11) and $CaO-TiO_2$ (12,13) binary systems were used to interpret the obtained experimental results.

RESULTS

Chemical composition of the starting mixtures as well as apparent and relative densities of the samples fired at 1600°C are given in Table 1. The results of the X-ray and EDX analyses are presented in Table 2. Compositions denoted in Table 2 with "*" were additionally examined using a high-temperature X-ray diffractometer. Phase composition of the mixtures determined on this way turned out to be identical with that obtained at room temperature for the samples fired at 1300 and 1600°C. The results of the phase analysis indicate that the calcium titanate phase which coexists with the chromium oxide in the compositions lying on the Cr_2O_3-CaO·TiO_2 tie line (samples 2.1-2.4) is able to dissolve 3-17% of chromium oxide. As it can be seen from the CaO and TiO_2 content found in the chromium oxide it is able to dissolve small amount of calcium titanate only. Mutual solubility was also found in case of calcium chromite and calcium titanate coexisting in the compositions lying on the corresponding tie line (samples 1.2-1.4). Only some compositions lying on the CT-K tie line reveal good sinterability and relative densities exceeding 90%.

---

* - Reagent-grade, E. Merck, Darmstadt (FRG)

** - Reagent-grade, Riedel de Haen, Seelze (FRG)

TABLE 1
Chemical composition of the examined mixtures and density of the samples obtained after firing at 1600 °C (N_2 atmosphere, 8h soaking period)

No	composition CaO (C) weight %	Cr_2O_3 (K)	TiO_2 (T)	molar ratio C	K	T	apparent density g/cm^3	relative density %
1.1	26.92	73.08	-	1	1	-	3.18±0.03	66.2±0.6
1.2	41.21	-	58.79	1	-	1	3.49±0.02	85.2±0.4
1.3	32.56	44.19	23.25	2	1	1	3.11±0.03	71.9±0.7
1.4	30.43	55.07	14.50	3	2	1	3.14±0.02	70.2±0.4
1.5	35.00	31.68	3.32	3	1	2	3.20±0.03	76.0±0.9
2.1	19.40	52.80	27.80	1	1	1	4.34±0.05	95.0±1.0
2.2	12.70	69.10	18.20	1	2	1	4.49±0.03	93.8±0.6
2.3	26.40	35.80	37.80	2	1	2	3.92±0.06	90.5±1.4
2.4	7.50	81.70	10.70	1	4	1	4.28±0.03	86.6±0.6
2.5	3.72	90.96	5.31	1	9	1	3.53±0.02	69.4±0.4
3.1	26.0	70.20	3.80	1	1	0.1	3.09±0.04	64.5±0.8
3.2	24.50	66.40	9.10	1	1	0.3	3.12±0.06	65.1±1.3
3.3	22.50	60.80	16.70	2	2	1	3.20±0.02	69.3±0.4
3.4	20.19	54.81	25.00	1	1	0.9	3.26±0.04	71.3±0.1
4.1	15.50	84.50	-	2	1	-	3.04±0.04	62.0±0.8
4.2	14.70	80.30	5.00	2	1	0.2	3.01±0.06	61.8±1.2
4.3	16.80	65.00	18.20	1.3	1.9	1	2.92±0.04	62.1±0.8

TABLE 2
Phase composition of the examined samples after firing at 1300 and 1600 °C according to X-ray and semiquantitative EDX-analyses

No	X-ray analysis phase composition after firing at 1300 °C	1600 °C	Semi-quantitative EDX-analysis chemical composition at the point weight % CaO	Cr_2O_3	TiO_2		presumed phases
1.1	CK	CK	24	76	-		CK
1.2	CT	CT					
1.3	CK, CT	CK; CT	26	63	11	(1)+	CK s.s.
			40	9	51	(2)	CT s.s.
1.4	CK, CT	CK, CT	39	9	52	(1)	CT s.s.
			21	65	14	(2)	CK s.s.
1.5	CK, CT	CK, CT					
2.1*	CT, K	CT, K	33	17	50	(1)	CT s.s.
			2	94	4	(2)	K s.s.
2.2*	CT, K	CT, K	2	95	3	(1)	K s.s.
			36	11	53	(2)	CT s.s.

TABLE 2 (cont.)

No	X-ray analysis phase composition after firing at 1300 °C	1600 °C	Semi-quantitative EDX-analysis chemical composition at the point weight % CaO	Cr_2O_3	TiO_2		presumed phases
2.3	CT, K	CT	2	95	3	(1)	K s.s.
			39	3	58	(2)	CT s.s.
2.4	K, CT	K, CT					
2.5	K, CT	K, CT					
3.1	CK	CK					
3.2*	CK,CT,K	CK,CT,K	11	77	12	(1)	CK s.s.
			39	11	50	(2)	CT s.s
			1	97	2	(3)	K s.s.
3.3	CK,CT,K	CK,CT,K	39	11	50	(1)	CT s.s.
			21	65	14	(2)	CK s.s
			3	95	2	(3)	K s.s.
3.4	CT,K,CK	CT,K,CK	4	91	5	(1)	K s.s.
			39	7	54	(2)	CT s.s.
4.1	CK, K	CK, K					
4.2	CK,CT,K	CK,CT,K	13	82	5	(1)	CK s.s.
						(2)	K
4.3	CK,CT K	CK,CT,K	2	95	3	(1)	K s.s.
			34	18	48	(2)	CT s.s.
			14	73	13	(3)	CK s.s.

CK - $CaO{\cdot}Cr_2O_3$, CT - $CaO{\cdot}TiO_2$, K - Cr_2O_3
* - Compositions subjected to high temperature X-ray analysis
\+ - Numbers denote the points on the samples' polished surfaces at which EDX-analysis was carried out. Corresponding SEM pictures of the analysed areas could not be presented due to space limit.

CONCLUSIONS

Under applied experimental conditions the internal area of the examined composition triangle CK-CT-K turned out to be mainly a three phase field.
Only in case of the compositions with TiO_2 addition lower than 4 weight % no other phase but β-$CaO.Cr_2O_3$ could be proven by X-ray analysis.
No ternary compound was found within the examined area. In the compositions lying on CT-K and CK-CT tie lines solid solution were found to be coexistent.
Only some compositions lying on the CK-CT tie line containing calcium titanate-chromium oxide solid solutions and no calcium chromite revealed after sintering at 1600 °C (N_2 atmosphere, 8h soaking period) relative densities exceeding 90%.

Acknowledgement: The financial support of the Deutsche Forschungsgemeinschaft (DFG) is greatfully acknowledged.

REFERENCES

1. Ford, W.F and Rees, W.J., The CaO-MgO-Cr_2O_3 Ternary System, Part I - A Partial Investigation of the CaO-Cr_2O_3 System. Trans. Brit. Ceram. Soc. 1948, **47**, 207-231.

2. Ford, W.F.and Rees, W.J., The CaO-MgO-Cr_2O_3 Ternary System, Part II - Further Experiments. Trans. Brit. Ceram. Soc., 1949 **48**, 291-312.

3. White, J. Appendix - The Interpretation of the Experimental Data on the System CaO-MgO-Cr_2O_3. Trans. Brit. Ceram. Soc., 1949, **48**, 312-321.

4. Ford, W.F. and White, J., The CaO-MgO-Cr_2O_3 Ternary System, Part III - The CaO-CrO_3-Cr_2O_3 Ternary and MgO-Cr_2O_3 Binary Systems. Trans. Brit. Ceram. Soc., 1949, **48**, 417-427.

5. Glasser, F.P and Osborn, E.F., Phase Equilibrium Studies in the System CaO-Cr_2O_3-SiO_2. J. Am. Ceram. Soc., 1958, **41**, 358-367.

6. Nadachowski, F., The Role of the Solid State Reactions in the Search for Novel Refractories. In Proc. 9th Symp. Reactivity of Solids, Elsevier, Amsterdam,1981, pp.747-765.

7. Zborowski, J.,. Krönert, W., Nadachowski, F., and Dhupia, G.S., Microstructural Changes of CaO·Cr_2O_3 Refractory under Slag Attack. In HighTech Ceramics, ed. P.Vincezini, Elsevier Science Publ. B.V., Amsterdam, 1987, pp.2415-2424.

8. Callister, W.D., Johnson, M.L., Cutler, J.B. and Ure, J.W. jr.. Sintering of Chromium Oxide with the Aid of TiO_2. J. Am. Ceram. Soc., 1979, **62**, 208-211.

9. Krivoruchko, P.P., Alpin, B.G., Lysak, S.V. and Degtyareva, E.V., Effect of Additions of some Oxides on the Sintering of Cr_2O_3 in Air and in Reducing Atmosphere. Transl. from Poroshkovaja Metallurgia,, 1976, **82**, 73-77.

10. Werner, H-D., Phase Relations in the System TiO_2-$CrO_{1.5}$. N. Jb. Miner. Mh., 1974, 218-234 (in German).

11. Anderson, S. and Jahnberg, L., Crystal Structure studies on the Homologous Series Ti_nO_{2n-1}, V_nO_{2n-1} and $Ti_nCr_2O_{2n-1}$. Arkiv f. Kemi, 1963, **39**, 413-426.

12. De Vries, R.C., Roy, R. and Osborn, E.F., Phase Equlibria in the System CaO-TiO_2. J. Phys. Chem.,1954, **58**, 1069-1073.

13. Roth, R.S., Revision of the Phase Equilibrium Diagram of the Binary System Calcia-Titania Showing the Compound $Ca_4Ti_3O_{10}$. J. Res. Natl. Bur. Standards, 1958, **61**, 437-440.

TRANSPORT PROPERTIES OF THE WUSTITE PHASE

F. ADAM, B. DUPRE, C. GLEITZER & J. NOWOTNY*

(Université de Nancy I, Laboratoire de Chimie du Solide Minéral, Associé au CNRS, BP 239, 54506 VANDOEUVRE CEDEX, FRANCE)

INTRODUCTION

It is known that a very high defect concentration in wustite ($Fe_{1-y}O$ with $0.045 \leqslant y \leqslant 0.11$) results in strong interactions and therefore in a complicated transport mechanism in this material.

Laub and Wagner (1, 2) have postulated that the chemical diffusion coefficient $\tilde{D}$ decreases with the deviation from stoichiometry y. This dependence is in an agreement with the results of Landler and Komarek (3) and of Sadowski et al. (4). In contrast to these reports, Rickert and Weppner(5), based on studies of the re-equilibration kinetics by using solid galvanic cells, have observed an increase of $\tilde{D}$ with y. The purpose of this work is therefore to clarify these discrepancies.

EXPERIMENTAL

The samples were obtained by zone-melting of a polycrystalline mixture of Fe_2O_3+Fe under controlled gas atmosphere. We used discs of 6 mm in diameter and 1 mm thick, with grain size about 0.05-0.1 mm.

---

* Present address : Max Planck Institute for Solid State Research, D-7000 STUTTGART 80, FRG

Thermogravimetric studies were performed in a $CO-CO_2$ mixture, with a linear flow velocity around 1 $cm.s^{-1}$. Oxidation and reduction runs were carried out within the wustite field.

ANALYSIS OF THE DATA

In the analysis of the equilibration kinetics, we assume that within the range under study there are a diffusion term, and a chemical reaction term corresponding to the surface re-equilibration :

$$t = t_D + t_S \qquad [1]$$

Assuming (6), that the surface reaction is of the 1st order :

$$d\alpha/dt = k\frac{A}{\Delta m_\infty}|c_i - c_f|(1-\alpha) \qquad [2]$$

α is the reacted fraction, $|c_i - c_f|$ the total change in oxygen concentration, A the surface and k the rate constant ; then :

$$\alpha = 1-\exp(-k\frac{A|c_i-c_f|}{\Delta m_\infty}t_S) \text{ and } t_S = -\ln(1-\alpha)/kA\frac{|c_i-c_f|}{\Delta m_\infty} \qquad [3]$$

As regards the diffusion step (7) :

i - for short times

$$\tilde{D} = (\frac{\Delta m}{A})^2 \frac{\pi}{t_D(\Delta c)^2} \text{ and } t_D = (\frac{\Delta m}{A})^2 \frac{\pi}{\tilde{D}(\Delta c)^2} \qquad [4]$$

ii - for long times

$$\tilde{D} = -\frac{d^2}{\pi^2 t_D}\ln\frac{\pi^2}{8}(1-\alpha) \text{ and } t_D = -\frac{d^2}{\pi^2\tilde{D}}\ln\frac{\pi^2}{8}(1-\alpha) \qquad [5]$$

where d is the sample thickness.

Therefore the numerical analysis of the experimental data allows to evaluate t_S and t_D at different stages of the reaction, and then k and $\tilde{D}$: for each couple of values (α_1,t_1), (α_2,t_2), equations [1] to [5] provide a couple of values t_S and t_D, and hence a couple of values k and $\tilde{D}$.

RESULTS AND DISCUSSION

The above-treatment allows to conclude that :

i - the ratio t_D/t_S increases steadily with α, so that the beginning of the reaction is mainly controlled by the surface reaction processes, and the end by the bulk diffusion

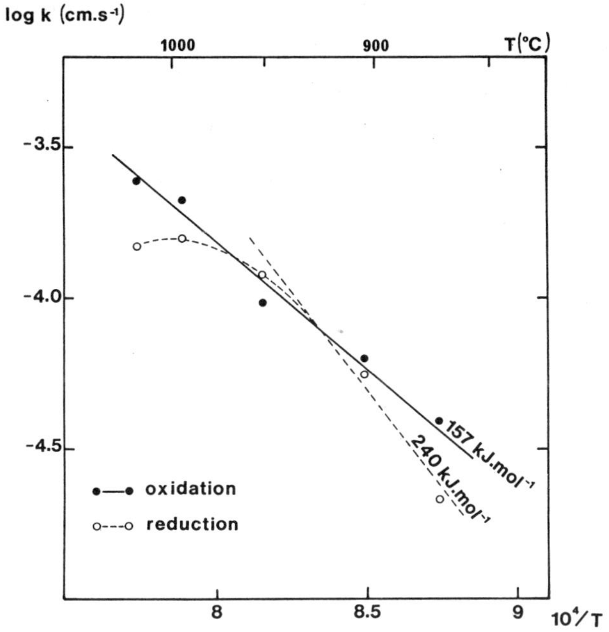

Figure 1. Variation of k with T.

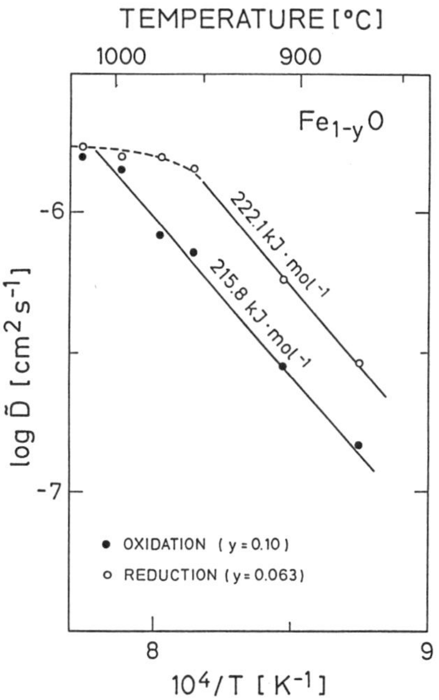

Figure 2. Variation of $\widetilde{D}$ with T.

ii - the precision of the data is not sufficient to derive well defined values of k and $\tilde{D}$ from any couple of experimental data, so that the most reliable values are obtained :

. for k, by using the initial slope of the α vs.t curves

. for $\tilde{D}$, by equation [5] applied to the last part of these curves.

The resulting variations of k and $\tilde{D}$ as a function of T are presented on figures 1 and 2.

The rate constant k does not display a simple thermally activated behaviour, in the reduction process, and the activation energies obtained at low temperature are rather high, in the range of 200 kJ/mol.

The chemical diffusion $\tilde{D}$ shows a normal trend in oxidation, whereas in reduction there is a kind of levelling of $\tilde{D}$ above 950°C. This may suggest that, at high temperature, there is some evolution of the basic processes. However, in the range 875-950°C, the activation energy for oxidation and reduction is quite comparable to the previous authors values. Secondly, the chemical diffusion coefficient slightly decreases when the departure to stoichiometry increases. This may be due to the clustering of point defects when y increases.

REFERENCES

1 - L.W. LAUB - PHD Thesis at Northwestern University, Evanston 1971

2 - L.W. LAUB & J.B. WAGNER - Oxid. Metals 7, 1 (1973)

3 - P. LANDLER & K. KOMAREK - Trans. AIME 236, 138 (1966)

4 - A. SADOWSKI, G. PETOT-ERVAS, C. PETOT & J. JANOWSKI, in preparation for publication

5 - H. RICKERT & W. WEPPNER - Z. Naturforsch. 29a, 1984 (1974)

6 - P. HAYES & P. GRIEVESON - Met. Trans. 12B, 319 (1981)

7 - P. CHILDS, L. LAUB & J.B. WAGNER - Proc. Brit. Ceram. Soc. 19, 29 (1971)

8 - E. RIECKE & K. BOHNENKAMP - Archiv. Eisenhütten. 40, 717 (1969)

MULLITE-FERRITE INTERACTIONS DURING THE SINTERING OF MgO BASED FERRITES

K.J. SIEVEWRIGHT and F.R. SALE
Manchester Materials Science Centre,
University of Manchester/UMIST,
Grosvenor Street, Manchester M1 7HS.

ABSTRACT

Reactions in mullite-ferrite mixtures have been studied because of stability problems of mullite saggers during the manufacture of ferrite yoke rings. Changes in the microstructure, chemical distribution and thermal stability of mullite have been investigated using S.E.M., T.E.M. and thermal analysis techniques. Results from these studies indicate that the cause of the microstructural changes is the formation of a low melting point liquid silicate as contamination occurs during use.

INTRODUCTION

After a number of passes through factory kilns mullite saggers used in the production of ferrite yoke rings have been found to stick to one another where sagger to sagger contact exists. In addition, the sloping edges, upon which sagger-ferrite contact occurred, increase in dimensions with the formation of layers which ultimately crack and break away. The result of such occurrences is a loss of productivity and increase in costs. In an attempt to solve this problem, different mullite sagger types have been tested. However, the problem was found to recur in each case regardless of whether the sagger was "normal sintered" (as usually used), "fused" or "low alkali" mullite. Experimental work has therefore been carried out in an attempt to identify the cause of these phenomena in terms of change in microstructure, chemical distribution and thermal stability as industrial usage occurs. This project has been carried out in conjunction with Philips Components, Southport, Merseyside, who have generously provided samples for study.

EXPERIMENTAL PROCEDURE

Microstructural Studies

Saggers, which had been contaminated in the industrial environment, were compared to unused materials. The microstructures were studied by taking a cross section from the sloping edge of the saggers, where sagger/ferrite contact had occurred. Such pieces were mounted, ground on metallographic paper and polished on 6 and 1 μm diamond impregnated pads respectively. Examination of the microstructures was carried out using Philips SEM 505 and Philips TEM 400 electron microscopes. Chemical analysis of the saggers was obtained using the EDAX analysis facility on both microscopes.

Thermal Stability Studies
Thermal stability investigations were carried out using a Stanton Redcroft DSC 1500 with a computerised data collection system (CETA). Powder samples were produced by taking thin slices from unused and contaminated areas of the saggers and then grinding them to powders. Powder samples of mass 40 mg were heated at 20°C/min in nitrogen to 1250°C, the minimum firing temperature used industrially. Powdered calcined pure alumina was used as the reference standard.

RESULTS AND DISCUSSION

The results presented are those obtained from studies of normal sintered saggers, however, these results are characteristic of each type of sagger investigated.

Figure 1 shows a S.E.M. micrograph of the microstructure of an unused mullite sagger. This consists of large, fairly dense grains within a matrix of smaller, finer grains. After use the saggers developed dense and interfacial regions at the outer edges, where contamination occurred. Such a microstructure is shown in Fig. 2. In this case the sagger was removed from service due to a large increase in its dimensions and spalling of the outer layer. As the figure shows the sagger microstructure consists of three distinct regions, a dense outer layer, the interface and a central area, which has a similar microstructure to the unused material.

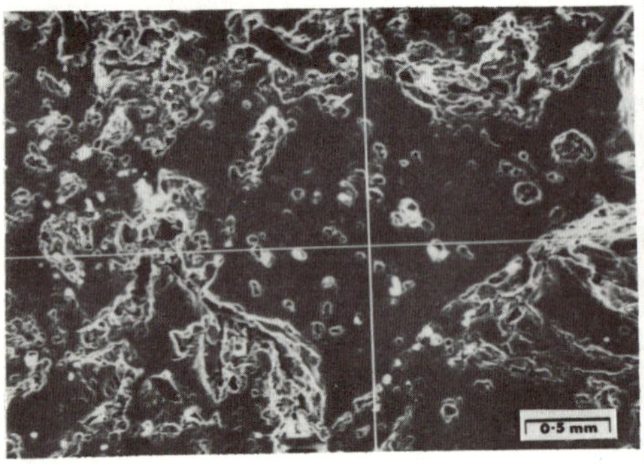

Figure 1. Unused sagger.

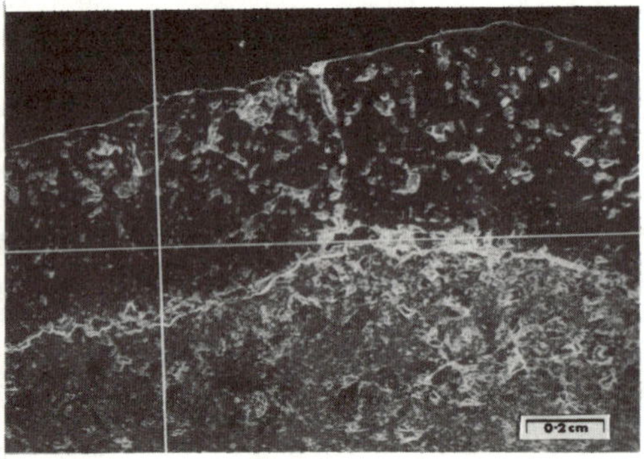

Figure 2. Contaminated edge of used sagger.

At higher magnification the TEM micrograph given in Fig. 3 shows cubic features which formed as a result of sagger use. These were found within a matrix of glassy material. These features were not seen in any of the unused materials but existed in groups between large grains within contaminated pieces. Chemical analysis of the severely contaminated sample shows that each of the regions previously mentioned has a different chemical composition. The dense outer region contains aluminium and silicon, in proportions within the range possible for mullite, Klug et al (1). In addition to these, this area contains potassium as an impurity at low levels. The interfacial area contains a high level of silicon and much less aluminium than is usual for mullite. This contains increased levels of other elements such as potassium, iron, titanium and magnesium. The inner region has a microstructure and chemical distribution typical of the unused material. The dense grains are pure containing mainly aluminium and silicon, again within the mullite composition range. The finer matrix region has a much higher impurity content. Figure 4 summarises the composition of these areas.

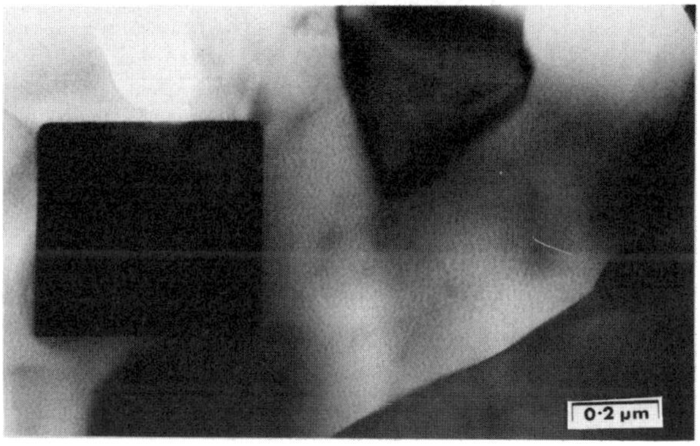

Figure 3. TEM micrograph of heavily contaminated region.

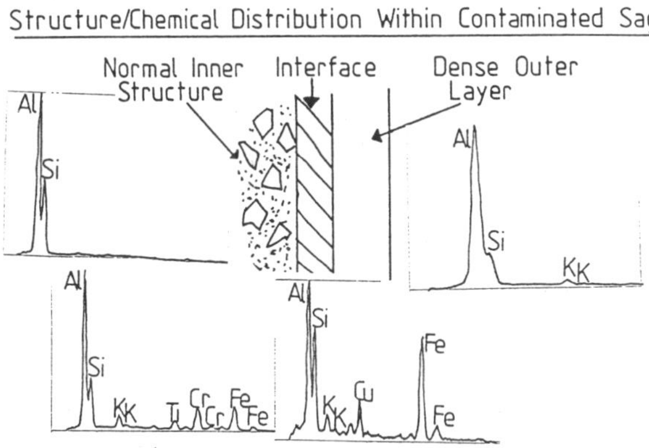

Figure 4.

Analysis of the cubic features, found using TEM, are summarised in Fig. 5. These features contain aluminium and silicon, within the range possible for mullite. Potassium was detected as an impurity within the material at low concentrations. The glassy matrix was found to contain a much higher level of silicon, a low level of aluminium and high levels of impurities such as iron, potassium and titanium.

The thermal characteristics of both the unused and contaminated samples are presented on Fig. 6 which shows the relationship between heat flow (mw) and temperature. Between room temperature and 1000°C both the contaminated and unused materials show an exothermic trend which is characteristic of the apparatus used. Above this temperature (~1000°C) there is a second exothermic trend probably caused by powder particles

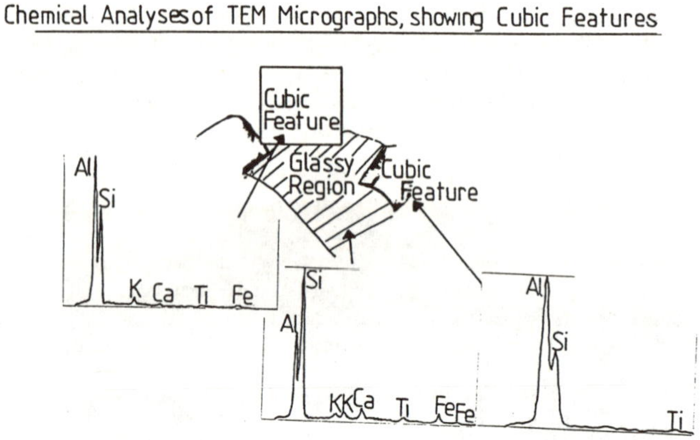

Figure 5.

sintering together. The contaminated material, however, undergoes a sharp endothermic event at approximately 1200°C. No such event occurs in the unused material.

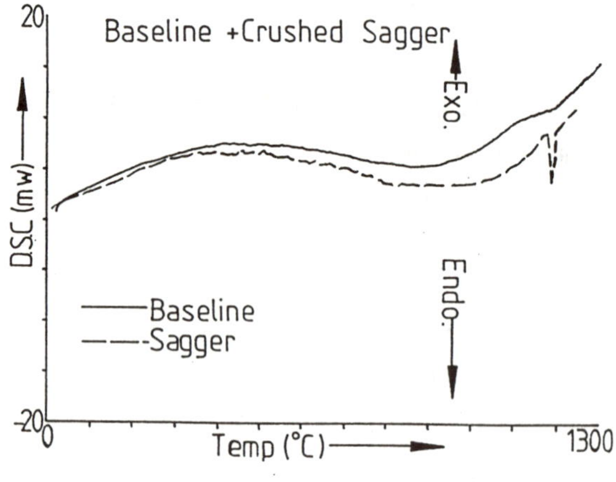

Figure 6. DSC traces showing baseline and contaminated sagger data.

It is clear from the microstructure and chemical composition data that some redistribution of elements within the mullite has occurred during use. Over many years a large number of workers have studied alumino-silicate refractories within blast furnaces. Some aspects of this work has relevance here because of the similarity in the major contaminants. Gad and Barrett (2) and Clews and Green (3) studied the alkali contamination of alumino-silicates in the presence of iron oxide and found a substantial redistribution of elements due to migration. More recently, Kennedy (4) found that molten alkali contact caused mullite to break down and form new compounds. On the basis of this previous work, it is proposed that an inward migration and hence concentration of potassium and sodium, from the original sagger, and iron and magnesium from the ferrite, occurs during the contamination of saggers during the firing of ferrites. It also appears that on repeated thermal cycling small cubic or rectangular grains form with an Al_2O_3 to SiO_2 ratio which suggests these are 2:1 mullite and which contain potassium as an impurity. Work by Harabi (5) has identified features which have a similar appearance. In this case the cubic appearance is thought to be caused by alkali impurities, in solid solution with mullite, causing the formation of a super-structure, which grows preferentially along the c-axis.

Chemical analysis indicated that increased amounts of silica were tied up with impurity elements within the interfacial region. Thermal analysis indicated, by the occurrence of the sharp endotherm at a fairly low temperature, that a liquid phase was formed. A combination of these data indicate the production of a liquid silicate, in small amounts, close to the outer edge of saggers within the first few passes through the factory kilns. This causes liquid phase assisted densification to occur at the outer edge. With an increased number of passes the dense layer increases in thickness and the liquid region moves inward. As the silicate migration inward proceeds the amount present increases by inclusion of free silica from within the matrix and further silica obtained by mullite decomposition. Thus a dense outer layer and interfacial layer, which consists mainly of silicates, form. Spalling of the outer layer may have occurred due to this dense layer having a different coefficient of thermal expansion from the bulk of the sagger. Kennedy (4) found similar occurrences in refractory kiln-ware. This results in a stress build up in the interfacial region when thermally cycling the material. Yamana et al (6) found similar stress build-ups in siliceous glass between mullite crystals. When solid, the stressed silicate has been seen to act as an easy path, along which microcracks may propagate, causing spalling of the outer layer, Prochauza and Klug (7).

Acknowledgements
The authors wish to thank Drs. W.M. Dawson, L. Boonstra and D. Tinsley of Philips Electrical Components, Southport, for advice and provision of samples. One of the authors, KJS, is grateful to the SERC for the provision of a maintenance grant for this work.

REFERENCES

1. Klug, F.J., Prochazka, S. and Doremus, R.H., J. Am. Ceram., 1987, **70**, 750.

2. Gad, G.M. and Barrett, L.R., Trans. Brit. Ceram. Soc., 1949, **48**, 352.

3. Clews, F.H., Richardson, H.M. and Green, A.T., Trans. Brit. Ceram. Soc., 1940-41, **40**, 426.

4. Kennedy, C.R., J. of Mat. for En. Sys., 1981, **3**, 27.

5. Harabi, A., 1989, Private Communication, Manchester Materials Science Centre.

6. Yamana, K., Tokonami, M. Nobugai, K., Morimoto, N. and Koizumi, M., Comm. Am. Ceram. Soc., 1981, C43.

7. Prochanzka, S. and Klug, F.J., J. Am. Ceram. Soc., 1983, **66**, 874.

THE ROLE OF THE REFRACTORY ON THE SINTERING OF Li CONTAINING MgZn-FERRITE.

Jack G. Boerekamp
Corporate Innovation Materials
Philips Components
Building BE-222, 5600MD Eindhoven, The Netherlands

ABSTRACT

If LiMgZn-ferrites are sintered on refractory support materials (e.g. mullite or alumina), a marked chemical reaction between ferrite and refractory is observed. The nature of these reactions is investigated. This study indicates that oxygen atmospheres seem to influence lithia loss. In mullite refractories crystallization of Zn-aluminate occurs which is enhanced by lithia present in the ferrite. It appears that a liquid phase, which results from the presence of lithia, accelerates the Zn reaction with mullite.

INTRODUCTION

LiZn and MgZn-ferrites are used for high ohmic, soft magnetic applications. In previous work we had observed that sintering of Li substituted MgZn-ferrite on mullite resulted in a deviation of magnetic properties as compared to those obtained for sintering on pure alumina refractory. A possible explanation for this deterioration is given by Ridgley et.al.[1]. It is ascribed to lithia disappearance due to evaporation of Li_2O [1,2,3]. Van Arkel et.al.[4] showed LiOH evaporation to occur when treating lithia at higher temperatures in water containing atmospheres. However, previous investigations have suggested that the quantity of lithia that is lost during sintering appears to be dependent on the specific refractory used. The role of the refractory support material has never been investigated, though Strickler and Roy [5], in studying the system $LiFeO_2$-Fe_2O_3, found no material loss when sintering samples which had been sealed in noble metal containers, not even up to 1500°C. This study considers the effects of various refractory materials and atmospheres, on the sintering behaviour of Li substituted MgZn-ferrite. Next the effect of ferrite firing on the used refractory is analyzed by SEM/EPMA.

EXPERIMENTAL

LiMgZn-ferrites were prepared according to the following compositions:

$$(Li_{0.5}Fe_{0.5})_x\ Mg_y\ Zn_z\ Mn_{0.1}\ Fe_{1.80}\ O_{3.85}$$
$$0 < x < 0.2,\ 0.35 < z < 0.50 \text{ and } x + y + z = 1$$

Raw materials of Fe_2O_3, ZnO, $Mg(OH)_2$, Li_2CO_3 and $MnCO_3$ were weighed out and mixed in an Eirich R02 mixer, and subsequently calcined on alumina plates at 850°C for 4 hours. After prefiring the powder was milled for 6 hours in a steel ball mill using demineralized water. The slurry was dried, broken down, mixed with binder and finally granulated. The granulate was pressed into toroids or discs of 5 mm thickness. Sintering experiments were done in a vertical tube furnace at 1150 - 1300°C.
As refractory material mullite ($Al_2O_3 \cdot 2SiO_2$), dense sintered Al_2O_3 and ZrO_2 were used. For electrical evaluation DC resistivity measurements were done by a two point method. The ferrite samples were analyzed for Li content using AAS, corrections for oxygen loss are made using Fe content analysis. Refractory samples, which were used for firing the Li containing ferrite, 1200°C / 150 hrs / air, were analyzed via SEM/EPMA.

RESULTS AND DISCUSSION

A. Influence of refractory and atmosphere on the ferrite

Previous studies have shown that the sintering of Li containing MgZn-ferrite on mullite supports results in inferior magnetic properties, similar to those documented by Ridgley et.al.[1]. The latter concluded that lithia evaporation (above 1000°C) leads to a ferrite containing excess iron; consequently, on cooling in O_2 containing atmospheres, hematite (α-Fe_2O_3) is precipitated as finely distributed second phase.
In this study, hematite was not detected via XRD, indicating it was not present at levels above 3%. Moreover, light microscopy did not reveal the hematite phase.
Table 1 shows that the extent of lithia loss during sintering is markedly dependent upon the refractory support material and upon the sintering temperature. For example, at 1200°C / 2 hrs / air, the amount of Li remaining in the ferrite is respectively 0.28, 0.23, 0.24 and 0.18 wt% Li for Pt, ZrO_2, Al_2O_3 and mullite substrates. The relatively large weight loss observed in the case of mullite suggest the lithia loss is enhanced by the presence of silica, possibly due to Li-silicate and at higher temperatures liquid phase formation. This is in accordance with thermodynamical data [6].
When lithia is lost, the ferrite changes from iron deficient (oxygen vacancies) to iron excess (cation vacancies):

$$Li^{+} + 1/2\ O^{2-} + 5/6\ V_o \rightleftharpoons 1/2\ Li_2O\uparrow$$

In an iron excess ferrite oxidation/reduction reactions play an important role. According to Reijnen [7] at higher temperatures cation vacancies disappear and oxygen vacancies are formed by reduction of Fe^{3+}:

$$Fe^{3+} + 3/8\ V_c + 1/2\ O^{2-} \rightleftharpoons Fe^{2+} + 1/4\ O_2\ (g)$$

Decreasing the partial oxygen pressure at a given temperature will also

TABLE 1

Li content analysis (AAS) on fired toroids of LiMgZn-ferrite.
The nominal Li-content is 0.32 wt% Li (x = 0.20).

Refractory	top Temp (°C)	top time (hours)	Atmosphere	Li-content (wt% Li)
Al_2O_3	850 (calcining)	2.0	air	0.31
Al_2O_3	1200	2.0	air	0.24
Al_2O_3	1250	2.0	air	0.16
mullite	1175	0.5	air	0.23
mullite	1200	2.0	air	0.18
mullite	1250	2.0	air	0.10
ZrO_2	1200	2.0	air	0.23
ZrO_2	1250	2.0	air	0.15
ZrO_2	1200	2.0	O_2	0.23
ZrO_2	1200	2.0	N_2	0.27
ZrO_2	1200	2.0	dry N_2	0.27
ZrO_2	1200	2.0	wet N_2	0.26
Pt	1200	2.0	air	0.27
Pt	1250	2.0	air	0.26
Pt	1200	2.0	N_2	0.29

serve to reduce the cation vacancy concentration. Furthermore, it can be assumed that the presence of cation vacancies in the ferrite will strongly enhance the Li^+ mobility and hence the lithia loss. This possibly explains the results obtained by Bandyopadhyay [8], who found higher irreversible weight (lithia?) loss for iron excess Li-ferrite (V_c) compared to iron deficient samples (V_0). The effect of the atmosphere on the lithia loss is demonstrated in table 1: firing of LiMgZn-ferrite on ZrO_2 at 1200°C / 2 hrs, in air and N_2, results in Li contents of 0.23 and 0.27 wt% respectively. However, as firing in O_2 is a prerequisite for obtaining good electrical properties, this detrimental influence of O_2 on lithia loss may be obviated by annealing in oxygen at a lower temperature.

The loss of lithia is also demonstrated by resistivity measurements on air fired LiMgZn-ferrite samples that contain various nominal Li contents, on alumina refractory at various temperatures. Increasing nominal Li content and firing temperature, gives rise to relatively more lithia loss during sintering, and results in a decrease in specific resistivity (figure 1). This indicates an increased Fe^{2+} concentration in the spinel lattice after sintering.

According to van Arkel [4] Li_2O is not volatile up to 1570°C, whereas the interaction of lithia with water results in the formation of LiOH which is volatile at high temperatures. However, in this investigation no effect of H_2O presence was shown; firing of the ferrite in dry and wet N_2 atmospheres resulted in similar Li contents (table 1).

Based on these results, the following mechanism for the lithia loss is proposed:

1. The lithia loss is mainly caused by diffusion and reaction with the refractory support material, especially with SiO_2 containing refractory, where glass phase formation seems likely.
2. Accelerated loss of lithia occurs in O_2-containing atmospheres at higher temperatures. Probably the oxygen atmosphere serves to increase the formation of cation vacancies which will enhance the Li^+-mobility.

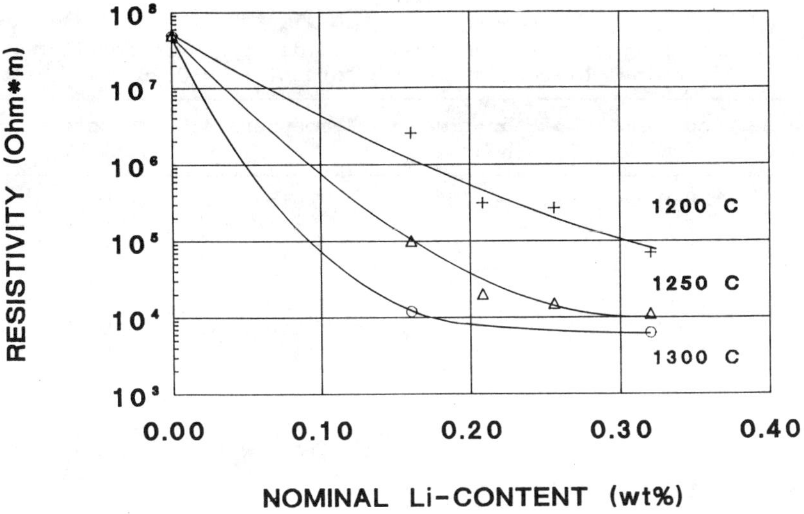

Figure 1. Specific resistivity of LiMgZn-ferrite fired on Al_2O_3 refractory as a function of nominal Li-content at various sintering temperatures.

B. Influence of the ferrite on the refractory

Given the problems with lithia loss, in which lithia seems to penetrate into the refractory, a more detailed study was carried out on the reaction products formed in the refractory. Figure 2 shows SEM/EDAX pictures with Zn mapping of mullite and Al_2O_3 refractory after interaction with a LiMgZn-ferrite (1200°C / 150 hrs / air). Crystallization of a Zn rich phase, identified as $ZnAl_2O_4$, is shown in mullite refractory. Interaction of the refractory types with a pure MgZn-ferrite is similar, however a smaller reaction layer is found in this case. No interaction is observed with the pure Al_2O_3 refractory for the two ferrites. Unfortunately Li mapping is not possible with SEM/EPMA. technique. We believe that a liquid phase formation in the presence of SiO_2 accelerates the reaction of zinc with the refractory, which is further promoted by Li_2O.

CONCLUSIONS

1. The sintering of LiMgZn-ferrites on mullite refractory yields inferior magnetic properties hence sintering on Al_2O_3, ZrO_2 or better still, Pt is preferred.
2. Higher lithia loss is observed on sintering in O_2 containing atmospheres, probably due to the generation of cation vacancies. Unfortunately, the magnetic and electrical properties require firing of LiMgZn-ferrites in an oxygen containing atmosphere.
3. Interaction of MgZn- and LiMgZn-ferrite with mullite leads to a crystallization of $ZnAl_2O_4$. No reaction is observed with pure Al_2O_3 refractory.
4. The reaction of Zn with the refractory is accelerated by SiO_2 and further promoted by Li_2O, possibly via a liquid phase mechanism.

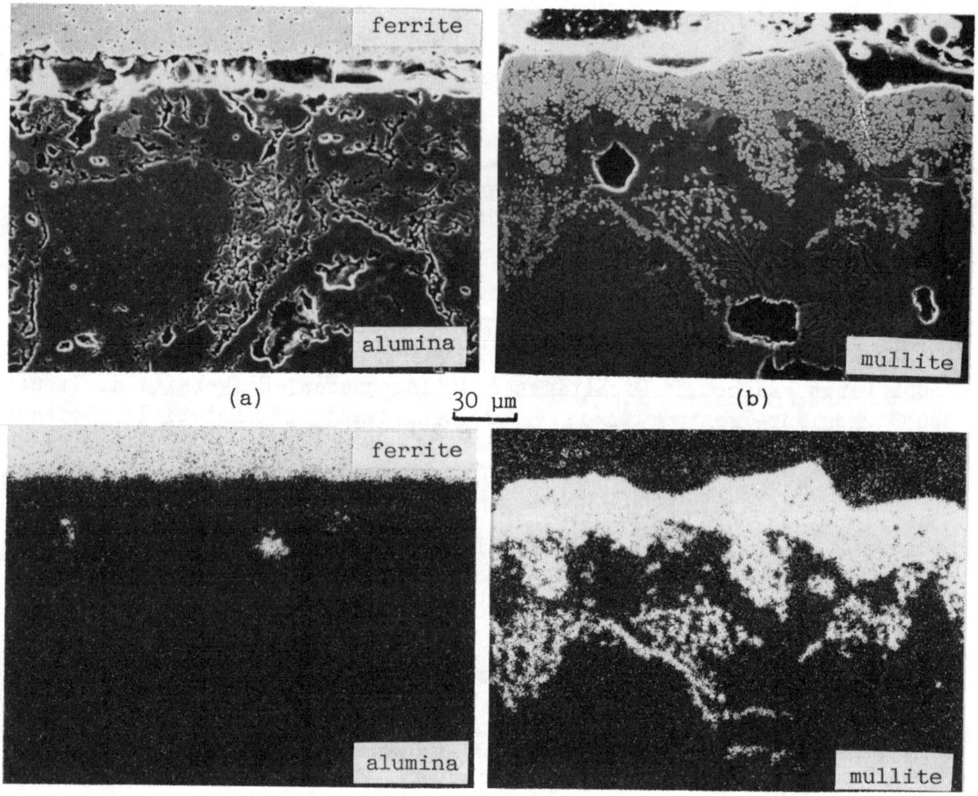

Figure 2. Top pictures: SEM. Bottom pictures: EPMA Zn-mapping.
(a) Al_2O_3 refractory; (b) mullite (Al_2O_3-$2SiO_2$) refractory.

LITERATURE

1. Ridgley D.H., Lessoff H. and Childress J.D., Effects of Lithium and Oxygen Losses on Magnetic and Crystallographic Properties of Spinel Lithium Ferrite, J. Am. Ceram. Soc., 1970, 53(6), 304-311.
2. Salmon O.N. and Marcus L., Note on Sublimation of Lithium from Li-Ni Ferrite, J. Am. Ceram. Soc., 1960, 43(10), 549-550.
3. Pointon A.J. and Saull R.C., Solid State Reactions in Lithium Ferrite, J. Am. Ceram. Soc., 1969, 52(3), 157-160.
2. Arkel A.E. van, U. Spitsbergen U. and Heyding R.D., Note on the Volatility of Lithium Oxide, Can. J. Chem., 1954, 33, 446-447.
5. Strickler P.W. and Roy R., Studies in the System Li_2O-Al_2O_3-Fe_2O_3-H_2O, J. Am. Ceram. Soc., 1961, 44(5), 225-230.
6. Reijnen P., Thermogravimetric Analysis Applied to Ferrites, Philips Tech. Rev., 1970, 31(1), 24-30.
7. Bandyopadhyay G and Fulrath R.M., Thermogravimetry of Lithium Spinel Ferrites, J. Am. Ceram. Soc., 1974, 57(11), 483-486.
8. Kubaschewski D. and Alcock C.B., Metalurgical Thermochemistry, 5th edition, Pergamon Press Ltd Oxford, 1979.

STRUCTURAL INVESTIGATION OF THE PYROLYSIS PROCESS OF COVALENT CERAMICS PREPARED BY POLYMER ROUTE.

FLORENCE BABONNEAU*[1], G.D. SORARU[2], J.D. MACKENZIE[3]
[1] Chimie de la matière Condensée, Université Paris 6, 4 place Jussieu 75005 Paris France. [2] Dipartimento di Ingegneria, Universita di Trento, 38050 Mesiano-Trento Italia. [3] Department of Materials Science, University of California, Los Angeles CA 90024 USA.

ABSTRACT

Advanced ceramics such as carbides and nitrides, have recently been prepared from metal organic polymers. This paper will present results concerning SiC and SiC/TiC systems. The precursor for SiC is a usual polycarbosilane while precursors for SiC/TiC composites were prepared by reacting a titanium alkoxide with polycarbosilane. The transformation of the polymeric precursors in final ceramics has been investigated by ^{29}Si MAS-NMR and XRD.

INTRODUCTION

The production of carbides and nitrides using a polymeric route is currently under investigation. Studies are mainly focussed on the preparation of new precursors or on the characterization and properties of final ceramics. Few work has been published on the conversion process of these polymeric precursors. The main results in this field concerned the most studied system, silicon carbide obtained from the pyrolysis of polycarbosilane [1]. However, the understanding of the structural evolution of these materials during the firing process appears quite necessary for any improvement of the properties of the final ceramics. The mechanical properties of polymer derived SiC fibers are known to deteriorate at temperatures above 1200°C, when grain growth occurs. Fibers in the Si-Ti-C system prepared from a polytitanocarbosilane seem to partially overcome this problem, and retain good mechanical properties up to 1400°C [2]. The difference of properties at high temperature seems closely related to the crystallization mechanism that appears different for the two systems. This paper presents a comparative study of the pyrolysis process of polycarbosilane and polytitanocarbosilane, using mainly ^{29}Si MAS-NMR and XRD.

EXPERIMENTAL

Commercially available polycarbosilane PC (Dow Corning X9-6348) was used in this study. The polytitanocarbosilane PTC, was obtained by refluxing PC and titanium n-butoxide in xylene according to a published procedure [3]. The chemical analysis gave a Ti/Si molar ratio of 0.15 for PTC. All the heat treatments were done under argon. MAS-NMR experiments were performed on MSL 300 and MSL 400 Bruker spectrometers. XRD patterns were recorded on a Philips diffractometer using Cu Kα radiation.

RESULTS

Characterization of the polymer precursors

PC and PTC were characterized by ^{29}Si CP MAS-NMR taking advantages of the presence of a large amount of protons in these polymers to use cross-polarization techniques (Figure 1). A structure for PC has already been proposed [1]. It can be mainly described with two kinds of Si units (1) SiC_4 with one silicon atom surrounded by 4 carbon atoms (δ=-0.8 ppm) and (2) SiC_3H with one silicon atom surrounded by 3 carbon atoms and one proton (δ=-17.6 ppm). The Si units can be more precisely described as $(CH_3)_2Si(CH_2)_2$ and $(CH_3)HSi(CH_2)_2$ although some bridging CH groups between polycarbosilane chains can exist.

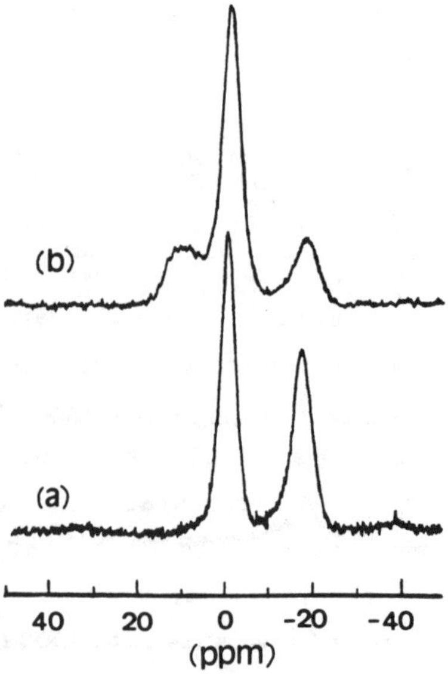

Figure 1: ^{29}Si CP MAS-NMR of PC(a) and PTC(b).

In the PTC sample, these two peaks are still present, but a new peak appears at 10 ppm that is assigned to the presence of SiC_3O units in which silicon atoms are surrounded by three carbon atoms and one oxygen atom. A comparison between the CP MAS-NMR spectra of PC and PTC clearly shows that the formation of Si-O bonds occurs with a decrease in the number of Si-H bonds. The reaction between PC and titanium alkoxide can be seen as a condensation reaction between Si-H bonds of PC and RO-Ti bonds of titanium alkoxide, and formation of Si-O-Ti bonds.

Pyrolysis process.

The pyrolysis process has been followed by XRD (Figure 2) for both systems. For the Si-C system, broad peaks due to crystalline SiC phases are present at 1200°C and sharpen by increasing the firing temperature. At 1500°C, the microcrystalline phase seem to be mainly β-SiC. For the Si-Ti-C system, the kinetics of crystallisation is different. At 1200°C, only very broad peaks are present reflecting a lower crystallinity than for the Si-C system. However, at 1500°C, the sample appears more crystalline than the fired PC sample, with peaks corresponding to β-SiC and TiC. The difference in the kinetics of crystallization could be due to the presence of titanium in the SiC matrix, but also to oxygen introduced via the titanium alkoxide. A diffraction peak assigned to graphite ($2\theta=26°$) is present in the fired PC samples while this peak seems absent in the fired PTC samples. In fact, the PC sample fired at 1500°C exhibits a C/Si ratio of 1.4 while no excess carbon seems to be present in the PTC sample fired at 1500°C that can be described as SiC, 0.15 TiC. In this last sample, the excess carbon certainly reacts with the oxygen present in the starting precursor ($0.5<O/Si<0.6$).

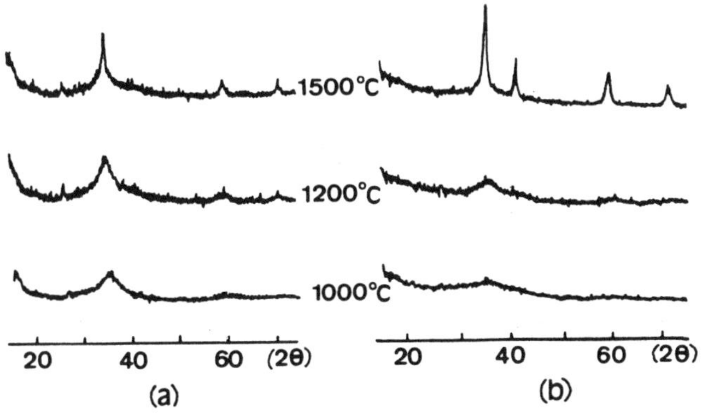

Figure 2: XRD patterns of fired PC(a) and PTC(b).

The evolution of the local environment of the Si atoms during the pyrolysis process has been followed by ^{29}Si MAS-NMR (Figure 3). As already mentionned [4], the first step of pyrolysis of PC is the comsumption of the Si-H bonds below 500°C. Above this temperature, only one peak is present in the NMR spectrum of the fired PC samples due to SiC_4 units. The position of the peak regularly shifts from -0.8 to -16.2 ppm, approaching values corresponding to crystalline SiC phases.

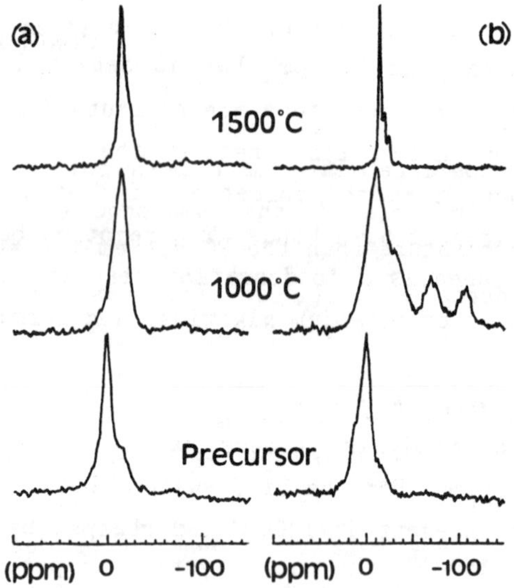

Figure 3: ^{29}Si MAS-NMR spectra of fired PC(a) and PTC(b) samples.

At the same time, the linewidth is decreasing, suggesting an ordering of the local environment of the Si units. In a first approximation, the main peak at -16.2 ppm can be assigned to β-SiC, even if this value is slightly different from published data [5]. The evolution of the NMR spectra of the fired PTC samples is different: at 1000°C, the spectrum exhibits a main peak at -11 ppm due to SiC_4 units, but also three additionnal peaks at -30, -61 and -109 ppm due to $SiC_{4-x}O_x$ with respectively x=2,3 and 4. The deconvolution of the spectra gave relative intensities of 52.8%, 24.8%, 13.9% and 8.6% for SiC_4, SiC_2O_2, $SiCO_3$ and SiO_4. The formation of $SiC_{4-x}O_x$ units with x>1 shows that the Si-C bonds of the polycarbosilane chains are not stable in the presence of oxygen below 1000°C. Above this temperature, the peaks due to $SiC_{4-x}O_x$ units regularly decrease: the oxygen is consumed by reaction with carbon that is present in excess at these temperatures. At 1500°C, three peaks at -16.3, -20.0 and -24.5 ppm are present due to SiC_4 units only. The main peak at -16.3 ppm is similar to the one found in the PC sample fired at 1500°C and can be assigned to β-SiC while the two others can be due to the presence of some α-SiC phases [5]. The difference in linewidth between the PC and PTC samples fired at 1500°C under the same conditions clearly shows a better crystallinity for the PTC sample, as already mentionned from the XRD patterns.

CONCLUSION

The pyrolysis processes of a polycarbosilane (PC) and a polytitano carbosilane (PTC) have been followed by ^{29}Si MAS-NMR and XRD. The PTC system retains the amorphous state up to 1300°C, at a higher temperature than the PC system for which microcrystallites of β-SiC are present at 1200°C. This has to be related to better mechanical properties of PTC fibers compared to PC fibers at high temperatures. It seems that an increase in the crystallite size is responsible for a loose in the mechanical characteristics of the system. However, if the amorphous state is retained at higher temperatures for PTC, the kinetics of crystallization is much faster in this case. These different behaviors seem to be due to the presence of titanium, but also to the presence of oxygen introduced via the titanium precursor in the PTC system. The ^{29}Si MAS-NMR experiments clearly show a deep modification of the local environment of the Si units in the fired PTC samples with the formation of $SiC_{4-x}O_x$ units that could prevent the formation of SiC crystallites. Above 1200°C, the oxygen is removed by reaction with the excess carbon, and the crystallization process is thus quite rapid.

ACKNOWLEDGEMENTS

NSF is greatly acknowledged for financial support (Contract No DMR 87 06379).

REFERENCES

1. Okamura, K., Sato, M., Matsuzawa, T., Hasegawa, Y. Ultrastructure processing of Advanced Ceramics, J.D. Mackenzie and D.R. Ulrich Eds, J. Wiley & Sons, 1988, 501.

2. Yamamura, T., Ishikawa, T., Shibuya, M., Okamura, K. J. Mater. Sci. 1988, **23**, 2589.

3. Yajima, S., Iwai, T., Yamamura, T., Okamura, K., Hasegawa, Y. J. Mater. Sci. 1981, **16**, 1349.

4. Soraru, G.D., Babonneau, F., Mackenzie, J.D. J. Non Cryst. Solids 1988, **106**, 256.

5. Hartman, J.S., Richardson, M.F., Sherriff, B.L., Winsborrow, B.G. J. Am. Chem. Soc. 1987, **109**, 6059.

PREPARATION AND CHARACTERIZATION OF SOME NEW NITROGEN GLASS CERAMICS

D.S.PERERA, P. KORGUL AND D.P. THOMPSON
Wolfson Laboratory,
Department of Mechanical, Materials and Manufacturing Engineering,
University of Newcastle upon Tyne, U.K.

ABSTRACT

Glass-forming compositions have been identified in M-Si-O-N and M-Si-Al-O-N systems which crystallize to give single-phase β"-magnesium sialon, Y_2SiAlO_5N, $MgYSi_2O_5N$ and $LnSiO_2N$ (Ln = La,Ce,Nd) glass ceramics. β"-magnesium sialon requires β'-sialon as a nucleating agent, and very rapid quenching is needed to avoid crystallization of Mg_2SiO_4; the resulting β" phase is stable up to ~1050°C. Y_2SiAlO_5N is an α-wollastonite which occurs over a range of compositions near the mid-point of the $YSiO_2N$ - $YAlO_3$ join, and is best prepared at slightly more SiO_2-rich compositions than that of the simple formula quoted above. It is unstable above ~1100°C. $MgYSi_2O_5N$ is a nitrogen pyroxene phase which was first observed during heat-treatment of Si_3N_4 ceramics densified with mixed MgO/Y_2O_3 additions. It has some range of homogeneity by substitution of Al + O by Si + N and is stable up to ~1450°C. $LnSiO_2N$ α-wollastonites are well known, but whereas $YSiO_2N$ melts incongruently and crystallizes to give mixed products, the La, Ce and Nd derivatives form single-phase glass ceramics, stable up to ~1700°C.

INTRODUCTION

Nitrogen glasses have been systematically explored during the last fifteen years and the extent of the glass-forming regions in a range of M-Si-O-N and M-Si-Al-O-N systems (M = Li, Mg, Ca, Y and Ln = La, Ce, Nd) has now been reasonably well established (1,2,3). Interest in nitrogen glasses is two-fold; firstly, they are important in their own right because of the improved refractoriness, higher viscosity and better mechanical properties compared with equivalent oxide glasses. Secondly, they occur in the grain boundaries of crystalline nitrogen ceramics where they are generally the dominant factor controlling high-temperature performance. In this context, numerous studies of the crystallization behaviour of nitrogen glasses have been carried out, but the systematic preparation of good quality, single phase nitrogen glass ceramics has not been attempted in any detail. Of course, suitable crystalline phases which occur in glass-forming regions cannot be predicted on a theoretical basis and this needs to be determined experimentally. In many cases these compounds are unstable at the high temperatures used for glass melting or for the preparation of crystalline nitrogen ceramics and they are therefore only observed during subsequent heat-treatment at a lower temperature. The present paper summarizes preparative and structural work on some nitrogen glass ceramic phases which have been observed during the last fifteen years of research on nitrogen ceramics but have only recently been systematically prepared and characterized as single phase glass ceramics.

EXPERIMENTAL

Samples were prepared by melting compositions made up from appropriate mixtures of binary oxide and nitride powders in either BN-lined alumina crucibles in a tungsten element resistance furnace in N_2, or in BN-lined graphite crucibles in an inductively heated graphite susceptor, both at 1650-1800 $^{\circ}$C. The melt was then either furnace-cooled (for high liquid viscosities) or poured out and quenched by compaction between graphite plates so as to retain a glassy product. The glass was then heat-treated in either SiC-element or Mo-element furnaces at various temperatures in the range 900-1500 $^{\circ}$C in nitrogen to crystallize the appropriate phase. Products were characterized by X-ray powder diffraction using a Hägg-Guinier focusing camera and CuKα_1 radiation. Scanning electron microscopy was carried out using JSM T20 and JSM35 instruments and TEM using a JEM 100CX microscope.

RESULTS AND DISCUSSION

β"-magnesium sialon

Extensive documentation already exists for the preparation of this phase (4-6). In the present work, glass compositions of $Mg_{20.6}Si_{19.4}Al_{2.8}O_{44.3}N_{12.9}$ and $Mg_{19.5}Si_{18.0}Al_{5.4}O_{44.4}N_{12.8}$ were used and glass cooling rates (which must be very high to avoid crystallization of forsterite) were 700 $^{\circ}$C/min. The best β" sialon was produced with about 5% of β'-sialon nuclei in the starting glass. The β' acts as a nucleating agent and crystallization takes place readily in the temperature range 900 - 1000 $^{\circ}$C. More details of this material have already been described elsewhere (7).

Y_2SiAlO_5N

This phase forms when the residual glass in commercial β'-sialon ceramics is cooled slowly, especially in the temperature range 900 - 1100 $^{\circ}$C (8). Y_2SiAlO_5N is part of a solid solution which extends from 50 - 70 m/o $YAlO_3$ in the $YSiO_2N$ - $YAlO_3$ system. Whereas the unit cell of $YSiO_2N$ is monoclinic with four layers of yttrium atoms and three-membered $Si_3O_6N_3$ rings, Y_2SiAlO_5N has perfect hexagonal symmetry and only two layers of yttium atoms and three membered rings per repeat. It cannot be prepared by melting and crystallizing glasses of precisely the quoted composition; purest samples are obtained at Y : Si : Al ratios of approximately 1.5 : 1.5 : 1.0 and an O : N ratio between 7 : 1 and 3 : 1. The viscosity of the Y-Si-Al-O-N liquid phase is much higher than that of the β" Mg-Si-Al-O-N glass and relatively slow cooling rates are sufficient to prevent crystallization on cooling. No satisfactory nucleating agents have been found which permit the controlled crystallization of a fine-grained microstructure. In the absence of a nucleating agent, crystallization generally takes place on nuclei within the glass and the microstructure is characteristically needle-shaped with residual glass clearly apparent (Figure 1).

In the final composition, Si and Al are disordered in the three-membered rings and the nitrogen atoms probably occupy bridging rather than terminal sites in the $(Si,Al)_3(O,N)_9$ rings (9). For some reason, Y_2SiAlO_5N always contains less yttrium than expected from the perfect 1 : 1 Y : (Si + Al) stoichiometry. The microstructure of this phase is always untwinned compared with the heavily twinned $YSiO_2N$ microstructure (Figure 2). More details of the preparation and microstructure of this phase have been given elsewhere (10).

$MgYSi_2O_5N$

This phase is observed during heat-treatment of Si_3N_4 ceramics densified with mixtures of MgO and Y_2O_3 (11,12). It is the first example of a nitrogen pyroxene and has the structure of diopside, $CaMgSi_2O_6$, but with yttrium replacing calcium and one in every six oxygens substituted by nitrogen. $MgYSi_2O_5N$ can be crystallized from glasses of similar composition by heat-treatment at temperatures in the range 1250-1400°C (see Table 1). This compound is more refractory than the other nitrogen glass ceramics described above and its melting point is in the range 1450-1500°C. This is consistent and even slightly higher than other pyroxene phases (see Table 2).

TABLE 1
Heat-treatment of $MgYSi_2O_5N$ glasses

Conditions T (°C)	Time (h)	Crystalline products
900	44	Glass
1050	24	
1100	24	$MgYSi_2O_5N$, tr. $Y_{10}(SiO_4)_6N_2$ (1050–1150)
1150	70	
1210	24	$MgYSi_2O_5N$, tr. $Y_{10}(SiO_4)_6N_2$ and tr. γ-$Y_2Si_2O_7$

TABLE 2
Melting points of some pyroxene phases

Phase	T (°C)
$CaFeSi_2O_6$	1130 (incongruent
$MnSiO_3$	1291 (incongruent)
$CaMgSi_2O_6$	1392
$LiAlSi_2O_6$	1425
$YMgSi_2O_5N$	1450 - 1500°C
$CaSiO_3$	1544
$MgSiO_3$	1557

N-α-wollastonites, $LnSiO_2N$ (Ln = La, Ce, Nd)

In the Y-Si-O-N system, it is impossible to prepare a glass of the $YSiO_2N$ composition; thus $YSiO_2N$ itself, which is difficult to prepare in high concentration by direct firing can be prepared in highest purity by heat-

Figure 1. Microstructure of Y_2SiAlO_5N heat-treated at $1070^{\circ}C$ for 50 hours.

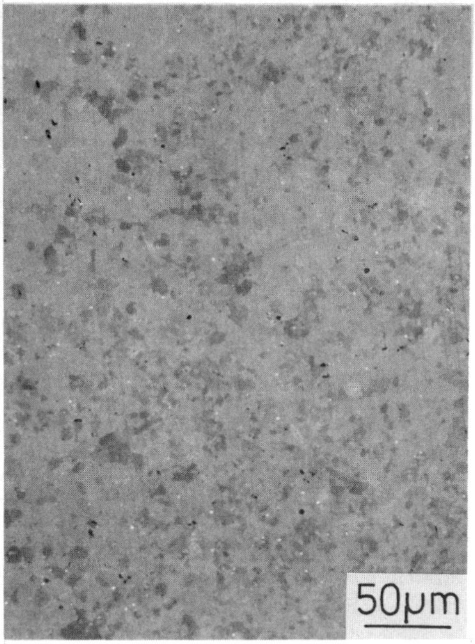

Figure 3. Microstructure of $NdSiO_2N$ after heat-treatment at $1300^{\circ}C$ for 6.5h.

Figure 2. Comparison of (a) Y_2SiAlO_5N, and (b) $CeSiO_2N$ showing the absence of twinning in the higher symmetry Y_2SiAlO_5N.

treating glasses on the silica-rich side of the perfect $YSiO_2N$ composition. However, in the La-Si-O-N, Ce-Si-O-N and Nd-Si-O-N systems, the liquid phase region is more extensive than in the yttrium system and $LnSiO_2N$ glasses and glass ceramics can be produced in all three systems (10). A typical microstructure is shown in Figure 3. Two types of X-ray powder pattern are observed depending on the heat-treatment temperature. At higher temperatures (>1450°C), all three wollastonites crystallize in the six-layer form, with well crystalline diffraction patterns. Below this temperature the structures are highly faulted and X-ray patterns apparently correspond to a two-layer modification which TEM studies show to be a result of the high frequency of faults. All three wollastonites are stable up to least 1700°C.

CONCLUSIONS

The range of nitrogen glass ceramics is steadily increasing and will be further extended as more detailed studies of glass crystallization are carried out in other M-Si-Al-O-N and M_1-M_2-Si-O-N systems. The compounds described in the present paper have nitrogen contents of up to 20 a/o, this being determined by the maximum nitrogen solubility in nitrogen glasses. An important area for further study is the identification of suitable nucleating agents to allow uniform, fine-grained microstructures to be produced.

REFERENCES

1. Loehman, R.E., J. Am. Ceram. Soc., 1979, **62**, 491-7.

2. Drew, R.A.L., Hampshire, S. and Jack, K.H., Proc. 2nd NATO ASI "Progress in Nitrogen Ceramics", Ed. F.L.Riley, pub. Nijhoff (The Hague), 1983, 323-30.

3. Hampshire, S., Drew, R.A.L. and Jack, K.H., Phys. & Chem. of Glasses, 1985, **26**, 182-6.

4. Drew, R.A.L., Ph.D. thesis, University of Newcastle upon Tyne, 1980.

5. Wild, S., Leng-Ward, G. & Lewis, M.H., J. Mater. Sci., 1981, **16**, 1815-9.

6. Leng-Ward, G., Lewis, M.H. and Wild, S., J. Mater. Sci, 1984, **19**, 1726-36.

7. Perera, D.S., Thompson, D.P. and Thorp, J.S., Brit. Ceram. Proc. **41**, 1989, 93-100.

8. Spacie, C.J., Jameel, N.S. and Thompson, D.P., Proc. 1st Int. Symp. "Ceramic Components for Engines", Eds. S.Somiya, E.Kanai and K.Ando, pub. KTK Scientific (Tokyo), 1984, 343-49.

9. Morgan, P.E.D. and Carroll, P.J., J. Mater. Sci., 1977, 12, 2343-5.

10. Korgul, P. and Thompson, D.P., Brit. Ceram. Proc. **42**, 1989 to be published.

11. Patel, J.K. and Thompson, D.P., Brit. Ceram. Proc. **39**, 1987, 61-8.

12. Patel, J.K. and Thompson, D.P., Brit. Ceram. Trans. & J, 1987, **87**, 70-3.

INFLUENCE OF THERMAL TREATMENT ON CRYSTALLIZATION OF HIGH-QUARTZ IN THE ZnO - MgO - Al_2O_3 - B_2O_3 - SiO_2 SYSTEM

E. TKALČEC, H. IVANKOVIĆ and *B. GRŽETA
Faculty of Technology, University of Zagreb and
*Ruđer Bošković Institute, 41000 Zagreb, Yugoslavia

ABSTRACT

Crystallization of high-quartz solid solution (h-Q s.s.) in the ZnO-MgO-Al2O3-B2O3-SiO2 glass system, with ZnO/Al2O3=1 and with variable quantities of MgO, in the presence of ZrO2 or ZrO2+TiO2 as nucleating agents has been studied. The effect of thermal treatment on the phase composition, on the stability and on the lattice parameters of h-Q s.s. has been examined by XRD. Crystallization of h-Q s.s. is accelerated and shifted to lower temperature if TiO2 is added as another nucleating agent. The lattice parameter a of h-Q s.s. is increased with TiO2 addition and decreased with heat treatment temperature, but not monotonously. Droplet-shaped separated phase enriched in Zn, Al, Zr, (Ti, Zr) is comfirmed by SEM and EDX as the first stage in the crystallization sequence of h-Q s.s. and gahnite.

INTRODUCTION

Transparent glass-ceramics with gahnite as predominant phase has recently received an increasing attention (1-4). Kisilev et al. (1) have reported about transparent gahnite glass-ceramics with ZrO_2+TiO_2 as nucleating agents, and with high-quartz solid solution (h-Q s.s.) present. However, Pinckley (4) considers that h-Q s.s. impairs transparency and crystallizes if more than 2 wt. % MgO is included in the glass , and that the best nucleating agent is TiO_2.

The aim of this paper is to compare the effectiveness of ZrO_2 and the mixture of ZrO_2+TiO_2 on the crystallization and thermal stability of h-Q s.s. in gahnite glass-ceramics with variable quantities of MgO.

MATERIALS AND METHODS

Quartz sand, used in crystal glass manufacturing, and reagent grade chemicals of $Al(OH)_3$, ZnO, H_3BO_3, $MgCO_3$, TiO_2 and ZrO_2 were used as raw materials. A batch of approximate 100g was melted in Pt crucible in electrical furnace for 2 hours at 1590-1600°C. The molten glass was poured onto the heated steel plate to form a circular patty about 5 mm thick, then placed in a furnace at 600°C for annealing: the furnace was turned off and the samples were slow cooled. Obtained bulk glasses were heated at the constant rate of $4^{\circ}Cmin^{-1}$ up to the maximum temperature (680, 720, 740, 760, 780 and 800°C), kept at it for 8 hours and slow cooled. The samples were also nucleated at 640°C for 4 hours and then crystallized at 950°C.

Differntial thermal analysis (DTA) was performed on the annealed and powdered glasses (grain size $<$ 40 µm), with the heating rate of $10^{\circ}C\ min^{-1}$. Crystalline phases prsent in heat treated samples, and in the samples previously subjected to DTA, were identified by X-ray diffraction analysis (XRD). The lattice parameters of h-Q s.s. were determined from the diffraction lines 100 and 101, since other lines were either superimposed or too weak to be measured. Scanning electron microscopy (SEM) and microanalysis using EDX system were performed on polished and/or fractured surfaces etched with 10 % HF.

RESULTS AND DISCUSSION

Glass composition is given in Table 1. DTA curves of glasses are shown in Fig.1 and the corresponding XRD in Fig.2. These data yield the following information about the crystallization behaviour of glasses: the crystallization stage in which gahnite $ZnAl_2O_4$, h-Q s.s., and low-quartz (l-Q) as main phases, and tetragonal and monoclinic ZrO_2 as minor phases could be identified, covers the temperature range from 900°C to 1000°C. This is characterized by appearence of one exothermic peak on DTA curves. The exception is the sample 1d (6.34 mol % MgO, ZrO_2), where DTA curve exhibits two distinct exothermic peaks: the first one is due to quartz, the second one to gahnite. The quantities of h-Q s.s. and l-Q are changing with the quantity of MgO and nucleating agent used.

TABLE 1
Glass compositions in mol %

Specimen No		SiO_2	Al_2O_3	B_2O_3	ZnO	MgO	TiO_2	ZrO_2
1	a	63.59	14.30	3.98	14.15	-	-	3.98
	b	62.54	14.06	3.90	13.91	1.67	-	3.90
	c	61.51	13.84	3.84	13.69	3.27	-	3.85
	d	59.56	13.39	3.72	13.25	6.34	-	3.72
2	a	63.59	14.30	3.98	14.15	-	1.60	2.38
	b	62.54	14.06	3.90	13.91	1.67	1.57	2.35
	c	61.51	13.84	3.84	13.69	3.27	1.53	2.31
	d	59.56	13.39	3.72	13.25	6.34	1.48	2.23

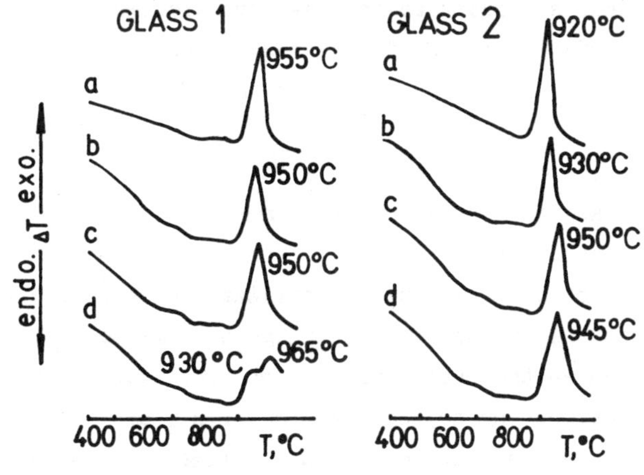

Figure 1. DTA curves of glasses: (1) with ZrO2 as nucleating agent, (2) with ZrO2+TiO2. Molar percentages of MgO in the samples are: (a) 0.0, (b) 1.67, (c) 3.27, (d) 6.34

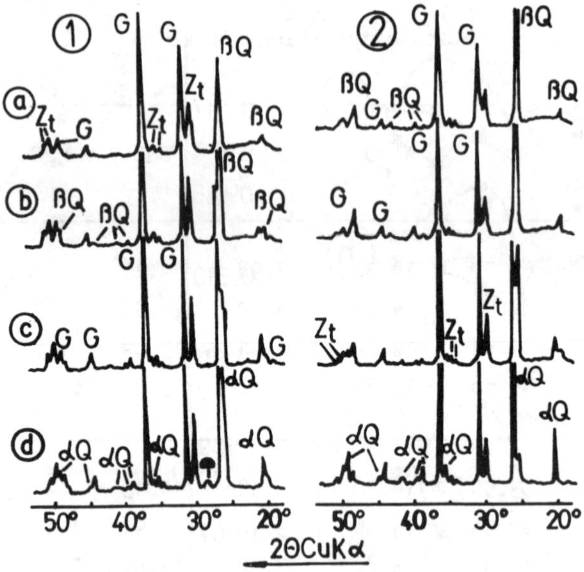

Figure 2. XRD patterns of glasses previously subjected to DTA: G-gahnite, Zt-tetragonal ZrO2, (●)-monoclinic ZrO2. Denotations of samples are same as in Fig.1

In order to determine the temperature range of transformation of h-Q s.s. to l-Q, glasses were subjected to single-stage treatment at 680, 720, 740, 760, 780 and 800°C for 8 hours and analysed by XRD. The content of the main crystalline phases formed during heat treatment was estimated as follows. Intensities of h-Q s.s. difraction line 101, l-Q line 101 and gahnite line 220 were measured and compared with the intensity of line 111 of Si, which was added as internal standard.

Obtained ratios I/I_{Si} are ploted vs. heat-treatment temperature in Fig.3. It can be notticed that the crystallization rate of h-Q s.s. is increased, whereas the crystallization of gahnite is not influenced by the addition of TiO_2 as another nucleating agent. The addition of MgO up to 3.27 mol % also favours the crystallization of h-Q s.s., but in greater amount it reduces the stability of h-Q s.s., likely by the exolution of Al^{3+} and the corresponding cations from $R_2O/RO\ Al_2O_3\ nSiO_2$ structure, what was examined by determination of h-Q s.s. lattice parameters a and c. Lattice parameter c of h-Q s.s. is slightly changing with temperature, therefore only a vs. heat-tretment temperature is shown in Fig.4. If ZrO_2

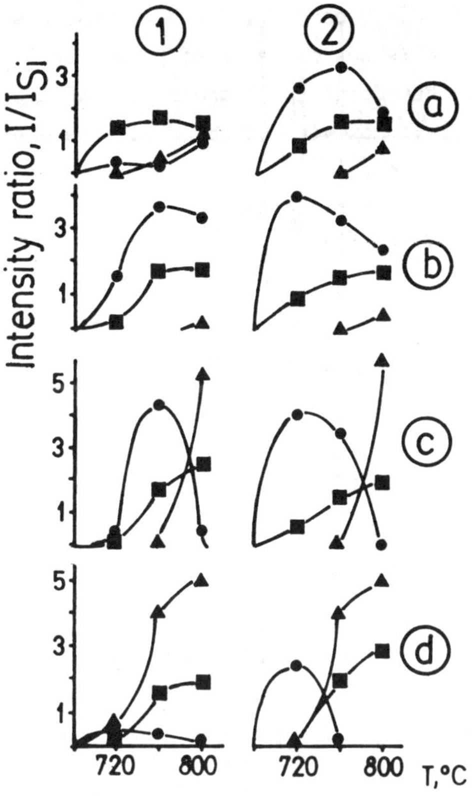

Figure 3. Relative content of h-Q s.s. (●), 1-Q (▲) and gahnite (■) in the heat-treated glass-ceramics. Denotations of samples are same as in Fig.1

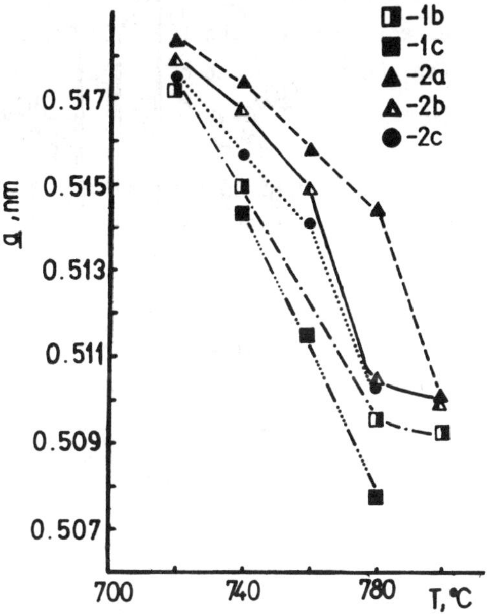

Figure 4. Variations of the unit cell parameter a of h-Q s.s. with heat-treatment temperature

is present, parameter a decreases uniformly with heat-treatment temperature. However, if both oxides are used (ZrO_2+TiO_2) there are inflexion on the curves for the samples 2a, 2b and 2c at approximate 0.5145 - 0.5150 nm. These inflexions ought to be explained in the light of different nucleating mechanisms of TiO_2 and ZrO_2 (5), and have to be further ivestigated. To get approximate SiO_2 content in h-Q s.s. the a values were fitted on the curve a versus SiO_2 for 2ZnO $1Li_2O$ $3Al_2O_3$ $nSiO_2$ system given by Beall et al. (6). Estimated quantity of SiO_2 in our samples varies between 80 - 90 mol %, what explains unstability of h-Q s.s. and its rapid transformation into 1-Q. The lattice parameters of

1-Q are slightly greater than given in JCPDS card No. 5 - 490, what has alredy been noticed (7). Slight but consistent decrease of parameter a for greater content of MgO at constant temperature is the result of MgO entering into the h-Q s.s. structure (6). However, it could also be due to exsolution of Al^{3+} and modifying cations.

SEM and EDX analysis (Fig.5) show the droplet-shaped phase separation with droplet about 3 μm in size, enriched in Zn, Al, Zr, (Mg, Ti) embeded into the silica glassy matrix. At about 720°C gahnite and h-Q s.s. as well as tetragonal ZrO_2,segragate from the droplets, and at 760°C crystallization in the silica glassy phase occurs. The samples heated at 800°C have rather coarse microstructure, but those nucleated at 640°C and crystallized at 950°C show fine microstructure of glass-ceramics.

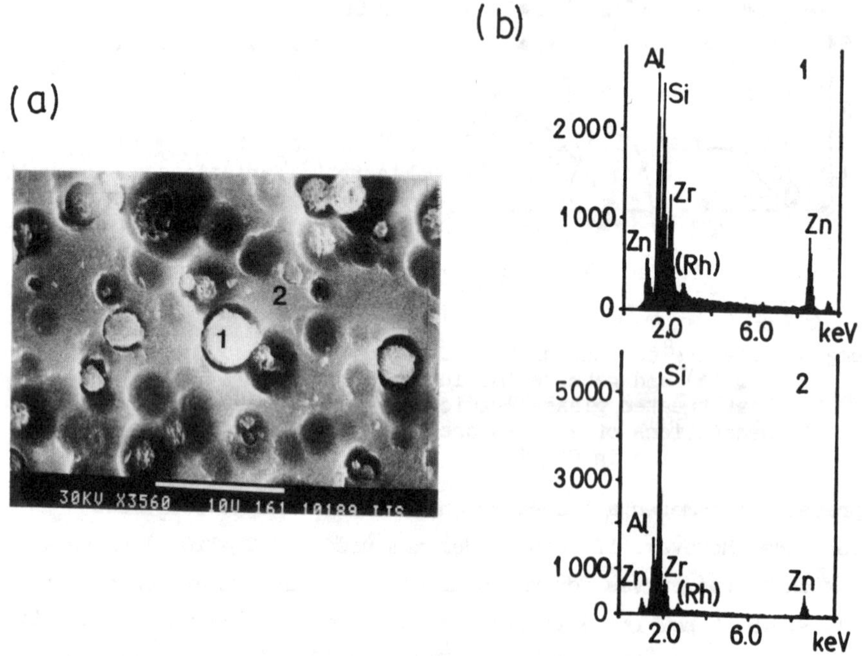

Figure 5. (a) SEM micrograph, (b) EDX microanalysis of sample 1b, at 720°C

CONCLUSION

Thermal stability of h-Q s.s. depends on MgO content and on the nucleating agents used. Up to 3 mol %, MgO favours the crystallization of h-Q s.s., whereas in greater quantities it accelerates transformation of h-Q s.s. into l-Q. TiO_2 added as another nucleating agent accelerates/or shifts the crystallization of h-Q s.s. to lower temperature. SEM and EDX analysis confirm droplet-shaped separation as the first stage in the crystallization sequence of gahnite and h-Q s.s.. The influence of TiO_2 on the inflexion point occurence on the curve a vs. heat-treatment temperature has to be explained.

REFERENCES

1. A. Kisilev, R. Reisfeld, A. Buch and M. Ish-Shalom, Chem. Phys. Letters **129**,1986, 450.

2. G. H. Beall, U.S.P. No. 3, 681, 102,1972.

3. R. Reisfeld, A. Kisilev, A. Buch and M. Ish-Shalom, J. Non-Cryst. Solids **91**, 1987, 333.

4. L. R. Pinckley, U.S.P. No. 4, 687, 750, 1987.

5. T. Dumas, J. Petiau, Springer Proceedings Physics, 1984, 2, 311.

6. G. H. Beall, B. R. Karstetter and H. L. Rittler, J. Am. Ceram. Soc. **50**,1967, 181.

7. M. F. Barker, P. F. James and R. W. Jones, J. Non-Cryst. Solids, **104**, 1988, 1.

PHASE TRANSFORMATIONS IN Y-TZP-CERAMICS IN BOILING WATER

P. LINTULA, T.T. LEPISTÖ, P. REUHKALA AND T. MÄNTYLÄ
Tampere University of Technology
Institute of Materials Science
P.O. Box 527 SF-33101 Tampere Finland

ABSTRACT

The degradation phenomenon of Y-TZP ceramics under humid conditions was studied by exposing the specimens to water vapour and boiling water. Phase transformation was observed to initiate without incubation time and extend deep inside the material. No change in the rate limiting process was observed. Neither yttrium depletion nor OH-group formation were found in the transformed layer.

INTRODUCTION

The degradation phenomenon means undesirable phase transformation from tetragonal to monoclinic crystal structure under humid condition at ca. 70-400°C. Tetragonal zirconia polycrystals with 2-3 mol-% Y_2O_3 and with relatively large grain size are especially sensitive for degradation. PSZ-ceramics (for example Mg-PSZ) are not sensitive because tetragonal precipitates are shielded from water vapour by cubic matrix.

MATERIALS AND METHODS

The materials having been tested were a commercial 3Y-TZP and uniaxially pressed and sintered 3Y-TZPs with varying grain sizes. The commercial TZP contained 2.9 mol-% Y_2O_3 with no monoclinic phase before treatment. Grain size was 0.45 µm. In the sintered materials yttria content was 2.8 mol-%, no monoclinic phase was found before treatment and grain sizes varied between 0.26 and 0.84 µm.

The exposure in boiling water was performed by immersing specimens into boiling distilled water. The materials were exposed to water vapour using the test configuration presented in Fig. 1. Argon, saturated with water, was passed over the samples at a flow rate of 4.2 l/min at temperature of 150°C. Argon gas (purity 99.998%) was bubbled through distilled water at temperature of T_2 (Fig. 1) and the argon-water vapour mixture was saturated leading it through a condenser at a lower temperature of T_1, in which the partial pressure of water in the mixture was fixed to 19.85 kPa. After passing the samples the gas mixture was condenced and the amount of water was measured to check the partial pressure of water.

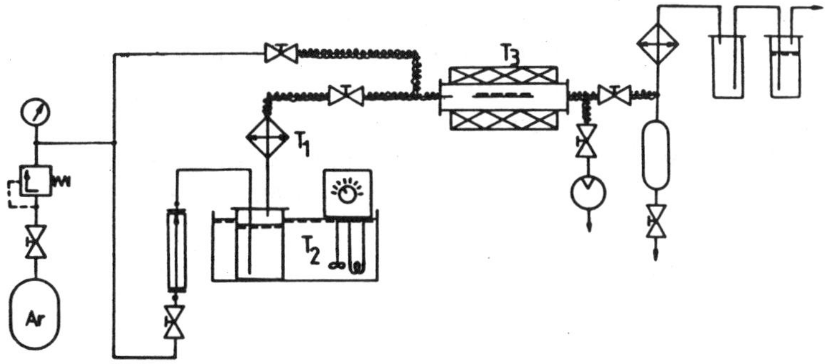

Figure 1. Schematic presentation of the test configuration used in the exposures in humid atmosphere.

The monoclinic phase content was measured by X-ray diffraction using tetragonal (+cubic) and monoclinic (111)-type peaks. The diffractometer was equipped with secondary graphite monochromator and copper tube. The lattice parameters of the tetragonal phase were measured with the same equipment using lanthanum hexaboride reference powder in the specimen holder for refining the peak positions. The lattice parameters were measured from the tetragonal (004)- and (400)-peaks because the accuracy increaces by using peaks at higher theta value.

The yttrium content in the transformed layer and untransformed material were measured by SEM+EDS. SEM was also used for measuring the transformation depth at the fracture surfaces.

ESCA was used for determining the existence of OH- groups at the material surface, inside the transformed layer and in the untransformed material. IR was also used for determining OH-groups at the surface.

RESULTS AND DISCUSSION

XRD studies revealed that the transformation initiated almost immediately when the material was exposed to boiling water or water vapour. Transformation in the uppermost grain layer can be observed by XRD because 20% of the total X-ray intensity comes from the 0.4 μm thick surface layer. Transformation was observed to proceed at constant rate, no change in the rate limiting process was observed. The propagating rate of phase transformation was 0.07 μm/h in boiling water for the commercial Y-TZP. In water vapour the monoclinic phase content in the transformed layer reached 10-80 % depending on the grain size distribution (Fig. 2). Grains which were larger than the critical size transformed to monoclinic structure during the exposure.

No significant lattice parameter variations of tetragonal phase were observed during the exposure. Slight growth of the c-value (Fig. 3) can be explained by the lattice strain induced by tetragonal to monoclinic phase transformation and unisotropic elastic modulus. Existence of tetragonal to rhombohedral transformation during the exposure can not be observed reliably by the XRD-studies performed due to peak overlapping and poor data statistic of the strained and twinned monoclinic phase (80% monoclinic content).

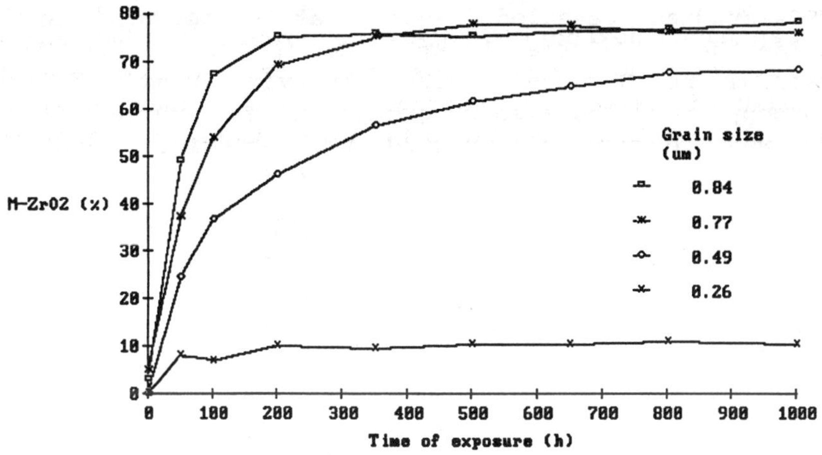

Figure 2. The monoclinic phase content as a function of time of exposure in water vapour at 150 °C. Four sintered materials with varying grain sizes were studied.

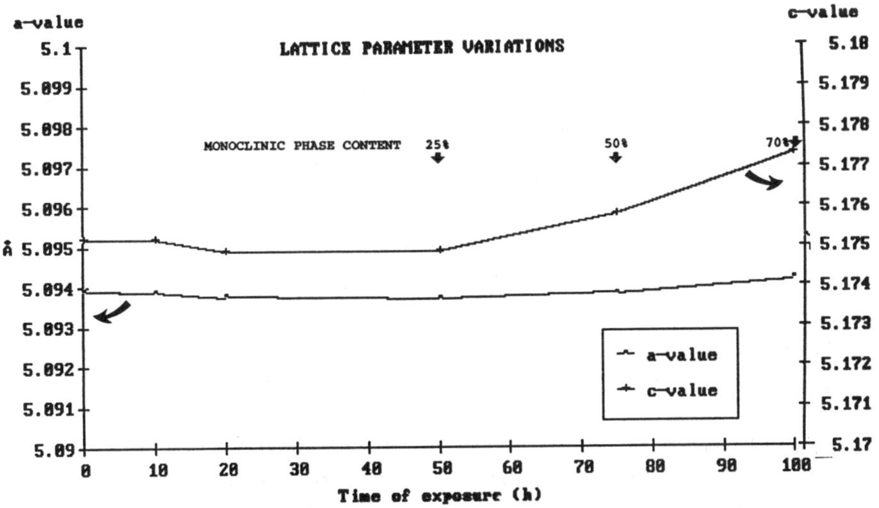

Figure 3. Lattice parameters of tetragonal phase as a function of time of exposure in boiling water.

The average yttrium content in the transformed layer was at the same level as in the untransformed material indicating that no yttrium depletion took place.

ESCA-studies revealed OH-groups at the material surface but inside the transformed layer (after mechanical scratching in vacuum) no OH-groups were observed (Fig. 4). IR-spectroscopy revealed also OH-groups at the material surface but it was not possible to study the material inside the transformed layer in vacuum by the IR used.

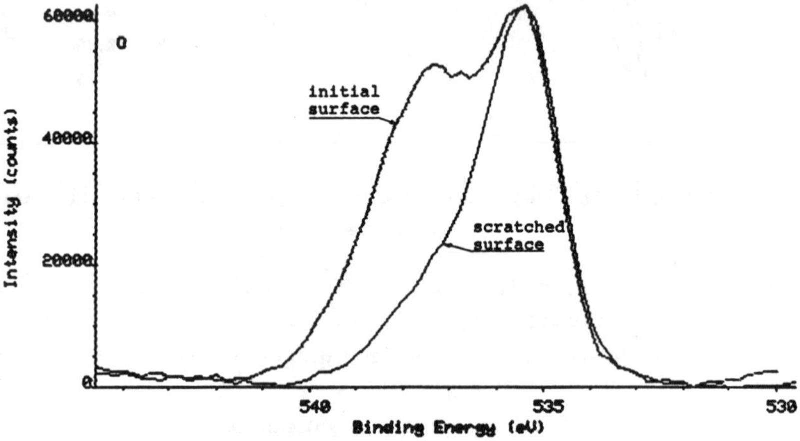

Figure 4. ESCA-pattern indicating the existence of OH-groups on the material surface but not inside the transformed layer.

CONCLUSIONS

1. Phase transformation initiates without significant incubation time and proceeds at constant rate.

2. Monoclinic phase content in the transformed layer saturates to a certain level depending on the grain size distribution. Grains larger than the critical size transform.

3. On the basis of the results of this study the degradation phenomenon is not caused by: yttrium depletion (distribution may change), ion penetration of depletion (practically constant lattice parameters) or hydroxide formation (no OH-groups inside transformed layer).

4. Degradation can be reduced by increasing yttria content or decreasing grain size.

THE ROLE OF ALUMINA IN ZIRCONIA-CERIA COMPOSITE ALLOYS

C. Schmid, H. Schubert* and S. Meriani
Istituto di Chimica Applicata e Industriale
Università di Trieste - Italy
*MPI Pulverlabor. - Stuttgart - BRD

ABSTRACT

Alumina acts as an effective grain growth inhibitor in Ce-modified tetragonal zirconia polycrystals. The two phases composite materials exhibited enhanced mechanical properties and one particular blend showed also a remarkable transformation plasticity behavior before failure in bending.

INTRODUCTION

Ceria-zirconia alloys receive increasing attention because of their high strength and toughness. Some difficulties may rise in the appropriate tailoring of their microstructure because of the tendency to an exaggerated grain growth. To overcome this problem we have introduced different amounts of alumina as grain growth inhibitor. For this work different ceria-zirconia alloys (CZ) containing 8.5 and 12 mol % ceria were mixed with 10 or 20 wt% alumina (A) to obtain composite samples: CZ8.5/A10; CZ8.5/A20; CZ12/A10 and CZ12/A20.

MATERIALS AND METHODS

Materials used were of commercial origin. Mixing was obtained either by attrition milling, with a zirconia lined camshaft in a plastic container using 2 mm zirconia balls, or via ball-milling with 10 mm zirconia pebbles over a period of 16 hours .in isopropanol. Slurries were dried, the resulting powders were sieved and pressed into both cylinders (12x12 mm) and bars (12x17x45 mm) via isostatic pressure. Some were uniaxially pressed as (2x7x35 mm) thinner slabs. Samples were fired at 1500°C for two hours and cooled in the furnace

Densities were measured by the water displacement method. Phases recognition and evaluation were made by XRD on as fired, polished and fractured surfaces to define the tetragonal vs monoclinic ratio, according to the formula proposed by Garvie and Nicholson (1). Microstructures were evaluated via SEM images of polished and thermally etched surfaces (80 min at 1400°C) using the lineal intercept method. Bend strengths were measured with a universal testing machine by the four points bend test using a jig with 30 mm outer, 10 mm inner span with a cross-head speed of 0.05 mm/min. The fracture toughness (K_{IC}) was calculated according to the theory of P. Chantiful et al. (2). Hardness was measured using the standard Vickers procedure with an indentation load of 98 N.

RESULTS AND DISCUSSION

The firing cycle allowed the attainment of densities in excess of 97% of the theoretical values calculated according to the rule of mixtures. Alumina behaves as an effective grain growth inhibitor in both matrices. The mean grain size of the matrices and of the composites is reported in Fig. 1.

The matrix microstructure of the less favorable case is reported in Fig. 2. There is evidence of some exaggerated grain growth within fairly regular grains with a mean size of 2.7 microns. The effect of a 10 vol % addition of alumina (CZ8.5/A10) could not effectively limit the grain growth, in fact the mean grain size did not change significantly. On the contrary, the (CZ8.5/A20) material showed a reduction of the mean size to 1.6 microns with a very limited number of oversized grains. It is quite evident that alumina particles exhibit a grain boundary pinning effect being mainly dispersed at the intergranular sites, Fig. 3.

Hardness values are collected in Fig. 4. They show a almost linear correlation with the alumina concentration

The response of the materials' surface to mechanical treatments is reported in Fig. 5 which collects data about the fraction of monoclinic phase present on both "ground" and 'fractured" surfaces with reference to their mean grain size. It can be easily pointed out that a mild mechanical action (open points=grinding) does not significantly change the previous phases ratio. In fact CZ8.5 matrix and CZ8.5/A20 alloy show the same monoclinic fraction even if their average grain sizes differs by a factor of two. Some dependency on the grain size is shown by the CZ12 series.

The effect of fracture on the t-->m transformation was significantly stronger, compare Fig. 5. It can be pointed out that the monoclinic content of the CZ12-/A10-/A20 shows a linear dependence from the mean grain size whereas the CZ8.5-/A10-/A20 group is in practice not affected. In the former case, where the ceria amount stabilizes the tetragonal phase, both grain size and mechanical stress can play their role on the t-m phase transition. Small grains restrain the t-m transformation, as it can be seen by the relative position of samples CZ12 and CZ12/A20 in Fig. 5. On the other hand, in the case of lower ceria concentration the chemical free energy dictates the t-m ratio and therefore for the tetragonal form much is left for further stress induced transformation (7).

Fracture strength data are reported in Fig. 6. Each group of samples appears twice under two separate headings: the first (UP) refers to "uniaxially pressed" powders directly into thin slabs suitable for fracture strength measurements, whereas the second group (IP) refers to the samples obtained by cutting down to the appropriate size, the large bars obtained by the "isostatic pressing" method. The reported data show that in general alumina increased the fracture strength of both composition with some exceptions. Materials based on the CZ12 matrix exhibited lower fracture strengths with a negative trend of the alumina role. In fact, IP-CZ12/A' samples decreased their strength as the alumina concentration increased, in contrast to all the other trends.

The UP samples behaved more properly in both series CZ8.5 and CZ12. The reasons for the different behaviors could be explained on the evidence of microstructural inhomogeneities which have been found as macro-pores in the samples cut from the IP bars. It was not unusual that IP samples broke far away from the indentation mark because of other internal defects acting as more effective stress concentrators.

During the 4 points bend tests, the cross-head displacement velocity was set at 0.05 mm/min and programmed to stop after a deflection corresponding to about 0.5 mm. By this procedure it happened that samples CZ8.5/A10 of both UP and IP origins, did not break under the applied load before the cross head had reached its maximum programmed displacement, whereas all the other samples did break before this instrument's set end-point. For this reason the CZ8.5/A10 data points are missing from the graphic in Fig. 6. On top of Fig. 7 two CZ8.5/A10 original samples are reported to be compared with two other pairs which survived the four-points bend test.

XRD diffraction patterns of the surface under tension exhibited a large fraction of monoclinic phase whereas that under compression showed only the tetragonal one. It is worth pointing out that "bent" specimens could be "counter-bent" without breaking for two more cycles provided that the cross-head displacement was properly programmed. On the tensioned bent surface the formation of shear bands was detected as an evidence of the presence of the dilatant monoclinic phase. Bent specimens maintained the strained configuration up to 250°C when they exhibited a full shape-recovery. This feature opens the very attractive area of "shape memory" effects in ceramics materials.

CONCLUSIONS

Plastic deformation of Ce-TZP before failure in bending has been previously reported (3-6). The phenomenon was presented as an evidence of slow, stable crack growth and it has been discussed in terms of transformation plasticity behavior. This latter provides a very effective stress field shielding and energy dissipation mechanism leading to a very remarkable toughness. Stress assisted transformation plasticity was found to develop in single phase tetragonal polycrystals

(Ce-TZP) being favoured by tensile conditions with a positive temperature dependance and, under certain conditions, exhibiting an autocatalitic trend. It has been argued by convincing arguments that this remarkable behavior becomes more difficult in the presence of a second phase (5). In our samples, despite the presence of a 10 weight % alumina, the basic features of transformation plasticity have been found. Furthermore, if we take into account the tensile strength (600 MPa) and the fracture toughness (12 MPa $m^{1/2}$) of this material we can expect it to tolerate surface mode I flaws in excess of 100 microns.

Acknowledgement: work performed with C.N.R. financial support and supported by the Federal Ministry of Science and Technology BMFT.

REFERENCES

1. Garvie R.C. and Nicholson P.S., J.Am.Ceram.Soc. 55(6) 303-07 (1972)
2. Chantiful P., Anstis G.R., Lawn B.R. and Marshall D.B., J.Am.Ceram.Soc. 64(9) 539-43 (1981).
3. Tsukuma K. and Shimada M., J.Mat.Sc. 20 (1985) 1178-84.
4. Chen I-Wei and Reyes Morel P.E., J.Am.Ceram.Soc., 69(3) 181-89 (1986).
5. Reyes Morel P.E. and I-Wei Chen, J.Am.Ceram.Soc. 71(5) 343-53 (1988).
6. Hannink R.H. and Swain M.V., J.Am.Ceram.Soc. 72(1) 90-98 (1989).
7. Schubert H. and Petzow G., Advances in Ceramics 24 p.21 (1988).

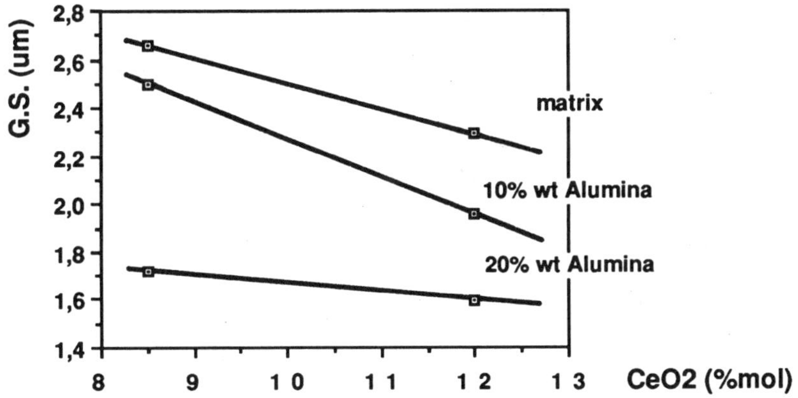

Fig.1 Grain size vs.ceria content.

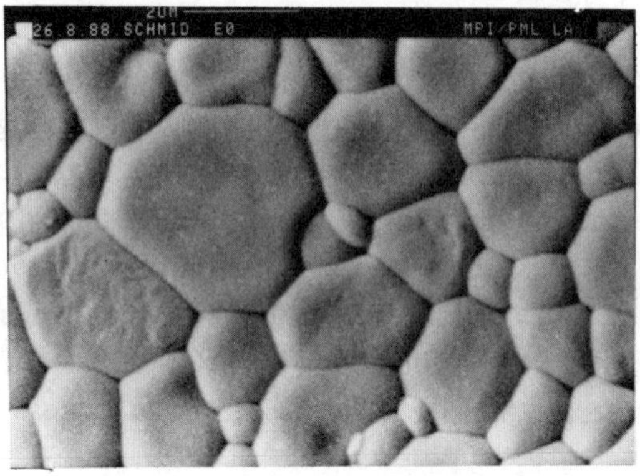

Fig. 2 Microstructure of CeTZP matrix.

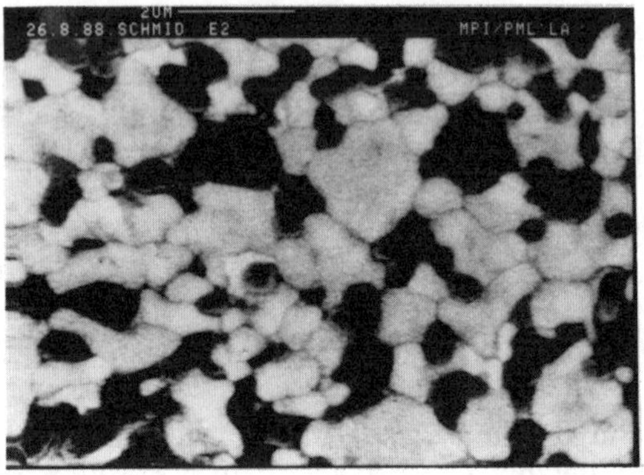

Fig. 3 Microstructure of CeTZP/A20 composite.

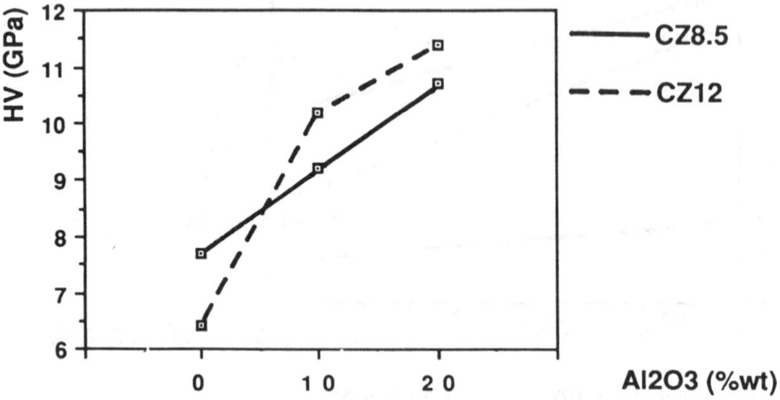

Fig. 4 Vickers hardness vs. Al2O3 content in CZ8.5 and CZ12 matrices.

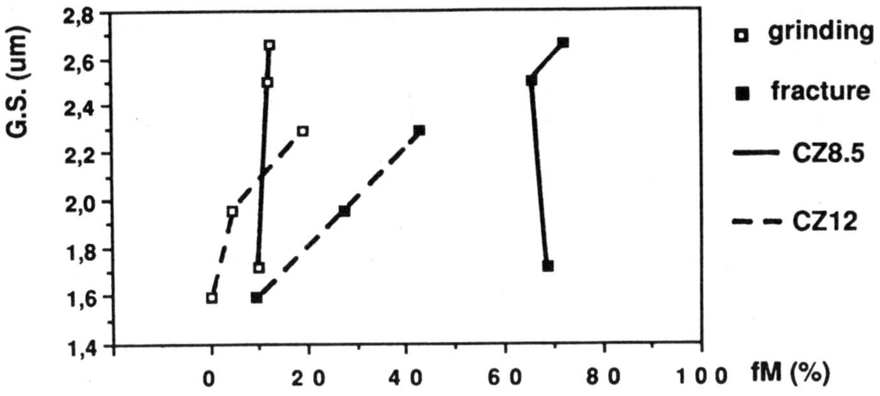

Fig. 5 Monoclinic faction on surface after grinding and fracturing vs. grain size for CZ8.5 and CZ12.

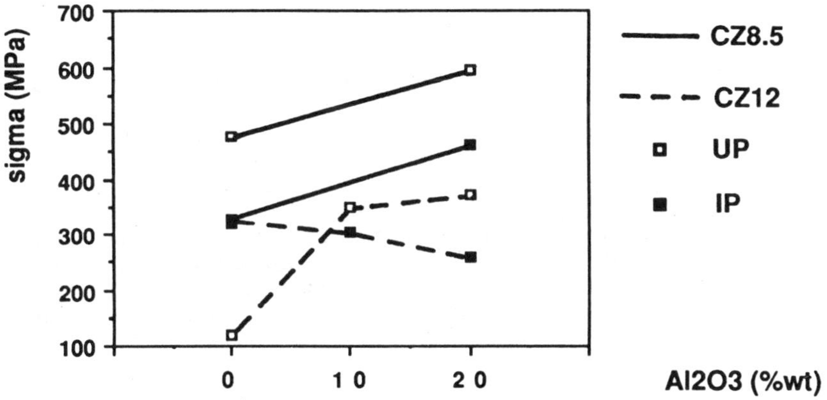

Fig. 6 Bend strength of uniaxially UP and isostatically IP pressed samples vs. Al2O3 content in CZ8.5 and CZ12.

Fig.7 a) original samples CZ8.5/A10
b) The same permanently bent — samples

THE ROLE OF MINERALIZERS ON THE DISSOCIATION OF ZIRCON

S.M.Naga x, I.S.Ahmed Farag + and D.M. Ibrahim x
x Department of Ceramics
+ Department of X-ray

National Research Centre, Tahrir Str., Dokki, Cairo, Egypt

ABSTRACT

The effect of small additions of different fluorides on the dissociation of zircon sand was studied by quantitative x-ray diffraction analysis. Thus Naf Lif, Caf_2, AlF_3 were added in the range between 2 to 12% by weight to zircon sand. Also mixtures of either NaF and LiF or CaF_2 and AlF_3 were tried. LiF and AlF_3 alone did not affect the dissociation reaction in the temperature range studied between 1200° to 1450°C. Best results were obtained by addition of either 8% NaF or a mixture of 6% NaF and 6% LiF. The amount of zircon dissociated was 59.25 and 65.5% respectively. The former was obtained at 1200°C while the latter at 1350°C.

INTRODUCTION

Mineralizers play an important role in many different ceramic processes. One of the essential economic factors concerning industrial production is the speed of the reaction. Accelerating the reaction takes place in several ways namely: the presence of a liquid phase, the introduction of lattice instability by diffusion of certain ions leading to Frenkel's mechanism or it may act on the surface of the individual grains. As a result crystallization of the liquid is enhanced forming new compounds.

The course of reaction of the mineralizer can be followed by thermal analysis. Locsei (1) studied the kinetics of the action of AlF_3 as a mineralizer added to kaolinite and to zircon sand (2,3).

Zircon and zirconia are becoming attractive in advanced ceramic technology. Zircon is used in structural products. Fused cast refractories are made from zirconia, mullite and or corrundum. The glassy phase in this material is a weak bond. It is readily attacked by molten matter. The increase in content of Zro_2 on the expense of SiO_2 in the mix will lead to an increase in the crystalline phases, improving thereby the performance of the structural material. This is achieved by getting rid of SiO_2 in the form of fluoride at a realtively low temperature to prepare the constituents for further reaction. ZrO_2-corrundum mix is advantageous basic material for the manufacture of abrasive grains.

Pigments based on zircon are another field for these products, [4,5]. The study comprises the reaction of different mineralizers namely AlF_3, CaF_2, NaF and Lif with zircon sand. The dissociation reaction is followed by thermal analysis and quantitative X-ray diffraction analysis. Activation energy is predicted for certain reactions.

MATERIALS AND METHODS

Materials used for the preparation of the different mixes were: pure zircon sand, and chemical reagents (BDH) NaF, Lif, AlF_3 and CaF_2 composition of the mixes prepared are present in TABLE 1. Mixes were thoroughly dry mixed in an agate mortar. Specimens in the form of discs 1 cm diameter and 0.3 cm thickness were processed under a pressure of $4KN/cm^2$, dried at 110 °C overnight then fired in an electric tubular furnace between 1200° to 1450°C with a soaking time of one hour at each maximum temperature.

Differential thermal analysis and thermogravimetric analysis were carried out for selected mixes up to 1300°C using an apparatus delivered by Netsch. The rate of heating was 10°C/min. Results obtained are present in FIGURE 1.

X-ray diffraction analysis was carried out for the

fired specimens using copper target with a scanning speed 0.5 °/min and time constant of 250 puls/sec. on a DIANO apparatus. Patterns were taken over the range of 2θ values between 42° to 48°. The unreacted zircon was quantitively determind using MgO as an internal standard. the spacing choen were d=2.066°A, hkl = 301, 1=20% for zircon and d=2.04°A, hkl = 400. I = 100% for MgO. Standard curve for zircon was obtained using glass as diluent and 10% MgO. Ratios of the areas under the speaks were taken.

TABLE 1

Composition of the different mixes wt%

Number of mix	AlF_3	CaF_2	NaF	LiF	Zircon
1	2				98
2	4				96
3	6				94
4		6			94
5		10			90
6	3	3			94
7			2		98
8			4		96
9			6		94
10			8		92
11			10		90
12			12		88
13				2	98
14				4	96
15				6	94
16			6	6	88
17			8	6	86
18			10	6	84
19			12	6	82

RESULTS AND DISCUSSION

The influence of type of mineralizer on the dissociation process of zircon is represented by the resuls of thermal analysis FIGURE 1. NaF, CaF_2 and the mixture of CaF_2 3% and AlF_3 3% show a gradual loss in weight with the rise of temperature starting approximately at 300° to 360°C and extending up to 1300°C. The amount of loss is about from 7 to 9%. It is represented in the DTA curve by a series of endothermic deflections at 350°, 580°, 750° and 930°C. According to Locsei (1) the reaction proceeds with the liberation of both HF and SiF_4. The first product is the result of the interaction of the added mineralizer with any moisture content present. Its duration lasts up to 900°C and is accompanied by loss in weight felt below 1000°C. The second product SiF_4 depends on the amount of fluorine available for the silica present. It manifests itself above 923°C but preferrably between 1067° and 1223°C.

$$2\ AlF_3 + 3H_2O \text{ ----- } Al_2O_3 + 6\ HF \quad (1)$$

$$4\ AlF_3 + 3ZrSiO_4 \text{ --- } 3ZrO_2 + 2Al_2O_3 + 3\ SiF_4 \quad (2)$$

The activation energies for both reactions were 17169 cal/mole and 2943 cal/mol respectively.

As it is evident from the results of TG the loss in weight was gradual showing no distinct reaction i.e. both reactions are overlapping. One cannot predict precisely the degree of dissociation of zircon in the early stages from the figures of loss in weight.

The results of x-ray analysis show that all concentrations of LiF or AlF_3 in the range of concentrations applied between 6 to 12% have no influence on the dissociation of zircon. Addition of mixtures of LiF and NaF or CaF_2 and AlF_3 have pronounced effects. The extent of dissociation reaction in terms of the amount of rest zircon after firing at 1200°C and 1350°C is shown in FIGURE 2. The effect of NaF concentration is more felt than that of increase in temperature. Thus the percentage of rest zircon at 1200° and 1350°C for concentrations 2 and 10% NaF were 27, 17% and 79, 87% respectively.

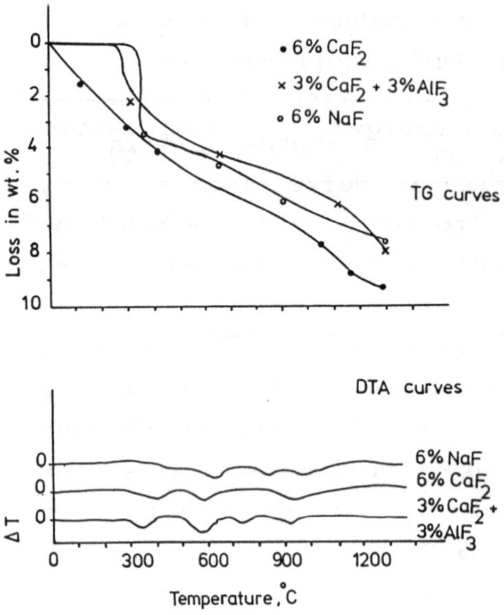

FIGURE 1. Thermal behaviour of the different mineralizers.

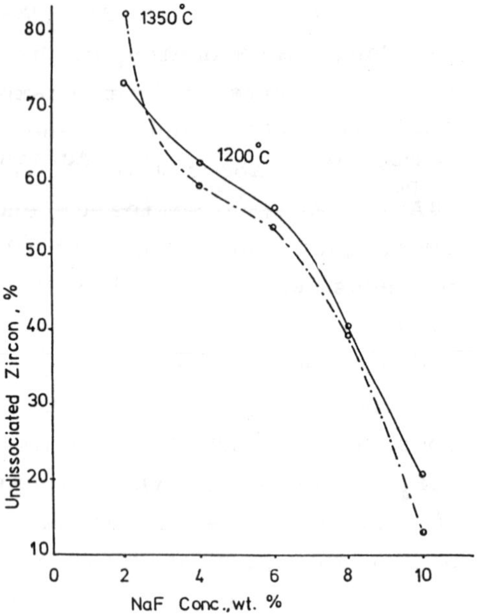

FIGURE 2. Effect of temp. on the dissociation of zircon.

Firing at a higher temperature of 1450°C did not favour the complete dissociation of zircon as indicated by the deflection in the curve FIGURE 3. The results indicate that in the temperature range studied and the percentages of fluorides added there is an optimum concentration, which is 8% NaF. Raising the amount to 10 and 12% cause the fusion of the specimens at 1450° and 1200°C, respectively.

The activation energy for the dissociation reaction for certain mixes was calculated using Arhenius equation, results are present in TABLE 2.

It is evident from FIGURES 3 and 4, that addition of 6% LiF to NaF enhanced the dissociation reaction. Higher percentages are not recommended as better results are obtained by NaF alone.

TABLE 2

Activation energy of the dissociation reaction of zircon in the presence of fluorides.

Number of the mix	Activation energy (cal/g)	Temperature (°C)
4 CaF_2 6%	3912.8	1200-1350
8 NaF 4%	668.5	
9 NaF 6%	381.8	
10 NaF 8%	392.8	
11 NaF 10%	675.6	
17 NaF 8% + LiF 6%	1814.7	1350-1450
18 NaF 10% + LiF 6%	20392.7	

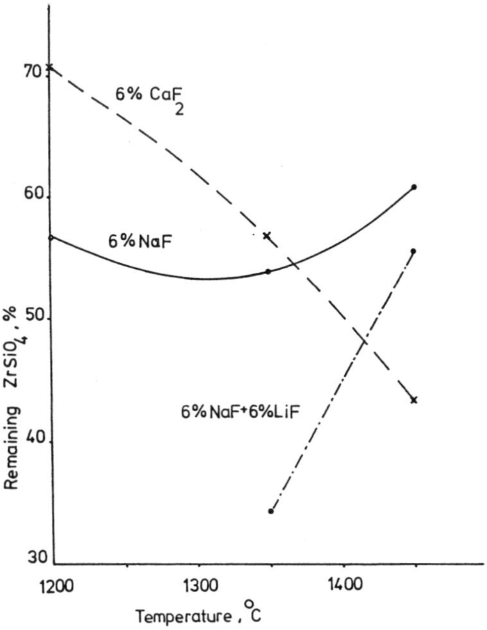

FIGURE 3. Fluoride effect on dissociation of zircon

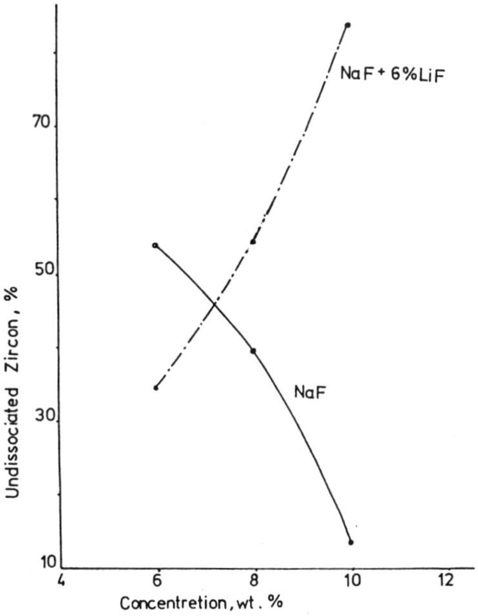

FIGURE 4. Effect of LiF addition on zircon

CONCLUSION

The dissociation of zircon is better followed by measuring the amount of zircon rest. Alkali flourides enhance the dissociation reaction when added as NaF or a mixture of NaF and LiF. Optimum dissociation is obtained by 8% NaF addition and a mixture of 6% NaF + 6% LiF.

REFERENCES

1. Locsei, B.P. and Umki, H., The role of mineralizers in some processes of silicate chemistry. Part 1: The reaction of AlF_3 and Kaolinite.
Interceram, 1980, 29, 392-397.

2. Ibid., part 2: On the Kinetics of the reaction of $ZrSiO_4$ with AlF_3.
Interceram, 1981, 30, 38-42.

3. Ibid., part 3: Improvement of resistance to corrosion of fused cast refractories containing ZrO_2.
Interceram, 1981, 30, 116-117.

4. Eppler, R.A., Mechanism of formation of zircon stains.
J. Am. Ceram. Soc., 1970, 53, 457-462.

5. Demiray , T., Nath, D.K. and Hummel, F.A. Zircon-Vanadium blue pigment.
J. Am. Ceram. Soc., 1970, 53, 1-4.

THE SOLID SOLUBILITY OF SILICA AND CALCIA IN MNZN-FERRITES

J.J. ROELOFSMA AND F.X.N.M. KOOLS
Philips Components Division, Corporate Innovation Materials,
Bdng. BE-2, 5600 MD Eindhoven, The Netherlands.

ABSTRACT

It is well known that silica and calcia have a significant influence on the properties of MnZn-ferrites. In this respect it is important to know the solid solubility of silica and calcia. Samples of a MnZn-ferrite with different amounts of SiO_2 and CaO were prepared, annealed to allow dissolution of Si and Ca, and finally quenched.
The solid solubility values were derived from XRD and EPMA measurements. It is found that the solid solubility of CaO decreases in the presence of SiO_2 and vice versa, being >0.49 wt% for CaO at low SiO_2 content and 0.07 wt% for SiO_2 at low CaO content, both at 1350°C.

INTRODUCTION

Silica and calcia in MnZn-ferrites are often present as impurities. Their presence during sintering may influence the final ferrite properties in several ways:

1) SiO_2/CaO may lead to the formation of second phases, thus influencing the microstructure and, hence, the microstructure dependend electrical and magnetic properties [1]. At usual sintering temperatures, a liquid phase is formed which promotes secondary grain growth.
2) SiO_2/CaO may dissolve into the main phase and partly segregate or precipitate during cooling, thereby affecting the grain boundary state. On the one hand this may lead to an increased grain boundary resistance and hence decreased eddy current loss [1]. On the other hand, grain boundary stresses will be built up, which may cause a deterioration in both permeability and hysteresis losses [1].
3) CaO/SiO_2 may dissolve into the ferrite lattice and remain partially present upon cooling. In this case the intrinsic properties may be affected. For instance it is expected that CaO replaces Fe^{2+} [1] resulting in a shift of the secondary peak maximum in the μ-T curve, which we observed.

It is therefore apparent that the dissolving of silica and calcia in ferrite plays a sifnificant role. Consequently it is important to know the solid solubility.
Unfortunately data about the solid solubility of CaO and SiO_2 in MnZn-ferrite are rather scarce. Taking into account the relevant ionic radii

and defect chemical effects ($CaO \rightarrow V_o$, $SiO_2 \rightarrow V_c$) a maximum solid solubility is expected in the region where the molar ratio $CaO/SiO_2 \approx 2$. Since the available data are not complete and not unambiguous, the present investigation has been carried out in order to determine the solid solubility in dependency of temperature.

EXPERIMENTAL OUTLINE

Preparation

The starting material was a calcined MnZn-ferrite powder with the formula $Mn_{0.47}Zn_{0.47}Fe^{2+}{}_{0.06}Fe_2O_4$, having a very low impurity level (0.011 wt% SiO_2 and 0.009 wt% CaO). Silica was added in the form of aerosyl and CaO was added in the form of $CaCO_3$. Three series of SiO_2 (S) and CaO(C) addition levels were applied (figure 1).

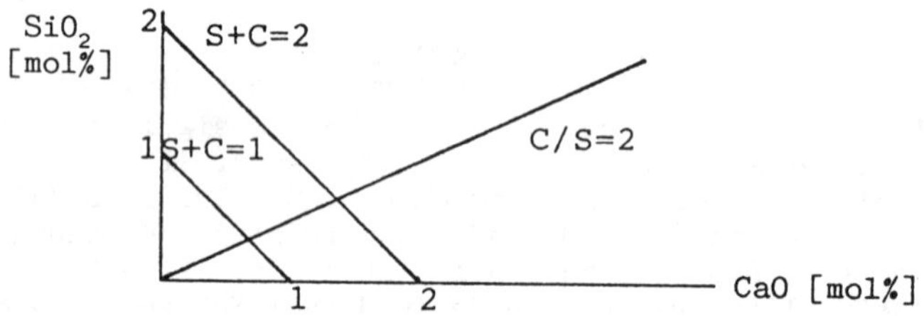

Figure 1. Three series of SiO_2 and CaO addition levels.
(see also tables 2 and 3)

The ferrite powder and the additions were wet mixed in a ball mill using demineralized water. After drying by evaporation, the resulting powder was granulated and pressed (50 MPa) to cylindrical compacts.
These compacts were presintered (8 hrs., 1200°C) in a controlled atmosphere to produce a grain size suitable for EPMA spot analysis in the grain (d_{grain} > beam diameter EPMA) and to adjust the Fe^{2+} content close to stoichiometry.
The presintered samples were encapsulated in platinum foil prior to reheating, in order to prevent an exchange with the ambient atmosphere during the annealing step while creating an equilibrium oxygen atmosphere inside this platinum foil.
After annealing at different temperatures (table 1) to allow dissolving of SiO_2 and CaO, the samples were quenched in water in order to avoid segregation/precipitation on cooling. The annealing time was varied to verify that equilibrium had been achieved.

TABLE 1
Annealing conditions

Temperature [°C]	times on top [hrs.]	
1350	4	1
1250	8	2
1300	4 (only for C/S=2)	

Analyses on quenched samples
In order to determine the solubility, local spot measurements with Electron Probe Micro Analysis were carried out inside the grains on safe distance from grain boundaries, 2nd phases and pores. For an analysis of the overall amount of SiO_2 and CaO integral measurements were carried out in which the electron beam was scanned over an area of 100 x 100 μm^2.
X-ray diffraction (XRD) analyses were carried out on the C/S=2 series in order to determine the lattice constant (a) as a function of the amount silica and calcia: it may be expected that a discontinuity in the lattice constant will be detected at the limit of the solid solubility region.
The Fe^{2+}-content was analysed by titrimetric analyses.

RESULTS AND DISCUSSION

Ferrous analyses
The Fe^{2+}-content of the samples was in the range of 1.40 to 1.63 wt%. Taking into account the limited measuring accuracy, this is close to stoichiometry.

X-ray diffraction
In figure 2 the XRD-results are presented, together with an indication of measurement errors. It appears that the lattice constant gradually increases, suggesting that the dissolved quantity also increases and, hence, neither the solid solubility of SiO_2 nor the solid solubility of CaO has been surpassed.

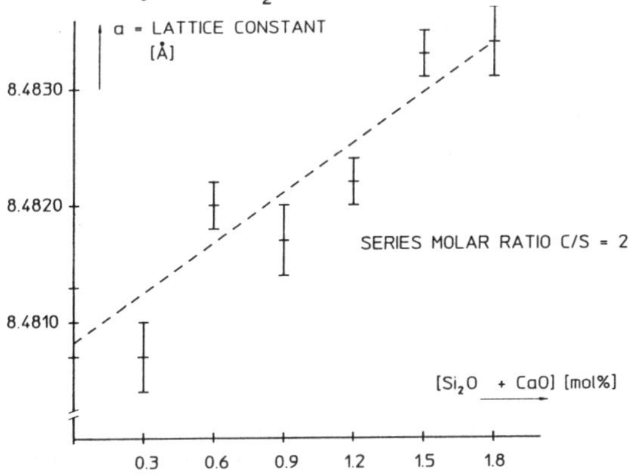

Figure 2. Lattice constant versus [SiO_2 + CaO] for the C/S=2 series.

EPMA-analyses
In table 2 spot and integral EPMA results are given for the C/S=2 series. This table shows that the concentration of SiO_2 within the grains, and hence the solid solubility, decreases in the presence of CaO and vice versa.
In table 3 results are presented for the extreme concentration levels of the series S+C=1 and S+C=2, this for various annealing conditions.
Since annealing for shorter times produced identical results, one may conclude that the equilibrium concentration has been reached (in table 3 only results for times ≥ 4 hrs. are given).
The CaO concentrations within the grains only depend on the addition level and do not vary with the annealing temperature. Therefore up to the

TABLE 2
EPMA-analyses series C/S=2, T_s=1300°C

SiO_2 [wt%]			CaO [wt%]		
spot	integral	nominal	spot	integral	nominal
.015	.015	.011	.006	.003	.009
.032	.024	.026	.042	.037	.048
↓	.055	.052	.075	.099	.096
	.066	.078	.101	.139	.144
	.084	.104	.128	.197	.192
	.147	.130	.160	.247	.240
.020	.133	.156	.193	.320	.288

TABLE 3
EPMA-analyses series S+C=1 and S+C=2

Annealing conditions	SiO_2 [wt%]			CaO [wt%]		
	spot	integral	nominal	spot	integral	nominal
1250°C 8hrs.	.015	.015	.011	.25	.23	.24
1350°C 4hrs.	.015	.015	.011	.25	.22	.24
1250°C 8hrs.	.015	.015	.011	.50	.46	.48
1350°C 4hrs.	.015	.015	.011	.49	.46	.48
1250°C 8hrs.	.04	.27	.26	<.005	<.005	.009
1350°C 4hrs.	.07	.25	.26	<.005	<.005	.009
1250°C 8hrs.	.04	.58	.52	<.005	<.01	.009
1350°C 4hrs.	.07	.52	.52	<.005	<.01	.009

maximum addition level which was investigated (0.48 wt%), CaO dissolves completely into the spinel lattice, which is clearly illustrated by the EPMA mapping of Ca in figure 3a. This consequently means that the solid solubility of CaO in "absence" of SiO_2 at $T \geq 1250°C$ probably is higher than 0.48 wt%.

The SiO_2 content within the grains varies only with the annealing temperature and is much lower than the integral content. The EPMA-mapping of Si in figure 3b shows clear enrichments. It may therefore be concluded that the solid solubility of SiO_2 is 0.04 wt% at 1250°C and 0.07 wt% at 1350°C.

A synthesis of all data is given in figure 4, representing 3 isothermal sections of the quasi ternary phase diagram in the MnZn-ferrite rich region with tie lines at 1300°C.

According figure 4 the line C/S=2 surpasses the solid solubility limit. The tie lines show that the amount of dissolved CaO increases with increasing total SiO_2 plus CaO content.

Since the ionic radius of Ca^{2+} is rather large this will lead to a gradual increase of the lattice constant, as is observed in figure 2.

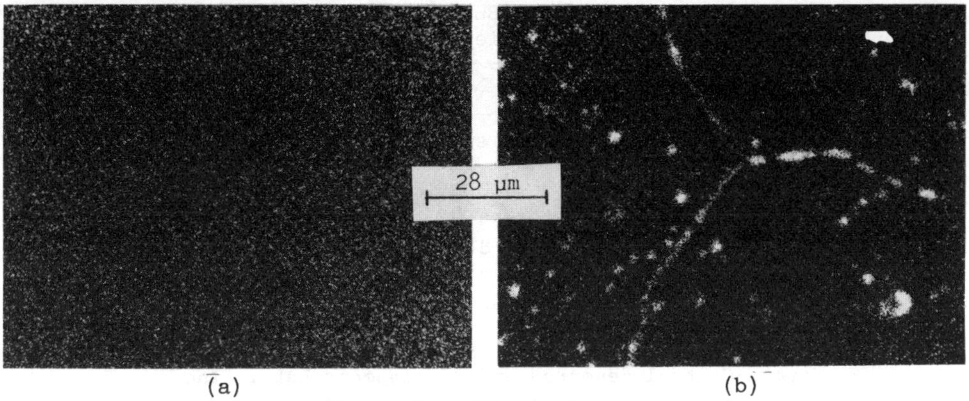

Figure 3. EPMA mapping.
(a) CaO=0.48wt% and SiO_2=0.011wt%; (b) CaO=0.009wt% and SiO_2=0.26wt%

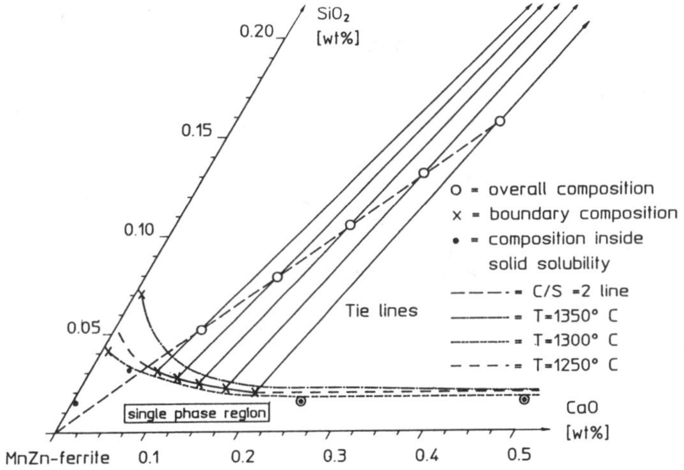

Figure 4. Ternary phase diagram of the SiO_2-CaO-MnZn ferrite system with tie lines at 1300°C.

CONCLUSIONS

The solid solubility of SiO_2 decreases with increasing CaO content, being 0.07 wt% at 1350°C for zero CaO.
There is no maximum solid solubility found for the molar ratio C/S=2.
The solid solubility of CaO decreases with increasing SiO_2 content, and is >0.49 wt% for SiO_2<0.02 wt%.

REFERENCES

1. Slick, P.I., Ferrites for non-microwave applications, Ferromagnetic Materials, ed. E.P. Wohlfarth, North-Holland Publishing Company, 1980, 2, pp.189-241.

FORMATION AND STABILITY OF Y-α'-SIALON CO-EXISTING WITH β'-SIALON

YOSHIO UKYO AND SHIGETAKA WADA
Toyota Central Research and Development Labs., Inc.,
Nagakute-cho, Aichi-gun, Aichi-ken 480-11, JAPAN

ABSTRACT

Si_3N_4 ceramics containing Y-α'-sialon were produced by hot-pressing the mixtures of Si_3N_4, Y_2O_3 and AlN powders. Hot-pressed bodies were composed of Y-α'-sialon and β'-sialon. The amount of α'-sialon was increased with increasing amount of Y_2O_3 and AlN. The lattice constant (solubility) of α'-sialon depended on the amount of Y_2O_3 and AlN, while that of β'-sialon did not. The amount of α'-sialon was decreased with increasing sintering time and temperature. This phenomenon was observed when Si_3N_4 ceramics containing α'-sialon were heated at higher temperatures for a long time. It is concluded that α'-sialon co-existing with β'-sialon is a meta-stable phase at higher temperatures.

INTRODUCTION

In an Y-Si-Al-O-N system, α'-sialon containing Y (Y-α'-sialon) forms in the limited range determined by Park et al. [1]. Mitomo et al. [2] obtained the Y-α'-sialon by sintering the mixtures of Si_3N_4, Y_2O_3 and AlN powders. In this case, the Y-α'-sialon is generated from the following reaction:

$$(4-3a)Si_3N_4 + a(Y_2O_3 + 9AlN) = Y_{2a}(Si_{12-9a},Al_{9a})(O_{3a},N_{16-3a}) \quad (1)$$

Therefore, the Y-α'-sialon is produced when Y_2O_3 and AlN in the mole ratio of 1:9 are added to Si_3N_4.

The characterization of sintered Si_3N_4 consisting of Y-α'-sialon and β-Si_3N_4 was done well by Mitomo et al. [2][3][4]. They reported that only Y-α'-sialon or both Y-α'-sialon and β-Si_3N_4 were obtained by sintering the mixtures of Si_3N_4, Y_2O_3 and AlN powders. In their reports, the effect of oxygen on the forming reaction of α'-sialon was not taken into account. However, oxygen usually exists in the Si_3N_4 powder as a surface oxide film. The formation of Y-α'-sialon seems to be influenced by the oxygen.

The purpose of this study is to investigate the formation of Y-α'-sialon by sintering the mixtures of Si_3N_4, Y_2O_3 and AlN powders with various compositions, the effect of oxygen on the formation of Y-α'-sialon and the stability of Y-α'-sialon at high temperatures.

EXPERIMENTAL PROCEDURE

Ten kinds of mixtures with different total amounts of additives, Y_2O_3 and AlN, were prepared. In all the mixtures, the mole ratio of Y_2O_3 to AlN was kept 1 to 9 as shown in TABLE 1. The powder mixtures were hot-pressed in a graphite die with BN coating. Hot-pressing was done in a N_2 atmosphere at 1750, 1800 and 1850°C under a pressure of 20 MPa for 1 to 10 hours. The phases and their lattice constants were determined by X-ray diffraction (XRD) analysis. The compositions of the grains and grain boundaries were analyzed by scanning transmission electron microscopy (STEM).

TABLE 1
Nominal compositions of powder mixtures (mol%)

No.	Si_3N_4	Y_2O_3	AlN	x *
1	91.0	0.9	8.1	0.077
2	89.0	1.1	9.9	0.096
3	87.0	1.3	11.7	0.115
4	85.0	1.5	13.5	0.135
5	83.0	1.7	15.3	0.150
6	81.0	1.9	17.1	0.174
7	77.0	2.3	20.7	0.215
8	74.0	2.6	23.4	0.255
9	67.0	3.3	29.7	0.355
10	50.0	5.0	45.0	0.622

* x: $Y_x(Si,Al)_{12}(O,N)_{16}$

EXPERIMENTAL RESULTS

Two sets of peaks corresponding to β-Si_3N_4 and α-Si_3N_4 were found in the XRD patterns of specimens 1 to 8. In the XRD patterns of specimens 9 and 10, on the other hand, peaks corresponding only to α-Si_3N_4 were found. All peaks were shifted towards lower angles, and the crystalline phases were, therefore, determined as the Y-α'-sialon and the β'-sialon. According to equation (1), all of Y, Al, O and N in Y_2O_3 and AlN should be dissolved in the Si_3N_4 structure, and the Y-α'-sialon must be formed. Actually, however, the β'-sialon was also formed. This means that some Al and O in Y_2O_3 and AlN were consumed for the formation of β'-sialon, instead of Y-α'-sialon.

Figure 1 shows the dependence of the amounts of Y-α'-sialon on the amounts of Y_2O_3 and AlN. The abscissa of this figure represents the x value in the general formula, $Y_x(Si,Al)_{12}(O,N)_{16}$. The x values were calculated from the nominal compositions of the mixtures shown in TABLE 1. As described before, SiO_2 contained in the Si_3N_4 powder as an impurity seems to affect the formation of α'- and β'-sialon. However, this effect of SiO_2 was not taken into the calculation of x in this experiment.

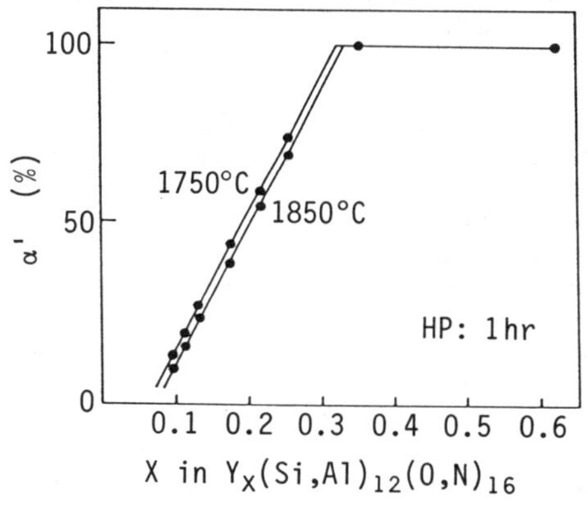

Figure 1. Dependence of the amount of Y-α'-sialon on x and hot-pressing temperature.

The amount of Y-α'-sialon was increased with increasing amount of Y_2O_3 and AlN, and the single-phase Y-α'-sialon was formed when the amount of Y_2O_3 and AlN was more than 30 mol% (x is larger than 0.33). The amount of

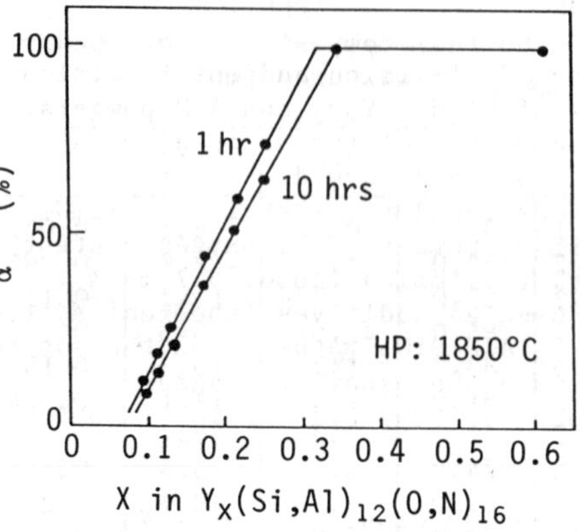

Figure 2. Dependence of the amount of Y-α'-sialon on the hot-pressing time.

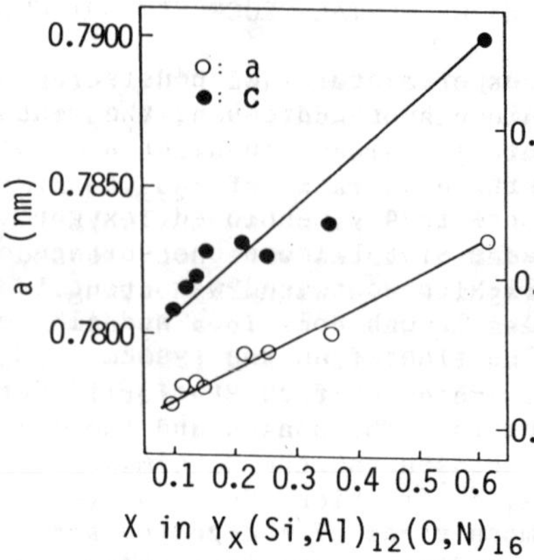

Figure 3(a). Relation between lattice constant of Y-α'-sialon and x.

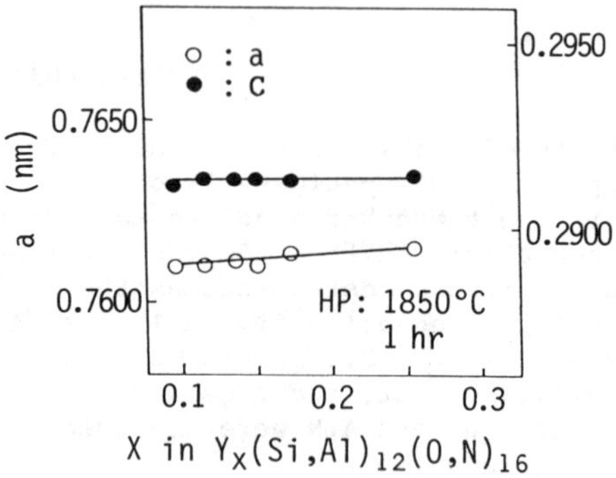

Figure 3(b). Relation between lattice constants of β'-sialon and x.

Y-α'-sialon was decreased with increasing hot-pressing temperature. Figure 2 shows the amount of Y-α'sialon in the specimens hot-pressed at 1750 and 1850°C for 1 and 10 hours. The amount of Y-α'-sialon is decreased with increasing hot-pressing temperature and time. These results lead to the conclusion that the Y-α'-sialon formed at the initial stage of sintering gradually changes into the β'-sialon.

The lattice constants of Y-α'-sialon and β'-sialon were calculated from the diffraction angles of the XRD peaks.

Figures 3(a) and (b) show the compositional dependence of the lattice constants of Y-α'-sialon and β'-sialon for the specimens hot-pressed at 1850°C. The lattice constants of Y-α'-sialon was increased with increasing x (the amount of Y_2O_3 and AlN) not only when only Y-α'-sialon formed but also when both Y-α'-sialon and β'-sialon formed. On the other hand, the lattice constants of β'-sialon were independent of x. The β'-sialon produced is expressed by the formula: $Si_{5.4}Al_{0.6}O_{0.6}N_{7.4}$.

The existence of two kinds of grains with different morphologies and compositions was revealed from the analysis by STEM/EDS. The grains with different morphologies have different compositions. TABLE 2 shows the compositions of the grains analyzed. The compositions were calculated on the assumption that the grains were composed of only Si, Y and Al, because O and N could not be analyzed by STEM. These results indicate that grains A and B are β'-sialon and Y-α'-sialon, respectively. The composition of triple point determined by the same method is also shown in TABLE 2. A larger amounts of Y and Al were found at the triple point than in grains.

DISCUSSION

In this experiment, Si_3N_4 consisting of the Y-α'-sialon and the β'-sialon was obtained by hot-pressing the mixtures of Si_3N_4, Y_2O_3 and AlN powders. Mitomo et al. [2] and Huang et al. [5] reported a different result that the α'-sialon and the β-Si_3N_4 (not β'-sialon) were formed.

As previously mentioned, oxygen is considered to exist in the Si_3N_4 powder as a SiO_2 film on the surface of Si_3N_4 powder. The oxygen content of the Si_3N_4 powder used was about 1.3 wt% (2.5 mol%) (about 5.7 mol% as SiO_2). Though only Y_2O_3 and AlN were used as additives, the real system must be considered as the system Si_3N_4-Y_2O_3-AlN-SiO_2 rather than the system Si_3N_4-Y_2O_3-AlN. Slasor et al. [6] determined the behavior diagram including the Y-α'-sialon and β'-sialon phases. The wide region in which the Y-α'-sialon and the β'-sialon are co-existing is found in their diagram.

Kamiya et al. [7] reported that SiO_2 easily dissolved in Si_3N_4 in the presence of AlN. Therefore, some part of AlN added will be consumed for the formation of β'-sialon as given by the following equation:

$$(2-0.5z)Si_3N_4 + zAlN + 0.5\,zSiO_2 = Si_{6-z}Al_zO_zN_{8-z} \qquad (4)$$

The result of XRD analysis shows that the lattice constant of Y-α'-sialon in specimen 1 with a small amount of Y_2O_3 and AlN is slightly larger than that of α-Si_3N_4 used, indicating that the solubility of Y-α'-sialon is very poor. However, the solubility of β'-sialon in specimen 1 corresponds to z=0.6. It is considered that the β'-sialon is formed in preference to the Y-α'-sialon at the initial stage of sintering, and that the residual AlN which is not consumed for the formation of β'-sialon dissolves in Si_3N_4, resulting in the formation of the Y-α'-sialon. Therefore, Y_2O_3 becomes to exceed AlN even if the nominal mole ratio of Y_2O_3 to AlN is 1 to 9. It is estimated that the excess Y_2O_3 reacts on SiO_2 and is left at grain boundaries.

The dependence of the lattice constants (solubility) of Y-α'-sialon is different from the results obtained by Mitomo et al. [3]. They reported that the Y-α'sialon co-existed with the β-Si_3N_4, that the dependence of the lattice constants of Y-α'-sialon co-existing with β-Si_3N_4 on x was not observed. It was observed in this experiment, however, that the Y-α'-sialon co-existed with the β'-sialon and that the lattice constants (solubility) of Y-β'-sialon increased with increasing amount of Y_2O_3 and AlN.

TABLE 2
Compositions of grains and grain boundaries

	Si	Y	Al
A (β'-sialon)	94.62	0.05	5.33
B (α'-sialon)	94.99	0.27	4.74
triple point	73.09	17.89	9.02

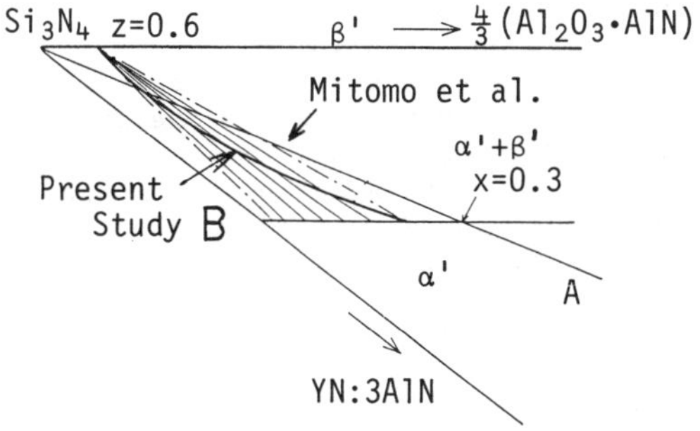

Figure 4. Compositional changes of specimen in the co-existing region of Y-α'-sialon and β'-sialon.

The reason for these differences has not been found. However, it is estimated that the difference of starting powder characteristics, such as purity and grain size, has effects on the forming reaction of Y-α'-sialon.

The tie lines were schematically drawn in the diagram determined by Slasor et al. [6] according to the lattice constants (solubility) obtained as shown in Figure 4. The compositions of the specimens should move along straight line A when the mole ratio of Y_2O_3 to AlN is 1 to 9. However, the compositions of the specimens obtained in this experiment deviated from straight line A and seemed to move along curve B. It is thought that this deviation is attributed to the existence of SiO_2 contained in the Si_3N_4 powder.

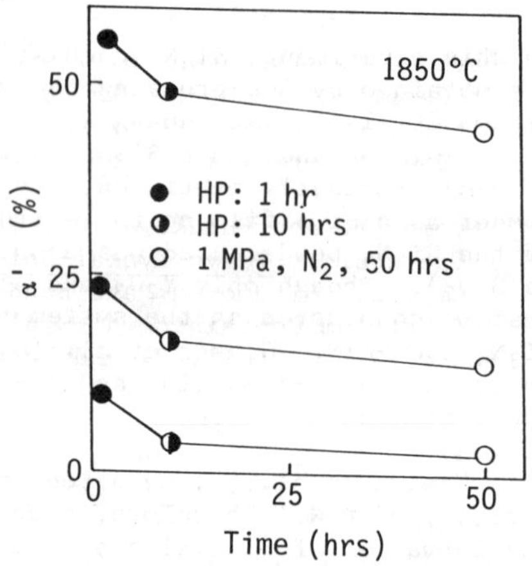

Figure 5. Dependence of the amount of Y-α'-sialon on the annealing time at 1850°C.

The amount of Y-α'-sialon was decreased with increasing hot-pressing temperature and time. This suggests that the Y-α'-sialon co-existing with the β'-sialon is unstable and changes into β'-sialon with time at high temperatures. The specimens containing various amounts of Y-α'-sialon fabricated by hot-pressing at 1850°C for 1 hour were annealed at 1850°C for 50 hours to investigate the stability of Y-α'-sialon co-existing with the β'-sialon. Figure 5 shows the change of the amount of Y-α'-sialon in the specimen annealed and in the specimen fabricated by hot-pressing at 1850°C for 10 hours. The amount of Y-α'-sialon was decreased with time as shown in Figure 5, and it is concluded that the Y-α'-sialon is less stable than the β'-sialon.

CONCLUSIONS

1. The Y-α'-sialon and the β'-sialon formed when the x value in $Y_x(Si,Al)_{12}(O,N)_{16}$ is smaller than 0.33. When the x value is larger than 0.33, on the other hand, a single Y-α'-sialon phase formed.
2. The lattice constant (solubility) of Y-α'-sialon depended on the amount of Y_2O_3 and AlN, while that of β'-sialon was consistent.
3. The Y-α'-sialon changed into the β'-sialon by heating at high temperatures for a long time. It is concluded that the Y-α'-sialon co-existing with the β'-sialon is a meta-stable phase at high temperatures.

ACKNOWLEDGMENTS

The authors thank Dr. T. Okamoto for his helpful discussions, and Messrs. T. Ishiguro and N. Suzuki for the STEM analysis.

REFERENCES

1. Park, H.K., Thompson, D.P. and Jack, K.H., α-Sialon Ceramics. Science of Ceramics, 1980, 10, 251-6.
2. Mitomo, M., Izumi, F., Bando, Y. and Sekikawa, Y., Characterization of α'-Sialon Ceramics. Proc. Int. Symp. on Ceramic Component for Engine, 1983, 377-86.
3. Mitomo, M. and Furukawa, O., The Stability of α'-Sialon at High Temperature. Yogyo-Kyokai-Shi, 1982, 89, 631-3.
4. Izumi, F., Mitomo, M. and Bando, Y., Rietveld Refinements for Calcium and Yttrium Containing α'-Sialons. J. Mater. Sci., 1984, 19, 3115-20.
5. Huang, Z-K., Greil, P. and Petzow, G., Formation of α-Si_3N_4 Solid Solutions in the System Si_3N_4-AlN-Y_2O_3. J. Amer. Ceram. Soc., 1983, 66, c96-7.
6. Slasor, S. andThompson, D.P., Preparation and Characterization of α'-Sialons. Proc. Int. Conf. on Non-Oxide Technical and Engineering Ceramics, 1987, 223-30.
7. Kamiya, N., Oyama, Y. and Kamigaito, O., Silicon Nitride Solid Solution in the Ternary System, Si_3N_4-AlN-SiO_2. Yogyo-Kyokai-Shi, 1975, 83, 553-7.

REACTION ENGINEERING PARAMETERS OF THE CARBOTHERMAL PRODUCTION OF β′-SIALON FROM KAOLIN

A.M. van Neerven, F. Blömer, R. Metselaar
Centre for Technical Ceramics
P.O. Box 595, 5600 AN Eindhoven, The Netherlands

ABSTRACT

The grain model has shown to be an appropriate reaction engineering model for the carbothermal production of β′-sialon from kaolin when pelletized mixtures of kaolin and coal are used as starting materials. When pellet size is below 1 mm in diameter (spherical pellets) pore diffusion can be neglected and the overall rate is controlled by the chemical reaction rate. Above this size, pore diffusion shows a significant influence. This holds for reaction temperatures of 1400°C. The apparent activation energy is about 400 kJ/mole.
When the CO partial pressure is kept below 0.05 atm., at 1400°C relatively simple reactor design calculations can be used. Under these conditions the formation of β′-sialon can be treated as an elementary reaction.
In general at least two steps of the complex reaction scheme are rate determining. This means that in order to achieve a reliable model, these steps should be distinguished. Experimental procedures should be developed to enable elucidation and separate modelling of these steps.

INTRODUCTION

The aim of this work is to achieve insight concerning the reaction engineering parameters of the carbothermal production of β′-sialon from kaolin. These parameters can be used for reactor design, upscaling and process control applications.
The overall reaction

$$3\ Al_2O_3.2SiO_2.2H_2O + 15\ C + 5\ N_2 \longrightarrow 2\ Si_3Al_3O_3N_5 + 15\ CO + 6\ H_2O$$

is known to proceed through a complex, reversible reaction mechanism with a reaction rate which depends on the partial pressure of carbonmonoxide.

Higgins and Hendry [1] subdivided the overall reaction.
At about 670°C, dehydration of kaolinite occurs giving amorphous metakaolin and at 1200°C mullite is formed:

$$3\ Al_2O_3.2SiO_2.2H_2O \longrightarrow 3\ Al_2O_3.2SiO_2 + 4\ SiO_2 + 6\ H_2O$$

At the reaction temperature, formation of sialon occurs in two steps:

$$4\ SiO_2 + 12\ C \longrightarrow 4\ SiC + 8\ CO$$

$$3\ Al_2O_3.2SiO_2 + 4\ SiC + 3\ C + 5\ N_2 \longrightarrow 2\ Si_3Al_3O_3N_5 + 7\ CO$$

Although quite some knowledge of the production of β'-sialon by carbothermal reduction of kaolin has been gathered, reaction engineering parameters, necessary for design, upscaling and process control applications, have not yet been elucidated.
In a first approach, one reaction step is assumed to be rate controlling, which enables a relatively simple treatment. Szekely's grain model [2] appears to be an appropriate reaction engineering model when pelletized mixtures of kaolin and coal are used as starting materials. This statement is based both on fundamental considerations and experimental results. The grain model can be regarded as a combination of the model for catalyzed reactions on porous solids and the shrinking unreacted core model. The first describes simultaneous pore diffusion and chemical reaction in porous pellets, the second is used to describe the chemical gas-solid reaction at the grains from which the pellets are made up [3,4].

A schematic diagram of a spherical pellet containing spherical grains of uniform size is shown in figure 1.

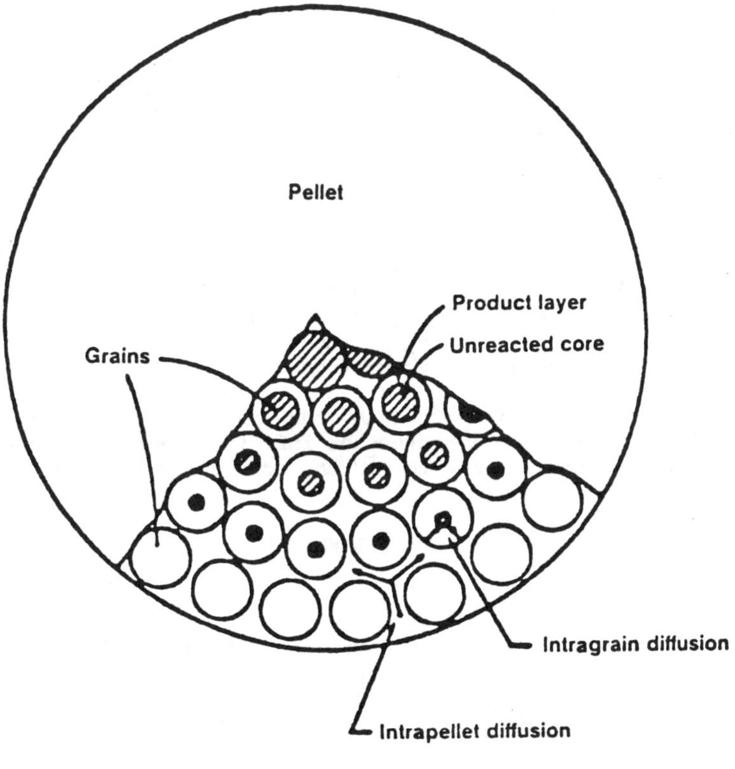

Figure 1. Sketch of the grain model [2].

The grain model combines equations for the conversion of the gaseous reactant A and product C with a mass balance for the solid reactant B, according to the reaction equation A(g) + b B(s) ---> c C(g) + d D(s).

$$D_e \nabla^2 C_A - r_{A,s} = 0$$

$$D_e \nabla^2 C_C + r_{A,s} = 0$$

$$b.r_{A,s} = -\rho_B \frac{dl}{dt}$$

where l is the characteristic coordinate of the shrinking core surface within the grains, R_s is the reaction rate at the grain surface, D_e is the effective diffusion coefficient and C is the concentration.
Analytical solutions can only be obtained for systems in which the surface reaction rate $r_{A,s}$ is a linear function of the reactant gas concentration C_A, i.e. first order kinetics. Also, the reacting surface area may not change during conversion, i.e. flat plate geometry for the grains and uniform grain size.

Szekely [2] showed that a simple and direct relationship exists by comparing numerical solutions for several cases, giving

$$t^* \approx \gamma_G(X) + \hat{\sigma}^2 \Phi_P(X)$$

where t^* is the dimensionless time t/τ^*, τ^* is the total reaction time when chemical reaction is controlling, Φ_P is the conversion function for pore diffusion control, γ_G is the conversion function for chemical reaction control, P, G are the pellet or particle, grain shape factors, X is the conversion and $\hat{\sigma}$ is the reaction modulus.

Film diffusion transport can also be incorporated, but appeared to be of negligible influence for the case at issue. The relative importance of pore diffusion is indicated by the reaction modulus $\hat{\sigma}$, which is equivalent to the Thiele modulus, a frequently used standard figure in chemical reaction engineering. When $\hat{\sigma}$ approaches zero, pore diffusion presents a negligible resistance to the progress of reaction. Under these conditions the reactant concentration is uniform throughout the pellet and is equal to that in the bulk gas phase. Reaction rate is then controlled by the chemical reaction rate at the grain surface. This situation holds as long as $\hat{\sigma}$ is below 0.3. As $\hat{\sigma}$ increases, diffusion through the porous pellet becomes the rate controlling factor. When $\hat{\sigma}$ approaches infinity, reaction occurs in a narrow zone between the unreacted core and the completely reacted layer.

EXPERIMENTAL

Experiments have been executed in a horizontal tube furnace. Szekely's grain model is a volume-element-model with a volume element of one pellet. Therefore, experiments should be carried out under so-called differential conditions. This means that reaction conditions should not be influenced by the reaction itself. In our case: the reaction rate should not be retarded by the reaction product CO. In case of a reversible reaction differential conditions can be achieved by limitation of the reactor content. It appeared that in our case, when the reactor content was below 0.3 gram, differential conditions were fulfilled.

All experiments have been carried out with approximately 0.25 gram of pellets. Conversion vs time curves were calculated from $P_{CO,outlet}$ vs time curves, taking into account the increase in gas flow, caused by the reaction.
Pore diffusion and chemical reaction were evaluated by converting pellets of 6.5, 3.5, 1 and 0.25 to 0.30 mm diameter, under differential conditions. The reaction temperature was 1400°C and N_2-gasflow 153 ml/min.

RESULTS

The conversion curves are given in figure 2. From these curves it can be calculated that the reaction modulus $\hat{\sigma}$ is below 0.3 when the pellet diameter is below 1 mm. This means that at a reaction temperature of 1400°C, pore diffusion can be neglected when the pellet size is 1 mm or less in diameter. Above this size, pore diffusion shows a substantial influence on the overall reaction rate.
Concerning the influence of the partial pressure of carbonmonoxide, it was shown that the assumption of only one reaction step being rate determining only holds as long as the CO partial pressure in the reactor is below about 0.05 atm. at a total pressure of 1 atm. and 1400°C reaction temperature. This is illustrated in figure 3.
Under these conditions, reactor calculations are relatively simple, as has

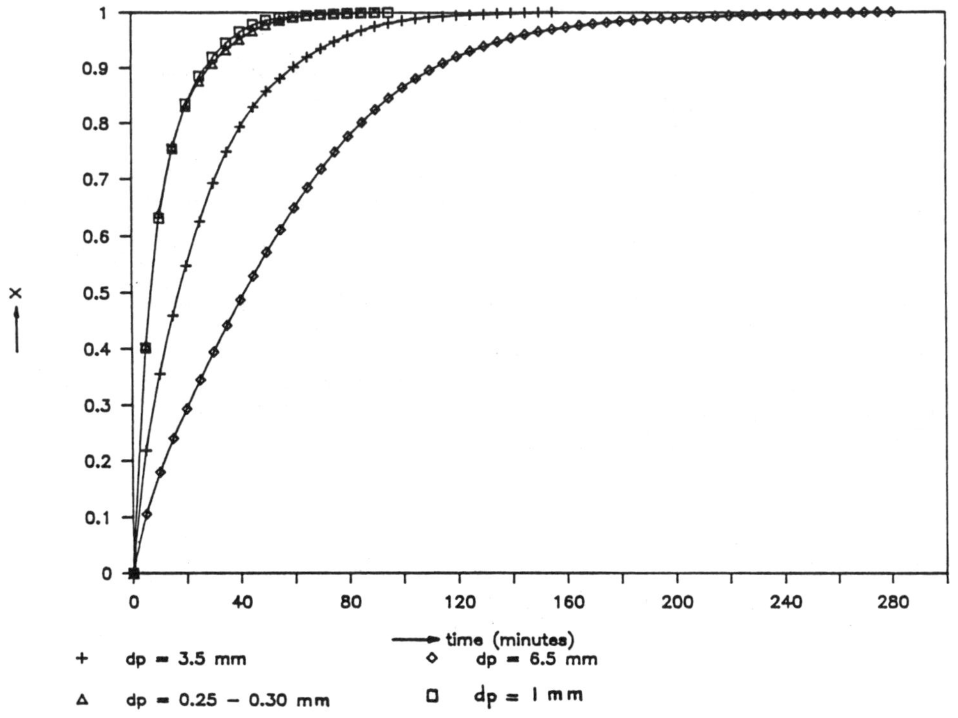

Figure 2. Conversion-time curves with various pellet sizes.

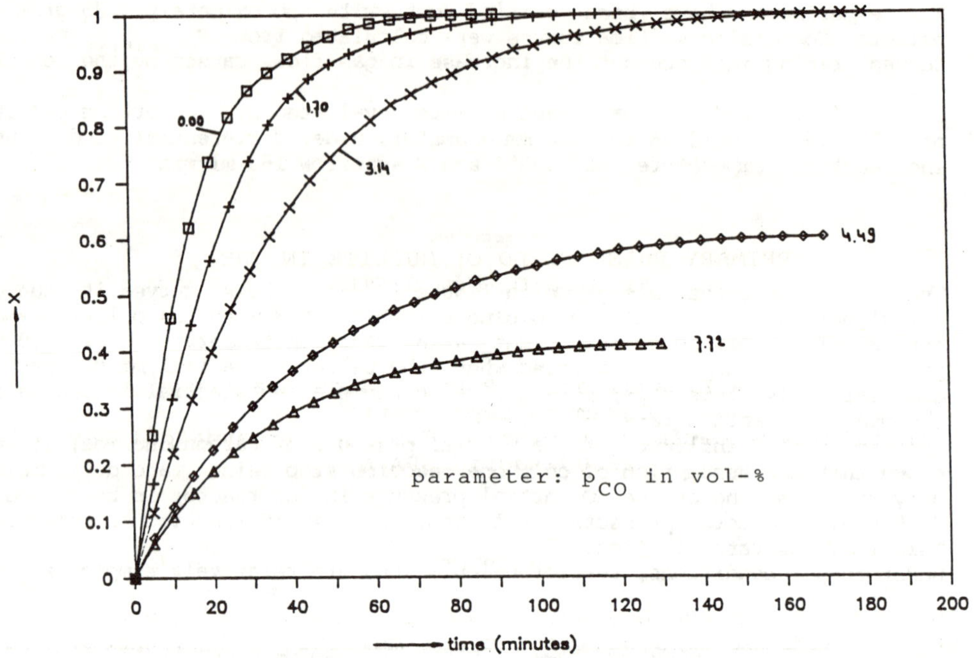

Figure 3. Conversion-time curves at various CO partial pressures. Differential conditions. Pellet size 500-800 µm. Reaction temperature 1400°C. Total gasflow 134-153 ml/min.

been illustrated for a continuous counter current plug flow reactor. At higher CO partial pressures, at least two reaction steps must be considered in order to obtain adequate models. The rate determining steps should be distinguished and experimental procedures should be developed to enable elucidation and separate modelling of these steps.

From conversion curves, determined at reaction temperatures ranging from 1350 to 1425 °C an activation energy of about 400 kJ/mole could be established. This value indicates that the conversion rate is controlled by the chemical reaction at the grain surface. However, this activation energy is only an apparent value as the temperature dependence of the equilibrium CO partial pressure has been neglected.

REFERENCES

1. Higgins I. and Hendry A., Br. Ceram. Trans. J., 85, 161-166, (1986)

2. Szekely J., Evans J.W. and Sohn H.Y., Gas-solid reactions, Academic Press, New York, 1976.

3. Levenspiel O., Chemical reaction engineering, John Wiley and Sons, New York, 1972.

4. Yagi S. and Kunii D., 5th Symposium (International) on Combustion, Reinhold, New York, 1955, p. 231.

PRIMARY PHASE FIELD OF MULLITE IN THE Al_2O_3-SiO_2-MgO SYSTEM

A.L.Cavalieri*, P.Pena, S.de Aza
Ceramics Department,
Instituto de Cerámica y Vidrio, C.S.I.C.,Arganda del Rey
28500-Madrid,SPAIN

ABSTRACT

The location of the boundary line which separates the primary phase fields of cristallization of mullite and corundum, in the system Al_2O_3-SiO_2-MgO, has been determined by establishing two isoplethal sections at 50 and 55 wt% alumina content. The tie-triangles where alumina, mullite and a liquid phase coexist at 1600 and 1650°C, were also determined by firing and quenching selected samples. The results are discussed in relation to the mullite melting behaviour.

INTRODUCTION

In view of the technological importance of mullite, a great deal of experimental works on the system Al_2O_3-SiO_2 at high temperature have been published (1-9). Neverthelešs, their contradicting results implies significant discrepancies concerning to melting nature of mullite and solid solution range of this compound.

The thermodinamical data allow reproduction of the contradictory experimental diagrams respect to the decomposition of mullite with minimal changes for the mullite data (10). They show either a peritectic decomposition or a congruent melting mullite phase. Thus, for the calculated diagrams, the limited accuracy of the data does not allow a clear decision and any diagram can be calculated to fit the experimental prejudices of the researches.

In a recent review of the present authors (11), on both melting nature and solid solution range of mullite, another route is proposed, to clarify the subjects in discussion, by establishing the boundary line which separates the primary

---

* Permanent adress in CETMIC-CONICET-CIC, La Plata,ARGENTINA.

phase fields of mullite and alumina in ternary systems of the type Al_2O_3-SiO_2-XO (XO = alkaline or alkaline earth oxides). After several considerations, the system Al_2O_3-SiO_2-MgO has been selected.

According it, the aim of this investigation is to establish the boundary line which separates the primary phase fields of crystallization of alumina and mullite in the Al_2O_3-SiO_2-MgO ternary system. Furthermore, the tie-triangles at 1600 and 1650 °C where alumina, mullite and liquid coexist, were also determined.

EXPERIMENTAL

Previously to select the isoplethal sections to be studied in the system Al_2O_3-SiO_2-MgO, taking into account the Osborn and Muan compiled diagram (12), a tentative boundary line which separate mullite and alumina primary phase fields of crystallization, in the mentioned system, was depicted considering the data of Aramaki and Roy(2,3) and Klug et al.(6) (Figure 1A) and those of Aksay and Pask (1) (Figure 1B) on the location of the eutectic and peritectic points in the sub-system mullite-alumina.

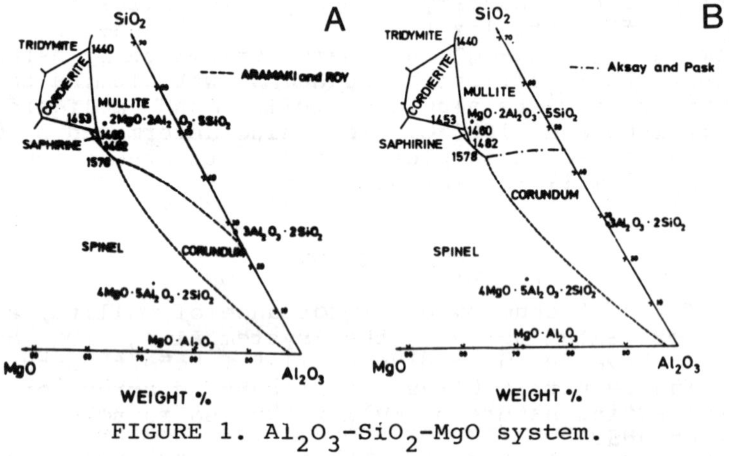

FIGURE 1. Al_2O_3-SiO_2-MgO system.

From Figure 1, theoretical isoplethal sections, with alumina constant content of 50 and 55 wt%, were drawn(Fig.2). All considered phase equilibrium relations below 1500 °C were those appear in the study published by Smart and Glasser(13).

The experimental compositions were prepared from dried low-sodium Al_2O_3 (99.998% purity), washed Belgian sand (99.9% purity), and AR grade MgO (1500 °C/2h calcined). The experiments were conduced by the very well know quenching technique which was described in earlier paper (14). Phase identification was made by ORLM and SEM-EDX.

Compositions with silica constant contents of 39 and 40 wt% , were additionally prepared in order to establish the tie-triangles mullite-alumina-liquid at 1600 and 1650 °C.

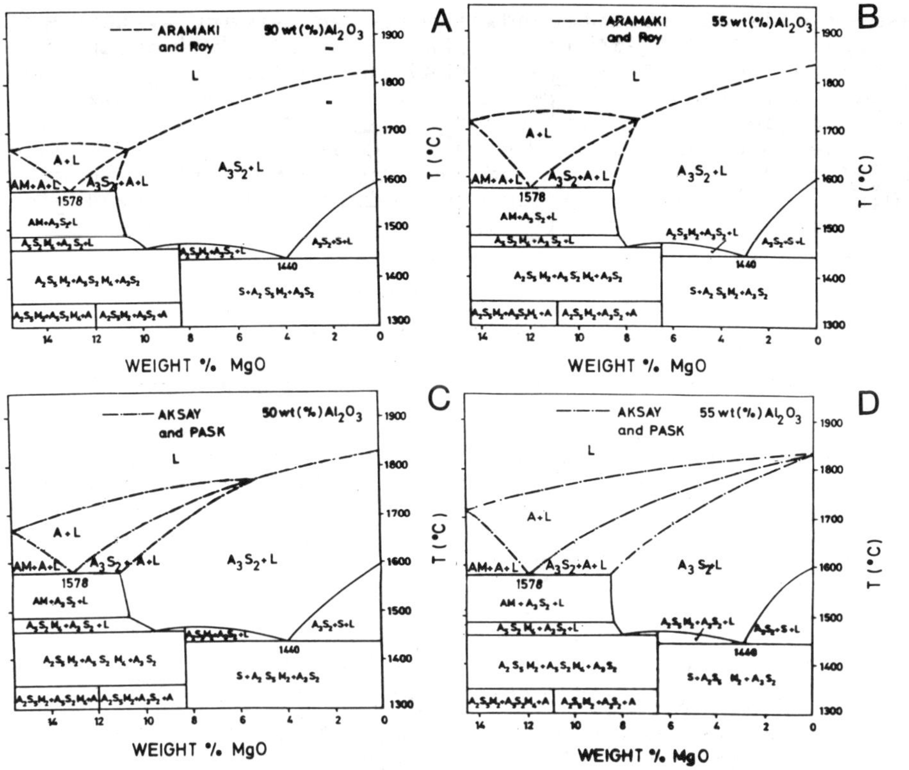

FIGURE 2. Theoretical isoplethal sections deduced from Figure 1A and 1B, in the system Al_2O_3-SiO_2-MgO.

RESULTS AND DISCUSSION

Studied compositions are located in Figure 3, which shows the isoplethal sections determined from the present experimental data. Heavy lines within sections represent boundaries determined from phase identification on the fired samples.

Considering the figures above, the obtained results are obviously in good agreement with the theoretical isoplethal sections obtained from the data of Aramaki and Roy or Klug et al., on the location of the invariant point of the subsystem mullite-alumina but not with those from Aksay and Pask.

Figure 4 depicts the tie-triangles mullite-alumina-liquid at 1600 and 1650^0 C respectively, determined from the experimental data on the isothermal section with a silica constant content of 39 and 40 wt%, and those corresponding to the isoplethal sections. The obtained results show that the tie-triangle at 1650^{0}C is within the tie-triangle at 1600^{0}C, which confirms that no alteration reaction occurs along the boundary line which separates the primary phase fields of mullite and alumina. This results together with those of the isoplethal sections confirm that the boundary is in accordan-

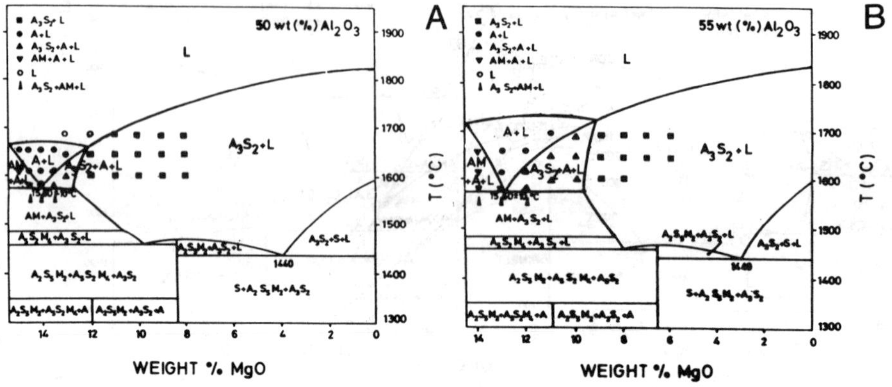

FIGURE 3. Experimental isoplethal sections for constant Al_2O_3 content of 50 and 55 wt%, in the system Al_2O_3-SiO_2-MgO.

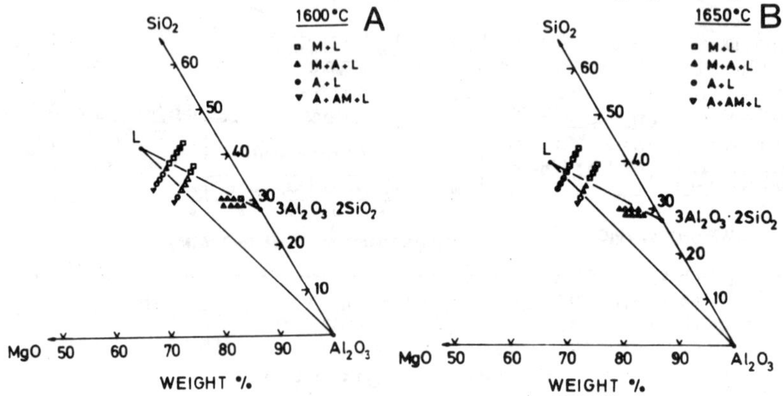

FIGURE 4. Mullite-alumina-liquid tie-triangles at 1600°C and 1650°C.

ce with that shown in Figure 1A.

The compositions of the liquid phases that coexist in equilibrium with alumina and mullite at 1600 and 1650°C were used to define more accurately the boundary between alumina and mullite primary crystallization phase fields.

From the obtained results and those previously reported in the literature (12), a modified projection of the liquid surface on the Al_2O_3-SiO_2-MgO system is proposed as is shown in Figure 5. This diagram differs slightly from that of Osborn and Muan on the location of the invariant point alumina-mullite-spinel-liquid as well as in its temperature.

Finally, from Figure 4, and considering that negligible quantities of MgO go into solid solution in mullite (15), can be deduced that compositions of mullite at 1600 and 1650° C are very close to 72 wt% Al_2O_3 with differences between them below the precision of the experimental method used.

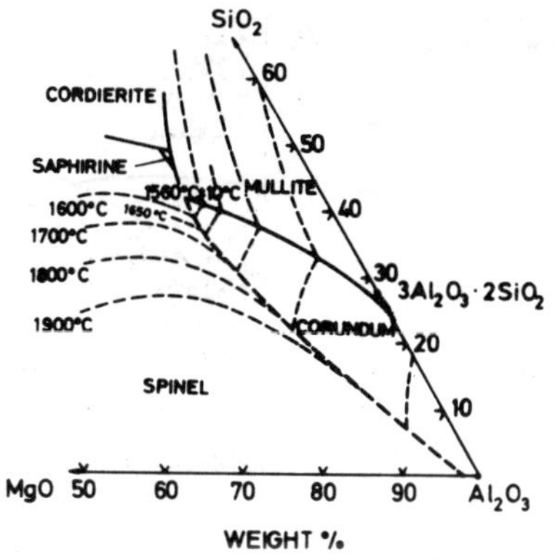

FIGURE 5. Revised projection of the liquid surface of the Al_2O_3-SiO_2-MgO system.

FINAL CONSIDERATIONS

The most important feature from the present results is about the location of the invariant point on the subsystem mullite-alumina, which can be infered to be located close to 78 wt% Al_2O_3 in the binary system Al_2O_3-SiO_2. The present experimental data do not allow to establish if mullite melts congruent or incongruently.

REFERENCES

(1) I.A.Aksay and J.A.Pask, J.Am.Ceram.Soc., 58(11-12)507 (1975)
(2) S.Aramaki and R.Roy, J.Am.Ceram.Soc.,42(12)644(1959)
(3) S.Aramaki and R.Roy, J.Am.Ceram.Soc.,45(5)229(1962)
(4) N.L.Bowen and J.W.Greig, J.Am.Ceram.Soc.,7(4)238(1924)
(5) R.F.Davis J.A.Pask, J.Am.Ceram.Soc.,55(10)525(1972)
(6) F.J.Klug, S.Prochazka and R.H.Doremus, J.Am.Ceram.Soc.70 (10) 750(1987)
(7) J.A.Pask, Ceram. Int.,9(4)107(1983)
(8) J.A.Pask, Proceedings of the 1st International Workshop on mullite, Japan(1987)
(9) S.Prochazka and F.J.Klug, J.Am.Ceram.Soc.,66(5)874(1983)
(10 R.S.Roth, J.R.Dennis and H.F.McMurdie, Phase Diagrams for Ceramists, Vol.VI, p.146-154, Ed. Am.Ceram.Soc.,Inc. (1987)
(11) A.L.Cavalieri, P.Pena and S.de Aza, Bol.Soc.Esp.Ceram.y Vidrio, (1988) (en prensa)
(12) E.F.Osborn and A.Muan, Comp. Temp. Phase Equilibrium Diagrams of the Refractory Oxides.Plate 3,ed.Am.Ceram. Soc., Ohio(1960)
(13) R.M.Smart and F.P.Glasser, Ceram. Int.,7(3)90 (1981)
(14) P.Pena and S. de Aza, J. Am.Ceram.Soc.,67(1)C-3(1984)
(15) H. Schneider, Ceram. Int.,13(3) 77(1987)

Financial support from CICYT,Spain (Project MAT88-0182)

PHASE RELATIONS IN THE $CaO-Cs_2O-Al_2O_3$ SYSTEM

Johan A.M. van Hoek, Frans J.J. van Loo, and Rudi Metselaar.
Centre for Technical Ceramics, University of Technology, Eindhoven, The Netherlands.

ABSTRACT

Phase relations in the system $CaO-Cs_2O-Al_2O_3$ were investigated. On basis of the experimental results, a reaction diagram for this system at 900°C is proposed. In this system a new compound was found and formulated as $Cs_2O.2CaO.4Al_2O_3$.
The phase relation results can be used to draw some conclusions about the cesium compatibility of calcia doped alumina. We found that all calcium aluminates react with Cs_2O forming a stable compound. From these results we conclude that calcium will have a negative effect on the cesium resistance of Al_2O_3.

INTRODUCTION

Only work on one quasi-ternary phase diagram involving Cs_2O and Al_2O_3 has been done in the past, namely on the $Cs_2O-SiO_2-Al_2O_3$ diagram (1). Yet there are some important applications of Al_2O_3 which merit such an investigation. In both thermionic energy converters and Magneto HydroDynamic generators, materials are exposed to cesium at temperatures in the range of 1000-1700°C. A number of research studies have been carried out in the past with the objective of finding materials with sufficient corrosion resistance (2,3). Alumina seems to be one of the best. The corrosion resistance is, however, largely determined by the sinter additives and grain boundary phases. We therefore examined the $CaO-Cs_2O-Al_2O_3$ system.

MATERIALS AND METHODS

The starting materials were Cs_2CO_3, $CaCO_3$ and Al_2O_3. Because Cs_2CO_3 is deliquescent, this material was handled in a glove box. The powders were mixed, pressed into pellets and subsequently heated to 900°C in air for 20 up to 200 hours. The pellets were taken out of the furnace while still hot and transferred to the glove box to cool down. Phases present in the pellet after this calcination procedure were identified by X-ray diffraction (XRD). In order to prevent reaction between the phases and water vapour from ambient air during XRD, the diffractometer was evacuated. During the heat treatment, the Cs_2O losses were often considerable. This means that the composition changed . By measuring the weight before and after the heat treatment, however, we could calculate the resulting composition.

RESULTS

In the first experiments (no. 1 to 3 in fig 1), we found from XRD measurements that together with Al_2O_3 and CsA[1] an unknown compound was always formed. We developed a new method to locate this compound. The principle of this new method is based on the fact that CsA decomposes upon heating at 900°C, leaving pure Al_2O_3. So when we heat mixtures of CsA, Al_2O_3 and the unknown compound for long times, Cs_2O will evaporate out of CsA until no CsA is present anymore. When no Cs_2O evaporates from the unknown compound, the resulting compositions will lie on a straight line between Al_2O_3 and the unknown compound. Once the composition of the compound was roughly located, we heated mixtures 19 to 27 (see figure 1) until their weights had stabilized. The diffraction pattern of mixture 23 showed only peaks of the unknown compound. This compound could be identified as $Cs_2O.2CaO.4Al_2O_3$ (CsC_2A_4). Details of this compound are published elsewhere (4).

The experiments discussed so far produced information about the Cs_2O-rich part of the diagram. In order to get information about the CaO-Al_2O_3-CsC_2A_4 triangle, we set up an other type of experiment. All the known calcium aluminates were synthesized in pure form by high temperature

[1] Cs = Cs_2O, A = Al_2O_3 and C = CaO

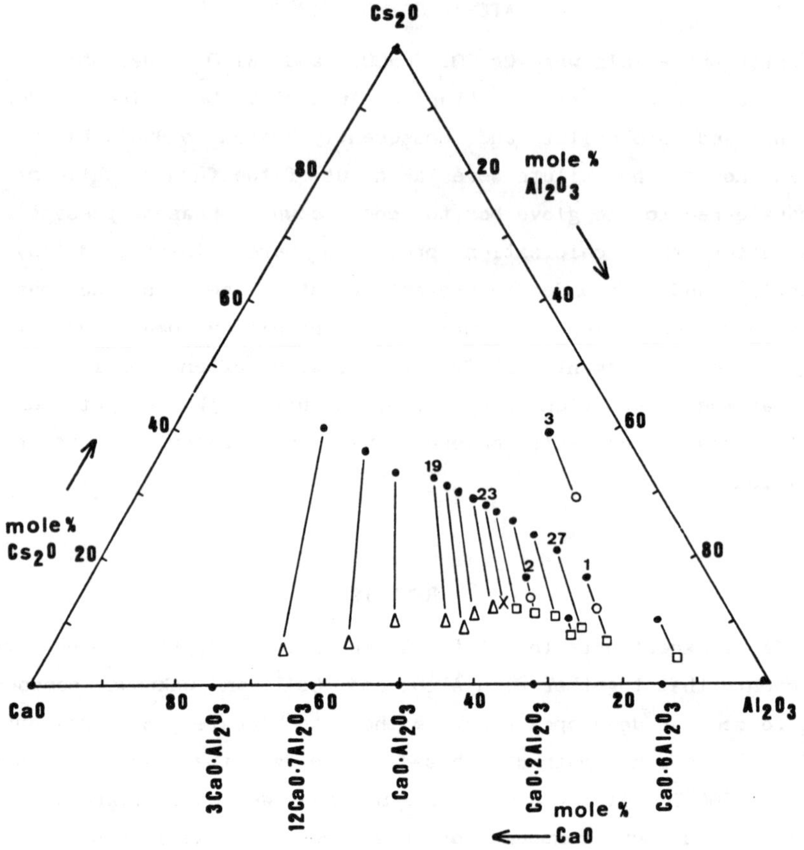

Figure 1. Representation of some illustrative experiments.

● starting compositions ($CsCO_3$-$CaCO_3$-Al_2O_3 mixtures)

○ $Cs_2O.Al_2O_3$, Al_2O_3 and $Cs_2O.2CaO.4Al_2O_3$

△ CaO and $Cs_2O.2CaO.4Al_2O_3$

□ Al_2O_3 and $Cs_2O.2CaO.4Al_2O_3$

X pure $Cs_2O.2CaO.4Al_2O_3$

syntheses and then heated with Cs_2CO_3 at 900°C. The results of these reactions can be written as:

$C_3A + 1\%\ Cs_2O \longrightarrow CaO + C_3A + CsC_2A_4$

$C_{12}A_7 + 1\%\ Cs_2O \longrightarrow CsC_2A_4 + C_{12}A_7 + C_3A$

$CA + 5\%\ Cs_2O \longrightarrow CsC_2A_4 + C_{12}A_7 + CA$

$CA_2 + 5\%\ Cs_2O \longrightarrow CA_2 + CsC_2A_4$

All the results are in good agreement with the diagram proposed by us in figure 2. The dashed lines between CsA and CsC_2A_4 are used to indicate that, although CsA is not stable at 900°C, CsA and CsC_2A_4 were found together at the beginning of the experiments. Cs_2O will evaporate out of CsA when heated for longer times in an open system, leaving Al_2O_3 and evaporated Cs_2O.

In the Al_2O_3-rich part of the diagram, some problems arose. The results of the CA_6 + Cs_2CO_3 experiment could not be understood on the basis of the diagram we constructed. When CA_6 was heated together with Cs_2CO_3, we found that an amorphous material had been formed. This experiment neither confirms nor contradicts the proposed diagram. This part of the diagram is not completely understood and is therefore shaded.

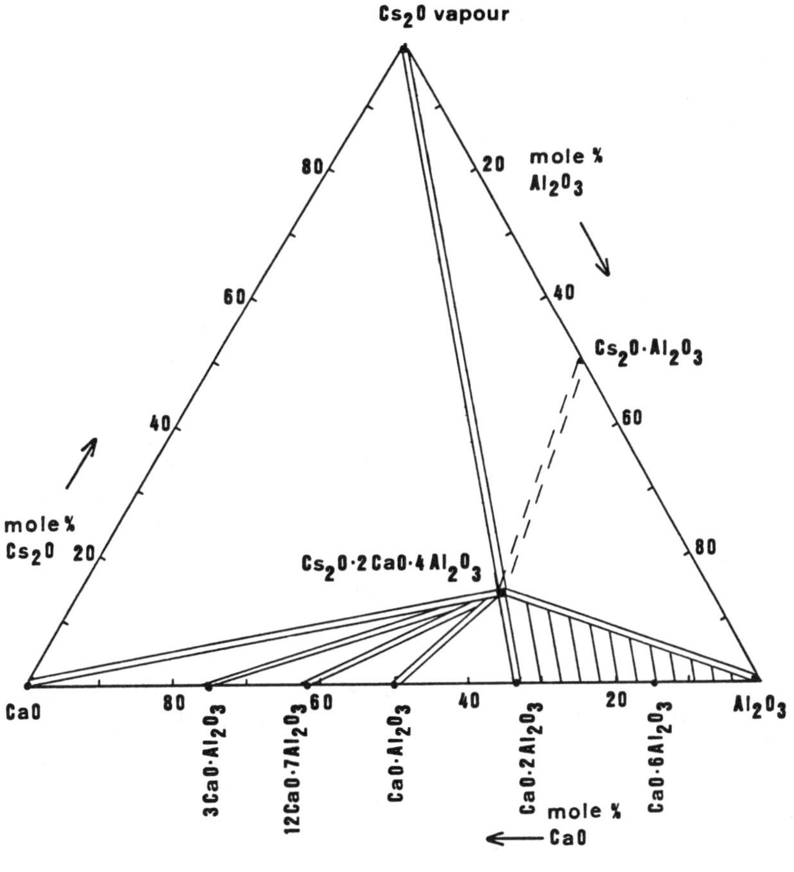

Figure 2. Reaction diagram at 900°C in the Cs_2O-CaO-Al_2O_3 system.

DISCUSSION

All the experimental results obtained during this study are in good agreement with the diagram proposed by us. In spite of this, we should also make some critical remarks. It has been reported that CsA_{11} can be formed after a year heating at 700°C and that this compound is stable up to 1050°C. We decided to leave it out of the diagram since the compound was only reported once and because it was never found during our own investigations.

The problem with CA_6 is well known. It is not formed at 900°C but it is believed to be stable at that temperature. It is known that CaO segregates on the grain boundaries of polycrystalline Al_2O_3 (5) probably leading to the formation of CA_6. This is the reason why it is important to know the behaviour of this compound when it is exposed to Cs_2O. For this reason we did not want to leave CA_6 out of the diagram, although we do not know the exact phase relations. For these reasons we prefer to call the diagram a reaction diagram instead of a phase diagram.

CONCLUSION

CaO segregates to the grain boundaries of Al_2O_3. As shown in the experiments, all the calcium aluminates react with Cs_2O to form CsC_2A_4. This means that even without knowing in which form (phase) CaO is present on the grain boundaries, we can predict from the diagram proposed by us that the grain boundaries will be corroded by Cs_2O. From these results one may conclude that it is better to avoid CaO as a sinter aid for alumina that is to be used in Cs containing atmospheres.

REFERENCES

1. Solov'eva T.V., Moroz I.Kh. and Vydrik G.A., "X-ray diffraction study of the synthesis of aluminum cesium disilicate", Russ. J. Inorg. Chem., 1970, **15**, 909-11.
2. Wagner P. and Coriell S.R., "High-temperature compatibility of cesium gas with some dielectrics", Rev. Sci. Instr., 1959, **30**, 937-8.
3. Hargreaves F., Mayo G.T.J. and Thomas A.G., "A study of the long-term compatibility of thermionic converter materials with caesium", J. of Nucl. Mater., 1966, **18**, 212-8.
4. Van Hoek J.A.M., Van Loo F.J.J., Wolff L.R. and Metselaar R., "Reactions in the CaO-Cs2O-Al2O3 system at 900°C", to be published.
5. Franken P.E.C. and Gehring A.P., "Grain boundary analysis of magnesia-doped alumina", J. Mater. Sci., 1981, **16**, 384-8.

THE $Y_3Al_5O_{12}$ REGION OF THE PHASE DIAGRAM $Y_2O_3 + Al_2O_3 + MgO$

M. Bolech, H.A.J. Oonk, A.J.H.M. Kock* and J.J.C. Oomen*.
Chemical Thermodynamics Group, University of Utrecht, Padualaan 8,
3584 CH UTRECHT; *Philips Lighting B.V., P.O. Box 80020,
5600 JM EINDHOVEN, The Netherlands

ABSTRACT

In order to obtain a better insight in the phase diagram of Yttria +Alumina +Magnesia results obtained with phase diagram analysis were combined with experimentally determined liquidus points. Since $Y_3Al_5O_{12}$ (YAG) is of interest for its application as optical ceramics, the region of the phase diagram near the YAG composition was investigated in detail. The calculated part of the phase diagram shows more detail than shown in the hitherto reported diagrams, while a good correspondence between calculated and experimental liquidus points was found.

INTRODUCTION

Yttrium aluminium garnet (YAG), sintered to translucency only a few years ago, appears to be a promising material for application as discharge envelope in high pressure sodium discharge lamps. While powder synthesis and subsequent powder processing were being developed, the knowledge of the phase diagrams of the systems $Y_2O_3 + Al_2O_3$ + sinter dope lagged behind. MgO being an effective sinter aid, a thermodynamic analysis was made of the system $Y_2O_3 + Al_2O_3 + MgO$, for compositions approximating the YAG composition. In the present analysis both literature data and new experimental liquidus data are incorporated.

AVAILABLE DATA

The binary phase diagram of $Y_2O_3 + Al_2O_3$ has been subject of a number of investigations [1]. Probably due to the experimentally difficult character of the system (high temperatures and three compounds), the various results do not agree in every respect. In particular the melting behaviour of the compound $YAlO_3$ is not well established. The ternary phase diagram of $Y_2O_3 + Al_2O_3 + MgO$ has been calculated by Udalov et al.[2].

OBJECTIVES

In order to obtain a better insight in the phase diagram of $Y_2O_3 + Al_2O_3 + MgO$, we decided to follow the combined approach of thermal analysis and thermodynamic phase-diagram analysis.

In this approach assessments are made of the phase diagrams of the binary subsystems, using phase-diagram data and the heats of melting of the three components and the various compounds. Subsequently, these assessments are used, in combination with a limited number of ternary liquidus data, to perform a computation of the ternary phase diagram. A full description of the applied procedures is beyond the scope of this paper, but can be found in Moerkens et al. [3]. Because our interest is mainly focused on YAG , the efforts were concentrated on the region of the phase diagram around YAG. In this paper, experimental efforts in order to measure the heat of melting of YAG are described and a number of experimentally determined ternary liquidus points are reported.

EXPERIMENTAL

Sample material was prepared by mixing $MgAl_2O_4$ and $Y_3Al_5O_{12}$ with either Al_2O_3 or Y_2O_3 powders. These powders had a purity of at least 99.9% and their average grain size was 4 μm. The powders were mixed in an aqueous suspension of pH = 3.5 (HNO_3), ultrasonically de-agglomerated and next stabilized by means of a 5 wt.% solution of methylcellulose to get a suspension containing 3 and 2.5 wt.% solid material and methylcellulose respectively. Subsequently the mixture was dried for three hours in air at 383 K.

Prior to liquidus measurements powdered sample material was pressed to rings, 3.5 mm in outer diameter, which were prefired at 1273 K in air to expell the binder. The liquidus measurements were made using an inductively coupled furnace with provisions for visual examination of the sample by means of an optical system (microscope Wild M650, CCD camera Philips LDH0600, TV-picture magnification 40-100). For each composition four measurements were made. At liquefaction the ring collapses to a drop-like shape. The thermocouple reading was interpolated to the readings of paladium (1821 K), alumina (2327 K) and polycrystalline YAG (2208 K). The experimental set-up is given in figure 1.

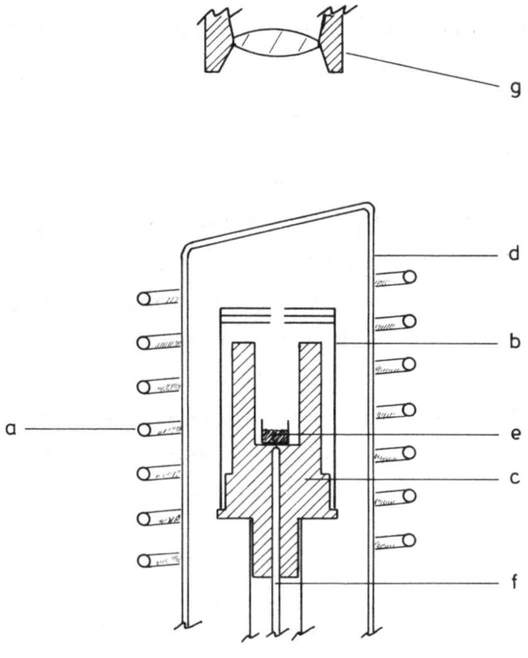

Figure 1 Cross-sectional image of the experimental set-up. **a)** primary coil (water cooled copper tubing) connected to a high frequency power source. **b)** tantalum heat shield (secundary coil). **c)** molybdenum body for temperature equalisation. **d)** quartz bell to contain an inert atmosphere. **e)** sample in molybdenum container. **f)** thermocouple. **g)** microscope and CCD camera.

To obtain an experimental value for the heat of melting of the compound $Y_3Al_5O_{12}$ some preliminary Differential Thermal Analysis (DTA) measurements were made. The measurements of the sample (cell loaded with YAG) and the reference (i.e. the empty cell) were performed in separate but identical heating cycles of this DTA-oven.

RESULTS

The above technique to determine the heat of melting of YAG appeared to be not suitable to measure a heat effect of this magnitude. Note that the heat effect is only some 20 J at a temperature of 2208 K. Therefore the thermodynamic phase-diagram analysis of the system $Y_2O_3 + Al_2O_3$ was performed with an estimated heat of melting of YAG, being 440 kJ mol^{-1}. With this value the experimental [1] and calculated liquidus points of the binary phase diagram of $Y_2O_3 + Al_2O_3$ showed the closest correspondence.

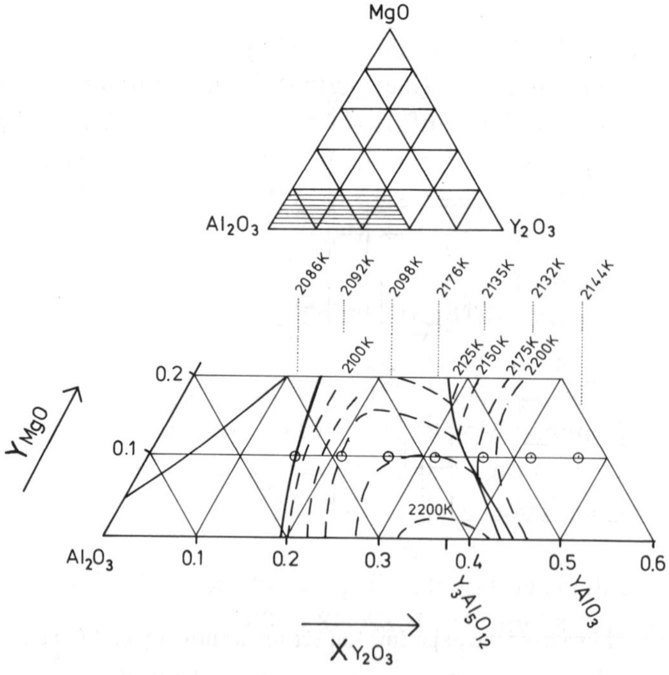

Figure 2 The region near YAG in the system Alumina-Yttria-Magnesia. The experimentally determined liquidus temperatures are indicated. Isothermal lines are represented as broken lines and eutectic lines are represented as heavy solid lines.

The subsystems Y_2O_3 + MgO and Al_2O_3 + MgO were subject of analysis too, and the results were combined to calculate the ternary phase diagram.

The calculated YAG region of the ternary phase diagram is presented in figure 2. The experimentally determined liquidus points are shown in this figure. The standard deviation in each of the measured liquidus temperatures is about 15 K. A limitation of the method used to determine the liquidus points -visual examination of the heated sample- is that the sample, when it is melting, may still contain a certain amount of solid matter. As a consequence of this heterogeneity on microscale, actual liquidus temperatures may be higher than the presently reported values indicate. Nonetheless the overall result -from a thermodynamic point of view- is satisfying. There is a good fit between the experimental and the calculated liquidus points. Compared with the work of Udalov [2], the presented part of the phase diagram shows more detail and has an improved link to physical reality.

CONCLUSIONS

The presented part of the ternary phase diagram Y_2O_3 + Al_2O_3 + MgO shows that the combined approach (thermodynamic phase diagram analysis and experimental work) can be fruitful. The quality of the calculated phase diagram is, however, governed by the quality of the thermodynamic input data. Further investigations are necessary to collect additional thermodynamic data for the system Y_2O_3 + Al_2O_3 + MgO.

REFERENCES

1) e.g. Cockayne, B., The uses and enigmas of the Al_2O_3-Y_2O_3 phase diagram. J. Less Common Metals ,1985, **114**, 199-206.
or Adylov, G.T., Voronov, G.V., Mansurova, E.P., Sigalov, L.M. and Urazaeva, E.M., Zh. Neorg. Khim., 1988, **33**, 1867-1869.

2) Udalov, Y.P., Calculated liquidus diagrams of the A12O3-Ln2O3-MO Systems, Zh. Neorg. Khim., 1977, **22**, 483-489.

3) Moerkens, R., Bouwstra, J.A. and Oonk, H.A.J., Thermodynamic assessment of binary data and calculation of ternary equilibrium, Calphad, 1983, **7**, 219-269.

The Aluminum Mono-Oxycarbide Al_2OC in the System Si-C-Al-O-N: some thermodynamic properties

J.-M. LIHRMANN, CNRS-LIMHP, Université Paris-Nord, 4 Av. J.B.Clément, F-93430 VILLETANEUSE; T. ZAMBETAKIS and M. DAIRE, EHICS, Département Science des Matériaux, BP 296 F-67008 STRASBOURG.

Abstract

Inside the quinary system Si-C-Al-O-N or Si,Al/C,N,O the high temperature behavior of the aluminum mono-oxycarbide Al_2OC is described in terms of its thermodynamic stability as a pure component and as a solid solution with aluminum nitride. It is shown that the solid solution $(Al_2OC)_{1-x}$-$(AlN)_x$ is always more stable than pure Al_2OC, whether stable or metastable equilibrium may occur. In view of complementary data recently published in the literature, a part of the phase equilibrium diagram of the pseudo-binary system AI_2OC-AlN is tentatively suggested.

INTRODUCTION.

The acronym "SiCAlON" has been first employed in 1978 by I.B. Cutler et al. (1) to characterize five component phases encountered in the quinary system Al,Si/C,N,O. Particular emphasis was given to materials with the hexagonal (2H) wurtzite crystal structure, and to probable combinations based upon them.
In this context silicon carbide (SiC) and aluminum nitride (AlN) synthetically processed into an intimate mixture at 1600°C were reported to form a continuous solid solution in the range 2 mol% AlN to 100 mol% AlN; as a similar solubility was mentioned for aluminum mono-oxycarbide (Al_2OC) in SiC, the article concluded with the probable existence of 5 component Si-C-Al-O-N phases that might be, among others, ternary solid solutions between Al_2OC, AlN and SiC.
An isothermal section of the quinary system Al,Si/C,N,O can be represented by a regular trigonal bi-pyramid with equal sides, as shown in Fig.1, where only condensed phases are taken into account. When equivalent-%

concentrations are used (e.g.Refs 2-3), the binary carbides, nitrides and oxides of aluminum and silicon turn out to be located at reciprocal corners of a regular prism with equal sides, known as Janecke concentration prism (Fig.1b); the latter encompasses all intermediate compounds as long as the valency states of the 5 base-elements remain constant.

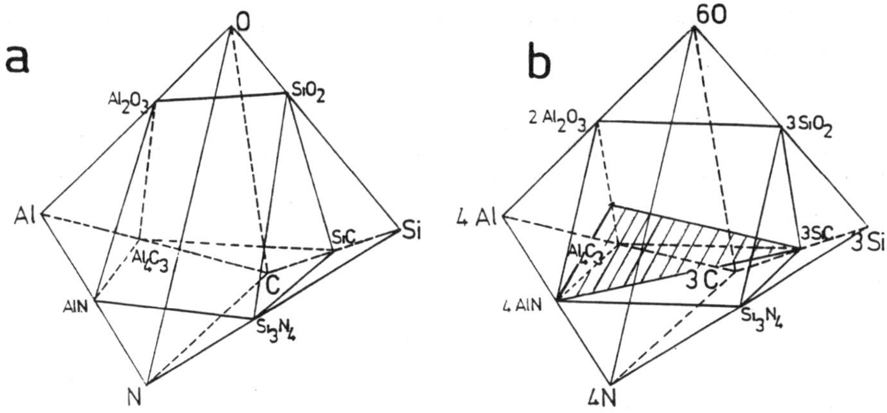

Fig.1 - Representation of the system Si,Al/C,N,O
a. in atom -% b. in equivalent -%

The system Al_2OC-AlN-SiC is hatched on Fig.1b. Materials generated within this triangle could possibly combine high mecanical properties of SiC and high functional properties of AlN (high thermal-low electrical conductivity), and the binary system AlN-SiC has been first investigated in this purpose (4-6). Despite excellent properties AlN-SiC materials require hot-pressing at quite high temperatures (up to 2300°) to densify and as such they appear technically limited, at least at a reasonable cost.

In contrast to aluminum nitride, the addition of aluminum mono-oxycarbide, Al_2OC to SiC (7) permits obtention of dense materials by pressureless sintering at reasonable temperatures (1850°C-1950°C), thanks to the eutectic liquid occurring in the system Al_2O_3-Al_4C_3 at 1840°C for the composition 86 mol% Al_2O_3- 14 mol% Al_4C_3 (8).

In this regard, the incorporation of Al_2OC to AlN-SiC materials may be interesting from a technological as well as from a fundamental point of view, since it also provides an opportunity to study the reactivity of some typical iono-covalent materials. To date, a small number of high temperature, isothermal sections are partially known in the pseudo-binary diagrams Al_2OC-SiC (7) and Al_2OC-AlN (9-11), and to the best of our knowledge a single composition has been tested within the system Al_2OC

-AlN-SiC (12).

The understanding of Al_2OC has long remained insufficient in that existing phase diagrams of the system Al_2O_3-Al_4C_3 were contradicting each other, and thermodynamic data on Al_2OC, inconsistent (13).

In previous studies (9, 14) we have attempted to clarify the high temperature behavior of Al_2OC. This compound has been found to undergo 2 solid state phase transformations in the system Al_2O_3-Al_4C_3, the one in stable equilibrium conditions and the other, in metastable equilibrium conditions (9). The purpose of this paper was to study more precisely the phase field of the solid solution $(Al_2OC)_{1-x}$-$(AlN)_x$ and to compare its high temperature behavior with the one of pure Al_2OC.

EXPERIMENTAL

Commercial powders of α-aluminum oxide(Alcoa A-16 SG) , aluminum carbide (Sofrem : Société Française d'Electrométallurgie) and aluminum nitride (Aluminium Pechiney) (Table I) weighed in appropriate amounts were dry-homogenized by vibratory milling, ball-milled in absolute ethanol and after drying were cold-pressed under 5 T/ cm^2 and without binder into cylindrical pellets 18 mm diameter X 25 mm height.

Green samples were heated in graphite crucibles in a graphite resistance furnace (Cristal 3000,La Physique Appliquée Industrielle) filled with Helium at a pressure very slightly exceeding 1 atmosphere. Beyond 1400°C heating was automatically conducted by a P.I.D. programmer ; the temperature was controlled by a pyrometer connected to the programmer, and was monitored by an additional optical pyrometer on top of the furnace.

The compositions discussed in this paper are listed in Table II. A first series (samples 1 to 10) were selected along the tie-line joining pure Al_2OC and pure AlN , and corresponding powder mixtures were prepared by addition of appropriate amounts of AlN to the binary composition 50 mol% Al_2O_3- 50 mol% Al_4C_3. After heating at 1980°C for 20 minutes they were furnace-cooled to room temperature; the cooling rates were of the order of 40°C/minute down to 1000°C and 10°C/minute below 1000°C. Three of these samples (with contents 5, 15 and 20 mol% AlN respectively) were further heat-trated at 1600°C for 20 minutes before cooling to room-temperature occurred. A second series of compositions (samples 11 and 12) were chosen in the aluminum oxide- rich corner of the diagram Al_2O_3-Al_4C_3-AlN.

After heat-treating and cooling, surface layers were ground off the samples on a depth of several millimeters, and central parts were either crushed into powder for X-Ray analysis or embedded into a resin and prepared for optical and scanning electron microscopy. Unless specified chemical etching was unnecessary to identify the different phases.

Table I Characterization of raw materials

	Purity	grain size from cumulative curve		
		d(10%)	d(50%)	d(90%)
Al_2O_3	0.1% max. Na_2O 0.07% max. SiO_2 0.03% max. Fe_2O_3	0.1	0.7	2.5
AlN	99.5%	0.2	1.0	4.0
Al_4C_3	/	18	50	120

Table II Compositions (in mol%) of powder mixtures, as-weighed

sample Nr.	composition (mol%)		
	Al_2O_3	Al_4C_3	AlN
1	47.5	47.5	05
2	45	45	10
3	42.5	42.5	15
4	40	40	20
5	35	35	30
6	30	30	40
7	25	25	50
8	20	20	60
9	15	15	70
10	10	10	80
11	81	13	06
12	83	11	06

RESULTS.

A. The solid solution $(Al_2OC)_{1-x}-(AlN)_x$ along the tie-line Al_2OC-AlN .

A1. At 1980°C.

Samples 1 to 10 exposed to 1980°C for 20 minutes and furnace-cooled were crushed into powder; X-Ray analysis using Cu-Kα radiation revealed the presence of solid solution $(Al_2OC)_{1-x}-(AlN)_x$ in all of them, with volume of lattice cell varying vs AlN content, as shown in Fig.2.

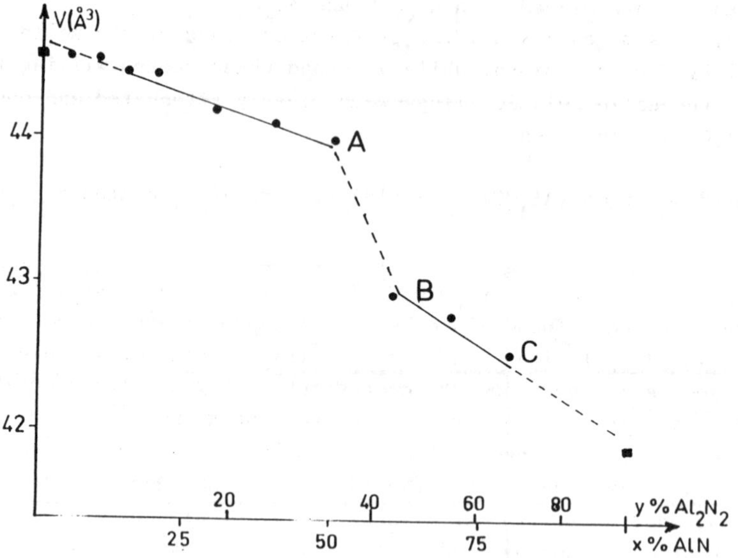

Fig.2 - Variation of the unit cell volume of solid solution $(Al_2OC)_{1-x}-(AlN)_x$ versus AlN content

In the range 15 mol% AlN - 75 mol% AlN the solid solution was the only phase detected by X-Ray Analysis. For lower AlN contents the four major lines of aluminum tetraoxycarbide Al_4O_4C could be distinguished at 5 mol% AlN, barely guessed at 10 mol% AlN and were absent above 10 mol% AlN, whereas traces of residual aluminum carbide were still present in sample 1. For AlN contents ⩾ 80 mol% , AlN no longer dissolved in the solid solution and it appeared as a separate phase.

For comparison purposes sample B1 (comp. 55 mol% Al_2O_3- 45 mol% Al_4O_4C-

0 mol% AlN) heated to 1980°C for 20 minutes and furnace-cooled was found to contain high amounts of pure Al_2OC and minor amounts of Al_4O_4C .

A2. At 1600°C.

After samples 1,3,4 and B1 (and only these) had reacted at 1980°C for 20 minutes, the temperature of the furnace was lowered to 1600°C and maintained at that temperature for 20 minutes before final cooling occurred. X-Ray analysis clearly revealed (9) that the solid solution $(Al_2OC)_{1-x}-(AlN)_x$ was still present in samples 1,3 and 4 after heat-treatment at 1600°C, whereas pure Al_2OC had all but disappeared in sample B1 and been replaced with Al_4O_4C and Al_4C_3.
Furthermore, in samples 3 and 4 the diffraction lines of the solid solution were still by far predominant while a second phase seemed willing to form. In sample 1 the solid solution lines were clearly attenuated and conversely those of Al_4O_4C, reinforced.

B. The solid solution $(Al_2OC)_{1-x}-(AlN)_x$ at high Al_2O_3 contents (≥ 80 mol%).

Samples 11 and 12 heated at 1970°C for 8 minutes and furnace-cooled to room-temperature were found to consist of 2 predominant phases: aluminum oxide and solid solution ($(Al_2OC)_{1-x}-(AlN)_x$, both of which exist in 2 different aspects as shown on the micrograph in Fig.3. The microstructure essentially consists in a primary phase appearing as white dendrites embedded in a fine-grained eutectic matrix. Semi-quantitative electron microprobe analysis identified the dendrites as consisting of solid solution, whereas the 2 light-grey, apparently out-of-equilibrium, acicular grains consist of aluminum oxide. The eutectic background is a very fine mixture of both Al_2O_3 and $(Al_2OC)_{1-x}-(AlN)_x$. The presence of dendrites attests cooling from a melt as was confirmed by thermal analysis recordings.
After annealing both samples at 1600°C for 65 minutes, the diffraction lines of the solid solution were only very slightly attenuated whereas small amounts of Al_4O_4C had begun to form; aluminum oxide was essentially unmodified. For comparison purposes sample B2 (composition 82 mol% Al_2O_3- 18 mol% Al_4C_3- 0 mol% AlN) was exposed to a similar thermal treatment . After exposure to 1970°C for 8 minutes it consisted of Al_2OC dendrites embedded in a matrix of Al_2O_3- Al_2OC eutectic; at the end of the annealing treatment at 1680°C Al_2OC was no longer present in the sample,having been replaced with Al_4O_4C and Al_2O_3.

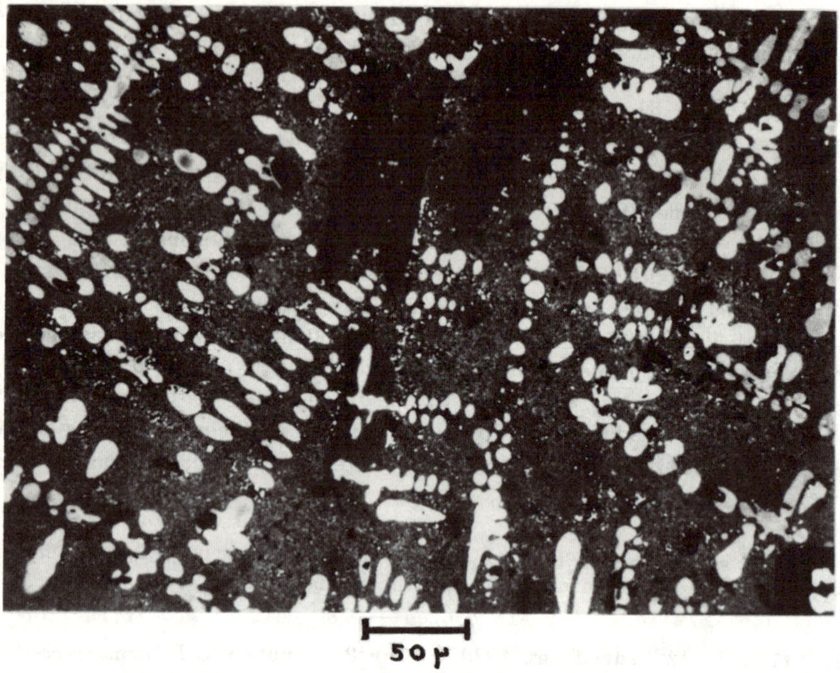

Fig.3 - Typical microstructures of samples 11 and 12 (composition of specimen shown is 89.5 mol% Al_2O_3 - 5.5 mol% Al_4C_3 - 5.0 mol% AlN)

DISCUSSION.

A. The solid solution $(Al_2OC)_{1-x}$-$(AlN)_x$ along the tie-line Al_2OC-AlN

A1. At 1980°C

Thermal analysis recordings that accompanied heating and cooling of samples 1 to 10 showed no liquid phase up to 1980°C, and micrographs indicated no trace of a transient liquid phase . It is therefore important to note that the solid solution $(Al_2OC)_{1-x}$-$(AlN)_x$ can be obtained by the heating and solid state sintering of powdered solid components. Equimolar mixtures of Al_2O_3 and Al_4C_3 are believed to form at first Al_2OC by the solid state reaction $Al_2O_3 + Al_4C_3 \rightarrow 3\ Al_2OC$, the latter being activated by the presence of AlN, a compound isostructural with Al_2OC ; then Al_2OC and AlN react to form the solid solution shown in Fig.2. All diffraction lines of this solid solution were quite satisfactorily indexed in terms of the normal wurtzite (2 H) symmetry.

It is also worthwhile noticing that the solid solution appears to be the only phase present at 1980°C in the range 15 mol% AlN- 75 mol% AlN. Fig.2 shows an abnormal variation (2.2%) of the volume of the unit cell of this phase between 50 mol % AlN and 60 mol% AlN, accounting for an abnormal variation of lattice parameters a and c within this range. This can be seen as characteristic of an ordering phenomenon leading to the stabilization of a new entity, which may be either Al_3CON, a wurtzite-structure component of composition 50 mol% Al_2OC - 50 mol% AlN (1), or $Al_{10}O_3C_3N_4$ of composition 42.86 mol% Al_2OC - 57.14 mol% AlN (9).In either case this intermediate compound is able to dissolve important amounts of both Al_2OC and AlN, since its solubility range at 1980°C approximately extends from 15 mol% AlN to 75 mol% AlN.

For AlN contents $\leqslant$ 15 mol% , the solid solution no longer exists as a single phase, since at least Al_4O_4C has been detected by X-Ray diffraction; this result is in agreement with the incongruent melting character of pure Al_2OC (8,15). Similarly for AlN contents $\geqslant$ 80 mol% , the diffraction lines of aluminum nitride clearly separate from those of the solid solution, indicating that AlN no longer dissolves in its lattice. It is therefore believed that the solid solution $(Al_2OC)_{1-x}$-$(AlN)_x$ is not complete between its end-members at 1980°C.

A2. At 1600°C

We have shown in a recent article (9) that the stable phase equilibrium diagram of the system Al_2O_3-Al_4C_3 initially proposed by Foster et al. (8) ought to be revised owing to the decomposition of Al_2OC below 1715°C according to the balanced reaction 4 $Al_2OC \rightarrow Al_4O_4C + Al_4C_3$ [1].

In this context it is expected that sample B1 annealed at 1600°C after heat-treating at 1980°C undergoes reaction [1], as is indeed the case. (It should suffice to remind that merely annealing sample B1 at 1700°C for 15 minutes guarantees the total decomposition of Al_2OC following reaction [1] (9,14)). Conversely samples 1, 3 and 4 heat-treated at 1600°C for 20 minutes are barely affected by reaction[1], clearly showing evidence of the stabilization of the solid solution $(Al_2OC)_{1-x}$-$(AlN)_x$ compared to pure Al_2OC.

A3. Tentative phase equilibrium diagram Al_2OC-AlN

Recently published data (11) on the tie-line Al_2OC-AlN mentions demixtion of the solid solution $(Al_2OC)_{1-x}$-$(AlN)_x$ below about 1800°C in the central part of the section Al_2OC-AlN. Additionally, pure solid solutions have been observed at 1950°C for 40 mol% AlN, and at 1900°C for 60 mol% AlN. In agreement with our results these data suggest the partial, stable phase equilibrium diagram of the pseudo-binary system Al_2OC-AlN shown in Fig.4. It should be stressed that this phase diagram is both incomplete (since our interest was limited to pure solid solution $(Al_2OC)_{1-x}$-$(AlN)_x$) and only tentative in that much more data would be necessary to ascertain it.

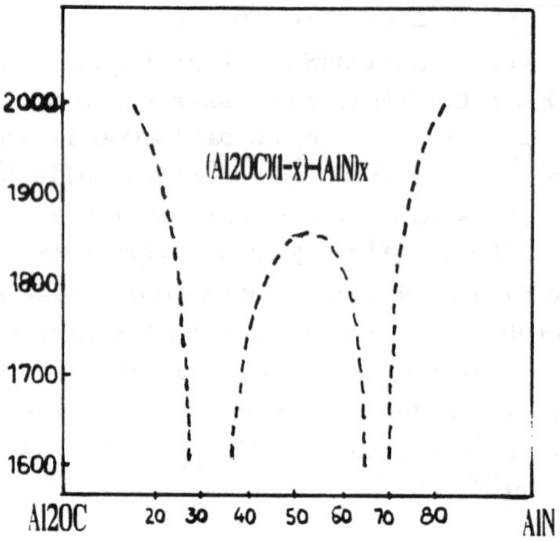

Fig.4 - Phase field of solid solution $(Al_2OC)_{1-x}$ - $(AlN)_x$ in the quasi-binary system Al_2OC - AlN. Demixtion is after Ref. 11

B. Compositions with high Al_2O_3 contents (≥ 80 mol%)

Compositions 11 and 12 were chosen after preliminary experiments performed along a line joining AlN and binary point B3 (of composition 86 mol% Al_2O_3 - 14 mol% Al_4C_3) had shown that $(Al_2OC)_{1-x}$-$(AlN)_x$ only formed for AlN contents ≤ 15 mol%, and that oxynitride phases started to form above 10 mol% AlN . Furthermore the proximity of points 11 and 12 with the eutectic liquid from the system Al_2O_3-Al_4C_3 was expected to promote densification (7). In this regard thermal analysis recordings have indeed indicated that a liquid phase formed at temperatures of 1875°C ± 20°C (sample 11) and 1890°C ± 20°C (sample 12). However the kinetics of formation of dendritic phase $(Al_2OC)_{1-x}$-$(AlN)_x$ appeared to be very slow below 1930°C, so that samples 11 and 12 have been heated to 1970°C for 8 minutes. (It may be worthwhile mentioning that sample 12 heated at 1910°C for 5 minutes contained no traces of solid solution but essentially Al_4O_4C as well as γ - and δ - oxynitride phases).

As regards the behavior of samples 11, 12 and B2, we have shown in a previous article (14) microstructural evidence for a metastable diagram in the system Al_2O_3-Al_4C_3, and justified its existence (9) by T.T.T. curves illustrating solid state reaction

$$Al_2O_3 + x\ Al_2OC \rightarrow (1-x)\ Al_2O_3 + Al_4O_4C \quad [2]$$

In view of these curves it is expected that Al_2OC in sample B2 completely transforms into Al_2O_3 and Al_4O_4C after 22 minutes at 1680°C. The reaction observed for sample B2 after 65 minutes at 1680°C is thus quite normal. On the other hand samples 11 and 12 which contained solid solution $(Al_2OC)_{1-x}$ - $(AlN)_x$ instead of pure AlN had barely started to undergo reaction [2] after 65 minutes at 1680°C: this result again ascertains that the solid solution $(Al_2OC)_{1-x}$-$(AlN)_x$ is less metastable than Al_2OC, in that the presence of AlN sustantially shifts toward the right the T.T.T. curves observed in the same conditions in the binary system Al_2O_3-Al_4C_3 (Fig.5).

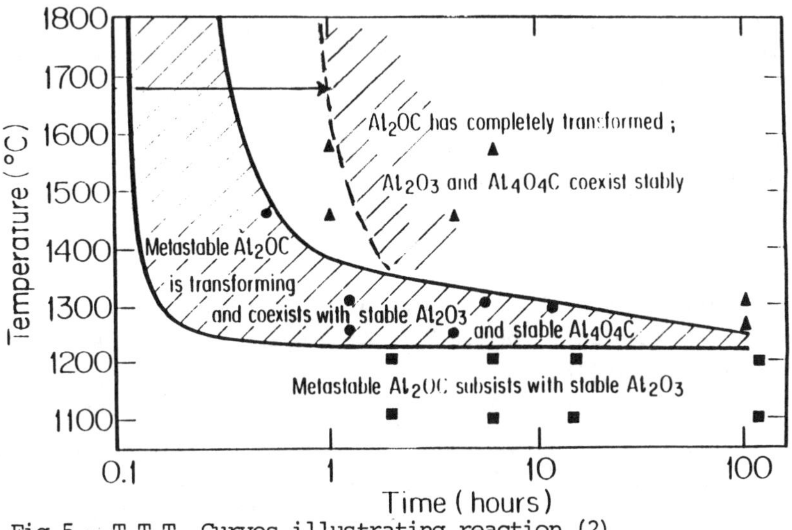

Fig.5 - T.T.T. Curves illustrating reaction (2) .
——— in the metastable system Al_2O_3 - Al_4C_3 at high Al_2O_3 contents (≥ 80 mol%)
– – – – with $(Al_2OC)_{1-x}$ - $(AlN)_x$ instead of Al_2OC, in the ternary system Al_2O_3 - Al_4C_3 - AlN at high Al_2O_3 contents (≥ 80 mol%)

Additional evidence to this stabilizing effect was given by microscopic observations of sample 11 after (fast) furnace-cooling and after slow cooling from 1970°C : dendrites of $(Al_2OC)_{1-x}$ -$(AlN)_x$ were equally present in both cases.
In contrast to this experiment the presence of Al_2OC dendrites in the metastable diagram Al_2O_3-Al_4C_3 and at high Al_2O_3 contents (⩾ 80 mol%) was only observed at room temperature after fast cooling (9,14).

SUMMARY AND CONCLUSIONS

The tie-line joining Al_2OC and AlN has been extensively studied at 1980°C. Within the range 15 mol% AlN- 75 mol% AlN, $(Al_2OC)_{1-x}$-$(AlN)_x$ exists as a single,2H wurtzite phase which extends about either Al_3CON or $Al_{10}C_3O_3N_4$. It is believed that the solid solution is incomplete at 1980°C because aluminum nitride starts to separate at 80 mol% AlN. Toward the other end, the solid solution may extend to pure Al_2OC but does not exist as a single phase due to the incongruent melting character of Al_2OC. This solid solution is interesting from the fundamental standpoint that it was obtained at 1980°C by pressureless, exclusively solid state sintering of powdered solid components.
The solid solution was also observed to form, but no longer as a single phase, at high Al_2O_3 contents.
In both cases, $(Al_2OC)_{1-x}$-$(AlN)_x$ offered the advantage of being thermodynamically more stable than pure Al_2OC, whether stable or metastable equilibrium may have occurred. Therefore on a thermodynamic point of view , dissolving AlN into Al_2OC has a beneficial effect on the lattice of aluminum oxycarbide. A later paper will describe some mechanical properties of Al_2OC and $(Al_2OC)_{1-x}$-$(AlN)_x$.

REFERENCES

1. I.B.Cutler, P.D.Miller, W.Rafaniello, H.K.Park, D.P.Thompson, and K.H.Jack, "New Materials in the Si-C-Al-O-N and Related Systems", Nature, 275 (5679) 434-35 (1978)

2. K.H.Jack, "Review. Sialons and Related Nitrogen Ceramics", J.Mater.Sci.,11, 1135-58 (1976)

3. G.Schneider, L.J.Gauckler, and G.Petzow, "Phase Equilibria in the Si,Al,Be/C,N System", in Energy and Ceramics pp 399-408, Ed. Vincenzini, Publ. Elsevier, Amsterdam, The Netherlands (1980)

4. W.Rafaniello, M.R.Plichta, and A.V.Virkar, "Investigation of Phase Stability in the System SiC-AlN", J.Am.Ceram.Soc., 66(4) 272-76 (1983)

5. R.Ruh and A.Zangvil, "Compositions and Properties of Hot-Pressed SiC-AlN Solid Solutions, J.Am.Ceram.Soc., 65(5) 260-65 (1982)
6. S.Y.Kuo, A.V.Virkar, and W.Rafaniello, "Modulated Structures in SiC-AlN Ceramics", J.Am.Ceram.Soc., 70(6) C125-C128 (1987)
7. J.L.Huang, A.C.Hurford, R.A.Cutler, and A.V.Virkar, "Sintering Behavior and Properties of SiCAlON Ceramics", J.Mater.Sci., 21(4) 1448-56 (1986)
8. L.M.Foster, G.Long, and M.S.Hunter, "Reactions between Aluminum Oxide and Carbon. The Al_2O_3- Al_4C_3 Phase Diagram", J.Am.Ceram.Soc., 39(1) 1-11 (1956)
9. J.M.Lihrmann, T.Zambetakis, and M.Daire, "High Temperature Behavior of the Aluminum Mono-Oxycarbide Al_2OC in the System Al_2O_3- Al_4C_3 and with Additions of Aluminum Nitride", J.Am.Ceram.Soc., xxxx (1989)
10. J.L.Henry, J.H.Russel, and H.J.Kelly, "The System Al_4C_3 -AlN - Al_2O_3. Powder Forming and Sintering Behavior, Phase Identification and Refractory Composition Properties", R.I.7320, Bureau of Mines, 24p. (1969)
11. S.Y.Kuo and A.V.Virkar, "Phase Equilibria and Phase Transformation in the Aluminum Nitride- Aluminum Oxycarbide Pseudobinary System", J.Am.Ceram.Soc., 72(4), 540-50 (1989)
12. S.Y.Kuo, Z.C.Jou, A.V.Virkar, and W.Fafaniello, "Fabrication, Thermal Treatment and Microstructure Development in SiC-AlN-Al_2OC Ceramics", J.Mater.Sci., 21, 3019-24 (1986)
13. see Ref.9,Introduction
14. Y.Larrere, B.Willer, J.M.Lihrmann, and M.Daire, "Stable and Metastable Phase Equilibrium Diagrams of the Binary System Al_2O_3 - Al_4C_3" (in Fr.), Rev. Int. Hautes Tempér. Réfract., Fr., 21, 3-18 (1984)
15. Prof.C.Petot, private communication, Université Paris-Nord (1989)
Louis Pasteur, Strasbourg (1985)

A PRELIMINARY STUDY ON CuO-BASED CERAMICS

A. DE PRETIS, D. MINICHELLI
Istituto Chimico, Università di Udine, Udine, Italy
L. PODDA
Istituto di Chimica Applicata, Università di Trieste, Trieste, Italy
F. RICCIARDIELLO
Dipartimento di Ingegneria Chimica dei Processi e dei Materiali
Università di Palermo, Italy

ABSTRACT

Several ternary phase diagrams based on high melting oxides, alkaline and earth-alkaline oxides and copper oxide were investigated. The considered oxide were: CeO_2, BaO, SrO, CaO, K_2O and CuO. The equilibrium phases were investigated by means of X-ray diffraction, thermal analyses and optical microscopy techniques in the temperature range 750-950°C.

INTRODUCTION

Interest in the search for compounds based on copper oxides suitable for non-traditional applications prompted an investigation on ternary phase diagrams based on high melting oxides, alkaline and earth-alkaline oxides with copper oxide. As some superconductors are CuO-based ceramics and literature does not report ternary phase diagrams in this field, the present investigation was undertaken.

The following phase diagrams: BaO-CeO_2-CuO, SrO-CeO_2-CuO, CaO-CeO_2-CuO and K_2O-CeO_2-CuO gave the more interesting results. In this field the literature reports data on the binary diagrams CeO_2-CaO, CeO_2-SrO, CeO_2-BaO (1,2,3,4) and CuO-CaO (5). No data appeared on the related ternary diagrams.

MATERIALS AND METHODS

The materials used are CeO_2, $CaCO_3$, $SrCO_3$ and $BaCO_3$ (all 99.9 %

purity) and $CuCO_3$ (99.5 % pure).
The weighted stechiometric mixtures were homogenized in agate mortars and in agate ball mills. The powders were fired in platinum crucibles at 950°C for 50 hours, then pressed into disks under 300 N/mm^2 and fired in oxidizing atmosphere (air) for different times at different temperatures in order to ensure the completation of the reaction. These specimens were subjected to X-ray diffraction analyses, optical microscopy investigation and thermal analyses (TGA, DTG and DTA).

TABLE 1
Composition of the sample tested (mol %)

Sample	CeO_2%	CuO%	CaO%	SrO%	BaO%	K_2O%
1	28.57	42.86			28.57	
7	16.67	50.00			33.33	
8	18.18	54.55			27.27	
31	30.00	50.00			20.00	
2	28.57	42.86		28.57		
9	16.67	50.00		33.33		
10	18.18	54.55		27.27		
32	30.00	50.00		20.00		
3	28.57	42.86	28.57			
11	16.67	50.00	33.33			
12	18.18	54.55	27.27			
6	33.33	50.00				16.67
16	16.67	50.00				33.33
17	18.18	54.55				27.27

RESULTS AND DISCUSSION

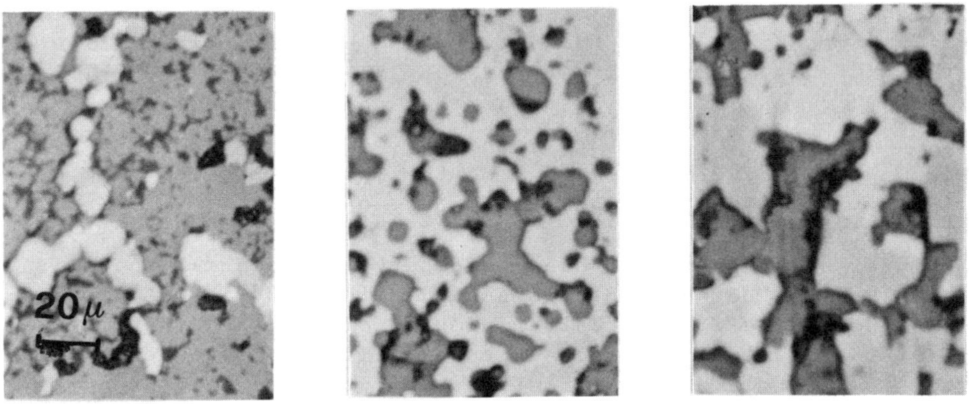

Figure 1 - Photomicrograph of the sample 1,7 and 8 at 950°C.

<u>System BaO-CeO_2-CuO</u> : For all the investigated compositions three phases were detected by microscopy as can be seen in Fig.1. These three phases can be indicated, after evaluation of XRD diagrams as $BaCeO_3$, a solid solution of CeO_2 in CuO and a compound belonging to the binary system CuO-BaO.
In Fig. 2a is reported the most probable ternary phase diagram.

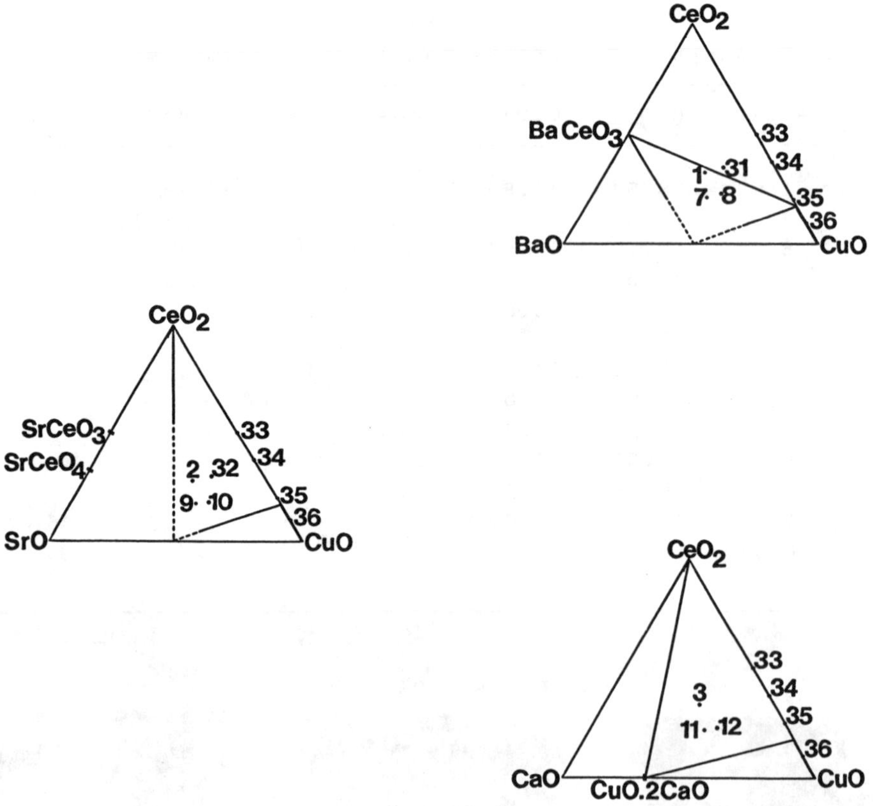

Figure 2 - Isothermal sections of the Systems BaO-CeO_2-CuO, SrO-CeO_2-CuO and CaO-CeO_2-CuO at 950°C.

An interesting phenomenon can be observed by thermal analysis data (Fig.3). Between the two endothermic peaks in DTA, corresponding to the loss of carbon dioxide of $CuCO_3$ (ca. 400°C) and of $BaCO_3$ (ca. 1000°C), another endothermic peak was detected : this is due to a melting process which promotes degasation of $BaCO_3$ at low temperature

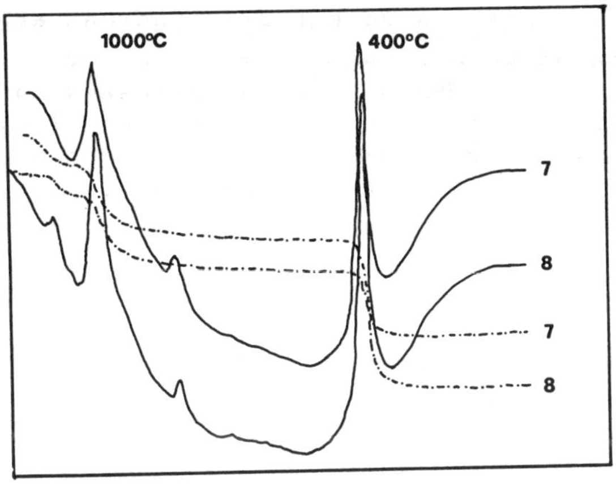

Figure 3 - TG (dotted line) and DTA curves for the samples 7,8.

System SrO-CeO_2-CuO : For all the investigated compositions three phases were detected by optical microscopy.(Fig. 4). These three phases can be indicated, after evaluation of XRD diagrams as CeO_2, a compound belonging to the SrO-CuO system and the same solid solution detected in the previous ternary system on the CuO-CeO_2 system. In figure 2b is reported the most probable ternary phase system. The behaviuor of DTA diagrams is similar to the one discovered in the previous system with the presence of a smaller endothermic melting peak (Fig. 5).

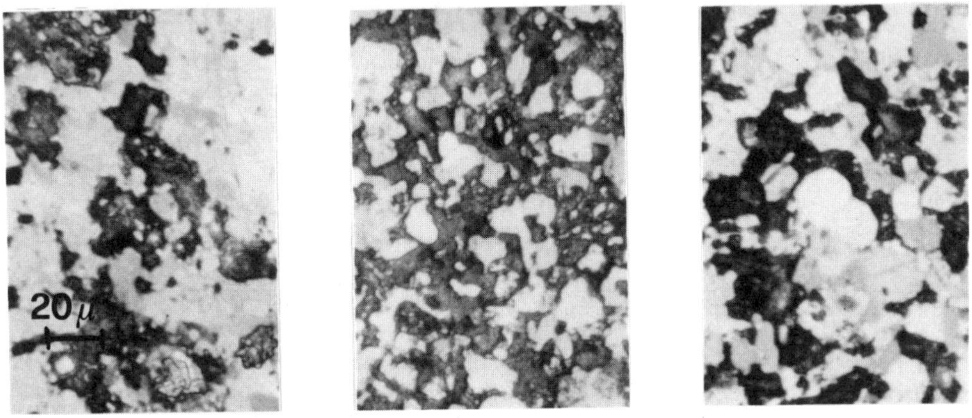

Figure 4 - Photomicrograph of the samples 2, 9 and 10.

System CaO-CeO_2-CuO : For all the investigated compositions three phases were detected by optical microscopy; after evaluation of XRD diagrams these phases can be indicated as CeO_2, the same solid solution in the system CuO-CeO_2 and the compound CuO.2CaO. In figure 2c is reported the most probable ternary phase diagram. A very small endothermic peak between the degasation peaks was detected in thermal analysis data (F.6)

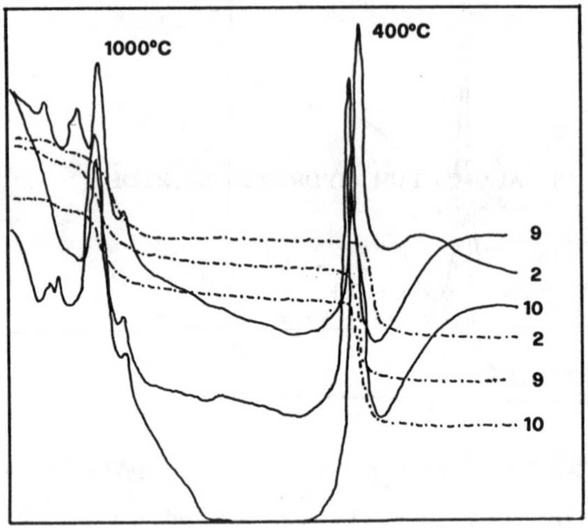

Figure 5 - TG (dotted line) and DTA curves for the samples 2, 9 and 10.

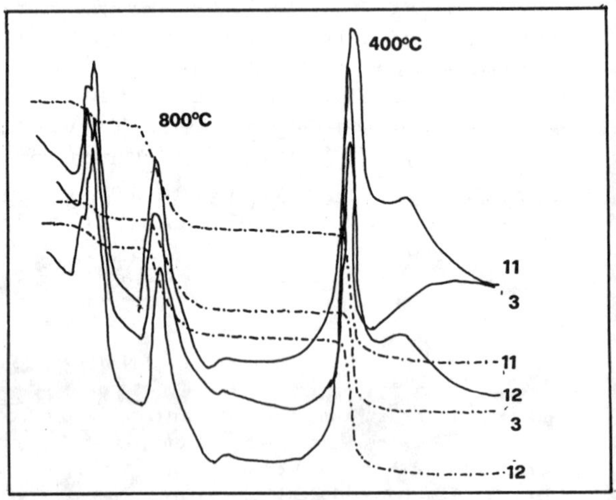

Figure 6 - TG (dotted line) and DTA curves for the samples 3, 11 and 12.

System K_2O-CeO_2-CuO : The presence of liquid phase was detected in the temperature range 850-950°C.

REFERENCES

1 .Longo,V.,Minichelli,D.and Ricciardiello,F.,Science of Ceramics Vol.11 ,1981 ,pp 171-6.

2. Longo,V.,Podda,L., Ceramurgia 1, 1971, 11.

3. Longo,V.,Podda,L., Ceramurgia 1, 1971, 83.

4. Danelon,M.,Barbariol,I.,Cocco,A., Annali Chimica 59,1969,465.

5. Gadalla,A.M.M.,White,J., Trans.Brit.Ceram.Soc.,65,1966, 185.

THERMAL DECOMPOSITION OF ALUMINIUM HYDROXYCHLORIDE

MAŁGORZATA SOPICKA-LIZER
Institute of Materials Science and Engineering,
Graniczna 16, 40-019 Katowice, Poland

ABSTRACT

Aluminium hydroxychloride has been obtained by mean of Al solvation in its chloride and solidification at room temperature. Its thermal decomposition have been studied using DTA, FT-IR, XRD and TEM methods. It has been found that decomposition took place in two main stages followed succesively. The first resulted in removal of coordination water from Al-13. Decomposition of Al-13 ion occured below 400°C however it has not been completed below 700°C. The porous, amorphous product of decomposition crystallised in the form of nanocrystalline γ-Al_2O_3 and then transformed into larger grains of α-Al_2O_3 with low angle boundaries and open pores network.

INTRODUCTION

Basic aluminium salts are of interest as precursor materials for the formation of aluminium oxide. Alumina obtained by the monolithic transformation from boehmite gels at low temperatures (below 1200°C) has been reported earlier [1]. However transformation produces large "wormy" grains enclosing a high proportion of intraganular pores, the elimination of which does not occur below 1600°C. Nonetheless an interesting results by M.D. Sacks, Tseung-Yuen Tseng and Soo Young Lee [2] indicate that densification at 1200°C is possible by alumina transformation from hydrated basic aluminium sulphate. Aluminium hydroxychloride can act in a similar way since it is known as a main component of sol-gel derived alumina fibres [3]. The process of hydrolysis in aluminium chloride solution leads to formation of several polymeric species with $[Al_{13}O_4(OH)_{24}(H_2O)_{12}]^{7}$ ion as a main component in the solution of 2-2.5 neutralisation ratio [4]. However, there are still some controversies in its exact formula [5]. The aim of the current investigation is to characterize the solidified hydroxychloride and the resulting products formed during calcination at temperatures up to 1300°C.

MATERIALS AND METHODS

Analytical reagent-grade chemicals and distiled water were used throughout the process. Aluminium powder was dissolved in 0.3 M solution of $AlCl_3$ with 2.5 neutralisation ratio at 70-75°C during 18 hrs. This solution was then evaporated, solidified at room temperature and ground. Resulted powders were calcined in the static air atmosphere for 5 hrs at temperatures up to 1300°C. Powder X-ray diffraction patterns were obtained using Ni-filtered CuK_α radiation. FT-IR spectra were determined using samples prepared with KBr. DTA and TG were run on uncalcined samples from room temperature to 1500°C in air with the heating rate of 8°C/min. The density of samples was determined applying pycnomethric technique with toluene. Samples for observation with a Philips EM-400T transmission microscope were prepared by spreading the powder on a copper grid covered with a thin carbon film.

RESULTS AND DISCUSSION

The density of a dried material is about 1.7 g/cm^3 and is even less than that of basic aluminium sulphate in the form of crystals - 1.86 [6] or precipitated from aqueous solution - 2.0 [2]. After 5 hrs of calcination the density rises progressively to 3.92 at 1300°C, close to that of α-Al_2O_3 .

The DTA curve shows that decomposition of aluminium hydroxy-chloride occurs in two stages , however mass losses are observed up to 1000°C. Both reactions occur succesively and they are not completed at 374°C with the end of the second endothermic reaction. Mass losses calculated from the decomposition of the compound with the hipothetical formula of $[AlO_4Al_{12}(OH)_{24}(H_2O)_{12}]Cl_7$ are close to that observed after calcination of tested specimens, assuming a loss of hydrated water in the first stage and decomposition of Al-13 ion in the second one.

The strong, broad absorption band in the region ~ 3150 to 3550 cm^{-1} and the absorption band which peaks at 1655 cm^{-1} (Fig.1) are evidences for molecular water presence in the structure. It cannot be determined from the FTIR spectrum if the materials contain hydroxyl groups coordinated with aluminium ions. However, significant decrease of the intensity of the two bands observed after heating at 140°C indicates the removal of that what can be assumed as the molecular water. The further decrease of the intensity of two mentioned bands have been noticed after calcination at 400°C and can be connected with decomposition of Al-13 ion though hydroxyl groups are still persistent up to 700°C. Two neighbouring bands centered at 2337 and 2358 cm^{-1} are due to chlorine bonds [7] and significant decrease of their intensity has been observed after calcination at 400°C, though, removal of chlorine ions has been completed at 1100°C.

The part of FTIR spectra in the region of 1000 - 400 cm^{-1} shows aluminium bonds (Fig.1). The absorption peak centered at

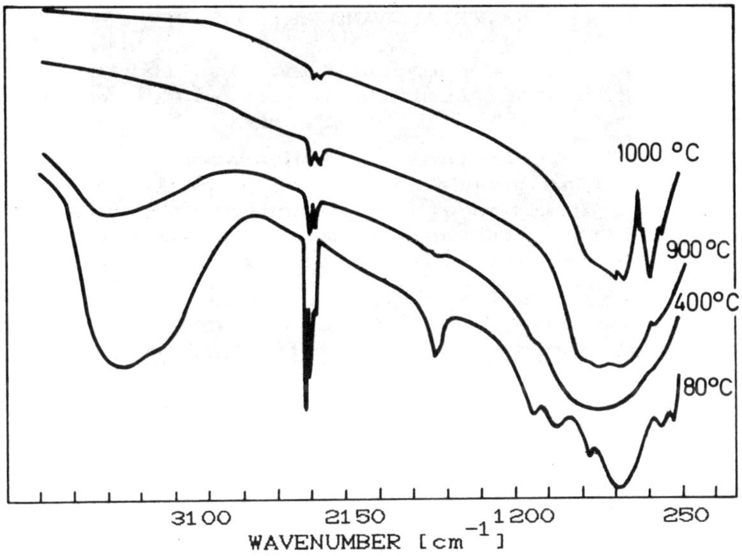

Figure 1. FT-IR spectra of aluminium hydroxychloride calcined at different temperatures.

310 cm^{-1} probably results from Al-OH vibrations [8-9]. The strong, broad absorption band, centered at 590 cm^{-1} may be attributed either to Al-O stretching vibrations in AlO_6 [10] or to Al-OH wagging librational mode of molecular water [9]. The peak centered at 775 cm^{-1} is seemingly due to either Al-O stretching [9] or valence vibrations in AlO_4 tetrahedra [10] or to Al-OH rocking librational mode of molecular water [9]. The bands at 970 cm^{-1} probably result from Al-OH deformation [10] and at 1083 cm^{-1} from Al-OH stretching [9] or deformation vibration [10]. Those bands ascribed to the frequency mode of coordinated water become fainter after calcination at 130°C. The 400°C FTIR spectrum shows only one broad absorption band which reflects a high degree of disorder in tested materials. Specimens remain amorphous up to 900°C but at 1000°C reordening of aluminium ions into corrundum structure takes place.

Another feature of interest in the FTIR spectrum is the development of the absorption band at 1083 cm^{-1}. This band ascribed to Al-OH vibration is still persistent in specimens calcined up to 700°C and disappears after heating at 800°C together with bands due to O-H and H-O-H vibrations. The mentioned band must result from Al-OH vibrations with "free" hydroxyl groups and it confirms that dehydroxilation is completed at 800°C.

The XRD pattern confirms amorphous structure of tested materials in the range from dried hydroxychloride up to 600°C. The first traces of crystalline phase have been found in specimens calcined at 700°C in the form of a single broad peak at d-spacing d=0.139 nm. Since it is a common diffraction peak to all transition alumina they cannot be distinguished.

50 nm

Figure 2. Powders calcined at 1000°C. TEM.

Calcination at 900°C leads to two other broad peaks at d=0.201 and d=0.238 nm which are probably due to γ-Al_2O_3, but material remains nanocrystalline. Further calcination at 1000°C causes the conversion to α-alumina. Material heated at 1100°C consists of far larger corrundum grains, considering the broadening of the lines.

The XRD does not indicate the presence of δ or θ -alumina which are common transition phases during the conversion $\gamma \rightarrow \alpha$ -alumina [1], but they were reported mainly in hydroxide -derived aluminas as well as in aluminium chloride, nitrate and sulphate hydrates [9]. In contrast the direct transformation from a low temperature transition alumina appears to be more common in basic salts-derived aluminas.

TEM studies of samples calcined at 900°C reveals that grains of alumina are about 10 nm in size and they develop into 10 times larger grains of corrundum after heating at 1100°C. Electron diffraction of specimen fully transformed to α-Al_2O_3 shows a spot pattern, Fig.2, indicating that the material is a monocrystalline one. These grains must contain low-angle boundaries. Similar results were reported by M.D.Sacks [2] and X.Yang [1]. It seems likely that the coarsening process is controlled by surface diffusional transport.

CONCLUSIONS

The present investigations of aluminium hydroxychloride shows that it solidified in the form of the low density amorphous material with significant amount of coordinated water and hydroxyl groups. The former can be removed below 200°C. This reaction followed by nonstoichiometric Al-13 ion decomposition below 400°C results in water and HCl removal completed at 800°C for dehydroxilation and at 1100°C for dechlorination. Amorphous highly porous material converts into nanocrystalline transition alumina at 700°C. Direct transformation from γ-Al_2O_3 to α-Al_2O_3 results in the density growth, the open pore network formation and the coarsening of the corrundum grains. This transformation changes individual particles from polycrystals of γ-alumina to the single crystals of corrundum indicating the formation of α-Al_2O_3 grains with the low angle boundaries.

ACKNOWLEDGMENT

The author is very grateful to the British Council for the stipend and Prof. A. Hendry at Strathclyde Universty, Glasgow for support and help during this work and dr R. Bailey for his kind assistance in the use of the FT-IR spectrometer.

REFERENCES

1. Yang, X., Pierre, A. C., Uhlman, D. R., TEM study of boehmite gels and their transformation to α-alumina. J. Non-Cryst. Sol., 1988 100, 371-377
2. Sacks, M. D., Tseung-Yuen Tseng, Soo Young Lee, Thermal decomposition of spherical hydrated basic aluminium sulphate. Am. Cer. Soc. Bull., 1984, 63, 301-310
3. Birchall, J. D., The preparation and properties of polycrystalline aluminium oxide fibres. Trans. J. Brit. Cer. Soc., 1983 82, 143-145
4. Akitt, J. W., Greenwood, N. N., Khandelwal, B. L., Lester, G. D., 27 Al NMR studies of the hydrolysis and polymerisation of the hexa-aquo-aluminium (III) cation. J. Chem. Soc. Dalton Trans., 1972, 5, 605-611
5. Bottero, J. Y., Tchoubar, D., Cases, J. M., Flessinger, F., Investigation of the hydrolysis of aqueous solutions of aluminium chloride. 2. Nature and structure by small-angle X-ray scattering. J. Phys. Chem., 1982, 86, 3667-3673
6. Johansson, G., On the crystal structures of some basic aluminium salts. Acta Chem. Scand., 1960, 14, 771-773
7. Nakamoto, K., Infrared and Raman Spectra of Inorganic and Coordination Compounds. Wiley, New York, 1978
8. Waters, D. N., Henty, M. S., Raman spectra of aqueous solutions of hydrolysed aluminium (III). J. Chem. Soc. Dalton Trans., 1977, 3,
9. Sato, T., Ozawa, F., Ikoma, S., Thermal decomposition of aluminium salts - hydrates of chloride, nitrate and sulphate and ammonium alum. J. Appl. Biotech., 1978, 28, 811-822
10. Schonher, S., Bertram, R., IR-spektroskopische Charakterisierung basischer Aluminiumkationen. Z. Chem., 1983, 23, 105

FORMATION OF TiN FROM TiO_2.

G.V.WHITE, K.J.D.MACKENZIE
Chemistry Division, DSIR, New Zealand.
J.H.JOHNSTON
Victoria University of Wellington, New Zealand.

ABSTRACT

The kinetics and mechanism of TiN formation by nitridation of TiO_2/carbon mixtures at 1050-1250°C have been studied by X-ray diffraction for a number of fine, pigment grade TiO_2 samples. The results show that TiN is formed by stepwise reduction to Ti_3O_5 which then nitrides.

INTRODUCTION

Titanium nitride is hard, abrasion resistant, and chemically inert. As a thin golden coating on metal it reduces friction and wear. It is used as a powder component of a varied range of advanced ceramics, cermets, and composites.

Large deposits of ilmenite on the west coast of New Zealand are a source of TiO_2 from which TiN can be prepared by carbothermal reduction. The chemistry of this reaction is the subject of this study.

EXPERIMENTAL

The Carbothermal Reduction

Mixtures of TiO_2 with fine powdered carbon (molar ratio 1:4) were heated in oxygen free flowing nitrogen in an electric tube furnace. Nine carbons and nine TiO_2 samples including both anatase and rutile forms, were reacted in batches.

The reaction products were analysed by quantitative X-ray powder diffraction (Philips PW 1700 diffractometer with graphite monochromator) using Al_2O_3 as the internal standard. The Al_2O_3 was added prior to heating because variable carbon consumption during the reaction militated against the normal procedure of adding the internal standard on a weight basis. Careful experiments indicated that no aluminium titanate was formed in the temperature range of the experiment, neither was nitridation of the Al_2O_3 to form aluminium nitride evident. The X-ray power diffraction pattern of the Al_2O_3 was also unchanged by the heating.

Analysis of the effluent reaction gas was performed by Infra-Red spectrophotometry, using a gas cell in a Perkin Elmer 337 grating spectrophotometer.

RESULTS AND DISCUSSION

The typical sequence of the phases detected during carbothermal reduction of TiO_2 is shown in Figure 2, which also includes the change in CO concentration in the gas atmosphere with increasing temperature.

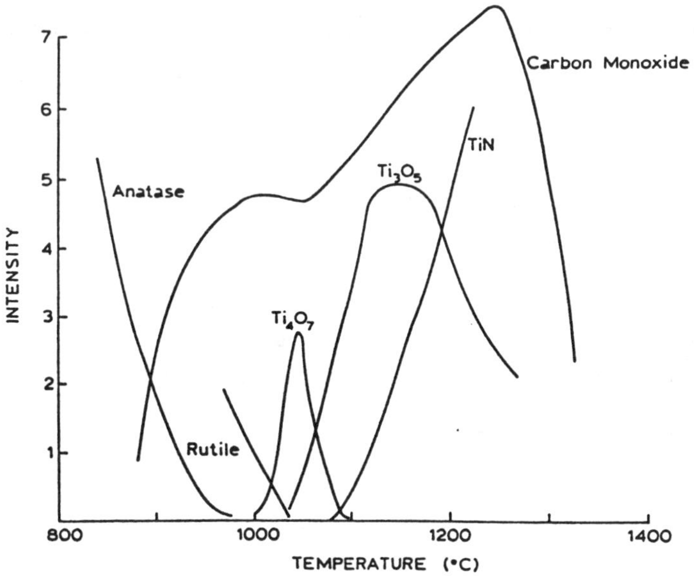

Figure 1 Semischematic Diagrams of Phases in Nitrogen

Evolved gas analysis shows that carbon monoxide is formed in two major steps. The first corresponds to the reduction sequence of TiO_2 to Ti_4O_7 via an homologous series of lower oxides of general formula Ti_nO_{2n-1} ($10 > n > 4$), first reported by Andersson et al [1].

$$4TiO_2 + C \rightarrow Ti_4O_7 + CO$$

The second step involves the formation of TiN from Ti_3O_5.

$$Ti_4O_7 + C + 2N_2 \rightarrow 4TiN + 7CO$$

This is consistent with the greater evolution of the carbon monoxide in the second step. The minimum in carbon monoxide evolution coincides with the transition from Ti_4O_7 to Ti_3O_5 suggesting that this is the rate determining step for the reaction. Also thermodynamic calculations show that this is not an energetically favoured step.

Kinetics

A typical series of growth curves of TiN from TiO_2 are shown in Figure 3.

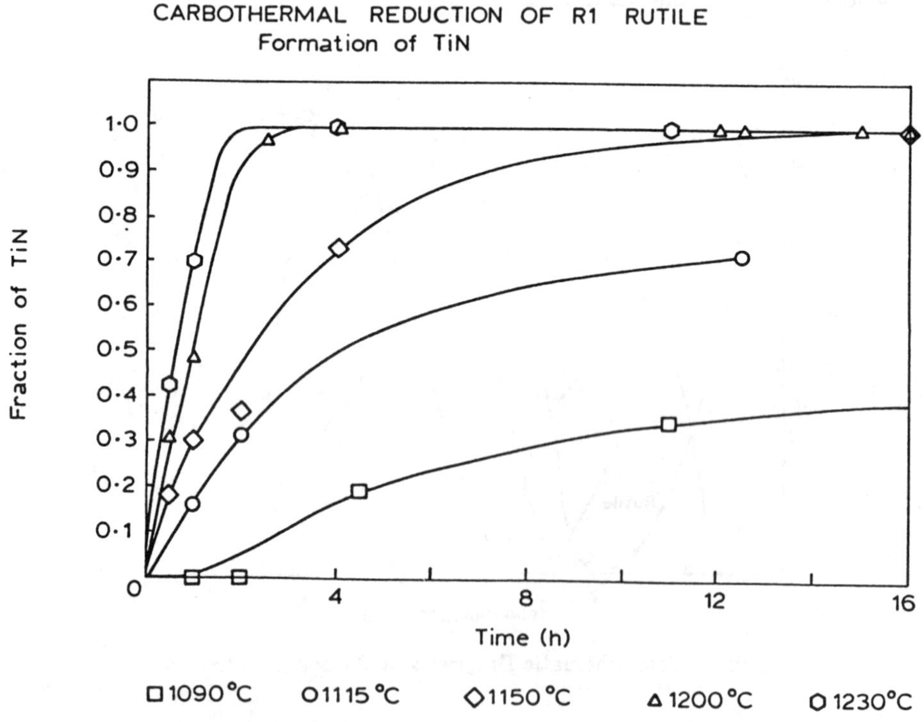

Figure 2 TiN Growth Curves from a Rutile.

The growth curves for TiN from a number of TiO_2 samples (anatases and rutiles) at 1090°C-1230°C are fitted better by a first order mathematical model than by diffusion or phase boundary model (Figure 4).

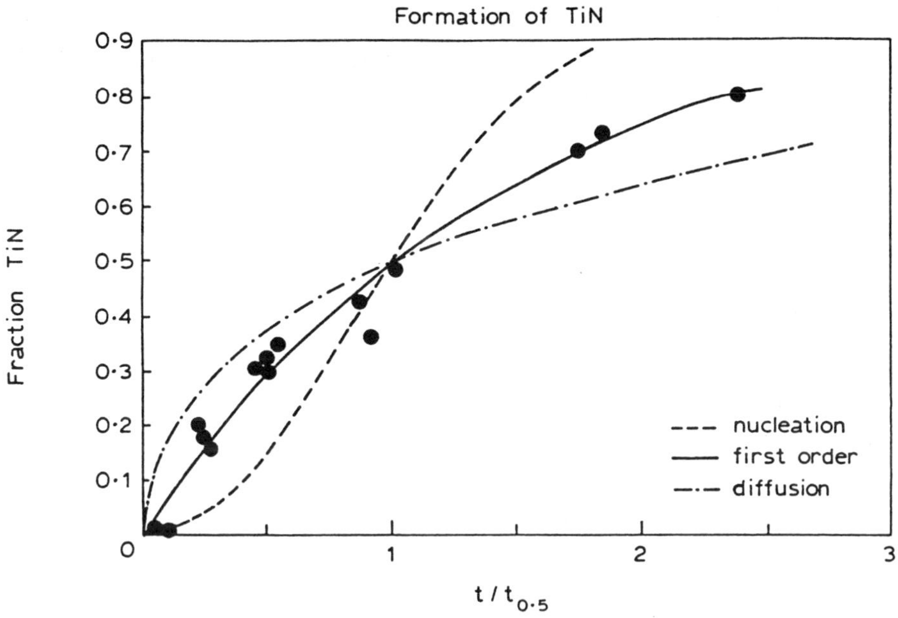

Figure 3 Reduced Growth Curves for Mathematical Models with Data Points from a Rutile.

Activation energies were calculated from the Arrhenius plots based on the first order rate equation, which consist of two straight lines. (Fig. 5)

Activation energies for the higher temperature process range from 260-390 kJ/mole. At the lowest temperatures a different mechanism appears to operate. These activation energies are significantly higher than those reported by Samsonov and Polishchuk [2] who worked with pressed pellets rather than the powders used here. This implies that for powder mixtures, the nitridation step involves both contact reduction and indirect reduction.

The kinetic parameters of nitridation are virtually independent of the physical and chemical properties of the TiO_2, starting Material because, as this work shows, the reaction sequence is extremely complex.

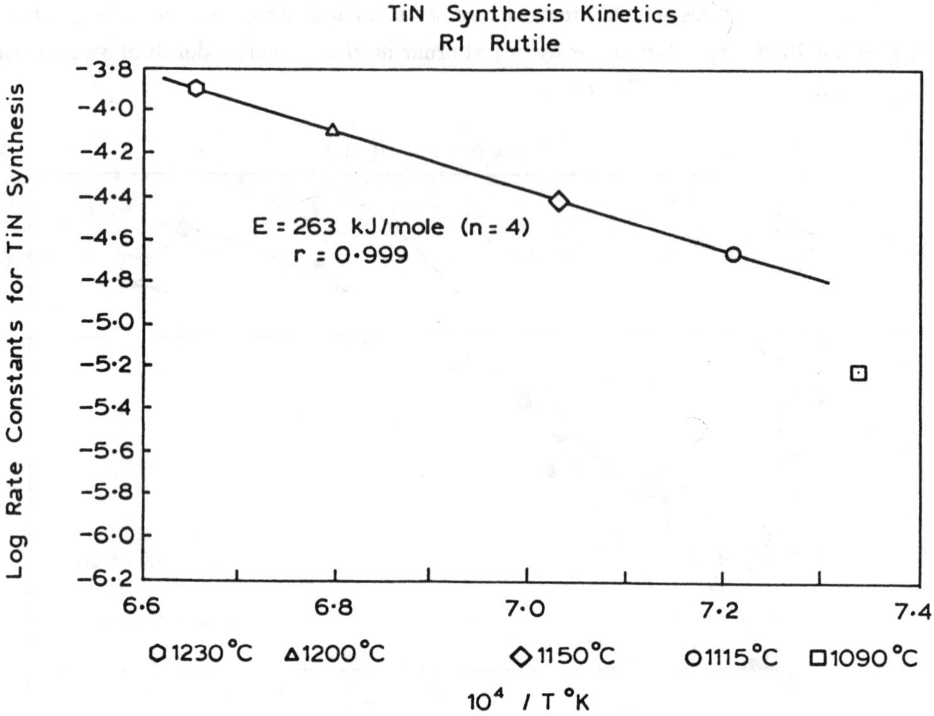

Figure 4 Arrhenius Plot for a Rutile.

REFERENCES

[1] Andersson, S., Collen, B., Kuylenstierna, U., and Magneli, H. Phase Analysis Studies on the Titanium-Oxygen System. Acta Chem. Scand. (1957) 11 1641.

[2] Samsonov, G.V. and Polishchuk, C.S.
Conditions for the Preparation of Titanium and Niobium Nitrides by the Carbon Reduction of Oxides with Simultaneous Nitridation.
Zh. Prikl. Khim. (1973) **46** (2) 241-6.

SYNTHESIS OF THE PEROVSKITE COMPOUND $Pb(Sc_{2/3}W_{1/3})O_3$

MARKO PETRIČ, DANILO SUVOROV and DRAGO KOLAR
"Jožef Stefan" Institute, University of Ljubljana, Ljubljana, Yugoslavia

ABSTRACT

$Pb(Sc_{2/3}W_{1/3})O_3$ was synthesised with a pyrochlore-perovskite conversion. The formation temperature of the perovskite phase, which is preceded by various reaction steps from the oxides, was approximately 1070 K. At 1245K the compound decomposed to Sc_2O_3 and a liquid phase. Dense and homogeneous samples for dielectric measurements were prepared by hot pressing. The temperature dependence of permittivity showed a maximum at about 225 K with $\varepsilon(max) = 2400$.

INTRODUCTION

Many complex lead-based perovskite compounds with the general formula $Pb(B_xB_y)O_3$ have received considerable attention in the area of electronic ceramics. These materials have good ferroelectric properties and low sintering temperatures. Examples are $Pb(Mg_{1/2}Nb_{1/2})O_3$ [1-4] and $Pb(Fe_{1/2}Nb_{1/2})O_3$ [5]. The synthesis of this type compound from the oxides is often difficult due to the appearance of an unwanted pyrochlore phase during the initial reaction steps. To reduce the amount of pyrochlore phase by employing an appropriate preparation procedure, knowledge of the formation process of the perovskite type compound from the oxides is necessary.

Formation of the ternary compound with perovskite structure in the PbO - Sc_2O_3 - WO_3 system was studied. The compound was already mentioned by Soviet authors in 1965 [6]. The following values were reported: cubic symmetry with the parameter of the subunit cell $a_0 = 4.067$ Å, $\varepsilon_{20} = 2300$ and a transition temperature $T_c = 261$ K.

The present communication is intended to describe the sequence of reaction steps leading to the $Pb(Sc_{2/3}W_{1/3})O_3$ compound with perovskite structure. The results of dielectric measurements are also reported.

MATERIALS AND METHODS

The raw material oxides PbO*, Sc_2O_3+ and WO_3# were reagent grade. Appropriate proportions of the oxides were mixed by alumina ball milling in ethanol for two hours. The material was dried, crushed in an agate mortar and sieved. The powder was then cold-pressed at 180 MPa and fired in alumina crucibles at different temperatures. The components in the fired samples were determined by means of X-ray diffraction analysis. To establish the process of formation of the perovskite compound, the high temperature diffraction method in the range between 773 K and 1373 K was used.

To prepare the perovskite $Pb(Sc_{2/3}W_{1/3})O_3$ compound for dielectric measurements, appropriate mixtures of starting oxides were fired four times for 45 minutes at 1100 K with intermittent crushig and pressing. The powder was then hot pressed at 20 MPa pressure for 30 minutes at 1100 K. The dielectric properties of ceramic discs, 8 mm in diameter and 0.5 mm thick were measured by an authomatic capacitance bridge (HP 4192A LF) with 1 kHz frequency.

RESULTS AND DISCUSSION

To investigate phase relationships in the PbO - Sc_2O_3 - WO_3 system several oxide mixtures were prepared. The mixtures were fired at temperatures ranging from 973 K to 1373 K for various soaking times. In the X-ray diffractograms, besides known peaks belonging to PbO, Sc_2O_3, WO_3, $PbWO_4$, Pb_2WO_5, Sc_6WO_{12} and $Sc_2W_3O_{12}$, a set of unidentified peaks was observed. It was determined by indexing that the reflections belonged to a compound with a face centered cubic unit cell with a_0 = 8.141(2) Å. Half of this value´fits well with the reported parameter of a subunit cell, a_0 =4.067 Å for $Pb(Sc_{2/3}W_{1/3})O_3$ [6].

TABLE 1

d-values, relative intensities and indices for $Pb(Sc_{2/3}W_{1/3})O_3$ of perovskite type

d/Å	I/I_0x100/%	h	k	l	d/Å	I/I_0x00/%	h	k	l
4.68	4	1	1	1	1.822	3	4	2	0
4.06	10	2	0	0	1.663	30	4	2	2
2.88	100	2	2	0	1.565	2	5	1	1
2.444	3	3	1	1			3	3	3
2.353	11	2	2	2	1.439	11	4	4	0
2.035	26	4	0	0	1.376	1	5	3	1
1.868	2	3	3	1	1.357	1	6	0	0
					1.288	6	6	2	0

Formation of $Pb(Sc_{2/3}W_{1/3})O_3$ from an initial mixture with PbO, Sc_2O_3 and WO_3 in the molar ratio 3:1:1 was followed by a sequence of intermediate reactions (Fig. 1).

* PbO, 60 mesh, 99 %, lot Nr. 031220402, Ventron

+ Sc_2O_3, 99 %, batch No. 47461, Koch-Light Laboratories LTD

WO_3, 99,7 %, lot Nr. 31961806, Ventron

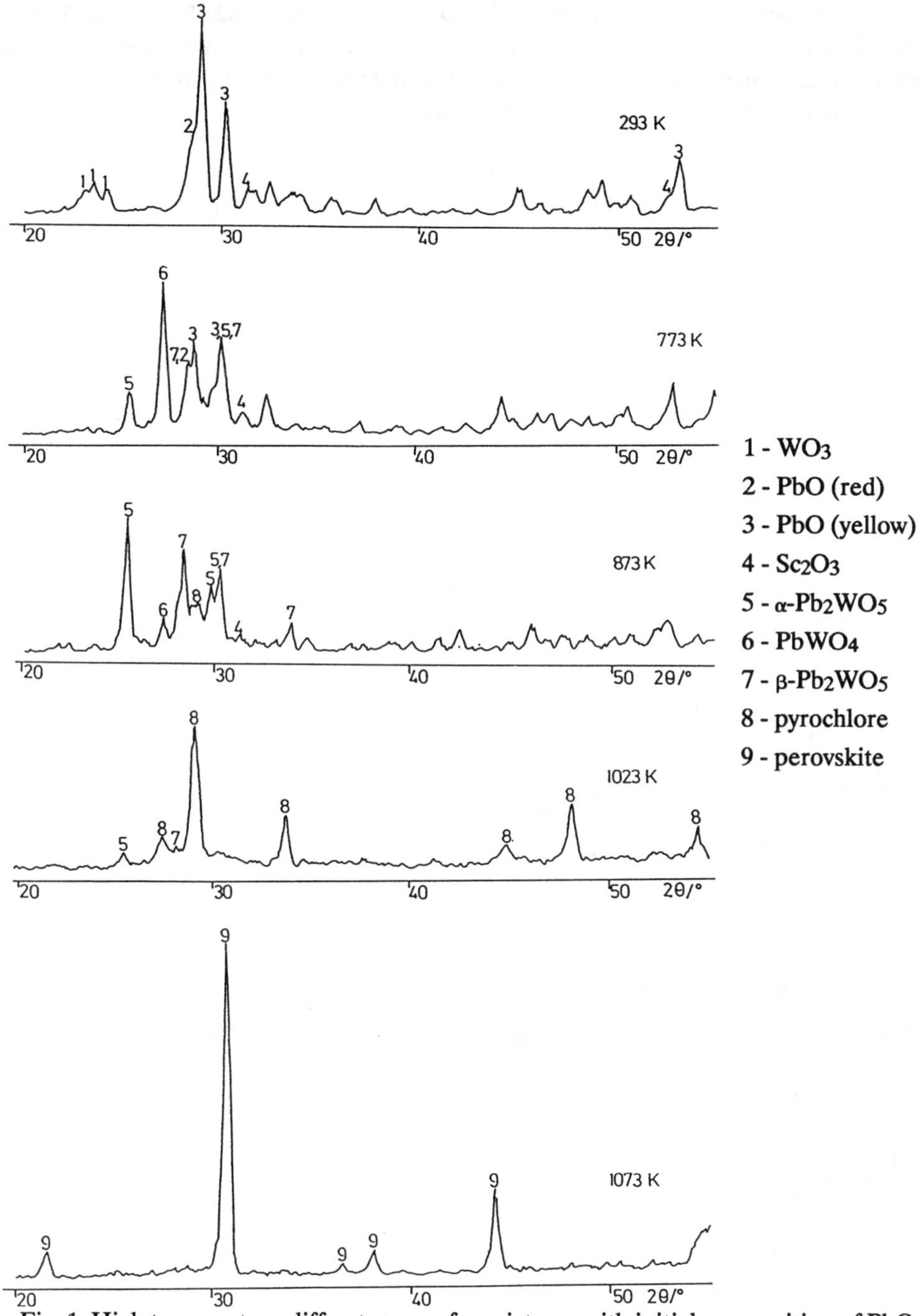

Fig. 1. High temperature diffractograms for mixtures with initial composition of PbO, Sc_2O_3 and WO_3 in the molar ratio 3:1:1.

X-ray analysis showed that PbO and WO_3 reacted at about 625 K to form $PbWO_4$. The following reaction at about 775 K was the formation of α - Pb_2WO_5, which was partially transformed into β - Pb_2WO_5 at approximately 825 K [7]. At around 875 K a compound with a pyrochlore structure began to form.

TABLE 2

d-values, relative intensities and indices for the compound with a pyrochlore structure

d/Å	I/I_ox100/%	h	k	l	d/Å	I/I_ox100/%	h	k	l
6.12	20	1	1	1	1.805	10	5	3	1
3.76	5	2	2	0	1.624	45	5	3	3
3.21	30	3	1	1	1.607	5	6	2	2
3.08	100	2	2	2	1.537	15	4	4	4
2.668	40	4	0	0	1.493	5	7	1	1
2.175	<5	4	2	2			5	5	1
2.053	5	3	3	3	1.387	10	7	3	1
		5	1	1			5	5	3
1.885	55	4	4	0	1.333	<5	8	0	0

The calculated parameter of the cubic unit cell was a_0 = 10.659(3) Å. The pyrochlore converted to a perovskite between 1025 K and 1075 K. The perovskite phase

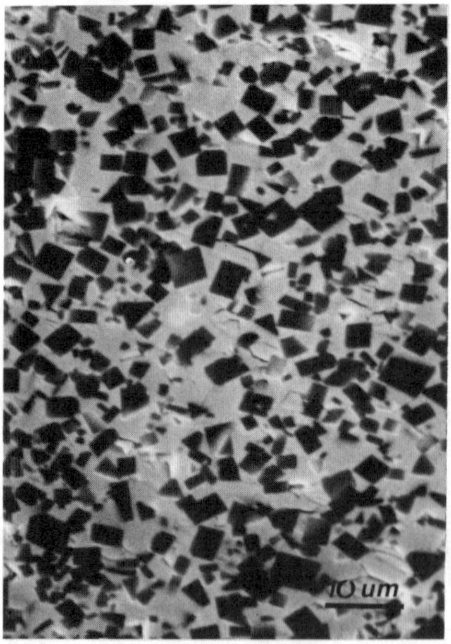

Fig. 2: Well shaped Sc_2O_3 grains in a liquid phase after decomposition of $Pb(Sc_{2/3}W_{1/3})O_3$ at 1245 K, 30'.

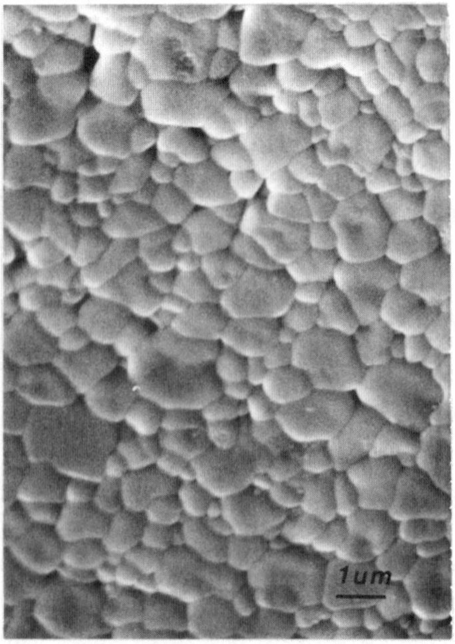

Fig. 3. Microstructure of hot pressed $Pb(Sc_{2/3}W_{1/3})O_3$ ceramic.
T = 1100 K, 30', P = 20 MPa

was stable up to 1245 K, where incongruent melting occurred giving a liquid phase containing well shaped Sc_2O_3 grains which were identified by EDS analysis (Fig. 2).

For the purpose of dielectric measurements, dense samples were prepared by hot pressing. The microstructure of a dense sample, prepared for dielectric measurements is shown in Figure 3.

The temperature dependence of the dielectric constant showed a maximum at about 225 K. Measured values of permitivitty and dielectric losses at room temperature were 2100 and 2×10^{-4}, respectively.

CONCLUSIONS

The compound $Pb(Sc_{2/3}W_{1/3})O_3$ was synthesised. The process of formation of $Pb(Sc_{2/3}W_{1/3})O_3$ from the oxides was as follows:

$$\begin{aligned}
PbO + Sc_2O_3 + WO_3 &\xrightarrow{\sim 625K} PbWO_4 + PbO + Sc_2O_3 + WO_3 \\
PbWO_4 + PbO + WO_3 &\xrightarrow{\sim 775K} \alpha\text{-}Pb_2WO_5 + PbWO_4 + PbO + WO_3 \\
\alpha\text{-}Pb_2WO_5 &\xrightarrow{\sim 825K} \beta\text{-}Pb_2WO_5 \\
(\alpha+\beta)Pb_2WO_5 + PbWO_4 + PbO + WO_3 + Sc_2O_3 &\xrightarrow{\sim 875K} \text{PSW(PYROCHLORE)} \\
\text{PSW(PYROCHLORE)} &\xrightarrow{1025K\text{-}1075K} \text{PSW(PEROVSKITE)} \\
\text{PSW(PEROVSKITE)} &\xrightarrow{\sim 1245K} \text{LIQUID} + Sc_2O_3
\end{aligned}$$

The dielectric properties of hot pressed $Pb(Sc_{2/3}W_{1/3})O_3$ ceramic ($P = 20$ MPa, $T = 1100$ K, 30') were:

$\varepsilon = 2100$ $tg\,\delta = 3 \times 10^{-4}$ (room temperature)

$\varepsilon_{max} = 2400$ (at about 225 K)

REFERENCES

1. Goo, E., Yamamoto, T. and Okazaki, K., Microstructure of lead - magnesium niobate ceramics. J. Am. Ceram. Soc., 1986, **69**, C-188 - C-190.
2. Guha, J.P. and Anderson, H.U., Preparation of perovskite $Pb(Mg_{1/3}Nb_{2/3})O_3$ using $Pb_3Nb_2O_8$ and MgO. J. Am. Ceram. Soc., 1986, **69**, C-287 - C-288.
3. Kang, D.H. and Yoon, K.H., Dielectric properties due to excess PbO and MgO in lead magnesium niobate ceramics. Ferroelectrics, 1988, **87**, 255 - 264.
4. Chen, G. and Fu, S., Fabrication of perovskite $Pb(Fe_{2/3}W_{1/3})_x(Fe_{1/2}Nb_{1/2})_{0.9-x}Ti_{0.1}O_3$ by double calcination process. J. Mater. Sci., 1988, **23**, 3258 - 3262.
5. Halmi, M., Desgardin, G. and Raveau, B., Improved lead perovskite compounds (PFN -PFT) for Z5U capacitor applications. Adv. Ceram. Mat., 1988, **3**, 32 - 37.
6. Tomashpol'skii, Yu. Ya. and Venevtsev, Yu.N., New lead containing perovskites. Sov. Phys. Solid state, 1965, **7**, 412 - 414.
7. Bosselet, F., Mentzen, B.F. and Bouix, J., Etude de la solution solide $Pb_2O(S_{1-x}W_xO_4)$ et structure cristalline de la variete alpha du monooxodiplomb (II) tetraoxotungstate (VI) $Pb_2O(WO_4)$ par diffraction X sur poudres. Mat. Res. Bull., 1985, **20**, 1329-1337.

LIST OF CONTRIBUTORS

INDEX OF FORMULAE

SUBJECT INDEX